W9-AOP-826

Operational Research and the Social Sciences

Operational Research and the Social Sciences

Edited by

M. C. Jackson and **P. Keys**
University of Hull
Hull, United Kingdom

and

S. A. Cropper
University of Strathclyde
Glasgow, United Kingdom

Plenum Press • New York and London

Library of Congress Cataloging in Publication Data

IFORS Specialized Conference on Operational Research and the Social Sciences (1989:
Queens' College, Cambridge)
  Operational research and the social sciences / edited by M. C. Jackson and P. Keys
and S. A. Cropper.
    p.   cm.
  Proceedings of an IFORS Specialized Conference on Operational Research and the
Social Sciences, held April 10-13, 1989, at Queens' College, Cambridge, United
Kingdom" — T.p. verso.
  Includes bibliographical references and index.
  ISBN 0-306-43149-1
  1. Social sciences — Methodology — Congresses. 2. Operations research — Congresses.
I. Jackson, Michael C., 1951–   . II. Keys, Paul, 1954–   . III. Cropper, S. A. IV.
International Federation of Operational Research Societies. V. Title.
H61.I38   1988   89-30284
300′.72 — dc19   CIP

Proceedings of an IFORS Specialized Conference on Operational
Research and the Social Sciences, held April 10-13, 1989,
at Queens' College, Cambridge, United Kingdom

© 1989 Plenum Press, New York
A Division of Plenum Publishing Corporation
233 Spring Street, New York, N.Y. 10013

PREFACE

        Twenty five years ago, in 1964, The Operational Research Society's
first International Conference (held at Gonville and Caius College,
Cambridge) took as its theme "Operational Research and the Social Sciences".
The Conference sessions were organised around topics such as: Organisations
and Control; Social Effects of Policies; Conflict Resolution; The Systems
Concept; Models, Decisions and Operational Research. An examination of
the published proceedings (J.R.Lawrence ed., 1966, Operational Research and
the Social Sciences, Tavistock, London) reveals a distinct contrast between
the types of contribution made by the representatives of the two academic
communities involved. Nevertheless, the Conference served to break down
some barriers, largely of ignorance about the objects, methods and findings
of each concern.

        In the ensuing twenty five years, although debate has continued about
the relationship between OR and the social sciences, mutual understanding
has proved more difficult to achieve than many must have hoped for in 1964.
There remains a feeling that the full potential of bringing together the two
sides to address current problems of organisations and society remains to
be realised. It was this belief which motivated the organisers to mount
another Conference on the theme of "OR and the Social Sciences" in 1989,
this time at Queens' College,Cambridge. This belief is reinforced by the
knowledge that both OR and the social sciences have developed in the past
twenty five years in directions which make the prospects for satisfactory
collaboration considerably better. These developments suggest several reasons why this Conference is particularly timely.

        OR as a community is more diffuse than in 1964. For one, it is larger;
and it now contains a significant number of academics based in OR or Management Science departments. Few, these days, regard OR as being simply
applied mathematics. The recognition that OR is a process of intervention
in organisational and human affairs is now wider and more explicit. There
has been a penetration and diffusion of ideas from the social sciences into
OR, reflected most strongly in the body of writing about soft OR methods
and soft systems thinking. Social and political skills are now recognised
as critical to the success of OR practice and this is particularly so as
operational researchers have sought to extend their client base outside
that conventionally served. The rise of computer technology, embraced by
OR, has required some thought to be given to its powerful impact on and
consequences for organisations, people and processes of decision. As a
consequence, perhaps, of broadening its methodological base and attempting
to extend its impact, OR has encountered the problem of competing 'paradigms'
- a condition long experienced by the social sciences.

        These factors suggest that a closer relationship with the social sci-
ences would be productive, and increasing numbers within OR recognise the
need to examine and potentially to draw from the bodies of knowledge that

make up the social sciences. Yet, there has been a reluctance to embrace what is seen as social science "jargon". As a consequence OR remains conceptually underpowered and practitioners who have integrated traditional OR and social science are relatively few and far between. Further, there is little sign of this being put to rights "naturally" -social science graduates are, in general, neither attracted to nor recruited to OR; and there is precious little social science taught on OR courses. The loss in 1985 of an institutional embodiment of the theme of the last 'OR and the Social Sciences' conference, the Centre for Organisational and Operational Research based at the Tavistock Institute for Human Relations, reflects the difficulties that have dogged the relationship.

Within the social sciences there has been an increasing uncertainty about role, purpose and identity. This is highlighted by the proliferation of alternative 'affiliations' and paradigms challenging conventional social science - particularly, perhaps, the growth of critical and post-modernist perspectives. The attempt to remove the status of "science" from social research, reflected in the change from Social Science Research Council to Economic and Social Research Council, is further evidence of a crisis of identity and legitimacy. Problems within and pressure from without make it imperative to demonstrate the relevance and applicability of the social sciences. The present climate is forcing the social science disciplines to become more pragmatic, providing more points of contact with operational researchers. Social scientists should find much of value in OR's ability to transform ideas into effective tools for practice.

The prospects for a fruitful debate between OR and the social sciences look much better, therefore, in 1989 than in 1964, and circumstances are more pressing. The conditions for initiating a period of creative and constructive synthesis between OR and the social sciences are now present. The time is right for a deeper consideration of the way in which the social sciences can be applied to problems of organisations and society and, in particular, how they can enter and inform OR practice.

The 1989 Conference at Queens' College, Cambridge, will hopefully provide a stimulus to this interaction. Plenary sessions, formal presentations and discussions at conferences have an impact upon those present. However, it is the written word which has a lasting impact on the wider community. Hence, this book also has a significant catalytic role to play. Its structure reflects the design of the Conference. The plenary addresses are first. Then come various sections which correspond to the parallel sessions around which the Conference was organised. They are:

* Intervention and Change
* Systems and OR
* Methodology
* Design and Planning
* Power, Conflict and Control
* Problems of Measurement
* Modelling Social Behaviour
* OR as a Social Science
* Information Systems
* OR and the Decision-Maker
* Post-Industrial Society
* Work

Some of these are not so different from 1964. The content and the spirit of the papers, though, reflects the changes described above; and it is clear from a comparison between these proceedings and those of the 1964

Conference that the debate has moved forward. Each section is introduced
by a brief sketch of developments in its area of concern. There is no
synthetic overview, deliberately: the body of material within this book
is intended as a starting point rather than a conclusion to cross fertilisation between OR and the social sciences.

M. C. Jackson,
P. Keys and
S. Cropper

CONFERENCE ORGANISATION

OR AND THE SOCIAL SCIENCES
CAMBRIDGE, 10-13 APRIL 1989

Organising Committee   M.C.Jackson (University of Hull, Chair)
                        P.Carter (University of Hull)
                        Dr.S.Cropper (University of Strathclyde)
                        Dr.N.Jackson (University of Hull)
                        P.Keys (University of Hull)
                        G.Mitchell (British Coal)
                        V.Symons (University of Cambridge)
                        N.Tobin

Programme Committee   P.Keys (University of Hull, Chair)

Stream Organisers   National Representatives

Dr.M.Alvesson (University of Lund)   Professor P.Agrell (FOA, Stockholm,
Dr.S.Cropper (University of   Sweden)
  Strathclyde)   Dr.S-K.Ahn (Kyung Hee University,
Dr.L.Davies (RMCS, Shrivenham)   Korea)
Dr.V.Degot (Ecole Polytechnique,   Professor A.Blumstein (Carnegie-
  Paris)   Mellon University, USA)
Dr.R.Flood (City University)   Professor J.D.Coelho (Universidade
Dr.D.Knights (UMIST, Manchester)   Nova de Lisboa, Portugal)
Dr.P.Ledington (RMCS, Shrivenham)   Mr.A.K.Datta (IEL Ltd., Calcutta,
Professor P.Manning (Michigan State   India)
  University)   Professor M.Despontin (VUB-CSOO,
Dr.J.Mingers (University of Warwick)   Brussels, Belgium)
Professor G.Peters (Open University)   F.G.Feichtinger (Technische
V.Symons (University of Cambridge)   Universitat Wien, Austria)
Dr.J.Tait (Open University)   Dr.R.Hall (University of Manitoba,
Dr.G.Walsham (University of   Canada)
  Cambridge)   Professor C.Jacobsen (Technicon,Israel)
Dr.G.Wright (Bristol Business   Professor S.Jorgensen (Copenhagen
  School)   School of Economics and Business
Dr.C.Yewlett (UWIST, Cardiff)   Administration, Denmark)
                                        Professor B.Kavanagh (Swinburne
                                          Institute of Technology, Australia)
                                        Professor A.A.Sissouras (University
                                          of Patras, Greece)
                                        Dr.A.Smith (Ministry of Energy,
                                          Wellington, New Zealand)
                                        Professor L.Streitferdt (Universität
                                          Hamburg, West Germany)
                                        Professor H.Takamori (Aoyama-Gakuin
                                          University, Japan)
                                        Professor J.Wallenius (University of
                                          Jyvaskyla, Finland)
                                        Dr.W.Y.Yeong (National University of
                                          Singapore, Singapore)

Administration

                    Ray Showell & June Hedge
                    (Operational Research Society)

CONTENTS

INFORMATION SYSTEMS

PROBLEMS OF MEASUREMENT

MODELLING SOCIAL BEHAVIOUR

POWER, CONFLICT AND CONTROL

POST-INDUSTRIAL SOCIETY

O.R. AND THE SOCIAL SCIENCES

PLENARY SESSIONS

A BEHAVIOURAL SCIENCE PERSPECTIVE ON OPERATIONAL RESEARCH PRACTICE

J. G. Burgoyne

Centre for the Study of Management Learning
University of Lancaster
Lancaster, UK

INTRODUCTION

The main aim of this paper is to propose a 'map' of the different ways in which behavioural science perspectives and the theory and practice of Operational Research may be thought about in relation to each other.

The main argument is that the relationship may be thought about at three fundamentally different levels.

At the first level behavioural science can be conceived as a 'tool' discipline to Operational Research, to be looked to for help, for example, in finding ways of persuading clients to value, accept and implement OR solutions.

At the second level behavioural science and OR can be thought of as intellectual and professional disciplines with overlapping interests, which can have a dialogue and debate about matters of mutual interest. Each discipline may, for example, offer alternative accounts of the same phenomena, particularly where they concern the behaviour of people in an organisational setting. There may be genuine intellectual frontiers between the two broadly conceived disciplines, of which artificial intelligence, managerial decision making and expert systems may be examples. Finally, to the extent to which either or both claim to be multi-disciplinary, it is possible to consider the extent to which each area already incorporates some of the insights of the other, or the extent to which both draw on common academic and professional sources.

At the third level OR, both as an academic discipline and an area of professional practice, can be considered as a phenomenon that behavioural science can look at, interpret, and raise critical questions about. Thus questions can be asked about the social significance of the instigation, evolution and growth of OR, and the eras, cultures and organisational settings where this has and has not happened. Questions can be asked about the values and ideologies that are built into and supported by OR. Questions can be asked about the implicit and explicit assumptions concerning the nature of both people and organised activity which underlie OR theory and practice. And to the extent to which reality is, at least in part,'socially constructed' (Berger and Luckman, 1967), as many behavioural scientists argue it is, then the question of the kind of personal and organisational realities created by OR practice can be asked. In addition, the behavioural scientist can pose questions about the kinds of people who become OR practitioners, the nature of their careers, how OR is organised within itself and within work organisations, and how its practitioners acquire and use power.

The stance underlying this paper is that each of these perspectives is valuable and useful. However it is the author's observation that operational researchers tend to prefer the first of these levels, and behavioural scientists the last, and that dialogue is less fruitful without agreement about the level or levels at which to conduct it. This is an extension of the phenomenon, observable amongst the protagonists of all disciplines claiming relevance to management, organisation or business, in which dialogue is based on the rival claims that 'my discipline includes/interprets/is "meta" to, your discipline'. Resisting the temptation to fall into this trap, and the very real difficulty in doing this, is a major challenge to inter-disciplinary cooperation, and suggests that the middle level may be the most fruitful starting point from which to reach out to the other two.

The main aim of this paper is therefore to map some of the issues, ideas and dilemmas at each of these three levels, as an aid to thoughtful dialogue between behavioural scientists and operational researchers.

The remainder of this paper will therefore consist of a discussion of issues at each of the three levels. To make such a discussion possible it will be necessary to make a number of assumptions about what is meant by both OR and behavioural science, both of which are likely to be contentious issues. To make the discussion possible these assumptions will be stated briefly before the discussion of the three levels.

ASSUMPTIONS ABOUT THE NATURE OF OPERATIONAL RESEARCH

The naive reader approaching the OR literature to see what it is all about gets a reasonably consistent picture of what, at its conventional core, OR is. It is concerned with being 'scientific' about problems that exist in a 'real' operational world, usually with some notion of optimisation being applied, through various techniques, to manipulate a model, usually mathematical, of the problem situation, and decide a 'best solution'. Such is the flavour of basic definitions offered by much quoted authors such as Churchman, Ackoff and Arnoff (1957).

The notion of what is 'scientific' seems to be, very much as Checkland (1983) says, based on an image of empirical natural science, giving much emphasis on interdisciplinarity, both among different fields of science and across to practitioners in the form of professionals and managers. Scientists were seen as providing method, with their substantive knowledge being incidental, while the professionals and managers knew and had influence over the problem. It is interesting to note,however, that the example quoted by Beer (1966) of a wartime group centred on Professor Blackett included 'another physicist, three physiologists, a surveyor, two mathematical physicists, an army officer and an astro-physicist', behavioural science being notable for its absence, although this was the era in which behavioural science, particularly in its more psychological forms, was growing rapidly in its applied usage in the military context. More recently the multi-disciplinary feature has faded out of accounts of OR, in favour of a view of OR as a discipline in its own right, with applied mathematics, modelling and optimisation as its substantive and methodological core.

The second point of historical interest is the early claim to find observable stable patterns of events even where these are 'rather unexpected in view of the large number of chance events and individual personalities

4

and abilities that are involved in even small operations' (Blackett, 1962). This statement, of observable mathemtical regularities in aggregated events involving people, seems to constitute the standard OR justification for not needing to grapple with the confusing variety of behavioural science thought.

ASSUMPTIONS ABOUT THE NATURE OF BEHAVIOURAL SCIENCE

Behavioural science presents no such unified central concepts, methods or history, though many strands of all of these are identifiable, and the history longer. The label is immediately problematical both in the terms 'behavioural' and 'science'. The safest statement to make may be the most common sensical: that behavioural science is the study of people.

The broad traditional categories of approach are probably still the psychological and the sociological, but with other long standing disciplines like anthropology in attendance, intermediate disciplines like social psychology, new disciplines like linguistics, hybrids like 'organisational behaviour', which has tended to become the standard label in management and business school curricula, particularly American ones, and areas of professional practice like organisational development (referred to as 'OD'), all making up a complicated and messy picture. The term 'behavioural' may tend towards the psychological end of the spectrum, and even bridge out into physiological explanations of behaviour. The term 'social', as in 'social science' may emphasise the other end of the spectrum - concerning people in the collective sense, and the meanings that individuals attach to themselves, their actions and institutions surrounding them.

One assumption or belief shared by much of behavioural/social science is, stated simply, that there is more to people and what they do than can be explained by their material make up and organic biological processes. It is because of this, primarily negative statement, that the issue of what it means to be 'scientific', and all the methodological and philosophical issues that follow from this, is more complicated and contentious in the behavioural sciences than it appears to be in OR. As an extension of this, a mutual non-acceptance of what the other regards as scientific, and on the behavioural side, not being able to present a clear consensus on methodology anyway, may be one of the major blocks to comprehension and dialogue.

The empirical natural science model adopted by OR represents one approach used in the behavioural sciences, characterised as the logical positivist approach when viewed from a philosophical perspective, and manifesting itself in experimental design approaches copied from natural science, biology, clinical trials and agricultural research. Strong arguments have been mounted against this approach, along the general lines, greatly simplified, that this degenerates behavioural science back to the biological and material, and that an appropriate methodology must allow for the reality of the non-material or biological, such as subjective experience, social process, patterns of meaning and symbolism (see Reason and Rowan, 1981, Morgan, 1983, Harre and Secord, 1972, for such arguments). However in some areas of the behavioural sciences, this argument has been won well in the past, in others it has scarcely been heard.

If a general statement about what behavioural science means by being scientific can be risked, it is perhaps that the method must be open to, and appropriate for, the kinds of phenomena that behavioural science theories can postulate, about the nature of what is human over and above the physical and physiological. This may involve a much broader conception of what is acceptable empirical evidence, and it may entail non-empirically based analysis of the nature of theories themselves and the ways in which it is possible to conceive human processes (e.g. Burrell and Morgan, 1979).

5

BEHAVIOURAL SCIENCE AS A TOOL FOR OR

The basic hypothesis of this paper is that OR is confident enough of its own essentially non-behavioural science general aim and approach to prefer to see behavioural science as a peripheral tool rather than a core inter-disciplinary contributor, or a radical critic.

There appears to be two main areas of contribution for behavioural science as a tool within the classical OR investigation: contribution of concepts, models and measures within the investigation, and dealing with the client system before, during and after the investigation.

Within an OR investigation it may seem desirable to include some be-havioural variables, while staying with the basic concept of a quantifiable model or a net of causally related variables, making possible the calculation of an optimal arrangement. Within a manpower planning model, for example, it may be necessary to use such variables as 'morale' and 'job satisfaction' as determinants of labour turnover, 'skill' as a determinant of performance, 'learning curve' as a predictor of growth of performance over time. In such an investigation behavioural science might be looked to first to suggest the variables, second for hypotheses about how they might fit into the model, and thirdly for empirical methodologies for measuring them.

A survey of the use of 'behavioural' variables in OR models is well beyond the scope of this paper, but two casual hypotheses might guide such an investigation: firstly that OR will avoid behavioural variables wherever possible, and second, where they are used they will be borrowed most readily from areas of the behavioural sciences that have stayed closest to the 'natural science' model of doing research.

Shilton (1982) provides an example of the former: in a model to pre-dict demand for train services he uses a variable Q for 'quality of service', which he defines operationally as a function of speed and frequency of ser-vice, rather than as a behavioural variable rooted in the experience of the railway passenger, and to be assessed empirically.

An example in support of the latter hypothesis is provided by Polding and Lockett (1982) in their attempt to study a phenomenon associated with the second part of this section: implementation of OR projects. This study does grasp the empirical nettle and attempts to assess the experiences of pract-itioners and clients through questionnaire survey. However the analytical methods of principle components and regression analysis are immediately called in to give a reassuring, if possible spurious, sense of order to this data, and present it in a form more compatible with the normal OR model. Such data-raking techniques are mainly used in those branches of the behavioural sciences that feel the need to convert their data into quantitative express-ion of quasi-causal relationships. It seems significant to this author that such an approach was used in preference to in depth qualitative case studies of a small number of OR studies from instigation to implementation.

Although, as stated above, the conventional core of OR seems clearcut, there is an apparent sense of 'crisis' in some people's minds (Checkland, 1983, Ackoff, 1979) which appears, in the first instance, to be a crisis of mismatch between theory and practice, rather than within the theory itself. To the extent to which the crisis is real, there seem to be two responses to it. The first is that the problem is, after all, a substantive and theoretical one, and that OR needs to shift its theoretical base in a systems/cybernetic direction. The second is that the OR core is basically sound, and the crisis derives from a mismatch between the rational logic of OR and the less rational psycho-logics of the processes involved in the acceptance and implementation of OR solutions. In this view it is the applied behavioural science of the consultant interaction with client

6

systems which holds the key to overcoming the crisis, and actually describes what the day to day work of the OR practitioner is like, in contrast to that which is implied by the OR curriculum. This is therefore the area in which OR practitioners may be most ready to look for help from behavioural science tools.

Such help is on offer from some branches of the behavioural sciences, particularly those concerned in the broadest sense with organisational development, see for example Lippitt and Lippitt (1978).

The general theme of such advice is that clients need to be regarded as people rather than rational decision making machines simply needing a correct technical solution. Some of the kinds of points that arise are:

(1) Understand the problem the client sees, and how they see it, and talk to them in their language.
(2) Recognise the nature and limits to the power and authority of the client.
(3) Be sensitive to the pressures on the client, and their needs to maintain their own self-esteem.
(4) Recognise that clients may be more able to take on new ideas at some times rather than others, and can sometimes do this only slowly.
(5) Recognise that clients can either be 'complacent' about a problem or in a quiet state of panic about it - the former need the reality bringing home to them, the latter need encouraging to see that there are solutions and ways out.
(6) In attempting to get a problem recognised and a solution accepted, consider the questions: who knows about the problem, who cares whether it is solved or not, and who can do anything about it. If an overlap between these categories exists or can be brought about, then this may define a fruitful client to work with.
(7) Try to put forward solutions and recommendations that are not only technically sound but also politically acceptable and administratively feasible.
(8) For a client to take action he or she needs to recognise a problem, having a vision of its solution, see a path to this solution, and know in detail the first step in this journey (this 'list' can provide a framework for a presentation).
(9) Clients are likely to grow to trust consultants only slowly, as they prove trustworthy, reliable, helpful and sensitive - initially over small matters.
(10) A Consultant can usefully give a client a balance of support through listening and sympathy about a problem and confrontation with evidence of new aspects of the problem, analyses or interpretations of it and ideas for ways to solve it.

Such 'practical guidelines', while being offered by some branches of the behavioural sciences may well be criticised by others as untested folklore, manipulative, or undermining the true role of behavioural science. However, since any consultant is likely to be a behavioural-scientist-in-practice, points like the ten above may at least provide a starting point for becoming more aware of some of the processes involved.

The underlying principle of most of these points is about understanding the psycho- and socio-logics of the client situation. Where this problem has been addressed from within the OR tradition (Churchman and Schainblatt, 1965), it has been on the basis of the concept of 'mutual understanding'. Polding and Lockett (1982) marshall a certain amount of evidence to support the view that when 'mutual understanding' is egocentrically interpreted to mean getting clients to think and understand solutions in OR terms, then this is counterproductive in terms of implementation. They conclude that 'there is possibly a need for mutual understanding more in a sociological than an intellectual sense.'

BEHAVIOURAL SCIENCE AND OPERATIONAL RESEARCH AS ADJACENT DISCIPLINES

Given the pluralistic nature of the behavioural sciences, the easiest areas for substantive dialogue appear to occur where there are similar methodological assumptions.

The 'natural science' OR model would see prediction and control as major aims of this form of science, aligning it closely with technical objectives of manipulating material entities to the aims of human kind. Much of behavioural science can be seen as concerned with explanation rather than prediction and control, and with such powers of prediction and control as do arise from explanation serving the ends of individual self-control by actors in a situation, rather than control by a centralised 'human engineering' function. Thus explanation serves illumination and emancipation rather than centralised control.

Areas of common interest between behavioural science and OR appear to be limited to:

(1) behavioural science concepts that can translate into variables of the kind acceptable to OR models
(2) methodological issues, at the level of research design and statistical, particularly multivariate, analysis
(3) the substantive area of decising making.

Concepts and variables like 'morale', 'job-satisfaction', 'skill' and 'learning curve' were given as examples of areas of common interest in the 'behavioural science as tools' section. Such concepts, their meaning and the interpretation of empirical observations about them can potentially also be an area of inter-disciplinary discussion, though examples are hard to find. Some branches of the behavioural sciences, like research on managerial effectiveness, and leadership style (Campbell et al., 1970) seek to develop models in the OR style which identify and measure independent variables, like 'participativeness', and relate them to, and evaluate them against, measures of 'productivity' and 'performance'. Even here, however, interdisciplinary debate seems more potential than actual.

Methodological common interests may be more obvious, but only seem to occur in the parts of the pluralistic behavioural science scene that overlap the home territory of OR theory and practice. Overlaps occur at the levels of research design, and of analytical technique. Some behavioural scientists and operational researchers see themselves as having similar problems in investigating the causal relationship between variables in a many variable situation which lends itself more easily to study in the field rather than in a controlled laboratory setting. Issues like those of experimental design (Campbell and Stanley, 1964) may represent common ground. Equally, at the level of analytical technique, some behavioural scientists, particularly psychometricians, would find much common ground with some operational researchers' on issues of regression analysis, factor analysis and so on.

Given that OR is in many ways defined by its methodology rather than the substantive phenomena that it investigates, and that it is presented as a practical, problem solving activity, it is clear that decision making itself, as a phenomenon of mutual interest, is in a special category. OR interest in decision making is rooted in the normative and rational, whereas the behavioural science interest is likely to be more descriptive and psychological. However the situation may be similar to the common but different interests of the computer scientist and psychologist in artificial intelligence. Both parties are interested in an operational clarification of the concept of intelligence, and both look to the other for hypotheses about processes and mechanisms underlying it.

Studies like those by Newell and Simon, (1972) Clarkson, (1962) approach decision making from the descriptive, behavioural side, but seek to develop models in the systematic form characteristic of the normative. With the emergence of 'expert systems' there appears to be a genuine area of common interest, and a practical incentive to explore it, between the computer scientist, operational researcher and psychologist.

It is possible that areas such as this will allow a renaissance of multi-disciplinary OR.

A number of strands of thinking in the behavioural sciences, at a more theoretical level, are converging on the issue of human agency - the sense in which people can create new meanings, make free choices, initiate actions, rather than act out parts in a deterministic story (Giddens, 1976). This suggests that the traditional OR view, that if there are regularities in aggregate events then underlying behavioural processes can be avoided, could usefully be reconsidered.

In summary, therefore, it seems that:

(1)   The multidisciplinary base of OR never had a high behavioural science content.
(2)   What there was has eroded away as OR has become a discipline in its own right.
(2)   Areas of common interest are:
       (i) methodological (ii) in decision making.
(4)   The emergence of 'expert systems' may provide the practical incentive and pressure to address behavioural issues directly.
(5)   The evolution of thought in the behavioural sciences suggests that this could be a good time to look at the question of individual 'free' choice and regularities in aggregate behaviour.

BEHAVIOURAL SCIENCE VIEWS OF OPERATIONAL RESEARCH AS A PHENOMENON

Behavioural science views of OR can be conveniently organised under two headings: views of OR as an institution and a social phenomenon, and views on operational researchers, as people, and the way they are organised as a profession.

Considering OR in a historical context, it can be asked why OR sprang up and grew when it did, and where it did, and what its significance is in social history. It seems possible to discern an era of 'scientific rationality' in the two or three decades following World War II, in the West at least, corresponding to a decline in religious and spiritual values, an increasing influence of formal organisation on everyday life, and a corresponding decrease in the influence of family and community, all supported by the belief that science would produce solutions to all human problems. The 'crisis' in OR (Checkland, 1983) can be interpreted as the beginning of the end of that particular era, and the beginning of the current one, of which the pattern is not yet clear, but which seems to contain a recognition that matters are decided by political process, reference to ideology and dogma, and various forms of fundamentalism, either religious or secular (Peters and Waterman, 1984), and politically oriented 'think tanks' displace detached scientific advisers as the shapers of policy. The interesting question is, how is OR reacting to this new era?

Looking at OR from a behavioural science perspective in less global terms raises a whole series of issues about the implicit and explicit assumptions that OR makes about the nature of organisations.

It seems clear that the OR approach fits best with functionalist/systems views about organisations, as opposed to interactionist/humanist ones (Burrell and Morgan, 1979, Silverman, 1970). Thus organisations are seen as 'designed mechanisms' with unitary objectives and structures independent of people, rather than evolving social structures emerging from human interaction, satisfying multiple needs and objectives for multiple stakeholders. Most OR studies involve the notion of a global, collective objective function which can be optimised for maximum payoff by the right arrangement of the sub-systems causing this payoff. Thus 'payoff' is seen as the property of the abstract global entity known as the 'organisation', rather than something that accumulates to a multiplicity of people in a multiplicity of forms.

It can be argued that functionalist/systems views of the nature of organisations are self-fulfilling, in the sense that organisations behave in this manner if enough of the people in them see them this way. OR practice can be seen to support this process, and add to the centralisation of power in as much as OR projects are sponsored by, and serve the ends of, those most powerful in defining objectives, functions and structures.

At a more micro level, OR can be seen to favour objective material criteria rather than possibly more meaningful subjective ones, and thus have an effect on organisational life. For example, OR studies of hospitals are more likely to address themselves to bed-occupancy rate, cure-rate, waiting list time, than to attempt to assess human suffering and wellbeing.

From a behavioural science point of view, therefore, OR, considered across the spectrum from macro to micro issues concerning its practice, represents a value position, has its own ideology, reinforces certain material, structural and power realities in organisations, and thus has a moral and ethical content.

Finally, behavioural science can offer interpretations of phenomena to do with operational researchers, and the ways they are organised. The personalities, careers, professionalisation and education of operational researchers, the way in which the profession is organised, and the way in which OR consultancies and departments are organised within themselves and within work organisations are examples of some of the issues that could be addressed.

The extent to which OR as an occupation attracts and selects people with certain personalities, the extent to which OR provides lifelong careers, and the 'career anchors' (Schein, 1978) of OR practitioners could be investigated. The extent to which OR is a profession, and exhibits the phenomena of professionalisation: claims to protect the public, maintenance of ethical standards, regulation (restriction?) of entry to the profession, exploitation of monopoly power, operation of a labour market, and the fixing of fees, could also be explored.

The organisation of OR consultancies and the extent to which they are organised as commercial companies or on the model of professional partnerships, must be of interest, as is the issue of how OR departments are organised and located within work organisations.

Finally, the education and training of operational researchers suggests itself as a topic of some interest. The question of a gap between theoretical and practical OR is pertinent here, as are questions of the role of education in licencing, formally or informally, practitioners as well as

developing appropriate skills. Finally, the education of operational researchers could be considered from the point of view of its role in creating
and maintaining the value systems and ideologies implicit and explicit in OR
practice.

SUMMARY AND CONCLUSIONS

The main aim of this paper has been to map some of the ways in which
behavioural science and operational research can be thought about in relationship to each other, rather than to draw any strong specific conclusions.

It has been argued, and hopefully demonstrated, that behavioural science
can be thought of across a broad spectrum from 'technical servant' through
'intellectual partner' to 'radical critic', in relation to OR theory and
practice.

The survey has suggested that the existing linkages at all three levels
are weak, and such linkages as were promised by the early interdisciplinary
nature of OR have eroded as OR has developed as a discipline in its own right.

It is suggested that the middle level, of 'intellectual partner', is the
best ground for fruitful debate, and that within this, methodology issues
arising from human free will and agency, and the study of human decision
making in relation to expert systems, possibly constitute the most fruitful
areas of discussion. Exploration of these issues is likely to establish
the ground from which to reach out to the other two levels.

REFERENCES

Ackoff R.L., 1979, The Future of operational research is past, J.Opl.Res.
        Soc., 30:93.
Beer S., 1966, "Decision and Control", Wiley, London.
Berger P.L. and Luckman T., 1967, "The Social Construction of Reality",
        Penguin, Harmondsworth.
Blackett P.M.S., 1962, "Studies of War: Nuclear and Conventional",
        Oliver and Boyd, Edinburgh.
Burrell G. and Morgan G., 1979, "Sociological Paradigms and Organisational
        Analysis", Heinemann, London.
Campbell D.T. and Stanley J.C., 1964, Experimental and quasi-experimental
        design for research in training, in "Handbook of Educational
        Research", Rand McNally, Chicago.
Campbell J.P., Dunnette M.D., Lawler E.E. and Weik K.E., (1970),"Managerial
        Behaviour Performance and Effectiveness", McGraw-Hill, New York.
Checkland P., 1983, OR and the systems movement: mapping and conflicts,
        J.Opl.Res.Soc., 34:661.
Churchman C.W., Ackoff R.L. and Arnoff E.L., 1957, "Introduction to
        Operations Research", Wiley, New York.
Churchman C.W. and Schainblatt A.H., 1965, On mutual understanding,
        Mgmt.Sci., 12:B40.
Clarkson G., 1962, "Portfolio Selection: A Simulation of Trust Investment",
        Prentice-Hall, Englewood Cliffs.
Giddens A., 1976, "New Rules of Sociological Method", Hutchinson, London.
Harre R. and Secord P.F., 1972, "The Explanation of Social Behaviour",
        Blackwell, Oxford.
Lippitt G. and Lippitt R., 1978, "The Consulting Process in Action",
        University Associates, San Diego.
Morgan G., 1983, "Beyond Method: Strategies for Social Research", Sage,
        Beverley Hills.

Newell A. and Simon H.A., 1972, "Human Problem Solving", Prentice-Hall, Englewood Cliffs.

Peters T.S. and Waterman R.H., 1984, "In Search of Excellence: Lessons from America's Best-run Companies", Harper and Row, New York.

Polding E. and Lockett G., 1982, Attitudes and perceptions relating to implementation and success in Operational Research, J.Opl.Res.Soc., 33:733.

Reason P. and Rowan J., 1981, "Human Enquiry: A Sourcebook of New Paradigm Research", Wiley, London.

Schein E.H., 1978, "Career Dynamics", Addison-Wesley, Reading, Mass.

Shilton D.C., 1982, Modelling the demand for high speed train services, J.Opl.Res.Soc., 33:713.

Silverman D., 1970, "The Theory of Organisations", Heinemann, London.

ACKNOWLEDGEMENT

This paper first appeared in "Further Developments in Operational Research", Rand G.K. and Eglese R.W. (eds.), 1975, Pergamon Press. The editors are grateful to the Operational Research Society for their permission to include it in this volume.

AN OPERATIONAL RESEARCHER LOOKS AT THE SOCIAL SCIENCES

A.D.J. Flowerdew

Faculty of Social Sciences
University of Kent at Canterbury
Canterbury, U.K.

INTRODUCTION

For eighteen years I have been a participant observer of the social
science scene as projected onto the University dimension in Britain. For
seven of these I was a covert observer, wearing the disguise of an economist
and therefore a genuine social scientist. After that I "came out" in my
true colours as a management scientist, otherwise an operational researcher,
but have still been permitted to retain the status of a social scientist as
a member of a Faculty of Social Science. Before being claimed by Academia,
however, I used to work in industry and for local and central government,
as an operational researcher, then a planner (another kind of social scientist), then an economist. Almost from the start of my first OR project (on
systematic activity sampling), I was having to consider ideas, theories and
data relating to the social sciences, but I don't think I was aware of this
before Cambridge 1964, the occasion on which we are now looking back. Not
that I attended this conference, but my boss at Richard Thomas & Baldwins.
Steve Cook, and three of his lieutenants, Mike Simpson, John Banbury and
John Lawrence, were heavily involved in its planning, and I slowly realized
that anything which was taking up so much time and intellectual energy had
to be important. It's not hard to see why. Almost all of the systems with
which operational researchers are concerned are social systems, involving
some or many human decision-makers and actors. If OR claims, as it does,
to represent the application of science, then the science to be applied is
likely to include social science. From then on, I have seen OR as almost
inseparable from the social sciences, sometimes some of them, and quite often
all of them. I wonder how many OR practitioners see OR like this, and how
many social scientists?

In the previous paragraph, I have exposed some of my credentials. However this paper is not intended as an extended curriculum vitae but as a
contribution to two debates that I suppose should take place at this conference. The first debate should, I submit, be concerned with what has happened
to the relationship between OR and the Social Sciences since 1964. Has OR
proved itself as an important component of the Social Sciences, or is it still
one of a number of candidate, non-voting members? Have the Social Sciences,
or some of them, learnt from OR, have they got more to learn, and if so,
what's stopping them from doing so? Mutatis mutandis, why haven't we learnt
more? Do OR practitioners all BELIEVE in the Social Sciences? As Sciences?
(Clap hands if you do.) Or, like the Prime Minister, do we believe that

studies is a more appropriate designation than Sciences, with the possible
exception of economics? (When the Social Science Research Council had its
name changed there was a proposal that the new title should be Social Studies
Research Council; the loss of scientific status was hotly contested by
economists which led to the compromise of Economic and Social Research Council). Having established that, where do we go next? Does OR still have the
missionary role that Stafford Beer and Russ Ackoff used to emphasize? Does
it still need a low profile to survive? (When OR entered the civilian part
of Government its earliest backers were insistent that it should be concerned
largely with management issues as opposed to policy issues with which the
Economist and Statistician grades concerned themselves. Like scientists,
the OR boffins were to be on tap and not on top.) Has OR done its job of
fertilising the Social Sciences so that they can produce decision-making
aids like one-armed economists or sociologists, and can it now safely be
eaten by its much larger mates? Where will we go and (not the same question
except to Marxist Social Scientists) where should we go?

These questions are difficult to answer and I would not be even attempting an answer if I didn't feel among friends. They (the questions, not the
friends) could do with some research, not very easy to do and likely to come
up with a range of answers depending on the preconceptions of who's doing
the research (and what does that say about how scientific the Social Sciences
are?) I hope the debate can be interesting enough to enliven the Conference
even without the evidence that is lacking. But I mentioned two debates and
it is the second of these which I am marginally better qualified to open. I
have even done a little (very simple) research.

The second debate relates to OR education and the Social Sciences. OR
is now taught in almost every conceivable educational location - in schools
(decision mathematics) - in low level in service management training such as
the Certificate in Management Studies - in Faculties of Engineering, of
Social Sciences, of Mathematics - in Departments of Statistics, of Economics,
of Business Studies - in Political Science, in Accounting, in Electronics.
It is taught in joint degree programmes with Physics, German, French, Social
Anthropology, Chemistry, Economic History and Sociology (and that is just at
the University of Kent). Probably somewhere there is a student studying
Divinity and Operational Research, Classics and Operational Research, Astron-
omy and Management Science. Yet the number of Operational Researchers who
are members of the UK Society (a low cost, high benefit proposition for
practitioners of the noble art) is (OR newsletter October 1988) a paltry
2971, which is only just over twice as many as there were 25 years ago!
(1327) Where are they all going? What kind of people take these courses?
Will the nature of the courses affect the work that is done? For instance,
it may be easier to find people to teach the mathematics of OR rather than
the methodology of OR. Will this result in OR ultimately becoming more of
a mathematical subject? Conversely, will the juxtaposition of OR in academic
courses with other subjects, especially perhaps Social Science subjects like
Economics, Social Psychology, Politics, Accounting, Law and Sociology, assist
in the production of broadly educated operational researchers who understand
in a way that my generation of operational researchers could not, the context of their work?

So those are the two debates I want to initiate, or contribute to if
they are already under way. The reader may not be aware that I have only
3000 words to write in this paper, and am writing to a deadline of October
15. However I have on occasions in my career found the truth of the old
McKinsey maxim: there are no problems, only opportunities. So, following
the Opportunist Theory of Management, I am proposing to write a 3000 word
paper now to introduce the questions described above, and, if possible, to
present a companion paper at Cambridge itself, which I hope will have benefitted from some feedback from the present draft as well as the opportunity

for a small additional amount of research. Having taken nearly 1000 words to get this far, I am left with about 1000 each to open up the two debates; so here goes.

OR AND THE SOCIAL SCIENCES

The last twenty-five years have seen some dramatic and some unobtrusive changes in the nature both of OR and of the Social Sciences. In the field of OR I have been most impressed at the applications end by the influence of computers and especially microcomputers. Decision Support Systems have provided an important opportunity for OR workers to show what they can do without threatening management; so to a lesser extent have multi-criteria methods and Expert Systems. Because user-friendly micro-computers can encourage management to use systems themselves, they begin to understand the OR models more easily and to contribute to their design. There needs to be less emphasis on optimisation, and evaluation of changes in practice can be done without going through a little understood accounting process.

At the theory end, and especially in recent years, it has been interesting to see the ever increasing emphasis on soft OR. Though this is not a field in which I have worked much myself, it is clear that there have been some major influences from the social sciences, such as Cognitive Mapping theories from Social (perhaps Individual) Psychology. It might be arguable that soft OR provides a bridge between Psychology and Econometrics or Social Statistics, in that the starting point for model-building in these subjects will often be an arrow diagram representing hypothesised causal relationships between variables. The importance of soft OR applications however are sometimes hard for the outsider to grasp, especially when papers are written which extol the virtues of felt-tip pens and flip charts as problem-solving methods (Huxham et al., 1988).

Two other areas which were of considerable influence in OR circles during the 1970s, Global Modelling and Cost/Benefit Analysis, both considerably influenced by economics, seem to attract rather little OR interest nowadays (and not just OR interest - for while both Global Modelling and Cost/Benefit methods are used today in serious forecasting and decision-making applications, in neither case are they used as much as in their heyday in the early 1970s). Yet I believe that in both cases it was the input from OR which made both of these techniques usable and perhaps because of this helped them to be perceived as a threat by busy or non-quantitative decision-makers. The case of Cost/Benefit Analysis has been extensively discussed, so I will refer here mainly to Global Modelling. As long as the Meadows' apocalyptic model could be seen to rest on various highly questionable assumptions, their work could be seen as a salutary warning but need not be taken too seriously when determining short or medium-term policy. However once these models had been developed to provide predictions based on assumptions that could not be rejected out of hand, the conclusions became less dramatic but less easy to reject on technical grounds; it became easier to find fault with the technique itself.

Organisational links between OR and social science research have had a significant influence over the period. The Institute for Operational Research had just been set up at the Tavistock Institute by 1964, and this proved a valuable channel for ideas from Social Psychology, although perhaps the IOR's best known innovation, AIDA (Analysis of Inter-related Decision Areas) has closer links with Political Science than with Social Psychology. Another organisation, not British but important enough to mention here, which linked OR with other Social - and Natural - Sciences in a fruitful way, was the International Institute for Applied Systems Analysis at Vienna, which provided a fillip for Multi-Criteria theory, reliability analysis and much else, as well as providing a much-needed link between East and West research. OR's involvement in this enterprise did not, however, prove so obviously successful as to persuade Lord Rothschild and others involved in setting up the Government's Think Tank in the 1970s that OR should be included (nor, in the early Thatcher years, was OR included in the Rayner exercise on Government efficiency).

Some comment on intellectual fashion changes in the Social Sciences and their relationship to OR over the period may also provoke some discussion. The most dramatic of these has probably been the death of Keynesianism and the rebirth of the older ideas of Monetarism. Strictly, the argument between Keynesians and Monetarists could be summarised as being whether a particular parameter occurring in an equation in a rather simple macroeconomic model of the economy should have the value 0 or 1, when any OR worker could immediately see that the value is likely to lie between them. In practice we have seen these theories of political economy being adapted to encompass families of vaguely related theories: for instance about public ownership, fiscal policy, industrial relations and even management style. Within Government, the replacement of Keynesian by Monetarist ideas does not appear to have been particularly inimical to OR (nor the other way round). Management efficiency drives can result in more use of OR methods, although sometimes management policies such as accountability and cash limits may impose suboptimisation.

A second occurrence, less important now because it has come and largely gone, possibly contributing a little to the death of Keynesianism before it left, was the (anti?-)intellectual idea sometimes referred to as ethno-methodology, but most famous or notorious for its claim that within the social sciences all research was inevitably subjective. Of course this claim has some substance and all social scientists have to be on their guard to avoid unconscious bias; but traditionally the social scientist would have claimed as an important part of their methodology the avoidance of bias, while the ethno-methodologists seemed to glory in it. Unless it is in some of the ideas in Soft OR, I think the ethno-methodological episode has not really affected OR very much one way or the other. We still claim to use scientific method with about the same mixture of justification and wishful-thinking. Perhaps one helpful new concept in model construction may be plausibility: some models such as long-term forecasting models can hardly be justified empirically, but the plausibility of their results can be tested; this concept has recently been applied to some of my own work in Land-Use/Transport Modelling. But it seems unlikely that the ethno-methodologists would be much impressed by such tests, since all that has been shown is that the preconceptions of the testers have been satisfied.

There is not room for more than one other example, but I think an interesting one. There have been three major areas of public policy in which substantial use has been made of large-scale computer models: economic policy, transport planning and global modelling. I have already referred to the role of OR in global modelling. The role of OR in transport planning has been significant and techniques devised or improved by OR researchers are widely taught and used. By contrast, the development of economic models has been largely carried out by economists without contributions by OR people - although Professor Jim Ball at the London Business School was interested in OR, his successors working on the LBS model and those concerned with other models did not apparently find any insights from OR helpful in their work. Nevertheless, these models appear to have found themselves a permanent niche in decision-making. How might their development have been altered with a greater OR involvement?

OR, THE SOCIAL SCIENCES AND EDUCATION

The second theme of this paper is intended to be the various ways in which OR education has developed in Britain and how this may have been affected by Social Science education or vice versa. This theme although more restricted in scope and hence perhaps more easily researchable, nevertheless presents me with similar difficulties in trying to move away from personal impressions to testable, let alone valid, generalisations. However let me start by looking at OR teaching. In 1964 very few University courses in OR/MS existed, and they were all, I think, 1-year postgraduate courses. Since then, the number of such courses has increased, but they cater for a rather small pool of British graduate students. By contrast, the undergraduate courses which have sprung up since 1964, including those at my University, attract large numbers of home students. Of course this is not surprising given the ungenerous treatment of graduate students in comparison with the relatively generous treatment given to undergraduates, but it has important implications for the impact of Operational Research on management in Britain.

Also the period since 1964 has seen the growth, now accelerating, of Business School graduate teaching, and of many levels of management teaching, such as the Diploma in Management Studies (DMS). Almost all such degree programmes include some element of Operational Research teaching. Won't the management of the 21st century be far better informed of the value of OR compared with the managers that we have known and (in some cases) loved?

When I changed from teaching economics to teaching Management Science, just over 10 years ago, I was struck by an article which stated that very few of the intake into general management positions in top companies or the Civil Service came from management or business courses. These organisations preferred to recruit from English or History, French or Mathematics; while the graduates of management programmes went into specialist service functions like Work Study, Production Control or Personnel. I wondered why this should be, and also felt that it could not be a good idea. (Having been forced to study Classics at school had left me with no illusions about the concept of some subjects having an educational value transcending their applicability while others did not.) I believed that the image of management courses as being boring, materialistic and only of vocational value, was wrong, and that within the study of management were issues as intellectually challenging as any to be found in the University curriculum. But of course as long as the image persisted both in schools and in recruiting companies, the best students would not come. The image was clear to read, e.g. on the UCCA application forms: "Pete Smith is a pleasant young man, not academic but a stalwart of the second 15. I am sure he will make a success of Business Studies." I am afraid Pete was not accepted at Kent.

But by the end of the 1970s things seemed to be changing. The standard of student entry became much higher and we resolved at Kent, no doubt in common with many other Universities to make the most of this opportunity so that those who had chosen Management Science for mainly economic, i.e. vocational reasons, would find their choice equally rewarding from an intellectual viewpoint. Having convinced the students of this is not enough, since recruiting firms and the schools that advise the applicants also need convincing, but it is a step in the right direciton.

Most undergraduate degrees are likely to allow students to study a range of topics alongside their main choice and in subjects like Operational Research this is undoubtedly a plus point. Those OR courses which are within a Social Science faculty can benefit from the ability to study many topics which relate closely to the environment within which they will be working, from Accounting and Business Law to Industrial Relations and Social Psychology, and can in return feed in some OR ideas to degree programmes in these subjects. Desirable as this undoubtedly is, it does not always turn out so well. An accounting course for accounting students will contain material which is also taught to OR students but in a more mathematical way as well as material which it is useful for OR students to know about accounting, and some subjects which, essential for accounting students, are not particularly relevant to OR. Either the OR students must take this course; or they must have another course specially designed for them; or some kind of modular approach must be adopted. The first course of action is unsatisfactory and will discourage OR students from taking accounting if it is optional; the second course is expensive in resources, which in this area are exceedingly scarce; and the third is not easy to arrange.

However these are minor worries. The link between OR and social science in teaching and research seems to be a strong one, and not only at Kent. 20 members of the Committee of Professors in Operational Research were kind enough to answer a small questionnaire I sent them (incidentally, in spite of the name of the Committee, it includes 9 Professors of Management Science and only 3 Professors of Operational Research, although 3 more have OR in their title). 17 out of 20 had members of their department doing joint research with social scientists: in 5 out of 20 all members of the department were doing such work. A more detailed report on the results of this questionnaire is to be prepared.

CONCLUSION

I should like to finish this paper with what I believe to be some of the major achievements of OR in relation to the Social Sciences over this period. It has seen the award of Nobel Prizes to some who have been active in OR, such as Herbert Simon and Sir Richard Stone. It has seen the concept of model, so strongly emphasized in the early accounts of OR methodology, forming a central part of the theory of all the social sciences, whether quantitative or not. It has seen the adaptation of OR techniques, unheard of by practitioners of social science in 1964, to form part of the curriculum for social sciences from accounting (linear programming, regression-based cost models) to political science (hypergames). And it has seen many of the social sciences, perhaps most importantly of all economics, reorient their research agenda towards problem solving and at the same time recognise that an axiomatic approach is only as good as its axioms. That seems to sound an optimisitic note: but unfortunately this progress within the social sciences has hardly been matched by a determination by Government, at least in Britain, to make good use of their insights and achievements. Solving that problem can't wait another 25 years.

REFERENCES

Huxham C., Eden C., Cropper S. and Bryant J., 1988, Facilitating Facilitators - a story about group decision-making, OR Insight, 1:13.

ON FLEXIBILITY AND FREEDOM

R.L.Ackoff

Interact
Philadelphia, U.S.A.

The conference on "Operational Research and the Social Sciences" held at Cambridge University twenty-five years ago was full of promise. That promise still comes through a rereading of its proceedings. Part of that promise was fulfilled for me through an extended collaboration with Eric Trist and Fred Emery. Their work stimulated and attracted many, including me, to studies of quality of working life, and organizational learning, development, and culture. However, it seems to me that their work and its offsprings had relatively little effect on OR and the systems sciences, and even less effect on the social sciences. Fortunately, it has had a great deal of effect on organizations and their work forces.

After the Cambridge conference several of my colleagues and I created a Ph.D. program and research center focusing on what we called the social systems sciences. The program and center were dedicated to improving the performance of systems in which people play the most important role. We hoped in vain for the involvement of behavioural and social scientists housed in our university. Fewer students from these areas came to our courses and research projects than from any other part of the university.

Those applied social scientists who engage in studies of quality-of-working life, organizational development, and human-resource planning and development have received no more support from traditional social scientists than those of us in applied systems science.

The university-based group of which I was a part until recently brought to light a number of issues that should have excited social scientists; and its many client-oriented projects should have been seen by them as providing unique opportunities to experiment with and test a variety of important social ideas.

I mention six of what I consider to be such exciting issues here. This sample is selected with a malice of forethought that will become apparent later.

Most of those organizations that must compete effectively in order to survive appear to have to reorganize totally or substantially with great frequency. Such reorganizations are adaptations to changing environments or internal conditions, but they are costly, time-consuming, and often demoralizing. Are they necessary? Is it possible to design an organization that can adapt without reorganizing?

Our national economies are organized as approximations to the free-market system. Autonomous business enterprises are supposed to be regulated only as much as is necessary to make their competition beneficial to consumers. Communist economies are seen as antithetical to the free-market system. They are centrally planned, controlled, and noncompetitive. Despite our aversion to such economies at the macro (national) level, we embrace them at the micro (enterprise) level. Most of our business enterprises have an internal economy that is like that of the Soviet Union, not the United States. What would a business enterprise organized with an internal free market look like and how well would it operate?

In that part of the western world that we call "free", nations operate democratically. By democracy I mean a system of government in which there is no ultimate authority and in which each person who has control over others is subject to their collective control. Most western countries are committed in both principle and practice to continuous pursuit of the democratic ideal. Despite this commitment at the national level, most of the public and private organizations in these countries whose purpose it is to provide goods or services are organized autocratically. Very few business enterprises or government agencies are operated democratically. Can they be, and what would be the consequences of doing so?

Is it possible to have a democracy that operates by consensus rather than by majority or plurality rule? Would the requirement for consensus preclude the ability to make decisions and thereby paralyze those organizations that require it? Are there ways of stimulating consensus? Would they prevent tyrannies of the majority?

Democracy reduces the power that officials have over others. But this does not necessarily reduce their ability to implement their decisions, their power to. How are power over and power to related, and how is this relationship affected by the educational level of those who are managed or governed?

In democratic governments there is usually a separation of legislative and executive powers. In the United States, for example, Congress has legislative but no executive powers. However, in virtually every public and private organization both types of power reside in managers. Why isn't their separation desirable in such organizations? If it were, how might it be accomplished?

These six issues did not arise out of academic reflections or discussions but out of projects directed at improving the effectiveness of real social systems in real situations. In response to them my colleagues and I have designed and implemented organizations that (1) virtually never have to reorganize, (2) have an internal free-market economy, (3) are completely democratic, (4) operate by consensus, (5) maximize power to but minimize power over, and (6) separate legislative and executive functions. Despite all this, we believe that these organizations are more effective, more adaptive, faster learning, and provide higher qualities of work life than do most organizations.

Now what amazes me is that even though all of our work has been published in easily accessible places it has attracted little attention of social scientists. I could forgive them for taking issue with it, but not for ignoring it. I cannot believe they are working on more important or challenging problems.

It is with the hope that I might attract some attention to these issues that I bring together here the principal ways by which my colleagues and I have tried to address them. The price I must pay for trying to do so in a paper-length presentation is that I must sacrifice depth for breadth. I will not be able to discuss the many variations of these designs that are in use. No two applications have ever been exactly the same because they are, and should be, adapted to the unique properties of the organizations and environments involved.

ELIMINATING THE NEED FOR REORGANIZATION

To organize is to divide labour and to coordinate that labour in order to produce a desired output. There is no need to organize where there is no need to divide labour. In a typical organization chart the horizontal dimension shows how labour (responsibility) is divided. The vertical dimension represents the flow of authority, how labour is coordinated.

No matter what the nature of an organization, its labour can be divided in only three different ways: (1) by functions (inputs) - for example, purchasing, finance, legal, and manufacturing; (2) by products or services (outputs) - for example, Cadillac, Buick, Oldsmobile, Pontiac, and Chevrolet; and (3) by markets (user characteristics) - for example, by their location as in North and South American, European, Asian, and African divisions, or by such other characteristics as end-users, retailers, wholesalers, and mail-order houses.

Organizations are normally designed from the top down. Their design process is usually started by drawing a box for the chief executive. In designing the first level down at which labour must be divided, one of or more of the three criteria are used. At each successively lower level, labour must again be divided, hence one or more criteria are used at each level. The resulting organizations always reflect the relative importance assigned to each criterion. The higher the level at which a criterion is used, the more important it is. For example, single-product companies are usually organized functionally at the top, multiproduct companies by products, and multinationals by country or region.

A change in or of an organization's environment or its role in that often requires changing the order of the criteria. For example, as a one-product company enlarges its product line, it may have to convert to a product-oriented structure. As a company expands its markets and operating locations it may have to convert to a market-oriented organization.

Almost all costly and disruptive reorganization involves reordering of the criteria for dividing labour, but this is not widely recognized. Once recognized, it can be avoided by creating a multidimensional (MD) organization, one in which there are functional, product- or service-, and market-defined units at all or most levels of the organization. (See Figure 1.)

Where the number of units at any level is too large to be handled by one manager, intermediate coordinating managers can be used, for example, see Figure 2. Every unit of an MD organization can itself be designed in three dimensions. Moreover, units at any level of an organization can be organized in this way whether or not higher or lower level units have been.

When every part of an organization is organized multidimensionally, the organization can be said to be a fractal (Gleik, 1987). Fractals are entities which appear the same whatever the scale at which they are examined. In view of the fact that fractals are currently considered to make possible major advances in our understanding of nature, one would think that fractal organizations would attract the attention of social scientists. Not so.

Many variations of an MD organization are possible. I show only one here (Figure 3). It is one in which two MD units have a common set of parts along any one of their three dimensions.

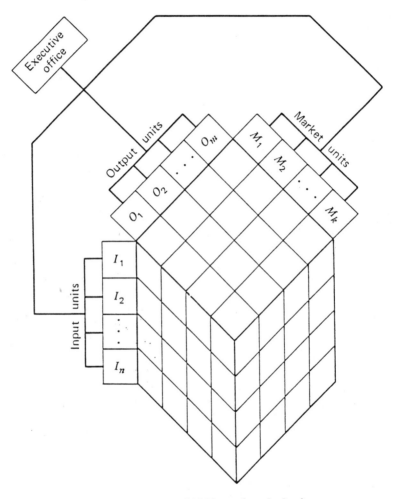

Figure 1. A multidimensional design

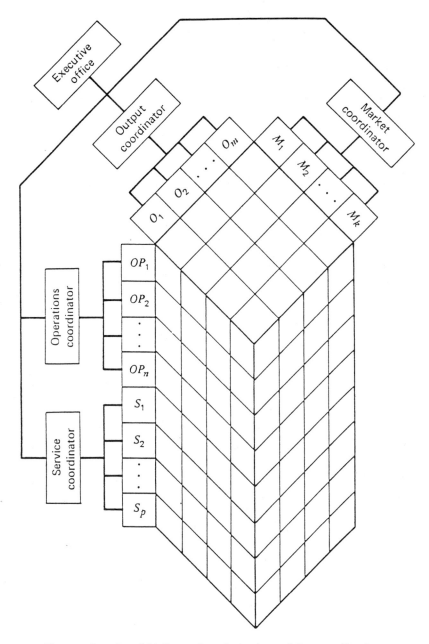

Figure 2. A multidimensional design with coordinators

23

In MD organizations every manager is a general manager no matter how specialized the unit managed. This simplifies succession planning and management development. Another advantage lies in the fact that units can easily be added or subtracted from such an organization without reorganizing it. It is equally easy to combine or divide units.

When all three criteria for dividing labour are used at a particular level of an MD organization, that level does not require reorganizing when the relative importance of these criteria change. All that is required is a reallocation of resources to the different types of unit.

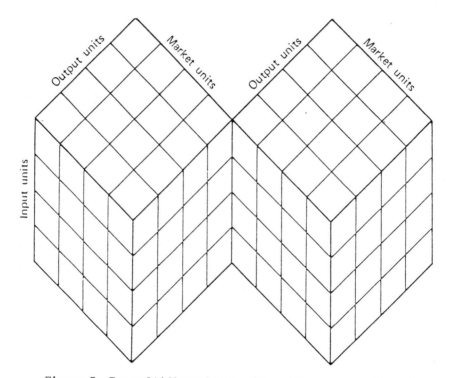

Figure 3. Two multidimensional units with a common dimension

Although advantages can be obtained from organizing multidimensionally without any other changes, even greater advantages can be obtained when such an organization is provided with an internal market economy.

INTERNAL MARKET ECONOMY

To create a market economy within the firm the following characteristics are designed into the multidimensional organization.

1. Each unit in the organization, including the executive office, operates as a profit centre.

2. Each unit also operates as an investment bank and government for each of its immediately subordinate units: it taxes their profits and receives payment (in the form of interest or dividends) for the capital it provides them. As a government each unit can enact laws that preclude specified activities of subordinate units. For example, it may preclude outsourcing of products that involve a "secret" formula or production process.

3. Each unit is free "within the law" either to supply itself with whatever goods and services it wants or to buy them internally or externally at a price acceptable to it. (This freedom is subject to an override which is discussed in paragraph 6 below).

4. Each unit is also free "within the law" to sell its services or products internally or externally to whomever it wants at a price it establishes (subject to the same override).

5. Each unit can retain up to a specified amount of its profit for use at its own discretion. Any amount above this must be turned over to the next higher level of management in return the unit is paid interest on this amount by the receiving unit.

6. Any manager can override a subordinate unit's decision to make, buy, or sell, but that manager must cover any additional cost or loss of profit to the unit affected.

The fact that every unit operates as a profit center does not imply that its principal objective is to maximize annual profit or return on investment. What is implied is that profitability be taken into account in evaluating each unit's performance, but this need not be annual profitability. Nor does this requirement imply that every unit must be profitable. An organization may maintain an unprofitable unit for nonfinancial reasons, for example, for the prestige it brings. Nevertheless, repeatedly unprofitable units should be reviewed periodically for possible discontinuation or divestment.

Note that this type of internal economy eliminates subsidized monopolies operating within an organization. All internal units that supply goods or services, and for which there are alternative external sources of supply, must compete even for internal business. This makes them more responsive and efficient suppliers.

An internal market economy is as applicable to a not-for-profit or governmental organization as it is to one for-profit. Even though an organization as a whole is subsidized by an external source, it can distribute funds internally solely for services rendered or products supplied. For example, students can be given freedom to choose their schools and the income of publicly supported schools can be restricted to a payment for each student enrolled. These are the essential parts of the voucher system originally proposed by Jencks (1970)and later varied by Ackoff (1974, Ch.8). It endows a publicly supported school system with an internal market economy.

Now we turn to a consideration of how organizations of any type, but especially MD organizations with internal market economies, can democratize and simultaneously increase their effectiveness.

ORGANIZATIONAL DEMOCRACY: THE CIRCULAR ORGANIZATION

The central idea involved in this design for democratization is that every person in a position of authority - each manager and supervisor - is provided with a Board. First we consider the composition of these Boards, then their responsibilities, and finally how they operate. (See Figure 4).

Composition of the Boards

The suggested minimal membership of Boards at every level of the organization except the top is

(1) the manager whose Board it is,
(2) his/her immediate superior, and
(3) his/her immediate subordinates.

Therefore, in the Board of any manager who has more than two subordinates, the subordinates may constitute a majority. In the case of the Chief Executive Officer, since his/her Board cannot contain a superior, it should contain representatives of each of the organization's major external stake-holder groups. However, the representatives of no stakeholder group should constitute a majority of this Executive Board.

Some of the Executive Boards have also included representatives of employees at all or a sample of subordinate levels of the organization.

Any Board has the right to add members drawn from within or outside the organization. For example, Boards of functionally defined units - like marketing or finance - commonly invite managers of other functional units at the same level of the organization to participate as either voting or non-voting members. Boards of units that have internal consumers of their products or services frequently invite some of them to participate. In some cases, even external customers have been invited to be part of Boards. In geographically dispersed organizations - for example, in multinational corporations where different units operate in different countries - repres-entatives of various external stakeholder groups have been invited to part-icipate. These have included representatives of relevant communities, and consumer or environmental groups.

Where a manager is supported by a staff, a separate Staff Board is usually established. If a manager has a Chief of Staff, then it will be that Chief's Board with the manager participating. If the manager acts as Chief of Staff, then the Staff Board does not generally include that manager's immediate superior.

In a unionized organization, union officials of the appropriate level are usually invited to participate on Boards - shop stewards at lower levels, department chairmen at a higher level, and union executives at the highest level.

Participation in Boards is generally made compulsory for managers and supervisors, but voluntary for others. There is obviously something wrong with Boards that fail to attract a majority of their manager's non-manager-ial subordinates. In such cases, the performance of the managers whose Boards have this characteristic should be evaluated by their superiors.

Managers other than those at the top and bottom of an organization will interact directly with five levels of management through their participation on Boards: two higher levels and their own level in their immediate super-ior's Boards, and two lower levels in their immediate subordinate's Boards. Such interaction makes it possible to achieve a degree of coordination and

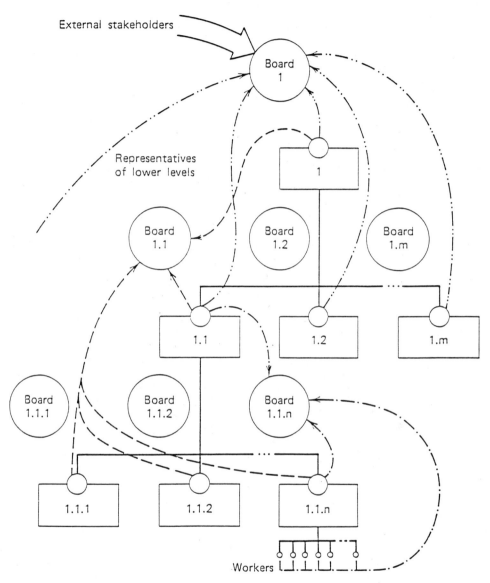

External stakeholders

Representatives
of lower levels

Board
1

Board
1.1

Board
1.2

Board
1.m

Board
1.1.1

Board
1.1.2

Board
1.1.n

1

1.1

1.2

1.m

1.1.1

1.1.2

1.1.n

Workers

Figure 4. A democratic hierarchy: the circular organization

integration of plans, policies, and activities that conventional organizations seldom attain.

Responsibilities of the Boards

Boards normally have the following responsibilities:

(1) planning for the unit whose Board it is,
(2) policy making for the unit whose Board it is,
(3) coordinating the plans and policies of units at the next lower level,
(4) integrating its own plans and policies and those of its immediately lower level units with those made at higher levels, and
(5) making decisions that affect the quality of working life of those on the Board.

In addition, most but not all Boards are given power:

(6) to evaluate the performance of the manager whose Board it is, and remove him/her from his/her position.

Consider these functions in more detail.

Each Board has responsibility for planning and making policy for the unit whose Board it is. A policy is a decision rule, not a decision. For example, "hire only college graduates for managerial positions" is a policy; hiring such a person is a decision. Boards do not make decisions other than those directly affecting the quality of working life of its members; except for these, managers make the decisions, not Boards. In general, Boards are analogous to the legislative branch of government, and managers to the executive branch.

Managers may, of course, consult their Boards about decisions they have to make. However, even if they take the advice given by their Boards, they must take responsibility for the decision made, not their Boards. The use of Boards in a consulting role is common, as is their use to facilitate communication up, down, and across. Because they are used in these ways, the need for other-than-Board meetings is frequently reduced, if not removed.

Each Board is responsible for coordinating and integrating the plans and policies made by units at the level immediately below it. Since the managers at this level participate in the Board and generally constitute a majority of it, coordination is to a large extent self-imposed with participation of two higher levels of management. Integration involves preservation of hierarchy: no Board is permitted to make a plan or policy that is incompatible with a higher level Board's output. However, it can request a revision of a higher level plan or policy through its members who participate in one and/or two higher level Boards. Moreover, because the members of every Board have an extended vertical view of the organization, they are not likely to make plans or policies that are either harmful to lower levels of the organization or incompatible with those made at higher levels.

No Board is permitted to implement a plan or a policy that directly affects another part of the organization over which it has no authority without either agreement of that part or, where such agreement cannot be obtained, approval of the lowest level Board at which the affected units converge going up the hierarchy.

Quality-of-working-life decisions are ones that directly affect the satisfaction employees derive from their work and their perception of its meaningfullness. Responsibility for the amount of work done and its quality remains with the manager. However, it has also been our experience that when employees are given power over their quality of working life, significant improvements in the quantity and quality of their work are usually attained, usually much more than has been obtained by programs explicitly directed at increasing productivity and quality (Ackoff and Deane, 1984).

The most controversial responsibility that Boards can assume is for evaluating and improving the performance of the manager whose Board it is. In these cases managers cannot retain or be appointed to their positions without approval of their Boards, hence, in most cases, a majority of their subordinates. Because this function is so controversial, only about half the corporations that have implemented the circular organization have initially given this responsibility to the Boards. Most have added it after running about a year. However, without this provision complete democracy cannot be attained in the workplace.

Note that even with this provision a group of subordinates cannot fire their boss; they can only remove him/her from his/her position. Only a boss can fire a subordinate. This means that managers cannot hold their positions without the approval of both their bosses and their subordinates. There have been very few cases in which Boards have removed their managers. It has not been necessary for them to do so because they can obtain most of what they want from the manager through constructive criticism. Once each year the immediate subordinates of each manager meet with or without a facilitator to discuss what their boss can do to make it possible for them to manage more effectively. The subordinates are precluded from telling their boss what he/she should not do; they must confine themselves to constructive suggestions covering what their boss might do that would enhance their, not their boss's, performance. When they have organized their suggestions, they meet with and present their suggestions to their manager.

"Receiving managers" are expected to respond to the suggestions in one of three ways. First, they can accept a suggestion. In our experience, they have accepted about 75 per cent of the suggestions. These suggestions usually involve things the managers had not previously thought of and the virtue of which they see immediately. For example, one such group suggested that their boss use his full vacation allowance so they would not feel guilty if they used theirs.   The boss was surprised to learn that his failure to use his vacation allowance was preventing his subordinates from using theirs. He accepted the suggestion and implemented it. Another suggestion involved more flexible budgets, particularly with respect to the purchase of minor equipment and external services. Again the boss agreed and appropriate changes in budget control were made.

Second, receiving managers can reject a suggestion but, if they do so, they are required to give their reasons for doing so. In most cases their rejections involve constraints imposed on them from above, constraints of which their subordinates were unaware. Their subordinates do not have to agree with, but should understand, the reasons for rejection of their suggestion. There have been very few cases in which such explanations did not produce the required understanding.

Finally, the boss can ask for more time to consider a suggestion and commit to a time by which he/she will make it.

These constructively oriented evaluative sessions have generally brought boss and subordinates closer together and bound them into a more collaborative relationship. They reflect the fundamental change in the concept of management that is implicit in the circular organization: the principal responsibility of managers is to create an environment and conditions under which their subordinates can do their jobs as effectively as their capabilities allow. It is not to supervise them. Put another way: the principal responsibility of a manager is to manage over and up, not down; to manage the interactions within their units and of their units with the rest of the organization and its environment, not to manage the actions of their subordinates. If subordinates require supervision beyond an initial break-in period, they should be replaced by persons who do not require it.

Managers who do not have the support of their subordinates have great difficulty in managing effectively. There is nothing a manager can want to have done that his/her subordinates cannot sabotage if they so desire. On the other hand, subordinates who approve of their superiors try to make them look as good as possible.

Modes of Operation

Each Board should prepare its own operating procedures, but most Boards choose to operate in similar ways; for example, they decide by consensus rather than majority rule. The advantage of consensus over divided opinions is apparent. What is not apparent is how consensus can be obtained where there is an initial divergence of opinions. Where consensus cannot be attained easily - and this occurs in less than 25 percent of the cases - the first "fall back" procedure consists of designing a test of the alternatives that all Board members accept as adequate and fair. For example, in one company, agreement could not be reached as to whether plant maintenance in a multiplant operation should be placed under plant managers or a corporate manager of engineerings. A multiplant test was designed which all Board members considered to be a good and fair one. The test was conducted and the findings were accepted and implemented across the system.

Differences of opinion are often based on different beliefs about "facts". The critical fact in the maintenance cases involved the relative effectiveness and efficiency of maintenance personnel under two different organizational arrangements.

The most dramatic case I have experienced was with a small Indian village in a very remote part of Mexico. The community could not reach agreement on whether to use capital punishment for capital crimes. One side argued that such punishment was necessary to reduce capital crimes; the other argued that it had no such effect. Therefore, the difference in attitude was based on a difference of opinion about a question of fact. A test was designed in the form of a retrospective experiment carried out across Mexico. (No one was killed in the test.) All accepted the design. The test showed that capital punishment had not decreased the incidence of capital crimes in Mexico, hence the community did not adopt it.

In some (but few) cases there is either not the time required to design and conduct a responsive test because a choice must be made quickly, or agreement on a test cannot be reached. Every Board must eventually decide how to handle such cases. Most Boards adopt a procedure like the following. When it is apparent that consensus cannot be reached through discussion and a test is not practical, the chairperson goes around the room asking each person to summarize briefly his/her position. Then after the chairperson states what decision he/she will make if consensus is not reached, he/she goes around the room for a restatement of opinions. If on this second go round consensus is reached, even if it is on a choice that differs from that of the chairperson, that consensus decision holds. Note that a lack of agreement on any decision other than that put forth by the chairperson is, in effect, consensus on the chairperson's decision.

Most Boards are chaired by the manager whose Board it is. Agendas for coming meetings are prepared by the chairperson's secretary using inputs from any of the Board members. After the initial "break-in" period most Boards schedule monthly meetings which usually last about three hours. Additional meetings are called when required either by the manager whose Board it is or by any of its other members.

Some Common Concerns about Boards

When one executive who had installed Boards throughout his company was asked when he found time to get his work done, he replied essentially as follows.

The ten Boards in which I take part meet once per month for three to four hours each. Therefore, they require no more than 40 hours per month. If I worked only 40 hours per week - like most managers, I work closer to 60 - this would be no more than 25 percent of my time. According to a number of studies of how managers spend their time, most managers spend only about 20 percent of their time managing. Therefore, the question that should be asked is: What do I spend the other 55 percent of my time doing?

There is a better answer. In my Boards I plan, make policy, coordinate the plans and policies at the level reporting to me, integrate these with those made at my level, and evaluate and guide the performance of my subordinates two levels down. What should I be doing that is more important than this? Moreover, our Boards have eliminated the need for most other meetings, such as those of staff. However hard it may be to believe, I spend less time in meetings now than I did before Boards were introduced, and I get a lot more done.

It takes some special skills to run Board meetings. For this reason most organizations that initiate Boards provide managers at all levels with training in group skills. Such training usually takes only two or three days. In many cases these sessions are made available to union and non-managerial personnel as well.

It usually helps to have a skilled facilitator at Board meetings, at least in the first few meetings. A facilitator can help establish the rules of procedure and keep the meetings flowing smoothly. Once the manager whose Board it is gains confidence in his/her ability to run meetings, he/she can take over. It often helps for the facilitator to remain as an observer for at least the first few meetings run by the manager. This enables the facilitator to make constructive suggestions to the manager in charge. In some companies - for example, Met Life - special training programs have been developed to provide the large number of facilitators required.

To the best of my knowledge only two organizations that have initiated Boards subsequently discontinued them, both for the same reason, a change of their chief executive.

The principal obstruction to the successful implementation of Boards has been a lack of commitment by middle managers who were not brought into the process at its beginning. For this reason, appropriate education and involvement of managers, particularly those in middle ranks is very important. Since there may be a large number of managers and they may be geographically dispersed, video tapes of relevant presentations and discussions have often been used to orient managers. Consultation with an experienced person helps and so do visits to organizations that have used a circular organization.

Where such groups as Quality Circles already exist, they are sometimes retained and run in parallel, but in most cases these groups are incorporated into Boards, thereby reducing the number of meetings required.

The best place to start the use of Boards is wherever the one who asks the question is located. Boards have been initiated at every level of an organization and have subsequently spread throughout. In some organizations, like Anheuser-Busch, Boards were initially established at the top. After the executives involved had experience with it, most started their own Boards. Then they moved down layer by layer.

At Kodak the first Boards were established in a unit whose manager was at the fifth level of the organization. Boards then spread up, down and across. At Alcoa's Tennessee Operations, Boards were simultaneously established at the top and bottom with the participation of the Union (Local 309 of the United Steel Workers) at both levels. Boards subsequently moved up and down the organization until they met at the middle. In some smaller organizations, all personnel have been indoctrinated at the same time, hence Boards were initiated all over the organization at the same time.

In some cases, managers below the top of an organization initiated Boards without the participation of their immediate superiors who begged off claiming to be too busy to be involved. However, as the value of the Boards became apparent, most of these superiors joined their subordinates' Boards.

Participating senior executives are not always able to attend all the Board meetings of their immediate subordinates because of other demands on their time. However, they usually attend critical meetings.

It has been both easier and more difficult to introduce Boards in a unionized company. Where unions have collaborated, they make it easier to involve the unionized workforce. Where they are opposed, they make it difficult but not impossible to introduce Boards. In some cases, unions that were initially opposed to the idea come around to embrace it fully.

Boards are not just another gimmick. Their use involves a radical cultural change. In a circular organization managers are no longer either commanders or supervisors, but are required to be leaders, facilitators, and educators. These are roles to which many managers are not accustomed. It takes time to make the necessary conversions, but the rewards for doing so can be very large. For example, at the time Boards were introduced in Alcoa's Tennessee Operations, in the early 1980s, these Operations were scheduled to be shut down because of their very low productivity and poor quality of product. In less than two years their productivity and product quality improved so much that corporate headquarters reversed its decision and initiated a modernization program. At the end of 1987, the most advanced aluminum sheet rolling mill in the world was opened at these Operations.

Earlier in this decade A&P closed all its supermarkets in the Delaware Valley, the metropolitan area of Philadelphia, because they were unprofitable. Subsequently, in a joint effort of A&Ps management with Locals 27, 56, 1357, 1358, and 1360 of the United Food & Commercial Workers, to which the A&P employees had belonged prior to losing their jobs, a new chain of supermarkets was designed. These involved the use of Boards at all levels. The new chain, called Super Fresh, was initiated mostly in the old facilities. It is now the fastest growing chain in the area and is profitable. This chain is being extended geographically, replacing traditional A&P stores.

Armco's Latin American Division (ALAD) was reorganized around the use of Boards and has experienced significant improvements in performance as well as morale. The same has been true of a number of other organizations including Metropolitan Life, Central Life Assurance, Clark Equipment, several new units of Alcoa, a variety of departments and divisions of Kodak, and Anheuser-Busch.

Democracy and efficiency are not inimicable.

CONCLUSION

In concluding I'd like to share with you a speculation that derives from my experience with the designs presented here. In the past the social sciences have tried desperately to adopt and adapt the methods of the physical sciences, but not with great success. The social sciences have failed to produce the kinds of dramatic progress that the physical sciences have. There has been a great deal of speculating about why this is the case, most of it focusing on the differences of subject matter and, in particular, on the difficulty of conducting controlled experiments on social systems. Unlike the objects of study in the physical sciences, those of the social sciences can "talk back", and react to investigations carried out on them.

These differences are real but their recognition has not led to any methodological breakthroughs with the possible exception of the development of what has been called "action research". However, it seems to me that the power in this concept has not been effectively exploited by the social sciences.

Because of the passiveness of the subjects of study in the physical sciences, physical scientists study them and engineers try to use the knowledge produced by the scientists to design better artifacts. Conversely, engineers find out how to build better artifacts without understanding the reasons for their superiority, leaving it for scientists to find out by research. The point is that in the physical sciences there has been a fruitful division of labour between physical scientists and engineers. My point about this is that the corresponding separation in the social sciences of basic from applied research may be the principal reason for its lack of progress.

I believe that effective social science can be carried out only by designing or redesigning social systems, implementing these designs, evaluating them in use, and adapting them to internal and external changes as well as what is learned about the system from its use. It is this need for continuous adaptation and learning in a rapidly and profoundly changing environment that prevents real social systems from remaining still long enough to be experimented on as physical systems are.

Put another way: I am arguing that in the social sciences the separation of so-called 'pure' and 'applied' science is a self-imposed constraint in the development of the social sciences. They must operate with a methodological paradigm in which design, research, and clinical treatment are synthesized, fused. In the paradigm required, social scientists would simultaneously act as architects, investigators, and clinicians. The few social systems scientists who are currently using this paradigm are having remarkable success in improving the effectiveness of the social systems on which they work. Such success, after all, should be the ultimate measure of success of the social sciences.

REFERENCES

Ackoff R.L., 1974, "Redesigning the Future", Wiley, New York.
Ackoff R.L. and Deane W.B., 1984, The Revitalisation of ALCOA's Tennessee Operations, Nat.Prod.Rev., Summer: 239.

Gleik J., 1987, "Chaos: Making a New Science", Viking Penguin, New York.
Jencks C., 1970, Giving parents money for schooling: education vouchers,
     Phi Delta Kappan, September :49.

OR AND SOCIAL SCIENCE: FUNDAMENTAL THOUGHTS

P.B. Checkland

School of Management
University of Lancaster
Lancaster, U.K.

INTRODUCTION

Operational Research has always been proud of its heritage in natural science. Many of its pioneers were Fellows of the Royal Society and they invented the subject by showing that their kind of scientific thinking could be usefully applied to one type of human situation, namely military operations. Since then most operational researchers have thought of themselves broadly as scientists, and it is noteworthy that the definition of OR which used to appear in every issue of the Journal of the OR Society (now wisely dropped) made three references to science and the method of science in five lines.

This means that the implicit claim made by the very existence of OR as a subject or discipline is that scientific thinking and the method of science, as exemplified in the natural sciences, can be applied not only to the phenomena of Nature but also to at least some problematical human situations. This bold claim is not one to accept lightly, not least because of the ongoing debate within the science of human situations, social science, about whether human and social phenomena are intrinsically the same as or different from those studied by natural science. (It is not completely obvious that a phenomenon such as "voting", for example, is the same kind of thing as, say, "cooling", susceptible to the same kind of investigation.) This debate in fact underlies that witnessed within OR in the last twenty years, namely the debate about the perceived gap between what the OR textbooks say and what the experienced practitioners do, about the so-called "crisis" in the subject, about whether or not the subject is "dead". (Dando and Bennett, 1981; Churchman, 1979; Ackoff, 1979; Machol, 1981; Checkland, 1983.)

The existence of this debate within OR means that it is very worthwhile to explore the scientific claims of OR and to examine in what way it can claim to be "scientific", and to ask what are the implications of those issues for the future practice of OR. This paper will examine these issues. It will examine briefly: the nature of the method of (natural) science; the implications of that for OR; the nature of "social science"; and the implications of that for OR practice.

THE NATURE OF NATURAL SCIENCE

We need to think of natural science not as a body of knowledge but as a mode of investigation. Particular bodies of results, such as the Newtonian picture of the physical universe, are replaced by new results (in this case those of Einsteinian physics) which pass more severe tests. What remains constant throughout these changes, however, is the method of science, the way of investigating the physical world. The method was elaborated in its modern form in the Scientific Revolution of the 17th Century and we may drastically condense it into a set of core principles which have to be followed if the results it produces are to be accepted, in our society and culture, as "scientific".

The principles are familiar to any educated person in Western culture. We know how scientists go about their work.

Firstly they do not try to investigate the whole of the universe: they select. They study separately, through controlled observation, the regularities of, say, magnetism, or heat, or the reactions of amino acids. They are reductionist in designing their experiments, and then reductionist again in two further senses. In explanation they accept the principle of Ockham's Razor (when faced with competing explanations accept the simplest) which makes for rigour in qualitative thought, and they seek, if possible, explanations which eliminate meta-concepts. Thus a natural scientist would be happy if psychology could be reduced to biology and biology to chemistry and chemistry to physics (Longuet - Higgins et al.,1972).

The second great principle of the scientific method is that the results of the reductionist experiments must be repeatable. If you discover the inverse square of magnetism in Basingstoke, then the results are accepted as scientific only if your experiments are repeatable in Basildon and Bombay.

Thirdly, scientific progress depends upon testing hypotheses to destruction, showing how hypotheses can be refuted, as Cardamus refuted the hypothesis that "the magnet feeds on iron" by putting a magnet and some iron filings away in a draw and weighing them periodically, over a period of months.

The extraordinary power of this method of investigation based upon reductionism, repeatability and refutation is precisely that it yields, not opinion, but testable public knowledge (Ziman, 1968). The public knowledge is at least in part socially defined, by the consensus which defines particular investigations as meaningful. And the public knowledge obtained by the scientific method is always provisional, waiting to be displaced by the results of more sophisticated investigations, as, for example, when the modern molecular-motion theory of heat replaces the phlogiston theory, and Einstein's physics replaces Newton's. But as long as repeatable experiments can be repeated and observed, public knowledge,not private opinion, will result. (See Checkland, 1981, for a more elaborate discussion).

The method of science is thus a very powerful way of investigating the natural world. What are the implications of trying to apply it, as OR does, to human situations? It is not immediately apparent that it can be used unchanged to investigate "voting" as well as "cooling".

THE IMPLICATIONS OF SCIENTIFIC THINKING FOR OR

People trained to design and carry out repeatable experiments in which hypotheses and conjectures are subjected to severe tests develop particular ways of thinking about the world in their professional work. They ask such questions as: what would constitute reliable evidence in a given situation, and by what criteria might one be judged? As human beings, of course, scientists are very likely to behave just as erratically and illogically as anyone else, making their contribution to the rich farce of human life in the tribe, but in professional work which establishes public knowledge which others can test, there will be more than usual pressure to argue rationally: the logic of the method of science will tend to filter out a good deal of casual opinion, unconscious bias and sheer illogicality.

At the birth of OR we had groups of scientists trying to apply scientific modes of thinking to military operations. Lest it appear completely "obvious" to us that this is manifestly the right thing to do, it is worth reminding ourselves that it is in fact a very particular, and unusual thing to do. Consider the battle of the Somme in the First World War. The Allies' "Big Push" was preceded by the biggest-ever artillery bombardment of the German defences. More shells were fired in a week than in the whole of the first year of the war, and these were directed at a 16 mile front (Middlebrook, 1971). The military supposition was that this would destroy the defences, making the Big Push a walkover. Anyone thinking scientifically about this would conclude that it was essential to obtain hard evidence that the defences had in fact been destroyed before sending the infantry into No Man's Land. But no one was thinking scientifically, the defences were intact, and the Allies suffered 60,000 casualties on the first day, two for every yard of the front.

By the Second World War the application of scientific thinking to military operations had made some progress, and with demonstrably useful results (see, for example, Waddington, 1973).

When the pioneers of OR undertook this work they were actually somewhat surprised at the manifest relevance of scientific thinking to situations involving human action. Blackett, one of the early protagonists, well aware that human operations entail "chance events" and "individual personalities and abilities" was surprised to find relatively stable repeatable results:

> ...many more useful quantitative predictions can be made than
> is often thought possible. This arises to a considerable extent
> from the relative stability over quite long periods of time of
> many factors involved in operations. This stability appears rather
> unexpected in view of the large number of chance events and
> individual personalities and abilities involved in even a small
> operation. But these differences in general average out for a
> large number of operations, and the aggregate results are often
> found to remain comparatively constant (Blackett, 1962; written
> in the 1940s).

This is one of the most important statements in the history of OR. It carefully sets out conditions under which the scientific method can apply to human operations: namely, there must be many similar examples of essentially the same operation, and the results must be taken to apply not to any one example but to the aggregate. In other words, scientific analysis of some human operations can reveal what follows from the logic of situations. Blackett had demonstrated that a testable scientific account of the logical entailments of some human situations could be given, even though any one example might legitimately be well outside the general findings.

Academic OR has acted thoroughly upon these foundations, and the resulting textbooks describe the algorithms which express the logic of many human situations which recur, such as queues, deciding where to place depots or when to replace capital equipment, etc. etc.. It is not atypical that a standard university textbook of OR devotes only 22 of its 1000 pages to the overall methodology and implementation of OR, the rest being devoted to the applied mathematics of the algorithms (Wagner, 1975). This proportion is probably reversed for practitioners in industry, and here lie the origins of the "crisis" debate in OR.

The application of the method of science to real-world operational problems obviously takes place in a human context, and this is likely to be much more significant than the laboratory context in which much natural science is carried out. OR is concerned with "articulate intervention" in human affairs (Boothroyd's happy phrase, 1978) and hence aims to help the formulation and realisation of human intentions, where natural science fundamentally aims at the acquisition of new knowledge of the world for its own sake.

The history of OR is littered with problems which at core stem from this crucial difference. Sir Bernard Lovell in his Blackett memorial lecture (1988) gives examples of Blackett's occasional very unscientific behaviour when it came to the meaning and application of OR results in a political context. And Freeman Dyson has told the story of what happened in some of his wartime work in Bomber Command. He showed in 1944 that, contrary to the official mythology, the established correlation between a bomber crew's chance of surviving a mission and their degree of experience no longer held. His findings were scientifically correct but he found his work suppressed because it was politically unacceptable (Dyson, 1979). More so than with natural science, the context in which OR is carried out will always be political.

And there is another characteristic of the manager's world which can also make problems for OR. The OR algorithms apply to the aggregate results of repeated operations, spelling out the logic of the situation. But most managers spend much of their time dealing with the unique idiosyncracies of specific situations rather than with their general logic. They are liable to be impatient when attention is given to what makes their problem general; they seek help with what makes it unique. This may be why a member of the Commission for the Future Practice of OR reported that he always hid new copies of the Journal of the OR Society "in case one of my client managers should see it" (Tobin, 1988).

These arguments suggest that OR should concern itself more with its human context and with what social science can teach about that. Unfortunately, this is not simply a case of absorbing an established body of accepted knowledge.

THE NATURE OF SOCIAL SCIENCE

Trained natural scientists who come late to the literature of social science are usually surprised by what they find. They expect it to be full of substantive contributions, as are the literatures of physics, chemistry and biology, but discover instead that fundamental questions concerning the nature of social science are still being debated. What is the nature of social theory? Is social science properly regarded as "science"? These are still live issues.

The ultimate reason for this is the special self-conscious, self-motivating character of human beings. How different studying the chemistry of the reaction of nitrogen and hydrogen to yield ammonia would be if the moleculres of nitrogen and hydrogen could decide capriciously whether or not to combine, doing so today but deciding not to next Thursday! But that is the situation the would-be social scientist is in. Human beings conscious that they are being investigated can in principle act deliberately to confirm, refute or make nonsense of the scientist's hypothesis. And if the experimental subjects are deliberately kept in ignorance of their status, tricky ethical problems are raised.

Social science offers no single agreed response to this major issue. Instead, two broad traditions have emerged. One, stemming from Emile Durkheim argues that social science should concern itself with the empirical study of "social facts". "Treat social facts as things" was Durkheim's famous dictum in 1895,and he wanted "the proper domain" of the new science of sociology to be the scientific study of social facts (Durkheim, 1895). A social fact is an observer-defined category believed to be relevant to a social group as a whole. For example, the proportion of school leavers entering higher education is a social fact which tells us something about the society in question. Studies of it over time could be made, and comparative studies of different societies could be conducted. Individual decisions about entering higher education will be taken for all kinds of logical (or illogical) reasons, or for no conscious reason at all, but the Durkheimian social scientist ignores that and chooses to define the aggregate of such decisions as a category worth investigating.

There are obvious echoes here of Blackett's version of OR, which focuses on the aggregate results of repeated operations, ignoring the idiosyncracies of individual cases. And hence the manager concerned with a particular individual case is unlikely to find the Durkheim tradition of social science relevant to his or her problems.

The alternative tradition, associated with Max Weber (1904) takes the unique interpreting ability of human beings as its starting point. Within this tradition the subject is concerned not with the objective study of social facts but with the actor's subjective understanding of social action. For Weber,human interactions were affected by something missing from the non-human world, namely the attribution of meaning to what is observed by human actors. Society was the result of a totality of intentional acts,and the task of social science was to analyse and explain social action by the study of the subjective meanings by which individuals direct their conduct.

The two traditions are based upon two very different Weltanschauungen; ultimately they stem from two different philosophical traditions. The Durkheimian tradition of social facts is underpinned, as is natural science, by the philosophy of positivism according to which all true knowledge is based upon empirical data, data which is in principle testable by all. The Weberian tradition links to the philosophy of phenomenology, in which the central concern is not the world as something independent of all observers of it, but rather the nature and context of our thinking about the world (Checkland, 1981, Chapter 8).

Durkheim's positivism clearly connects to Blackett's kind of OR - the study of the aggregate results which reveal the logic of situations - and there the operational researcher can import from natural science the techniques of statistical analysis which scientists have developed. But the manager concerned with a unique situation involving particular people might well hope for an OR based ultimately upon Weberian concepts, since he or she will be very much concerned with how individuals attribute meaning to, and interpret, the problem situation being faced. No such fully-fledged version of OR has been developed, but in the late 1980s we are probably witnessing its birth. It is noticeable that since the two conferences on the relation between OR and systems thinking in 1983 and 1985 (which led to the issues of the Journal of the Operational Research Society for August 1983 and September 1985) the phrase "soft OR" is creeping into currency. The phrase has no generally agreed meaning; rather it is gradually coming to be seen to be meaningful! I would argue that, technically, "soft OR" refers

to an OR based upon phenomenology rather than positivism, with the important addendum that any piece of work based upon phenomenology can always
encompass within itself "hard" studies based on positivism, so that hard
and soft OR are complementary to each other, not rivals, the former being
a special case of the latter (Checkland, 1985).

We reach a point, then, at which (a) we see the need for OR to pay more
attention to its human context and (b) we find a tradition within social science
(the interpretive tradition of Max Weber) apparently relevant to OR's aim
of helping the formulation and realisation of human intentions. Unfortunately the interpretive strand of thinking in social science has not so far
developed methodological approaches which could be imported intact into OR
practice.

A way forward, however, is potentially available from developments during
the last two decades in the practical use of systems thinking in real-world
problems (Checkland 1981, 1983, 1985, 1988; Wilson, 1984). Soft systems
methodology is a systems-thinking-based approach to tackling messy real-
world problem situations dominated by different attributions of meaning.
It provides structured guidelines for examining the social context of any
would-be scientific intervention in real-world affairs. It offers ways of
examining the intervention itself, as well as the problem situation, as both
a "social system" and "a political system", using those phrases in their
everyday language sense (Checkland, 1986). These analyses help selection of
a number of systems of purposeful activity relevant to the problem situation.
These systems are modelled and compared with the real world of the problem
situation, that comparison serving to structure a debate in which meanings
and intentions are coherently explored (Checkland, 1981). The approach thus
provides a meta-level framework in which the social context of OR intervention can be explicitly examined. Had it been available to Freeman Dyson in
the 1940s he would not have been taken aback by the political suppression
of his study because his analysis would have included examination of the
social context of his work. He would have prepared for the rejection his
work received and had a strategy ready for that eventuality.

What is being argued in general is that whenever an algorithm is used
or a model built of some part of the real world of human action, with a view
to alleviating a problem situation, the analyst should also explore - from
several different points of view - the purposeful activity systems which
will be concerned with that intervention. Recent developments in systems
thinking can provide help in that.

CONCLUSION

We may summarize the argument of this paper in the following terms.

1.   OR has been rightly proud of its scientific heritage, but can be truly
scientific in the mode of natural science only when it works out the logic
of problem situations which recur, and produces results applicable to the
aggregate of empirical results gleaned from such situations.

2.   However, most client managers will for much of the time be concerned
not with aggregate results deriving from situational logic but with the
idiosyncratic features of a unique situation.

3.   Also, because OR aims to help realise human intentions rather than
establish new public knowledge, the context in which it is done will always
be political.

4. We might hope that social science could provide some help in enabling the operational researcher to pay more attention to the social context of his or her studies.

5. The positivist (Durkheimian) tradition in social science is not helpful here, but the alternative interpretive (Weberian) tradition is more relevant. Unfortunately Weberian social science is not well developed methodologically.

6. However, recent work in systems thinking does provide an approach which could usefully enrich OR as it faces these issues.

REFERENCES

Ackoff R.L., 1979, The future of operational research is past, Jnl.Op.Res. Soc., 30:93.

Ackoff R.L., 1979, Resurrecting the future of operational research, Jnl.Op. Res.Soc., 30:189.

Blackett P.M.S., 1962, "Studies of War", Oliver and Boyd, Edinburgh.

Boothroyd H., 1978, "Articulate Intervention", Taylor and Francis, London.

Checkland P.B., 1981, "Systems Thinking, Systems Practice", John Wiley, Chichester.

Checkland P.B., 1983, OR and the systems movement: mappings and conflicts, Jnl.Op.Res.Soc., 34:661.

Checkland P.B., 1985, From optimizing to learning: a development of systems thinking for the 1990s, Jnl.Op.Res.Soc., 36:757.

Checkland P.B., 1986, The politics of practice. International Roundtable on the Art and Science of Systems Practice, IIASA, Laxenburg, Austria.

Checkland P.B., 1988, Soft systems methodology: an overview, Jnl.Appl.Sys. Anal., 15:27.

Churchman C.W., 1979, Paradise regained: a hope for the future of systems design education, in "Education in Systems Science", B.A. Bayrakter et al., eds., Taylor and Francis, London.

Dando M.R. and Bennett P.G., 1981, A Kuhnian crisis in management science?, Jnl.Op.Res.Soc., 32:94.

Durkheim E., 1895, "The Rules of Sociological Method", translated by S.A. Solovay, J.H.Mueller, 1964, Free Press, New York.

Dyson F., 1979, "Disturbing the Universe", Harper and Row, London.

Longuet-Higgins H.C. et al., 1972, "The Nature of Mind", University Press, Edinburgh.

Machol R.E., 1981, OR/MS in Europe - an American's impressions, in "Operational Research '81", J.P.Brans, ed., North-Holland, Amsterdam.

Middlebrook M., 1971, "The First Day on the Somme", Allen Lane, London.

Tobin N., 1988, OR Newsletter, 18(2):1.

Waddington C.H., 1973, "OR in World War 2", Elek Science, London.

Weber M., 1904, 'Objectivity' in social science and social policy, in E.A. Shils and H.A.Finch, eds., "Max Weber's Methodology of the Social Sciences", 1949, Free Press, New York.

Wilson B., 1984, "Systems: Concepts, Methodologies, Applications", John Wiley, Chichester.

Ziman J.M., 1968, "Public Knowledge: an Essay Concerning the Social Dimension of Science", University Press, Cambridge.

OPERATIONAL RESEARCH AS NEGOTIATION

Colin Eden

Strathclyde Business School
Glasgow
Scotland

INTRODUCTION

I start from the view of Operational Research as a formal and scientific model building activity aimed at helping people tackle 'real-world' problems. It is thus an activity which addresses itself to the choices and actions of persons other than the OR practitioner. Its success can only be measured by the extent to which it influences the thinking and action of other people. Put most succinctly OR is client oriented not solution oriented. At its best, OR practice must therefore be the collusion (sic!) of the dispassionate and 'objective' activities of science combined with those aspects of social science which reflect the passion of interaction in organizations. By this I mean the parts of social science that relate to interaction and engagement between people rather than that part which is concerned to study large groups of societies. In recent writing I have referred to this collusion as aiming for a multiplier effect between skills of process management and content management (Eden, 1987a), where the multiplier comes from treating the two skills as intimately and continuously intertwined rather than separately applied.

The above view of OR implicitly argues that a part of the body of concepts residing in the social science literature should be useful to the OR practitioner. They should be useful because they should be capable of informing the OR process and practice. The crude demand we might make of social science concepts is that they should "earn their living" by being obviously useful to us.

In my experience of training OR practitioners and working with OR academics, most of them hold the belief that the process management skills of the good OR practitioner are best developed intuitively. Indeed those who thoughtfully argue the case will suggest that process and content skills can only become systemically intertwined if they are not bombarded by two separate streams of concepts and ideas - those of formal modelling and those of social interaction.

In this paper I intend to discuss some of the concepts from social science that have had a profound effect on my practice. I hope to demonstrate that these concepts are non-trivial, are unlikely to be fully utilised if they grow solely from intuitive experiences, and that a conscious understanding of them enriches the ways in which we can make sense of our organizational experiences. This is patently not an easy task. OR Society conferences are bereft of explicit social scientific debate; this conference is twenty-five years on from the last. At the same time we should not punish ourselves too much: in the field of social policy we might expect a greater admission of relevance, and yet in a book titled "Social Science and Public Policy" Martin Rein (1976) suggested that "social science does contribute to policy and practice, but the link is neither consensual, graceful nor self-evident". An apt description of the relationship between OR and the Social Sciences?

In 1986 the UK OR Society Commission on the Future Practice of OR reported. Of many important findings, there were some that implicitly suggested a role for social science ideas. For example social and political skills were believed to be of more consequence in the practice of OR than in most apparently related activities; the successful OR worker was seen as adaptive, opportunistic, articulate, and sensitive to the political and social environment around him or her. The findings also suggested that the successful OR consultant must understand the nature of power and influence. Most significantly the inquiry clearly revealed that it is the way in which OR is practiced which makes for success or not. This focus on the "way" of practice is not intended to imply the need for better project management, but rather to focus on skills of bringing formal analysis into the arena of organizational and thus personal action.

THE ORGANIZATIONAL SETTING FOR OR PRACTICE

An organization is most often described in terms of its task with respect to the environment, and its members described in terms of the roles they are expected to enact. That is, for example, "the purpose of the organization is to make and sell widgets at a profit" and there is "a production manager who will ensure that enough widgets are made of a satisfactory quality to meet the needs of the market", and so on.

This view of an organization is useful because it tells us something about what should happen. Using the play as an analogy, it is the equivalent of the playwright's instructions. But, as a matter of course, we do not expect any group of actors and director to present the same experience to the audience (the customer) or to one another. In the same way the above description of the organization tells us little about how things will come about in the organization. It is not that a structuralist view of the organization does not impinge on what happens it is more that the people in the organization do things within a subtly negotiated order.

Organizations as Negotiated Order

Social order depends for continuity and change upon continuous social interaction between people renegotiating their relationships. They have to continue to enjoy social intercourse with one another. For an OR intervention to be successful it must account for the need of all members of the client group to accommodate or accept a part of the reality projected by OR analysis, including the projected social consequences of decisions made. That is they are all participants in what social scientists call the social construction of reality and the negotiation and management of meaning.

For the Operational Researcher this process can be remote for he or she rarely "lives with" the newly negotiated order. Such remoteness is often an advantage because the OR consultant is able to, and often asked to, take an "objective" outside view of a situation. However, what is at issue here is not whether the analysis is or is not objective but whether the analysis resides outside of the social negotiation that determines action. I shall argue that the need to understand organizations as negotiated order is non-trivial because it implies the need to clearly explore and understand that order so that OR models and the ORer influence (rather than direct) the negotiation of new order.

Putting this social scientific view more formally, I am adopting the view that an organization can usefully be seen as "the interaction of motivated people attempting to resolve their own problems [and that] the environment in which an organization is located might usefully be regarded as a source of meanings through which members defined their actions and made sense of the actions of others" (Silverman, 1970 p.126).

Conceiving of an organization as negotiated order is most powerfully sustained by considering how organizations manage to keep going when, as many have argued, the material world is in such flux (Toffler, 1970, Schon, 1971). In order to do so the people within the organization become involved in "the processes of give-and-take, of diplomacy, or bargaining" (Strauss et al., 1963). Order exists because "numerous agreements are continually being terminated or forgotten, but also continually being established, renewed, reviewed, revoked"..."policies and rules serve to set the limits and some of the directions of negotiation" (Strauss et al., 1963, reporting on the hospital and its negotiated order). This negotiated order includes all levels of an organization. The power of the CEO is set within the context of a clear understanding of current order. Chester Barnard recognized this aspect of power when he argued that "the decision as to whether an order has authority or not lies with the persons to whom it is addressed.....there is no principle of executive conduct better established than that orders will not be issued that cannot or will not be obeyed" (1938, pp.163-167).

Organizations with a Purpose?

It is surprising how much of OR presumes that the objectives of parts of the organization are both known and unambiguous. Objectives that are enduring and can be agreed upon are usually only those that are 'all things to all men', what we normally call 'motherhood and apple-pie'. Once we get down to detailed everyday goals and objectives it becomes importantly the case that agreement is difficult if not impossible to achieve. It is the negotiated settlement to these that govern action. Almost all decisions can be explained with respect to the enduring statements of purpose, including those rationalistically implied by the OR person as if they were the only interpretations. If all decisions are legitimated in this way, then influence over real action can only result from a continued and contingent negotiation by the consultant with the client and members of the client group (Eden and Sims, 1979) rather than by reference to the objectives of the organization. I think it was Max Weber (a sociologist) who noted that the top of organizations are never bureaucratized, rather they always belong to somebody. There is nobody in whom the objectives of the organization reside, there are many bodies who will each have their unique interpretation of what the needs of the (reified entity they label) organization (as if it were an individual) are.

Perrow (1972), also a sociologist, significantly goes on to argue that: "organizations must be seen as tools...for shaping the world as one wishes it to be shaped. They provide the means for imposing one's definition of the proper affairs of men upon other men" (p.14). Classically OR aims to impose the 'objective' logic of science on the affairs of men; however it often fails because of its commitment to being right rather than to a commitment to negotiation. Operational Researchers are no less certain about the way in which the world should be shaped, they are rather often pathetic in their ability to see their so-called objective (that is, non-evaluative) definition of the world as one of many other legitimate and rational definitions.

45

Thus, debate about facts is sterile, it is only the debate about the facts as they relate to values/objectives/goals of individuals that make the facts significant. Purpose gives meaning to events, it gives individuals the will to bargain for action. As ORers we are unable to investigate the facts without explicitly acknowledging their relationship to our own values. Similarly we cannot inquire with intent to negotiate without acknowledging the meanings others will attribute to our analysis.

The moment we do this we discover that there is debate about values and about facts and about their relationship one to another. There is debate because any assertion about the nature of the world is seen as having action implications for disturbing the social order. A potential disturbance to order creates ambiguity, ambiguity creates discomfort, and individuals will seek to reduce the ambiguity. As an example consider the time and energy exerted on the resolution of the apparently trivial issue of company car policy!

This need to understand individual purpose is two fold. Firstly the interests of the organization simply do not explain or inform the sort of decisions that OR attempts to influence. Secondly, "men do not exist just for organizations they track all kinds of mud from the rest of their lives with them into the organization and they have all kinds of interests that are independent of the organization" Perrow (1972, p.5). Indeed as Vickers (1965) suggests "the goals we seek are changes in our relations or in our opportunities for relating; but the bulk of our activity consists in 're-lating ' itself...the most important aspect of activities, the ongoing main-tenance of our ongoing activities and their ongoing satisfactions" (p.33), in other words, the maintenance of social order.

Organizations as Political Cauldrons

The traditional view of power is relatively unhelpful to the OR practitioner. That is the element of power that comes from the ability to control scarce, critical resources that are required by others is relevant to an analysis of the client problem situation but not particularly relevant to managing the consultancy intervention. The OR consultant very rarely has the power to use coercion as a method of persuasion, he or she is more dependent upon psychological or negotiative devices or persuasion. "Politics concerns the creation of legitimacy for certain ideas, values, and demands - not just action performed as a result of previously acquired legitimacy" (Pettigrew, 1977).

That the attainment of one's own purpose demands the cooperation of fellow workers is almost obvious. Even those groups that are traditionally conceived of as having little power can so strongly determine the success of the organization by their lack of cooperation and by working without energy and commitment - by using the rules as tools to impose their own will. Participants in an organization can withhold information that is relevant, and sometimes critical, to decisions and do so without risking accusations of sabotage of the 'should's' or 'ought's' of the bureacracy. Consider for example the significance of the relationship between the ORer and the client's secretary. Nothing decrees that this shall be fruitful, rather it depends upon successful and subtle negotiation between one individual and another. And yet this relationship is usually crucial to the consultant having data which will importantly inform an understanding of the problem situation. Perrow calls this "particularism" and realistically argues that to abhor such interactions as disorderly is merely to express an ideal and not accept the reality of organizations.

Within OR practice there is one overriding consideration - the power to manage meaning, as the ability to "mobilise bias" (Schattschneider, 1960).

"Key concepts for analyzing meaning are symbolism, language, belief, and myth. Language is not just a means of expressing thoughts, categories and concepts: it is also a vehicle for achieving practical effects" (Pettigrew, 1977).

It is this more psychological view of power that is most difficult to understand. We are accustomed to Machiavellian views of power and politics where careerism, ambition, and sheer bloody-mindedness are the focus of attention. The manoeuvreing of people along Machiavellian dimensions is relatively easy to identify, but it is, in my experience, much less common than the politics that result from the wish to define reality. This latter form of politics is the essence of human life, it derives from honest people believing they know what is best for the organization. This belief means that meaning will be managed by the formation of coalitions, by trading in support of different versions of reality.

EARNING A LIVING?

The above has been an attempt to bring together a number of social science perspectives (of "negotiated order", the "social construction of reality", and the "management of meaning") that all support the notion of negotiation. While the central idea of organizations as "negotiated order" comes from Strauss et al. (1963) and their empirical research, it is most powerfully supported by the writing of Berger and Luckmann (1966), Perrow (1972), Silverman (1970) and the symbolic interactionist school of Blumer (1969) and Mangham (1978). Each of them support the important social psychological view that problems are ultimately addressed through a social not logical process.

What does this perspective imply for the process of doing Operational Research? It does not argue against the consideration of structure, and control. Neither does it argue against the analysis of the interaction between concrete things. For example, we may need to know how service time varies for supermarket queues according to different till layouts and queue disciplines. This demands a proper piece of analysis that will be informed by the use of a mathematical model, possibly combined with simulation. And yet for all that the notion of negotiated order would demand that the ORer be involved in a number of other important activities. The meaning of different options will be different for each of the individuals who are party to the currently negotiated way of doing things. Thus if we consider who the people are who have reached an accommodation with others in contriving ways of working they will be those for whom the knowledge of an OR study will release uncertainty and ambiguity. We might guess that they will, in order of highest anxiety level, be the till supervisor, till operators, store manager, and several categories of customer. The analysis is thus based upon an understanding of power which stems from the ability to create legitimacy for certain ideas.

Let us take the example further. In the initial model building the options we might wish to consider in our analysis will need to be dependent upon our expectation about the process of negotiation that will occur amongst the identified actors if an option was implemented as action. Included within the negotiated order is the conception of 'self' for each of the actors - the supervisor will actively seek to negotiate a supervision conception of self with the till operators regardless of the action taken and the intentions supporting it. The operators will collude in the negotiation because their relationship with the supervisor as person has real and lasting consequences for maintaining social order. They will jointly devise order in ways that are little different from the serious anecdotal evidence we all have of the ability individuals have to break almost all imposed control systems by what is often immensely creative action. They can do so with all of them aware of, and colluding in, cheating, theft, or simply more efficient problem solving on behalf of their interpretation of the purpose of the organization.

Thus far the implication for OR is that actors need to be identified. They should not be identified through the broad sweep of "stakeholder analysis" that is implied by CATWOE within the "Soft Systems Methodology" (Checkland, 1983) or Ackoffian liberalistic notions of "Idealised planning" (Ackoff, 1974). The socially constructed reality of each of these actors needs to be understood and preferably modelled. While the intention is to get close to what Checkland unhelpfully calls the "Weltanschauung" of the actors, a negotiated order perspective would put greater emphasis on world-taken-for-granted and patterns of interaction that maintain social order. The tools of "rich pictures" (Checkland, 1983) and "cognitive maps" (Eden, 1988a) have a role to play at this investigative level.

An additional implication is that we have acknowledged the significance of considering implementation as a part of initial problem construction rather than as the last stage of a project.

However if the analyses we undertake are to have a more profound impact on change within the organization then we need models (and therefore model building methods) that are more transparent so that they can be used to facilitate the management of new meanings. This may mean an increased use, and development, of visually attractive interactive models that all members of a group can work with together. Visual interactive simulation has been making continuous strides in this direction but it has been designed solely as an analytical device rather than as a device to guide negotiation in the group. In our example we need a model which can act as a negotiative device (Eden, 1988b) to facilitate the involvement of key actors in gradually constructing an alternative social reality. Current research and developments in Group Decision Support Systems (Zigurs et al., 1987, Ackermann, 1987) are helping us understand the potential of decision aiding as a part of social activity.

Mintzberg (1973) has firmly argued that executive problem solving does not follow the logical sequential steps suggested by most OR based problem solving methods. The OR process is likely to be most effective when it can be a part of a cyclical and non-linear process. In addition it must recognise the role it has to expand the "boundedness of rationality" by becoming more centrally a part of the psychological negotiation that is going on as individuals "change their mind" about problematic situations. I have argued elsewhere (Eden, 1987b) for attending to the nature of problem finishing in contrast to problem solving, where a problem is finished for a large number of psychological and social psychological reasons other than a solution having been attained. However the essence of the case is that a person thinks, reconstrues, and socially interacts and so changes their mind. It is a gradual and subtle process within which OR models could play a more significant part only if they can be a processual "toy" that can be played with in an engaging and enjoyable manner.

In previous papers (Eden, 1982, 1987b) I have depicted the social business of working on problems as a cycle of "presenting a portfolio of solution/options", "problem construction", pondering upon and "making sense of the situation" and "defining the situation". My own methodology for helping groups negotiate action within complex problems - SODA (Strategic Options Development and Analysis) - places a significant emphasis on problem construction because of the above assertions implying that negotiation is most likely to take place at the problem construction stage. This means that the method is aimed at 'solution' falling out from the "making sense" and "definition" that naturally follows "construction". Indeed as I noted above, implementation is not a stage in a process of problem solving, but rather is embedded in the negotiation about the nature of the problem. People do not construe problems without also considering the practical world of getting things done.

The SODA method (Eden, 1988a) also prescribes the above cycle, and many so-called "soft" OR methods (Decision Conferencing - Phillips, 1987, Soft Systems Methodology - Checkland, 1981, Strategic Choice - Friend and Hickling, 1987)and interestingly well established approaches to international concil-iation (Fisher & Ury, 1982) follow a similar cyclic process. Most particu-larly they all follow a dialectical process of engaging the client group in a cycle of negotiation. However few of them use their methods as entry to the more formal quantitative analysis that typifies OR, but which could also take place in a negotiative framework. Notably the negotiative methods all seek to move the group through a process of identifying new and creative options which fall from members of the group seeking to generate a new social order. (It is notable that the recently published views of Fisher on nego-tiation are that the role of relationship building is absolutely crucial to the formal negotiation process (Fisher and Brown, 1988)).

Vickers (1983) argues that acts of creativity are "integrative" solutions which are "attained only by changing the way in which the situation is re-garded or valued (or both) by some or all of the contestants, a change which enlarges the possibilities of solution beyond those which existed when the debate began" (p.208), he goes on to say that this depends upon changes in the ways in which the minds concerned are predisposed to see the realities of the situation" (p.209). It is, perhaps, important to comment that the need for cyclical processes to be used for negotiating a new order relates to the complexity of the issue being addressed (Mintzberg,1973).

CONCLUDING REMARKS

Ultimately, if OR methods are to play any significant role in aiding negotiation they must also extend the bounds of rationality of problem solv-ing and open the psychological constraints on the way in which the problem is formulated (Reitman, 1964). To a very limited extent the so-called 'soft' approaches do this at the level of qualitative thinking, however it is possible to conceive of quantitative modelling offering similar possibilities. It is interesting to note the reflections of Walton on a linear programming project (Bryant, 1988). Here the opportunity to thoroughly reflect on an apparently straightforward mathematical programming project led Walton and Bryant to identify key process elements, indeed one of the section headings in the paper is "mainly process management". In effect I am arguing for progress in OR practice via two developmental thrusts. First, the modelling activity must be much more intertwined with the clients progression of thinking about the problem, and therefore we must devise modelling techniques that can collect views and data on a continuously interactive basis and be capable of rewarding the client group with analysis continuously. Second, and it is a plea I have made before (Eden and Sims, 1979), we must find ways of better understanding the social science of good OR practice by providing appropriate frameworks for reflection (Argyris and Schon, 1974).

REFERENCES

Ackermann F., 1987., SODA as Group Decision Support through negotiation,
        presented to International Symposium on Decision Management, Toronto.
Ackoff R.L., 1974, "Redesigning the Future", Wiley, New York.

Argyris C. and Schon D., 1974, "Theory in Practice: increasing professional effectiveness", Jossey-Bass, San Francisco.

Barnard C.I., 1938, "The Functions of the Executive, Harvard University Press, Cambridge, Mass.

Berger P. and Luckmann T., 1966, "The Social Construction of Reality", Doubleday and Co., New York.

Blumer H., 1969, "Symbolic Interactionism", Prentice Hall, Englewood Cliffs.

Bryant J., 1988, Frameworks of inquiry, J.Op.Res.Soc., 39:423.

Checkland P.B., 1981, "Systems Thinking, Systems Practice", Wiley, Chichester.

Eden C., 1982, Problem construction and the influence of OR, Interfaces, 12:50.

Eden C., 1987a, PxC: Finding the multiplier, presented to International Symposium on Decision Management, Toronto.

Eden C., 1987b, Problem-solving or problem-finishing, in Jackson M.C. and Keys P. (eds.), "New Directions in Management Science", Gower, Aldershot.

Eden C., 1988a, Cognitive mapping: a review, Eur.J.Opl.Res., 36:1.

Eden C., 1988b, Cognitive maps as a visionary tool: strategy embedded in issue management, Long Range Planning, forthcoming.

Eden C. and Sims D. 1979, On the nature of problems in consulting practice, Omega, 7:119.

Fisher R. and Brown S., 1988, "Getting Together", Houghton-Mifflin, Boston.

Fisher R. and Ury W., 1982, "Getting to Yes", Hutchinson, London.

Friend J. and Hickling A., 1987, "Planning Under Pressure", Pergamon, Oxford.

Mangham I.L., 1978, "Interactions and Interventions in Organizations", Wiley, London.

Mintzberg H., 1973, "The Nature of Managerial Work", Harper and Row, New York.

Perrow C., 1972, "Complex Organizations", Scott Foresman, Glenview.

Pettigrew A., 1977, Strategy formulation as a political process, Intl.St. Mgt.Org., 7:78.

Phillips L., 1987, Decision Analysis for Group Decision Support, presented to International Symposium on Decision Management, Toronto.

Rein M., 1976, "Social Science and Public Policy", Penguin, Harmondsworth.

Reitman W.R., 1964, Heuristic decision procedures, open constraints and the structure of ill-defined problems, in Shelley M. and Bryan G. (eds.), "Human Judgements and Optimality", Wiley, New York.

Report of the Commision on the Future Practice of Operational Research, 1986, J.Opl.Res.Soc., 37:829.

Schattschneider E.E., 1960, "The Semisovreign People", Holt, Reinhart and Winston, New York.

Schon D.A., 1971, "Beyond the Stable State", Temple Smith, London.

Silverman D., 1970, "The Theory of Organizations", Heinemann, London.

Simon H.A., 1957, "Administrative Behaviour", Macmillan, New York.

Strauss A., Schatzman L., Ehrlich D., Bucher R. and Sabshin M., 1963, The hospital and its negotiated order, in Friedson E., (ed.), "The Hospital in Modern Society", Macmillan, New York.

Toffler A., 1970, "Future Shock", Random House, New York.

Vickers G., 1983, "The Art of Judgement", Harper and Row, London.

Zigurs I., Poole M. and DeSanctis G., 1987, A study of influence in computer-mediated decision-making, Working Paper MISRC-WP-88-07, University of Minnesota.

THE VISIBILITY OF SOCIAL SYSTEMS

Robert Cooper

Department of Behaviour in Organisations
University of Lancaster
Lancaster, U.K.

INTRODUCTION

The role of the senses in social life is a neglected theme in social analysis. Despite the philosopher Whitehead's (1938) early call for systematic study of the 'organization of the sensorium' in human communication, it is only recently that this has been taken up as a serious challenge by the human sciences. Much of this work has focused on the dominance of vision, especially in modern social systems subjected to 'bureaucratization' and the processes of formal organization more generally. As Rorty (1980) has reminded us, modern thought has privileged the eye as the sense organ by which we may represent the world to ourselves most effectively. Yet it was not always like this. The historian Lucien Febvre (1982), in a study of sixteenth century social life, comments on the underdevelopment of sight in that period: 'Like their acute hearing and sharp sense of smell, the men of that time doubtless had keen sight. But that was just it. They had not yet set it apart from the other senses. They had not yet tied its information in particular in a necessary link with their need to know' (p.437). Consequently, they lived in a fluid world 'where nothing was strictly defined, where entities lost their boundaries and, in the twinkling of an eye, without causing much protest, changed shape, appearance, size....'(p.438). The stabilization of the world required a specialized training of the visual sense and its elevation over the other senses. The process of stabilization called for quantification and calculation and this necessarily involved the eye: 'The passage from the qualitative to the quantitative is essentially linked to advances in the predominance of visual perception' (Rey quoted in Febvre, 1982, p.432).

It is customary to view the careers of societies in terms of the division of labour where the emphasis is placed on the specialization of skills and occupations in an hierarchical framework. The social analysis of the human sensorium suggests an alternative way of viewing this process: instead of the 'division of labour', the differentiation of the senses draws attention to the complex processes at work in actively shaping the human agent as a perceiving organism in the social system. This differentiation process we may more accurately call the 'labour of division' since it not only highlights the act of division itself (as opposed to the specific agents of 'labour') but it also suggests that the 'division' in this context is significantly bound up with the act of 'seeing' - that is, 'vision' is an intrinsic component of 'division'. We can take this analysis further through the work of the philosopher Jacques Derrida (see, for example, Norris, 1987) who views the logic of 'division' in terms of (1) <u>hierarchy</u> and (2) <u>interaction</u>. (These are not Derrida's terms - I have retained his ideas but used a more familiar terminology). In hierarchy, it is recognised that systems, social or otherwise, are structured around binary oppositions (e.g., good-bad, male-female) in which one of the terms dominates the other. As Derrida (1981) writes: '...we are not dealing with the peaceful co-existence of a <u>vis-a-vis</u>, but rather a violent hierarchy. One of the two terms governs the other (axiologically, logically, etc.), or has the upper hand' (p.41). In interaction, as Derrida reminds us, there is a continuous double movement <u>within</u> the binary opposition so that the positively-valued term (for example, good) is defined only by contrast to the negatively-valued second term (for example, bad). In fact, the relationship between the apparently opposing terms is really one of mutual definition in which the individual terms actually inhabit each other. In other words, the separate, individual terms give way 'to a process where opposites merge in a constant <u>undecidable</u> exchange of attributes' (Norris, 1987, p.35). Interaction describes precisely the fluid, changing forms of perception characteristic of the human agent in the sixteenth century, as noted by Febvre (1982). The conversion of this state of perpetual ambiguity into a more determinate structure necessitated the hierarchical step of raising the status of sight over the 'primitive' senses of touch, taste and smell in favour of the <u>visualization</u> of perception. In this way, the general process of the 'labour of division' enables greater control over the social and material world through enhanced clarity, transparency and visual certainty <u>at a distance</u>.

VISIBILITY AND THE DEVELOPMENT OF MODERN SOCIAL SYSTEMS

The first widespread attempts to organize and systematize the social world occurred in the sixteenth century. As Foucault (1979) points out, the sixteenth century saw the emergence of an increasing concern with the 'art of government' and political economy, that is, with the management of people, territory and raw materials. Perhaps for the first time, people began to be defined in terms of their relationship with 'wealth, resources, means of subsistence, the territory with its specific qualities, climate, irrigation, fertility, etc.'...as well as in their relation to 'accidents and misfortunes such as famine, epidemics, death, etc.'. (Foucault, 1979, p.11). In short, the management of social-economic systems began to take shape in men's minds. The creation of visibility, the harnessing of a precise visual bias to the control of people and things, was a necessary component of this development.

In his study of hospitals and medical practice at the end of the sixteenth century, Foucault (1973) identifies a new phenomenon of organized perception - the 'gaze' - which subjected the individual to the definition of a new social practice, i.e., medicine, that assumed the authority of a 'science'. The concept of the gaze as a technique for ensuring the maximum visibility of individuals is further elaborated in Foucault's (1977) study of the prison system in which the gaze finds its special pedigree as a power (here the significance of the French 'pouvoir', with its inclusion of the verb 'to see', underlines the imbrication of power, knowledge and visibility) that 'must see without being seen' (p.171). The necessary role of surveillance in the construction and maintenance of visibility is perhaps best exemplified in Jeremy Bentham's architectural concept of the Panopticon, 'the polyvalent apparatus of surveillance, the universal optical machine of human groupings' (Miller, 1987, p.3). The Panopticon was a circular building with a central tower from which continuous surveillance could be unilaterally exercised on inmates housed in perimeter cells. The Panopticon represented a more general idea - the Panoptic principle - in which Bentham dreamt 'of a transparent society, visible and legible in each of its parts' (Foucault, 1980, p.152) which would expurgate 'the fear of dark-ened spaces, of the pall of gloom which prevents the full visibility of things, men and truths' (p.153) that haunted the second half of the eight-eenth century. Hence the need to elaborate that compendious theoretical system of administration which absorbed so much of Bentham's life in the pursuit of the criteria of visibility: transparent knowledge <u>at a glance</u> through surveillance <u>at a distance</u>.

The pursuit of visibility was advanced, as Foucault (1977) notes,through the introduction of the 'examination', a technique which combines the power of an 'observing hierarchy' with that of a 'normalizing judgement'. The examination 'is a normalizing gaze, a surveillance that makes it possible to qualify, to classify and to punish. It establishes over individuals a visibility through which one differentiates them and judges them' (p.184). In the eighteenth century, the hospital emerged as one of the first modern examples of an 'examining apparatus'. It regularized the form and frequen-cy of the medical inspection and made it 'internal' to the hospital. Form-erly a religious hospice, the hospital became subject to medical authority and the physician, hitherto an external element, displaced the religious staff into a subordinate role in the technique of the examination. The hospital became a place of training and of the development of a form of knowledge which had as its object the subjection of the patient through techniques of surveillance and the normalizing gaze. The medical examination exemplified the logic of division: it hierarchized individuals into the categories of 'normal' and 'abnormal' and, just as importantly, it served to prevent the 'interaction' or intermixing of these categories. Ultimately, it enabled everyone <u>to see more clearly</u> the difference between 'good' and 'bad' physical and mental health.

In the same way, the early schools became a sort of examining apparatus. 'The Brothers of the Christian Schools wanted their pupils to be examined every day of the week: on the first for spelling, on the second for arith-metic, on the third for catechism in the morning and for handwriting in the afternoon, etc.' (Foucault, 1977, p.186). The school examination made the pupil subject to a 'discipline' of continuous surveillance and visibility: 'it guaranteed the movement of knowledge from the teacher to the pupil, but it extracted from the pupil a knowledge destined and reserved for the teacher' (p.187).

There are three specific ways, according to Foucault (1977), in which the examination promotes visibility while at the same time linking the latter to knowledge and power. First, '<u>the examination transformed the economy of visibility into the exercise of power</u>' (p.187). In contrast to traditional forms of power whose efficacity rested on the full display of the power holder (e.g., king, general) and the passive invisibility of its servitors, the new examination-based disciplinary power reverses this process by im-posing a 'compulsory visibility' on its servitors while it itself remains invisible. 'In discipline, it is the subjects who have to be seen. Their visibility assures the hold of the power that is exercised over them. It is the fact of being constantly seen, of being able always to be seen, that maintains the disciplined individual in his subjection' (p.187). Second, '<u>the examination...introduces individuality into the field of documentation</u>' (p.189). The examination not only locates individuals in a 'field of sur-veillance' but it 'also situates them in a network of writing: it engages them in a whole mass of documents that capture and fix them' (p.189). Through administrative writing it became possible to describe, identify and record in precise form the visibility of individuals according to the logic of division. Through what Giddens (1985) has called 'textually mediated organ-ization' (p.179), the examination made it possible to develop a system of accountancy for 'normal' and 'abnormal' behaviour: it enabled 'the constit-ution of the individual as a describable, analyzable object, not in order to reduce him to "specific" features, as did the naturalists in relation to living beings, but in order to maintain him in his individual features, in his particular evolution, in his own aptitudes or abilities, under the gaze of a permanent corpus of knowledge' (Foucault, 1977, p.190). Third, 'the examination, surrounded by all its documentary techniques, makes each individual a "case"' (p.191). Through writing, the examination documented the individual as a 'case', made him visible as an object of 'normalizing' management or, more precisely, 'an object for a branch of knowledge and a hold for a branch of power' (p.191). The management and control of social systems finds its leverage increasingly at the level of the bio-scopic individual 'case'.

Against this background of the visible insinuated within the lisible, Hoskin and Macve (1986) have traced the history and development of accounting out of the technology of the examination. Accounting is that branch of modern management which has capitalized most on those two characteristic features of writing's visibility - instantaneity and distance; in other words, writing enabled information to be made available at a glance and in a depersonalized form (i.e., free from the possibility of 'contamination' by social 'interaction', in Derrida's sense). Writing within the general context of Foucault's analysis of 'knowledge-power', Hoskin and Macve argue that 'examination, discipline and accounting are historically bound together as related ways of writing the world (in texts, institutional arrangements, ultimately in persons) into new configurations of power' (p.107). These new ways of re-writing the social world date conceptually from antiquity but their influence is limited until medieval times when a 'new knowledge elite appears, centred around the nascent universities' (p.109). The new elite of clerks and masters 'produce a vast new range of pedagogic re-writings of texts, i.e., techniques which grid texts both externally and internally in the service of information-retrieval and knowledge-production' (p.109). Hoskin and Macve note that 'the scholars of the cathedral schools and universities begin to use visualist metaphors both to denote reading (e.g., videre, inspicere) and composition (scribere for dictare)' (p.110). Above all, the gridding of texts - the use of alphabetical order, a visually-oriented layout, systems of reference - and the substitution of read-easy arabic numerals for the more clumsy roman numbers created a new ordering of knowledge and understanding based on visibility. The new techniques were first perfected in the universities, especially those of Paris and Bologna but later the exponents of these methods took their places as 'professionals' either in the church or at court. They included men like Thomas à Becket, who began his career, after qualifying in Paris, as a clerk-accountant in London in the 1140s. Double-entry book-keeping appeared in the thirteenth century as a particularly sophisticated form of the new examination-based writing. Significantly, as Hoskin and Macve point, 'double-entry' was based upon the visualist metaphor of the Mirror - it was a mirror-book that reflected the 'equal and opposite signs of debit and credit' (p.121). This, too, owed its emergence to a university teacher and cleric - Pacioli. Despite its relative sophistication, 'double-entry' was limited to 'financial examinatorial control' and therefore could not make 'human accountability' more visible. Hoskin and Macve argue that the accounting of human behaviour in terms of the debit-credit system had to await the introduction of the examination mark into the educational system in the nineteenth century (pp.129-130).

VISIBILITY, KNOWLEDGE-POWER AND PROFESSIONALIZATION

The point of Hoskin and Macve's analysis is to reveal the power of accounting to create a specific form of knowledge (knowledge-power) which subjects individuals to a fixed and determinate visibility. As such, it is a specific example of a more general trend in modern society - the development of knowledge-power by means of professionalization. In fact, Hoskin and Macve characterize accountancy as a profession that is squarely founded on the knowledge-power of accounting technology. The implication here is that the professions are those groups in society that are accredited with the task of creating and maintaining the appropriate 'visibility' of social agents through such techniques as the examination. This view of the professions is clearly at odds with the prevailing general and academic understanding of the professional's contribution to society as being both benevolent and selfless. As Goldstein (1984) points out, the sociological definition of a profession rests on four criteria: (1) it is a body of highly specialized knowledge which has to be formally mastered before it can be practised; (2) it is a monopoly, i.e., the body of knowledge is recognized as the exclusive competence of those who practise it; (3) it is autonomous, i.e., the professionals control their own work and how it should be done, and (4) it embodies a service ideal, i.e., 'a commitment of ethical imperative to place the welfare of the public or of the individual client above the self-interest of the practitioner, even though the practitioner is earning a living through the exercise of the profession' (p.175). This is essentially a 'theory' about certain kinds of occupational groups. When Foucault's ideas about 'disciplines' and knowledge-power are applied to the professions a totally different picture emerges. In fact, as Goldstein argues, the sociologist's professions turn out to be none other than Foucault's 'disciplines'. Like disciplines, the professions are now seen to be not only bodies of esoteric knowledge but also social practices. As social practice, disciplines/professions institute and maintain procedures for 'discipling' individuals in the sense of subjecting them to the new forms of theoretical knowledge that epitomize the major professions. Human subjects become 'framed' in a 'picture' that is meticulously drawn in the practice of the professional's knowledge-power. Hence they can be made operationally visible.

Another version of this argument has been presented by Frug (1984) who has identified professionalism with bureaucratic legitimation. In an exemplary analysis which uses Derrida's idea of 'interaction', Frug shows that the central problem of bureaucracy (as studied by organization theorists and scholars of corporate and administrative law) is that of reconciling the relationship between subjectivity and objectivity. Derrida's notion of 'interaction' means essentially that no division can ever realistically be made between these concepts since they are 'confused' within each other. But the visual imperative that orders the social world means that (somehow) they must be kept apart - otherwise they could not be 'seen'. 'All the stories of bureaucratic legitimation...share a common structure: they attempt to define, distinguish, and render mutually compatible the subjective (and objective) aspects of life. All the defences of bureaucracy have sought to avoid merging objectivity and subjectivity - uniting the demands of commonness and community with those of individuality and personal separateness - because to do so would be self-contradictory. Moreover, it has never been enough just to separate subjectivity and objectivity; each must also be guaranteed a place within the bureaucratic structure' (Frug, 1984, p.1287). In this case, the process of making bureaucratic structures clear and legible is self-deceiving for while the 'objective' qualities of professionalism are seen 'as something outside the individual to which he must adapt, they are qualities that the professional himself helps to define' (p.1331). Such interweaving of objectivity and subjectivity necessarily undermines the 'transparent' or 'obvious' features of bureaucratic logic - the lack of division or distinctness between the two making it difficult to see either. The answer to the problem is the creation of a 'fiction' of objectivity. The suspicion that this 'fiction' may be vision-based is suggested in a study by Weinstein and Weinsten (1984) of what they call 'the visual constitution of society'. Briefly, Weinstein and Weinstein base their study on Sartre's famous analysis of the 'look' (le regard) - essentially, Foucault's gaze - which functions to objectify the social world of fixing or freezing individ-

uals in a determinate framework from which they cannot escape since they are
held (and, of course, beheld) by 'eyes that must see without being seen'
(Foucault, 1977, p.171). Because it cannot be seen, the 'look' cannot be
questioned - this is its lethal, invisible power; vision and objectivity
are thus mutually-defining. The significance of the 'look' is not only that
it objectifies but also that it is the prototypical form of visual interaction
in formal social systems such as bureaucratic organizations.

It would be myopic to assume that the visual bias of the 'objectifying
gaze' is limited to the examples of the disciplines/professions - medicine,
accountancy, etc., - we have discussed here. The definitive feature of
modern society is not so much the presence of 'disciplined vision' in form-
ally-recognised institutions but its unobtrusive diffusion in the least expected places. For example, Bové (1986) has recently shown how the institution of modern Anglo-American literary criticism was 'professionalized' by
the famous Cambridge critic, I.A.Richards, beginning in the 1920's. Significantly, Richards was a follower of Jeremy Bentham and Bové shows in some
detail how Bentham's Panoptic principle - Foucault's gaze again - shaped
Richards' model of literary worth as well as his teaching practice. Literary
criticism became less concerned with novels and poetry as an expression of
human endeavour and emotions but was turned into a 'project for the production of knowledge, the exercise of power, and the creation of careers'
(Bové, 1986, p.48) on the model of 'the other positive disciplines such as
economics, psychology, medicine, and anthropology' (p.48). Richards' project
was concerned with the training (i.e., 'discipling') of teachers and readers
as part of a wider programme which would 'manage (the) larger forms of socio-
cultural and political difference between men and women, various classes,
and competing ideologies and nations' (p.53)...'reducing the complex function
of modernist literature to an ahistorical training school for teacher education in cultural management' (p.55). His purpose was to develop a way of
seeing the world and its literature according to a 'normalized' and 'transparent' perception. Like the early constructors of the objectifying gaze
studied by Foucault in the fields of medicine, prisons, education and the
human sciences, Richards succeeded in applying the same normalizing vision
to a body of expression traditionally valued for its exploration of precisely
those areas of human sensibility which lie beyond the 'gaze of discipline':
the spontaneous, the erotic, the uncanny, etc..

The analysis of knowledge-power and the 'labour of division' in social
systems has two important critical functions: (1) it reorients analysis away
from the static picture of social structures produced by the division-of-
labour perspective to reveal 'division' as a central force in the social
production of 'visibility', and (2) it underlines the cardinal role of the
discipline-professions as social practices which create and maintain
'division' through 'visibilization' techniques.

REFERENCES

Bové P.A., 1988, "Intellectuals in Power: a Geneology of Critical Humanism",
    Columbia University Press, New York.
Derrida J., 1981, "Positions", University of Chicago Press, Chicago.
Febvre L., 1982, "The Problem of Unbelief in the Sixteenth Century", Harvard
    University Press, Cambridge.
Foucault M., 1973, "The Birth of the Clinic", Tavistock, London.
Foucault M., 1977, "Discipline and Punish", Allen Lane, London.
Foucault M., 1979, Governmentability, Ideology and Consciousness, 6:5.
Foucault M., 1980, "Power/Knowledge", Harvester, Brighton.
Frug G.E., 1984, The ideology of bureaucracy in government law, Harvard Law
    Rev., 97:1276.
Giddens A., 1985, "The Nation-State and Violence", Polity Press, Cambridge.

Goldstein J., 1984, Foucault among the sociologists: the "Disciplines" and the history of the professions, History and Theory, 23:170.
Hoskin K.W. and Macve R.H., 1986, Accounting and the examination: a geneology of disciplinary power, Acctg. Org. and Soc., 11:105.
Miller J-A., 1987, Jeremy Bentham's panoptic device, October, 41:3.
Norris C., 1987, "Derrida", Fontana/Collins, London.
Rorty R., 1980, "Philosophy and the Mirror of Nature", Basil Blackwell, Oxford.
Weinstein D. and Weinstein M., 1984, On the visual constitution of society: the contributions of Georg Simmel and Jean-Paul Sartre to a sociology of the senses, Hist.Eur.Ideas, 5:349.
Whitehead A.N., 1938, "Modes of Thought", Macmillan, New York.

POST MODERNISM: THREAT OR OPPORTUNITY?

G. Burrell

School of Industrial and Business Studies
University of Warwick
Coventry, U.K.

In this paper, it will be argued that the current debate between mod-
ernists and postmodernists opens up issues which are rarely if ever considered in the O.R. literature and that a position needs to be taken with
regard to a number of choices facing us in a self reflective way rather
than merely out of habit. Whilst postmodernism is seen as offering new and
innovative approaches to some dilemmas facing social scientists, it may be
that we turn our back upon modernism at our peril.

The state of social science in the late 1970s was characterised in a
number of ways by observers on the Anglo-American scene. For some (Giddens,
1979) it was a Tower of Babel in which the biblical description of Genesis
was a good metaphor for the diaspora of theorists to widely spaced encamp-
ments in which they spoke differing languages and viewed each other with
thinly disguised hostility. For others, the fragmentation that we witnessed
was of crisis proportions in that Parsonian hegemony was breaking down in
the face of new 'satellites' such as ethnomethodology (Gouldner, 1979). OR
too saw similar developments as conventional methodology came under attack
(Jackson, 1987). Whilst today there is still some concern at this level of
fragmentation, the social science journals by and large are not replete with
such issues. Partly this reflects a small body of opinion which believes
that the theory of structuration in the form developed by Anthony Giddens
has solved some of the problems posed by the clash of phenomenologically
inspired approaches with structuralist and functionalist orientated theories.
Partly the quieting of the debate reflects the political pressures upon
social theory in which sociology in particular has faced a variety of threats.
The natural response to these pressures has been a closing of ranks and the
underplaying of internal differences within the discipline. And partly the
internal dissension has been overlaid with the new concerns raised by the
rise of postmodernism.

Postmodernism is exceptionally difficult to define involving a movement
in architecture, in theatre, in art as well as in social theory. The gurus
of modernism of course have attempted to suggest that postmodernism is itself
an inchoate, fragmented and transitory set of approaches which does not bear
the weight of the epithet "movement" - and in this, as in other things, they
have a point. Nevertheless, the postmodernists in social theory share a
common concern which is based upon a mistrust of the notion of rationality,
truth and progress. They find the notion of a unity of theory, of history
and the subject as very difficult to justify. In place of such notions

dependent as they are upon a belief in the achievement of the Enlightenment, postmodernists suggest that the reversibilities of history, the importance of the contingent, the superficial and the relativities of the world are the key features which social theory needs to address. Postmodernism stands for post 1968 decentralisation, the dominance of the image and appearance over 'reality', of parody and nostalgia over realism. The importance of this shift in orientation has very recently been highlighted in a very insightful set of articles in Marxism Today. In these articles there is a discussion of modern times and the cultural artifacts of that period. This is juxtaposed to New Times in which the features of postmodernism are well highlighted. The differences between these two cultural systems are evident in the following:

MODERN TIMES	NEW TIMES
Fordism	Post-Fordism
Modern	Post-modern
Steinbeck	Pynchon
Le Corbusier	Venturi
Sarte	Foucault
Futurism	Nostalgia
Marlon Brando	William Hurt
Production	Consumption
Mass Market	Market Segmentation
Ford	Toyota
Self-control	Remote control
Depth	Surface
Elvis	Michael Jackson
Interpretation	Deconstruction
Butlins	Theme Parks
Relationships	White Weddings
The Beatles	Bros
Determinism	The arbitrary
Maxwell House	Acid House
Concrete	Holographic Glass
Liberalism	Libertarianism
Mass Hysteria	Fatal Attraction
Humanism	Post structuralism
Raspberry Ripple	Hedgehog crisps
Lady Chatterley	Blue Velvet
World Wars	Terrorism
Angst	Boredom
Roosevelt	Reagan
In/Out lists	New times guides
Newspapers	Colour supplements
Z Cars	Miami Vice
Conservation	Thatcherism
Emotion	Affectation
Dow Jones	Nikki Index
Stalinism	Glasnost
Free Love	The Free Market
The Titanic	Challenger
The Cabinet	The Prime Minister
Bingo	The Big Bang

A COMPARISON OF MODERN AND NEW TIMES (Source: MARXISM TODAY)

Now it may be that the reader is unable to relate to the New Times column at all. Many of the items listed may well be ephemera but the table does catch the mood of cultural transition - and this is the important point.

60

One can see from this table that, for <u>Marxism Today</u>, postmodernism is subsiduary to post-Fordism. That is, culture and changes within it are to be explained in relation to the rise of Post-Fordism. Post-Fordism includes all or some of the following characteristics: more decentralised forms of labour process, the contracting out of functions and services, an emphasis on marketing and design and upon product differentiation, a rise in the service and white collar classes with an attendant feminisation of labour, the globalisation of financial markets and a new spatial organisation of multinationals autonomous in large measure from nation state control. In John Urry's somewhat questionable terms (Lash & Urry, 1986) we are leaving the stage of organised capitalism and entering its 'disorganised' phase where class, city, region, nation and party - even the word - are melting into air. Notice here that 'new' technology is not attributed a key determining role. Its subservience to developments in capitalism is assured.

Now whether post-modernism is a reflection of these economic changes at the level of superstructure or whether there is some structural autonomy for the contemporary cultural apparatus is not easy to answer. However there is some acceptance of the fact that 'All that is solid melts into air'. In other words, we are supposedly in the midst of major changes which are transforming our society, its economy, its culture and its politics. We must proceed, however, with caution. It seems to me that almost every generation must have believed that it was in the last throes of some terrible upheaval the like of which the world had never seen before. Life and death, growth and decay are a feature of human existence so change is something that even the 'coldest' societies (in which change is very slow) must have imagined that they lived with. Today, such is human egocentrism that to believe that one's life is not marked by major disruptions is a negation of self-importance and individuality. Therefore such beliefs are belittled.

For the sake of argument, however, let us admit that post-modernism is reflective of changes underway within 'advanced capitalism'. Of course it does not represent a transcending of capitalism but rather post-modernism is part of Jameson's new 'cultural logic of capital'. It is the expansion of capital into hitherto uncommodified areas. Nevertheless, I would submit that this period of 'new times' needs to be taken seriously by those in O.R. In this, let us echo the words of Rosenhead and Thunhurst (1982) "that in its history and in its current practice O.R. is deeply rooted in wider social processes". Developments in OR can only be understood in terms of the development of capitalism as a whole.

POSTMODERNISM AND O.R.

In this section, I will consider some of the interrelated processes locked up in post-modernism. Each theme has to be seen as closely connected with the others and as forming part of a totality. What then are the cultural and economic forces likely to bear upon O.R. and its theory and practise if 'postmodernism' is seen as a descriptor of some utility?

<u>The decline in a belief in rationality and progress</u>

The growth of powerful mathematical techniques marks the history of O.R. development (Jackson, 1987). But so too does a belief that solutions and problems are possible, that 'performativity' can be enhanced by better problem solving techniques. And what if problems are <u>not</u> tractable? What if mathematics cannot answer the basic political issues of in whose interests is the solution being developed? But worse still, what if there is no progress - merely advances and then reversals, reversals and counter-reversals-cyclicality or randomness rather than unilinear development? What threat does this concentration or uncertainty pose. Of course it may be pointed out that soft systems thinking deals with such 'messes' (Ackoff, 1981, Jackson, 1987) but even in the work of thinkers such as Checkland (1981) there is clear commitment to learning and the reduction of uncertainty. Rationality is still assumed to be the goal.

The rejection of history as unilinear

We may reject 'history' as being far from unilinear but the temptation is not to do so. We are tempted to say we arrived here having journeyed deliberately. But what if the future is pluralistic and random. What does that say for long range planning, forecasting and the like? The 'predict and prepare' paradigm, of course, may no longer be pre-dominant but it appears to me to be the accepted form of 'normal science' in the discipline. The sophisticated critiques of this predict and prepare approach developed by leading figures may have had the desired impact but if this is the case, why do they continue to develop the critique?

The move towards decentralisation and sub-contracting within large organisations

Yet on the one hand, there appears to be optimism that O.R. is progressing, helped largely by developments in computing technology. Yet at a recent meeting of O.R. specialists which I attended the atmosphere was pessimistic. The large centres of O.R. expertise in public and private corporations were seen as being broken up by the 'need' of the organisation to be flexible and to be fashionable in using the notion of core and periphery (Atkinson, 1986). This declaration was seen as a force which might encourage the peripheralization of the O.R. function. The role for O.R. as a separate function with its own staffing located in one department in this 'post-Fordism' phase is less clear than it was. Of course, decentralisation and centralisation are not antonyms. It is possible to have both processes simultaneously at work. Sub-contracting may be an empowering as well as weakening development.

Consumption rather than production

The emphasis on the superficial packaging and appearance of an object rather than its construction and production (Baudrillard, 1988) has shifted the balance of power in many organisations in the direction of marketing and away from traditional O.R. fields. Here again a shift in O.R. techniques towards marketing and image might be what seems like a 'logical' step but have we not just suggested that logic is not a post-modern concept? Moreover has O.R. in the USA not been involved in the marketing areas for many years? Baudrillard's horrifying dictum of the post-modern world 'I shop, therefore I am' encapsulates this movement away from production to consumption.

Pastiche and parody, irony and colour

In post-modern architecture, Miami is held up as an advanced city. Styles there are mixtures of historical periods in juxtaposition with no sense of unilinear developments but in place of this sense of time and progress they exhibit pastiche and parody with heavy use of colour and irony. I may be wrong but I see little of these forces of humour and lightness in contemporary O.R. Who is constructing a post-modern O.R. in which these superficialities are given a high profile? The response to this point may well be that who should construct such a set of superficialities? Again, the issue turns upon disciplines which are in tune with change and which can reflect dynamic forces against those which are unaware of their contexts.

No legitimatory role for O.R.

According to Rosenhead and Thunhurst (1982) a major fun
in the (modernist) phase of capitalism now receding from viet
atory. O.R., it is said believed 'leave it to the expert' an
the same boat, don't rock it'. But legitimation requires a co
the times and a role in which people can believe for themselvet
a discipline seeks to continue this role which, after all, keep   , of us
in business then it needs to reflect New Times rather than Modern Times. -
Or does it?

The rise of O.R. in the Post-World War II period is difficult to explain but clearly its development was related to the growth of the nationalised industries and a belief in state involvement in production. Moreover,
O.R. continued into the 1970s to play a role in national and local government. The emphasis upon analysis and detailed study marked O.R. off from
Computer Departments which were seen as mere technical outfits in which
'number-crunching' was carried out but not deep analysis of problems per se.
Little recognition seems to have been given to the huge budgets which those
in charge of main frame computers would be able to claim and scant regard
was paid to the consequences this access to financial power would create.
So as the demand for centralised systems in the Health Service, Local Government and Nationalised Industries grew it created a functional position for
O.R. which it might have more completely grasped. Nevertheless, O.R. is
associated, whether it likes it or not, with the post-war consensus, with
Keynesian economics, with a corporatist orientation and with the large scale
nationalised industries.

O.R. is a product of the 'Butskellite' policies pursued by most Governments until the spring of 1979. O.R. then is essentially modernist and
Fordist but may grow again for all that. Who is to say that Thatcherism
will be with us in the 1990s? Who is to say that the post-Fordist concentration on high quality, high price luxury goods is a viable strategy for a
Britain in a post 1992 Europe? Already Britain is seen by some of those in
the Pacific Rim as a low wage tax-incentive economy. Geopolitical moves in
terms of economic shifts and political realignments may mean that a return
to liberal corporatism, mass production and state welfareism yet again produce a climate in which the old O.R. might flourish.

TOWARDS A POSTMODERNIST O.R.?

What if modern times do not return, however? Well, first of all, there
is the belief that OR is changing. Community OR, Soft Systems Thinking and
so on are seen by some as escaping from the traditional perspectives. But
my point is that even these progressive views do not realise the full force
of New Times. If it is decided that O.R. should embrace The New Times,
what sort of things might be on its agenda. Let me suggest three topics.

1. Formal organisations are the present expressions of power that mas-
    querade as rational constructions. In Michel Foucault's analysis,
    organisations are episodic and unpredictable manifestations of a play
    of dominations (Smart, 1983:76). In other words, O.R. would need to
    get political!

2. Foucault also suggests in his 'genealogical method' the following ad-
    vice to researchers:-

    (i)   Record the singularity of surface events looking at the meaning
          of small details, minor shifts and subtle contours.
    (ii) There are no fixed essences or underlying laws. There is dis-
          continuity and arbitrariness. Since the world is at it appears
          seek out the 'superficial secrets'.

(iii) Act as the recorder of accidents, chance and lies. Oppose the search for depth and inferiority.

Would it be possible for such an O.R. to be developed in which analysis was given up in favour of image and superficiality? This may be very much against OR's self image as the revealer of underlying truth but what if such a geneological method were tried? What would it look like?

3. Finally would it be possible to develop an ironic, humorous O.R. in which there was a lively, colourful, sardonic, sarcastic, shocking language. What linguistic devices could be developed to bring O.R. into this world of a new vocabulary - a brave new word world. Only half tongue in cheek, one suspects that the mathematical and systems engineering origins of O.R. have left a mechanistic legacy and in the terms of mechanical failure we find elements of alternative language. When commodities or products cease to work effectively the language of sexuality is invoked (They're buggered, it's knackered etc.etc..) Perhaps here is the basic position from which a new colourful linguistic code for O.R. might be developed. In any event, New Times require New Language.

But for those whom all this is mere philosophical affectation there is a note of optimism. Habermas has accused post-modernists of being irrational and of turning their backs on all the achievements of the Enlightenment. For him, progress is notable; for has West Germany not forgotten the gas chambers and made great political and economic strides since 1945? Without a belief in progress, human endeavour and critical human agency, Habermas sees a dark descent into irrationalism. For many O.R. personnel this may be precisely the way they see the world. Postmodernism is a force threatening and antagonistic towards the world in which O.R. grew up - the modern world. But if we make a decision to turn our backs on postmodernism in its economic and cultural forms let us do so in the self reflexive glow of having thought about our beliefs - and not simply with them.

REFERENCES

Ackoff R.L., 1981, The art and science of mess management, Interfaces, 11(1):20.
Atkinson J. and Meager N., 1986, Is flexibility just a flash in the pan?, Pers.Mgt., September.
Baudrillard J., 1988, "Selected Writings", Polity Press, Cambridge.
Checkland P.B., 1981, "Systems Thinking, Systems Practice", Wiley, Chichester.
Giddens A., 1979, "Central Problems in Social Theory", Macmillan, London.
Gouldner A., 1970, "The Coming Crisis of Western Sociology", Heinemann, London.
Jackson M.C., 1987, Present positions and future prospects in management science, Omega, 15:445.
Lash S. and Urry J., 1986, "Disorganised Capital", Polity Press, Cambridge.
Rosenhead J. and Thunhurst C., 1982, A materialist analysis of O.R., Jnl.Op.Res.Soc., 33:111.

OR AS A PROBLEM SITUATION:

FROM SOFT SYSTEMS METHODOLOGY TO CRITICAL SCIENCE

H. Willmott

School of Management
University of Manchester Institute of Science and Technology
Manchester, U.K.

During the past 25 years, the discipline of OR has become increasingly self-conscious and self-critical of its foundations (Dando and Bennett,1981: Jackson and Keys, 1987). Whereas problems were thought to exist out there, in the world, a growing number of OR academics and practitioners now believe that traditional, 'hard' OR is itself as much a source of problems as it is a means of resolving them. Specifically, it has been suggested that these problems arise from an incapacity to appreciate the reflexive, socially constructed nature of problem situations. As faith in the assumptions of 'hard' OR has been shaken, alternative, 'softer' methodologies have emerged to challenge its supremacy. In turn, these efforts to break out of the traditional mould of OR have stimulated a critical evaluation of the adequacy of their challenge.

The architects of 'softer' methodology as well as their critics, have sought inspiration from critical thinkers in the social sciences (e.g. Habermas, 1972) who, during the same period, have been engaged in a major intellectual struggle over the question of what is to count as 'scientific' in the study of the social world. Reviewing the developing fragmentation of methodologies within the social sciences, Giddens (1979) observes how, during the 1950s and 1960s, there was a widespread belief in the adequacy of functionalist thinking in which biology was seen to provide 'the proximate model' for the study of the social world', a view underpinned by a faith in logical empiricism as a unified (positivistic) methodology for all forms of science (Giddens, 1977).

In the past two decades, however, reflection upon the claims of the 'hard' sciences has been inspired above all by Critical Theory (Adorno et al., 1976; Bernstein, 1976). Critical studies of science have questioned whether logical empiricism actually does underpin natural scientific enquiry (e.g. Barnes, 1974); and philosophers of social science have doubted the (exclusive) legitimacy of logical empiricism as a foundation for the social sciences (e.g. Keat and Urry, 1977; Johnson et al., 1984). As a consequence, there has been a rather dramatic fracturing of the orthodox consensus within both the social sciences and the management disciplines (e.g. administrative science, accounting as well as OR), so that it is now increasingly difficult to deny that 'the logical empiricist view of science represents only one possible philosophy of science among other available philosophies' (Giddens, 1979: 238).

In the vanguard of this challenge, 'softer', more qualitative methodologies present an alternative, and increasingly legitimate, to the logical
positivism of 'hard' systems thinking (Dando and Sharp, 1978). Although a
scan of the journals reveals that the credentials of 'soft' methodologies,
both in practice and in theory, are not universally accepted, 'softer'
approaches are in the ascendant. Many 'hard' OR practitioners are clearly
unconvinced of their value. At the same time, the theoretical foundations
of these methodologies have been subjected to a critical examination by
those who are skeptical of its theoretical and practical claims. In general, the critiques of the theoretical foundations of 'soft' methodologies
are not made by defenders of 'hard' thinking who, it appears, seem content
to ignore the challenge. Rather, the most penetrating critiques have been
developed by those who draw upon Critical Theory to argue that 'soft' methodology is <u>insufficiently penetrating</u> in its critique of orthodox, 'hard'
systems thinking. More specifically, it is insufficiently reflexive about
its own values and problem-solving capacities, a deficiency which is at once
a condition and a consequence of its lack of a social theory necessary to
grasp why problem situations become defined in particular ways (e.g.,
Mingers, 1980; Jackson, 1982; Tinker and Lowe, 1982; 1984; Ulrich, 1983).

The purpose of this paper is to review and advance the debate about the
adequacy to the 'soft' systems challenge to 'hard' systems thinking. More
specifically, the paper explores the use of Critical Theory (Held, 1980;
McCarthy, 1978; Roderick, 1986) in assessing the theoretical foundations
of 'soft' methodology before considering its practical application. In the
course of the paper, three uses are identified. The first is Checkland's
use of Habermas' work to interpret and legitimate the commitment and contribution of his 'soft' system methodology (SSM). I have not paid much attention to this as I believe it to reflect and expose the very limitations of
SSM which are exposed by those who used Critical Theory to critique 'soft'
methodology. Much of the chapter is taken up with the debate on SSM pro-
voked by Jackson's (1982) critical review of the work of the major proponents
of 'soft' methodology: Ackoff, Churchman and Checkland. Focussing upon the
debate between Jackson and Checkland, the paper seeks to shed light upon the
strengths and weaknesses of their respective arguments. In essence, Jackson
is criticised for failing to provide a more explicit and convincing rationale for his critique of SSM, while Checkland is criticised for using this
failure as a defence against the force of Jackson's argument. In the remainder of the paper, a third application of Critical Theory is briefly considered in which Habermas' work provides the inspiration for an alternative
'soft' methodology for OR developed by Ulrich.

THE CHALLENGE TO 'HARD' METHODOLOGIES OF OR

Three of the most influential figures in OR - Ackoff, Churchman and
Checkland - have urged, and made major contributions to, the development of
an alternative, 'softer' methodology (Ackoff, 1974; 1979a; Checkland, 1981;
Churchman, 1979a; 1979b). Questioning the ontological identity of natural
and social reality, their arguments reflect a greater awareness of the
presence of reflexivity in producing, defining and resolving problem situations.[1] This awareness they seek to develop among practitioners and users
of OR in order that the complexity and dynamics of these situations may be
more adequately appreciated and addressed. Before examining the case for
'soft' methodology more closely, however, it may be helpful to give a definition of OR in which the basic contours of the 'hard' approach are traced.
The following is the official definition of OR offered by the U.K. Society
in <u>Operational Research Quarterly</u>:

'Operational Research is the application of the methods of
science to complex problems arising in the direction and
management of large systems of men, machines, materials and
money, in industry, business and defence. The distinctive
approach is to develop a scientific model of the system, incorporating measurements of the factors such as choice and
risk, with which to predict and compare the outcomes of alternate decision strategies or controls. The purpose is to
help management determine its policy and actions scientifically'. (quoted in Dando and Sharp, 1978 :940)

Critics of 'hard' methodology have challenged a number of assumptions
contained in such definitions. For example, they have questioned the
appropriateness of scientific methods based upon logical empiricism for
studying phenomena whose identity is contested; they have criticised the
treatment of 'men' as if they were no different from other elements of the
system; and they have doubted the assumption that OR practitioners do,and
should, simply observe and predict, rather than actively participate in and
produce, the design of systems. With the possible exception of Churchman
(1979a), the most rigorous presentation of a 'soft' methodology has been
made by Checkland (1981). Checkland argues that its capacity to develop
models which are analytically sound harbors and conceals a failure to
appreciate idiosyncratic features of the social world which are of practical
significance 'in most real-world problem situations' (Checkland, 1983a; 670).
For, in this 'real-world', practitioners are 'dealing with a unique situation with its own possibly crucial chance events and its own individual
personalities; to such situations the aggregate results do not apply' (ibid.
:668).

In contrast to 'hard' methodologies which assume that the reality of
situations can be 'captured within systemic models', 'soft' methodologies
regard

'system models as models relevant to arguing about the world,
not models of the world' (Checkland, 1985 :765, emphasis
added).

Checkland illustrates this relevance when commenting upon the studies at
the Airedale Textile Company and in Cordia Engineering which were so influential in the rejection of 'hard' systems thinking in favour of SSM. He
writes

'Had it been possible to use hard systems methodology in soft
problem situations, the implication would have been that social
reality is indeed systemic, and that the problem for the analyst
is simply that of finding an appropriately sophisticated systems
model.

In fact, the dramatic - not to say traumatic - finding in our
formative studies...was that it simply was not possible to
take any system as given...[because]...there were potentially
as many relevant descriptions of the system as there were concerned actors in the problem situation. The early experiences
have been multiplied many times since then, and the general conclusion has to be that this research experience does not support
either the positivist assumptions of hard systems thinking or the
positivist account of the nature of social reality' (Checkland,
1981 :278)

From Checkland's perspective, the basic limitation of 'hard' systems
methodology is that, in reducing the singularity of all situations to the
common, aggregated logic of the situation, it disregards the way in which

each situation is uniquely constructed by those who are engaged in its (re)
production. By overlooking the constructed nature of reality, 'hard' systems thinking is seen to equate and confuse the modelling of problems with
the complex situations which these models seek to represent. Contrasting
their respective orientations, Checkland (1981 :279) stresses that SSM

> 'declines to accept the idea of "the problem". It works with
> the notion of a situation in which various actors may perceive
> various aspects to be problematical....The methodology...offers
> the chance of taking action in the real world by means of some
> systems thinking about the real world. Its emphasis is thus not
> on any external "reality" but on people's perceptions of reality,
> on their mental processes rather than on the objects of those
> processes'

Only in comparatively rare cases where interconnections between systems
elements are understood to derive from a universal logic is the use of a
'hard' methodology   deemed appropriate. As examples, Checkland gives
problems where there is a clear, shared objective - such as arrangements
for the manufacture or assembly of products (Checkland, 1983a :670). What
differentiates these problems, Checkland argues, is the ease with which
consensus can be formed on the definition of the problem : 'what at the
operational level may be agreed problems quickly become, at higher levels,
issues created by clashing norms, values and Weltanschauungen' (Checkland,
1985 :765). As a further illustration of his argument, Checkland contrasts
the problem of cataloguing library stock which, he suggests, is amenable
to a 'hard systems' engineering solution, with the problem of how the budget of the library should be allocated, an issue which is said to demand a
'soft systems' approach.

In the majority of problem situations, SSM is said to be more appropriate because the interconnectedness of phenomena is dominated not by 'the
logic of the situation', as in the case of arranging for the assembly of
products or the cataloguing of library stock, but, rather, by 'the meaning
attributed to their perceptions by autonomous observers' (Checkland, 1983a;
670, original emphasis). As a response to this interpretation of the 'real
world', SSM is presented as 'a systems-based learning system, one providing
guidelines for coping with real-world complexity' (Checkland, 1988 :28).
Stressing that there can be no uncontested, authoritative representation of
most problem situations since, to use another of his examples, one person's
'terrorist system' is another person's 'freedom fighting system',Checkland
(1983a :671) stresses the importance of remembering that 'systems thinking
is only an epistemology, a particular way of describing the world. It does
not tell us what the world is'.

In sum, whereas 'hard' OR assumes a unified reality containing well
defined objectives which can be more effectively fulfilled through the
application of its techniques, SSM is oriented to learning about the complex, multi-faceted constitution of social reality so that desired and
feasible objectives may be agreed and attained.

THE CRITIQUE OF 'SOFT' SYSTEMS METHODOLOGY

In this section, I consider the critical assessment of 'soft' systems
methodology by those who have drawn upon Critical Theory to reveal its limitations. Since Checkland's position has been examined in most detail above,
and has also received the most attention from the critics of 'soft' methodology, it will be convenient to take the debate stimulated by his work as
a focus. Checkland's response to his critics will also be considered, as
will the rejoinder to him. By examining the arguments closely, it is hoped
to clarify the key differences between the 'soft' and the 'radical' positions

and to offer an assessment of the strengths and the weaknesses in respective arguments. In doing so, it is perhaps worth noting that Habermas' work has been a common influence upon both positions, though the significance of his work is interpreted and applied in rather different ways. This influence is also reflected in the willingness of both parties to engage in a dialogue about their differences to which the present paper is intended as a contribution.

In what is perhaps the most ambitious examination of the work of the most influential 'soft' systems thinkers, Jackson has criticised the work of Ackoff and Churchman, as well as Checkland, for its 'conservative, regulative bias' (Jackson, 1982 :24; c.f. Mingers, 1984). Drawing heavily upon Burrell and Morgan's (1979) discussion of the paradigmatic location of varieties of social scientific enquiry, Jackson challenges the adequacy of Checkland's self-understanding concerning the problem-solving potential of his methodology. For whereas Checkland argues that SSM can promote either conservative or radical forms of change, Jackson argues that this contention reflects a fundamental misunderstanding of the inherent bias within SSM towards the former, to the exclusion of the latter.

What Checkland claims is that SSM is

> 'neutral in itself but contains the potential to be as reactionary, as emancipatory, as radical, as revolutionary as the user makes it' (Checkland, private communication, cited in Jackson, 1982 :24).

According to Checkland, if conservative, regulative outcomes are generated from applications of SSM, this reflects the purpose of the analyst, not the nature of the methodology. Against this claim, Jackson contends that the regulative bias in SSM stems from its phenomenological commitment to attend exclusively to 'people's perceptions of reality, on their mental processes rather than the objects of those processes' (Checkland, 1981 :279, emphasis added, also cited above). It is this commitment, Jackson argues, which renders SSM incapable of appreciating the 'particular social arrangements and institutions' (Jackson, 1982 :25) within and through which Weltenschauungen emerge and are negotiated to produce a consensus. And, as a consequence, SSM overlooks and therefore colludes in preserving a silence about the role of social arrangements in constraining and distorting the process of consensus formation.

To support this critique, Jackson refers to seven stages which comprise the process of analysis within SSM. Specifically, he focusses upon stages four and five during which models of competing Weltanschauungen, constructed during the earlier stages of SSM, are discussed. During the stages, different perspectives on a problem situation are compared and contrasted, judgments about their empirical and normative content are examined and perhaps revised. Out of this process, a consensus gradually emerges about change that is both desirable and feasible. Drawing upon Habermas' (1970a; 1970b) theory of communicative competence, Jackson (1982 :25) argues that, in the context of 'societies characterised by great inequalities' conditions free of domination are rarely, if ever, realised. In these circumstances, communication will express structural asymmetries of power between participants producing a consensus which, in part, will be the product of force, and therefore is deceptive or 'false'.

> 'the kind of unconstrained debate envisaged [by Checkland] cannot possibly take place. The actors bring to the discussion unequal intellectual resources and are more or less powerful. The result...is that the "ideologies" of the powerful are imposed upon other actors who lack the means of

recognising their own true interests.   The result of the inequalities of power is that the existing social order, from
which power is drawn, is reproduced' (ibid :25, emphasis added)

This assessment directly contradicts Checkland's (1981 :20) own claim
that SSM 'is a formal means of achieving the "communicative competence" in
unrestricted discussion which Habermas seeks'.   For Jackson, SSM is defic-
ient because it accepts uncritically the agreement which emerges from stages
five and six of SSM. Not only does it disregard the inequalities of power
which may serve to contrive or 'force' agreement for want of the resources
necessary to challenge it.   But, in failing to explore this possibility,SSM
serves to legitimise the understanding that agreements reached through its
application are the product of free and frank discussions through which prejudices are aired and dissolved, and a consensus about desired and feasible
change is attained.   The phenomenological foundations of its methodology
deny the very possibility of speaking about objective social arrangements
which enable and constrain agreement.   Or, as Burrell (1983) has phrased
this critique, SSM is unable to connect its identification of meanings or
Weltanschauungen to 'the structure of interests' which underpins both the
diversity of Weltanschauungen and the forcing of consensus within assymmetr-
ical relations of power.

When responding to Jackson's critique, Checkland (1982) attacks its very
basis : namely, the claim that the adoption of a radical perspective provides
knowledge of the objective structure of domination and the possibility of a
true consensus.   These claims are seen to rest upon no more than an implicit,
unsupported and therefore unjustified faith:

'Jackson takes as given an objective social reality characterised by "structures" which put "constraints" upon "groups"; he
believes (again without offering evidence) that there could be
a true consensus, as opposed to the "false consensus" which is
all that Churchman and Ackoff and Checkland might achieve. The
reader may feel it significant that when Jackson writes of Habermas'
view that the social world takes on constraining objective features,
man being "in the grip of unconscious forces", he writes not that
Habermas believes this to be the case but that he "recognises" it.
(ibid :37)

Instead of awaiting the arrival of a utopian situation in which there
is an equality access, resources, etc., Checkland defends the immediate and
proven value of his approach as a learning process in which 'people negotiate and renegotiate with others their perceptions of the world' (ibid :38).
Instead, elsewhere Checkland (1983b :129) suggests that the extent to which
'debate is constrained by societal structures and dispositions of power'
could be investigated experimentally by applying SSM.   Further, Checkland
reaffirms his view that, in principle, the outcome of this process of negotiation cannot be known in advance, and that it may either confirm the
status quo or it may radically challenge it, depending on the kind of change
of perception which emerges from discussion.   Conceding that in most applications 'defining changes which are "culturally feasible"' (Checkland, 1981
:281) has led to rather conservative use of the methodology', he nevertheless insists that this is not inevitable.

'It would be legitimate to argue that soft systems methodology
inevitably props up the status quo if, and only if, it were
possible to demonstrate that problem-owners' Weltanschauungen
were immovable. All of the experience with the methodology, as
well as much observation of everyday life, suggests that
Weltanschauungen do change, sometimes incrementally, sometimes
radically' (Checkland, 1982 :39)

How, then, does Jackson defend his critique? To Checkland's observation that his criticisms are informed by an unquestioned and therefore unstated Weltanschauung, Jackson (1983) states only that he is 'prepared to view the world through the radical sociological paradigms identified by Burrell and Morgan'(ibid : 111). However, quite apart from the inconvenience of the fact that, in the latter's assessment, the incommensurability of these paradigms denies the coherence of this claim, the only justification offered by Jackson for privileging this perspective is his interest in seeing what can be learnt by doing so. That is to say, he presents no argument for taking seriously what this perspective reveals against what may be learnt from the adoption of other approaches, such as SSM. Without qualification, he simply repeats his assertion that 'it is possible to identify and study persistent structural features in such systems' (ibid: 111) and that 'the "real" nature of social organisation' (ibid :112) may be concealed when the Weltanschauungen of more powerful actors are adopted. Despite a modesty which leads him to clothe his priviledged access to 'reality' in scare quotes, Jackson remains coy about how this access is gained and justified.

In sum, SSM stands accused of silencing examination of what they take to be the objective conditions which frustrate, as well as facilitate, the discussion of Weltanschauungen and the process of consensus formation. In this sense, SSM is understood to 'confine itself to working within the constraints imposed by existing social arrangements' (Jackson, 1982 :27). In his defence, Checkland attacks as unjustified or dogmatic the faith of his critics in the existence of the real structures whose objectivity SSM is accused of failing to recognise. The potential for SSM to promote regulation or radical change, he argues, would only be closed off if it could be demonstrated that Weltanschauungen are immovable, a position which he finds both implausible and paradoxical when adopted by proponents of radical change. Responding to this defence, Jackson repeats his claim that the structural features of social systems can be identified when the world is viewed through a radical paradigm. But, beyond that, he offers no justification of this claim or of the value of adopting this perspective.

DISCUSSION

It is difficult to resist the force of Checkland's arguments that Jackson is insufficiently reflective about the foundations of his critique. Jackson's critique is seriously weakened by a failure to anticipate and head off the dismissive response that its 'reference to "objective social reality"...needs no kind of justification or defence' (Checkland, 1982 :37). The existence of institutions and social arrangements which constrain discussion are simply asserted, without any attempt to justify either the claim itself or the deduction that their presence means that, as a consequence of inequalities of power, many actors 'lack the means of recognising their own true interests' (Jackson, 1982 :25). While Checkland can be criticised for exploiting this weakness in order to avoid any serious consideration of Jackson's argument, it is clearly necessary to provide a more adequate justification of the Weltanschauung which informs Jackson's critique. Simply stating that, when viewed through Burrell and Morgan's radical sociological paradigms, social systems 'exhibit "objective" properties' (Jackson, 1983 :111) and that SSM 'has an intrinsic bias towards regulative action' (ibid :112) does not suffice.

The basic difficulty with Jackson's argument is that it lacks any justification for privileging the radical paradigm of analysis beyond the understanding that the choice is made 'to see what I can learn about the [the social world] by doing so' (ibid :113). Presented in this way, Jackson's self-understanding of his decision appears to be highly voluntaristic and, indeed, subjectivistic. That is to say, the adoption of the paradigm appears to be directed either by personal curiosity or by an interest in knowledge for knowledge's sake. This is unfortunate since Jackson's subsequent references to 'life chances', systemic constraints upon action and the association of "truth" with a demoncratic consensus suggest that his choice of methodology is fundamentally associated with (undisclosed) moral and political concerns.

However, in Jackson's defence, it is relevant to point out that, by exposing the dependence of his critique upon an absolute, mystical knowledge of the existence of 'societal constraints, domination, etc.', Checkland (1982 :39) evades Jackson's efforts to draw attention to the incapacity of SSM to stimulate reflection upon the <u>possibility</u> that the material and ideological conditions do shape the content and negotiation of Weltanschauungen. Despite weaknesses in its presentation, a rational kernel does lie buried in Jackson's mystical, absolutist shell in which he encases his critique. The chief deficiency in 'soft' systems methodology, identified by Jackson but then unconvincingly critiqued, is not that it fails to recognise 'objective structures'. Rather, its major shortcoming lies in its <u>unnecessarily limited</u> capacity to promote reflection upon the <u>possibility</u> that the content and negotiation of <u>Weltanschauungen</u> are expressive of <u>asymmetrical relations of power</u> through which they are constructed and debated. <u>Because it is unexaminable within SSM, the question of how situations come to be defined in particular ways and why one definition of a situation is more widely accepted than others, albeit that this acceptance may be contested and temporary, is left untheorised.</u>

Phenomenology, and SSM in particular, simply lacks a <u>social theory</u> capable of accounting for why particular sets of perceptions of reality emerge, and why some perceptions are found to be more plausible than others. Although SSM seeks to elucidate different typifications of the problem situation as a basis for a debate in which a consensus on agreed changes may be reached, it is unable to shed any light upon the question of how or why different typifications exist, why some typifications are more widely held than others, or what circumstances facilitate of impede the process of consensus formation.

Although Checkland (1981) allows that SSM may be 'inhibited by society's structure' (ibid :283), this inhibition is understood as a <u>universal</u> condition of social existence. Or, as he puts it, 'it is the nature of society that this will be so' (ibid). In support of this argument Checkland cites Popper who also universalises this inhibition of rational criticism when he observes that 'at any point in time, we and our values are products of existing institutions and past traditions' (ibid., citing Popper; c.f. Popper, 1976 :293-4). This understanding simply reflects the lack of a social theory - that is to say, a theory which seeks to explore how the <u>particular features</u> of contemporary social reality are objectified and reproduced. The limitation of phenomenology is that it universalises the processes of reality construction through which particular social realities are reproduced or transformed; it cannot account for how any particular social world is maintained.[2] In order to develop such an explanation, it is necessary to reveal the role of <u>power</u> in defining and transforming the boundaries of what it is possible to think, and to be, in the social world.

Of course, this is precisely what the critics of SSM are driving at when they argue that 'it attributes to ideas...too much efficacy in relation to other aspects of social systems' (Jackson, 1983 :112). As Jackson goes on to note, ideas - or perceptions of problem situations, as Checkland might prefer to say - are not produced in a social or historical vacuum. They arise, and gain plausibility, within particular relations of power. Or, as Jackson expresses this understanding, in 'some political and economic circumstances ideas generally will develop better than in other political and economic circumstances' (ibid). However, while Jackson identifies the inherent limitations of a phenomenologically based methodology, he assumes that a radical perspective provides unproblematical access to 'the "real" nature of social organisation' (ibid) in a way which is denoted by phenomenology.

TOWARDS A CRITICAL METHODOLOGY IN OR?

Precisely how the presence and role of power is revealed and critiqued is a matter of considerable theoretical debate (e.g. Lukes, 1974; Miller, 1987). For Habermas, the answer lies in an appeal to the theory of communicative competence, incorporating a model of an ideal speech situation, which serves as a standard for assessing the degree of distortion within empirical arenas of consensus formation. However, as Ulrich (1983) has argued, Habermas' preoccupation with the theoretical justification of the claims of critical science has been at the expense of developing a methodology which could illuminate relations of power by revealing the normative content of problem definitions and methodologies. It is this challenge which Ulrich takes up in the development of Critical Systems Heuristics (CSH).

Taking the project of advancing reasonable practice as its mission, CSH acknowledges its incapacity to provide a theoretical defence of its categories. Its categories are self-consciously metaphysical in the sense that they contain unjustified hermeneutic judgments about the world and that these judgments are recognised as sources of possible deception. But, against this formal deficiency is posed the substantive lack of critical theory which CSH seeks to fill. Namely, 'a conceptual framework that could transform critical theory from a mere research program into a practical tool of critical social inquiry and design' (ibid :155).

There is insufficient space here to explore the details of Ulrich's critical science. Two features only will be highlighted. The first of these he shares with 'soft' systems methodology: the concern to appreciate the purposeful nature of human systems. However, whereas Checkland applies this insight to learn about the plurality of Weltanschauungen which constitute/interpret 'problem situations', Ulrich (1986) argues for the need to go beyond the exploration of (competing) frames of meaning to appreciate how definitions of problem situations necessarily involve what he terms 'justification break-offs'. The concern to render transparent the implicit normative significance of breaking off the process of problem definition from further critical reflection upon the plausibility of definitions of the situation is the second, most distinctive, feature of CSH.

At the heart of Ulrich's approach is the understanding that any definition of the situation, whether developed by the expert or the lay person, necessarily involves a field of taken-for-grantedness which is 'unjustified'. For such a definition to have stability or continuity, there must be a point at which critical reflection upon the good sense of a definition of the situation is suspended. And, at this point, for all practical purposes, the definition assumes the status of an "objective necessity". The purpose of CSH is to reveal the normative content of the boundary judgments associated with 'justification break-offs'. So doing, it is intended to enhance the rationality of systems design by opening up debate upon the normative underpinnings of both descriptive and prescriptive definitions of problem situations.

Built around a checklist of twelve questions which are intended to enable both those involved in the developing systems to interrogate 'a design's normative content and challenge the "objective necessities" by which the other side may seek to justify or dispute the underlying boundary judgments' (ibid :6), the objective of CSH is 'to render transparent, and reflect on, the normative content of systems designs, problem definitions, or evaluations of social programs' (ibid :3). Having exposed the boundary judgements, they are then subjected to critical examination through an application of polemic. The term 'polemic' is used in a somewhat ironical way to recall that all definitions and statements necessarily rely upon 'justification break-offs' and, in this sense, share a common limitation. Ulrich contends that this shared characteristic empowers the lay person/expert to challenge the definitions favoured by the expert/lay person without any special obligation to justify the rationality of such a challenge. So long as she/he does not privilege the validity of her/his own definition of the situation, it is possible to impose the burden of proof, or justification upon those who do (implicitly) make positive validity claims about their own definitions.

> 'The polemical employment of reason secures to both sides an
> equal position for reasonable dialogue...by advancing its own
> good grounds, i.e. "facts" and "values" capable of consensus
> such as existential needs of all individuals, ecological know-
> ledge....The fact that ultimate justifications remain impossible
> provides no sound reason for renouncing any effort to bring in
> one's own good grounds. Nor does it provide the other side
> with a good ground to refuse entering upon such an effort :
> skepticism, turned into an argument against any argumentative
> effort, is no less dogmatic than the expert's reference to
> "objective necessities"' (ibid :13)

It remains to be seen how fruitful Ulrich's methodology is in eliciting a rational justification of the normative contents of boundary judgements. In common with Checkland's SSM, it offers a methodology for learning about the complex construction of the social world; it makes no claim to tell us what the world really is. In this respect, it escapes the criticism levelled against Jackson's critique of 'soft' systems methodology. However, Ulrich's Critical Heuristics also goes beyond the confines of SSM when it seeks to expose the normative content of the 'justification break-offs' associated with boundary judgements. For it attends directly to the way in which perceptions or definitions of problem situations are founded, more or less explicitly, upon normative boundary judgements upheld within particular relations of power. By revealing their foundations, CSH is able, in principle, to 'expose the dogmatic character of the expert's "objective necessities"' (ibid :12), and thereby elicit from the proponents of these definitions a justification of their judgements.

However, there are some grounds for doubting the efficacy of Ulrich's methodology. As Ulrich himself observes, 'an equal position for reasonable dialogue' can be achieved only if each side 'renounces the dogmatic or cynical employment of boundary judgements' (ibid :13). Absent is any consideration of how such a renunciation is to be achieved, the conditions which promote it, or the steps which need to be taken to create such conditions. Instead, it is simply argued that the successful operation of the methodology requires this renunciation to occur. The basic problem, then, with CSH is its assumption of the availability, if not the presence, of the very conditions which are implicitly denied by the recognition that justification break-offs associated with boundary judgements are very often dogmatic or cynical. To remedy this weakness, it is necessary to complement CSH with a theory which accounts for why, in the contemporary context, dogmatism is substituted for reasonable dialogue; and, secondly, a theory which may enable individuals to challenge and transfer the asymmetrical relations of power which promote and are protected by dogmatism and cynicism. Only by developing an analysis of why justification break-offs occur together with a theory which facilitates an effective challenge to their powerful continuation may it be possible to construct a methodology which facilitates a change in the relations of power which undermines the very foundations of dogmatic and cynical justification break-offs. Such a development will require a 'positive' theory of power in the sense that it must go beyond the universalistic observation that the privileging of definitions of reality is accomplished through relations of power which they frequently serve to reproduce.

When formulating such a theory, it is important to acknowledge its provisional nature and, in particular, to resist the suggestion that the adoption of a radical paradigm yields access to the political and economic circumstances. In other words, it is necessary to acknowledge that any analysis of power - Marxist, Critical Theoretic, Foucauldian or whatever - is simply a (provisional) means of exploring the understanding that privileging of particular definitions of problem situations is not arbitrary or beyond rational analysis. And that this concern or orientation to the social world has not arisen out of intellectual curiosity but, rather, is guided, like all other forms of enquiry, by an implicit normative (i.e. moral and political)concern. In this case, the concern is not to improve control over existing systems using 'hard' methodology; nor is the concern to employ 'soft' methodology to gain a more appreciative understanding of the diversity of perceptions of problem situations. Rather, the normative concern is to identify and transform the conditions which impede or distort their rational analysis and transformation.

CONCLUSION

The paper has explored the contribution of Critical Theory to the development of alternatives to 'hard' systems thinking. From the perspective of Critical Theory, processes of social reproduction cannot be adequately understood either in terms of the functional needs of the system or by reference only to the Weltanschaunngen which guide social action. Where the former, 'hard' approach disregards the reflexively constructed nature of social reality, the latter 'soft' orientation can offer no explanation of why particular Weltanschauung are favoured. Critical Theory invites the construction of a methodology which acknowledges how the privileging of Weltanschauungen is associated with the organisation of power in society, an organisation which inhibits and distorts the process of identifying rational solutions to practical problems.

When 'soft' systems methodology (SSM) is accused of failing to locate this process of negotiating Weltanschauungen within 'the constraints imposed by existing social arrangements' (Jackson, 1982 :25), the phenomenologist's response is to be expected : namely, that the existence of these constraints is an expression of a Weltanschauung whose relativity his critics fail to acknowledge. To deny this relativity, it is argued, is to slip back into an objectivism where 'certain knowledge (concerning the stability of societal constraints, domination, etc.) is known absolutely' (Checkland, 1982 :39). While the proponents of 'soft' methodology are right to attack the absolutism inherent in Jackson's critique, this itself provides no answer to the question of why particular Weltanschauungen are favoured or privileged. What such critiques both reflect and obscure is the absence of a social theory in phenomenology/SSM capable of addressing this question.

For an application of Critical Theory intended to provide a critical alternative to 'soft' methodology, it was necessary to turn to the work of Ulrich. Through the use of 'polemic', Critical Systems Heuristics seeks to interrogate the normative basis of the definition of problem situations. In doing so, it intended that a dialogue will occur between the 'experts' who design systems and the lay persons who use them in which both parties challenge the other to explicate the unspoken basis of what they each take for granted. The intention of this methodology is to expose the presence of cynical and dogmatic judgements so that these will no longer be designed into systems. Though commendable in principle, it was argued that this methodology makes the idealist assumption that these parties are both willing and equipped to engage in such a dialogue. To the extent that these conditions are absent, or only partially established, it was concluded that this methodology needs to be complemented by a theory which provides a (provisional) analysis of resistances to communication sustained by the contemporary organisation of power within ideologies and institutions. Only by drawing upon such an analysis, it was suggested, could a critical methodology of OR be capable of identifying the actions most likely to be effective in challenging and transforming the relations of power which currently support justification break-offs based upon cynicism or dogmatism and stubbornly resist idealist appeals for their rational removal.

Notes
1.   It is relevant to note that the 'soft' critics of 'hard' systems thinking do not speak with one voice. It seems that Ackoff, for example, accepts the basic frame of reference of the traditional, 'hard' definition of OR, including its management-centredness, even though he is highly critical of the ineffectiveness of 'hard' methodologies where problems are treated as if they were an aggregation of tidy, stable problems that are susceptible to modelling, measurement and optimum problem-solving. Seeking to remedy this weakness, Ackoff recommends a 'softer' approach in which problem situations are perceived as untidy and changing : that is, are recognised as 'messes' (Ackoff, 1979 :99). Ackoff's distinction between the reality of 'messes' and the identification of 'problems' indicates an awareness of social reality as purposeful and negotiated - an insight which is reinforced by his concern to develop a methodology of OR which is actively engaged in construction of reality, not mere passive adaptation to it (Ackoff, 1979b :197). However, this awareness of the socially construction of reality is not explicitly thematised. His critique of 'hard' systems methodology is directed at its practical limitations, and not at its theoretical foundations. For a more systematic presentation and defence of the theoretical foundation of a 'soft' methodology it is necessary to turn elsewhere.

2.   When appealing to the authority of Popper, Checkland appears to believe that, in this instance, the person who is 'usually regarded as the implacable enemy of the Frankfurt School',is actually in agreement with Habermas. Leaving aside the issue of whether it is sensible to view Popper as an unequivocal enemy of the Frankfurt School (Giddens, 1977), Checkland completely overlooks the basis of their disagreement when he associates their respective recognition of the role of values and traditions in limiting the powers of critical reason. This disagreement is directly relevant for the discussion of SSM since, in a number of key respects, Checkland's methodology is vulnerable to the same attacks made by Habermas (1976) against Popper's critical rationalism. Namely, that it is insufficiently reflective about the historical conditions which define the limits of the rationalism upon which science is constructed. More specifically, it fails to theorise social phenomena in relation to the circumstances which they reflect and reproduce. To put this another way, criticism is restricted to the methods of social science to the exclusion of a critical examination of the specific conditions which support the historical appearance of its 'objects' of investigation. Yet, as Habermas argues, these 'objects' derive their appearance of objectivity from the relations of power which suppress or displace alternative possible objectivities, and are fully intelligible only when situated within these relations.

REFERENCES

Ackoff R.L., 1974, "Redesigning the Future", John Wiley, New York.

Ackoff R.L., 1979a, The Future of Operaitonal Research is Past, Jnl.Op.Res. Soc., 30:93.

Ackoff R.L., 1979b, Resurrecting the Future of Operational Research, Jnl.Op.Res.Soc., 30:189.

Adorno T.W., Alpert H., Dahrendorf R., Habermas J., Pilot H. and Popper K.R., 1976, "The Positivist Dispute in German Sociology", Heinemann, London.

Barnes B., 1974, "Scientific Knowledge and Sociological Theory", Routledge and Kegan Paul, London.

Bernstein R.J., 1976, "The Restructuring of Social and Political Theory", Basil Blackwell, Oxford.

Burrell G., 1983, 'Systems Thinking, Systems Practice': a review, Jnl.Appl. Syst.Anal., 10:121.

Burrell G. and Morgan G., 1979, "Sociological Paradigms and Organisational Analysis", Heinemann, London.

Checkland P.B., 1981, "Systems Thinking, Systems Practice", John Wiley, London.

Checkland P.B., 1982, Soft Systems Methodology as process: a reply to M.C.Jackson, Jnl.Appl.Syst.Anal., 9:37.

Checkland P.B., 1983a, OR and the systems movement: mappings and conflicts, Jnl.Op.Res.Soc., 34:661.

Checkland P.B., 1983b, 'Systems Thinking, Systems Practice': a response to Burrell's review, Jnl.Appl.Syst.Anal., 10:127.

Checkland P.B., 1985, From optimising to learning: a development of systems thinking for the 1990s, Jnl.Op.Res.Soc., 36:757.

Checkland P.B., 1988, Soft systems methodology: an overview, Jnl.Appl.Syst. Anal., 15:27.

Churchman C.W., 1979a, "The Systems Approach", Dell, New York.

Churchman C.W., 1979b, "The Systems Approach and its Enemies", Basic Books, New York.

Dando M.R. and Bennett P.G., 1981, A Kuhnian crisis in management science? Jnl.Op.Res.Soc., 32:91.

Dando M.R. and Sharp R.G., 1978, Operational Research in the UK in 1977: the causes and consequences of a myth, Jnl.Op.Res.Soc., 29:939.

Giddens A., 1977, Positivism and its critics, in: "Studies in Social and Political Theory", A. Giddens, Hutchinson, London.

Giddens A., 1979, "Central Problems in Social Theory", Macmillan, London.

Habermas J., 1970a, On systematically distorted communication, Inquiry, 13:205.

Habermas J., 1970b, Towards a theory of communicative competence, Inquiry, 13:360.

Habermas J., 1972, "Knowledge and Human Interests", Heinemann, London.

Habermas J., 1976, The analytical theory of science and dialectics, and, A positivistically bisected rationalism, in: Adorno T.W. et al. "The Positivist Dispute in German Sociology", Heinemann, London.

Held D., 1980, "Introduction to Critical Theory: Horkheimer to Habermas", Hutchinson, London.

Jackson M.C., 1982, The nature of 'soft' systems thinking: the work of Churchman, Ackoff and Checkland, Jnl.App.Syst.Anal., 9:17.

Jackson M.C., 1983, The nature of 'soft' systems thinking: comment on the three replies, Jnl.Appl.Syst.Anal., 10:109.

Jackson M.C. and Keys P. (eds.), 1987, "New Directions in Management Science", Gower, Aldershot.

Johnson T., Dandeker C. and Ashworth C., 1984, "The Structure of Social Theory", Macmillan, London.

Keat R. and Urry J., 1977, "Social Theory as Science", Routledge and Kegan
            Paul, London.
Lukes S., 1974, "Power: a Radical View", Macmillan, London.
McCarthy T., 1978, "The Critical Theory of Jurgen Habermas", Hutchinson,
            London.
Miller P., 1987, "Power and Domination", Routledge and Kegan Paul, London.
Mingers J.C., 1980, Towards an appropriate social theory for applied
            systems thinking: critical theory and soft systems methodology,
            Jnl.Appl.Syst.Anal., 7:41.
Mingers J.C., 1984, Subjectivism and soft systems methodology - a critique,
            Jnl.Appl.Syst.Anal., 11:85.
Popper K.R., 1976, Reason or revolution, in: Adorno T.W. et al., "The
            Positivist Dispute in German Sociology", Heinemann, London.
Roderick R., 1986, "Habermas and the Foundations of Critical Theory",
            Macmillan, London.
Tinker T. and Lowe T., 1982, The management science of the management
            sciences, Hum.Reln., 35:331.
Tinker T. and Lowe T., 1984, One-dimensional management science: the making
            of a technocratic consciousness, Interfaces, 14(2):40.
Ulrich W., 1983, "Critical Heuristics of Social Planning", Haupt, Bern.
Ulrich W., 1986, Critical heuristics of social systems design, Working
            Paper 10, Dept. of Management Systems and Sciences, University
            of Hull, Hull, UK.

CRITICAL HEURISTICS OF SOCIAL SYSTEMS DESIGN

W.Ulrich

Department of Public Health and Social Services
Bern
Switzerland

APPLIED SCIENCE AND THE PROBLEM OF JUSTIFICATION BREAK-OFFS

The stuff of applied disciplines such as OR/MS is what epistemologists call the 'context of application', in distinction to the so-called 'context of justification'. Epistemologists such as Karl R. Popper (1961, 1968, 1972) have claimed that the context in which science is applied is relatively irrelevant for the justification of its propositions. In distinction to this position, I propose to understand - and indeed define - applied science as the study of contexts of application. Of course this definition renders the distinction between the two contexts obsolete. From an applied-science point of view, the distinction is really quite inadequate: To justify the propositions of applied science can only mean to justify its effects upon the context of application under study. The key problem that makes applied science, as compared to basic science, so difficult to justify lies in the normative content that its propositions gain in the context of application.

By 'normative content' I mean not only the value judgements - the normative premises - that inevitably flow into practical propositions such as recommendations for action, design models, planning standards or evaluative judgments, but also their normative implications in the context of application, i.e., the life-practical consequences and side-effects of the 'scientific' propositions in question for those who may be affected by their implementation.

Speaking of the 'context of application' is a scientifically neutral way to say that applied science, whenever it really gets applied, tends to affect citizens that have not been involved in the scientific justification of the propositions. What does it mean to be scientific, or to 'justify' the propositions of applied science, in view of the uninvolved being affected?

Basically, the answer is to understand 'justification' no longer as the business of the involved only, but as the common task of both the involved and the affected. Hence a dialogical concept of rationality must replace the conventional 'monological' understanding of rational justification. Whereas the latter relies on deductive logic and empirical corroboration of falsification attempts on the part of the involved, the former must be grounded in a model of rational discourse that would explain the conditions for reaching 'rational'(as opposed to merely factual) consensus among all the involved and the affected in regard to the 'rightness' (acceptability) of a design's normative content.

The problem of how rational discourse can redeem the validity claims of practical propositions - their claim to secure improvement and to be rationally justifiable - is known as the problem of practical reason. The branch of philosophy dealing with this problem, practical philosophy, has recently experienced a considerable renaissance. Contemporary practical philosophers such as Paul Lorenzen (1969); Lorenzen and Schwemmer (1975) and Jürgen Habermas (1971, 1973, 1975, 1979; see also McCarthy, 1978) have developed 'ideal' models of practical discourse. They give us essential insights into the conditions that would allow us to justify disputed validity claims. The problem is only that these models, because they are ideal designs for rational discourse, are impractical (not realizable): They assume ideal conditions of rationality that will always remain counter-factual. In fact they presuppose what they are supposed to produce, namely, rational argumentation - the ability and will of all participants to argue cogently and to rely on nothing but the force of the better argument. Most importantly, they do not take into account the inevitability of argumentation break-offs. In practical discourse, just as in conventional 'monological' justification strategies, every justification attempt must start with some material premises and end with some conclusions that it cannot question and justify any further. In other words, every chain of argumentation starts and ends with some judgments the rational justification of which must remain in open question.

CRITICAL HEURISTICS, OR HOW TO DEAL CRITICALLY WITH JUSTIFICATION BREAK-OFFS

From what has been said it follows that the crucial problem for any applied scientist seeking to justify his propositions is the question of how to deal critically with the justification break-offs that inevitably flow into these propositions. As long as he does not learn to make transparent to himself and to others the justification break-offs flowing into his designs, the applied scientist cannot claim to deal critically with the normative content of these designs.

Critical Heuristics (or by its full name: Critical Heuristics of Social Systems Design) is a new approach to both systems thinking and practical philosophy, an approach that aims to help the applied scientist in respect to this task. It does not seek to prove theoretically why and how practical reason is possible (as do all presently known 'schools' of practical philosophy) but rather concentrates on providing planners as well as affected citizens with the heuristic support they need to practice practical reason, i.e., to lay open, and reflect on, the normative implications of systems designs, problem definitions, or evaluations of social programs.

In order to achieve this purpose, Critical Heuristics takes three requirements to be essential:

First, to provide applied scientists in general, and systems designers in particular, with a clear understanding of the meaning, the unavoidability and the critical significance of justification break-offs;

Second, to give them a conceptual framework that would enable them to systematically identify effective break-offs of argumentation in concrete designs and to trace their normative content; and

Third, to offer a practicable model of rational discourse on disputed validity claims of such justification break-offs, that is to say, a tool of cogent argumentation that would be available both to 'ordinary' citizens and to 'average' planners, scientists, or decision takers.

For each of these three basic requirements, Critical Heuristics offer a key concept. I can only give a brief introduction here; for a more complete explanation, the reader is referred to the main sources (Ulrich, 1983, 1984); for a helpful review see Jackson (1985). The application of these concepts has been illustrated in two earlier-published case studies (Ulrich, 1981; 1983, Chapters 7 and 8).

Key concept no.1: Justification break-offs as boundary judgments (whole systems judgments)

Systems science offers a concept that is helpful to understanding the meaning of justification break-offs, though unfortunately its critical significance is not always adequately understood. I mean the well-known concept of boundary judgments.

Whenever we apply the systems concept to some section of the real world, we must make very strong a priori assumptions about what is to belong to the system in question and what is to belong to its environment. The boundary judgments representing these assumptions can therefore be understood in a twofold, and complementary, way:

- as whole systems judgments, i.e., the designer's assumptions about what belongs to the section of the real world to be studied and improved and what falls outside the reach of this effort;

- as justification break-offs with regard to the demarcation of the context of application that is to be relevant when it comes to justifying the normative implications of a design for those affected by its effects.

In contemporary systems science, the problem of boundary judgments is either entirely ignored (typically in textbook exercises and case studies) or else it is discussed in terms of formal criteria of modelling, rather than in terms of the normative content of whole systems judgments and corresponding justification break-offs. Frequently, models of 'systems' are presented as if the boundaries were objectively given, and the model itself does not tell us whether the boundaries in question have been adequately chosen. If the problem is discussed at all, it is seen merely from a modelling point of view: so as to facilitate the modelling task, boundaries are determined according to the availability of data and modelling techniques. But even from a merely technical modelling point of view, this way of dealing with the problem of boundary judgments is inadequate. First, the implicit criterion is that everything that cannot be controlled or is not known falls outside the boundaries of the model, so that the model itself looks neat and scientific. In point of fact, the reverse criterion should be applied: we cannot understand the meaning of the model (and hence the systems in question) if we do not understand the model-environment. Hence aspects that are not well understood ought to be considered as belonging to the system in question rather than to its environment, at least until their significance has been studied. Second, such studying of boundary questions must not be restricted to the 'is' (or 'will be') but must always include the 'ought'. Whether or not a certain boundary judgment is rational depends less on what boundaries are presently established than on what the boundaries should be, given the purpose of the model (the systems map or design). The normative content of the answer to the question of what the boundaries should be cannot be justified by referring to data availability, to presently accepted boundaries, or to the success of purposive-rational action. The normative content can be justified only through the voluntary consent of all those who might be affected by the consequences. Hence all the citizens affected, be they involved in the process or not, ought to be regarded as being part of the context of application.

Such an openly and critically normative understanding of boundary judgments has far-reaching implications for systems science and systems design. To mention but two of them:

(a) Systems science will have to employ a new, 'critically-normative' concept of what represents an adequate definition (map, design) of a system: We shall say that a definition (map,design) of a system is 'adequate' if it makes explicit its own normative content.

(b) The designer (applied scientist) will have to aim not at an objective but at a critical solution to the problem of boundary judgments. That is to say, it is his responsibility to secure the transparency of the boundary judgments on which he relies and to trace their possible normative consequences; but he cannot delegate to himself the political act of positively sanctioning these consequences - only the affected can. No standpoint, not even the most comprehensive systems point of view, is ever sufficient to validate its own implications. The rationality of a systems design is to be measured not by the degree to which it fulfills the impossible role of providing an elitary justification of its own normative implications, but rather by the degree to which it renders explicit the underlying justification break-offs and thus enables both those involved in and those affected by the design to reflect and discourse on the validity and legitimacy of these break-offs.

Key concept no.2: A priori concepts of practical reason

In order to facilitate the systematic identification and examination of justification break-offs, Critical Heuristics has developed a check list of twelve boundary questions. They aim at boundary judgments that inevitably flow into any systems design. By means of these questions, both the involved and the affected can question a design's normative content and challenge the 'objective necessities' by which the other side may seek to justify or to dispute the underlying boundary judgments.

The twelve questions are:

1. Who ought to be the client (benificiary) of the system S to be designed or improved?
2. What ought to be the purpose of S, i.e., what goal states ought S be able to achieve so as to serve the client?
3. What ought to be S's measure of success (or improvement)?
4. What ought to be the decision taker, that is, have the power to change S's measure of improvement?
5. What components (resources and constraints) of S ought to be controlled by the decision taker?
6. What resources and conditions ought to be part of S's environment, i.e., should not be controlled by S's decision taker?
7. What ought to be involved as designer of S?
8. What kind of expertise ought to flow into the design of S, i.e., who ought to be considered an expert and what should be his role?
9. Who ought to be the guarantor of S, i.e., where ought the designer seek the guarantee that his design will be implemented and will prove successful, judged by S's measure of success (or improvement)?
10. Who ought to belong to the witnesses representing the concerns of the citizens that will or might be affected by the design of S? That is to say, who among the affected ought to get involved?
11. To what degree and in what way ought the affected be given the chance of emancipation from the premises and promises of the involved?
12. Upon what world-views of either the involved or the affected ought S's design be based?

The twelve boundary questions are organized into four groups of boundary judgments, each group comprising three kinds of categories:
- The first group asks for the sources of motivation flowing into the design in question: Who contributes (ought to contribute) the necessary sense of direction and 'values'? What purposes are to be served? Given a tentative planning purpose, whose purpose is it?
- The second group is to examine the sources of control built into a design: Who contributes (ought to contribute) the necessary means, resources, and decision authority, i.e., 'power'? Who has (ought to have) the power to decide?
- The third group of questions is to trace the sources of expertise assumed to be adequate: Who contributes (ought to contribute) the necessary design skills and the necessary knowledge of 'facts'? Who has (ought to have) the know-how to do it?
- The fourth group, finally, helps reflect on the sources of legitimation to be considered: Who represents (ought to represent) the concerns of the affected? Who contributes the necessary sense of self-reflection and 'responsibility' among the involved? How do the involved deal with the different world-views of the affected?

In short, the first group of boundary questions asks for the value basis of the design; the second for its basis of power; the third for its basis of know-how; and the fourth for its basis of legitimation.

The three questions of each group refer to the following three kinds of categories: the first question refers to social roles of the involved or the affected, the second to role-specific concerns, and the third to key problems or crucial issues in determining the necessary boundary judgments relative to the two previous categories.

The critical relevance of these categories and corresponding boundary questions may best be seen by contrasting each 'ought' with the pertaining 'is' judgment:

1. Who is the actual client of S's design, i.e., who belongs to the group of those whose purposes (interests, values) are served, in distinction to those who do not benefit but may have to bear the costs or other disadvantages?
2. What is the actual purpose of S's design, as being measured not in terms of the declared intentions of the involved but in terms of the actual consequences?
3. What is, judged by the design's consequences, its built-in measure of success?
4. Who is actually the decision taker, i.e., who can actually change the measure of success?
5. What conditions of successful planning and implementation of S are really controlled by the decision taker?
6. What conditions are not controlled by the decision taker, i.e., what represents "environment" to him?
7. Who is actually involved as planner?
8. Who is involved as 'expert', of what kind is his expertise, what role does he actually play?
9. Where do the involved seek the guarantee that their planning will be successful? (E.g. in the theoretical competence of experts? in consensus among experts? in the validity of empirical data? in the relevance of mathematical models or computer simulations? in political support on the part of interest-groups? in the experience and intuition of the involved? etc.) Can these assumed guarantors secure the design's success, or are they false guarantors?
10. Who among the involved witnesses represents the concerns of the affected? Who is or may be affected without being involved?

11. Are the affected given an opportunity to emancipate themselves from the experts and to take their fate into their own hands, or do the experts determine what is right for them, what means quality of life to them etc.? That is to say, are the affected used merely as means for the purposes of others, or are they also treated as "ends-in-themselves" (Kant), as belonging to the client?
12. What world-view is actually underlying the design of S? Is it the view of (some of) the involved or of (some of) the affected?

Contrasting 'is' and 'ought' boundary judgments provides a systematic way to evaluate the normative content of planning while at the same time laying open the normative basis of the evaluation itself: The 'is' questions aim at determining a design's effective normative implication in the light of the 'ought' answers, that is to say, without any illusion of objectivity.

It remains to be explained why the title to this section speaks of a priori concepts of practical reason: First, the suggested boundary questions represent mere 'forms of judgments', that is, they are in need of being substantiated with respect to both their empirical and normative content. Second, they can help to fill critically-heuristic categories such as 'client', 'purpose', etc. with empirical and normative content, but not to justify this content. The boundary judgments identified or postulated remain dependent for their justification on a discursive process of consensus formation - a rational discourse - among the involved and the affected.

As previously suggested, contemporary models of ideal discourse do not provide a practicable way to redeem disputed validity claims of justification break-offs; it remains to be shown how practical discourse can secure at least a critical solution to this problem.

Key concept no.3: The polemical employment of boundary judgments

The concepts that have been introduced thus far are to provide a tool of reflection for tracing the normative implications of systems designs. But they cannot guarantee such reflection. How then can affected citizens cause the involved decision takers, planners, and experts to reflect on a design's normative content if the involved are not willing to do so on their own? On the other side, how can the conflicting demands of democratic participation (of the affected) and of cogent argumentation (on the part of everybody involved, including the witnesses of the affected) be reconciled so that ordinary citizens can bring in their personal concerns without being convicted of lacking rationality or cogency?

Critical Heuristic's basis conjecture in this regard may by now seem familiar: Any use of expertise presupposes boundary judgments with respect to the context of application to be considered. No amount of expertise or theoretical knowledge is ever sufficient for the expert to justify all the judgments on which his recommendations depend. When the discussion turns to the basic boundary judgments on which his exercise of expertise depends, the expert is no less a layman than are the affected citizens.

It follows that every expert who justifies his recommendations, or the "objective necessities" he may disclose in the name of reason, by referring to his expertise without at the same time laying open his lay status relative to the underlying boundary judgments can be convicted of a dogmatic or cynical employment of boundary judgments. Dogmatically he employs them if he fails to recognize his lay status in respect to boundary judgments and hence asserts their objective necessity; cynically, if he very well sees through their character as justification break-offs but against his better judgment conceals them behind a facade of objectivity or pretends other than the true ones to be his boundary judgments.

Now anybody who is able to comprehend the unavoidability and the meaning of boundary judgments in general can also learn to see through - and to make transparent to others - the dogmatic or cynical character of the 'objective necessities' disclosed by experts in specific contexts of application. To this end, concerned citizens and professionals will have to master two tasks of argumentation:

1.   They must be in a position to demonstrate that the boundary judgments of the involved cannot be justified rationally, i.e., cogently.

2.   They should be able to translate their own subjective way of being affected by the boundary judgments in question into rational, cogent argumentation.

How can ordinary citizens without any special expertise or 'communicative competence' (as required by the ideal models of rational discourse) accomplish this apparent squaring of the circle? My answer is: by means of the polemical employment of boundary judgments.

Immanuel Kant, in his discussion of the 'polemical employment of reason'(1787, B767), calls 'polemical' an argument that is directed against a dogmatically asserted validity claim and which does not depend for its cogency on its own positive justification. A polemical argument has only critical validity; but in regard to this merely critical intent, it must be rational, i.e., cogent. Thus the polemical employment of reason, as understood by Kant, has nothing in common with 'polemics' in the contemporary, vulgar meaning of the term; it aims at the cause rather than at the person, and it must be logically compelling.

The use of boundary judgments for merely critical purposes almost ideally fulfills this condition: Boundary judgments that are introduced overtly as personal value judgments entail no theoretical validity claim and hence do not require a theoretical justification. Hence no theoretical knowledge or any other kind of special expertise or 'competence' is required. Indeed, it is not even necessary to pretend that a boundary judgment used polemically may not be false or merely subjective. What matters is only that no one can demonstrate the objective impossibility (and hence, irrelevance) of a polemical statement any more than its proponent can demonstrate its objective necessity.

Now the crucial point is this. So long as affected citizens employ their boundary judgments for critical purposes only, i.e., without asserting any positive validity claims, they can secure for themselves an advantage of argumentation by imposing the burden of proof upon the involved experts: As against the expert's boundary judgments, they can with equal right and with overt subjectivity advance their own boundary judgments, thereby embarrassing the expert for being unable to prove the superiority of his boundary judgements by virtue of his expertise. In this way, they can demonstrate three essential points:
    (a) that boundary judgments do play a role in the expert's propositions;
    (b) that his theoretical competence is insufficient to justify his own boundary judgments or to falsify those of his critic;
    (c) that an expert who seeks to justify his recommendations by referring to his competence or by asserting 'objective necessities' argues either dogmatically or cynically and thereby disqualifies himself.

Thus the polemical employment of boundary judgments enables ordinary people to expose the dogmatic character of the expert's 'objective necessities' through their own subjective arguments, without even having to pretend to be objective or to be able to establish a true counterposition against the expert. Therein, I believe, lies the enormous significance of Kant's concept of the polemical employment of reason for a critically-heuristic approach to planning, an approach that would actually mediate between the conflicting demands of democratic participation (of all affected citizens) and those of rational, cogent argumentation (on the part of the involved planners and experts).

To be sure, as soon as the affected claim positive validity for their boundary judgments, they lose their advantage of argumentation and disqualify themselves no less than the experts do. But it would be to mistake the situation if the polemical employment of boundary judgments were considered to produce a mere 'symmetry of helplessness'. The polemical employment of reason secures to both sides an equal position for reasonable dialogue: each side, if only it renounces the dogmatic or cynical employment of boundary judgments, can now argue its case and work towards mutual understanding about the premises and consequences of planning, by advancing its own good grounds, i.e., 'facts' and 'values' capable of consensus such as existential needs of all individuals, ecological knowledge, ethical principles, principles of constitutional democracy, and so on. The fact that ultimate justifications remain impossible provides no sound reason for renouncing any effort to bring in one's own good grounds. Nor does it provide the other side with a good ground to refuse entering upon such an effort: scepticism, turned into an argument against any argumentative effort, is no less dogmatic than the expert's reference to 'objective necessities'. Hence a 'symmetry of helplessness' arises only with respect to dogmatic or cynical argumentation attempts; for the rest, an essential condition of rational discussion is secured, namely, the possibility of 'competent' participation of affected citizens in the process of unfolding the normative implications of planning.

Opening up the applied disciplines for such a process of unfolding is certainly no royal way to solve the problem of practical reason; but it might be an important step towards dealing critically with the justification break-offs that inevitably flow into the definition of specific contexts of application to be considered. It is only thus that applied science can hope to fulfill its mission - to secure improvement of the human condition, by studying contexts of application of human knowledge and design.

REFERENCES

Habermas J., 1971, Vorbereitende Benerkungen zu einer Theorie der kommunikativen Kompetenz, in: "Theorie der Gesellschaft oder Sozialtechnologie-Was leistet die Systemforschung?", Habermas J. and Luhmann N., eds., Suhrkamp, Frankfurt.
Hanermas J., 1973, Wahrheits theorin, in: "Wirklichkeit und Reflexion", Fahrenbach H., ed., Neske, Pfullingen.
Habermas J., 1975, A postscript to Knowledge and Human Interests, Phil.Soc. Sci., 3:157.
Habermas J., 1979, What is universal pragmatics?, in: "Communication and the Evolution of Society", Habermas J., ed., Beacon Press, Boston.
Jackson M.C., 1985, The itinerary of a critical approach, Jnl.Op.Res.Soc., 36:878.
Kant I., 1787, "Critique of Pure Reason", translated by N.K.Smith, 1965, St. Martin's Press, New York.
Lorenzen P., 1969, "Normative Logica and Ethics", Bibliographisches Institut, Mannheim.
Lorenzen P. and Schwemmer O., 1975, "Konstruktive Logik, Ethik und Wissenschaftstheorie", Bibliographisches Institut, Mannheim.
McCarthy T., 1978, "The Critical Theory of Jurgen Habermas", MIT Press, Cambridge.
Popper K.R., 1961, "The Logic of Scientific Discovery", Basic Books, New York.

Popper K.R., 1968, "Conjectures and Refutations: The Growth of Scientific
           Knowledge", Basic Books, New York.
Popper K.R., 1972, "Objective Knowledge", Clarendon Press, Oxford.
Ulrich W., 1981, A critique of pure cybernetic reason: the Chilean exper-
           ience with cybernetics, Jnl.Appl.Syst.Anal., 8:33.
Ulrich W., 1983, "Critical Heuristics of Social Planning: a New Approach
           to Practical Philosophy", Haupt, Bern.
Ulrich W., 1984, Management oder die Kunst, Entscheidungen zu treffen,die
           andere betreffen, Die Unternehmung, Schweizerische Zeitschrift
           für betriebswirtschaftliche Forschung und Praxis, 38:326.

Note: This paper first appeared in European Journal of Operational
      Research, 1987, 31:276-283.

OR : SOCIAL SCIENCE OR BARBARISM?

J. Rosenhead

Department of Operational Research
London School of Economics
London, U.K.

SOCIAL SCIENCE OR BARBARISM?

Neither the future of operational research nor that of social science is knowable. Statements about their future inter-relationship therefore have the status of conjectures. But there are certain preconditions before one can even develop and support plausible conjectures. One is a clarification of concepts - without which general confusion can all too easily prevail. Another is that conjectures about the future need to be firmly based on an analysis not just of the present (a mere snapshot in time), but of the past.

Not everyone is agreed that the mutual involvement of OR and the social sciences is to be encouraged. Some may even deplore the tendency. Others, perhaps Conference participants, may base their support on particular assumptions about OR, social science and the form of their interaction, which others might not accept. The task of this Conference should surely be, not just to attempt an undifferentiated and uncritical boost to all OR practice with some arguable social scientific component. We should try, rather, to generate a continuing dialogue about what versions of mutual involvement hold out promise for a future operational research we would be eager to participate in.

WHICH SOCIAL SCIENCE?

There are a multiplicity of disciplines within the social sciences. My own academic institution has departments of sociology, social psychology, anthropology and economics. But it also has departments of geography, government, economic history, and international relations, among others. The standing of the various subjects is different. Some have a fairly well-defined subject matter and set of methodologies; some have only a defined subject matter, and borrow widely from the methods of other disciplines. Can OR make statements, or blank declarations of intention, concerning such diverse domains?

Another factor we need to encompass is the diversity within particular disciplines. There are divisions by subject matter - sociology of religion, sociology of ideas, sociology of development....But within one discipline there are also differing schools - functionalist, structuralist, neo-Weberian

... - which may lay competing claims to particular subject matters. Their approaches are so at variance (for example, in the type of explanation sought for particular phenomena) that OR cannot propose a marriage, even of convenience, with all of them simultaneously.

This is by no means the end of the complexity. There is also the time factor. Operational research has experienced changes over the past quarter century, and so, perhaps more dramatically, have the social sciences.

Sociology can serve as one example. The topography of the subject bears little resemblance now to that of a generation ago. Much of what was then in place has been swept away, and the usurpers in turn usurped. Mathematical sociology, for example, appeared poised for great advances in the later 1960s, but its moment never came, and it is now quite marginal. Ethnomethodology seemed to have swept the field in the 1970s, but now merits only one question on the undergraduate Theory examination. And so on. Sociology has been, and doubtless will be, a moving target.

A similarly fluid and fragmented pattern is evident in anthropology. If any unity is to be found, it consists largely of a studied eclecticism of theory and method. Economics is in a not dissimilar condition, split between the warring tribes of monetarists and marxists, with only a few embattled neo-Keynesians stranded in the middle ground.

Certain changes which seem to be more enduring can however be observed. In 1964 sociology, like the other social sciences, was most often presented as neutral and value-free. Under the influence of feminist and marxist criticism few would hold to that posture today. The feminist influence in particular has been pervasive, both in its critique of gender bias in mainstream work, and in its construction of a feminist sociology - analysing, for example, women's (unwaged) contributions to work and economic development. The same is true of anthropology, where gender-insensitive work in any field is regarded as out of touch with modern developments.

Is this new feature of the landscape one which operational research can orientate itself to? Not, one would have thought, without quite radical changes within OR itself. The issue of the future relationship between OR and the social sciences raises not only the question of 'which social science?' but also 'what OR?'

NATURAL SCIENTIFIC OR?

Not all operational researchers think that social science content in OR is to be applauded. A distinguished academic, ex-President of the Operational Research Society of America, some years ago savaged a published case study dealing with personal perceptions and group dynamics in the following terms: "didn't say anything about anything...time for someone to stand up and shout that the emperor isn't wearing any clothes...no content in any of this...lots of words, some very impressive...self-styled operational research" and so on (Machol, 1980). The emotive force of such attacks stems from the sniffing out of sin, or at least the shocked discovery of a social science cuckoo egg in the OR nest.

Generally those who oppose a social scientific emphasis within OR do so because they favour a natural science model for our subject. Thus Rivett (1983) writes of the need to bring the world of problems where repeated experiment is impossible, that is the world of OR, "within the classical mainstream of science." "Operational Research" he continues "must always be part of that mainstream, and the task of the OR scientist is to remain a scientist while tackling problems which the traditional scientist has traditionally avoided." There is no room here for the multiple world views, problematic nature of reality, analysis of sectional advantage and ideological hegemony etc. which a social scientific perspective might import to the subject. It should not be thought that these unreconstructed attitudes to operational research are necessarily only those of unrepresentative individuals. My own experience is that such an approach is widespread in the OR community, though possibly declining in prevalence.

In fact I believe that there are a number of, so far mostly good-tempered, low-key disputes in progress which roughly map onto each other. This is a dispute about social science versus natural sciences. There is a dispute about 'hard' versus 'soft' methods. There is a dispute about a managerialist versus a broader role for operational research. And there is a divergence of political commitment. When any one of these arguments erupts into visibility there tends to be a good deal of shadow-boxing. Punches apparently aimed at particular targets actually land on the propositions behind those ostensibly at issue.

Understanding what is going on is made more difficult by the fact that on several of these dimensions there are three rather than two main positions. (This is consistent with Dando and Bennett's (1981) finding of three competing OR paradigms - 'official', 'reformist' and 'revolutionary'). Thus there are those who would embrace only that subset of the social sciences known as the behavioural sciences; and those who wish to break out of the rather limited managerial remit of operational research only or principally in order to find a broader role with management. Those who combine together on one dimension may find themselves in opposing camps on others.

The outcome of these arguments will be extremely important for the future of the subject. It is a battle for the soul of operational research, or (less emotively) for which is to be the dominant paradigm. Some of the issues are usefully explored by Jackson (1987a). Here I will only address their relevance to the debate about OR and the social sciences.

MANAGERIALIST OR?

A denial of the relevance of social science to OR is an assertation that the systems which OR analyse can be treated as if they were natural systems, not human activity systems. (See Checkland 1981). By implication all the elements of these systems can be regarded, for practical purposes, as inanimate objects without volition. Evidently this is consistent with a belief in a strong version of managerial prerogatives - in the unfettered right to command the labour power for which the organisation is paying,and to expect obedience.

This view of control mechanisms in work environments is rather simplistic, directly applicable perhaps in the buccaneering era of the industrial revolution, and in present day non-unionised sweat shops round the world. Where the workforce has some countervailing power, more sophisticated control strategies are encountered. The corresponding OR approach recognises the contribution which behavioural science can make to its work. Thus "it is no longer sufficient for OR practitioners to propose a complete change in the method of production-scheduling within a company without paying due regard to the corresponding implications for social change. Managers are becoming increasingly aware that any advantage accruing from the introduction of this type of technical change may be far outweighed by the cost of social disruption....The widespread application of quantitative techniques, with the aid of computers, is rapidly reaching a barrier because of our failure to know and understand the behaviour of people and systems." (Heald, 1970).

The quotation is from the early days of such perceptions - and in the current restructuring of our economic system, management has often decided to face down the social disruption rather than attempt to evade it. (Managerial strategies, like the social sciences, are diversified at any point in time, and over time evolve, recycle and respond to fashion). However versions of this position are still much in evidence. Thus Müller-Merbach (1988), in a statement about the present Conference, bases his support on the light which the social sciences can shed on individual motives, group dynamics, informal structures of social systems and on the mechanisms of leadership - which can in turn improve the success of "deciding how to best design and operate man-machine systems".

So far I have sketched in two 'managerialist' postures on OR and the social sciences - rejection, and the acceptance of social science only in the form of behavioural science. There is a third more recent stance,based not on the humanity of workers but on the individuality of managers. In this view sensitivity to individual psychology, group dynamics etc. should inform, not our attempt to control the work process, but our attempt to help management structure its problems. 'Soft' OR methodologies, aiming to assist this process by a variety of participative means, have burgeoned (Rosenhead, 1989a). Some,but by no means all, have an explicit reliance on social scientific theory.

This third version of managerialism is paradoxically compatible, in the short-run, with the priorities of those operational researchers who adopt a position towards the left on the political dimension. Traditionally it has been held to be bad taste to bring politics into a discussion about operational research. However the politics will not go away because we decline to talk about it. (Indeed socialists, and socialist ideas, were formative at the very origin of our subject fifty years ago, Rosenhead, 1989b). Politics, of both the left and the right, is present within OR and helps to determine the work that people do and how they do it.

A thought-out socialist perspective cannot but include a rejection of, indeed an attack on, behavioural science as inherently manipulative, and manipulative on behalf of the dominant interests in society. Such research, into the behaviour and motivations of the weak on behalf of the powerful, can be seen as tantamount to stealing knowledge from the former to bolster the power of the latter. Much social science beyond the confines of behavioural science is also, from this viewpoint, highly suspect; its function is to provide conceptual or practical tools whose effect is to maintain, or justify the maintenance of, an unequal and exploitative society.

However there are aspects of the current condition of the social sciences which make it an attractive model for radicals. The social sciences exhibit a vigorous argumentation in which both analytic approaches and the structures of society are subjected to criticism, both separately and in conjunction. There is no such tradition of critical discourse in operational research (Bevan, 1976).

Political conservatives and methodological traditionalists, given the mainstream dominance of technical, managerialist OR, stand to gain little from any attack of self-reflective debate - and generally fail to respond to criticisms which are made of their approach. The natural science culture also militates against polemical engagement. So generating a debate which might advance consciousness among operational researchers is hard conceptual work. Borrowing from the more developed critical tradition of the social sciences provides a leg-up which makes it less of an uphill struggle. (See, for example, Jackson, 1982).

Where radicals converge with more sophisticated managerialists is in their advocacy of 'soft' operational research methodologies. One reason for this is the appeal of methods which reject the 'dictatorship of the optimum' - the underlying assumption that there is a consensus on objectives, and hence only one solution. But there is also the potential of soft methodologies for the alternative OR clientele of relatively disadvantaged 'community groups', currently under development in Britain at least (Jackson, 1987b, Thunhurst, 1987). Community organisations are only as strong as their members' commitment and mobilisation behind a shared view of their problematic situation, and the strategies which this view implies. Participative methods for the transparent structuring of problems therefore have an evident appropriateness for such groups.

The movement to broaden the role of operational research is not supported only by radicals, nor is it restricted to work with community groups. Müller-Merbach (1986) , while President of IFORS, pointed out the "huge variety of problems waiting for feasible solutions to which the operational research community can contribute" - regional and civil wars, East-West confrontation, the North-South divide, unemployment, hunger, crime, etc.. Others during the past quarter-century have urged a wider social role for OR (Cook, 1973). I hope that it is not wishful thinking to hold that such a move now seems more likely to succeed. (Certainly in Britain an Operational Research Society initiative to foster work on AIDS has met with a good response.) OR's mainstream algorithmic approach and natural science perspective has in the past been a conceptual albatross. It has largely unfitted us for contributing to the formulation, argumentation and reformulation of such issues, with their conflicting world views and high content of social interaction. Here too problem structuring, as opposed to problem solving, methods offer a possible way forward.

WAYS FORWARD

At the end of a conference such as this it is appropriate to consider ways forward. It would be impertinent to suggest how the social sciences might gain from OR. Our remit is, rather, to find hopeful directions for operational research.

The discussion in the previous sections of this paper indicates the implausibility of making a value-free statement about the future of the OR - social science relationship which is of any value. It is a relationship of indeterminate form with problematic terms at both ends.

However the situation changes if we can fix one of the terms. Once we select a desired future for operational research, certain categories of social science will rule themselves out as ingredients or partners, as will certain forms of relationship. My own pretence at neutrality must therefore end here, or I will have next to nothing to say.

The nexus of attitudes to the philosophy and practice of operational research identified above may without too much violence be simplified as follows. There are perspectives, with variants, which support the maintenance of the status quo - in the sense of a continuation of key relationships of dominance in society. I will call this the control paradigm of operational research. There are other perspectives which are concerned to promote change in society in ways which will empower those who are currently the objects rather than the subjects of relations of domination. I will call this the emancipatory paradigm of operational research.

For the remainder of these remarks I will address myself to the implications of the emancipatory paradigm. It is the one which I adopt and attempt to practice; and it is an approach which attempts to make a contribution to the creation of a less exploitative society. As such it is certainly consistent with the aspirations of the socialist pioneers of operational research, who were motivated not just by a penchant for rationality, but also by a vision of a fairer world (Rosenhead, 1989b).

Both paradigms, control and emancipatory, may make calls on explicit 'knowledge about society' in some form - though this is less crucial for the control paradigm. Those who seek purposeful change have the more difficult task, and not only because the resources of society are unequally stacked. It is more difficult because work which does not take account of the social mechanisms of change and immobilism will be vitiated; and because OR intervention must occur in more problematic, less pacified environments.

Operational researchers operating within the emancipatory paradigm need, therefore, to be more sophisticated in their practice. Conversely their practice, if conducted with openness and lack of dogma, can help to generate the necessary sophistication. Sensitivity will be required, in particular, to the process of problem resolution within groups, and how it can be enriched rather than impeded by formal analysis. Social science will have a role to play in informing this practice. But it will certainly not be exclusive. We will need to learn also from common sense, folk wisdom, bitter experience, apprenticeship to skilled practitioners, etc.. Scientific knowledge is only one of the inputs to an effective practice with and on behalf of the less privileged groups in society.

What branches of social science might be thought to have the greater direct contribution to make in this venture? Certainly not behavioural science - manipulation and emancipation are incompatible. The most plausible candidates, perhaps, are economics and sociology rather than psychology. The predicament of the have-nots will virtually always have a strong, even dominating economic component. To at least alleviate this predicament by purposeful structural change is necessarily a collective achievement. Psychology, with its individual basis, seems likely to be less helpful in representing the mechanisms of transformation, though its insights into group decision processes may well be of value.

Emancipatory operational research can draw sustenance from the social sciences in other ways than by direct borrowing of content. We should also emulate the more developed self-critical tradition to be found in the social sciences. One aim would be to generate deeper and more widespread awareness of the narrowness and partiality of mainstream operational research (Rosenhead, 1986). However there is a danger of such critique being seen as an end in itself, becoming ever more isolated from real world concerns as it becomes more refined. Vigilance will be necessary to ensure that this self-sustaining but crippling introspection, all too evident in some branches of social science, is kept at bay. Or perhaps vigilance will not be necessary. Pragmatism is so deeply engrained in operational research that there is little risk of theory running away with itself. A rigorous self-reflection, if it can be grown in the operaitonal research community, should encounter conditions for a healthy relationship of theory and practice.

As to fields of application of emancipatory operational research, there is no shortage of candidates. The unifying characteristic is that of development - enabling the disadvantaged not only to secure a more equitable share of life's resources and chances, but also assisting them to the knowledge and self-confidence required to continue that process. The 'social OR' identified and advocated by Ravn and Vidal (1986) to promote development in less developed countries falls squarely within this description. But under-development is not limited to the Third World. In developed countries also there are extensive groupings - geographical, ethnic, class- or gender-based - characterised by deprivations and disadvantage. Community Operational Research is an initiative, originating and currently thriving in Britain, which is addressing the predicaments of such groups. The developed OR communities of the developed world could well copy.

I was one of the participants in the 1964 Conference, and shared in its optimism, its belief in progress, and its fundamentally uncritical ethos. Rereading its proceedings is, twenty five years on, an instructive but not rewarding experience. Predominantly operational research is perceived as managerialist, and the relevant science is seen as behavioural science. I must have heard the closing address of Kreweras (1966), with its conclusion that he "had never seen so many prominent scientists devote several days to questions so closely related to the key problems of modern society". Looking back from 1989 on the record, both of the proceedings of that Conference and of the paucity of practical consequences, I must register my belated dissent. We should affirm the determination that, when our proceedings and their impact on practice are reviewed with the hindsight of 2014, our record will be more substantial.

ACKNOWLEDGEMENTS

I am indebted to Michael Edwards, Graham Ives, Peter Loizos and Leslie Sklair for insights into the development of their disciplines.

REFERENCES

Bevan R.G., 1976, The language of operaitonal research, Jnl.Op.Res.Soc., 27:305.
Checkland P.B., 1981, "Systems Thinking, Systems Practice", Wiley, Chichester.
Cook S., 1973, Operational Research, social well-being and the zero growth concept, Omega, 1:647.
Dando M.R. and Bennett P.G., 1981, A Kuhnian crisis in management science? Jnl.Op.Res.Soc., 32:91.
Heald G., 1970, Preface, in "Approaches to the Study of Organisational Behaviour: operational research and the social sciences", G. Heald, ed., Tavistock, London.
Jackson M.C., 1982, The nature of soft systems thinking: the work of Churchman, Ackoff and Checkland, Jnl.App.Syst.Anal., 9:17.
Jackson M.C., 1987a, Present positions and future prospects in management science, Omega, 15:455.
Jackson M.C., 1987b, Community Operational Research: purposes, theory and practice, Dragon, 2:47.
Kreweras G., 1966, Closing remarks, in "Operational Research and the Social Sciences", J.R.Lawrence, ed., Tavistock, London.
Machol R.E., 1980, Comment on 'Publish or Perish', Jnl.Op.Res.Soc., 31:1109.
Müller-Merbach H., 1986, quoted in J.Rosenhead, 1986, op.cit.
Müller-Merbach H., 1988, OR and the Social Sciences, Letter from the President, IFORS.
Ravn H.F. and Vidal R.V.V., 1986, Operational Research for developing countries - a case of transfer of technology, Jnl.Op.Res.Soc., 37:205.
Rivett P., 1983, A world in which nothing ever happens twice, Jnl.Op.Res. Soc., 34:677.
Rosenhead J., 1986, Custom and practice, Jnl.Op.Res.Soc.,37:335.
Rosenhead J. ed., 1989a, "Structuring Decisions", Wiley, Chicester.

Rosenhead J., 1989b, Operational Research at the crossroads: Cecil Gordon
            and the development of post-war OR, Jnl.Op.Res.Soc., 40.
Thunhurst C., 1987, Doing operational research with the community,
            Dragon, 2:143.

ON TECHNICAL, PRAGMATIC AND ETHICAL ACTION. A CHALLENGE FOR OR?

H. Müller-Merbach

Universität Kaiserslautern
Kaiserslautern, Germany

This paper is a plea for comprehensive thinking, comprehensive planning, comprehensive leadership - and comprehensive education. This plea covers three layers,

(i)   science and the humanities,
(ii)   the interdependence between technological progress, economic growth, and social change as well as
(iii)   the unity of technical, pragmatic, and ethical action (according to Kant) - leading to the "diamond of thought" after being extended by a fourth dimension of knowledge, the field of methodology and formal procedures.

THE DIVORCE OF INTELLECT - SCIENCE AND THE HUMANITIES

In his short story "Geteiltes Wissen" (Split Knowledge), Kurt Kusenberg (1895-1962) reported about a father who educated his twin sons by means of a dictionary. Peter had to learn everything from A through L, and Paul everything from K through Z. In a quite witty way, Kusenberg discussed the consequences. The common base of mutual understanding was limited by K and L.

This is a perfect model of our academic system, on the one hand science (S = Paul's domain) and on the other hand the humanities (H = Peter's domain). What do they have in common?

There seems to be a deep canyon between science and the humanities. Do humanists not consider the scientists as ignorant because they produce new technologies without reflecting the consequences for society? And do scientists not consider the humanists as arrogant because they talk about consequences of new technologies without understanding the technologies? The tiny bridge of K and L is too narrow to understand each other, to talk to each other, to argue with each other; and they would not trust each other, they would not appreciate each other; they might even prefer to ignore each other.

There really does seem to exist "two cultures", a split of academia. This split between the cultures of thought was impressively described by Snow (1964).

97

Sometimes, it seems as if science and the humanities are not even brothers anymore like Peter and Paul. Instead, their mutual ignoring gives the impression of a divorced couple. There is very little intellectual interaction across the canyon.

The divorce is dangerous since the understanding of the world and the guidance of society into the future would require an intensive cross-fertilisation between science and the humanities.

THE INTERDEPENDENCE BETWEEN TECHNOLOGICAL PROGRESS, ECONOMIC GROWTH, AND SOCIAL CHANGE

The development of industrial systems can only be understood through the interdependence between technological progress, economic growth, and social change. These three fields stick together like strands in a rope. One can not understand $e=mc^2$ if one only has knowledge of energy (E = Peter's domain) or only of matter (M = Paul's domain). Similarly, one can not create deep insight into the development of industrial systems, if his knowledge is restricted to technology or to economics or to sociology etc..

Interdisciplinary Understanding, the Traditional Virtue of OR

Any responsible role in an industrial system requires an interdisciplinary understanding of the whole.

This is the reason for the frequently repeated plea for interdisciplinarity in Operational Research (see among many others: Ackoff 1973, Müller-Merbach 1984). Even if this traditional virtue seems to be forgotten in many OR groups, it is still alive in others. Interdisciplinarity has been spelled out with respect to science on the one hand and to the humanities on the other hand, for example, at a conference on "Education in Systems Science" (with "Systems Science" as a synonym for "Operational Research") in 1978. It was particularly emphasised that education programmes in this field should cover - economics and business administration,
- behavioural and social sciences, and
- hard sciences and/or engineering (Bayraktar et al., 1979, p.41, p.68, p.84).

The argument for such educational breadth was not only to provide an understanding of the development of industrial systems; it was particularly stressed that science and the humanities
- have to deal with different kinds of data and different data qualities,
- are based upon different kinds of theory, and
- are different in their characteristic research approaches.

Education for responsible positions, it was concluded, should be broad enough to provide an understanding and appreciation of both, science (engineering included) and the humanities (i.e. the social sciences or liberal arts, including fields like economics, sociology, psychology etc. which deal with human beings as individuals as well as in groups).

The same intention can be found in a recommendation for university education in the Palatine (Rheinland-Pfalz), one of the 11 states of Germany (see Expertenkommission 1985, pp.107ff.). A "compensatory extension" of all the courses of study was recommended, that is the courses in science and engineering ought to be extended by foundations of the humanities, and the courses in the humanities ought to be extended by foundations of science and engineering. The driving power behind this recommendation was again the anticipated necessity of understanding the processes of technological progress, economic growth, and social change in their independent totality.

This applies to the three large groups of objects of machinery: <u>material</u> and its processing, <u>energy</u> and its processing, <u>information</u> and its processing.

The Necessity of Integration

Today, science and the humanities house distinct <u>cultures of thought,</u> and the three strands of industrial development are more or less being studied in mutual isolation.

The traditions of university education link
- science (engineering included) with technology and its progress on the one hand
- and the humanities with economics and economic growth as well as society and social change on the other hand, see table 1.

These traditions of university education would not be as crucial if former students made up for their deficiencies in their years of higher maturity - five or ten or 20 years after their degree. But do they really have the chance to overcome the deficiencies? What the university system provides is not only knowledge - it is to a great extent a culture of thought, a cognitive style, a Weltanschauung; and there is some evidence that this would not be subject to much change after a person has past his degree; an engineer remains an engineer, an economist remains an economist, a psychologist remains a psychologist etc.. A narrow university education would more or less result in a remaining narrow understanding of the world by the graduates. In contrast, interdisciplinary breadth of education would eventually result in a comprehensive, holistic understanding of the world - a chief qualification for responsibility and leadership in our technological world.

KANT'S THREE KINDS OF ACTION

The plea for breadth of education does not only include <u>science</u> and the <u>humanities</u> or <u>technological progress</u>, <u>economic growth</u>, and <u>social change</u>. Rather, there is another important layer, derived from a classification by the German philosopher Immanuel Kant (1724-1804). He distinguished between three kinds of action,

Table 1. Traditional links (from left to right) and barriers (up and down) of today's university system

Cultures of Thought	Strands of Industrial Development
science	technological progress
humanities	economic growth
	social change

- <u>technical</u> action,
- <u>pragmatic</u> action, and
- <u>ethical</u> action.

Kant's numerous statements on this classification were put together by Hinske (1980). Kant's classification was a basis for the author's suggestion to organise the university courses of study similarly (Müller-Merbach 1987a, 1987b, 1988).

Technical Action

Technical action, according to Kant, has to deal with material objects, i.e. <u>matter</u> and <u>machine</u>. Today we should extend this list by <u>capital</u>, <u>energy</u>, and <u>information</u>.

Technical action, according to Kant, requires <u>skill</u> (Kant: <u>Geschicklichkeit</u>). Technical action is the domain of scientific endeavour. Most educational programmes concentrate on technical action, i.e. they provide <u>skill</u>. This is not fully occupied by science (in the sense of natural science and engineering); it also includes parts of economics and other fields of the humanities. For instance, economics provide skill for technical action as long as they use the fiction of the <u>homo economicus</u>, a machine-like man.

Pragmatic Action

Skill in <u>technical</u> action is not sufficient for anybody in society, particularly not for leaders at any level. Leadership would also require individual understanding of all those with whom direct and indirect communication takes place. For this category, Kant introduced the term of <u>pragmatic</u> action. It means dealing with people.

Pragmatic action requires <u>prudence</u> (Kant: <u>Klugheit</u>). This includes the ability to understand individuals and groups, to deal with them, to cooperate with them, to respond to their desires, to motivate them, to form teams of excellence and guide them etc.. Pragmatic action, in Kant's terminology, is the domain of practical philosophy, but not that of science.

Ethical Action

General objectives of technical and pragmatic action are <u>efficiency</u> and <u>effectivity</u>. Both are derivative measures and depend on "given" goals. The goals have to be defined at a higher level.

This is the level of Kant's <u>ethical</u> action, the level of morals and benevolence, of social responsibility. Ethical action requires <u>wisdom</u> (Kant: <u>Weisheit</u>). Ethical action, according to Kant, is again the domain of practical philosophy, as is pragmatic action. It has to deal with questions of goodness, of final virtues, of basic values. There is no measure of ethical action from a higher level, since the level of ethical action is the highest itself. There is nothing beyond it.

Kant's Hierarchy of Imperatives

The different levels of action find consideration in Kant's hierarchy of imperatives. Ethical action is subject to his <u>categorical imperative</u>, while technical and pragmatic action are subject to <u>hypothetical imperatives</u>. Hypothetical imperatives require highest efficiency and best effectivity with respect to given goals.

In contrast, the categorical imperative deals with the goals themselves. Kant requested any actor to take into consideration that his goals become the principle of a general law. Thus, any goal setter should ask for the consequences if the same goals were chosen by others as well. In Kant's terminology: "Act only according to a maxim by which you can at the same time will that it shall become a general law." Or: "Act as if the maxim of your action were to become through your will a general natural law." (Russell 1988, p.683).

The difference becomes clear through Russell's (1988, p.683) words: "There are two sorts of imperative: the hypothetical imperative which says 'You must do so-and-so if you wish to achieve such-and-such an end'; and the categorical imperative, which says that a certain kind of action is objectively necessary, without regard to any end."

The Unity of Action

Technical, pragmatic and ethical action can be considered separably. But in practice, they form a unity, a non-separable totality. This has two consequences:

- While setting goals, i.e. at the level of ethical action, one has to see the consequences for technical and pragmatic action, i.e. one takes responsibility for the efficiency and effectivity measures as well. The goal setter's responsibility includes the consequences.

- And while acting according to efficiency and effectivity measures for technical and pragmatic action, one has as well to reflect the goals themselves, i.e. one has to be aware that any action is ethical action at the same time. Nobody is completely free of goal responsibility.

The Comprehensiveness of Education

Kant's three kinds of action can be immediately used to evaluate the university courses of study:

- Most courses of study have their emphasis in scientific endeavour, i.e. they provide skill in handling matter, machines, capital, energy, and information.

- Much less educational attention is paid to prudence, the requirement for pragmatic action, the dealing with people. Even the humanities seem to have their centre of gravity in technical, rather than in pragmatic action, and they provide more skill than prudence.

- It seems as if ethics get even less attention in most courses of study. A weakness of the contemporary educational programmes? Kant's classification between technical, pragmatic, and ethical action, his qualifications of skill, prudence, and wisdom, his hierarchy of imperatives can be used as a measure of our education programmes. Are not those systems to be aspired which combine the training in technical, pragmatic as well as ethical action?

Can we be satisfied to live in a society with highly advanced skill only? Would we be satisfied to live in a society with highly advanced skill and professional prudence only? Or do we not long for a society with highly advanced skill and professional prudence, both combined with a wisdom of deep morals?

Good education should be comprehensive in that it provides foundations for skill, prudence and wisdom.

THE DIAMOND OF THOUGHT

What about courses of study in mathematical· and other fields of methodology? They do not find a space within Kant's three kinds of action. This is not of a surprise since any methodology itself does neither deal with objects nor with people, let alone ethical values.

Since quite a few courses of study mainly emphasise mathematics and other methodologies, it seems to be useful to extend Kant's system of three kinds of action into a system of four. However, there are no four kinds of action; instead there are four dimensions of <u>knowledge</u> and <u>understanding</u>.

The four dimensions of knowledge and understanding can be arranged in a diamond (Figure 1). The base is <u>methodology</u>. This includes any kind of method, be it language (and its grammar), mathematics, logic, computer programming, up to philosophy of science. All the other three corners of the diamond depend on methodology. <u>Technical knowledge</u> (referring to Kant's technical action) covers all kinds of skills to deal with objects, be it matter, machines, capital, energy, information etc.. These skills can be based upon science as well as the humanities, including economics, management science, sociology, psychology etc.. <u>Pragmatic understanding</u> (referring to Kant's pragmatic action) covers all kinds of prudence to deal with people as individuals as well as in groups. They can be teachers or students, employees and employers, sellers and buyers, voters and politicians or any other kind of subsets in social systems. <u>Ethical consciousness</u> (referring to Kant's ethical action) is based on general wisdom. It is necessary for defining and accepting any kind of goal or end.

A university course of study should provide the students with some competence in all the four dimensions of knowledge and understanding. Some actual courses of study,however, concentrate mainly on methodology, particularly the programmes in mathematics, partly in informatics (computer science and partly in operational research and management science. This is very narrow.

THE COVERAGE OF AN IDEAL EDUCATION

    From the three layers,
    (i)   science and the humanities,
    (ii) the interdependence between technological progress, economic
          growth, and social change and
    (iii) the unity of technical, pragmatic, and ethical action and the
          "diamond of thought" covering the four dimensions of knowledge
          and understanding,
a frame for an ideal university education can be derived.

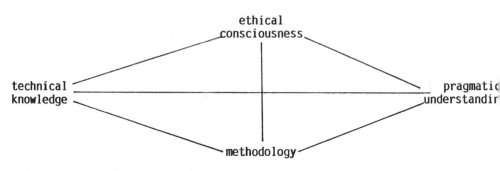

Figure 1. The "Diamond of Thought" combining the four dimensions of knowledge and understanding (see Müller-Merbach 1987c)

The foundation of an education should be broad, while any specialisation can be as narrow as it is practised today - if it is based on the broad foundation.

The breadth can be based on the frame of Table 2. The student should get enough insight into this frame such that he can localise the field of his speciality and that he can link his speciality with the totality of knowledge and understanding. Such a frame should become a solid component of each student's individual culture of thought, his cognitive style, his Weltanschauung.

THE CHALLENGE FOR OPERATIONAL RESEARCH

From this frame comes the challenge of OR education and OR practice. Certainly, specialisation is necessary, and one would not come far if he remains a generalist only. However, any specialisation without a general frame is without orientation. If the old virtue of interdisciplinarity of OR had been maintained over the five decades of OR, the plea in this paper would not be necessary at all. The situation is, however, that in many schools the idea of interdisciplinarity has been abandoned.

In contrast, there was a time that OR was understood as the only interdisciplinary approach to problems, contrasting all the traditional monodisciplines. Any attempt to restore the interdisciplinary approach of OR seems to be worth any effort. Interdisciplinarity would not mean that somebody understands everything. It only means that he has a basic understanding of anybody's role in society and of anybody's Weltanschauung. Thus, the operational researcher becomes an agent of communication between specialists. Table 2 is only one frame among many other possible frames for interdisciplinarity.

Table 2. The Frame of an ideal education

Culture of Thought	Strands of Industrial Development	Dimensions of knowledge and understanding			
		Methodology	technical knowledge	pragmatic understanding	ethical consciousness
science	technological progress				
humanities	economic growth				
	social change				

REFERENCES

Ackoff R.L., 1973, Science in the Systems Age: beyond IE, OR and MS, Opns.Res., 21:661.
Bayraktar W.A. et al. (eds.), 1979, "Education in Systems Science", Taylor and Francis, London.
Expertenkommission, 1985, "Wettbewerbsfähigkeit und Beschäftigung", Mainz, Rheinland-Pfulz.

Hinske N., 1980, "Kant als Heransforderung an die Geganwart", Alber,
        Freiburg-Munchen.
Muller-Merbach H., 1984, Interdisciplinarity in Operational Research - in
        the past and in the future, Jnl.Op.Res.Soc., 35:83.
Muller-Merbach H., 1987a, Immanuel Kant: drei Arten des Handelns,
        Technologie and Management, 36:60.
Muller-Merbach H., 1987b, Kant's Three Categories of Action, Letter from
        the President, IFORS.
Muller-Merbach H., 1987c, The Diamond of Understanding, Letter from the
        President, IFORS.
Muller-Merbach H., 1988, Ethik ökonomischen Verhaltens, in
        "Wirtschaftswissenschaft und Ethik", H. Hesse (ed.).
Russell B., 1988, "A History of Western Philosophy", Unwin, London.
Snow C.P., 1964, "The Two Cultures", Cambridge University Press,
        Cambridge.

O.R. AS A SOCIAL SCIENCE

OPERATIONAL RESEARCH AS A SOCIAL SCIENCE

Lynda Davies

Management Development and Organizational Research Division
The Management School
University of Lancaster, United Kingdom

INTRODUCTION

This conference is the second of its kind with the first being some twenty-five years ago. The proceedings from the first conference did not deal with operational research as a social science but rather showed a concern for showing how operational research may be used as a form of applied social science (Lawrence, 1966). It is a sign of growing self-awareness that this second conference sees the notion of operational research as a social science to be one which is worthy of consideration and debate. This session is particularly concerned with that debate and that is why the title is posed as a question. Can operational research be considered as a social science? If so, what mode of scientific inquiry is it pursuing, either implicitly or explicitly? It seems that operational research has become more aware of its social attributes, particularly in relation to the effects of changes which it may introduce into organizations and societies. This may not be acceptable to all as a mode of social scientific inquiry. The distinction between restricting usage to seeking to know (science) and the application of knowledge to the purposeful acts of change (technology) is one which is seen to be of particular relevance in this session. Any reflection upon operational research as a social science should be willing to challenge its status as such considering the intentions of operational research as a method of inquiry into social life, or as a means of creating change in groups, organizations, or societies.

SCIENCE OR TECHNOLOGY?

If we are to take operational research as a social science then it is necessary to ask what knowledge of the behaviour of individuals, groups, organizations, and societies has been gained specifically through operational research inquiry techniques. The first problem which becomes apparent is that it is uncertain whether there are such inquiry techniques. What specifically does operational research ask about the nature of social life? How does it go about developing research frameworks for challenging current knowledge about such? It seems that such knowledge is gained as a diversification of operational research techniques, as in such areas as the study of decision-making, of conflict resolution, or of organizational development and change. It is the application of operational research to more traditional theoretical areas of social science such as psychology, sociology, economics, or politics which has rendered knowledge about human behaviour rather than any direct development of operational research. From this recognition, which it is possible may be found highly contentious by some individuals, the realisation is made that operational research is not a social science in the old sense of the term. It does not merely seek to inquire, through theory, models, data collection, etc into the nature of human behaviour as operational behaviour. There are further perspectives which are often adopted from other disciplines to enhance the notion of operational research as a social science.

So far the model of social science which this paper has implied is only one of many currently available. Modern debate in the social sciences is having to deal with issues regarding the validity of traditional approaches to inquiry into human behaviour. The argument has been put forward on many occassions that much of the experimental knowledge gained from psychology, positivistic sociology and, to a lesser extent, economics has little external validity, i.e., it is severely limited in its transferability to the everyday world of human action. This has led to the development of research which is involved in collecting data and inferences from the everday world. Approaches such as ethnography, participant observation, introspective recording, and other qualitative approaches have moved from a notion of social science as objectively observable, accountable, repeatable, and test-able to one of social science as intersubjectively constructed, reflexive, of unique occurrences and experiential. This distinction represents a range of polar opposites in social scientific inquiry of the present but examples of both extremes are found in the liter-ature. The movement of change has broadly been from measuring generalistic data to interpreting contextually-dependent meaning constructs. The scientific inquiry into human actions is no longer confined to the Popperian experimental view.

A further challenge which has arisen is the realisation that social inquiry interferes with and changes social action. This can be seen as a nuisance to be guarded against or as a focus of inquiry in itself. The former approach is found in many qualitative sociology modes of inquiry whilst the latter has given rise to a phenomenon known as action research. Action research purposefully seeks to change with that change procedure becoming part of the object of inquiry. With action research, the researcher is concerned with understanding changes in the context of the inquiry and also understanding the approach to change being used. The interaction between these two (context and method) is also considered to be of prime importance. It is action research which changes the more formal model of social science into one of social technology.

This purposeful nature of change as both a process and object of inquiry is the one which bears great resemblance to much of operational research as seen in organizational change procedures. However, it is difficult to see this in many of the key journals of operational research which gives the impression that the chief concern is with the creation of algorithms to deal with complex but highly closed problems of logistical control. However, there is a recent indication that a more action research style of operational research is coming to the fore. The growing interest in the 'soft' areas of operational research implies this. Writers such as Eden, Ackoff, Checkland, Friend, and Rosenhead have shown that purposeful change can be accepted as an approach to social scientific inquiry. The approach does concern itself directly with the application of knowledge to the changing of social life – be this in groups, organizations, or societies. This is a form of technology whilst also a form of action-based social science. This newer form of operational research does seem to bridge the gap between science and technology whilst not losing its validity as both a form of inquiry and a means to change.

A MORE MORALLY AWARE OPERATIONAL RESEARCH?

One of the key questions being addressed by modern writers on operational research is that concerned with the responsibilities of agents for change. Users of operational research do change situations but until recent times they have not explicitly considered the moral effects of their actions as important in that change procedure. The recent concern with the development of community O.R. has shown a dramatic shift in that concern. It seems that revisioning operational research as an action-based social science/technology can lead to a more morally aware generation of future users of operational research. Changes in operations do change the human curcumstances, be it at the individual, group, societal, or global level. Action researchers have been concerned with that interventionist mode of change. More recent challenges have meant that even this new wave of 'soft' operational researchers must go beyond intervention per se to a consideration of how to create changes which allow for the protection of the freedom of the individual. A morally aware future seems to be possible for operational research and this has been helped by the recognition that operational research is a form of social science, albeit one which is concerned with purposeful change. This 'technology' of operational research does still allow for knowledge generation regarding social life but also allows for intervention and change into social situations. This can lead to a better future for the inquirers/intervenors using operational research and the beneficiaries of the actions of operational research. Such is the vision of operational research as a morally responsible social science and one which it is hope the next twenty-five years of O.R. will witness in abundance.

REFERENCE

J.R. Lawrence, ed., "Operational Research and the Social Sciences" Tavistock, London, 1966.

SOCIAL SCIENCES FOR OR - AN OVERVIEW

Per S. Agrell

FOA
S-10254   Stockholm   Sweden

This is not going to be an original paper with a brilliant conclusion. It is to be an overview of how the OR field is appreciated by different writers in the field. From this overview it will then be seen to what extent the different social sciences are relevant for the different kinds of OR. Some kinds of OR will appear as a more or less conventional social science, others will appear using the social sciences.

In the overview not only will be seen how the different kinds of OR borrow pieces of social science, but also how the different ways of slicing the field of OR will make the social sciences and their importance appear differently; that is, different perspectives on OR yield different aspects of the social sciences.

CLASSIFICATIONS

An important attemt to distinguish different kind OR is made by M. Dando and C. Eden 1977. They introduced the expression "reflect on our own studies" and they contrasted a reflecting work against the simple direct application of a specialized technique. They state that "reflection" is needed for impacts of OR on the real world.

Even the simple applications need to use a social sciences knowledge when the systems studied are social to some extent. In any case such knowledge is needed in elaborating the process of the applied project and in the reflections described by Dando and Eden. This elaboration is basically social engineering it can be said, but in that case including a perceptive work of an ethnographical character. The work milieu must be understood for the initiatives in the project procedures to work. Dando and Eden state that reflection is needed for a good application.

Attempts towards objective observation are more necessary than easy here. "Contretransfer" (Devereux 1980) is a Freud inspired methodology in this direction explicating and explaining subjective inclinations. Practically this is done by different kinds of teamwork. By revealing the nature of an observer's subjectivity, objectivity becomes more possible.

Another fundamental and similar distinction is the trichotomy offered by J. van Gigch (1989). At least three levels of analysis are important for him: 1)application, 2)methods development and 3)paradigmatic elaborations. The first case of Dando-Eden is very similar to the van Gigch case one. Their second one is mainly the union of the vG levels 2 and 3. There is the similarity between the two slicings that the latter categories are considered necessary for good applications.

The vG case two, as the others, contains subcases. The first subcase, creativity methods development needs normally more of a social sciences knowledge than the development of methods for critics and analysis. There is a facet of psychology and group dynamics which De Bono calls lateral thinking (1973). There are also issues of organizational inertia well enough described by for example Schon (1972, before he changed his name to Schön) and Hatchuel et al (1987).

Methods of critics and analysis, the other subcase, are often treated by some kind of pure logic, without human concerns. However, as M. Berry points out (1983), there are social values of all methods and of all their different applications. That is, not all methods work equally well in different social milieus, and all of them have both expected and unexpected social effects. This really should count for the OR methods developer. Implementation works differently in different contexts.

Transfer of experience is no more easy for OR methology than for other fields; each experience is unique and the generalizations are uncertain. Here again some kind of anthropology of modern organizations is needed, anthropology rather than sociology, their epistemologies being fundamentally different. The former tries to make the best out of obviously subjective observations the other leans rather on repeated observations, experiments and statistics. The epistemological maturity is naturally enough smaller, but the epistemological vitality is greater whith the anthropologists as far as the author has been able to judge it.

For the OR methods development the anthropological logic is of an autstanding importance since so many applications are unique, and generalized knowledge must be drawn from those whether this is scientifically possible or not. The time is out for those who believe that normative methods reserach can be done without a concern for the shifting local human logics and cultures concerned (van Gigch 1989, Grunwald & Fortuin 1989).

The third level of van Gigch (the more superiour and also the more fundamental one) deals with the individuals characters in thought and research and with stable patterns of culture. This level explains and controls the other two. Still Churchmans book "On the design of inquiring systems" (1971) is the outstanding drawing of such paradigmatic differences. In this book archetypes of modern research styles are identified. The main idea is to see how classical philosophy may serve to describe and to influence modern applied reserach.

Research and analysis on this third level imports to the applied
OR projects a better and a more flexible implementation as well
as an improved cooperation within the OR-teams.   To methods
development and other more theoretical research it provides a
clarification of important assumptions.   This level is basically
philosophy, but there is also psychology in it, for example in
finding out about each others' reigning paradigms.

For normative methods research for example, the paradigmatic
elaborations specify the domains of validity for its resulting
methodological advice. M.C. Jackson for example (1988) specifies
three paradigms, or here degrees of coherence, in an organization
"unitary, pluralist and coercive" which demand clearly different
methodologies.   It is not surprising that this succesful author
works in a strictly client oriented environment, community OR.
Such client oriented organization of OR in conferences, working-
groups and institutes is what OR needs today rather than an
organization on techniques or issues. Such a reorganization
would mean a greater importance for all the so far defined and
undefined social sciences.

A further, one and the same slicing of the OR- field to be
considered seem to have been developed independently by three
different researchers, Samuel Eilon 1974, Steen Hildebrandt 1982
and Werner Ulrich 1988. A first reading of their respective
texts displays differences, but these would mainly be referred to
supplementary dimensions in their texts. Their basic
trichotomies are very much alike.

The first case in this slicing is about OR for statements on the
adequacy of a system itself (not the control of it). Hildebrandt
says "thinking" and Ulrich says "operational systems management".
Here we have the puzzle solving expert, the empiric evaluator,
the organization diagnostician and the systemizer. We have also
all methods development connected. Social sciences concerned are
both descriptive and normative, both for application and for
generalized theory building. This vast class of OR deals with
all sides of human life and thus with all social sciences.

The second case then is about OR for improvments in the behaviour
that controls a system. Hildebrandt says "communication" and
Ulrich says "strategic systems management". Here we have the
neutral communicator, the not diagnosing process consultant, the
sparring partner, the iconoclast (changes beliefs and behaviour),
the decision theorist mentioned below and the empiric evaluator
already mentioned above under case one. Scientifically this is a
smaller niche than the case one, it is behaviour. It is both
individual and organizational behaviour. It is not only social
sciences however. Cybernetics and control theory for example are
also relevant.

The third case is "search-learning" or "normative systems
management" with a purpose of dismantelling the controling
behaviour into its ideals and other stable features. If one
speaks about paradigms in this case it is about the paradigms of
management, not the paradigms of research as does van Gigch in

his case three above. In a pure case we have something like a classifier, who organizes information and influences ideas as to what a system is and can and will do under certain circumstances. A dialectician and a change agent could take a role of this kind as well as the earlier mentioned iconoclast. The kind of deep psychology implied here should not be considered odd by the OR man. Many more than Ackoff (1979) have found that participation and influence on the level of ideal-formation is good OR especially if you want to avoid the usual implementation problems. In negociations and other competitive situations you must understand the other parties in depth to win. P. Bennett has explained this in terms of a "hyper game" theory. See for example his paper together with C.S. Huxham 1982. K.C. Bowens' methodology for problem formulation in conflicts (1983) works by a group of actors who learns in a structured way about each others ideals and normative systems.

Attributes in all the three dimensions are quite possible to combine in the same project. For example, yes or no, effort or not in the respective dimensions, make dichotomies, and each OR project falls or falls not into each one of those. Eilon in fact is the most elaborate author on this point describing seven of the eight combined archetypes possible. What he has missed is the normative decision theorist and thus the oportunity to make the complete 2x2x2 matrix. The latter insight belongs to Ken Bowen (1984).

PARAMETRIC DESCRIPTIONS

Explicit parametric descriptions of OR-types are less frequent. Grunwald and Fortuin however (1989) draw a space on three rationality dimensions: "social, system and quantification" qualities. All facets need social sciences. "Quantification" least so, but still from time to time psychological and social theories with their quantifying rationalities. What they call a "social" rationality becomes a continuous parameter by indicating to what degree research and executive behaviour are integrated. The range of this variable is between a take it or leave it analysis and the complete end user modelling. The third dimension, the "systems" rationality one goes from cybernetics with boxes, arrows and mathematics to the so called autopoietic systems, e g systems which preserve themselves and their identity by continuous renewal. A good decision support system and good OR, they say, should have all three rationalities.

This parametric description of different types of OR brings us towards scientific activities which are no doubt social science but not very strictly in any of its defined subdisciplines. The anthropology of using information for example is an important issue worth more academic attention.

Another parametric description of OR types could be presented by a reference to the author's own works. There is generally so much misuse of these auto-references that I hesitate to include them, but this one reference is unavoidable for the present scanning attempt, it is a co-authored paper and it is just one out of many treating the same theme (P. Agrell and R. Vallée 1986).

There are basically four dimensions in the descriptor: "matter, method, programme and paradigm". For "matter" there is the duality between a focus of primary interest and an attention to the actors concerned including an attention to the researcher himself. For "method" there is a similarly symmetric duality between imagination and analysis. For "programme" the duality is between thinking and communication and for "paradigm" it is between beliefs and standards. By this taxonomy all applied OR, in fact all research applied for a client, has got attributes in all the four dimensions. It is even claimed that good OR demands a certain equillibrium in all the four dualities.

The scanning by this four level taxonomy confirms the impression that all applied projects need the social sciences, but not only social sciences taken from the shelf. New research problems have to be formulated, solved and applied. Such issues are the following, taking at random a few urgent ones from the four dimensions mentioned above:
o How do you make the OR people understand, that the knowledge of "matter" they need is not only what they can learn from their clients but also a systems theory (with and without social elements) not known to those?
o How can the analysts and the scientific OR community ensure that the systems theory presented contains and provides something that is new enough to the specialists of the "matter" concerned?. (Systems theory too often only restates the obvious.)
o What is a good description of the "actors", one which helps in the project work?
o How do you specify "actors" and other social parts of the "matter" so that such specifications may make conditions and domains for a respectable methods choice and a normative methods research?
o What creativity "methods" are possible in the different kinds of bureaucracy? This is difficult - explicit creativity methods are not very respected generally.
o What reactions do you find to the use of different "methods" of analysis in different social contexts?
o How can different epistemological "programmes" be expressed in terms of bureaucratic action, contacts and communication?
o What "paradigmatic" beliefs and standards about OR are common in different cultures? Which help OR and which do not help; and how do you use the helpful ones and avoid or change the more troublesome ones?

RESULTS

The essence of this overview is that OR needs much of the social sciences. An anthropology of modern organizations for example is needed to enable general conclusions to be drawn from case studies. OR moreover needs a very flexible social science which is not bound by the existing traditions and the existing research organizations. Without discarding any of its mathematical, systemic or interdisciplinary aspects, OR needs to establish itself as a social science, a SOCIAL OPERATIONS RESEARCH perhaps.

REFERENCES

Ackoff, R. Resurrecting the Future of OR, J Op Res Soc 2, 1979.
Berry M. Une technologie invisible, Ecole Polytechnique, Paris 1983.

Agrell, P. and Vallée R. Different concepts of Systems analysis, Kybernetes Vol 12, 1985.

Bennett P. and Huxham C.S. Hypergames and what they do, J Opl Res Soc 33(1) 1982.

de Bono E. The Use of Lateral Thinking, Harper, New York 1973.

Bowen K. C. A Methodology for Problem Formulation, Thesis, Royal Holloway College, London University 1983.

Bowen K.C. unpublished memo, London University 1984.

Churchman, W. The Design of Inquiring Systems, Basic Books, London 1971.

Dando, M and Eden C. Reflections on Operational Research. A report from the Euro II Congress, Omega 5 no 3, 1977.

Devereux, G. De l'angoisse à la methode, Flammarion, Paris 1980.

Eilon, S. Seven Faces of Research, Omega 2 no 1, 1974.

van Gigch, J. The Potential Demise of OR/MS: Consequences of Neglecting Epistemology, EJOR, to appear 1989.

Grunwald, H.J. and Fortuin, L. DSS and ES in the "Information Organization" - Back to the Roots of OR, EJOR to appear 1989.

Hatchuel A. et al, Innovation as System Intervention, Systems Research 4 no 1, 1987.

Hildebrandt S. The Operational Research Process as a Cooperative venture, Bedriftsekonomen 5, 1982.

Jackson M.C. Some Methodologies for Community Operational Research, J Opl Res Soc 39 no 8, 1988

Schon, D. Blindgångare mot framtiden, Pan Stockholm 1972.

Ulrich, W. Systems Thinking, Systems Practice and Practical Philosophy, in Systems Practice, Plenum Press 1988.

O.R. IMAGES IN ORGANISATION THEORY: A CRITIQUE

Haridimos Tsoukas

Manchester Business School, University of Manchester
Booth Street West,
Manchester, M15 6PB   UK

THEORIES, METAPHORS AND IMAGES...

Any phenomenon, especially if it is a social one, can be viewed in a
number of ways depending on the purpose and the viewing position of the
observer. Yet there is an important distinction to be drawn amongst all
the potential descriptive accounts. Phenomenological descriptions are
common sense accounts of the observable phenomena of the world, while
scientific descriptions deny (or transcend) these phenomenological accounts
in the name of (usually directly unobservable) more profound ontological
truth-claims (Davis, 1971; Maki, 1985). In that sense it can be said that
scientific theories re-describe the world in order to lay bare the mechanisms responsible for the observable phenomena we experience. It is because
our phenomenological accounts fail to expose these generative mechanisms
that scientific discourse is necessary and possible (Bhaskar, 1978).
Showing the hitherto unknown mechanisms requires painstaking thought and
creative imagination. We usually attempt to transcend our ignorance by
using metaphors, analogies, models and images which will enable us to make
inferences "about one of the things, usually that about which we know least,
on the basis of what we know about the other" (Harre, 1985; see also Morgan,
1980, 1986).

Of course any metaphor or model is one-sided and partial, and that is
why scientists have not been yet forced out of business! This however does
not imply that all of our theoretical schemata are equally illuminating and
useful as Morgan (1980, 1986) and Checkland (1985, 1987) seem to suggest.
As a matter of fact the very process of creative production of metaphors is
distinctly different from their epistemological status. A certain theoretical
redescription is superior to some others if it provides an ontologically
deeper description of the "ontic core" of the phenomenon under study, that
is if it demonstrates that what the rival redescriptions have shown are
derivative phenomenological manifestations of the ontic core as described
by the new theory or model (Maki, 1985). In other words, if the object of
study redescribed by a certain perspective is what the perspective claims
it to be, this particular perspective should be able to explain all the
empirical manifestations and their interrelations connected with its object.
If it does not, it needs to be refined or rejected in favour of some other
perspective which is, provisionally at least, problem-free. Viewed in this
way, metaphors and models do not only possess a heuristic value, but constitute also the hypothesised mechanisms which might really be responsible for

the phenomena under study. To satisfy our insatiable Aristotelian "desire to understand" we need to go deeper and deeper to new strata of knowledge.

...IN SEARCH FOR THE OPTIMUM ORGANISATION DESIGN

Operations Research (O.R.) has been a repository of images used in organisation theory in (chiefly) designing organisational institutions. More particularly, the optimisation image, which has historically been the defining property of O.R., has taken the notion of best fit between, or optimisation of, usually two 'variables', subject to a number of predominantly environmental constraints, to be a necessary and sufficient condition for the maximisation of total organisational performance. The socio-technical systems theory (STST) in particular has conceived of organisations as consisting of people who produce desirable products or services using a certain technology. The social sub-system includes the relationships among the organisational participants, their attitudes and expectations. The technical sub-system is based upon the tasks to be performed, including the equipment, the tools, the operating techniques and the lay out. As the STST argues, technological demands constrain the operation of the social system through the allocation of work rules and the technologically given dependence relations between tasks, thus shaping the behaviour of the social sub-system. However, the social sub-system has certain socio-psychological properties of its own, independent of technology. Thus, it is argued, an organisation will function optimally only if the social and technical sub-systems are designed to fit the demands of each other, in a given environmental context, namely if they are jointly optimised. Every other solution may be optimal for a sub-system but not for the system as a whole (Trist et al, 1963; Herbst, 1974; Pasmore and Sherwood, 1978; Trist, 1981; Emery and Trist, 1981; Pasmore et al, 1982).

In other words what the socio-technical systems theorists seem to have been saying is that organisation designers, like the operational researchers, can somehow be viewed as the 'independent' optimisers of social systems via the construction of evaluative models which will enable them to compare alternative work designs and to select the 'best' one for the situation under study (see Lupton, 1975 for an illustration of such an exercise). Naturally, such a view of organisation design presupposes (a) commonly agreed, unambigously defined and temporally stable 'objective' (that is performance goal), 'variables' (that is technical and social sub-systems) and 'constraints' (namely environmental restrictions); (b) that the evaluative model for arriving at the 'optimum' design is a perfect representation of the social system; (c) that the 'variables' to be 'optimised' can be viewed independently of their social, cultural and political context; and (d) that theory or model-building always precedes practice, or to put it differently, design formulation and implementation are two temporally and analytically distinct phases in which more than one course of action is first vetted, and the 'best' one is put into practice respectively. Furthermore, it is assumed that theory can have primarily instrumental value, namely in helping practitioners to intervene in social situations, without any concern for providing explanations too. In fact it is tacitly assumed, or sometimes even explicitly postulated, that there is a disjuncture between "the explanation of structure" and "the construction of a theory of organisation design" (Donaldson, 1985:142). Somehow, explanatory and interventionist knowledge are viewed antithetically.

A CRITIQUE OF THE OPTIMISATION IMAGE

There has been a growing body of empirical literature reporting socio-technical applications, without however being able to assess conclusively the STST claims. There still remain ambiguities calling for a substantive refinement of the model, although as I shall argue later there are more problems with its paradigmatic underpinnings which render its theoretical usefulness questionable.

In a substantive sense, even leaving research design problems aside (cf. Pasmore et al, 1982), establishing the direction of causality in the reported successful socio-technical designs has been problematic. A socio-technical intervention is a multi-faceted process and it is difficult to know "whether the higher productivity resulting from reorganization of work is derived from increased efforts because people are better motivated or whether the effects are due to the fact that work is organized better" (Stymne, 1985:99). Kelly (1978, 1985) has argued that extrinsic motivational aspects (e.g. payment) and common rationalisation measures (e.g. elimination of idle time and unnecessary movements, lay out changes etc) are mainly responsible for positive economic results rather than the socio-technical theorists' emphasis on the satisfaction of the social sub-system's needs. To be sure, better research designs will alleviate these problems, but at the same time there is a need for STST to refine its concepts and their consequent operationalisations.

More importantly, the most central tenet of the optimisation image, namely the principle of joint optimisation, has rarely been implemented literally in organisational practice. Empirical evidence suggests that almost invariably a more appropriate social system is sought to be designed around a given technological configuration (Lupton, 1975; Kelly, 1978). For instance, Pasmore et al (1982) in their review of 134 North American socio-technical experiments found that technical changes were accomplished in only 16% of the studies. More often than not the 'rigid' logic of efficiency supersedes the 'malleable' logic of satisfaction. This finding underscores a major lacuna in the optimisation image, that is the failure to include the wider socio-economic milieu in its analytical focus. 'Optimisation' exercises are not executed against a socially neutral background, but rather the latter is marked by certain organising principles which need to be conceptually incorporated into any theoretical organisation design model. In case of capitalism for instance, "the activities of business - in so many ways the very epitome of independently conceived and purposefully motivated action - are perceived as obeying imperatives that originate below the surface of economic life. The structuring effect that this netherworld casts over the course of business activity is never precisely revealed in the pattern of economic events, any more than the design produced by a magnet in the arrangement of iron filings exactly reveals the arcs and loops of its field of force. Yet in the one case as in the other, the persisting presence of an order-bestowing influence is felt" (Heilbroner, 1987:17-8). The use of capital as a means in an endless process for amassing more wealth is the most central characteristic of the nature of capitalism (see Heilbroner, 1985; Berger, 1986) which has certain repercussions for business organisations, and throws the joint optimisation principle into question. Efficiency and satisfaction simply do not stay on an equal footing, since social formations are not tabula rasa but already set up agendas.

The latter remark leads me to the second, paradigmatic part of my critique. Applied theories or models for intervention in human affairs tend to be inadequate if they are not simultaneously concerned with broader explanatory questions. For to improve current work arrangements one must ask why concrete work configurations arose, were developed and reproduced, as well as how they can be modified (Silverman, 1983; Whitley, 1984). In turn this implies going beyond surface patterns of events to the mechanisms of their production and reproduction. Epistemological questions with regard to the mode of linkage between the "nether-world" of a social formation and the surface of organisational manifestations become crucial, especially if the Scylla of determinism and the Charybdis of voluntarism are to be avoided. Yet the optimisation image remains remarkably reticent and crucially unable to handle these issues.

The optimisation image tends to underestimate the malleability, change-ability and the purposfulness of the 'variables' of the 'objective function' to be 'optimised'. Performance standards for instance, are not given but chosen by the organisational power-holders, who are in a dialectic relation-ship with their environment (Child, 1972). Even the survival of a firm is susceptible to multiple realisations - and this is a crucial distinction between biological and social systems - which do not necessarily imply uniform organisational designs. Furthermore, even if there is agreement on the relatively stable imperatives of the technical sub-system, it is hard to see how such a unanimity can be reached on the changeable needs of the social sub-system. The logic of satisfaction mutates over time, it is influenced by the accumulating experience of the people, and more import-antly it can be moulded by powerful organisational members. The current burgeoning literature on organisational cultures precisely underscores the fact that people are social products whose needs, aspirations and desires are not immutably given, but socially (re)created (Berger and Luckman, 1967). Thus, as I remarked earlier on, it is not surprising to find that most of the socio-technical changes amount to a reorganisation of the social sub-system to suit the purposes of technology. People are, or can be made, infinitely more flexible than machines!

The optimisation image is found wanting on two further points. Even if organisation designers could in principle reach optimal work designs, empirical evidence suggests that the latter would be inevitably modified (and thus they would cease to be optimal) at the operational level, via the "unofficial manipulatory devices" (Warmington et al, 1977:159) of all those involved, in order to further their interests and preferences (see also Ackoff, 1979; Child, 1985). In that sense 'optimal' work designs are always intended strategies which are invariably diluted (i.e. 'de-optimised') to a lesser or greater extent, en route to their realisation (Mintzberg, 1978, 1987). This points to an additional trait of social phenomena, which has been underplayed by the optimisation image, namely that social institu-tions have an irreducibly emergent character. Formulation and implement-ation are not necessarily dichotomically distinct, but quite frequently they are entwined in a process of learning. It is noteworthy that in one of the first socio-technical studies (Trist et al, 1963) it was the miners themselves who had initially applied what was later called composite work groups. Work configurations usually emerge as a variable combination of intended designs and non-intended (from the designer's point of view that is) responses and modifications of those engaged in the designs' implement-ation. Emergent work patterns can in turn be fed back on new designs which give rise to new work patterns and so on, in a continuous loop of adaptation and learning (Mintzberg and Waters, 1985). The role of the organisation designer is better captured not by the image of a planner but by the image of a craftsman who works in a dialectical interactive relationship with his material and his environment (Mintzberg, 1987). If this is accepted then the role of organisation design becomes different. It is not so much the content of designs which is important as the form of meta-design, that is the way design itself is conceptualised and practised. The organisational theory orthodoxy has viewed organisations primarily as machines or organisms. It is time we made our images more anthropocentric, thus being appreciative of the uniqueness of social reality. This of course requires a different conception of social reality itself, and of the role of knowledge in inter-vening in human affairs. In short, a different meta-theory is needed.

120

REFERENCES

Ackoff, R. L., 1979, The future of opperational research is past,
      Journal of the Operational Research Society, 30:2
Bhaskar, R., 1978,"A realist theory of science", The Harvester
      Press, Hassocks
Berger, P. L., 1987, "The capitalist revolution: Fifty propositions
      about prosperity, equality and liberty", Gower, Aldershot
Berger, P. and Luckman, T., 1966, "The social construction of reality:
      A treatise in the sociology of knowledge", Penguin Books,
      Harmondsworth
Checkland, P., 1985, Form optimising to learning: A development of
      systems thinking for the 1990s, Journal of the Operational Research
      Society, 36:9
Checkland, P., 1987, The application of systems thinking in real-world
      problem situations: The emergence of soft systems methodology,
      in: "New directions in management science", M. Jackson and P. Keys,
      eds., Gower, Aldershot
Child, J., 1972, Organisation structure, environment and performance:
      The role of strategic choice, Sociology, 16:1
Child, J., 1985, Managerial strategies, new technology and the labour
      process, in: "Job redesign: Critical perspectives on the labour
      process", D. Knights, H. Willmott and D. Collinson, eds., Gower,
      Aldershot
Davis, M. S., 1971, That's interesting! Towards a phenomenology of
      sociology and a sociology of phenomenology, Philosophy of Social
      Science, 1:1
Donaldson, L., 1985, "In defense of organization theory: A reply to the
      critics", Cambridge University Press, Cambridge
Emery, F. and Trist, E. L., 1981, Sociotechnical systems, in: "Systems
      Behaviour", Open Systems Group, eds., Harper and Raw, London,
      3rd edition
Harre, R., 1985, "The philosophies of science: An introductory survey",
      Oxford University Press, Oxford, 2nd edition
Heilbroner, R. L., 1987, "The nature and logic of capitalism", W. W. Norton
      & Company, New York
Herbst, P. G., 1974, "Socio-technical design: Strategies in multidisciplin-
      ary research", Tavistock, London
Kelly, J., 1978, A reappraisal of sociotechnical systems theory, Human
      Relations, 31:12
Kelly, J., 1985, Management's redesign of work: labour process, labour
      markets and product markets, in: "Job redesign", D. Knights,
      H. Willmott and D. Collinson, eds., Gower, Aldershot
Lupton, T., 1975, Efficiency and the quality of working life: The
      technology of reconciliation, Organizational Dynamics, 4(2)
Maki, U., 1985, Issues in redescribing business firms, in: "Proceedings
      of the first summer seminar of the group on the theory of the firm",
      K. Lilja, K. Rasanen and R. Tainio, eds., 7-8 August 1984, Espoo,
      Finland, Helsinki: Helsinki School of Economics, Studies B-73
Mintzberg, H., 1978, Patterns in strategy formation, Management Science,
      24:9
Mintzberg, H., 1987, Crafting strategy, Harvard Business Review, 87:4
Mintzberg, H. and Waters, J., 1985, On strategies, deliberate and
      emergent, Strategic Management Journal, 6
Morgan, G., 1980, Paradigms, metaphors and puzzle solving in organization
      theory, Administrative Sciences Quarterly, 25:1
Morgan, G., 1986, "Images of organization", Sage Publications, Beverly
      Hills
Pasmore, W. A., Francis, C., Haldeman, J. and Shani, A., 1982,

Sociotechnical systems: A North American reflection on empirical studies of the seventies, Human Relations, 35:12

Pasmore, W. and Sherwood, J., eds., 1978, "Sociotechnical systems", University Associates, San Diego

Silverman, D., 1983, Socio-technical systems, in: "Organisations as systems", M. Lockett and R. Spear, eds., Open University Press, Milton Keynes

Stymne, B., 1985, Reorganization of work: Causes and effects, in: "New challenges for management research", A.H.G. Rinnooy Kan, ed., Elsevier Science Publishers B.V.

Trist, E. L., 1981, The sociotechnical perspective, in: "Perspectives on organization design and behaviour", A.Van de Ven and W. F. Joyce, eds., John Wiley & Sons, New York

Trist, E. L., Higgin. G., Murray, H. and Pollock, A. B., 1963, "Organisational choice", Tavistock, London

Warmington, A., Lupton, T. and Gribbin, C., 1977, "Organisational behaviour and performance", MacMillan, London

Whitley, R., 1984, The scientific status of management research as a practically-oriented social science, Journal of Management Studies, 21:4

SOCIAL SCIENCE AND OR TECHNOLOGY

Paul Keys

Department of Management Systems and Sciences
University of Hull
Hull, U.K.

INTRODUCTION

This paper addresses some issues in the relationship between OR and
the social sciences which are raised by accepting the view that OR is a
technology. As such it is seen to be fundamentally different from the
social sciences which seek to establish a body of understanding about
certain phenomena whereas OR is more concerned with processes of practical
value. The view that OR is a technology has a history dating back thirty
years since when Lathrop (1959), Barish (1963), White (1970), Dando,
Defrenne and Sharp (1977), Raitt (1979), Malin (1981) and Rosenhead (1986)
have consistently reminded the OR community of this possibility. In the
following the validity of the case for seeing OR as a technology is taken
as given. What is of concern are the consequences for understanding OR
and its relationship with the social sciences which follow from this
perspective.

In the first half of the paper the traditional model of the science-
technology relation is interpreted and assessed for the specific case of
OR and the social sciences. It is argued that the expected nature of
this relationship is not found to exist. Consequently an alternative
model of the science-technology complex is explored in an attempt to
provide a better explanation. Some observations are made about this model
and issues raised by the analysis discussed.

TRADITIONAL SCIENCE AND TECHNOLOGY

Barnes and Edge (1982, pp 147-54) see the traditional model of the
science-technology relation to be hierarchical in structure and see
technology as applied science. A key distinction is made between science,
which is concerned with the production of new knowledge, and technology,
which applies the products of science in a routine and mechanical fashion.
Technological developments are therefore triggered by scientific discovery
and each discovery will eventually play a part in technological progress.
To understand a technology requires an appreciation of the sciences on
which it is based and the processes by which it applies those sciences.

There have been several attempts to understand OR using this model.
Flood (1962) visualises OR as an applied science built upon a theoretical
base of behavioural and systems sciences and using the methods of the

the experimental sciences to apply such knowledge to problematic situations. Dando, Defrenne and Sharp (1977) recognise OR to be a technology which needs a science of decision-processes. Bonder (1979) calls for an operational science to be established which provides knowledge about those operational and managerial phenomena which are associated with OR practice. It is clear from these observations that the social sciences have a role to play in providing a theoretical base for OR. They provide knowledge about the context in which OR operates as well as about some of the processes which OR might have to confront directly in its work. The difficulty with this model lies not in the matching of social sciences and OR with science and technology but in the failure of experience to show any of the expected consequences of its adoption.

First, the model implies that an important distinction exists between the roles of OR and its underlying body of knowledge. Thus the only mechanism by which understanding of social, organisational and individual behaviour can be acquired is by work done within the social sciences. This is then transferred to OR and used as appropriate. However many operational researchers are involved in activities which have consequences for the knowledge base. This knowledge might be specific to one organisation but it cannot be denied that operational researchers hold a reservoir of knowledge about the behaviour of organisations which acts as a theoretical base, however informally defined, for their work. Many would argue that successful OR cannot be carried out unless such knowledge is acquired and used. Thus the presence of uni-directional flow of knowledge from the social sciences into OR is debatable.

Second, it is implicit in this model that a technology is under-pinned by a unified science. Yet the social sciences, to a greater extent than the natural sciences, are characterised by differences in paradigm and attitude which make them difficult to reconcile into a unified body of knowledge. Operational researchers looking to use elements of social sciences will find basic differences between them. Such differences will also be observed to exist between the social sciences generally and the natural sciences which use different methods of enquiry, modelling and validation. Hence the existence of a unified body of knowledge supporting OR is brought into question.

Third, this model suggests that technology cannot exist without a science to support it and that this science triggers technological change. In the current context this is interpreted to mean that OR would not be able to function in organisations unless it received direct and useful support from the social sciences on how to understand and behave in such settings. However the experience suggests that operational researchers receive little formal education in organisation theory and managerial behaviour and develop the necessary knowledge in an experiential manner. Although there is a growing awareness of the importance of the social sciences to OR, historically they have paid little attention to developments and these seem to have had little impact on OR practice. Thus OR seems to have survived with little dependence on and has not altered in response to developments in the social sciences.

Fourth, the traditional model suggests that OR consists of routine activities. Yet, in many cases, OR involves a creative element which is often central to its success. The model sees the culture of OR to be determined by its underlying science base and does not allow for a separate technological culture to exist. It can again be noted that OR has developed its own professional attitudes, language, networks and structure independently of any sciences, social or natural.

These difficulties with the traditional model are not unique to OR and Barnes and Edge (1982) and Mayr (1976) have noted that similar problems arise when other technologies have attempted to be understood by this model. In response to these difficulties an alternative model of the science-technology relation has been developed. This symmetrical model will now be explored in connection with the relationship between OR and the social sciences.

SYMMETRICAL SCIENCE AND TECHNOLOGY

The modern understanding of the science-technology relationship emphasises science and technology as separate bodies of activity which support each other in a mutually beneficial way. The view that science drives technology is replaced by the view that science and technology are self-maintaining. Within each party progress is made on the basis of previous work within that sphere of activity. Science builds new knowledge on the foundation of old and technology develops new mechanisms and systems on the basis of existing devices. The influences between the two strands is much weaker than the internal momentum. Hence technology is attributed here with a much greater degree of autonomy than in the traditional model. Science, consequently, is seen as not being as influential in technological progress as is the case in the previous model. Technological developments can occur without prior, causal developments in science.

The evolution of technology is influenced by two forces (Price, 1969). The strongest of these is that set of aims which a technology seeks to achieve. As the aims change then so the technology responds and adapts in such ways as to move towards them. A major influence upon OR is therefore the set of issues which it addresses in organisations. If these remain static and of a similar form then this will be reflected by OR being slow to develop. When novel issues and tasks are addressed then OR will evolve new skills and concepts to tackle them.

A weaker and less direct influence comes from those sciences and technologies which form an environment for OR. As new concepts and understanding are gained in these sciences they will slowly enter OR and influence the work undertaken there. Examples might be the use of social psychological notions by Eden et al. (1983) or cybernetics by Beer (1985) in their efforts to tackle issues in new ways. Thus the social sciences influence OR slowly and indirectly and this appears to correspond to what has happened over the recent past. OR has drawn upon the social sciences but in an almost random fashion. It has certainly not been the case that the social sciences have been strongly associated with OR.

This model presents the relationship between OR and the social sciences as being weak. The major forces of progress in both OR and the various social sciences are seen to be driven internally rather than externally. Only rarely will needs of OR result explicitly in change in the social sciences and vice versa. This is in contrast to the traditional model where developments in the social sciences should lead to progress and change for OR.

The symmetrical model of the science-technology relationship leads to several issues of concern for the future interaction of OR and the social sciences. Some of these are now considered.

One issue concerns the way in which knowledge and methods are transferred between the social sciences and OR. At the present this flow appears to act in a haphazard and sparse fashion. If OR is to take advantage of the insights generated by the social sciences suitable mechanisms need to be set up to facilitate the exchange of information. These mechanisms must also act as interpreters of what social scientists are doing in order that this work is made relevant and compatible with OR practice.

A second issue concerns the papyrophobic nature of technologies (Price, 1969). The tradition of open publication of new results in the sciences is not found in the technologies, partly due to a need to maintain commercial competitiveness. It is also not seen as being appropriate to technologies where a criterion of success is the achievement of goals set rather than the production of new knowledge. This situation poses a problem for the education of technologists, and hence of operational researchers. The absence of a growing and contemporary literature means that technologists receive a formal education which is based on second or third-hand experience. The informal education received on-the-job then sharpens this up by providing direct exposure to situations of relevance. Technologists receive some education in appropriate sciences, natural and social, but these tend to be the safe, proven and uncontroversial parts of those sciences. Hence the social science to which operational researchers are exposed are the mainstream, conventional theories of organisational behaviour and not the more radical and novel approaches. There is a generation gap between what the social scientists are doing and what is perceived of the social sciences by operational researchers. This gap needs to be overcome if the continually expanding level of understanding of organisations and management held by the social sciences is to be accepted and made practical by OR.

CONCLUSION

This short paper has provided an overview of two models of the science-technology relationship. Some difficulties associated with the traditional model are overcome by adopting the more recent, symmetrical model. This, in turn, opens up issues of importance concerning the tasks which face OR if it is to absorb and become more open to the advances made in the social sciences.

REFERENCES

Barish N.N., 1963, Operations research and industrial engineering: the applied science and its engineering, Ops Res., 11:387.

Barnes B. and Edge D. (eds.), 1982, "Science in Context", Open University Press, Milton Keynes.

Beer S., 1985, "Diagnosing the System for Organisations", John Wiley, Chichester.

Bonder S., 1979, Changing the future of operations research, Ops. Res., 27:209.

Dando M.R., Defrenne A. and Sharp R.G., 1977. Could OR be a science? Omega, 5:89.

Eden C., Jones S., and Sims D., 1983, "Messing About in Problems", Pergamon, Oxford.

Flood M.M., 1962, New operations research potentials, Ops. Res., 7:423.

Malin H., 1981, Of kings and men, especially OR men, Jnl. Op. Res. Soc., 32:953.

Mayr O., 1976, The science-technology relationship as a historiographic problem, Tech. and Cult., 17:663.

Price D.J. de S., 1969, The structures of publication in science and technology, in "Factors in the Transfer of Technology", W.H. Graber and D.G. Marquis, eds., MIT Press, Cambridge.

Raitt R.A., 1979, OR and science, Jnl. Op. Res. Soc., 30:835.

Rosenhead J., 1986, Custom and practice, Jnl. Op. Res. Soc., 32:953.

White D.J., 1985, "Operational Research", John Wiley, Chichester.

O.R. - TECHNOLOGY

René Victor Valqui Vidal

The Institute of Mathematical Statistics and
Operations Research (IMSOR), The Technical University
of Denmark, DK-2800 Lyngby, Denmark

INTRODUCTION

O.R. has been conceptualized in different ways: as a scientific activity, as a craft, as a social process, as an ideology, as management's
action research, etc. In this paper we introduce a new dimension in these
discussions about O.R. epistemology, we conceptualize O.R. as technology.
This short paper outlines some of the main conclusions of a forthcoming
report: An assessment of O.R.-technology and the practice of O.R. with
social responsability, where a complete list of references is also available.

Our research work complements the global and general work of for instance Rosenhead (From management science to worker's science) and Jackson
(Some methodologies for community Operational Research). We look at O.R.-
technology as a process closed related to the production process. We obviously lose generality but we gain deepness in the understanding of the
reasons for using the technology and the consequences for the different
social groups involved in such a process.

TECHNOLOGICAL DEVELOPMENT

The historical development of market-oriented societies can be characterized by the following trends:

- a long-term relative increase of investments in machinery and raw materials compared to investments in labour,
- a growing concentration and centralization of capital,
- a growing division of labour among firms in different sectors,
- a cost price reduction in a single firm's strategy to survive competition, this can be achieved by using new technology in the production
  process. That is technological development of the production process
  is a system-immanent demand in a market-oriented economy.

Cost price reduction in a single firm can in principle be achieved in
three different ways: by economy of labour, i.e. reduction of labour costs
per product; by economy of machinery, i.e. reduction of machine costs per
production; and by economy of raw materials, i.e. reduction of raw material costs per product. Table 1 illustrates some of the methods used
to achieve costs price reduction. Remark that cost price reduction can

127

Table 1. Cost price reduction - some methods

Economy of	Methods
Labour	- introduction of machinery requiring less manpower - division of labour - increasing the intensity of work
Machinery	- introduction of work shifts - reduction of machine stop times - using better qualities of raw materials to increase machine speed
Raw Materials	- raw material substitution with cheaper products - better use of raw materials and reduction of production waste - recycling of waste

be achieved either by purchasing or by better utilization. Furthermore, technological changes can be done by technical means or by organizational means.

A technological change usually will have an effect on all these ways to reduce cost price, some will be positive other negative in economic terms. It is management's task to select projects that will maximize cost reduction from many technical feasible alternatives.

The reasons for using this theoretical framework is twofold:

a) the methods of cost price reduction that a firm chooses is strongly conditioned by the business cycle. Thus, depending on the market conditions, prices, labour supply, etc., one or other method will be preferred. For instance, in periods of prosperity and full employment introduction of labour-saving machinery is the common method, while in times of recession economy of raw materials is the prevalent method.

b) the social assessment of technological changes in a given plant is conditioned by the method used. Thus the non-economic consequences of technical changes are of great concern to the working people (changes in working conditions) and to the community as a whole (employment, pollution, etc.).

O.R. AS TECHNOLOGY

Our conception of O.R. - technology consists, as it is the case for any technology, of two elements:

- O.R. techniques, usually highly related to computer facilities,
- the way how they are used in production to achieve cost price reduction.

Table 2. depicts O.R. - technoly and its application in production at all levels of the firm. At the strategic level the technological changes are achieved by modifying the structure of the production system. At the tactical and operational level the technological changes are achieved by using "optimaly" the given structure. Let us, for the sake of concreteness, see briefly three case studies.

The Danish pork industry

This industry has experienced during the last 20 years a radical reduction of producing firms (from 62 to 18). Complicated location studies have been performed with the following two purposes:

a) the large producers used them in their struggle against the small producers arguing for the necessity of centralization/concentration due to economical reasons
b) to dimension and locate the new modern plants.

It is without any doubt that the first purpose was the major reason for the use of O.R., that is as a tool for legitimation of a pre-decided development. The "objectivity" and "neutrality" of the mathematical models are used by those powerful groups to reach their own goals. Consentration of plants has also meant the use of new techniques in production.

The consequences of such development have been:

- a reduction in employment in spite of a tremendous increase in production,
- an increase in division of labour and the rate of work,
- serious regional problems due to the closing of plants,
- the increase in transport distances gives problems with pork quality.

In this industry linear programming has been extensively used to solve production planning problems. One of the biggest plants has a large computer to elaborate 3-month plans. The model determines the optimal use of raw materials and machines. No work-force variables are included in the model. This means that fluctuations in work-force level are very high. Nobody in the firm dares to change the model, to do it more flexible, maybe using fourth generation hardware and software that are already avaiable in the firm. Moreover nobody in the firm is able to run the plan without the computer model. This is a good example of O.R. - technology fetichism.

The Danish brewery industry

While the technological development of the pork industri did not receive major opposition from the trade union, this was not the case for the brewery industry. Although having achieved a high degree of centralization (one firm controls nearly 85% of the market), the concentration of plants and automatization of production have not gone as fast as the firm expected mainly due to the opposition of the trade union.

During the sixthies and seventies the brewery trade union was very active in the class struggle. They have achieved good wages, relative good working conditions and a great deal of influence in the use of new techniques in production. Trade union strategy and demands have been developed in close cooperation with students and researchers from different universities in Denmark.

This was not a satisfactory situation for the firm, especially because during the seventies the international competition was very hard, and the forecast for the future showed that competition in the EEC was going to be even harder. Cost price reduction could be achieved by

Table 2. O.R. - technology in production

O.R. techniques	Economy of		
	Labour	Machinery	Raw materials
1.			
Location	x	x	x
Lay out	x	x	
Job-design	x	x	
Investment Planning		x	
2.			
Production Planning	x	x	x
Project Planning	x	x	
Inventory Control		x	x
Quality Control	x	x	x
Distribution Planning	x	x	
3.			
Sequencing		x	
Process Control		x	x

1. Strategical Level, 2. Tactical Level, 3. Operational Level.

reducing transportation costs, for instance by locating new plants abroad. Another method was to increace productivity and reduce labour costs, this was achieved by building a modern new plant in 1979 in Jutland (far away from Copenhagen) that needed only 1/3 of the workforce that was used in the plant of Copenhagen. Moreover, the workforce did not need to be skilled therefore it was cheaper. Due to the economical crisis, that means massive unemployment, it was impossible for the trade union to avoid such a development.

The design, location, construction and actual function of the plant has been done by a massive use of O.R. techniques. Working conditions have been redically change as a result of a highly automatized factory. This case illustrates the political role of O.R.-technology.

The Danish steel industry

This industry consists of one steel mill using primarly scrap as raw material. It is located in an undeveloped region in Denmark and has therefore a major impact on the environment.

The steel mill has, like all the steel industry in Europe, during the sixties and early seventies invested in new technology. In the seventies a serious economic crisis affected the steel industry. The result of this crisis was that all over the world there were unutilized production capacity for producing steel. The competition on the steel market today is vere hard, and if the EEC and the US Government did not control the steel market, many steel mills would soon close. This means that the steel mill in Denmark is limited to producing only 450.000 tons p.a. while actual capacity is 1 million tons p.a.

In the seventies the market price of oil and labour have increased at a much higher level than the market price of steel. To meet this increase in cost price the steel mill must be able to produce steel at a lower price. This is only possible trough optimizing the existing production system, because investments in new technology do not pay off in short terms.

To meet the hard competition and the increasing production costs the steel mill has chosen to use "quality" and "just-in-time" as parameters in the market. Thus the steel mill has succed to survive during the seventies. But in the eighties this is not enough any longer, the competition is much harder today. This causes a wish to optimize the production system in order to get lower cost price.

This the background for a serie of projects (MSc thesis) carried out at IMSOR dealing with inventory problems, production planning, simulation to find bottlenecks, cutting stock problems and booking systems using personal computers. Cost price reduction have been achieved by economy of machinery and economy of raw materials. Since the production process is highly automatized no major changes have ocurred in working conditions and employment. On the contrary, the massive use of computer techniques have created new jobs.

This case study illustrates the fact that new investments will give higher cost price reductions but this alternative was not feasible. The second best alternative was the massive use of O.R. and computer facilities.

O.R. WITH SOCIAL RESPONSABILITY

Rosenhead has supplied in the above mentioned article the methodological criteria to characterize an "alternative O.R.". I want to add one more criterion, this is the question of the practice of O.R. with social responsability. This practice will mean working together with local trade unions, local authorities, communities and grass-roots organizations, to help to desmystifying traditional O.R., critize established plans and contruct alternative plans.

Let us describe, for the sake of concreteness, some of the projects we have been working on:

- DUE-project (DUE is an acronym for: Democracy, Development and EDP):
  The main objective was to contribute to the efforts of the local trade unions to gain influence on the use of computer system in production. Methodologically, our approach has many similarities with the sociological method known as "action research",

- Assessment of technological development within the food sector: The main objective was to evaluate the effect (environmental, food quality, regional development) of future technological development within this industry in cooperation with consumer organisations.

- A decision support system for industrial solid waste management: The main objective is to develop a dynamic tool (using advanced computer systems and O.R.) for the local Office of Environmental Protection, The City of Copenhagen, to steer/evaluate the consequences of the different strategies for industrial waste management,

- A computer based system for environmental control in working places: The main objective is to use advanced information technology for the proactive control of the environment af a small firm,

- Methods for assessment of rural energy projects in developing countries: The main objective was the development od a new approach to project analysis for the Department of Energy in Zambia, implemented on a personal computer being modular and user oriented,

The practice of O.R. with social responsability is not an easy task mainly due to the following reasons:

- traditional O.R. does not provide methods/techniques that can be used in the above mentioned projects,

- the establishment is no always willing to support economically such "alternative projects",

- the process of cooperation between working people and university people is rather slow,

- the difficulty of finding experienced people to carry out such projects demanding both technical and sociological expertise,

- the project by definition are of important political relevance to society, therefore there are many endogeneous and exogenous conflicts that have to be tackle while implementing the achieved results.

FINAL REMARKS

Our future work will be focusing on more broad social problems. More specifically the use of the systems analysis approach to environmental and natural resources management. Our first step has been the organization of a conference in Poland dealing with the Baltic region (September 1988). The next conference will be already in November 1989 in Leningrad, USSR. In this connection an international network dealing with questions of analysis, monitoring, modelling and managerial solutions for national and international problems is being established.

The future of our societies will be to a high extend conditioned by the way how we solve environmental and natural resources problems. O.R. has a lot to contribute to these areas and it is our social responsability to participate to the solution of such serious problems.

AUTO-REFLEXIVE ALGONDONIC LOOPING

OF PEDAGOGIC OPERATIONAL RESEARCH

Allan L. Ward

Department of Speech Communication
University of Arkansas at Little Rock
Little Rock, Arkansas

Operational Research, inwardly directed, feeding back into itself
the environmental information gained as a result of its applications,
can effect its own future by creating an autoletic educational
environment for the pedagogic descendants of its current practitioners.
The present generation of operational research specialists encountered
it in their adult years of advanced study, superimposing it over a
variety of other often educationally-static behavioral layers. During
the developmentally sensitive years, these concepts can be internalized
to impact on personal developmental interaction with the environment.
Our contemporaries have the capability of operationalizing a system of
education that could apply a self-reflexive and self-rewarding process
to the succeeding generation. Beginning with the earliest, and
continuing with each subsequent level of education, the individual
student could be provided an opportunity of utilizing operational re
search in the analysis of systems for personal application, for group-
classroom application, a school-wide application, and hypothetical
extensions to the levels of city, region, nation, and planet, involving
all of the social sciences. This opportunity would make the student not
just the recipient of static concepts, but a participant in a process
whereby he would exercise a degree of control in the shaping of his own
environment. The classroom setting would become part of the "real
world" to which he would apply increasingly complex degrees of systems
analysis and institute self-rewarding alterations in his environmental
sub-systems. Such training could develop a generation of participants
for whom operational research was not a superimposed theory but a
behavioral reality. Their impact on all of the social sciences could a
major factor during the next quarter century.

"To me," wrote Varela, "the chance of surviving with dignity on
this planet hinges on the acquisition of new mind (1986:6)." The
implications of this statement are reflected in the attitude of
conferring about Operational Research (OR) and the Social Sciences, in
setting the agenda of that relationship for the next quarter century.

There are several implications inherent in such an approach:
first, that OR has something to offer that can positively impact on the
social sciences; second, that the current degree of that impact can be
accelerated, augmented, and improved in the future; third, that a broad
theoretical construct at present can give perspective for detailed study
implementation and revision in the future.

As Estep summarized in a related context, "On the level of social urban analysis and the analysis of world dynamics representing the interactions of population, pollution, industrialization, natural resources, and food, simulation models are possible to show the dynamic consequences of choices made or not made relative to these indices effecting the future (1988:591)."

In the current situation, by simple modeling, we have (A) a body of current OR knowledge, (B) people who have varying degrees of that knowledge and who to varying degrees apply it in a variety of situations, (C) an array of fields of study dealing with human welfare commonly categorized as the Social Sciences, (D) a spectrum of outcomes from social science concerns when those individuals apply or do not apply that knowledge, and (E) the value assumption that the outcomes tend to be more positive when using OR constructs.

The concern of the present is to facilitate the means of increasing the competence of OR practitioners applying that knowledge in the future. "The main interest," wrote de Zeeuw in a related context, "is excited by the introduction of new ways of ordering existing resources or of recognizing still unused resources. In fact, the term problem solving is often reserved for the latter kind of change. Where such change relates not only to problem solving on a single occasion, but also the ability of being able to solve such problems in the future, and hence, even to prevent them for occurring, we can also speak of increased competence (1988:288)."

If part of the energy exerted by the current generation of OR practitioners, generally aimed at current problems, is looped back into the educational system, so that OR principles are utilized to greatly influence that educational system, then the next generation of OR practitioners would, within a quarter century, exert a profound influence on every facet of the social sciences.

Today's practitioners often encountered the complex of concepts related to OR in their advanced undergraduate eduation or more often in their graduate or professional experience. By that time in their lives, their actions and perspectives had long since been influenced by the problem-solving behaviors current in the social and educational environments in which they chanced to find themselves during their learning-sensitive years. OR practices, superimposed over these ingrained habits may be, in consequence, diluted in much of personal behavior.

Utilizing the best of general systems concepts, the current generation of OR practitioners can pull out patterns of problem solving in layers of decreasing-increasing complexity, determining which can be incorporated at various levels of a child's development. At his earliest socialization at home and among peers and at preschool, the child formulates the ways of coping with the people and other elements in his environment. These are the most rudimentary patterns of problem solving, over which are layered subsequent encounters, and therefore, not too early for the introduction of the most productive problem-solving behaviors.

At each educational level, there could be built into the school activities, projects which would dramatically depart from the old pattern of lecture, memorization of patterns of words, and words repeated on tests judged by how closed they came to the teacher's original words. The student in the old model of education learned to sit in ordered rows for twelve to twenty years and repeat phrases spoken or written by authority figures. The student was expected to exert little change over the environment as an infant, child, youth, and young adult, and then miraculaously, was expected to graduate and begin to exercise a significant degree of change on complex global problems. To this simplistic retrospect must be added that there are numerous experiments with alternatives but these are often isolated and without continuity in the traditional system. Individuals may be innovative in spite of, not because of, the educational framework.

OR must now turn its ability to relate to complex problems of the educational system itself, pulling together the notable experiments underway, and purposing viable alternatives at every educational level, from early pre-school through university training, for creating participant behaviors in the next generation. To do this, positive feedback to the individual is essential. In each class problems need to be addressed, using OR processes at the appropriate level of complexity, which will impact favorably on the individual, the class as a whole, and the institution. Hypothetical situations must be introduced similarly that related to the larger environment of the city, region, nation, and world.

The thinking must become systemic. A water pollution problem, for instance, will no longer be seen as what exits a school faucet, but the pipes, purification centers, sewer workers, bureaucracy, water tables, river-lake sources that make up the physical and social interconnections that bring the water to the faucet and deal with it after its departure down the drain.

A meta-awareness must be imparted. The student, as a biological entity, must come to see himself as a complex of chronobiological cycles. As Axelrod noted, "Biologic periodicity has been demonstrated in virtually all plant and animal species, from unicellular organisms to humans. Oscillation of a number of biochemical and behavioral functions in man are of diagnostic and therapeutic importance (1971)." And, the student must note, they are of educational importance. The types of seating, lighting, colors, temperatures, the amount of movement permitted, the degree of novelty and change in the environment, the effect of various nutrients introduced into the digestive tract, are only some of the variables over which the participants in the educational system must begin to share increasing decision-making responsibilities, using OR methods to relate to the complex system of issues involved. Similarly, there must be incorporated in behavioral practice, information about the function of the brain, itself, including the seemingly divergent systemic and linear abilities of the right and left lobes; the sequence of learning-sensitive periods during human development; the relation of change and stress and the means of coping with stress produced by educationally-induced change; the uses of modeling, including verbal, mathematical, computerized, and three - dimensional; cultural literacy; and others which the OR process of educational renovation can identify.

Innovative experiments exist, from which general systems may be extrapolated as applicable to other educational areas, including varied subject matter and age groups. As an instance, the Hawkesbury Experience, discussed by Woog and Bawden, apply "a weaving together of experiential learning/problem solving activities with systems theory and practice (1988:1254)" to an agricultural facility. They cite Lewin as describing experimental learning "as a cycle of activity which involve the learner in concrete experiences, reflective observation, abstract conceptualization and active experimentations. The experiential learning process combined with sytems methodology can provide a tool for problem solving." Organizations, as represented by the Association for Integrative Studies, foster a variety of creative approaches to be utilized.

Global modeling by current experts must produce their counterparts in classroom modeling, institution modeling, the local environment modeling for the student who will become the expert of the future. In this way, global modeling will not come as an educational culture shock, but natural transition of maturation in dealing with the plural environments. Jay W. Forrester is credited for being the creator of the first world model (Pecci, 1982:101). He takes into consideration the life cycle of growth, economic long-wave or Kondratieff cycle, and the short-term business cycle, in human development. The Mesanovic-Pestel model was intended to be a "planning and decision-making tool related to managing human affairs during the transition of a new global order (Pecci, 1982:139)."

As a current group of global modelers, the Club of Rome is identified by its members as "a body concerned with the well-being of mankind as a whole, is future-oriented in its thinking and must necessarily take into account the incongruities of the human condition, its values and goals, both actual and desirable, if the species is to survive (Laszlo, 1977:vii)." The group devotes its attention to the gathering of facts from the past and present, building a model of conditions as they were and are, including the dynamics of movement from the past to the present, and then projecting additional possible change-agents to project theories for the future. "Only by charting his (humankind's) further course intelligently," the Club indicates, "has he a chance of advancing into new territories offering immense opportunities for his continued enterprise (Laszlo:vii)." The current impact of groups such as the Club of Rome, can be intensified by preparing a future generation in which these perspectives have been internalized from childhood.

The efforts of such concerned groups, as valuable as they may be, deal with their benefits of model building primarily with adults rather than with children in their formative years. Just as the efforts to teach a second language are much more readily absorbed during the language-sensitive period than in later years, so the OR, systems, cybernetic, model-building principles are more readily assimilated and utilized when introduced during the social-communication sensitive periods of development. The need in the next quarter century is not only for new research findings in an advanced field of study. We need first, the application of these principles to the total renovation of educational systems, utilyzing the finest that current experimentation has to offer. Second, through the introduction of the most creative problem solving methodology available to students at every level of education, we need to develop a generation who practice these complex methodologies as facilely as they speak their native language.

Bruce (1988:372), described how, in the summer of 1874, Alexander Graham Bell "let a kaleidoscope of miscellaneous notions swirl about in his mind." Then, "suddenly the jumble fused into a great insight: the fundamental principle of the telephone." Consider that it was the process of perception as a system, not the production of new information that occured: "His magnificently simple inspiration depended on no scientific discovery or natural phenomenon that had not been well-known for nearly half a century. Why had no one thought of it before Bell did?"

136

In this example, we have diverse segments of information, separated by specialized compartmentalization, suddenly integrated into a systemic whole through a process occuring in the brain of a perspective, interdisciplinary-oriented person. The world utilized the instrument and generally overlooked the implications of the process by which it came to be: the latter is the greater of the two. OR concepts, used to serve as a integrative agent, can bring together already existing expertise into a new, user-participative framework of education, where, in part, the content and process may be perceived as one. A dictum of cultural wisdom notes, "Give a man a fish and he will eat for a day; teach him to fish and he will eat for a lifetime." For our purposes of future projection, we would then alter the dictum to read, "Apply OR principles to a given social science concern, and limited conditions may improve for a time for the people involved; teach the next generation to behaviorally utilize OR concepts, and they will positively change the condition of the social sciences continuously." They can, as it were, acquire a new perspective of a "new mind," if the implications are thoroughly pursued.

References

Axelrod, Julius, 1971, "Chronobiology," paper distributed at International Society for the Study of Biological Rhythms, Arkansas Medical Center, Little Rock.
Bruce, Robert V., 1988, Alexander Graham Bell, "National Geographic," 174, 3:358–384.
de Zeeuw, G., 1988, Modeling for Applied Social Science, "Cybernetics and Systems," Kluwer Academic Publishers, Dordrecht.
Estep, Myrna, L., 1988, Systems Analysis and the Concept of Power, "Cybernetics and Systems," Kluwer Academic Publishers, Dordrecht.
Laszlo, Ervin, 1977, "Goals for Mankind," The Research Foundation of the State University of New York, New York.
Pecci, Aurelio, 1982, Global Modeling for Humanity, "Futures," 14, 2:92–94.
Varela, Francisco J., 1986, Laying Down a Path in Walking: A Biologist's Look at a New Biology and Its Ethics, "Cybernetic," 2, 1:6–15.
Woog, Robert and Richard Bawden, 1987, Systems Education – The Hawkesbury Experience, "Problems of Constancy and Change," Society for General Systems Research, Budapest.

O.R. AND SOCIAL SCIENCES:

A STORY OF ON-OFF RELATIONSHIPS

Ronald G. Stansfield

The Conference in Cambridge on Operational Research and the Social Sciences - "The 1964 Conference" - with the Conference book containing the text of the papers and a very adequate account of the proceedings and discussions (Lawrence 1966) - in retrospect marks the high point in the relationship between O.R. and social science. The theme was chosen for the Operational Research Society's first international conference because of widespread recognition of the need for better integration of the closely related interests, needs and resources of O.R. and various social sciences (see especially "Editor's Preface" in Lawrence 1966). John Lawrence's account of these needs is as valid and important now as when it was written. The calibre of the participants was outstanding; so was the quality of the papers and discussions. Today, most of the papers are well worth re-reading; several continue to be of real importance. The present paper owes much to that Conference, as background and still more as stimulus.

O.R. studies problems in organizations and systems, in all of which people form part and take part. So from its earliest days, knowledge and expertise from the field of the human sciences have been relevant. The ways, and extent, to which they have been used have fluctuated. This paper reflects on the history of such use, and on possible lessons to be learnt from the reasons for such fluctuations, particularly as seen from the author's own experience. Care in nomenclature is needed, in partic- ular regarding the overlapping categories of human sciences and social sciences, especially since labelling sciences and groups of sciences is a matter of convenience; all boundaries are in some degree arbitrary and are fuzzy. The author normally thinks in terms of a main grouping into physical, biological and social sciences, one which is very widely used; but for some purposes it is useful to speak of human sciences, meaning those sciences which directly study human beings (in particular, as distinct from other animals). For the author, the criterion of a social science is that it studies situations in which several creatures interact; - usually two, several or many human beings, but more generally, possibly other animals, e.g. ants, bees, chimpanzees. To illustrate the dividing line: a study in psychology of the reaction-time of a car-driver when a traffic-light turns red is human, but not social, science; study of the change in reaction-time when a uniformed policeman is nearby is both human and social science (the example is taken from old research in experimental psychology). As it happened, the author first met the need for human science expertise before he had to pick up some acquaintance with social science. His first operational research-type assignment, in 1939 and early 1940, was a study in War conditions of failures which turned on errors about the positions of aircraft. It brought in the physics of radio waves (affecting radio bearings), school-level geometry and statistical error-theory; more important were the mistakes made by people reading maps, map-grids and other displays of information. These mistakes clearly followed regular patterns, patterns depending on circumstances; but no one seemed to know anything useful about these patterns or the reasons for them. The upshot was that the author found himself making field studies in immediate collaboration and mutual support with an experimental psychologist making parallel laboratory experiments. This experience, its context and what it led to, is described more fully in Stansfield 1981a. The scope of that paper is wider than its title suggests; it contains substantial previously-unpublished historical information. The 1940 work on legibility continues to be relevant today; e.g. Stansfield 1981b.

A theme underlying Stansfield 1981a was that O.R., because of direct observation of and involvement in problems of organization, has been fertile in bringing out new ideas and new lines of approach and in this way has been important in helping the start and the growth of new areas of science. The re-shaping and expansion of U.K. Government provision for supporting human and social science research which took place during the decade after 1945 was greatly helped by the prestige enjoyed by O.R. among influential people who were well-informed about its Wartime importance, and by O.R.-style thinking about what was needed. Stansfield 1981a traced the history of these developments as far as the creation of the Social Science Research Council. In particular, the growth of sociology (especially industrial sociology) benefited from the development.

On the human sciences side, ergonomics is another important example. It is now established as that discipline where experimental psychology, physiology, anthropometry and allied subjects meet with physics and some engineering in dealing with what has been dubbed "fitting the job to the worker", studying the relation between the man or woman and his or her tools and working equipment as well as immediate working environment. The aim in design is to take proper account of the way in which human beings perceive and process information, as well as the size and strength of their bodies and the characteristics of their mental and physical skills. The 1940 study of legibility happened to be a very early, and quite minor, strand which led after the War to the formation in 1949 of the Ergonomics Research Society. Other roots in O.R. were more important; Sir Charles Goodeve, Steve Laner and Alan Welford, main figures in the 1964 Conference, also are prominent in the early history of ergonomics.

Reference to ergonomics is relevant here, not only to show the contrast between human and social sciences, but also to draw attention to the way in which the 1964 Conference dealt with social sciences to the nearly complete exclusion of non-social aspects of human performance. In retrospect this seems surprising, in view of the reasons given above why almost inevitably these aspects also are relevant to O.R. But ergonomics has itself shown as much "on-off" as social science in its use of O.R. practitioners; the author wonders, for example, how many O.R. studies have failed to notice important matters such as sound-cues provided by machines to their operators, or serious health hazards from noise or dust. Quite recently, many manufacturers and users of computers and similar equipment have learnt "the hard way", at great expense, the consequence of ignorance of the motor and perceptual abilities and limits of the people who would use them.

Relations between experimental psychology and social psychology, and more generally between ergonomics and social science, have also fluctuated greatly. The author has on various occasions noticed instances when researchers studying human performance under varying conditions have not pursued the reasons for differences when these appeared to lie in changes in social, rather than in material, factors. It would seem to be important to seek causes for these on-off variations; possibly there are common reasons why researchers should shy off study of what - on the face of it - are interesting problems, and moreover ones of considerable practical importance. This question came to the author's attention especially in the late 1950s when he was both on the Executive Committee of the Ergonomics Research Society and a member of the Committee of the Social Section of the British Psychological Society; and also at the same time Recorder of the Sociology Section of the British Association for the Advancement of Science. Reasons appeared to be complex, often involving the prestige of being "scientific". Psychologists seemed to be particularly sensitive about this, and concerned to avoid the danger of being seen as "not really scientific"; so they preferred to work in the laboratory on experiments carefully-designed and subject to mathematical analysis. There was widespread belief that wider studies of human behaviour were bound to be "not really scientific" and woolly. The author's initial research training happened to be in experimental physics; it seemed to him that the images held of what constituted "scientific method" and the criteria for "being scientific" were often greatly idealised. Clearly there were also cogent reasons why researchers should avoid problems calling for the use of more than one discipline or membership of a cross-disciplinary team; such work would be of doubtful future help towards a secure job or career. Even by 1960, signs had appeared of a tendency for research workers and research groups to pull back from novel and unfamiliar fields into work nearer the established core of their own discipline; this was a precautionary defensive response. Later on it became more evident as research staff ceased to grow or were cut, and competition for resources increased.

The 1960s interest by O.R. people in the social sciences before the 1964 Conference led the O.R. Society to set up a Discussion Group On Operational Research and Social Organization. It first met on 3rd April 1963 and for a number of years held regular meetings. Most were in London at 5.30 p.m., with a single paper followed by substantial discussion. For some time it was one of the Society's best-regarded of the varied Discussion Groups. It saw itself as drawing on and carrying forward the success of the 1964 Conference; the book of that Conference was published, and sent to all participants, in mid-October 1966 and on 8th November the Discussion Group held at The Olney Club, Cookham, Berks. a full-day Seminar to discuss it. The list of expected participants shows 62 names, a good representation from England and Wales of the men and women, many quite senior, then engaged in O.R. or closely concerned with it. The 1964 experience seems to have encouraged half a dozen people centrally involved in running the O.R. & S.O. Study Group (as it was often called) or in close touch with it to set up an ad hoc Committee which organized a series of two-day pre-Christmas conferences in London. The leaflet giving details of the second conference, on 18th and 19th December 1969, was headed as reproduced at the top of the next page.

There were four such Conferences on O.R. and the Behavioural Sciences, each announced by a similar leaflet showing the same sponsorship. The second line of the topic was the same for the 1968 Conference as in 1969; for a December 1971 Conference it became "Approaches to the Study of Industrial Relations & Manpower Organisation", and for the December 1972 Conference, "Planning and the Individual". Each Conference was attended by between 300 and 350 people. In the author's opinion, at the time and

O.R. AND THE BEHAVIOURAL SCIENCES:

APPROACHES TO THE STUDY OF ORGANISATIONAL BEHAVIOUR

SECOND JOINT CONFERENCE

sponsored by

Operational Research Society (Social Organisation Study Group)
British Psychological Society (Occupational Psychology Section)
British Sociological Association
Ergonomics Research Society

now in retrospect, the conferences were both good in themselves and most valuable in forming contacts between the varied subjects of study and groups of people involved, as well as eliciting good and important papers (Heald 1970). The author was never clear about where, and in what ways, responsibility lay for these Conferences (the present account amplifies and to some extent amends Stansfield 1981a p.277). Gordon Heald had become secretary of O.R. & S.O.G. in 1967; in January 1973 he was writing to members as if the Conferences were annual conferences of the Group. There are signs that the effort of organizing them had slowed down attention to other aspects of the Group's activity.

Serious disruption of the Group began late in 1972 when a number of members, many newly-joined, made a determined effort to redirect it to discussion of the social responsibilities of O.R. workers. Sharp, even personal, exchanges were followed by irenic efforts by senior O.R. people. These led the A.G.M. in February 1973 to distinguish two aspects of the Group's operation: one to be a bridge between O.R. and the Behavioural Sciences, the other to examine the social impact of operational research. As a compromise it was agreed that both aspects should be incorporated into the Group's programme, with neither dominating it. The March 1974 A.G.M. recognized that this had not worked well, and by 1975 there was a serious suggestion that the Group be wound up. In retrospect, the whole episode seems to have brought about a period of lively interest in the Group, at the cost of losing, immediately or rather later, active members needed to keep it viable. No more conferences were held; the Group gradually lost vitality. The last meeting of which the author has record, attended by 14 people, was on 1st December 1981.

The 1989 Conference, "Operational Research and the Social Sciences", repeats exactly the title of the 1964 Conference. The word "and" leads to the immediate thought that O.R. is itself a social science, if what was said above is correct, since nearly all O.R. studies turn on the behaviour of people, two or more. This is so although it may well not be made explicit in the formulation of the problem or the presentation of the data.

It will be interesting to see how the social sciences represented in the 1989 papers compare in range and balance with those in 1964. Curiously, the 1964 Conference never took up the criterion for being a social science, nor what areas of social science it did not touch on (for example, social anthropology; also social history, whose importance had hardly been recognized in 1964). In retrospect, the backgrounds of the participants and the content of the papers would now classify the conference as one based mainly on O.R., sociology, social psychology (its relationship to individual psychology was not explored), managerial and organizational science, together with some economics. The rapid growth of economics since the early 1960s has been well described by Smith 1987.

As it happens, the author's contribution in discussion (clearly reported by Tom Burns in Lawrence 1964 p.547) was about ambiguities of words and other symbols. It helped to sensitize him to the various alternative meanings included within the dictionary definitions of science, such as "pure science, one depending on deductions from self-evident truths, as mathematics, logic", contrasted with "natural, physical, science, one dealing with material phenomena and based mainly on observation, experiment and induction, as chemistry, biology" (The Concise Oxford Dictionary 1964). The different uses are long-established; Langland (c. 1370) made Lady Study counsel Peter the Plowman about the importance of diligent study and of learning science from teacher and book. The other approach she put firmly in its place: "The men of old used carefully to note all the wonders that they saw ... yet no one was ever saved by this natural science, nor did their books lead a single soul to heaven". The author was concerned also about the development of mathematics for the social sciences without adequate understanding of what the mathematics could stand for, a concern later developed into fuller discussion of quantophrenia, the abuse of inadequately-specified numbers to give a spurious appearance of precision to quantitative findings (Stansfield 1981a, p.278 and 1986). Stansfield 1986 gave reasons why O.R. workers and economists may be especially vulnerable to this hazard, as they tend to use numbers drawn at second-hand from observation, and why such data are likely to have been "massaged" in transit from the primary source.

The 1989 conference is to be held in Queens' College, Cambridge, so it is appropriate, as a small illustration of the post-1964 decline in active interest in social science, to conclude by reference to a 1978 O.R. Society seminar, a lively one-day conference held in the Cambridge Graduate Centre a few hundred yards further up the River Cam. A short note handed out by the author (Stansfield 1978) said: "in operational research I have often met problems on which social anthropologists have been the scientists providing the most useful help. For example, when I studied typing pools in a Civil Service Department I needed rapidly to gain a working understanding of the power-structure within and around each pool; and I found that weaknesses in bricklayer training were to be understood only in terms of the social values and moral code of the building industry, values and code not that of the educational world to which I was adjusted." (c.f. Stansfield 1979). The note opened by observing that the recent O.R. Society Conference in York had noticeably lacked awareness of the relevance of cultural issues and seemed ill-informed about available social science and skills. To end by returning to the original: "I asked a few people ... and found that we appear to have the not-unfamiliar situation in which an area of interest is dealt with independently by people in different university Departments, perhaps in different Faculties, people who use different terminologies for the same concepts, are mutually unaware of each others' interests, and make no use of the relevant related knowledge and expertise held by the other group. The data from my informants, a small and strictly haphazard sample from Cambridge and elsewhere, suggests strongly that University anthropologists and operational research teachers are alike in that they have a vague and much-distorted idea of the nature of the other group's subject, if they have heard of it."

References

Heald, Gordon, ed., 1970, "Approaches to the Study of Organizational Behaviour", Tavistock Publications, London.

Langland, William, c. 1370, "Piers the Plowman", translation into modern English by J.F. Goodridge, Penguin Books, Harmondsworth, Middx.

Lawrence, J.R., ed., 1966, "Operational Research and the Social Sciences", Tavistock Publications, London.

Smith, Cyril S., 1987, Networks of influence: the social sciences in Britain since the War, in "Social Science Research and Government: Comparative Essays on Britain and the United States", M. Bulmer, ed., Cambridge University Press, London.

Stansfield, R.G., 1978, O.R. and anthropology, note for O.R. Society Seminar, O.R. in Developing Countries, 27 October 1978, The Society, Birmingham.

Stansfield, R.G., 1979, Typing Pools: a study in satisfaction in work, in "Satisfactions in Work Design: Ergonomics and Other Approaches", R.G. Sell and Patricia Shipley, eds., Taylor & Francis, London.

Stansfield, R.G., 1981a, Operational research and sociology: a case-study of cross-fertilizations in the growth of useful science, Science and Public Policy, 8:262.

Stansfield, R.G., 1981b, Quis custodiet?, OR newsletter, 10:8.

Stansfield, R.G., 1986, The place of observation in OR methodology or, when and where do OR workers observe?, paper given at Eighth European Conference on Operational Research, Lisbon, Portugal, deposited at British Library Document Supply Centre, Boston Spa.

THE TOTAL ORGANISATIONAL SYSTEM (TOS)

WHY OPERATIONAL RESEARCH AND THE SOCIAL SCIENCES CANNOT MATCH

Teddy Weinshall

Professor Emeritus
Tel Aviv University
Israel

Abstract

A quarter century after the first Cambridge Conference on Operational Research and the Social Sciences, I feel that it is high time to summarize the mutual contribution between these two areas. In this paper OR are the quantitative sciences and tools, while SS (the Social Sciences) represents the qualitative approach or what we would like to refer to as real life.

The area of SS relevant to our theme is that of the actual dynamics of organization and management, represented by the so-called Total Organizational System (TOS) described in the first part of the paper. This part includes a discussion of why the TOS cannot benefit from any contributions of OR. Obviously the SS do derive from the quantitative sciences statistical testing tools for verification of findings based on large numbers of data (or such statistical applications as quality control testing), as well as the obvious utilization of electronic data processing (EDP).

The second part of the paper considers the question why OR has, on the one hand, only contributed marginally to and could never match the SS, and on the other hand, has developed and expanded over the last 25 years in spite of the halo of mathematical (and simplistic!) solutions. The relative ease with which mathematically oriented publications could be evaluated by scientists from different disciplines for the purpose of nominating, promoting, and tenuring faculty members is discussed.

The last part of the paper presents a few examples of the harm done by OR to the healthy and prosperous development of SS. I subsequently describe the damage that OR in its wider sense has caused Israel, especially in the areas of management and public health.

Comparing the TOS "Real-Life" Contingency System with OR and other
 Modelistic Systems

In 1983 I summed up my work after having converted, a quarter century earlier, from mechanical and industrial engineering to general management and organizational behavior. Introducing the TOS, I first described then compared it with OR and other mathematical models as follows (Weinshall and Raveh, 1983):

The dynamics of management and organizations over time are described and analysed by way of the TOS (Total Organizational System). This notion of a total organizational system assumes a dynamic relationship between management of the organization and the environment, both of which are affected by changing conditions of size, place, and human nature. Any change in one part of the total system affects the other parts and the survival of any organization is continually threatened by the changes brought about by growth, which in itself is essential for survival.

The TOS (see Figure 1) is composed of the immediate environment, the wider environment, the organizational strategy, the scope of decision-making, the managerial structure, and the managerial characteristics. The immediate environment includes the organizations competing for cooperation of the stakeholders (management, labor, trade unions, bankers, shareholders, suppliers, customers, government, etc.). The wider environment includes the systems of the employment market, the money market, the supply and demand market for materials and products, as well as the technological-scientific and socio-cultural systems. The scope of decision-making is the total amount and complexity of the decisions imposed upon the management by their own organizational strategy, influenced by both the immediate and wider environments. Managerial structure refers to the actual way in which decision-making is practiced, formally or informally. Managerial characteristics are the leadership and followership characteristics of the hierarchy from the chief executive down.

Consequently, organizational systems are actually contingency systems; that is to say, the different components are contingent upon each other. A major change in one subsystem may not only affect what is happening in other subsystems, but also alter the rules by which these systems are governed. Therefore, the principles which until quite recently have governed management education and literature are based on false assumptions. For example, all of the following principles turn out to be absolutely wrong:
   --There exists a 'good' and desirable organizational structure within which an organization should operate at all times;
   --there are 'good' and 'bad' managers, i.e., a good manager will always be good and a manager who has completely failed in one organization could never succeed elsewhere;
   --it is desirable to have people continue to work in the organization as long as possible, and the organization should do whatever it can to hold on to employees who are currently doing well:
   --there are rules which should govern the establishment of organizational structures; one of these rules is the so-called 'span of control', which represents the number of subordinates that a manager can control. (pp. 1-3)
'System' is one of those words used loosely in different disciplines and has recently penetrated the fields of organization and management. Systems, as referred to in organization and management, are either 'descriptive' or qualitatively analytical on the one hand, and 'modelistic' or quantitatively analytical, on the other. Operations Research scientists have been using exclusively 'modelistic' systems, while organizational scientists have, on the whole, been using 'descriptive' ones, but with some distinct exceptions.

The TOS is definitely not a quantitative model. Indeed, we do not believe that in the foreseeable future, a useful quantitative model can or will be developed for it, the reason being that the contribution of many of its variables is, in the present state of the art, incapable of quantitative characterization.

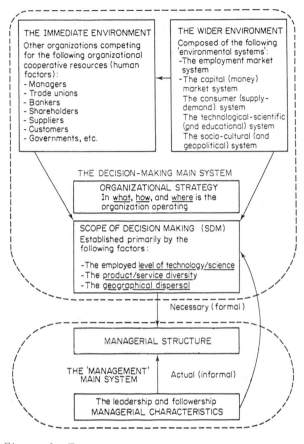

Figure 1   Two main systems and six principal
           systems of the TOS

The validity of a mathematical model is usually conditional upon
specific variables not included in it, not being part of it, and appearing
under the overall caption 'all other things being equal'. The validity of
a quantitative model is directly proportionate to the verifiable scope of
the variables included in it, and should be reasonably high. Otherwise,
its relevance quickly diminishes. If a quantitative model can cover, say
80 per cent or more of the phenomena it claims to describe, then it may be
regarded as useful. If, by contrast, a model covers only up to, say, 20
per cent of the phenomena, it is worthless.

Unfortunately, enormous amounts of effort and means are wasted on
constructing models of the latter type, and then in collecting,
processing, interpreting, and even acting upon data and conclusions which
were irrelevant from the outset.

It is indeed the level of abstraction of OR which accounts for the
usefulness of mathematical models. In those systems where one usually
concentrates on only a small part of one component of the TOS to the
exclusion of the rest, the scope of variables covered by the model can be
easily reduced to 20, 10, 5, and even to 1 per cent. This limited domain
is regarded in a dangerously self–deceiving way, as isolated from the
effffects of other phenomena. However, only within such defined boundaries
of the problem can a quantitative model expect to become useful by
increasing the scope of variables to, say at least 80 per cent.

Why did OR Prosper at its Outset, Survive and Thrive since World War Two, when it cannot Match the SS

The explanation as to the success of OR at its initial stages during World War Two is quite simple. In wartime there exists what Mike Jackson calls a "mechanical-unitary problem-context," namely "the methodology of improvised OR (and the same can be argued for systems analysis and systems engineering) is useful as long as the problem solver can establish the objectives of the system in which the problem resides and represent that system in a quantitative model which will simulate its performance under different operational circumstances" (Jackson, 1987, p. 61)

And as to how OR managed to survive and thrive for over forty years since the War--the answer is to be found in the "publish or perish" dictum of the academic world, and in the vested interests of publishers and university presses. When considering promotions in their faculties, universities do something which is contrary to practice of any other type of organization--industrial, business, government, etc: they seek the opinion of members of other universities, preferably on the other side of the globe; in other types of organizations this would be tantamount to seeking advice from competitors. In order to express their opinions, learned colleagues (sometimes not too proficient in the English language) have to understand the content of the papers. This they best do if the papers are written in the language common to them all--modelistic mathematics. Publishers, too, in order to increase their international sales among those who understand the modelistic language but are not necessarily interested in the clinical problems of a certain country or group or organizations, etc., also urge authors to write modelistic books.

Additional Examples of the Damage Caused by the OR way of Thinking and Acting

I am, of course, referring to OR in general terms--to the modelistic approach in teaching, research, consulting, and action in society. One of the main reasons for individuals choosing a more quantitative career is their desire to avoid, as much as possible, interpersonal relationships within and outside organizations. This is true not only of such fields as OR, mathematics, or physics, but also of the majority of those who take up professions like engineering and accounting. Mathematics, however, has usually a "quicksand" effect on its votaries. Once one starts to play around with mathematics, he is in danger of being absorbed by it. Take, for instance, Herbert Simon, whose first book (1947) was an effort to rewrite and explain Chester Barnard (who together with the Hawthorn researchers, embarked on organizational behavior and was the father of managerial behavior). This work of Simon's could have developed into a most significant contribution to organization and management, but over the years he shifted further and further into the modelistic area and ended up receiving the Nobel Prize in Economics which had hitherto been awarded only for achievements in econometrics. Two others, Paul Samuelson and Milton Friedman, who are world renowned for their contributions--the first for his Econometrics book which popularized economics for millions of students, and the second for helping governments all over the world to improve their economic performance, were likewise Nobel Laureats, honored for some obscure (at least to others than modelistic econometicians) mathematical formulations, which probably manifested one side of their genious, certainly not the one which really contributed to SS and to humanity.

Why are these examples of the harm that OR inflicts on SS? The answer is that by encouraging scientists to engage in modelistic work,

thousands of capable and brilliant OR and affiliated scientists are away
from descriptive work. No wonder that even for people like Churchman
(1967), Ackoff (1986) , and Checkland (1984), as well as Rosenhead(1986a
and b), and Jackson (1982 and 1987), it is difficult (perhaps impossible)
to break completely away from OR. After all, who wants to start his
doctorate over again? And all the while universities continue to hatch
ORers, who in turn continue to bring the gospel to organizations
(including communities), in most cases wasting their time and money.

The extent of this harm is clearly exemplified by what happened to
management education and to public health care in Israel. Let me first
explain that the modelistic "publish or perish" dysfunction is graver in
Israel than elsewhere. The reason is that the age-old tradition of the
Jewish mother desiring her son to become a doctor has turned into a
veritable obsession of Israeli parents who wish both their sons and
daughters to become university professors!!! Let us now consider what has
actually happened in Israel.

Management Education. There are graduate programs in business
administration and industrial management in the following five
Israeli universities: Bar Ilan, Ben Gurion, Hebrew, Technion, and
Tel Aviv. All these programs are to a lesser or greater degree
modelistic ones. Their doctorates resemble OR ones more than management or business administration ones, and they take double the time
to complete compared to their counterparts, say, in the U.S. or the
UK. True, in the U.S. there also exist such modelistic graduate
business administration programs (Chicago is such an example), but
among over 400 business administration and management schools they
probably constitute one percent. When, on the other hand all five
Israeli programs are modelistic, this is a scandalous situation, both
because Israel is not sufficiently rich to afford this waste, and
mainly because Israeli universities (overwhelmingly government-
financed), do not supply the economy with any worthwhile managers.

Health Care. At this very moment a State Inquiry Committee (chaired
by a Supreme Court judge, and manned by two senior professors of
medicine who are heads of the two most prominent hospitals, and two
professors of manpower and of labor economics) is investigating the
deplorable situation in the Israeli health care system. I believe
that the worst aspect of this situation is that the majority of
doctors in Israeli hospitals strive to rise in the academic
hierarchy, while attending, treating, and healing patients as an
objective only insofar as it furthers their academic career. There
are about 5,000 doctors in Israeli hospitals, and about 2,500 on the
faculties of the four medical schools. This means that 50% of the
doctors in Israel are mainly concerned with the papers they have to
publish in order to survive and thrive in the academic system. Then
there are all those doctors who aspire to an academic career but have
not yet found their way in; they too write papers... . When a famous
professor, the head of a hospital department (i.e., the consultant)
makes the rounds in an Israeli university training hospital, the
behavior of his entourage of assistants and medical students differs
completely from that of their British counterparts: while the latter
carefully watch and listen how the chief attends the patient, in the
Israeli scene both the professor and entourage copy the entries on
the patient's chart to be used as data for their papers! It is also
said that while in Britain only one out of every ten general
hospitals is a university training facility, in Israel the figure is
eight out of ten.

Acknowledgements

I am mainly grateful to Mike Jackson and Jonathan Rosenhead, who while aware that I disagree with their views, encouraged and helped me to prepare this paper. My thanks go to Derek Pugh who directed me to the two main sources of material—Anthony Hopwood and (again) Jonathan Rosenhead of the LSE, both of whom I also thank. Last, I am indebted to my erstwhile Technion colleague Eliezer Goldberg for helping me with the final draft of this paper, just as he had done with its predecessor in the Lawrence volume 25 years ago.

References

Ackoff, R. L., 1986, Management in Small Doses, John Wiley and Sons, N.Y.

Checkland, P., 1984, Rethinking a systems approach, in: Tomlinson and Kiss, 1984)

Churchman, C.W., 1967, Wicked problems, Management Science, 14:B-141.

Jackson, M. C., 1982, The nature of soft systems thinking: the work of Churchman, Ackoff and Checkland, Applied Syst. Anal, 9:17

Jackson, M. C., 1987, Community operational research: purposes, theory and practice, Dragon, 2:47.

Rosenhead, J., 1986a, Custom and Practice, JORS, 37:335.

Rosenhead, J., 1986b, From the president (of the Operational Research Society), OR newsletter, Vol. 15, No. 9.

Tomlinson, R. and Kiss I., editors, 1984, "Rethinking the Process of Operational Research and Systems Analysis," Pergamon Press, London.

Weinshall, T. D. and Raveh, Y. A., 1983, "Managing Growing Organizations," John Wiley and Sons, Chichester.

O.R., THE SOCIAL SCIENCES AND POLICY-MAKING IN A NETWORK OF PROFESSIONS

Erik Anders Eriksson[1]

Department of Defence Analysis
Swedish Defence Research Establishment
S-102 54   Stockholm

INTRODUCTION

The OR tradition is certainly an exponent of the "social engineering" quest for increased rationality in social policy that also motivated the post-war growth of social science research in countries like the U.S. and Sweden.

This quest has been harshly questioned in the last couple of decades - certainly in the general public debate and also in the more specialized scientific one. As a consequence, many social scientists of today do not hold in high esteem the idea of contributing, in any more direct sense, to increased rationality of policy-making. Also, I have a feeling that many of them see us OR analysts as naive and old-fashioned, or perhaps as dangerous technocrats.

Although this view may be justified in many individual cases, the critics fail to observe the ability of analysts to learn - not least from the disciplines of social science - and change. Majone (1985) convincingly demonstrates this to be a main characteristic of the OR tradition (including also labels like systems analysis and policy analysis).

In this paper, I shall argue that the social sciences could also benefit from the OR tradition, and in favour of social engineering - although in a sense much different from the classical Popperian (Popper, 1945).

THE CRITICISM AGAINST "SOCIAL ENGINEERING"

Examining the criticism against "social engineering" more closely, one substantial point is the question - rational for whom? Issues of conflicting interests and values were high-lighted e.g. by the left of 68 and by the ecologists.

---

[1] Thanks are due to Per Agrell and Jan-Erik Svensson for commenting on earlier versions of the paper.

On the scientific side, the work of Kuhn and his follow-
ers (and predecessors like the French conventionalists) threw
doubt on the received view of rationality in science itself
by pointing out its social aspects. And findings of behav-
iourily oriented researchers like Simon and Lindblom
challenged the models of rational individual and societal
decisionmaking prevailing in economics, political science -
and OR.

In spite of these findings, there is also a strong
tendency in social science of today, spreading from neo-
classical economics to other disciplines, applying very
strong rationality assumptions to modelling human interaction.

AN ANSWER: TOWARD PROCEDURAL CRITERIA OF RATIONALITY

Both the former - _arational_ - and the latter -
hyperrational - tendency, if sufficiently strongly or care-
lessly interpreted, cast grave doubts on the program of
"social engineering" and OR: How does one increase the
rationality of decisionmaking if it is either non-existent or
already perfect?

In my view, the arational tendency gives - by far - a
more realistic picture of the world at large than does the
hyperrational. Thus, for the following argument, I shall use
arationality as the starting-point with hyperrational
fundamentalism being simply an environmental factor along
with e.g. religious fundamentalism. This does in no way
restrain me from choosing hyperrationality as my starting-
point in other situations - I argue below for O.R. as an
"open" multi-disciplinary field free to choose the approaches
from various sources deemed suitable for a particular
situation.

Faced with the findings of the arationality tendency,
the OR tradition can respond along two different lines -
apart from throwing up our hands. One is sticking to the
empirically refuted rationality assumptions as norms and
trying to implement then as standards for decisionmaking.
This is the route chosen by orthodox decision analysts and
cost-benefit analysts.

I think there are situations where that response is
quite adequate. However, in other contexts it has only a very
limited relevance. In a world of decision-making rational in
a classical sense, analysts - or social scientists - could
expect to be asked well-formulated questions by decision-
makers. These in turn would then put the analysts'/scien-
tists' answers to appropriate use. In a world of policy-
making processes, it is hard to conceive of analysis being
put to use, unless it takes an active part of the process,
prepared to explain its findings and to reconsider them in
the light of the resulting discussions.

From that perspective, rather than a relatively mechanic
procedure for finding an efficient (in the Pareto sense)

trade-off between fixed preferences, given a set of alternatives, 'analysis should be an innovative constituent of change processes involving strong elements of interaction and social learning - this learning includes also the invention of new alternatives and the formation of preferences. In a catch-phrase, the analyst is seen as a <u>change-agent</u> rather than a <u>methods specialist</u>. Judging from Swedish analysts and such standard texts as Miser and Quade (1985) the process-view is nowadays state-of-the-art in the analytic profession.

I argue that - as the second response to arationality findings - one may conceive of a broader rationality also rendering possible the evaluation of problem solving through social interaction. In my view, this should be sought along the lines of <u>procedural rationality</u> (as opposed to <u>substantive rationality</u>) discussed by Simon (1976) - although his discussion is in the context of individual rather than group decision-making.

I think there is no hope of getting such clearcut conceptual tools for the study of procedural rationality of problem-solving by social interaction as the SEU hypothesis of decision theory or Pareto efficiency of welfare economics. Also, since preferences are not taken as given, the relationship between individual and collective rationality will become muddled as compared to the view taken by decision theory and game theory. One consequence of adopting the model of procedural rationality would be that such conflicts between individual and collective rationality as the "prisoners' dilemma" become less acute than they seem from the point-of-view of standard economic theory. On the other hand, the indicated model of procedural rationality of social interaction is not applicable in situations with radically opposed interests and views. In spite of the social learning, of course, differences of interests and values will remain to highlight the question - rational for whom?

Below I shall indicate the application of the notion of procedural rationality to policymaking as well as to scientific progress. It should be clear that operational criteria of procedural rationality could not encompass the actual outcome - knowable perhaps only several years after the decision. Instead they must deal with the quality of the deliberations preceding the decision: in the context of societal policymaking (i) the likelihood that they disclose potentially available knowledge relevant to the situation; (ii) how well these are transmitted to relevant constituencies; (iii) how their, by now hopefully well-informed, interests are traded off; and (iv) how well the deliberation process prepares the various actors for dealing with subsequent surprises.

The classical example of procedural rationality - in a quite specific and limited context - is provided by legal proceedings. The discipline of jurisprudence has a longstanding tradition of blending the normative and the empirical, which could serve as a paradigm for scientific studies of procedural rationality also in a broader sense.

KNOWLEDGE UTILIZATION RESEARCH - THE ROMANTICISM OF THE ENLIGHTENMENT MODEL

In the last decade, the study of social science knowledge utilization in policymaking has emerged as a scientific discipline, largely within the arationality tendency (see e.g. Weiss, 1977 and the journal _Knowledge_). The findings go against the direct "social engineering" model of knowledge utilization and favour the indirect "englighten- ment" model. There, social scientists and policymakers are seen as widely separated (the two-communities hypothesis) with insights spilling over only in an uncontrolled and haphazard manner.

Webber (1983) is a paper within the above-mentioned tradition of particular interest when discussing the relationship between OR and the social sciences. Webber considers three possibilities of better knowledge utilization from the use of policy scientists - as _advisers_, as _lobbyists_, and as _research brokers_.

The first option, according to Webber, calls into question whether the policy researcher remains a policy researcher: "Outside of the policy research matrix the policy researcher may not be any more insightful, thorough, or credible than any other policy adviser..."

The second option is more highly regarded by Webber. So, he contends, it is compatible with the enlightenment model.

The third option is the one most acutely criticized by Webber: "Research brokers - who are, in a sense, quasi- researchers because they are asked to consider and evaluate policy research - would not be subject to either community's [policy researchers' or policymakers'] evaluation. Further- more, the tension between bureaucratic loyalty and research integrity poses serious problems."

I largely agree with Webber in his criticism against the policy researcher as policy adviser. As for the second option, although I accept the right of the professional organizations of economists and sociologists etc. to engage in lobbying, I do not think that is a very rational model as I shall develop below.

Although I find Webber's discussion of the broker option worthy of attention, it manifestly overlooks a possibility: the establishment of the research brokers as an _intermediary profession_ in between policymakers and researchers subject to its own evaluation - and also that of scientists and policy- makers. This is the answer of the analytic tradition.

Here I think it may be enlightening with a brief digression on the relationship between physical science and industrial practice. During Enlightenment, science seems to have sought its motivation entirely from practical utility. As a matter of fact, though, the impact of science on the emerging industry is very much questioned by modern history of technology (see e.g. Lindqvist, 1984). During Romanticism, utilitarian science was criticized. Theoretical and practical knowledge were seen as fundamentally different entities - cf. Webber's practical policymakers as opposed to theoretical policy researchers. Romanticism was not necessarily opposed to science, but practical utility was banned as a drive and motive for it.[1]

However, in fact during Romanticism, the engineering profession emerged, fusing theory and practice in its own way. In my view, a key attribute of engineers is their accepting practical (economic) rather than scientific evaluation of their work, although this has frequently been questioned by some of them. Generally, modern history of technology stresses the differences between science and technology, the formula "technology as applied science" is dismissed as a myth, and the existence of an engineering knowledge tradition distinct from science is recognized (see e.g. Hughes, 1979).

For the present discussion, we note the possibility of successfully establishing new professions. This, of course, is not sufficient to demonstrate the practicability of an analytic profession, but invalidates Webber's argument against knowledge brokers. In other words, we note that the social engineering model of knowledge utilization research is burdened with the outdated maxim of "technology as applied science". Accepting the possibility of intermediary professions, we get the "real" social engineering model, entailing an independent analytic profession, as an alternative to the enlightenment model with its romantic dichotomy between theory and practice.

However, just as I think that physical science benefitted from having utilitarian demands removed by Romanticism - and, in the long run, probably society too, benefitted from this, even in a utilitarian sence, I think it is beneficial to lift demands of short-range policy relevance off social science. In this respect I sympathize with the opponents of "social engineering" knowledge utilization. However, it must be discussed what claims such academic social scientists - relatively free with respect to policymakers but subscribing to a disciplinary tradition - can justifiably make for the attention of policy-makers, be it as advisers or as lobbyists.

PROCEDURAL CRITERIA OF RATIONALITY IN SCIENCE

Given the insights of modern studies of scientific knowledge, we can hardly expect to find the formal, universal criterion to distinguish true scientific knowledge from the claims of pretenders necessary for Popperian social engineering.

Further, rather than covering separate fields of social activity, the various disciplines of social science represent different points-of-view of the same arena. Their answers to given questions are sometimes entirely determined by

---

[1] My picture of romantic science is largely based on a talk by Professor Gunnar Eriksson, March 1985.

fundamental assumptions (the paradigm, research tradition, or whatever label one may attach). This calls for the judgement of "open" multidisciplinary scholars and analysts. In a similar spirit - though entirely intrascientifically - members of the French Annales school assign to the methodologically heterogeneous discipline of history the role of a laboratory for determining the ranges of applicability of the various homogeneous social science disciplines (de Certeau, 1974).

Therefore, there is need for debate, and craft knowledge, as to the applicability of various traditions - e.g. standard economic theory with its strong rationality assumptions, or perhaps knowledge utilization research - to a given problem situation. Also, there is little probability that such craft knowledge will ever emerge into anything like scientific laws, mainly due to the fact that man is a learning subject; this includes effects of social science knowledge on the actual game in the arena studied by social science.

Along the lines of such a debate on the effectiveness of various research traditions in coping with problems of the real world or of other research traditions, I believe that we can develop procedural criteria of good scientific knowledge and of scientific progress (these ideas are partly inspired by Laudan, 1977, although he does not see his criterion as a social one).

I believe the notion of a network of professions to be a suitable starting-point for such a development. In contrast, influential theories of professions seem to stress the autonomy and exclusiveness of them (see e.g. Larson, 1977). This may be very true for e.g. law or medicine, but there are other settings where different groups of people, some or all required to have completed a very advanced training and displaying a strong group identity, work in the same field without clearcut subordination - in a network. One case in point would be the performing arts with authors, various performers, theorists, critics and so on.

I think that this type of situation is more typical for "modern" professions, than is the situation of law or medicine. Still, I think it is highly warranted to use the term profession also in the "modern" settings. It primarily alludes to the transfer of craft knowledge taking place in a profession.

If a "modern" profession cannot conceivably claim the exclusive power to judge its own results, what autonomy could it have? Laudan distinguishes between two modalities of appraisal: acceptance and pursuit. I suggest that a "modern" profession operating in a network of professions, should have relatively great powers to pursue what it judges promising lines of activity, whereas the acceptance of results should be mainly for other, normally adjacent, professions or groups to decide.

Further, if the social interaction in science is reasonably rational, research traditions that do not have much to offer their neighbours, in due course will fade away for lack of new students, funds, etc.

CONCLUSION: THE USEFULNESS OF O.R. TO THE SOCIAL SCIENCES

I am afraid that many social scientists see the OR tradition as a candidate for such fading away. In my view that would be highly irrational since OR analysts have accumulated a lot of knowledge and ethical awareness on bringing (social and natural) science knowledge to bear in the context of societal change processes. Given the inter-active nature of knowledge and practice in this arena, intermediaries like OR (and problem-oriented, multi-disciplinary fields like studies of R&D) should be expected to have more profound contributions to offer the fundamental social sciences than has engineering with respect to physical science.

REFERENCES

Certeau, M. de, 1974, L'Opération historique, in Le Goff, J., and Nora, P., eds, "Faire de l'histoire III", Gallimard, Paris.

Hughes, T.P., 1979, Emerging themes in the history of technology, Technology and Culture, 20: 697-711.

Larson, M.S., 1977, "The Rise of Professionalism", University of California Press, Berkeley and Los Angeles

Laudan, L., 1977, "Progress and Its Problems," University of California Press, Berkeley.

Lindqvist, S., 1984, "Technology on trial: the introduction of steam power technology into Sweden, 1715-1736," Almqvist & Wiksell International, Stockholm.

Majone, G., 1985, Systems analysis: a genetic approach, in Miser and Quade, op.cit.

Miser, H.J., and Quade, E.S., eds, 1985, "Handbook of Systems Analysis," Wiley, Chichester.

Popper, K.R., 1945, "The open society and its enemies. Volume I: the spell of Plato", 5th ed 1966, Routledge & Kegan Paul, London.

Simon, H.A., 1976, From substantive to procedural rationality, in: "Method and appraisal in economics," Latsis, S., ed., Cambridge University Press, Cambridge.

Webber, D. J., 1983, Obstacles to the utilization of systematic policy analysis: Conflicting world views and competing disciplinary matrices, Knowledge: Creation, Diffusion, Utilization, 4: 534-560.

Weiss, C.H., ed., 1977, "Using Social Research in Public Policy Making", D.C. Meath, Lexington, MA.

EFFECTIVENESS VERSUS EFFICIENCY :

OPERATIONS RESEARCH IN A NEW LIGHT

Sasan Rahmatian

Department of Information Systems
California State University, Fresno
Fresno, California, 93740

I. INTRODUCTION

Operations researchers have traditionally been concerned with well-structured problems which have one best solution or a range of acceptable solutions. Social scientists, on the other hand, deal with hard-to-structure problems in which it is very difficult to define what "best" means let alone find the best solution. Stated differently, operations research attempts to find the most efficient means of reaching a goal given a particular problem formulation. In contrast, social sciences tend to question the validity of the goal itself by treating the problem formulation not as a given but as open to challenge and revision.

The above difference can be partially traced to the difference between mathematical problem solving and real-world problem solving. In solving mathematical problems, from elementary school all the way to the postdoctoral stage, we are used to situations in which the basic facts are given, the assumptions are unnegotiable, and the problem is already structured and formulated. The only valid formulation of the problem is the one provided. To solve the problem based on a different set of givens becomes tantamount to solving a different problem. Not so in the real world. As I have argued elsewhere (Rahmatian, 1985), real-world problems exhibit a distinctly hierarchical nature. This is because problems are based on objectives, and objectives form a hierarchical ends/means chain. With assumptions questioned at one level, we move to a higher level where real goals become more manifest. We thus minimize chances of correctly solving the wrong problem; of reaching an efficient but ineffective solution.

Effectiveness and efficiency are indeed two interesting concepts which can help us understand the interplay of problems and solutions in the real world. In the next section, a conceptual analysis will be offered, challenging some of the traditional wisdom surrounding the two concepts. In section three, the outline of a theory will be sketched, while in the last section the theory will be illustrated in the domain of information technology.

II. EFFICIENCY VS. EFFECTIVENESS

Effectiveness has to do with "what" is accomplished; efficiency with "how" it is accomplished. As Peter Drucker (1974) put it, effectiveness has to do with doing the right things whereas efficiency has to do with doing things right. Another conception equates efficiency with the ratio of output over input, while defining effectiveness in terms of the difference between the actual and the desired output. With this last definition of efficiency, an interesting situation arises. Since

$$\text{Efficiency} = \text{Output/Input} \ ,$$

it has always made sense ("common sense"!) to regard as equivalent the following two ways of increasing efficiency :

A. By obtaining a higher level of output with the same amount of input;
B. By obtaining the same level of output with a smaller amount of input.

Are these two modes of increasing efficiency equivalent? Mathematically yes, but practically no.

First consider situation A. Take a "production process" (in its most general form) in which 10 units of input produce 100 units of output. Suppose due to technological innovations, the efficiency of this production process is increased such that with the same 10 units of input 105 units of output are now produced. This may mean slightly higher profitability and nothing more. Now suppose due to a fundamental technological breakthrough, the same 10 units of input can produce 1000 units of output. This time the qualitative impacts of this quantitative change need to be examined. Going from 100 to 1000 units of output may mean

* a larger sales force;
* a different organizational structure;
* a revised pricing/advertising strategy;
* etc.

These are all the qualitative impacts of a quantitative change, which indicate that "more" is more than merely a simple quantitative notion.

Now take situation B where we can obtain the same level of output with a smaller amount of input. With the same technological breakthrough still underway, we now consider producing the same 100 units of output but this time with only 1 unit of input. This may imply:

* less raw material;
* fewer production runs (fewer production shifts, lower
  labor cost);
* smaller goods-in-process inventory;
* etc.

   To be noted is the difference between situations A and B
in terms of goal-setting. With situation A, there is a
tendency to take advantage of the higher efficiency in terms
of increased levels of production. With situation B, the
higher efficiency tends to affect the cost of production.

   To be noted is also the similarity between situations A
and B. In both we have efficiency increase to such an extent
as to begin to impact effectiveness. This is again due to the
difference between abstract mathematical problem-solving and
concrete real-world problem-solving. In purely mathematical
terms "6 times 2" and "12" are the same. In practice, buying
two six packs of beer has consequences different from buying
twelve individual cans of beer. With the two six packs, one
ends up with two cartons, which can either translate into a
problem (how to get rid of them) or an opportunity (making a
toy for a child).

   Effectiveness and efficiency can be separated for
analytical purposes. In practice, however, they are
inseparably linked. The relationship between the two will be
explored in what follows.

III. TOWARDS A THEORETICAL FOUNDATION

   The ideas explored in the previous section can be
formalized into a theory consisting of two basic axioms.

A X I O M   1 :   For every production process, there
                     is a critical threshold point such
                     that when the efficiency of the
                     process increases beyond that point,
                     then the effectiveness of the process
                     is impacted (usually positively).

A X I O M   2 :   For every production process, there
                     is a critical threshold point such
                     that when the efficiency of the
                     process falls below that point, the
                     effectiveness of the process is
                     impacted (usually negatively).

An example may help clarify the meaning of these axioms.

   Imagine the very common situation in which an important
document (such as a proposal for winning a competitive
contract) is to be produced under considerable time pressure
by a certain deadline, say Friday 4 PM. Suppose based on the
expected length of the document it is estimated that it will
take eight hours to type, proofread, edit, correct, and
revise it. Typing must therefore begin around 8 AM.

Efficiency can be defined here in terms of the number of hours of typing (labor input) needed, whereas effectiveness can be defined in terms of the quality of the report (i.e., the extent to which it will be instrumental in winning the contract).

Suppose the regular typist (who would take eight hours to type the document) is replaced by a more efficient one who would be able to do the job in seven hours. The time (one hour) thus freed up may not be long enough to let the writers of the proposal (who wrote it under considerable time pressure) improve its quality significantly. This could be true for both operational and psychological reasons. Now suppose due to a technological breakthrough (such as word processing) efficiency increases dramatically, such that it will take only two hours to produce the same document. This means the six hours thus freed up can be spent improving the quality (i.e., effectiveness) of the proposal. This is precisely what Axiom 1 means.

Axiom 2 is easily understood in the context of the above example too. Suppose the efficiency of the process drops from its original level of eight hours (due to the typist badly injuring three fingers) to the new level of twelve hours. This means that typing must start Thursday at 1 PM (presupposing an 8-to-5 workday), which implies that Thursday evening cannot be used by the authors for working on the proposal, further exacerbating the time pressure they are under. The adverse impact this will have on quality (effectiveness) is obvious.

The proposed axioms are perhaps in some way related to the basic tenet of catastrophe theory, namely that for a certain class of events, continuous changes beyond a threshold point produce discontinuities. It would be challenging to explore the similarities between the threshold point in catastrophe theory and the one in the above axioms.

IV. APPLICATION DOMAIN : INFORMATION TECHNOLOGY

The proposed theory perhaps finds its clearest application in computer-based information technology. This type of technology was originally conceived as entirely efficiency oriented. Even today, it is being marketed primarily as a productivity enhancing force, whether in the office, the factory, or at home. To regard computers as merely (or even primarily) efficiency oriented without any consideration of their impact on effectiveness (along the lines explained in the previous section) would be just too simplistic. Computers have increased efficiency in many areas to unimagined levels. Indeed, in some domains, the productivity gains have gone beyond the critical threshold points hypothesized earlier. That is why computers are an excellent vehicle for allowing us to see the application of the proposed theory to the real world.

Computers are mind-amplifying tools (Rahmatian, 1987). But this is not to say that they are merely more efficient manual systems. With computers allowing dramatic increases in efficiency, many qualitative changes are bound to occur. As clerical tools, computers have found a secure place for themselves. The more recent development is using computers for strategic purposes. The two axioms explicated in the previous section can perhaps shed some light on how this shift (from computers as clerical tools to computers as strategic tools) came about. An example may help clarify the explanation.

Consider a bank which is trying to attract new patrons. Service is an important factor affecting the decision of a prospective client whether or not to open an account in this bank. An element of service is the length of waiting lines which, in turn, is partially determined by computer response time. Imagine computer response time being such that on the average six (n = 6) people wait in the line.

With computer efficiency improving such that n becomes equal to five, the perception of a shorter waiting line is not likely to be created. However with dramatic increase in efficiency causing n to equal 1 the situation may become qualitatively different. It may make the difference between "a pretty long line" and "practically no line at all". This shift in perception can make the difference in the prospective client's mind between opening and not opening an account in that bank. And <u>this</u> is a strategic, not merely clerical, issue.

It also works the other way around. With computers facing unresolved maintenance (or other) problems, the response time can significantly increase, thus making the line much longer, and the bank as a whole appear less service-oriented.

Many other examples can be provided illustrating the ways in which the effects of information technology have gone beyond pure efficiency considerations and have begun to spill over into the domain of effectiveness.

It may be advantageous to bring the paper to a conclusion by going back to the issue raised at the beginning. Social sciences, in their widest scope, attempt to understand human behavior. Human behavior can be explained in terms of purposefulness (Ackoff and Emery, 1972). Understanding human goals, their variety, and their relationships is thus an important part of social sciences. In practical problem-solving situations, it becomes even more important to understand the exact nature of the results being sought. Failure is as likely to occur due a limited or erroneous understanding of the problem as it is likely to occur due to adopting an incorrect solution. Effective problem solving then becomes a goal to strive for.

Operations research, in its widest scope, attempts to impose a mathematical structure on a given problem formulation, thereby identifying the most efficient solution to it. Armed with the latest techniques and technologies available, operations researchers are tempted to make efficient problem solving a goal to strive for.

If this paper has succeeded in explaining the relationship between effectiveness and efficiency as foci of action, then perhaps it has helped shed some light on the relationship between operations research and the social sciences as different approaches to the investigation of human behavior.

REFERENCES

Ackoff, R. and Emery F.E.,1972, "On Purposeful Systems", Aldine Atherton, Chicago.

Drucker, P., 1974, "Management: Tasks, Responsibilities, Practices", Harper & Row, New York.

Rahmatian, S., 1985, The Hierarchy of Objectives : Towards an Integrating Construct in Systems Sciences, Syst. Res. 2:3.

Rahmatian, S., 1987, Design for Automation : The Mind Enhancing the Mind, Proceedings of the International Society for General Systems Research.

METHODOLOGY

METHODOLOGY

Paul Ledington

The Management School
University of Lancaster
Bailrigg
Lancaster, U.K.

Methodology is an undervalued area of interest. The term itself has different meanings within the Social and Management Sciences. In the Social Sciences the term is normally applied to the approach adopted in order to generate acceptable knowledge concerning whatever area of the Social world which is being scrutinised. In the Management Sciences the term is more usually applied to the framework of action which is adopted in order to bring about some change in an area of concern. Clearly there is a major difference between a Research Methodology and a Problem-Solving Methodology. However there are also links between the two because a Problem-Solving Methodology may also be employed as a means of inquiry into the nature of that which is being changed. Implicit in any Management Science Methodology is some model or theory concerning the nature of change and the nature of that which is being affected. If such a methodology is indeed effective in bringing about change, and hence solving problems, then there is the possibility of drawing 'weak' inferences between the ideas implicit in the approach and the reality being grappled with, but this assumes that there are adequate means of assessing and evaluating the results of employing a methodology.

This is not a simple problem. Checkland (1972) stated an inherent 'dilemma' in this area. If a methodology is asserted to work then the reply can be made 'how do you know that some other methodology would not have been more effective'; on the contrary if the methodology is asserted not to work then equally the reply even be made 'how do you know that the failure does not stem from you incompetent use of the Methodology'. In the sciences the outcome, the knowledge generated, is evaluated on the basis of the methodology employed and the competence of its application. Essentially the relationship between a methodology and its underpinning philosophy is first established and then accepted so that valid statements, valid within the philosophical framework, can be developed. In the sciences the evaluation of the methodology is based upon its logical veracity with respect to its underlying philosophy, whereas the knowledge generated is evaluated by considering the 'Methodology In-Action'. The distinction between Methodology as a set of principles for sensible action and 'Methodology In-Use' has not been so clearly delineated in the Management Sciences and much more emphasis on this latter area is required. Secondly the nature of the Management Science enterprise needs further consideration. The validity of change generated through problem solving activity is not evaluated with respect to the methodology employed; in most cases the results of management science activity are judged independently by the stakeholders in the problem situation. The real question being addressed in the Management Sciences is not 'how can we solve this problem efficiently' but rather 'what is the nature of the Management (problem-solving) process'. In this way methodology represents a model, or theory, of action which can be used to both guide real world activity and analyse that activity. This separates the role of user and management scientist. The user is concerned to solve a problem whereas the management scientist is concerned to understand the nature of the problem-solving activity. The user may adopt a methodology and integrate it into his own problem-solving activities and quite rightly may be concerned as to whether the employment of the methodology enhances or diminishes his problem-solving effectiveness. Once again questions are raised as to how this adoption process operates, how does the user learn to use the methodology, how is the methodology integrated into the user's natural propensities for problem-solving, and how and why all of this is sometimes judged acceptable and sometimes unacceptable. It is these questions which the Management Sciences must address.

If our understanding of the nature of methodology is to develop then we need to begin to consider it from a much broader framework than has been used to date. Problem-solving activity needs to be considered within a framework of, at least, three elements. The Content of the Problem-Solving. The context of the Problem-Solving, and the Process of the Problem-Solving Methodology is currently considered as a definition of the Process, whereas perhaps it would be better to consider how the adoption of a methodology impacts upon the Problem-Solving Process.

Methodology, as a set of principles for sensible action, has generally been based upon the idea of an external hard reality, little attention has been paid to how methodology fits within human problem-solving as that problem-solving activity itself helps to create the social reality it is trying to change.

A case has been made that there are potentially many areas concerned with methodology which require further elucidation. It is often said that 'if all you have is a hammer then all problems will look like nails' but if instead you have a large and sophisticated set of tools the problem of how to use them effectively comes into focus. We know little of the various strategies, management techniques and operating skills which are required by various approaches, nor how they combine together to provide levels of problem-solving performance. Understanding the nature of problem-solving performance is at the heart of Management Science, and methodology is surely a key concept within a framework with allows us to understand that performance. Perhaps in twenty-five years time, if OR and the Social Sciences is once more considered, there will be much more content concerning the nature of methodology than is available today.

Reference

Checkland P B (1972) "Towards a Systems based Methodlogy for Real-World Problem-Solving". Journal of Systems Engineering 3,2.

DUALISM IN SOCIAL THEORY AND SYSTEMS PRACTICE

Gilbert Mansell

The Polytechnic
Huddersfield

METHODOLOGY

A methodology is a body of knowledge that enables action to be taken, either to increase substantive knowledge of the world (as in the natural sciences), or to bring about some other desirable state of affairs, such as the solution of a practical problem (as in engineering or management science). Successful action depends on reliable theories both about the world, and how to intervene in it. Such theories can only be established, however, by action. In the history of a mature engineering discipline, theory and practice combine to progressively enhance the theory and increase the likelihood of successful practice. Not all engineering disciplines are equally mature. Management science, however, has more intractable problems to solve than even the most junior engineering discipline. There is no unified body of theory, and successful practice of management is possible without knowledge of any of it. This is due not to the incompetence of management scientists, but to the fact that management science is a form of applied social science. There is disagreement as to how to generate substantive knowledge in social science, and less agreement as to how to intervene effectively in social systems. One contribution to the methodology of management science has come from the systems movement.

SYSTEMS METHODOLOGY

Systems thinking (as presented for example by Kast and Rosenzweig, 1970) provided an ontology for the analysis and design of organizations. Organizations were conceptualized as systems, imbued with a life of their own, that enabled them to endure over time even though individual members left or were replaced, and to take collective action beyond the capacity of any individual member or indeed the whole unorganized membership. Talk about organizations treated them as entities that could act, own things and pursue objectives. Systems theory explained these attributes in terms of the emergent property resulting from organization, and explained the persistence of organization in terms of cybernetic control and adaptive goal setting in response to the environment. The role of systems methodology was to enable organizations, or subsystems within them, to be designed in an optimal way, as if organizations were complex machines. An important source of theory in this respect derived from systems engineering (operations research made similar theoretical assumptions).

171

SYSTEMS ENGINEERING

The essence of systems engineering methodology (classically described by Jenkins, 1969), is that it requires the clear definition of system objectives, and then designs a system that meets objectives and can be proved to do so. Proof of the correctness of design is achieved by the manipulation of a mathematical model that represents the system. More realistically, several designs would normally be produced, each of which represented the outcome of different, but defensible, trade-offs between alternatives, and therefore different cost/benefit profiles. Systems engineering methodology has been applied successfully to systems of machines, (defined at some appropriate level of abstraction), but was also recommended for man/machine systems, and even thought appropriate for systems of predominantly human activity. Historically, early accounts of systems engineering saw no particular problem in designing human activity in much the same way as the behaviour of machines. For this to be a sustainable proposition, human activity had to be regarded as systemic.

The relevance of systems engineering methodology to human activity can be challenged. Firstly the setting of system objectives is problematic. There may be difficulties in gaining agreement amongst the stakeholders in a system as to appropriate objectives. The objectives to be sought may be in conflict with one another, and fundamental contradictions may mean that it is impossible to design a rational system. It may prove impossible or undesirable to quantify objectives in such a way that it could ever be established that they had been met. There are serious practical and ethical objections to subjecting human beings to measures of supposedly objective performance in many areas of activity. For example such problems arise in the assessment of teachers and academics, and quite conceivably systems engineers as well.

A second objection to the application of systems engineering methodology to human affairs is that it treats human activity as systemic. Human activity is assumed to be sufficiently like the behaviour of a machine, to make it feasible to design human activity and socio-technical systems as if they were machines. Human activity is arguably systemic if it is the same tomorrow as it was today. Persistent activity over time, if co-ordinated and directed to a common end, can achieve results beyond the capacity of any individual agent. Nevertheless human beings are unlike the components of mechanical systems in that any individual may simultaneously belong to many associations or organizations, is conscious of this membership, and can strive to influence group or organizational goals. The notion of a human activity system, therefore, is metaphorical in a way that the notion of a mechanical or electronic system is not. Undue stress on the structure of human activity systems overlooks the fact that structure is created by process, and the process is influenced by the free, if constrained, choice of intelligent agents. The structure of human activity is maintained by conscious action as well as programmed behaviour, but also by exchange of information. Electronic information systems exist, but they are capable of exchanging or displaying only pre-determined messages. Human agents use their gift of language in a constant struggle to influence their fellows, and the nature and outcome of discourse is not always predictable.

SOFT SYSTEMS METHODOLOGY

Soft systems approaches can be viewed as the transformation of systems engineering (and operations research) methodology to ensure success in the area of human activity, rather than the hard engineering application area (Checkland, 1981). Systems engineering "worked" where there was agreement about systems objectives and no problem about routinizing human tasks. With fewer constraints on human thought and action, issues of subjectivity and complexity arose that were recognised and addressed in the work of, for example, R L Ackoff, C W Churchman and P B Checkland. Checkland's SSM, for example, does not depend on the paramount necessity of setting system objectives, does not presuppose the uncontentious objective existence of a system whose functioning is to be optimized, and allows for multiple alternative perceptions of a complex reality on the part of different stakeholders. What makes SSM a systems methodology, however, is the fact that an abstract idealized system is conceived of that could achieve some purpose if that purpose were agreed to be desirable. The idealized system is a human activity system, and its definition and elaboration is a subjective process. The world-view of the creator determines the nature of the system that is perceived to be relevant to some situation.

METHODOLOGY AND THEORY

A methodology that claims to be useful in assisting intervention in the social world is necessarily underpinned by social theory. This theory is concerned with what exists, for example with social structure and the way in which it can be said to exist, and with the way in which knowledge can be gained about what exists. Social theory is riven by dichotomy between subjectivism and objectivism (Rubinstein, 1981, Manicas, 1987). The mission of objective social science has been to make good a claim for parity of esteem with natural science. Objective social science has therefore been characterized by empirical enquiry, commitment to quantification and a search for predictive laws. A fundamental tenet is that knowledge is based on facts that can be discovered independently of any theoretical perspective. Subjective social science rejects the positivist doctrine of a unity of scientific method, and claims that the social world is qualitatively different from the natural world, because men have minds and human actions express intentions and have meaning for the actors. An objective account of human action is impossible because action is a reflection of an interior reality that can only by subjectively interpreted. Knowledge of the social world is based on a subjective understanding of it rather than objective explanation.

Objective social science is historically exemplified by behaviourism in psychology and structural functionalism in sociology. Behaviourists either deny that action exists or redefine action to exclude a mental component, or treat action as deterministically caused by mental phenomena. Structuralists believe that social life is explained not by the ideas of participants but by objective features of social organization. Expressions of subjective social science, such as ethnomethodology, regard social reality as consisting of the ideas of social actors, and reject scientific explanations that cannot be grounded in the ideas of actors. Both subjective and objective social science, although completely opposed on many theoretical issues, have a similar approach to the human mind. The mind is essentially private and inaccessible to objective study. This attitude stems from mind-body dualism.

CONSEQUENCES OF DUALISM

Objective social science must claim to have a neutral language with which to discourse about society, and also claim to discover facts about the social world. A claim to objectivity fails if what is said in the supposedly neutral language is derived from ideas in society, and if the "facts" discovered only count as such from some pre-determined theoretical perspective. Empirical research in the social sciences can be criticized for embodying an implicit process of interpretive understanding. The fundamental error in objective social science is to reify social institutions and to treat them as having an existence distinct from men's ideas about them.

Subjective social science suffers from its inherent idealism. It has no
means of distinguishing between correct and incorrect accounts of reality.
It has no theory to explain alienation and false-consciousness. It cannot
analyse the unintended consequences of human action. It cannot account
for properties of collective as distinct from individual human action.

SOCIAL THEORY OF SOFT SYSTEMS METHODOLOGY

Systems engineering is implicitly based on objectivist social theory.
Soft systems methodology is informed by subjectivist movements within
social science, for example phenomenology. Nevertheless soft systems
methodology cannot completely reject its objectivist origins without
losing its orientation to bringing about desirable change in the world.
In the case of SSM, for example, the subjectivist components are the
ontological status of human activity systems, and the role of world-view in
colouring perceptions of reality. The objectivist origin of SSM is shown
by the location of human activity systems in a systems typology. Two
sorts of subjectivist claim arise with respect to human activity systems.
Firstly there is the view that accepts the existence of systems of human
activity but denies that objective knowledge about them can be possessed,
the essential problem being human agency and the privacy of mind.
Secondly there is the position that rejects the the objective existence
of systems, but wishes to claim that mental modelling of human activity
as systemic can lead to valuable insights. The insights would not be
valuable, however, unless human activity could, in principle, be systemic.
Either view point allows for alternative models of relevant systems argued
to derive from alternative world-views. The origin of world-views is not
enquired into, and they are treated as private mental phenomena.

The relationship of human activity systems to social systems is an uneasy
one in SSM. In an early typology a distinction was maintained between the
two, with social systems being treated as natural systems and systems of
human activity as potentially open to social engineering. To distinguish
between the natural and the social world is subjectivist, but to treat
the social world as partly natural and partly designed is problematic.
Since man is part of nature society is natural, but constructed by man.
Not all men are equally free to contribute to the construction, however,
and to many society appears immutable. So mankind can become alienated
from his own construction.

A PREFERRED THEORY

The social theory underpinning both systems engineering and soft
systems methodology has been argued to be inadequate. The laws of society
(or rather the rules that define meaningful action in some society) are
not like the laws of physics, because they are constructed by social agents
in their everyday practices, and could be other than they are. The laws
that enable an engineer to confidently design bridges have no equivalent
in the social world to enable a social engineer to design systems of human
activity. This insight informs soft systems methodology, which is however
haunted by the spectre of idealism. The social world is created by agents,
but soft systems methodology has no means of distinguishing between correct
and incorrect accounts of the social world. The consequences of action may
be unintended by the agent and not properly understood. Going to work
reproduces the system of capitalism whether or not this is desired or
understood. Systems of human activity exist (as abstractions) because if
they did not a good deal of human action would be incomprehensible.
Objective accounts of social reality can be given because human activity
has an economic base that is independent of culture. A preferred theory
to underpin systems methodology would be realist or materialist,
therefore.

REFERENCES

Checkland, P B 1981, "Systems Thinking, Systems Practice," Wiley

Jenkins, G M 1969, "The Systems Approach", Jnl Sys Eng i:i.

Kast, F E, and Rosenzweig, J E 1970, "Organization and Management: A Systems Approach", McGraw Hill.

Manicas, P T 1987, "A History and Philosophy of the Social Sciences," Blackwell.

Rubinstein, D 1981, "Marx and Wittgenstein," Routledge.

A SCIENTIFIC APPROACH TO METHODOLOGY

Brian Lehaney

Centre for Accountancy Studies
Luton College of Higher Education
Luton, U.K.

THE SUBJECT MATTER OF SCIENTIFIC METHODOLOGY

Scientific methodology is the buffer between the philosophical and the practical aspects of any form of investigation, and is often neglected, but yet forms the very essence of any scientific discipline. Despite the importance of the subject, most economists and operational researchers will have had, at the most, a fleeting brush with methodology, and would probably not have more than a passing interest in it. Yet, the importance of methodology must be emphasised, particularly because its importance is not always immediately or obviously apparent.

Consider the practical and immediate problems facing a business economist wishing to forecast demand in a fast changing world, or the problems of an operational researcher developing a telecommunications network. Before they can approach their problems in anything other than an ad hoc fashion, theories and methods must have been developed for their use, and it is this development process that is a major part of the subject matter of methodology. This is quite distinct from methods, which are the techniques the practictioner uses to approach practical problems. For example, the method used to tackle an oil refinery problem might be linear programming, but the way the algorithm was developed is the concern of methodology. Another, very important, aspect of methodology is the sponsor/researcher relationship. How does a researcher decide which project to undertake, and how are results effected by the need to obtain further research funding? These latter aspects are referred to briefly, but the main concern of this paper is theory and technique development.

In order to understand any problem more fully, it is always useful to step back a stage, and consider some wider issues, and, in the case of methodology, the step back is philosophy, and the wider issues are questions such as, "What is science?", "What is social science?", etc. Only when these questions have been addressed, is it possible to consider the questions concerning scientific methodology. To attempt the latter without the former can be likened to an English person attempting to read untranslated German literature, without first having learnt some German.

The order of the paper will be to firstly discuss what is meant by the various terms used, then to outline the different approaches to methodology,

and finally, to propose how these views can be encompassed within a
pluralistic framework.

Science and Social Science

Why is it that physics is often considered unequivocably scientific,
astrology not to be taken seriously, and economics somewhere in between? The
answers may be found by addressing the questions posed in the last section.

To be considered scientific, a theory must be able to explain and to
predict. In the realm of physics, the archetypal example of natural science,
the symmetry is rarely seriously disturbed, for the following reasons:

    i)   The subject matter has no feelings of its own, and there are
         consequently no mental or emotional interaction between the
         observer and the observed.

    ii) The subject matter is entirely quantitative, and there is no
         attempt to quantify qualitative factors.

    ii) The possible shocks are finite, and, to use a simple
         example, the electrician can rely on a fixed relationship
         between current, voltage, and resistance, provided the
         temperature and humidity are reasonably constant.

Thus, the electrician can explain why a current was so many amperes
yesterday, and he can predict the current tomorrow, subject to certain finite
and coherent conditions. Why is social science not like this? For the very
opposite of the three reasons listed above. By definition, social science is
concerned with subject matter that is qualitative to some degree, which has
feelings, and which does not easily lend itself to predictions subject to
finite, coherent provisos. Why then is it called social science, rather than
something else? Why does economics consider itself above astrology? It is
because social sciences ostensibly adhere to scientific methodology. They
seek to explain and predict, and to do so using similar methodology to that
of a physicist, whereas an astrologer does not. In order to be considered
scientific, whether in the natural or social sciences, methodology must be
logical, repeatable, and a different scientist should achieve the same
results with the same data.

Scientific Methodology

Economics is one branch of social science, and its concern is the use of
scarce resources, given infinite wants. Whilst physicists can carry out
controlled experiments, governments cannot change just a single variable and
hold all others constant for a year or two. Consequently economists model,
as opposed to carrying out controlled experiments. Therefore, in the natural
sciences, if a competing theory arises, it is either rejected, or the old
theory is rejected, and the new becomes dominant, so it is usual to have only
one theory to explain a particular event in the natural sciences. In the
social sciences it is unusual to have a single theory to explain a particular
event, and the various branches of social science each tend to have at least
two current competing theories, which are equally matched, both in terms of
numbers of supporters, and on internal logic. Each theory will explain
something the other does not, but each would have problems of internal logic
were they to attempt to incorporate those aspects found in the alternative
theory.

The question must be asked, "Is there virtue in scientific method as an
end itself?". The answer must be no. For example, if reasonably correct
predictions for demand per month have been made for the last ten years, based

on a tea lady's ability to read tea leaves, the only reason a firm would want a more scientific method, in the short run, would be if it could produce better predictions. In the long run, however, the firm may be interested in someone else forecasting demand, once the tea lady retires. If the prediction process does not follow clearly defined, logical steps, then it would be difficult to train someone else for the job. In which case, a more scientific method would eventually have to be used. The point then being that it is the end which determines the means, and it is one thing to achieve a wonderful methodological process, and another to convince people, particularly practical businessmen, of its merits.

Why should the scientist/researcher be concerned to convince others? Amongst other things, the research that can be done without funding is limited, and, in order to obtain that funding, those in charge of finances have to be persuaded to support projects. Thus politics enters the arena, and the emphasis of research is then, to a large extent, determined by funding. This is particularly true in the social sciences, where, for example, in recent years, the Social Science Research Council has had to change its name to the Economic and Social Research Council, and its funds have been cut by government. Using the previous argument, if your aim is to obtain research funding, you will do so in the best way possible. If you don't obtain the funding, but feel you have been very scientific in not doing so, will you be consoled by this?

METHODOLOGICAL APPROACHES

Popper

Karl Popper's works (e.g. 1959, 1963) form the basis of modern debate about methodology in science and social science, and if economists or operational researchers were asked their general methodological approach, it would be ostensibly Popperian. Blaug (1980), though, argues that whilst economists have learnt to mouth the right methodological words they have yet to put them into practice.

Pre-Popperian methodology was inductive and empiricist in nature, observation being the keyword, with the basis of induction being to establish an existing pattern or trend, and then use this to extrapolate and theorise. Popper argued that induction was not a suitable method for science, and he used the example that no matter how many white swans have been observed, this does not mean that the next swan to be seen will also be white. He suggested that falsifiability is the key to scientific methodology, and argued for the deductive approach, the latter being to formulate a hypothesis, and to then try to disprove this hypothesis by looking at fresh data. If the hypothesis cannot be rejected, it eventually becomes a theory. Ad hoc adjustments would automatically reduce scientific status, and Popper rejected probablistic explanation in favour of nomological explanation. That is, scientific laws as opposed to trends.

In addition, Popper was keen to distinguish between science and pseudo-science, arguing that science requires risky predictions, which should be explicit, and not too general in nature. In particular, Popper was critical of Marx, Freud, and Adler for the latter reason, arguing that Marx could explain everything in terms of class struggle, without fear of being disproved.

Popper argued that the method of formulating hypotheses is not important, but the method of testing them is, and that science is a method for testing theories. In general, the Popperian method is as follows:

a) Formulate hypothesis;

b) Collect new data;

c) Test hypothesis on the basis of prediction;

d) Reject or do not reject hypothesis.

Thus, in the Popperian view, prediction is seen as the most important test, but since it is a hypothetico deductive nomological procedure, there should be a symmetry between explanation and prediction.

The rigorous nature of falsifiability has undoubtedly improved methodology, but it is not without critiscism, and there are four main problems with Popper's arguments. Firstly, they are created inductively themselves. Secondly, the criticisms of Marx etc, are of the axioms as opposed to the theories themselves. Thirdly, the idea that it is unimportant how hypotheses arise completely ignores the problem of the relationship between sponsor and researcher. Fourthly, positive existential statements can have no scientific status. That is, it is not scientific, in the Popperian sense, to state that the abominable snowman exists, since, by not finding any, this does not prove conclusively that they do not exist, so the statement cannot be falsified. However, the statement that abominable snowmen do not exist, can be falsified by finding one. In addition, whilst it is possible to have controlled experiments in natural science, this is not the case in social science, and consequently probablistic, as opposed to nomological explanation, would have to be used for any studies concerning people.

Post & Anti Popper

Hempel argued for hypothetico deductive probablistic procedure, which allows for the statistical testing of theories. This, of course, means that rational decision between competing theories becomes far less straightforward, as there may be equal evidence for and against each hypothesis. Hempel's ideas of probablistic explananation provide the link between natural and social scientific methodology.

Kuhn largely agreed with the Popperian view, but argues that paradigms are the normal course of events in science. Scientists work within the framework of knowledge (paradigm), until a scientific revolution replaces this framework with another. Thus, Kuhn argues that science has a puzzle-solving tradition within the existing paradigm, and that Popper's ideas are more relevant to scientific revolutions.

Lakatos is another who, in the main, is a disciple of Popper, but he argued that a main hypothesis can have auxiliary hypotheses, which can be adjusted ad hoc without losing scientific status. The change from one hypothesis to another can therefore be gradual, and need not have the instant refutation proposed by Popper.

Both Kuhn and Lakatos agreed with Popper that it is rationally possible to decide whether theories improve on others or not, but as mentioned previously, this is not straightforward when probablistic explanation is employed, and the social sciences are typified by competing theories.

Lessnof argued that social science is neither nomological or probablistic, but distinct from natural science altogether, because mental phenomena are involved in social science. This follows in the Weberian tradition of the subjectivity of social science.

Winch also distinguished between science and social science, but further argued that human behaviour is so varied that general theories are not possible in social science at all.

Hesse considered that theories are not determined by facts, but constrained by them, and since controlled experiments are not possible in social science, tests of prediction are very difficult. Hesse argued that social theories are more like the arguing of a political case than scientific explanation, and that facts cannot decide between competing theories, so conscious, or unconscious value judgements play a part in the researcher's choices. She did, however, consider that not all social science theories depend on value judgements.

WHICH METHODOLOGICAL APPROACH?

Each approach to methodology prescribes how scientists should carry out their research, but methodologists seem above such prescriptions for their own researches. In other words, scientific methodologists are not particularly scientific in their own approaches. Why then should a researcher choose one methodological approach as opposed to another? Because most other people do? Because its easiest?

In fact, the researcher must choose the most rigourous methodology available, for the current research, or lose scientific status by not so doing, and researchers should challenge the work of their peers on these grounds. Has the methodology been made explicit, and would it have been possible to use a superior approach? This differs from Caldwell's (1982) approach, in that Caldwell proposes how existing research programmes can be considered in the light of different methodologies, whereas what is suggested here is that there is a pluralistic approach to determining one suitable method for a particular study. There will doubtless be arguments as to which approach is superior to which, that some approaches are not suitable for some purposes etc. This is perfectly reasonable, provided that these arguments are explicitly expounded, allowing others to see exactly how the research was conducted. The important point is that the researcher can use any of the approaches, and is not limited by conditioning or dogma to force the research to fit a methodology which is unsuitable.

Undoubtedly, hypothetico deductive nomological method must rank highest of the approaches, on the basis that if controlled experiment is possible, then it is unequivocally possible to determine cause and effect. It is often argued that this is totally unsuitable for social science, but whilst this is true today, it is only true within our limited framework of knowledge. If knowledge of the brain became such that it were possible to determine the physical and chemical processes caused by certain sounds or actions then controlled experiment would lead to scientific laws of behaviour being formulated, as opposed to trends of behaviour.

If controlled experiment is not possible, then the researcher must consider hypothetico deductive probablistic method. The hypothesis should be formulated, and fresh data collected with which to test it. In this case, research has entered the realms of model building, and consequently results must then be ascribed probabilities. A recent trend has been for some weather forecasters to attach probabilities to their forecasts of rain, whereas they would previously forecast either rain or not. The newer approach is much more sensible, and the status of weather forecasters has improved as a consequence.

What if it is not reasonably possible to collect fresh data, for example, because a large enough sample size would take too long? In this case, hypothetico inductive probablistic method has to be used, and the researcher must be honest about doing so. This is much better than fooling oneself, and possibly others, that hypothetico deductive probablistic method has been used, when in fact, the sample sizes were really too small to do so, or the same data which was used to formulate the hypothesis was used to carry out the testing.

If none of the above are possible, then the researcher has to resort to argument as used in a legal case, or in putting forward political ideas. Again, it is surely better to be explicit and admit that the data is insufficient, or is of a qualitative nature, such that even hypothetico inductive probablistic method is not suitable. Attempts to overcome such situations have been made, particularly by the use of cost-benefit analysis, where qualitative variables are quantified. What then happens is that opposing protagonists build their models based on their assumptions, and interpret the results accordingly. This then means the arguing of the equivalent of a political or legal case, which might as well have been done without the aid of the models, which simply embodied the arguments of the researchers anyway. A good example of this is the argument between G. Stern (1976, 1978) and S. Abrahams et al (1978), regarding the Roskill Commission report on the siting of an airport. Putting forward arguments in the manner suggested, of course means that the debate is never clearly and completely won, but if it is the case that the reality does concern qualitative aspects, and is reasonably complex, then the building of a model will not magically resolve the situation.

Finally, it may be that there are certain areas of human behaviour which vary so much as to not be suitable for some generalised theory. Again, if this is the case it nust be accepted, and not squeezed into the category above, to gain that extra bit of status.

In conclusion, it must also be restated that the researcher should be explicit about methodology and should use the most rigorous possible. To argue as per a legal case when it is possible to formulate a hypothesis, and collect fresh data with which to test it, is to lose the status of being scientific. To argue as per a legal case, when it is not possible to do otherwise, is to be as scientific as it is possible to be under the circumstances.

REFERENCES

Abrahams, S., Flowerdew. A., & Smith, J., 1978, Correspondence on "Sosiping, or Sophistical Obfuscation of Self-Interest and Prejudice", Operational Research Quarterly, Vol. 29.

Blaug, M., 1980, The Methodology of Economics: Or How Economists Explain, Cambridge University Press, Cambridge.

Caldwell, B., 1982, Beyond Positivism, George Allen & Unwin, London.

Popper, K., 1959, The Logic of Scientific Discovery, Hutchinson, New York.

Popper, K., 1963, Conjectures and Refutations: The Growth of Scientific Knowledge, Routledge & Kegan Paul, London.

Stern, G., 1976, SOSIPing, or Sophistical Obfuscation of Self-Interest and Prejudice, Operational Research Quarterly, Vol. 27.

Stern, G., 1978, Clarification of Official Obfuscation Language?, Operational Research Quarterly, Vol. 29.

THE APPLICATION OF THE THEORY OF DIRECTED GRAPHS

IN THE SOCIAL SCIENCES

L.R. Foulds

School of Management Studies
University of Waikato
New Zealand

ABSTRACT

Network programming and more generally, the concepts of directed graphs (digraphs) have become a legitimate and very useful area of operational research (OR). As OR is being applied to more and more problems of society, it is apparent that digraph models and algorithms have the potential to be of great use in the social sciences. The purpose of this paper is to discuss some of the existing and possible uses of digraphs in the social sciences.

1. INTRODUCTION

As is well-known, operational research (OR) was first applied in military and then commercial settings. During the last three decades, there has been an increasing tendency to apply the OR approach to other problems of society, especially to what has become known as the social sciences. Coinciding with this more recent application area for OR, has been the increase in the power and utility of network programming as a part of OR methodology. As a network is a special case of what is known in combinatorics as a directed graph (digraph) it has been natural to attempt to use the theory and algorithms of digraphs to solve problems in the social sciences. These attempts have been quite fruitful and it is evident that digraphs have significant existing and potential use in societal problems.

The purpose of this paper is to discuss some of these uses. We begin with a brief introduction to the basic digraph concepts and then cite some useful literature on some of their social science applications. We then discuss some known applications in this area.

2. BASIC DIGRAPH CONCEPTS

A **digraph** is an ordered pair (V,A), where V is a finite, nonempty set, whose elements are called **points**, and A is a set of ordered pairs of distinct elements of V, whose elements are called **arcs**. From the definition of A, it is evident that neither: (i) **loops** of the form (v,v) where $v \in V$, nor **multiple arcs** of the form (u,v), (u,v), where $u,v \in V$, are allowed. However it is sometimes useful to allow such possibilities in modelling various phenomena in the social sciences. A structure which is similar to a digraph, except that loops are allowed is called a **pseudodigraph**, and if multiple arcs are allowed, it is called a **multidigraph**.

It is often useful to represent a digraph (V,A), in diagrammatic form with the points of V drawn as points in the plane and a directed line drawn joining point u to point v in the plane if and only if (u,v) is an arc of A.

A digraph (V,A) is said to be **symmetric** if (u,v)∈A if and only if (v,u)∈A. Thus any symmetric digraph has the property that its arcs occur only in oppositely-directed pairs between pairs of its points. In this case it is often convenient to replace each such pair: (u,v), and (v,u), by the undirected pair {u,v}. When this process is performed on any symmetric digraph, the resulting structure is termed a **graph**. Thus a **graph** is an order pair (V,E), where V is a finite, nonempty set whose elements are called **points** and E is a set of unordered pairs of distinct elements of V, whose elements are called **lines**. **Pseudographs**, and **multigraphs** can be defined analogously to their digraph counterparts.

The concepts that we have introduced all have great utility in analysing various societal problems. Throughout this paper we have adopted the notation and terminology of Robinson and Foulds (1980) where these and related concepts are developed and exploited. We now discuss some reports on the application of the above mentioned notions.

3. LITERATURE SURVEY

Due to limitations of space, only an extremely limited number of reports on social science applications of digraphs can be given. These should be taken as giving a mere flavour to this area and as providing signposts to finding descriptions of more specialised applications.

Harary et al. (1965) in writing the first book solely on digraphs, pointed out digraph applications in a wide variety of the social sciences, especially that of balance and clusterability. Roberts (1978,1979) has presented detailed studies of selected digraph concepts and their application to problems concerning: traffic flow, radio frequency assignment, indifference, measurement, seriation, independence, domination, balance theory and social inequities, social networks, data organisation, group decision-making, social choice, decision-making models, influence, power, and status. Busacker and Saaty (1965) introduced digraph models of: social hierarchy, communication patterns, sentence reconstruction, and disarmament. Hall (1971) has discussed digraph models for organisation structure, consistency of choice and urban studies. Finally, Robinson and Foulds (1980) have used digraphs to study: food webs, partial orderings, project management, tournaments, and classification.

4. DIGRAPHS IN THE SOCIAL SCIENCES

In this section we present a representative sample of the application of digraphs in the social sciences. Typically, a digraph or graph is used to model the relationship between a given set of objects. Each object is represented by a point and the relationship between various pairs of objects by one of:

(i)   an arc, if the relationship is ordered (e.g. **is taller than**), or
(ii)   a line, if the relationship is ordered (e.g. **is married to**).

Further the arcs or edges can be **signed** by assigning each of them one of two symbols, usually "+" and "-". These signs can be used to indicate a different type of relationship, such as : allied with/allied against, attracted to/repelled by, and so on.

4.1 Organisational Structure

Consider an organisation whose members and rankings are represented by the points and arcs respectively, of a digraph. This abstraction of the **corporate structure chart** will contain an arc (u,v), if and only if the individual represented by point v is inferior (within the organisation structure) to the individual represented by point u.

The social science concept of **social status** can be quantified with such a digraph. For most organisations involving human beings, the associated digraph is acyclic, as individuals cannot be their own superiors. (Cycles may occur in digraphs modelling the organisational structures of certain flocks of birds, which often have circular pecking orders). The level of subordinacy can be established for any individual, i say, in the organisation by calculating the length (in terms in the digraph as the number of arcs) of the shortest path from the point representing the superior to the point representing i. A collection of points with the same level of subordinacy represents a peer group of individuals. It is possible to quantify the **status** of an individual by defining a measure based on the number of subordinates the individual has at each possible level. Efficient shortest path algorithms for calculating the required paths are well-known in the theory of digraphs.

4.2 Social Hierarchy

Digraphs are often used to model kinship in anthropological groups. As an example consider a group of individuals who are represented by points in a digraph D_1 which contains arc (u,v), if and only if the individual represented by point u is a parent of the individual represented by point v. Suppose that a second digraph D_2 is similarly constructed, which models the parent-child relationships of individuals which are children in D_1. In D_2, all possible relationships involving the children of these individuals are modelled. The product of the adjacency matrices of D_1 and D_2 is the adjacency matrix of a further digraph which models the grandparent-grandchild relationships between these individuals.

4.3 Consistency of Choice

In many social science experiments, one is asked to rank a number of given objects by pairwise comparison, especially when numerical measurement is difficult. For instance, a dog may be presented with each different pair of dog foods from a set of n foods, one pair per day. The food out of the pair that the dog selects first is noted. The object is to rank the n foods in order of preference. The outcomes of each day's selection can be modelled as a tournament digraph D, in which each food is represented by a point and an arc (u,v) is present if and only if food u is preferred to food v. If D is acyclic, a clear order of preference can be established. If D is cyclic, digraph techniques involving ranking with the minimum number of violations can be employed to establish the best order according to certain optimality criteria.

4.4 Game Theory

Game theory has developed into an important OR tool as it can be used to find the best way to perform a set of tasks in a competitive environment. The concept of a **kernel** of a digraph is useful in finding a wining strategy. A set of points K, in a digraph D, is termed a **kernel** if: (i) no two points in K are joined by an arc of D, and (ii) every point v, of D, not in K, is joined by an arc vk, to some point k∈K. Kernels have been used by von Neumann and Morgenstern (1944) in the theory of games to define winning strategies. **Game digraphs,** with points representing game positions, and arcs representing effective preferences, provide a useful approach, if not a complete analysis for, two-person, perfect information, deterministic, finite games. As such, they have been applied to analyse bargaining, voting, market forces, oil cartels, deterrence, and disarmament.

Each game digraph has a unique **starting point,** representing the opening position, and at least one **closing point,** representing a position at which the game is terminated. The closing points are classified according to the outcome that they represent, such as: **player A wins,** **player B wins,** or a **draw.** Each player tries to find a path from the starting point to a closing point representing a win for that player. What is of interest is to establish, once the game has reached a certain point, whether or not either player can force a win.

We assume that player A moves first and that the player who makes the last move wins. The following theorems are relevant.

Theorem

Every acyclic digraph has a unique kernel.

Theorem. If the starting point is not in the kernel K, of a game digraph D, then player A can force a win in the game represented by D by always selecting points in K.

4.5 The Directed Chinese Postman's Problem

Consider a letter carrier who begins delivering mail in a network of one-way streets, starting from a post office. The carrier must traverse each street at least once (in the correct direction) and finally return to the post office, travelling the least possible distance. Models of this scenario application abound, including: refuse collection, milk delivery, inspection of power, telephone, or railway lines, the spraying of salt on roads, office block cleaning, security guard and snow plough routing, and even museum touring.

A relevant optimisation problem can be formulated as a digraph model with an arc-weighted digraph D, with its points, arcs, and weights, representing the relevant intersections, streets, and distances of the network.

A cycle in a digraph is termed **Euler** if it contains each arc of the digraph exactly once. Clearly, the problem is equivalent to finding an Euler cycle in D. If D contains such a cycle it is optimal. There exists an efficient (polynomially-time bounded) algorithm for identifying such a tour in a graph which is due to Fleury, as reported by Kaufmann (1967). It can easily be extended to digraphs. If D is not Euler, some arcs will have to be covered more than once. An efficient algorithm for this case has been presented by Edmonds and Johnson (1973). Basically,

the algorithm adds duplicated arcs of minimal weight so as to make D
Euler and then uses Fleury's rule.

4.6 The Location of Centres

Consider an urban network of one-way streets on which a number of
emergency centres, such as hospitals, police and fire stations are to be
located (not necessarily at the nodes). The optimality criterion is the
distance of the furthest node to a centre. There is a second, related
problem. For a given critical distance d, locate the smallest possible
number of centres so that all nodes lie within d from at least one
centre. These problems can be formulated using a digraph D as for the
Chinese postman's problem. The problems involve a set of points, some
existing in D, and possibly some new points inserted into the arcs of D,
which satisfy the above criteria. Christofides (1973) has called these
new points **absolute p- centres**, and has devised an efficient algorithm
which can solve either problem.

5. CONCLUSIONS AND SUMMARY

For many decades, the theory of digraphs has provided useful tools
for areas of the natural sciences where sometimes other branches of
mathematics and OR have not been of much utility. The success of
digraphs can, in part, be attributed to the special physical structures
present in the problems analysed. For a lesser time it has been apparent
that the same digraph modelling approach can be applied with fruitful
results in the less structured social sciences.

One of the major contributions of digraph theory has been the
provision of a precise, mathematically-based language and concepts by
which the social scientist can formulate, hypothesise, conjecture,
communicate, and analyse societal problems. What has not happened, to
any great extent, is the gaining of major breakthroughs in social science
by the application of the mathematical theory of digraphs.

However as more and more social scientists use the language and
techniques of digraphs they will be able to analyse their problems and
develop new theories more precisely than before. Conversely, the just
mentioned endeavours have lead to many important results in digraph
theory and to new unsolved theoretical digraph problems.

In this paper we have defined the concept of a digraph. We have
given a brief survey of some of the major reports of digraph applications
in the social sciences. We have outlined a representative sample of some
of these applications. Finally, we have come to the conclusion that, so
far, digraph theory has contributed little more than a precise language
for the social scientist. However there is an excellent prospect that
various problems in the social sciences, when expressed in digraph terms,
will stimulate new theoretical results which can be applied to provide
new insights into these societal problems.

REFERENCES

1. Busacker, R.G., and Saaty, T.L., 1965, "Finite Graphs and Networks:
   An Introduction With Applications", McGraw-Hill, New York,
   pp 209-214.

2. Christofides, N., 1975, "Graph Theory: An Algorithmic Approach",
   Academic, New York.

3. Edmonds, J., and Johnson, E., 1973, "Matching, Euler tours and the Chinese Postman", Math. Programming, 5:88-124.

4. Hall, C.W., 1971, "Applied Graph Theory", Wiley-Interscience, New York, Chapter 9.

5. Harary, F., Norman, R.Z., and Cartwright, D., 1965, "Structural Models: An Introduction to the Theory of Directed Graphs", Wiley, New York.

6. Kaufmann, A., 1967, "Graphs, dynamic programming and finite games", Academic, New York.

7. Roberts, F., 1978, "Graph Theory and its Applications to Problems of Society", Regional Conference Series in Applied Mathematics, Society for Industrial and Applied Mathematics.

8. Roberts, F., 1979, "Graph Theory and the Social Sciences", in "Applications of Graph Theory", R.J. Wilson and L.W. Beineke, editors, Academic, London, Chapter 9.

9. Robinson, D.F., and Foulds, L.R., 1980, "Digraphs: Theory and Techniques", Gordon and Breach, London.

10. von Neumann, J., and Morgenstern, O., 1944, "Theory of Games and Economic Behaviour", Princeton U.P., Princeton.

APPLYING THE DATA DICTIONARY CONCEPT

TO MODELS OF SOCIAL AND INDUSTRIAL SYSTEMS

Frederick P. Wheeler

Management Centre
University of Bradford
Bradford, BD9 4JL, U.K.

INTRODUCTION

The central idea in the present work is that of a data dictionary. Usually a data dictionary is thought of as a tool for documenting the development and maintenance of a database system. Here the concept of data dictionary is being proposed for the development of a simulation model. Our contribution is to suggest specific classes of dictionary entry which reflect the requirements of a continuous-time simulation and to suggest how to use the dictionary in the development of models of social or industrial systems. It is our opinion that the OR analyst of today has, at her or his disposal, an increasing number of software products which allow the organization of text and other data, based on their internal relationships, as well as easy-to-use modelling software, all of which may be integrated and available on a personal computer. The effective use of such tools requires a methodology which is independent of specific implementations. The present paper addresses this requirement.

From the perspective adopted here, the behaviour of an organization is seen to be largely determined by the interlinked results of human actions. We assume that these actions are the result of reproducible decision processes, so that essentially the same actions would be taken by an individual acting under similar initial conditions. We also assume that it is the task of a model to reproduce the outcomes of human actions. The model simulates human activity using computational processes which require input data flows and which generate output data flows. The inputs to these processes attempt to simulate the data actually used by individuals, bearing in mind that the data used are likely to be some subset of available data and may be out of date, filtered or corrupted (Simon, 1976). The outputs from these processes are the changes in observable variables of the system under study.

The roots of our approach are based in the methodology of System Dynamics (Forrester, 1961.) The approach is to model the behaviour of organizations by simulating the actions of decision-makers. Particular attention is paid to the influences which these actions have on the organization itself. Often, the causal linkages between actions form closed loops within the system so that the outcomes of early decisions feed into later decisions. These feedback loops dominate the qualitative behaviour of the system.

OBJECTS FOR MODELLING

We are interested in modelling evolving systems and in implementing the models on digital computers. This immediately affects our method of working because although the systems of interest evolve in continuous time, the models evolve by discrete computational iterations.

In such models each iteration corresponds to a step forward in time of the system. Although the magnitude of this timestep is under the control of the modeller, it is always finite. The effect of this is that the system is modelled as a sequence of 'still frames' in the same way as a cine film pictures a changing scene. In any frame it is only possible to record what is observable at that instant. Observable quantities include the state variables of the system. This means that observers must compare successive frames in order to make inferences about quantities that are not directly observable. Among the quantities which must be inferred in this way are the outcomes of human actions, since they are responsible for changes in the state of the system.

This brief overview of the domain shows the existence of three sets of of objects. We describe these as 'levels', 'policies' and 'increments' and discuss them in turn. First is the set of observable quantities which determine uniquely the state of the system. We adopt System Dynamics terminology and refer to these state variables as levels. The term 'level' implies that its values can be ranked in order on either a continuous or discrete scale. Certain observable variables cannot be ranked but can be classified. An example might be some variable, colour, let us say, whose domain is: (red, yellow, blue). Colour could be represented in the model by the disaggregation of all relevant levels into three sub-levels (i.e. red, yellow and blue sub-levels.) There is a practical limitation to the degree of disaggregation since the number of state variables is increased by a factor equal to the number of classes (i.e. three in the example of colour.) Disaggregation will not be discussed in detail in this paper.

The purpose of the model is what determines its policies. Morecroft (1988) makes this clear in his use of Seymour Papert's concept of 'microworld' to describe these models. Sometimes a model is constructed to test out hypotheses about policies, such as when exploring the mental constructs of individuals within an organization, and in such a case the hypothetical policies will be built into the model. The purpose of a model may be to understand the collective effect of the actions of the individuals in an organization so that policies should replicate individuals' actions, based on observations and interviews of them, in this case. When the purpose is to reproduce historical behaviour with a view to prediction, particularly in the case of macroscopic models of economic or social behaviour, it may be possible to achieve the required level of precision without reproducing individual human reasoning processes, but instead the desired outcomes may be generated by using laws of collective behaviour.

The role of all policies, so far as the model is concerned, is to generate changes of state. In other words, policies output the changes in levels that take place after each timestep. We call these changes 'increments' and we associate each policy with an increment, or a sub-set of increments in the case of policies which change the values of disaggregated levels. Thus, policies are simulated actions and increments are the simulated outcomes of these actions.

The objects described so far are unconnected components and it is necessary to have a means of connecting them. We do this by creating a 'master' process within which all policies are sub-processes. The master process updates levels by calling on policies to generate the increments needed for updating. In other words, the data inflow to the master process consists essentially of the old values of levels, whereas the data outflow consists of their new values.

With large models it may be sensible to break up the master process into 'sectors.' The separation into sectors makes sense from the point of view of model development, since the development effort can be broken into manageable chunks in this way. Sectors can be treated as separate processes that are all at the same level of the process hierarchy, namely above the level of policies. Although the dividing lines between sectors are arbitrary, they are drawn to simplify the modelling task. Thus sector boundaries are chosen such that sectors are largely self-contained and this implies that there is a minimal number of physical and informational flows across any sectorial boundary when this is mapped onto the system. The sectors are specified in a 'master-summary.'

We have not yet defined parameters or functions; these are required for all practical modelling. (Certain functions are intrinsic to many programming and simulation languages and do not concern us here because they do not need to be developed by the modeller). We will find it convenient to group parameters and functions together as similar objects, since we will assume that functions can be specified by a table of fixed values and a look-up procedure defined in terms of the independent variables, whereas parameters are single fixed values. Thus, functions will be referred to as 'tables' and parameters will be called 'constants,' in general. However, the definition of policies is simplified if one distinguishes parameters whose dimensions involve time (such as an average delivery delay.) These will be called 'time-dependants,' since their values are measured in units of the model's timestep, and their values will change as the timestep is changed.

Finally we should mention exogenous time series. It is easy to envisage models which are not entirely self-contained so that data may need to be supplied in order to simulate the external environment of the system, or sub-system, under study. There is no need to introduce new objects to represent these exogenous data. The data in tables, constants and time-dependants are supplied exogenously so these objects already provide the necessary channel for exogenous data input. Thus an exogenous time series can be supplied in a table whose look-up procedure is to select values sequentially as determined by the elapsed time of the model.

THE DICTIONARY

The dictionary is a repository of definitions for the objects which together comprise the model in its current state of development. These definitions can differ considerably depending upon the type of object being defined. Some objects, such as constants, can be defined very briefly while others, such as policies, require lengthier definitions. Figure 1 indicates suitable formats of dictionary entries for the different objects.

Dictionary Entries

Master Process.   In Figure 1, the master process is shown separated into a summary and sectors. The summary of the master process is a list of its sectors, while each sector definition is a list of its levels and the policies which cause them to be incremented. Sometimes there is the need to multiply an increment by a conversion constant, for example to retain dimensional consistency, and this is also indicated as necessary.

Fig. 1.   Formats for dictionary entries.

Key:-   Dummy names appear <u>underlined</u>; suggested reserved
words of structured English are in CAPITALS; optional items
are in <...> and repeated items are in [...]

```
MASTER PROCESS:
SUMMARY
                Model Name:        model
                Model Status:      version #.#
                Description:       <[introductory comments]>

                [sector     version #]
;
[
SECTOR
                Sector Name:    sector
                Status:         version #
                Process:        <[comment]>

                [level      [INCREASE_BY policy <CONVERT increment USING constant>]
                            [DECREASE_BY policy <CONVERT increment USING constant>]
                            <comment>]
;].
[
POLICY      Policy Name:    policy
            Purpose:        TO_SET increment
            Description:    <[... details of the procedure ...]>
.]
[
INCREMENT Inc. Name:        increment
            Composition:    fields in data record
            Values:         allowable values
            Dimensions:     units of measure
.]
[
LEVEL       Level Name:     level
            Composition:    pattern of fields for records in data store, e.g:
                            [lag, level-value|[sub-level-value]]
            Values:         allowable values
            Dimensions:     units of measure
.]
[
TABLE       Table Name:     table
            Composition:    pattern of fields in data store, e.g.:
                            [dependent_value,[independent_value]]
            Look-up:        notes to explain interpolation procedure
.]
[
TIME_DEPENDANT
            Name:           time-dependant
            Value:          allowable values
            Dimensions:     units of measure
            Time-Scaling:   proportional | inverse | other power
.]
[
CONSTANT  Name:             constant
            Value:          allowable values
            Dimensions:     units of measure
.]
```

Policy. Each policy has as its purpose to set the next value of an increment and the procedure that determines the size of this increment is described in the policy definition.

Increment. The definition of an increment lists its allowed range, its dimensions (i.e. its units of measure) and its composition. Increments are data flows, whose composition needs to be specified, in order to allow for the possibility of sub-increments to disaggregated levels.

Level. Levels are similarly defined and also require their composition to be specified. Levels are data stores and they will hold values for sub-levels in the case of disaggregation. It is also possible that the past values of levels may be required. Each past value has associated with it a lag, this is an integer which identifies the number of timesteps before the current timestep to which the past value refers. Both the lag and the corresponding values need to be recorded in the data store of the level.

Table. Tables are fixed data stores composed of arrays that relate values of a dependent variable to one or more independent variables. The way to interpolate the table, and also the procedure to follow when the independent variable values are outside the domain of the table, both need to be specified in a look-up procedure.

Time-Dependant and Constant. The dimensions and allowable values of exogenously fixed parameters are recorded in their definitions and the time-scaling of time-dependant parameters is also noted. For example, a business simulation may use an internal timestep corresponding to half a week of real time. In such a model, a real delivery delay of four weeks corresponds to eight timesteps, so the delivery-delay parameter scales inversely to the timestep.

Developing the Dictionary

The dictionary is in a state of flux as the model is being developed. Definitions are constantly added, revised and expanded as more knowledge of the system is included. It is a document that lends itself to structured model development, particularly in relation to policies, which are best developed in a structured manner. A typical development might proceed by sketching the sectors of an organization into the master-summary. At this early stage the master-summary would include little more than a title for the model, a reference to its status (e.g. version 0.1), a brief description, in the form of comments on the purpose of the model, and a list of the planned sectors to be developed. One such sector, in a manufacturing industry, might define the flow of materials. Investigation of the plant would reveal the existence of stocks of goods for despatch, work in process, raw materials in stock, raw materials on order, and so on. These would be identified as levels in the material flow. Each level would be named and its definition would be added to the dictionary. Next, the mechanisms which alter these levels would be identified. One such could be the raw materials purchasing policy. A definition for purchasing would be added to the dictionary, but at this stage it would be incomplete, stating only its purpose, namely, to set a value to materials purchased. This, in turn, would require a definition of the increment, materials-purchased, and this also would be added to the dictionary. Further enquiries would attempt to determine how materials are ordered and what information is used when purchase orders are placed. This purchasing policy would be investigated and statements about it might be obtained from interviews with purchasing staff. This information would be recorded under the description of the policy and would appear as unstructured text. Later, this text would be made more precise and ultimately it would be expressed in structured English (Caine and Gordon, 1975.) The policy's input requirements, in the form of the values of one or more levels, constants, time-dependants or tables, would become clear during this process of refining its description.

The dictionary that would result is the logical definition of the model and is clearly distinguished from the details of any particular implementation. This makes the dictionary valuable because it allows development of a logical model to proceed independently of the implementation. The decision on implementation will then rest with technical considerations such as the use of simulation software versus a programming language, the matching of model size to computational resources, and so on, which are incidental to the purpose of the simulation.

DISCUSSION

Our view of social and industrial systems is that their behaviour is determined by the interlinked results of human actions. It is evident that maps of the linkages between actions in social systems can rapidly lead to complex structures (Eden, 1988). When such structures are the basis of computer models, serious questions of model validity must be answered if faith is to be placed in the results of simulations. A systematic approach to model development has been proposed here which keeps a running documentation of the model in development, removes redundancy and separates the logical simulation from its implementation. In sum, the present approach applies existing ideas about software development (Yau and Tsai, 1986) and, in line with emerging environments for the development of software (Waters, 1985, Martin, 1985), focuses effort on problem definition rather than program development.

An additional benefit of the structured design of the documentation is that it can form the basis of a dynamic document, and can utilize current developments in interactive text processing (Meyrowitz and van Dam, 1982) for interrogating, viewing and editing the model.

REFERENCES

Caine, S.H. and Gordon, K.E., 1975, PDL - A tool for software design. AFIPS National Computer Conference Proceedings, 44:271-276

Eden, C., 1988, Cognitive Mapping, Eur. J. Opl Res., 36:1-13.

Forrester, J., 1961, "Industrial Dynamics," MIT Press, Cambridge, MA.

Martin, J., 1985, "Fourth Generation Languages," Prentice Hall, Englewood Cliffs, NJ.

Meyrowitz, N., and van Dam, A., 1982, Interactive editing systems, ACM Comput. Surv., 14:321-415.

Morecroft, J.D.W., 1988, System Dynamics and microworlds for policymakers, Eur. J. Opl Res., 35:301-320.

Waters, R C., 1985, The programmer's apprentice: A session with KBEmacs, IEEE Trans. Software Engng, SE-11:1296-1320.

Yau, S.S. and Tsai, J.J.-P., 1986, A survey of software design strategies, IEEE Trans. Software Engng, SE-12:713-731.

PARTICIPATION, PRODUCTIVITY, AND STABILITY:

EXPERIMENTS ON FAIR DEMOCRATIC INCOME POLICY DECISIONS

Norman Frohlich

Department of Public
Policy
University of Manitoba
Winnipeg, Canada

Joe A. Oppenheimer

Dept. of Government and
Politics
University of Maryland
College Park, Md. USA

INTRODUCTION

We analyze results from laboratory experiments which bear on the relationship between participation in redistributional decisions and subsequent attitudes and behavior. In particular, we examine the impact of democratic participation on the subsequent acceptability of a redistributive principle and on the productivity of participants. We believe the experiments have implications for the role of worker participation in economic enterprises and for the relationship between income distribution policies, economic incentives, and political processes.

The theoretical inspiration for the experiments is drawn from John Rawls who, in his A Theory of Justice, (1971), attacked (among many other problems) the question of a "fair" rule for distributing income in any society. In particular, he concluded that maximizing the primary goods of the worst off member of a society (the difference principle) was the fairest principle of distributive justice.

Rawls described a set of conditions under which he believed the difference principle would be chosen as the fairest principle for distributing income. Fundamental to those conditions was the notion of impartial reasoning. He argued that "fairness" could be evoked when one made rational decisions without knowing one's own self-interest. Along with a variety of other arguments, he sketched a situation (an original position behind a "veil of ignorance") in which individuals were to decide upon a principle for economic distribution without knowing which share they would receive. The analogy was to the familiar rule for dividing a piece of cake between two individuals. One cuts; the other chooses. In the cake cutting problem the divider can be presumed to be concerned about being as fair as possible. In Rawls' original position, ignorance regarding which share one would get was used to induce.

Although Rawls argued the case abstractly and with regard to a highly stylized, hypothetical, and idealized situation, his conclusions are implicitly based upon empirical assumptions. Recent experiments [Frohlich, Oppenheimer and Eavey (1987a, 1987b) and Lissowski et al. (1988)] explicitly test the validity of some of Rawls' empirical assumptions.

They tested whether the difference principle would indeed be chosen under approximations of the conditions Rawls specified. Those experiments showed that it was possible to obtain consensus on a single principle of distributive justice as the "most fair" using the methods Rawls outlined. But they also demonstrated an utter lack of support for the difference principle. Instead, groups generally chose a "mixed principle". The groups set a floor below which the worst off individual in the group would not be allowed to fall. Given this constraint, however, the remaining incentives were set to maximize production and hence maximize average income.

Subsequent experiments [Frohlich and Oppenheimer (1988)] showed that support for the group's chosen principle exhibited a fair degree of stability when producers in an ongoing production setting experienced its effects. Nor did the acceptance of the principle seem to have a detrimental impact on production despite the fact that the taxes and transfers were - at times - sizeable. Production rose and both taxpayers and recipients increased their output. The redistribution to which the subjects agreed behind the veil of ignorance did not appear to dampen incentives.

In those experiments, subjects freely chose the taxation system under which they were to work earn and be taxed (benefitted). Here we explore and contrast the effects of a similar regimen when the redistribution and taxation system is imposed by the experimenter.

Specifically, we examine whether individuals who are taxed (and commonly thought to feel imposed upon) have their incentives to work diminished. We also consider the potentially ameliorating impact of democratic participation in the choice of the redistribution and taxation scheme upon this negative tendency. In particular, we address three questions:

1. Does experience with redistribution and taxation affect acceptance of the principle of redistribution?

2. Does experience with the redistribution and taxation system affect productivity?

3. Does participation in choosing the redistribution and taxation system affect the answers to the two preceding questions?

EXPERIMENTAL DESIGN

Individuals were familiarized with principles for redistributing income and the implications of their realization for the actual distribution of income. They knew they would have to earn their income in a subsequent part of the experiments. However, since the nature of the task to be performed was not made known to the subjects prior to their decision, subjects were effectively unable to estimate their likely future productivity and economic status in the production economy. They were never permitted to know what position they would hold in the subsequent income distribution, and in that sense they were behind a veil of ignorance.

For purposes of this paper two main treatment groups were established. In one, the subjects had to decide, after discussion, by a secret ballot, on principles and policies of income distribution and taxation to govern their future income. Two variation of the choice procedure was used: unanimity and majority rule. For the purposes of this paper, however, these two conditions are collapsed into a single category representing participation in choice of the principle. The second treatment group was not allowed to choose the principle and associated taxation scheme. It was imposed by the experimenters. The distributive policy which was imposed was maximizing the average income with a floor constraint, implemented via a proportional tax. Taxes needed to raise individuals above the floor income were assessed proportionally against he earnings of those who earned more than the floor income. This tax rule was suggested by the experimenters and could have been modified by the subjects. In practice, it wasn't. The actual floor imposed ($9900) was the average of the floors chosen in the experiments cited above. Nineteen groups of five individuals constituted the first treatment group while the second consisted of ten groups of five. All were students at the University of Maryland.

After the governing policies were in place, subjects were assigned a task. The task was the correction of spelling errors in excerpted (and altered) texts of Talcott Parsons. Each excerpt contained about 20 spelling errors. Each subject received the same text, and received wages for his/her individual production. The marginal pay rate had increasing returns to scale.

Individuals' outputs were checked, their earnings, taxes, and take home pay were calculated and reported to them along with the equivalent yearly income flows implied by the earnings. The process was repeated three times in each experiment. Measurements of subjects' preference rankings of principles, and degree of certainty regarding the ranking of principles were administered at each stage of the experiments. This latter measure was designed to elicit the degree of conviction subjects had regarding their rankings to see whether their experience increased or undermined their conviction. They were asked:

"How do you feel about your ranking of these principles?

Very unsure; unsure; no opinion; sure; very sure."

Weights of 1 - 5 were assigned to the categories. The scores for their answer directly after the group choice (or imposition) were compared with the scores after each task and redistribution period to identify the impact of the production period on their conviction.

EXPERIMENTAL FINDINGS

Stability of Preferences

One of our concerns was the stability of individuals' acceptance of the taxation principle after economic experience with redistribution. An examination of first place rankings of principles reveals that there is a great deal of stability of first place rankings. Among the 138 subjects who recorded first place rankings of principles both before production began and after the third period, there were only 21 gross changes in first place rankings. Moreover, considerable stability was evident in both treatment groups. The gross number of changes in first place rankings among all principles was 12 for the 94 subjects in the choice experiments and 9 in the imposed rule experiments. Net, there were only 4 changes in first place rankings in the choice experiments and 8 in the imposed rule experiments: indicating marginally greater stability in the choice group.

The differential effect of participation in choice of principle versus non-participation was evident in the initial degrees of certainty exhibited by subjects in the two groups. For those who voted on the rules to govern production, their original degree of certainty regarding their rankings of principles was 3.988 (n = 94),for those who had the rule imposed their initial certaintyl was significantly lower: 3.460 (n = 50). The t statistic for this difference is 2.982 significant at a p of .003. Thus discussion and participation had a positive affect on subjects confidence in their rankings of principles. And the production process reinforced this tendency. For the subjects as a whole, the production experience had a positive impact upon the certainty of their convictions. They were more secure about their preferences after production and redistribution than they were before that experience. Security of conviction rose from 3.764 at the start of production to 3.984 at the end of the third period for an F score of 6.510 significant at a probability level of .012 with an n of 123. In the choice experiments, certainty rose from 3.932 before production to 4.189 at the end of production for an F value of 7.105, significant at a probability level of 0.009 (n = 74). In the group which did not participate in choosing the principle, the effect is evident but the increase is not significant. In that group the rise was from 3.510 to 3.673: an F value of 1.033 significant only at the 0.315 level (n = 49).

Impact of Production and Redistribution on Productivity

We also were concerned about the effect of the taxation / redistribution policy on productivity and whether any such effects would be a function of participation in the taxation decision.

A direct measure of productivity was available: the number of spelling errors correctly identified in each production phase. Experience with the task and redistribution INCREASED productivity for the group as a whole. The number of errors found rose on average from 5.886 in the first period to 6.711 in the third: an F statistic of 7.445 significant at the 0.007 with an n of 114. But again there was a difference between the groups which chose their own principle and those upon whom the experimenters imposed a taxation scheme. In the former group average productivity rose from 6.031 to 7.250 for an F of 8.330 significant at the 0.005 level for an n of 64. In the imposed experiments the story was different. Subjects in those experiments experienced only a marginal and statistically insignificant increase in productivity: from 5.700 to 6.020 for an F of 0.577 significant only at the 0.451 level (n = 50). Whatever effect experience with the task in question may have had on increased productivity, it was virtually wiped out by working under an imposed regime. Conversely, it was evident under a regime of choice. Thus, participation in a group discussion and choice of the appropriate taxation scheme for the group may actually have enhanced group productivity. Moreover, the effect was not uniform even on subjects within the treatment groups.

We divided the data into high producers (who as high income earners had to pay taxes) and low producers (who as low income earners received transfers). Most strikingly, only in the choice experiments did the transfer recipients account for a disproportionate amount of the increased productivity. When the group discussed and chose, poor producers improved; otherwise they didn't. Thus, in choice experiment, transfer recipients increased their production from 12.3% to 16.2% of the total between the first and third production periods [with an F of 4.422, significant at the .042 level]. The comparable analysis of taxpayers and recipients in the imposed rule experiments shows a very different picture. Under imposed rules, the recipients decreased their relative share of production while taxpayers actually increased their burden. The percent of group production transfer recipients accounted for actually FELL (from 12.7% to 10.5%) in the imposed experiments! But this trend was not statistically significant.

DISCUSSION

The results of these experiments offer tentative answers to the questions posed above. Subjects are, in general, more confident of their rankings of the principles after experiencing production. Such confidence is reinforced by participation in the choice of policies.

As to the impact of participation in discussion and choice on productivity, here again it appears to have a positive affect. Productivity increases, but differentially in the two treatment groups. Where discussion and choice takes place there is a strong and clear positive gain in productivity over the course of the experiments. This increase is disproportionately by the poor and thus benefits the taxpayers. In the imposed experiments no such significant rise in productivity occurs. Participation in discussions of taxation and input into the choice of a rule to govern redistribution has a clear and positive impact on subsequent productivity: especially for the poor. Participatory democracy (at least in an experimental setting) aids productivity. Rather than acting as a disincentive, the taxation and redistribution system which guarantees a floor when agreed to in a participatory fashion seems to act as a production incentive to those at the bottom of the income distribution.

In a short paper we cannot expand, at length upon our analysis and interpretation but a few caveats and comments are in order. The laboratory setting required a compression of time, income, and stakes. Subjects were University students. Perhaps the positive impact on productivity of participation in the decision process stems from group solidarity created during the discussion. But even if the gains were due to this "esprit du corps" rather than the combined effects of the principle and participation in the selection process, it still would seem that meaningful participation in a democratic process can have beneficial implications both for economic production and satisfaction with the system. Moreover, these results are consistent with the emerging arguments in labor relations that participation at the shop floor level can have a positive impact on productivity [Dolan and Schuler, (1987), and Theirs, (1987)]. Generalized experiments to that point might shed light on the exact nature of the operative variables in those effects.

REFERENCES

Dolan, S. L., and Schuler, R. S., 1987, "Personnel and Human Resource Management in Canada," West, St. Paul, Minnesota.

Frohlich, Norman, Oppenheimer Joe A., and Eavey, C., 1987a, Laboratory Results on Rawls' Principle of Distributive Justice, British Journal of Political Science 17:1 - 21.

Frohlich, Norman, Oppenheimer, Joe A., and Eavey, C., 1987b, Choices of Principles of Distributive Justice in Experimental Groups, American Journal of Political Science 31:606 - 636.

Frohlich, Norman, and Oppenheimer, Joe A., 1988, A Test of the Stability of a Political Conception of Justice in Experimental Groups. Mimeo.

Lissowski, Grzegorz, Okrasa, Wlodzimierz, and Tyszka, Tadeusz, 1988, Principles of Distributive Justice: Preferences of Polish and American Students, Mimeo, Warsaw.

Rawls, John, 1971, "A Theory of Justice," Harvard University Press, Cambridge.

Thiers, Richard J., 1987, Bradford-White: A Study in Employee Relations and Productivity, Personnel 64:74 - 77.

INSTITUTIONS AND INTELLIGENT SYSTEMS

Rosaria Conte

Istituto di Psicologia
Consiglio Nazionale delle Ricerche
Viale Marx 15, 00137 Rome (Italy)

INTRODUCTION

Within the structural-functionalist tradition, (comparatively formal) organizations like trade-unions or corporations, and institutions such as the Market, the Family etc. have been defined as "purposeful systems" (Barnard, 1938).

Thereafter, this view has been receiving a great deal of criticism, the most important of which are the following:
-ratiomorphism: apparently human characteristics (such as "goals" and "needs") are assigned to institutions by structural-functionalists;
-reification: institutions are treated as if they had a life of their own (for a discussion, see Donaldson, 1985).

However, three circumstances deserve special attention:
-some social scientists have recently argued for a collective understanding of order (Alexander and Giesen, 1987), thus favouring a revival of interest in defining institutions as well as other macroterms;
-system theory (ST) still claims that a systemic level of description is needed for collective actions to be explained, and that social systems are self-regulating systems;
-some literature on institutions describes them as doing cognitive activity (problem-solving, and the like).

Now, what is the meaning of these notions, and what is their relation to cognitive systems? In this paper, the question of ideal-type institutions as intelligent systems is addressed. It is to be noted that what will be referred to, here, as institutions is simply any legitimate and more or less organized group of actors sharing some goal(s).

The paper is organized as follows. First, the notions of self-regulating and problem-solving systems as used in the literature on institutions will be examined. Then, the requirements of a cognitive model of action will be presented. Finally, some answers are illustrated; the application of a cognitive model of action to the study of institutions is discussed.

INSTITUTIONAL SELF-REGULATION AND PROBLEM-SOLVING: ARE THEY COGNITIVE NOTIONS?

As stated above, ST, as well as the organization theory and more generally the literature on institutions, describe institutions as self-regulating (SR) systems, and sometimes as characterized by the doing of cognitive activities.

However, in ST, SR and goal-pursuing (Buckley, 1967) are cybernetic notions: goals are viewed as states-to-be-maintained and not as new-states-to-be-set-up. Action is then seen as a process of restoring those states, and not as a creative process.

In ST, social systems are not meant to modify reality, but to simply maintain a given level of integration and adaptation; consequently they are not thought of as cognitive systems, and if they are, this is due to a very poor notion of cognition: in Maturana and Varela (1980) a cognitive system is defined as a homeostatic one (for a detailed analysis, see Castelfranchi, 1985).

On the other hand, in the relevant literature institutions are often described as doing categorizing (Douglas, 1986), decision-making (Schotter, 1981), and problem-solving activities (Ullman-Margalit, 1977).

To clarify, let us focus on norms as problem-solving mechanisms (Ullman-Margalit, 1977). In this view, norms derive from interest conflicts among social actors and provide them with solutions such that the resulting overall net benefit is greater than what would be obtained if social actors acted in the absence of norms.

Sometimes, norms are simply customary, spontaneous solutions worked out by interacting social actors. Their origins are inscrutable since these norms do not derive from any deliberate human act. They point to the paradoxical phenomenon of "extramental" regulation: these norms impinge upon social actors, and yet no-one devised them. Also, this shows that social theory badly needs functional explanation, which now much displeases many social scientists.

In other cases, norms may be explicitly devised by a legislating system. This could be viewed as an intelligent system since:
-the outcome is a rational choice: norms result in a more rational state of the world;
-the outcome modifies reality: norms set up a new, hither-to unrealized state;
-the outcome has a regulatory power: norms regulate the actors' choices.

However, in the following it will be argued that these features prove unsatisfactory. Whilst they contribute to a seeming rationality of institutions, they do not allow for the claim that institutions are truly rational entities. Institutions cannot be said to be really intelligent unless their structure is looked into.

SIMPLE MACHINERY AND INTELLIGENT SYSTEMS

It is quite some time since Artificial Intelligence students began to debate about what is really needed for a machine to be defined as human-like, or at least intelligent (cf. the Dartmouth Conference, dated 1956). What begins now to be acknowledged is not only that machine outcomes and performances are not "good" criteria, but also that even complex, "planful" activities may be just a *seemingly* intelligent behaviour (cf. Agre and Chapman, 1987). In other terms, a system may behave in an adaptive, planful and effectful manner with no capacity for reasoning and planning. Thus, the following are not good criteria:

-effectiveness: simple machinery (like Agre and Chapman's PENGI) which exploits regularities in its interaction with the world may do well at performing certain tasks;
-complex and plan-like behaviour: PENGI engages in rather complex activities: however, it does not do any planning. Its moves are the direct result of the environment changes that it registers;
-flexible answers: PENGI's behaviour is highly flexible, and yet it does not reason about its actions. Its routines are triggered, stopped, and iterated as a consequence of registered changes and obstacles;
-acting on relevant aspects of the situation: PENGI perceives pragmatically relevant entities and events (features of the situations that may favour or damage it), without instantiating classes represented in its database. Thus, it cannot distinguish individual aspects: it only "knows" that-something-is-favouring/damaging-it-now. These relevant aspects select routines.

Thus, for a system to act and solve problems in a really intelligent way, one needs showing that it has:
-a representation of its goals (and not only a capacity for answering aspects of a situation);
-a model of the world, which allows for the representation of the future (and not only a registration of present aspects);
-a capacity for reasoning about actions and domain-knowledge (and not only a capacity for applying routines).

In the following, we will try to show how these features may be related to the study of institutions as intelligent systems.

INSTITUTIONS AS INTELLIGENT SYSTEMS? THREE POSSIBLE ANSWERS

Let us now get back to our central question: is it possible to say that collective ends are reached thanks to intelligent problem-solving performed by collective minds? Indeed, what is a collective mind, if it exists?

The Multi-Agent-Solution: A Negative Answer

One possibility is to deny that collective minds exist. Now, this answer comprises two different views:

1. one that what does exist is a plurality of interacting rational actors. These are capable of accepting immediate costs for future higher rewards, and can act cooperatively, if this increases the probability of obtaining a share of the rewards which compensates the cost of cooperation. In this view, institutions turn out to be either:
-shared intentions, plans, and beliefs, or
-collective intentions, plans and beliefs.
This is so, on one hand, for instrumental reasons: since social actors share goals, they may also have collective, e.g. cooperative, intentions. On the other, out of reciprocal adaptation: in social life, thanks to their day-to-day interaction, people "squeeze each other" into both having collective intentions and sharing intentions and beliefs (Douglas, 1986).

Now, this kind of answer seems to raise the following question: if institutions are not disengaged from individual minds, where do such things as norms and roles come from, how do they get into people's minds?. In the writer's opinion, this answer fails to acknowledge and to describe explicitly the role of institutions in regulating human minds. In addition, it fails to acknowledge and account for what was defined above as the extramental power of institutions, and possibly the functional mechanisms of their reproduction.

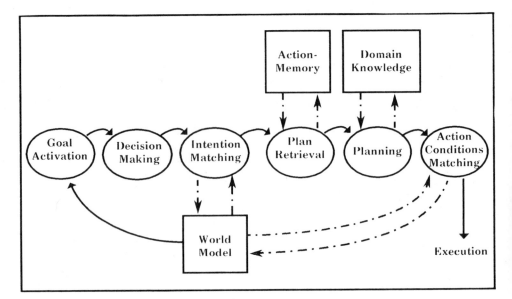

fig.1

2. A second view maintains that human minds do not provide exhaustive answers to the question of what an institution is and how it works; yet, it denies that such a thing as a collective mind exists. Basically, this view aims at solving the problem of what an institution is in terms of functional mechanisms. The advantage of this view is twofold: it disengages institutions from human minds and paves the way to answering the question of how norms, roles, and the like come to influence social actors, and at the same time gives an account of the paradox of extramental regulation. However, such a view is bound up with advances in functional theory, which is currently found wanting (but see Alexander and Colomy, 1985, for a "neo-functionalist" version of social theory).

A Weak Criterion for Collective Minds

Another possibility is that of saying: as long as institutions *act as if* they were intelligent, providing rational solutions and the like, they *are* intelligent. No matter what their real structure is, what counts is their outcome, and perhaps the emerging properties of social organizations (for an interesting as well as extreme version of emerging intelligence, see Steels, 1988).

Consequently, intelligent behaviour is the emerging property of, say, corporations, as well as of the Market: this is not far from the Hidden-Hand type of explanation. It is the emerging property of even lower-level animal organizations. This ratiomorphic view seems to conceal a vital question: what are the specific causes of different types of emerging seeming rationality? We would say that in the case of PENGI, the cause is its programmer's intelligence; in that of an ant-row, the "intelligence" of evolution; and in the case of market, a non biological functional mechanism.

A Strong Criterion for Collective Minds

A strong positive answer to our question states that collective minds do exist, but not all-institutions are such. More generally, not all supraindividual entities behaving in a rational-like manner can be defined as collective minds. Let us imagine a fictitious situation like the following: given a defined cognitive model of action (see fig.1), let us assume that:

-in situation X, final actions are carried out according to this model;
-each single agent, or group of agents, acts as a substep of the cognitive model, as a processing unit;
-none of the agents, included one acting as a central processing unit, if any, has a representation of each specific processing (see, Castelfranchi, 1985).
Could we say, in such a situation, that X is a collective mind? The answer seems controversial. Let us try to figure out what a discussion between a supporter and a sceptic would look like:

Sc: "Ok, you have a cognitive model of action carried out by a bunch of people. So what, it is just an example of multiagent cooperation."
Su: "Yes, but it is the procedure which is intelligent, and not only the guys involved in it."
Sc: "Well, yes, but a procedure is not more intelligent than a book. The agent who designed it is intelligent."
Su: "I don't care: human beings are designed by God, perhaps. Still, you say they are intelligent."
Sc: "Ok, but look: what people do is not in God's mind, it is in theirs. Now, what these X fellows do is also in someone else's mind: namely, the mind of the projector."
Su: "Wait: that is not true. The projector is not known of any specific problem-solving and decision-making. Just like God."
Sc: "I agree. But then I would say that we've got a plurality of cooperative agents. The organization being intelligent is just a metaphor, don't you see?"
Su: "It is not in the least more metaphorical than saying that some machines are intelligent. Listen, in your view, there are no intelligent machines, as long as they are programmed by someone."

To be sure, the discussion is an everlasting one. Two central points should be clarified, however:
-The relationship between "programming" (the institutional design) and the "machine running the programme" (the organization producing a specific output);
-The difference between a collective mind and a plurality of cooperative agents.

CONCLUSIONS

Two major points are stressed in this paper:
-Neither shared nor collective intentions and beliefs provide exhaustive answers to the question of what an institution is and how it works;
-if institutions are to be defined as intelligent systems, their internal structure and functioning are to be highlighted. A promising direction is to examine whether a cognitive model of problem-solving, decision-making, etc. could be applied to an institutional organization.

REFERENCES

Agre, P.E., and Chapman, D., 1987, Pengi: An implementation of a theory of activity, Sixth National Conference on Artificial Intelligence Proc., Morgan Kaufmann, Los Altos, 1:269.
Alexander, J. and Colomy, P., 1985, Towards neofunctionalism: Eisenstadt's change theory and symbolic interactionism, Sociologic.Th., 2:11.
Alexander, J.C., and Giesen, B., 1987, From reduction to linkage: The long view of the micro-macro debate, in: "The Micro-Macro Link," J.C. Alexander, B. Giesen, R. Muench and N.J. Smelser, eds., The California Press, Berkeley.
Barnard, C. I., 1938, "The Functions of the Executive," Harvard University Press, Cambridge.
Buckely, W., 1967, "Sociology and Moderns Systems Theory," Prentice Hall, Englewood Cliffs.

Castelfranchi, C., 1985, "Sulle Tracce della Mente," Technical Report Istituto di Psicologia del CNR, Rome.

Donaldson, L., 1985, "In Defence of Organization Theory. A Reply to the Critics," Cambridge University Press, Cambridge.

Douglas, M., 1986, "How Institutions Think," Routledge and Kegan Paul, London.

Maturana, H.R. and Varela F.J, 1980, "Autopoiesis and Cognition. The Realization of the Living," Reidel, Dordrecht.

Schotter, A., 1981, "The Economic Theory of Social Institutions," Cambridge University Press, Cambridge.

Steels, L., 1988, "A Reaction-Diffusion Model of Computation," VUB A.I. Lab. Memo, Brussels.

Ullman-Margalit, E., 1977, "The Emergence of Norms," Clarendon Press, Oxford.

LIMITATIONS OF MANAGERIAL COMMUNICATION RESEARCH:
NO SIGNIFICANTLY DIFFERENT METHODS EXPLORE STATUS
AND SEX OF MANAGERS

Suzanne Walfoort

Metropolitan State University
St. Paul, MN

This is an abbreviated report of a two phase research project. The goal of the first phase of this research was to describe and analyze relationships among status, sex, and use of communication control behaviors of corporate middle managers. The second phase explored the relationship of the findings to communication accuracy measures. The manager is the linchpin of the organization and as such his/her job is to communicate. Because of the centrality of the manager in communication networks, much can be learned from researching effective and efficient managerial communication action.

This study, however, merged interest in effective managerial communication with a second contemporary managerial trend: women in management. According to Blum and Smith (1988), over the last fifteen years the percentage of American women in executive, administrative and managerial occupations has doubled from approximately 18% to nearly 36%. Corporate management, therefore, offers a context in which women recently have attained enough higher status positions that a significant number of women managers now serve in positions superior to men in the organizational hierarchy. Beyond offering a rich context for the study of sex differences in managerial behavior and in effective management as well, management studies may also allow some unraveling of the variables of sex and status, variables almost hopelessly confounded in the American culture where men are in power, women almost always out of it.

Historically, the preponderance of studies of managers, and of sex differences, has explored perceptions such as: how managers are perceived to respond, to lead; how female and male superiors are perceived to behave or to communicate--differently or not; and how female managers are perceived in the business world. (For review of the literature, see Stogdill, 1981.) For each study in which sex differences were discovered, another produced no significant sex differences. But methods used to examine these perceptions reflect little imagination or diversity. What one finds is a litany of survey studies often not even using employees as subjects. Research on perceived differences through self and other report methods abounds, but the correspondence of such perceptions to behavior remains largely unexplored. Even within studies that do attempt to describe managerial behaviors are embedded validity problems. The same pattern holds in communication studies. Regardless of context, the research reflects the same over-reliance on student subjects, surveys, and perception studies. It is a dismal fact that in studies on managers and beyond, "there is a plethora of studies describing portions of what leaders and their subordinates say they do, could do or should do, but only a smattering of studies describing what they actually do..." (Lombardo & McCall in Husband, 1985, p. 105). From the sex differences in communication research end as well, many have called for but few have chosen field research to address this intriguing issue, leading researchers like Fisher to consider whether the findings of sex difference in communication remain mired mostly because of "the confounding of the research question with the observation of perceptions of communicative behavior rather than a direct observation of communication itself" (1983, p. 225).

Literature on superior/subordinate communication can boast more diverse methods than the research on sex differences and management can. Standard survey methods are still the norm, but two early exemplar superior-subordinate studies suggest the movement toward more behaviorally oriented research (Wager, 1972; Dansereau, Graen and Haga, 1975). These two studies, although not observational research, foretell the move by incorporating more than one perspective in assessing the impact of a superior-subordinate relationship within the chain of command.

Wager (1972) used surveys, but attempted through them to assess the amount of time managers behaved in certain roles during conferences, thus moving toward more objective measures of managerial behavior. The results of Wager's work will be discussed later, but one noteworthy design feature was the contrast of the managers' behavior in two contexts--in conferences they attend as subordinates and in conferences they attend as superiors. Dansereau, Graen and Haga (1975) also built in a contrast feature when they developed the "vertical dyad approach," interviewing actual superiors and their subordinates. Although their study also focused on perceptions, at least the method used interviews rather than surveys, allowed the employees to discuss their actual work relationships, and most importantly, included the goal of ascertaining the correspondence of perceptions between the superior and subordinates. They found subordinates divided into two camps: those who felt and were treated as an "in group" by the supervisor, receiving preferential treatment, and those who felt like and were considered "hired hands" for whom the relationship with the superior remained perfunctory. Unfortunately, no sex difference analysis was possible in either study because samples consisted exclusively of male managers.

The present study was designed as an innovative response to calls for departure from shopworn perceptual and self-report research investigations on both superior-subordinate (status) issues and sex difference issues. The hope was to account for observed communication patterns based on status differences as well as sex differences. So, it attempted to unravel the confounded variables of sex and status by studying them in one of the few available contexts in American culture in which women have power--in corporate management. This is, however, an abbreviated report of the study. For full description and analysis of the data, see Walfoort, 1987.

The design of the present study aimed to analyze managerial behavior on one aspect of superior-subordinate communication, specifically, in messages of control and instruction-giving. The control dimension was chosen as a typical and important aspect of the manager's job (Bartolome & Laurent, 1986), as a much considered issue affecting women in management, (Schein, 1982), and one that applies to both upward and downward managerial communication (Katz & Kahn, 1966). Giving instructions--both upward and downward--has been identified as an important managerial skill. The quality of this communication responsibility of managers has been judged by at least one expert to partially discriminate more productive from less productive managers (Redding, 1964). Human communication problems have great impact on production systems, and vice versa. In the volume resulting from the last conference on operational research and the social sciences, S.L. Cook (1966) claimed such misunderstandings were one of the greatest problems operations researchers meet in the course of work, and called for study of improved communication flow.

The research question in the present study revolved around sex and status differences in managers' communicating of control: Do female and male managers exert equal amounts of control in their communication with subordinates and their superiors? The operational definition of communication control used here was Putnam and Skerlock's referring to, "the communicative behaviors which restrict the type, direction, frequency, and amount of participation of the other person (1978, p. 6).

METHOD

Design

Seventy-two subjects, 36 males and females recruited from five large corporate organizations in the U.S. midwest agreed to participate in the study. The group consisted of 24 first-line managers, their superiors and subordinates. At each organizational level, males and females were equally represented.

A modified field experiment design was constructed as a methodology that hopefully capitalized on the merits of field and laboratory methods without compromising either. The design therefore incorporated three tiers of corporate status to allow comparisons of control communicated upward, downward, as well as between and within sex groupings. Status and relational authenticity seem particularly critical validity issues in communication research. Attempts were therefore made to overcome two major methodological problems referred to earlier, classic problems Weick (1965) labelled:
1. resemblence, an over-reliance on lab and attributional studies failing to resemble the managerial context, and
2. embeddedness, producing interaction that is not embedded in the existing relationships of the subjects.

Procedure

A direction-giving interaction exercise was employed which involved two trials, one a manager performed with an immediate superior and one with a subordinate. Essentially, in this exercise a designated "sender" must convey instructions to a receiver, a situation which typifies everyday managerial interaction. No explicit instructions were given, leaving the individuals to negotiate participation in this task. The order of trials was varied and no feedback allowed between trials. All trials were audiotaped and transcribed. Following the exercise, the participants completed self and other report questionnaires indicating perceived control during the exercise, hypothesis guessing, and their opinion of the typical communication style of the other person.

Coding

Frequencies were calculated on three operations frequently defined as interaction control mechanisms:
1. amount of one-way versus two-way communication—relative talk time (or control of the floor).
2. the number of statements and questions (and tag questions) employed.
3. the amount of interruption: successful, resisted, and attempted.
Occurrence of all except talk time were reported in occurrence per minute to control for the individuals' total talk time dictating all frequencies.

Teams of two trained coders analyzed the audiotapes and transcripts, but were checked for inter-coder reliability and inter-team reliability as well. Multiple methods thus allowed triangulated comparison of results among observed behavior, perceived (other) behavior and self-reported behavior, providing convergent validity measures.

RESULTS

As Table 1 shows, no pattern of significant sex differences was observed in the amount of communication control exerted by managers with subordinates or superiors. A pattern of significant differences did emerge, however, when the data were analyzed by status.

With the exception of questions, the operations indexing communication control reflected status differences. Results were not in the direction expected, however, in that managers overall directed more control messages toward their superiors than their subordinates. Concerning within group comparisons, examination of the variance in standard deviation terms presented in Table 1 does not reveal significant differences in the control behavior among female or male managers as superiors or subordinates.

Table 1. Communication Control Exerted by Managers (Displayed by Sex and Status of the Manager, in Frequencies Per Minute)

	Components				Interruptions		
	Statements	Questions	Tag Quest	Talk Time	Successful	Resisted	Attempted
BY SEX							
FEMALE MANAGERS							
M	9.95	.762	.582	.717	.309	.168	.158
SD	3.16	.513	.601	.171	.359	.178	.249
MALE MANAGERS							
M	9.03	.591	.377	.748	.272	.136	.139
SD	3.21	.531	.328	.128	.448	.279	.239
$p <$	.24	.76	.71	.81	.51	.36	.87
BY STATUS							
SUPERIORS							
M	10.91	.61	.42	.66	.43	.22	.23
SD	3.61	.58	.50	.15	.71	.37	.29
SUBORDINATES							
M	8.01	.74	.54	.80	.15	.08	.06
SD	2.51	.48	.5	.11	.32	.17	.16
$p <$	.002	.44	.30	.005	.03	.12	.03

n = 24 in each group

A total of 9,384 utterances were coded. Rough reliability estimates were reasonably high when calculated on inter-coder unit agreement. Acceptable levels of inter-coder reliability were set at 80%, on a simple percentage of agreement, and reached 83% for one set of coders; 85% for the second set of coders. As for inter-team reliability, with the exception of one tape, chi-squares performed on five randomly chosen tapes proved statistically significant at or beyond the .01 level. Concerning validity, in reference to how typically the manager was perceived to have behaved, both self and other reports reflected a relatively high degree of validity, 79% for the self-ratings of the managers, 83% for the receivers reporting the manager's communication behavior as typical.

DISCUSSION

In this case, no significant sex differences were detected in the amounts of communication control exerted by female and male managers. The relationship here between the two variables sex and status on communication control would more accurately be described as unfounded rather than confounded. The results point to the importance of examining status differences in sex difference studies.

One possible explanation for the unexpected status direction result is that in their eagerness to impress and perform successfully with their superiors, managers take charge. These results are consistent with Wager's 1972 study mentioned earlier which surveyed managers in their roles as both superiors and subordinates in conferences. Wager's expected result was that managers as superiors gave more information than they sought. The striking result was the degree to which managers as subordinates gave and sought information. Wager found the two major communication roles—information seeking and information giving—both more related to the subordinate than superior status. Yet questions are common information seeking strategies, and proved not to discriminate by status in the present study. Further research should aim to identify patterns of questions used in conversation to draw others out or up, to gain information, and to grant or gain control.

In terms of limitations, it must be registered that the present study was not true observational field research, and although the surveys done afterward did generally confirm control and communication patterns, the fact that self and other reports did not perfectly coincide must be noted. True too, coding is more an art than a science, and only gross measures of communication control were measured here.

There are several potential research outgrowths of this study. First, the study must be replicated in some fashion eliciting natural conversation from employees, over extended periods, using more sophisticated coding schema now being developed. Second, the nature of the dramatic status differences in control mechanisms should be pursued. Status analyses should be conducted where possible on the body of attributed sex differences this study's results dispute to reveal more concerning the nexus of status and sex. Finally, and most importantly, field study should be encouraged on applied communication research topics such as this one, for what is not heeded in most communication studies on managers is the common sense notion that to best learn about male/ female or superior/subordinate managerial communication, one must venture into the context of the business world and study individuals in those relationships.

Turning to methodology, what is most needed was prescribed years ago by Weick for organizational research in general. "We typically need multiple methods, or techniques which are imperfect in different ways" (Weick, 1969, p. 21).

PHASE II: PRELIMINARY ANALYSIS OF DATA

The second phase of the study explored the relationship of these findings to accuracy measures, since clear communication is a measure of managerial competence.   Groups of eight rated the resulting designs for accuracy.
Preliminary analysis of the accuracy measures on how successfully the managers communicated instructions to their subordinates and superiors revealed
high control correlated poorly with accuracy.   The situations in which managers exhibited high communication control yielded low accuracy in the product,
indicating that control of the conversation did not control the results of
it.   Low correlations of accuracy to control mechanisms ranged from a low for
interruption (.07 for resisting interruption and .10 for successful interruption) to highs of only .25 for control of the floor and .28 for number of
statements used.

This addresses the importance of feedback loops in any system.   According to Wager (1972), downward communication flow predominates in management.
It may, however, not be strictly the downward but also the one-way flow of
communication that causes substantial accuracy problems.   Perhaps managers
would reap production benefits by investing time through sharing talk time
while giving instructions.   And just as managers should consider investing
time in communicating, so should researchers consider investing time observing managers so doing.   This too could reap impressive production benefits.

References

Bartolome, F. and Laurent, A., 1986, The Manager: Master and Servant of
    Power, Harvard Bus. Rev., Nov.-Dec.:77-81.
Blum, L. and Smith, V., 1988, Women's Mobility in the Corporation: A
    Critique of the Politics of Optimism, Signs, 13:528-45.
Cook, S.L., 1966, Discussion and Commentary, in "Operational Research and the
    Social Sciences," J.R. Lawrence, ed., Tavistock Publ., London.
Dansereau, F., Graen, G., and Haga, W.J., 1975, A Vertical Dyad Linkage Ap-
    proach to Leadership with Formal Organizations: A Longitudinal Investi-
    gation of the Role Making Process, Org. Beh. and Human Perf., 13:46-78.
Fisher, B.A., 1983, Differential Effects of Sexual Composition and Interac-
    tional Context on Interaction Patterns in Dyads, Human Comm. Res.,
    9:225-38.
Husband, R.L., 1985, Toward a Grounded Typology of Organizational Leadership
    Behavior, Quar. J. of Speech, 71:103-18.
Katz, D. and Kahn, R.L., 1966, "The Social Psychology of Organizations," John
    Wiley and Sons, Inc., New York.
Putnam, L.L. and Skerlock, L., 1978, Verbal and Nonverbal Determinants of
    Dominance in Sex-Typed Female, and Androgynous Dyads, Paper presented at
    the International Communication Assn. Convention, Chicago, IL.
Redding, W.C., 1964, The Organizational Communicator," in "Business and
    Industrial Communication: A Sourcebook," G. Sanborn, ed., Harper and
    Row, New York.
Schein, V.E., 1982, Power and Political Behaviors in Organizations: A New
    Frontier for Research on Male/Female Differences, in "Women and the
    World of Work," A. Hoiberg, ed., Plenum Press, New York.
Stogdill, R., 1981, "Stogdill's Handbook of Leadership: A Survey of Theory
    and Research," B. Bass, ed., The Free Press, New York.
Wager, L.W., 1972, Organizational 'Linking Pins': Hierarchical Status and
    Communicative Roles in Interlevel Conferences, "Hum. Rels.," 25:307-26.
Walfoort, S., 1987, Status and Sex Links of Communication Control in the Cor-
    porate Chain of Command, Paper presented at Speech Comm. Assn., Boston.
Weick, K.E., 1969, "The Social Psychology of Organizing," Addison-Wesley,
    Co., Reading, MA.
Weick, K.E., 1965, Laboratory Experimentation with Organizations, in
    "Handbook of Organizations," J.G. March, ed., Randall Pub., Chicago, IL.

POVERTY AND NON-FARM EMPLOYMENT IN LDCs

Meera Tiwari and Alan Mercer

Department of Operational Research and Operations Management
University of Lancaster
Lancaster, UK

ABSTRACT

Growth has not benefited all sections of community in the Less Developed Countries. Nearly half the population of LDCs lives in extreme poverty. Causes for the deteriorating economic situation of poor people in rural and urban areas are examined using systems methodology. It is seen that shortage of non-farm employment opportunities in the rural area contributes to the increasing poverty amongst the landless, some of whom immigrate. The illiterate and unskilled migrant remains underpaid in the urban sector. The situation can be changed by increasing non-farm employment in rural area and improving the skills of migrants.

INTRODUCTION

The available income per person in Less Developed Countries (LCDs) has gone up nearly 75% during the last twenty five years. Yet there has been only a slight increase in the daily calorie consumption (about 20%) – quite inconsistent with the enormous rise in the income. The extra income has failed to reach the very poor who remain undernourished and underclothed. A vast majority of these people are the rural landless.

The affected individuals either migrate to the urban area in search of work or become poorer by staying in the rural area. Does migration to urban area improve the quality of their lives? The squalid conditions of the ever expanding urban slums indicate that the poor continue to suffer.

Migration in LDCs has been of major interest to the social scientists since the late sixties and seventies. More recently the problem of unemployment/underemployment within the urban informal sector has received attention. This paper examines whether these can be positively influenced by changes in the rural variables.

THEORETICAL FRAMEWORK

The rural landless of most LDCs depend on non-farm employment and seasonal agricultural work in the fields of big farmers for their livelihood. The very small farmers are in a similar situation because the yield from their farms is inadequate.

213

During the lean agricultural period (which may last up to eight months) the income generating opportunities are few and non-farm employment is scarce. The rural landless and the small farmer are grossly underemployed and their economic circumstances begin to decline. They consider migrating at this stage. There is another group "... the very poorest, who cannot afford the initial cost of movement" (Lipton, 1976). Trapped in a dire and deteriorating economic situation it is forced to borrow money at very high rates and remains in debt for generations. Over a period of time there emerges a class of rural poor which lives in sub-human conditions. Illiterate and unable to help itself it becomes even poorer.

The unskilled migrant on the other hand is unable to compete and find a job that significantly improves his economic status. He is relegated to the lowest strata of the urban informal sector and remains underemployed and underpaid. The higher cost of living has to be met from his meagre income. Like his country cousins he learns to survive in an unacceptable environment.

A small percentage of migrants who remain unemployed for longer than they can afford are forced to return. The very few whose circumstances improve stay and provide inspiration to others at home.

Clearly non-farm employment is of central importance in determining not only the economic status of landless people in rural area but also the migrants' employment prospects in the urban sector.

It is against this background that the dynamics affecting the rural poor in LDCs are analysed. It is essential to look at the problem not in isolation but within its total environment. Systems methodology makes it possible to simulate the existing situation. Behaviour of the system is observed in terms of its structure and policies for ten years. Variables responsible for 'poverty amidst plenty' are identified. Changes are then induced in these variables to study their effect on the system. Strategies are suggested to improve the situation.

The study looks at data for the state of Bihar in India. One tenth of Indians live here, 87% in the villages, 48% below the 'poverty line'.

METHODOLOGY

Data for this study is taken from the 1981 census of India.

The model developed for the analysis expresses the situation in terms of two broad components – levels and rates. Levels indicate the number of people belonging to a particular category and rates depict the dynamics of the system by indicating the movement of people from one category to the other. This is influenced by different parameters.

There are four levels in all. The first two include deficit farmers/ landless labours with low income and the poorest rural landless. The remaining two comprise workers in the urban informal sector and those who are grossly underemployed. The urban poor below the poverty line are members of the latter level.

The parameters may be (a) exogenous e.g. net birth rate and literacy which are taken as constants or (b) intrinsic to the system – the auxiliaries, estimated by the index method. In this method a curve based on qualitative data shows the relationship between dependent and independent variables. Existing values of each are represented by the point (1,1) and subsequent values are extrapolated from the trend.

The system is observed in stages. At first the independent variables

retain current values. Later some of these are altered and the effect on
system behaviour is analysed. The data selected in this study is
representative of the dynamics affecting the rural poor in most LDCs. The
model does not aim to predict actual numbers that can be achieved upon
changing the variables. Instead it highlights the trends caused by
intervention.

A third order delay is incorporated in the model to allow for the
observation that the process of finding a job in the urban area which meets
demands of urban life is slow for the migrant. Delay functions have been
introduced between the migrant's perception of the time required to find an
urban job/income which will improve his economic situation and the actual
time taken in finding such a job.

THE MODEL

Influence diagram of the model is given in figure 1. Dysmap has been
used to translate the model into the systems language.

The level of the deficit farmers/landless workers is depleted by the
rates at which some of its members migrate and some others become poorer.
The net birth rate of this sector and the migrants who are forced to return
increase the level. Those deficit farmers/landless workers who become poorer
join the level of the poorest rural landless people.

In the urban sector, the flow of rural migrants and the constant net
birth rate increase the level of the informal sector. Those migrants who
fail to acquire a job here join the level of underemployed persons. The rate
at which some of its members return to the rural area depletes the level of
underemployed persons.

In the rural area, the measure of the economic status (LEST) is a
parameter which affects both the rate of migration and transition to a poorer

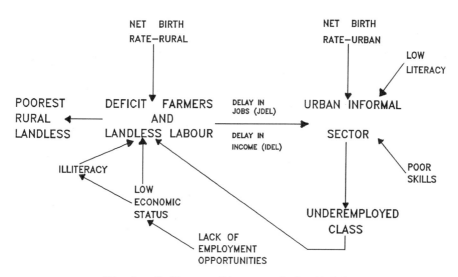

Fig.1. Influence Diagram of the Model

class. This parameter is a function of the agricultural work and the non-farm employment. The former is taken as constant in this model. The independent variable (REMP) therefore denotes the current status of non-farm employment and a constant amount of agricultural work. To study the effect of increase in non-farm employment on the system, values of the variable REMP are unchanged.

The high illiteracy rate (77%) also contributes to the declining economic status of the poor.

Other parameters affecting migration are the migrants' perception of an urban job (PJOB) and income (PINC) that will improve his economic situation. Both auxiliaries are functions of the actual jobs (AJOB) and the actual income (AINC) in the urban area.

The variable AJOB is a derivative of the informal sector growth. It retains its present growth rate in the model. Variable AINC is a derivative of the actual expenses incurred in urban living (AEXP). To study the effect of rising cost of urban living on migration, values of the variables AEXP are changed.

Parameters of low literacy and skill (SKL) affect the migrants' transition to the underemployed class in the urban area. Number of migrants leaving the informal sector is a variable dependent on their poor skills for any urban job. The value of variable SKL is changed to examine its effect on migrant's joining the underemployed group.

It is estimated from the available data that for a short duration a constant proportion of the urban population living below the poverty line will return to the rural area.

Results of the model are shown in TABLE - 1. STATES 0-3 give values of the rates at the end of model length of ten years.
STATE 0 - all parameters retain the present growth rates.
STATE 1 - all parameters except REMP (measure of non-farm employment) retain their present growth rates.
STATE 2 - all parameters except SKL (measure of migrants' skills) retain their present growth rates.
STATE 3 - the parameters REMP, SKL and AEXP are changed simultaneously.

TABLE - 1

VALUES OF SYSTEM RATES   (in '000)

	PRESENT VALUE	STATE 0	STATE 1	STATE 2	STATE 3
TRANSITION TO POORER CLASS	124	109	91	109	92
MIGRATION TO URBAN AREA	323	284	237	284	219
TRANSITION TO UNDER-EMPLOYED CLASS	11	26	24	19	18

STATE 1 reveals that the increase of non-farm employment in the rural area can reduce both the economic decline of the rural poor and their migration to urban area over a period of time. STATE 2 indicates a positive influence of increase in skills of the migrant on his potential to acquire an urban job. STATE 3 combines the above two changes with the increasing costs of living in the urban area given by the variable AEXP and shows an optimal reduction in the three rates when the system parameters are altered simultaneously. Increase in non-farm employment improves the economic status of the rural poor by providing more income generating opportunities. This leads to a decline in the rate of migration for employment and also the rate at which they become poorer. In the urban sector the increase in living costs is expected to discourage some immigrants, and cause further decrease in migration over a period of time. Finally when the skills of the migrants are changed, the prospect of finding an urban job improves and fewer people join the group of underemployed urban poor.

CONCLUSION

Systems dynamics is an important tool for examining the forces of poverty (rural and urban) in Less Developed Countries. The trends resulting from positive changes in (a) the rural non-farm employment and (b) the skills possessed by rural migrant are notable. A substantial fall is observed in the rate of transition to poorer/unemployed class and migration to the urban area.

The problems of urban decay and rural stagnation need to be viewed together. Both appear to be triggered and sustained by inadequate provision of employment opportunities. The unskilled migrant is unable to compete for suitable jobs in the urban sector.

A vicious cycle exists where poverty, lack of non-farm employment and insufficient skills perpetuate each other. This can be broken by intervening in the rural sector of the system through promotion of skill development and non-farm employment.

REFERENCES

1. Census of India - Series 1, 1981, Office of Registrar General, New Delhi.
2. Chambers, R., 1983, Integrated Rural Poverty, in: "Rural Development - Putting the Last First", Longman, London.
3. Chuta, E., and Lidholme, C., 1979, "Rural Non-Farm Employment : A Review of the Art", Michigan State University, East Lansing, Michigan.
4. Connell, J., Dasgupta, B., Laishley, R., and Lipton, M., 1976, "Migration from Rural Areas : The Evidence from Village Studies", Oxford University Press, Delhi.
5. Coyle, R.G., 1977, "Management Systems Dynamics", John Wiley and Sons, New York.
6. Forrester, J.W., 1968, "Principles of Systems", Wright-Allen Press Inc., Massachusetts.
7. Lipton, M., 1976, Migration from Rural Areas of Poor Countries : Impact on Rural Productivity and Income Distribution, in: "Migration and the Labour Market in Developing Countries", R.H. Sabot, ed., West View Press, Boulder.
8. Lipton, M., 1978, "Why Poor People Stay Poor", Maurice Temple Smith, London.
9. Muthiah, S., 1987, "A Social and Economic Atlas of India", Oxford University Press, Oxford.
10. Nayyar, R., 1977, Poverty and Inequality in Rural Bihar, in: "Poverty and Landlessness in Rural Asia", AWEP study ILO, Geneva.
11. Sethuraman, S.V., 1981, "The Urban Informal Sector in Developing Countries", ILO, Geneva.

12. Tiwari, M., thesis to be submitted shortly at the University of Lancaster, U.K.
13. Todaro, M.P., 1969, A Model of Labour Migration and Urban Unemployment in Less Developed Countries, AER 59 : 138.
14. Todaro, M.P. and Harris, J.R., 1970, Migration, Unemployment and Development : A Two-Sector Analysis, AER 60 : 126.

SYSTEMS AND O.R.

OR AND SYSTEMS

'MANAGEMENT OF MESSES'

R.L.Flood

Department of Management Systems and Sciences
Hull University
Hull HU6 7RX

INTRODUCTION

It is entirely appropriate that a session on 'OR and Systems' be titled 'Management of Messes'. The title reflects a relationship between the two 'disciplines'. This condition has both negative and positive features.

(a) Negative feature

The now widely used term 'mess' orginates from Ackoff's writings (Ackoff, 1974, for example), but was a feature of his transitionary thinking and from being noted as an OR scholar to one whose central 'problem solving' interests were based on pluralistic and systemic issues. His despair of OR is recorded in the two articles 'The Future of Operational Research is Past' (1979a) and 'Resurrecting the Future of Operational Research' (1979b). These were powerful criticisms from one considered to be a 'father' of OR (see for example Churchman, Ackoff and Arnoff, 1957, which was an important early text). And what was the plea in 'Redesigning the Future' (Ackoff, 1974)? It was for recognition that a 'Systems Age' prevails (replacing a 'Machine Age').

(b) Positive feature

OR and Systems are most often compared and contrasted as if they are of the same family. The genetic link in my view is that recent 'sorts' were conceived in the act of 'management' (and management of messes I interpret as 'problem solving' which is also a commonality of interests).

In these terms OR and Systems are linked by a tradition in 'management' (and 'problem solving'), but repel each other on methodological grounds. This might not have been so if Ackoff had been successful in his criticism of OR. That he has made little impression on the majority is, however, evident from the contents of the 'Journal of the Operational Research Society' where Ackovian influence is hard to find. (I am sure that this is of no surprise to Ackoff bearing in mind his pessimism expressed in the 'Postcript' of 'The Future of Operational Research is Past'.) True, it is impressive to see so many practitioners (from industry and commerce)with their names against papers – but does the content of their papers really reflect what they are doing? And if so, are these limited approaches the totality of the contribution that OR has made to Organisations and Society?

At this point, of course, I should expect to hear bitter complaints from the 'Soft OR' community, but what has been achieved here that is different? There certainly cannot be a claim of paradigm shift, as called for by Ackoff, and so the limited positivistic view (in social contexts) prevails, and this is coupled with a neglect of the utility of systemic concepts despite whiffs of interest apparent in qualitative causal diagrams and so on...

That is not to say that Systems is innocent on all these accounts since evidently it is not. Substantial amounts of social research which satisfy the criteria that it is also systems research (ie, the abstract organising structure 'system' is explicitly drawn upon) crumbles from its positivistic foundations upwards (this argument is well rehearsed elsewhere, see Checkland 1981, for example). Nevertheless, a healthy population of, at least, inter-pretivistic Systems thinkers and practitioners has emerged with Ackoff, as well as Checkland and Churchman, being key figures (Jackson, 1982).

Now, having argued that a methodological rift characterises the relationship between OR and Systems, it might be a little surprising to hear me call for a complementary relationship between the two. This is not so surprising, however, if we remember their general relationship in 'management' (and 'problem-solving'). Complementarity can be achieved if we bring these (and other) methodological approaches to the fore at once, yet guide method-ology choice according to situational context and the relevance of methodologies thereto. Differentiation of situational contexts has been discussed by Jackson and Keys (1984), Jackson (1987), Keys (1988) and Flood (1989) amongst others, and interestingly for this writing, in part in terms of Ackoff's differentiation between 'Systems Age' and 'Machine Age'.

The outcome of applying this notion is concilliation and openness, as opposed to the frictional climate that would otherwise emerge if the earlier words of this article were left to stand on their own. The single point of this introduction, then, is a call for the recognition that there are both limitations and applications for OR and Systems approaches. Honesty on this account is vital. Let us bear this in mind as we consider the following contributions.

References

Ackoff, RL, (1974). Redesigning the Future. New York: J.Wiley and Sons.

Ackoff, RL, (1979a). The future of Operational Research is past. J Opl Res Soc, 30(2), pp 93-104.

Ackoff, RL, (1979b). Resurrecting the future of Operational Research. J Opl Res Soc, 30(2), pp 189-199.

Churchman, CW, Ackoff, RL and Arnoff, EL (1957). Introduction to Operations Research, New York: J.Wiley and Sons.

Checkland, PB, (1981). Systems Thinking, Systems Practice, Chichester: J.Wiley and Sons.

Flood, RL, (1989). Six scenarios for the future of systems 'problem solving'. Systems Practice, 2(1), pp (in press).

Jackson, MC, (1982). The nature of "soft" systems thinking: The work of Churchman, Ackoff and Checkland. J Appl Syst Anal, 9, pp 17-29.

Jackson, MC, (1987). New directions in Management Science. In, (eds) Jackson, MC, and Keys, P, New Directions in Management Science Aldershot: Gower.

Jackson, MC, and Keys, P, (1984). Towards a system of systems methodologies. J Opl Res Soc, 35(6), pp473-486.

Keys, P, (1988). Methodology for methodology choice. Systems Research, 5(1), pp65-76

METHODOLOGIES FOR MESSES

R.L. FLOOD

Department of Management Systems and Sciences
Hull University
Hull HU6 7RX

INTRODUCTION

In the introductory article to this section 'Management of Messes' I made my view quite clear, that complementarity between approaches is of paramount concern, I noted that methodology choice could be guided according to situational context and some understanding about the assumptions on social reality inherent in methodologies (theoretical underpinnings and so on ...). In this paper the concern is with messy situations (where context is messy), and so I shall first discuss this in terms of complexity as understood through a number of paradigmatic interpretations. Once the notions of 'mess' and 'complexity' have been illuminated I shall move on to consider the availability of theories and methodologies for dealing with such 'sets'of circumstances.

PARADIGMATIC INTERPRETATIONS OF COMPLEXITY (AND MESS)

Ackoff's notion of 'mess' " .... is a system of external conditions that produces dissatisfaction. It can be conceptualised as a system of problems (which)...are abstract subjective concepts." (Ackoff, 1974, p21). I find similarities between this and some thinking of my own on complexity.

In two earlier papers (Flood, 1987, 1988) I drew up a definition of complexity by construction of a conceptual framework, this was not explicitly put into normative terms, ie preferred theory and value statements were not explicitly made. Let us consider some possibilities then.

If complexity is characterised by 'anything we find difficult to understand' (ie, problems which are abstract subjective concepts), then we might usefully think about it in terms of subject/object (eg,'subject = we or people' and 'object = anything'), and whether this is a dichotomy or an equality, and so on ...

Considering such theoretical (ontological and epistemological) issues is vital if DeVries and Hezerwijk (1978) are to be believed. They were concerned that systems theory has dealt mostly with 'what is', or so called essentialistic, questions that at best lead to classifications and taxonomies, but never lead to deeper explanations. The disassembly of complexity referred to above could be labelled as yet another such status

225

representation if I were not to highlight the normative aspects of this work, i.e. social reality status questions are normative questions in disguise. Of course, Kant has shown that thought and observation cannot be separated (Jung, 1926) and Ackoff (1979, p 102/3) discussed the implications of this forcefully for OR. In this text I am introducing the same line of thinking for complexity.

Holding to an ontological and espistemological debate, let us outline four possible theoretical groundings through which complexity as a concept can be considered. The combinations ontological realism and epistemological positivism (OREP), ontological realism and epistemological anti-positivism (OREAP), ontological nominalism and epistemological positivism (ONEP), and ontological nominalism and epistemological anti-positivism (ONEAP) may be drawn-up. In each case both the ontological and epistemological views influence each other, as can be seen in Table 1.

Table 1. Contrasting ontologies (with a 'geographic' analogy and epistemo-logies (with a 'visual' analogy relating to the 'geographic' analogy) of four possible combinations

COMBINATION	ONTOLOGY "on the social world"	EPISTEMOLOGY "on knowledge of the social world"
OREP	It is real, immutable, external and 'close'	Once known it is immutable, external and 'resolution can be chosen'
OREAP	It is real, immutable, external and 'distant'	Because of 'poor resolution' it is largely internal and transmutable
ONEP	It is assumed to be real although unknow-able as such; there-fore immutable and external, yet 'beyond our horizons'	What is known is immutable and external, reflecting the unknowable, whose 'resolution cannot be known'.
ONEAP	There is no external social world as such, but an internal and transmutable 'non-geographic' type of reality	What is 'known' is transmutable and internal and 'non-visual'

Let us now explore each of the four combinations in terms of complexity and by reflecting on the contents of Table 1. This analysis outlines a number of normative surfaces.

OREP suggests that complexity is of a real tangible world; that we can know it, represent it accurately and disseminate concrete knowledge of it. All we need to do is to develop isomorphic systemic representations of the complex real-world. We might alternatively choose to represent it with simpler models that maintain an identity, assessed through empirical observation, yet, which are more manageable. In this sense 'complexity' and 'system' are synonymous and the world is accepted as a complex of systems.

OREAP suggests that complexity is formulated in our perceptions, we conceptualise a real-world through systemic abstractions. The real-world is therefore thought of in terms of complexity, however, these abstractions are merely related to a world of reality, of real objects. This reality is, however, somewhat 'distant'. It is preferred not to think of complexity as if it were a real property of those 'distant' objects, only that it might be useful to think about them in this subjective way. Objects are assumed to exist in a real-world independent of a human observer, but it is the human observer who owns the concepts 'complexity' and 'system'. In this sense 'complexity' and 'system' are synonymous in abstract terms only.

ONEP assumes that complexity is of our perceptions, yet it is extremely difficult to know about that world, and impossible to know it as it is. Language structures, and labels and systemic representations can be used to describe the real-world where complexity lies. With these reflective forms we can learn about the structure and function of the real-world, that exists beyond our horizons, and hence the complexities of it. Therefore, real tangible immutable knowledge may be ascribed to a real-world which, nevertheless, remains beyond our absolute knowing of it. In this sense 'complexity' and 'system' might be synonymous, but only in abstract terms.

ONEAP accepts that complexity is formulated in our perceptions. A systemic structure may be used to organise these perceptions, yet no direct reference is made to a real social world because it is accepted as unreal beyond our consciousness. Complexity is only partly explained through systemic structures, being more clearly associated with our psychological and cultural being. 'Complexity' and 'system' are therefore not synonymous at all. Thus our knowledge is subjective, it is not immutable but is up for negotiation and reappraisal with our social partners.

Of the four combinations it is the last 'filter' through which I prefer to conceptualise the definition of complexity in Flood (1987). Other approaches to complexity to be found in the literature more typically are of the remaining 'filters', resting more comfortably with the scientific approach as derived in the natural sciences (but in various ways).

Now, it seems to me, that ONEAP would best characterise Ackoff's thoughts with respect to 'mess'. Further, he stated that "Objectivity is not the absence of value judgements in purposeful behaviour. It is the social product of an open interaction of a wide variety of subjective value judgements". (Ackoff, 1979, p 103). This implies that ideology is as intricately involved as theory. More broadly, on objectivity, Ackoff (1977) talks of objectivity as a social product of an open interaction of a wide variety of individual subjective judgements (encompassing, I feel, both ideas of theory and ideology). Objectivity is therefore immanent within the confines of the open interaction.

'Mess' is associated with plurality therefore. And plurality is characterised by conflict (by definition there are opposing views) or by coercion (where an opposing view is given a dominant position through the play of political forces).

'Mess' is also associated with a 'Systems Age' by Ackoff (1974, 1981, for example), which has, in turn, been associated with complexity (see Jackson and Keys, 1984).

So methodologies for messes must be able to deal with systemicity, and conflict or coercion; i.e. these features of messes must be reflected in methodological assumptions about the nature of the social world for a methodology to hold any utility in dealing with such situations.

METHODOLOGIES FOR MESSY-COMPLEX SITUATIONS

Jackson and Keys (1984, p 79-81) discussed in some detail the appropriateness of Ackoff's (1981) Interactive Planning approach and Checkland's (1981) Soft Systems Methodology (SSM) in conflictual-systemic situations. The inappropriateness of SSM in coercive situations is also discussed in some detail by Flood (1989), and there is little reason to believe that Interactive Planning would escape from that line of criticism. In essence, interpretivistic approaches are an inadequate model of social science because: they leave no room for an examination of the conditions which give rise to the actions, rules and beliefs which it seeks to expli- cate, neglect to explain the pattern of unintended consequences of actions, a feature of social life which, by definition, cannot be explained by refer- ring to the intentions of the individuals concerned; offer no method of analysing the contradictions which might exist between certain actions, rules, and common meanings, or between these and their causes and results; and neglects to explain historical change - (Fay, 1975, p 83-9). The interpretive model is inadequate as an account of how social theory is related to practice because: it promises an increase in communication but fails to take into account the conditions under which such a communication would occur; it assumes an inherent continuity in a particular society - (Fay, 1975, p 89-91).

Perhaps the most obvious feature of SSM (as one brief example) relating to the last point and thus to conservatism, which is open to such criticism, is the notion of cultural feasibility, i.e. is SSM not merely a 'tool' for management to the possible detriment of the workers? In general, the points above suggest that interpretivistic methodologies and coercive contexts are not well matched.

We will now consider briefly the essential ideas of critique which, I propose, are suitable for coercive situations (largely drawn from the accessible, concise and summarising work of Fay, 1975, with the intended meaning of course being my own responsibility).

The critical model is critical because it views theories as analyses of social situations in terms of those features which can be altered so as to eliminate certain frustrations which actors in it are experiencing. Therefore, the critical model needs: (a) to accept interpretive reasoning as a starting point; (b) to recognise that many actions actors perform are caused by social conditions over which they have no control; and (c) to understand that social theory and practice are inseparable so that 'truth' is in part determined by the methodological means through which theory is supposed to relate to practical action.

Critical theory takes on board the need to explain contradictions in social life which underlie tensions and conflicts that might be observed. So there is a need to know how and why actors have the particular purposes and needs they declare, and how they are unsatisfied, i.e. there may be structural conflicts in social order according to the phenomenological experience of actors.

Freud's psychoanalysis is often associated with the need to overcome feelings of frustration by actors coming to understand themselves in their situation. Perhaps as the product of certain inherent contradictions in their social order. Removal of the frustrations therefore means a change in social order.

Critical Systems Methodologies must inform and guide activities of dissatisfied actors, and lead to specific action which the theory calls for. The theory must be translatable into ordinary language in which the experience of the actors may be expressed. Therefrom it is necessary to demonstrate false consciousness, i.e. in what ways ideologies of social actors related to their frustrations are illusions.

CONCLUSION

Through the notion of 'mess' and 'complexity' we have come to understand that coercive contexts are not yet serviced in systems 'problem solving', but that the essential ideas are available and it is now our challenge to undertake some methodological developments from them.

REFERENCES

Ackoff, R.L., (1974). Redesigning the Future
    New York: J Wiley and Sons
Ackoff, R.L., (1977). Optimisation and objectivity = opt out.
    Eur J Opl Res, 1, pp 1-7.
Ackoff, R.L., (1979). The future of Operational Research is past.
    J Opl Res Soc, 30(2), pp 93-104.
Ackoff, R.L., (1981). Creating the Corporate Future.
    New York: J Wiley and Sons
Checkland, P.B., (1981). Systems Thinking, Systems Practice.
    Chichester: J Wiley and Sons.
DeVries, R.P. and Hezerwijk, R.Van, (1978). Systems Theory and the
    philosophy of Science. Annals Syst Res, 7, pp. 91-123.
Fay, B., (1975). Social Theory and Political Practice
    London: George Allen and Unwin.
Flood, R.L., (1987). Complexity : A definition by construction of a
    conceptual framework. Systems Research, 4(3), pp. 177-185.
Flood, R.L., (1988). Situational complexity, systems modelling and
    methodology. Trans Inst Meas Contrl, 10(3), pp 122-129
Flood, R.L., (1989). Liberating Systems Theory: Towards Critical Systems
    Thinking. Human Relations, (forthcoming)
Jackson, M.C., and Keys, P., (1984). Towards a system of systems
    methodologies. J Opl Res Soc, 35(6), pp. 473-486.
Jung, C.G., (1926). Psychological Types.
    New York: Harcourt and Brace.

"WHAT A MESS!" THE CASE OF THE NECESSARY MYTH

Lynda Davies

School of Management
University of Lancaster
Bailrigg
Lancaster, U.K.

INTRODUCTION

Operational research approaches have been severely criticised in the past for their reliance upon unambiguous statements of problems with the argument being presented that organisational life is not so clearly explicable. This criticism was led by Ackoff (1979a; 1979b; 1981) and follows criticisms made by Churchman (1967) with the metaphors of messy and wicked problems being generated to replace those of difficulties and structured problems. The view of organisations which follows on from these metaphors is not one of clear structures with well-managed functional operations but rather one of ambiguity, change, doubt, and disorder. This may seem a move from a positive and desired view of organisational performance to one of· negative, critical and undesirable. This is not the attitude taken by the author of this paper who views the <u>description</u> of organisations as chaotic as being richer and more consistent with experience than that of organisations as smooth-running machines. This attitude does not concern itself with a <u>prescriptive</u> view of how the author feels organisations ought to be but rather seen as a useful framework through which to inquire into organisations. This distinction between description and prescription is one which appears to be confounded in much of the writing on organisations. The systems theorists who deal with organisational analysis are as much guilty on this account as any others. The functional model of the organisation which is inherent in many systems documents is often spoken of as an ideal (descriptive) alongside addressing it as a model of inquiry (prescriptive). This confusion needs to be consciously avoided if the distinction between the analysis of organisations and the designing of organisational change is to be purposefully managed.

Approaches to the analysis of organisational life which are currently being considered worthy of debate have tended to move away from the model of the organisation as well-structured and clearly described to a model of the organisation as a puzzle which is seemingly chaotic. Pondy et al (1988) have argued that organisational models should create a vision of the organisation as a source of ambiguity. There are many different perspectives which could be taken to describe organisations and the incoherence between such models should be viewed positively rather than from a standpoint of failure of representation. This is more consistent with a pluralistic view of organisations. Pondy et al (op cit) argue that organisational behaviour demonstrates continual ambiguity and change with conflicting descriptions abounding and so any descriptions should expect to find interpretations to be ambiguous and changing. Martin and Meyerson (1988) argue that the intellectual construct of the notion of a paradigm may be useful to help show how different perspectives display ambiguous and contradictory interpretations of organisations. Such a view is becoming more commonplace in organisational analysis, particularly with the advent of the current interest in organisational culture.

In this paper a case study is presented describing a study where the author analysed an organisational event which was deemed to be problematical by members of that organisation. The purpose of the analysis was to intervene in this problem situation using Soft Systems Methodology (Checkland, 1981, 1988). The author was working in descriptive mode when she carried out this study and found it highly problematical. She was working under the assumption that the organisation could be <u>described</u> as a chaotic puzzle, full of ambiguity. It was not recognised that it was possible that the organisation was being <u>prescribed</u> to be interpreted as chaotic and that this was a purposeful act on the part of certain senior members within that organisation. How this appeared to be occurring, and the purposes associated with it, is presented in the next section.

THE CASE STUDY

PROVISO

In this case study the names of the institution and its members have all be avoided so that any possibility of contention regarding the 'truth' of the situation can be avoided. The author recognises that interpretation of organisational life is highly subjective and an objective true account cannot be found. The law is yet to recognise such as a counter-argument to libel and so greater precautions must be taken. This is a common problem to all accounts of organisational research, investigation or intervention and the reader is warned to bear that in mind when reading such accounts. At times it may be extremely difficult to present one's own interpretation for fear of reprisal, despite the validity of it from a pluralistic interpretive stance. As the behaviour uncovered in the analysis presented in this study would technically by possibly considered as embezzlement, any public presentation of that study must, of necessity, be extremely guarded. This exposes one of the dilemmas in organisational research, that of weighing the rights of the researcher to present the material uncovered against the rights of the access-providers to cause withholding of such accounts. This is the dilemma upon which all organisational research is founded.

THE CONTEXT

The study was centred around problems arising from the performance of a financial management information system (FMIS) and the author was asked to conduct an analysis which demanded intervention into that situation and the management of change to an improved situation. A participative approach to such change management was to be taken. The organisation was in the service sector and was a section of a larger corporate body. It was mainly concerned with training and some R&D functions of a specialised applied nature. This organisation had a rather unusual relationship with its major customer as they actually shared the premises where the organisation was situated. The customer was also a major active decision-maker on the committees and boards of this organisation. They were the supplier of the largest part of the finance responsible for the organisation's survival and were also involved in the decision-making as to how that finance should be used and disseminated. The organisation's corporate body also controlled the finance in that profits were regularly removed from that organisation to the coffers of the parent corporate body. This meant that an FMIS to aid management accounting control had to serve many masters as well as cope with the requirements of a variety of internal users.

The internal structure of the organisation was of three major groups which were strategically separable. Across the groups there were many functional similarities thus showing that the administration could be standardised to a large extent with only the specific content addressed being changed. Within these groups there were a series of departments which were also strategically different but functionally highly similar. The organisation had been in existence for nearly fifty years, with the only major recent change being that of a takeover from the new parent corporate body which had happened some three years before the start of the study. In this takeover, the organisation had moved from a public sector one to a private sector one.

The FMIS was used by all departments in all the groups in that its output was the means for the management of all internal accounting procedures. The actual information technology of the FMIS was managed centrally by a finance office which was headed by a very senior administrator. The finance office also had a senior accountant and a series of various accounts clerks.

THE PROBLEM SITUATION AND THE INTERVENTION

The way that the problem situation was presented to the author was under the assumption that the issue was really a technical one and that a new computer would solve all the problems. The author was very aware that this was often suggested as a solution but rarely the sole action necessary to deal with organisational issues surrounding IT related changes. The author decided to carry out a functional systems analysis of the paperflow surrounding the use of the IT and an analysis of the work structures associated with that. She then intended to look at the more social, organisational aspects of the FMIS and look for how the administrative issues could be dealt with alongside a possible IT change. She adopted a participative approach of stakeholder analysis followed by team management of designed change. From past experience, she fully expected there to be inconsistencies in accounts of the issues, ambiguities in interpretations, issues of a moral and political nature and anomalies in the overall picture presented. She felt that she was prepared for this and adopted Soft Systems Methodology to help her manage the dilemmas she felt that she was likely to uncover.

The first functional analysis uncovered a duplication of paperwork and recommendations were made to rectify this. The recommendations were accepted without any dissent. Eighteen months later they have still not been implemented. The easier level of 'difficulty' had, however, been dealt with and the issues surrounding the use of the FMIS output was next tackled. The author visited all departments and groups to hold interviews with people who had been put forward as the potential major stakeholders by the client. These were all managers. The author soon found that it was not the managers who were dealing with the output but their secretaries who were, invariably, left with the problems of reconciliation. A communication issue soon became apparent as the author was informed that the managers simply handed over the accounts and demanded action. No training in reconciliation was given nor was any feedback from the committees and boards on which the managers sat to discuss the issues surrounding the FMIS. This communication issue was presented to the client and a selection of the managers who agreed to a design and development team being set up between the secretaries to design their own FMIS requirements. The accountant and some of the managers were to sit on this team, in theory, although in practice they rarely did. The team dealt with the requirements and so a participative design could emerge. The study appeared to be going almost like a textbook case but the author was extremely perturbed without being certain why.

Above is an account of one perspective on the problem situation. However, this was not the only one which arose from the conduct of the study. During the study the senior administrator in the finance office was very positive about what was being uncovered but continually retorted that it was the IT of the FMIS which was to blame. The author had noted that the secretaries were all extremely concerned by the monthly accounts which were outputs from the FMIS to the departments and groups. They found them, without exception, irreconcilable. At first it seemed that this could be because the secretaries had had no training in reconciliation but further enquiry found that not to be the issue. The secretaries had become extremely competent at reconciling their own accounts but those sent from the central accounts office were constantly changing. The author looked at samples of the accounts and could see why reconciliation was such an issue. The balances carried forward were often altered without explanation. Debits from one set of accounts were sometimes transferred across departments and even across groups without any explanation. These debits were sometimes portions of amounts, rather than full amounts, and these portions did not even link to amounts for specific goods on invoice. Payments were duplicated without explanation and credit notes were issued with charges made against them which were spasmodic and non-relatable. The obvious source of linkage of tracing in such a situation is the coding of items. This was where the 'faults' with the FMIS came into the spotlight.

The FMIS was run centrally from the accounts office and was controlled by the accountant who was directly responsible to the senior administrator. The accountant continually expressed his concern regarding the FMIS. He said that its capacity was too limited and that the technology was out of date for his requirements. The author tried to explain that way that the accounting system was administrated appeared to be a greater issue than the technological constraints. He tended to agree but not listen in that he would say "Yes, that's fine" and then continue to blame the computer for everything. The author was getting rather frustrated at this point but realised that her role was merely to mediate and not to take control. The author explained that many complaints were coming from the secretaries who had to carry out the reconciliation and she explained the nature of these. The accountant explained that there had been major staffing difficulties in the accounts office, with a resultant one hundred per cent of staff turnover over a twelve month period. The accountant had been employed during this turnover period. This meant that only the senior administrator had been around for longer than one year and had had first hand knowledge of the changes occurring in the administration of the FMIS within the finance office.

The author's suggestions for change were accepted when they dealt with functional issues, such as duplication of actions, and when they dealt with participative design of requirements but still the accountant and the senior administrator continually blamed all issues on the computer. When the author remarked that reconciliation was extremely difficult because of the discrepancies between accounts' records from month to month she was told again that this was because of changeovers regarding the computer and asked whether she wouldn't rather look at the documentation of the FMIS. She did look at this but felt that other more relevant issues were being bypassed. She decided to still look at the reconciliation problems presented to her by the secretaries. The realisation that the discrepancies could be traced by following item codes led onto a tracing of these. This proved extremely difficult as it soon became obvious that the codes were constantly being changed and that there appeared to be little of a logical path to show how code changes were being recorded and monitored. The author went back to the accountant to query these code changes and was told that these occurred because the accountant was trying to find the most suitable coding for his database management. Again, he blamed any discrepancies on issues arising from the computer.

At this point the author was beginning to wonder how these accounts were ever cleared by auditors. The client did not seem to want the accounting issues to be looked at too deeply but wanted the dissatisfaction of the users to be dealt with by changes to the computing aspects of the FMIS and the use of participative approaches to designed change. The author did what the client required and the secretaries did express some measure of satisfaction as they were able to have a voice in relation to the IT changes. They continued to set up and run their own independent accounts that they were able to reconcile and started to file the accounts delivered from the FMIS away without a great deal of reference to them. The organisation was coping by setting up a duplication of an entire process so that the functions intended of that process could be carried out meaningfully independently of the finance office and the computer-based FMIS. The finance office continued to blame everything on the computer, to have high staff turnover and to present accounts which were irreconcilable. The senior administrator told the author that he was extremely happy with the work which she had done and offered more work in looking at major IT strategy for the whole organisation. She was extremely uneasy with the situation, gave thought to the further work and eventually declined the offer.

THE POSTSCRIPT

The author continued to have some contact with individuals inside the organisation. She had been rather perturbed by the nature of the study and had seen instances in the accounts analysis which appeared to indicate transferences across accounts to accounts which were only opened temporarily. The study had indicated that the people in the situation had considered the FMIS to be inadequate and the problem to be one of computer inadequacy and the need for participative design. The author had followed the lead of client and had acted upon the spoken issues rather than upon what appeared to be far more problematical ones. She felt distraught because she felt that she had been guided away from acting upon these other issues but recognised that her job as intervenor was not to take control and demand certain changes above others. She was, however, particularly perturbed that her more easily implemented recommendations had not been carried out and felt that her intervention as a participation guide had been rather cosmetic. She felt that she had been rendered impotent.

It is difficult to know when issues are constructed by the investigator as being meaningful or when they are recognised as highly significant by the organisational members even when the investigator is not present. This is one of the major problems in organisational research where an interpretive stance is taken; the interpretation is always mediated through the expectations of the investigator.

All accounts of events, no matter how publically declared, are only interpretations and images and the 'actual truth' is something which is ultimately unattainable.

The author was privy to rumours which occurred some time after her study had been completed. It is important to stress that these can only have the status of rumours because no public action was taken which could openly challenge them. However, certain organisational changes did make them appear to have a rather high level of credence. The rumour which occurred regarding the management accounting of the organisation was that a form of black economy had been operating which was controlled by the senior administrator. The rumour stated that the parent corporate body uncovered certain dealings in this economy and that the senior administrator was warned that if they did not stop immediately he would find himself prosecuted for embezzlement. The organisational actions which followed this were that many private accounts which paid for extra-curriculum activities were suddenly expressed as having run dry. Certain individuals even expressed that they thought they were going to find themselves in dire financial circumstances because of the need to repay certain hefty loans to the finance office. All sections of the organisation found themselves in a severe economy drive with many orders for development and new equipment being cancelled. If the black economy did exist, it seemed to have been drawn to a rather abrupt halt. Other activities regarding the passing of information to the customer who was on the premises were also spoken of, including one individual doing this in order to gain political protection so that his group would not be made to give up certain benefits. As his group was the only one which did not suffer changes this may have been true. On the other hand, it may have been a rather vindictive rationalisation of an effect upon an innocent party. At this point the computerised FMIS was replaced by a more easily standardised type of system.

CONCLUSION

No singular interpretation of the above events is adequate. The functional analysis did uncover administrative problems which were easily rectified, although action to do this was not carried out. The social issue of using participative design procedures to appease the complainers was carried out although the leader of that process, the author, acted in good faith believing that such an exercise would benefit the secretaries. It did help them to cope with a situation where they could not change the FMIS by allowing them to develop ways of setting up an alternative way of operating. This was not the intention of the participative exercise, but it had been the result. Had the situation been one requiring redesign of work and administration these two sets of activities would probably have sufficed. The author felt very dissatisfied with the result of this study and found it difficult to interpret why, apart from the obvious inability to explain the terrible states of the accounts.

Despite looking long and hard at the computerised FMIS, the author could find little wrong which some bolt-on further power could not solve adequately. It was the way that the FMIS was being used, or rather abused, that was most problematical. It appeared to be being used as a way of creating mess rather than as a way of alleviating it. Furthermore, this mess creation appeared to be purposeful rather than accidental. Without direct accusation, the author could not find any interpretation of this other than her own incompetence to uncover the source of the mess. Despite this, her client, the senior administrator, continually stated that he was happy with her work. The only conclusion she could come to was that she was being used as part of a smokescreen but this seemed rather fantastical. It was only as the rumours unfolded, after the completion of the study that a far more understandable interpretation was possible.

The interpretation that the author has come to it that the FMIS issues were a smokescreen for illicit activities. All the faults which were associated with it did not follow through when she looked closer at them. The faults appeared to be administrative and not just technological. It seems that one feasible interpretation is that the FMIS was a convenient artifact for propagating and sustaining a mess which could then attain mythical status. The author had recognised that it was acceptable to use the concept of 'mess' to <u>describe</u> the situation but it did not occur to her during the study that the concept may have been applied <u>prescriptively</u> as a way of designing a foil for certain organisational activities. The author, as interventionist, had provided a means for supporting that myth so long as she could be manoeuvred to carry out certain activities and avoid the pursuit of others. As the client had the control of where the author's activities should concentrate, this was possible.

The case study indicates that any attempt to manage messes should not take the unitary perspective that all those involved in the situation wish the mess to be cleared away. It is wiser to challenge why such a mess is reinforced when some actions to rectify it may be obvious and show little difficulty of implementation. Discussions of systems approaches to the tackling of messes still act as if it is wise and necessary to expel the mess. If it is serving an ulterior purpose then the intervenor may find her/himself continually frustrated in any expulsion attempts. It is necessary to stop and ask why this may be occurring and whether the intervenor wishes to be involved in such circumstances. This is an ethical decision which only the individual with her/his own conscience can decide upon.

The politics of the intervenor are rarely discussed in the world of operational research or systems (though see Checkland, 1986 for a discussion of this in relation to Soft Systems Methodology). The literature dealing with organisational behaviour and organisational development does give greater consideration to this, eg see Kakbadse and Parker, 1984. It seems that systems approaches to intervention should take greater account of this political dimension so that the intervenor may more easily recognise when they are being used as political pawns to maintain a rather dubious status quo. Because organisations do appear to be highly ambiguous and in constant change, the investigator may be led to believe that this description indicates a nature of the organisation which emerges non-purposefully from human interactions. That ambiguity needs to be challenged in case there are indicators of the creation and maintenance of that ambiguity for more purposeful reasons. The complexity of human actions is so rich that the intervenor needs to be aware that new deviations can constantly occur and the model or theories at hand may not be able to provide adequate interpretation of the events. It seems that an interpretation of messes as necessary myths rather than as accidental or as merely problematical could be a useful addition to the systemic practice of intervention in organisational problem situations.

REFERENCES

Ackoff, R. L., (1979a), "The Future of Operational Research is Past", J. Opl. Res. Soc., 30:93-104.

Ackoff, R. L., (1979b), "Resurrecting the Future of Operational Research", J. Opl. Res. Soc., 30:189-200.

Ackoff, R. L., (1981), "The Art & Science of Mess Management", Interfaces, 11:20-26.

Checkland, P. B., (1981), "Systems Thinking, Systems Practice", Wiley, Chichester.

Checkland, P. B., (1986), "The Politics of Practice", paper presented at the IIASA International Roundtable, 'The Art & Science of Systems Practice', Intl. Inst. Applied Systems Analysis, Vienna.

Checkland, P. B., (1988), "Soft Systems Methodology: An Overview", Jnl. Appl. Syst. Anal., 15:27-30.

Churchman, C. W., (1967), "Guest Editorial: Wicked Problems", Management Science, 14(4):B-141-B-142.

Martin, J., & Meyerson, D., (1988), "Organisational Cultures & the Denial, Channelling & Acknowledgement of Ambiguity", in Pondy et. al. (eds) (op. cit.).

Pondy, L. R., Boland, R. J. Jnr., and Thomas, H., (eds), (1988). "Managing Ambiguity & Change", Wiley, Chichester.

ON THE CONCEPT OF MESS

Paul Ledington

The Management School
University of Lancaster
Bailrigg
Lancaster, U.K.

The idea of the world as 'messy', and problems as 'wicked' and 'ill-structured' is a 'quiet' recurrent theme in the literature of Management Science, broadly defined, and in some degree the literature of the Social Sciences. It is to be sure a poorly defined idea but one which has of some value in understanding the development of Management Science. It is also an idea which, it will be argued, has profound significance and also presents a major challenge for future intellectual developments in the Management Science arena.

This paper begins an exploration of the ideas inherent in the concept of 'Mess' by sketching out some of the thinking which underpins the 'quiet' revolution of 'messiness'.

TWO THREADS IN THE DEVELOPMENT OF OR

OR, Systems Analysis, Management Science, whichever name you prefer, is a relatively new area of endeavour which emerged out of the efforts of scientists to tackle operational problems which arose in the conduct of World War II.

The past fifty years have seen OR develop rapidly its technical competence yet also face a continuous crisis of identity. The sources of the crisis seem to lie in two interlinked areas. The first area is the 'challenge of Relevance' and the second the 'challenge of Messiness'. In the opening to a recent article, Eden and Huxham (1988), draw the two threads together quite well.

"In recent years there have been two major concerns of practitioners working on strategic plans: first that the planning activity has had little impact on managerial action, and secondly, that it has taken scant notice of the dynamics created by the strategic options open to competitors.

241

The first of these concerns is of importance to all analysts whose roles are concerned with bringing about organisational changes. The rational approaches have frequently had less impact on organisational decision-making than their proponents would wish. This has been particularly true in their application to complex, 'messy' problems".

The 'challenge of relevance' notes the lack of impact of OR approaches and suggests that, at least the practice of OR requires further development. It is interesting to note that at the 1988 annual conference of the Operational Research Society, held in Sheffield, the most popular and well attended sessions were devoted to 'improving the practice of OR' and concentrated upon the skills and techniques required for 'client-handling'. There is little doubt that we can all improve our abilities in that area but is it the only area for development; is the other challenge, the 'challenge of Messiness', one only of degree or is it somewhat more fundamental.

In 1948, Warren Weaver suggested that there are three types of situation which confront the application of Science. These are Organised Simplicity, in which analytical techniques are effective, Disorganised Complexity, in which statistical techniques are effective and Organised Complexity, an area which presents major challenges to current techniques. One can, perhaps crudely, interpret the development of OR by using this model. On the one hand OR has made great strides in the development and application of analytical and statistical techniques and, one might argue, has extended the realms of Organised Simplicity and Disorganised Complexity in this way; on the other hand, there has been a continuing debate concerning the applicability of these techniques in the area of Organised Complexity. The debate, at its extreme, is a paradigmatic one in which there are two responses to the 'challenge of Messiness'. The Paradigm I response is that 'Messiness' is essentially a matter of degree. The application of Science in these areas is very new but has made remarkable strides. It is only to be expected that there are problems which have not yet been effectively tackled. The continued disciplined application and development of the methods and techniques we use will eventually come to handle those area not yet within our boundary.

The Paradigm II response is somewhat different. The 'challenge of Messiness' is not one merely of degree but is also one of quality. There are aspects and properties of the world which cannot be captured by the approaches relevant to Organised Simplicity or Disorganised Complexity. The extension of these approaches into these new areas is but a chimera, the need is to re-think the very nature of the endeavour and the approaches which are employed within it.

Characterised in this way, Paradigm I takes the World as consisting of 'Continuous Complexity', whereas Paradigm II views the World as consisting of 'Discontinuous Complexity' or in Schumacher's (1977) phrase that there are 'Ontological Gaps', or in Boulding's view that there is a hierarchy of complexities that forces the 'Skeleton of Science' (1956).

It is evident in the rise of Systems Thinking, (Checkland 1981), that there is a community which takes the idea of Organised Complexity seriously and is attempting to develop concepts which are useful in this area, but there is also a question of whether organised complexity is Synonymous with 'Messiness' and whether ideas developed in the one area are relevant in the other. The Scientist is concerned to understand the nature of organised complexity, whereas the Management Scientist is concerned, in Checkland's phrase, with 'Engineering' problem situations. Given that there is a need for at least considering the idea of 'Mess' itself as a Management Science concept, then is it possible to go further and examine the nature of 'Mess'. The next section begins such an exploration.

EXPLORING 'MESS'

It was suggested at the outset that there is a 'quiet' revolution of messiness, Ackoff (1979), is of course well known for his advocacy of the concept of 'Mess' and his disillusionment with 'classical' OR. However, the struggle to develop new ideas, which is the core of any intellectual revolution, is 'quiet' because there is still no commonly articulated language and set of ideas which adequately characterise the area. There is but isolated evidence of attempts to come to terms with something beyond the accepted view of the world. Forrester has argued that many areas of the World, and in particular, Social Systems, exhibit counterintuitive behaviour. Essentially a view that the observer is unable, without further aid, to comprehend the behaviour of complex systems. He further suggested the panacea of systematic analysis and computer-based simulation. Supported by Churchman, (1967), Rittel and Webber, (1974), and echoed later by Mason and Mitroff, (1981), suggest a different idea; that of the 'wicked' problem and enunciate the differing characteristics of 'wicked' and 'benign' problems. Emery (1981) recognises the problems for organisations created by a turbulent environment. Beer (1979) explores the Problem of 'viability' in a world of high variety.

Different expressions, languages and concepts yet all trying to express aspects of a highly complex, dynamic, slightly mysterious world. Yet is this expression sensible; are there 'wicked' or 'benign' problems in the World, is it possible to say that Social Systems have counterintuitive behaviour, or even that some situations are 'Messes'. Should we not say that these efforts and expressions all suggest that there are circumstances in which we do not have suitable models, or theories, which allow us to act sensibly in such situations. In both Management Science and Social Science our concern is with human action; not only "What is" but also "What ought to be". Both are but 'images', important 'images' none-the-less, which when they are not in tune seem to give rise to a feeling of Messiness. It is through the framework of "What is" and "What ought" that we make sense of our actions. When in a 'Mess', the discrepancy between "What is" and "What ought" creates major uncertainty about our actions.

THE IMPLICATION OF 'MESS FOR THE SOCIAL SCIENCES

At the heart of all the Social Sciences lies the idea of human action as based upon rational choice. In the use of the rational choice model there is a concern to explain action in terms of its relationship to desired goals in a known situation. The action chosen being the most effective means to achieve the desired end. Such a model implies a mechanistic closure between "What is", "What ought" and the 'action'. It is this closure which must be challenged if the concept of 'Mess' is to be taken seriously. If such closure is challenged, then the idea of rational action itself is challenged. Thus the idea of 'Mess' presents a challenge to rationality. A challenge which presents different aspects to Social and Management Science. Social Science must explain the way in which the meaning of action is constructed, but the challenge to Management Science is to change the way in which, in a particular situation, particular meanings of action are constructed. This challenge itself has a number of dimensions;

To what extend is change possible? - The Technical Problem.

What is the extent of 'legitimate' change? - The Ethical Problem.

How can change be achieved? - The Practical Problem.

How can we know if the direction of change is 'desirable'? - The Philosophical Problem.

Implicit in all of this is an acceptance that we are dealing with a 'World' which, at least in part, is constructed by and through human beings; a social reality rather than a natural reality.

CONCLUSIONS

Three interpretations of the concept of 'Mess' have been surfaced. The first is based on the idea that the complexity of the world is continuous but beyond the unaided capability of man to comprehend. A Mess is a psychological notion which can be tackled by enhanced information processing. The second suggests that the nature of the World is one of discontinuous complexity and thus that different 'models' of the World are required at different levels. A Mess arises due to the inadequacy of a model, or its incorrect use and is thus a product of the intellect. A situation which can be remedied by further intellectual development.

The third view is one of a mess arising from social construction. Through physical action man co-creates the World and through communicative activity constructs understanding of the World. If the relationship between these two processes becomes dislocated, then a Mess ensues. To handle this form of 'Mess', do we bring understanding in line with action, or action in line with understanding, or do we have to deal with both plus the inter-relationship of the two. In mitigating such a mess, the way forward is through the facilitation of purposeful action.

Perhaps only one thing can be said with any conviction and that is that the challenge of Messiness has yet to be mitigated.

REFERENCES

Ackoff, R. L., (1979), "The Future of Operational Research is Past", J. Opl. Res. Soc., 30:93-104.

Boulding, K. E., (1956), "General Systems Theory - The Skeleton of Science", Management Science, 2 (3).

Beer, S., (1979), "The Heart of Enterprise", Wiley, Chichester.

Checkland, P. B., (1981), "Systems Thinking, Systems Practice", John Wiley, Chichester.

Churchman, C. W., (1967), "Wicked Problems", Management Science, Vol. 14, No 4, December.

Eden, C., and Huxham, C., (1988), "Action Oriented Strategic Management", J. Opl. Res. Soc., 39:889-899.

Mason, R. O., and Mitroff, I. I., (1981), "Challenging Strategic Planning Assumptions, New York, Wiley.

Rittel, H. W. J., and Webber, M. M., "Dilemmas in a General Theory of Planning", Chapter 12 in R. L. Ackoff (Ed), Systems and Management Annual 1974, New York, Petrocelli, pp219-33.

Schumacher, E. F., (1977), "A Guide for the Perplexed", Jonathan Cape.

Weaver, W., (1948), "Science and Complexity", American Scientist, 36.

Emery, F. E., (1981), "The Emergence of Ideal-Seeking Systems", in F. E. Emery Systems Thinking, Vol Two, Penguin.

THE LIMITS TO HUMAN ORGANISATION:

THE EMERGENCE OF AUTOPOIETIC SYSTEMS

Fenton F. Robb

The University of Edinburgh
50 George Square, Edinburgh EH8 9JY

CONSTRUCTS OF MANAGERIAL ORGANISATIONS

Much contemporary thinking about organisations is directed towards creating and managing the unmanageable. Managerial structures, procedures, vocational training and the institutionalisation of functional knowledge are manifestations of a collective perception by groups of managers of a world in which they have the illusion of control. Managerial organisations so specified are "first-order" constructs in the everyday language of description (Bittner, 1965; Calder, 1977).

A second-order construction is needed, framed within the theoretical apparatus of science, with which to investigate the first-order constructs and to provide descriptions, even if not causal explanations, of managerial organisations. A construct has been offered by the cybernetic approach which has seen organisations as analogous to purposeful machines or organisms whose structures determine their behaviour.

The cybernetic tradition rests on the assumption that there is no distinction of "natural kinds" between entities in the physical, biological and social worlds. The same "laws" are thought to govern all systemic processes. What is of particular interest here is that we must now include new insights into the behaviour of systems far from equilibrium in those "laws".

SYSTEMS FAR FROM EQUILIBRIUM

Prigogine (1980) and Tritton (1986), among others, have demonstrated that if linear systems are over-driven and pushed far away from the influence of their stabilising attractors, they move into a state in which their processes become exposed to many small forces. These forces are not observable using the language of representation which described the system's self-regulation. They are too small, too many and operate in too many directions. Furthermore, the sequence in which they operate on the system may be an important determinant of its behaviour and the logic of the language of description does not accommodate the irreversibility of the state changes so induced. There are several possible outcomes from such a state. The system may remain in "stable chaos", it may be destroyed or it may produce a new dissipative structure of higher logical order which

247

requires a new language of description. In complex systems there are many degrees of freedom and the precise trajectory "selected" from the many available, cannot be anticipated. After each transition, other than that into stable chaos or to destruction, the system acquires a greater ability to resist intervention from external forces or from within itself.

We are now coming to see that concepts such as coherence, macroscopic order, spontaneous organisation, adaptation and growth, previously thought as features of living systems and solely in the provenance of biological sciences, are to be observed in the physical world and indeed permeate all nature, including, we offer, our own artifacts, managerial organisations.

COMPLEXITY IN MODERN ORGANISATIONS

The information technology "revolution" sets organisation in a new environment in which interconnections are being made piecemeal between managers at different levels and in different functions within the same organisation, and between many levels of managers in different organisations. Variety is growing almost exponentially, but human capabilities are unchanged. Only the increase in organisational complexity, and the consequent acquisition of many more degrees of freedom, has prevented collapse. Modern organisations may be likened to "fly-by-wire" aircraft which are inherently unstable and are maintained in flight only by their information and control systems. In these days they are often driven to the limits of their self-regulatory capabilities. The limits to the growth of organisational hierarchy appear to be set when supra-human autopoietic organisation emerges.

Maturana (1981), Beer (1980) and Pask (1981), have pointed to "autopoietic", "viable" and "potentially conscious" systems respectively. These are very similar concepts. Human social organisations may, when their processes interact very closely indeed, produce and maintain conscious, viable, autopoietic processes which hold their own unity homeostatic by organisational closure. Such an organisation need no longer serve human purposes. Instead it need serve only the purpose of maintaining its own unity. Intervention, in an attempt to design or adapt such systems by humans, themselves systems of a lower logical order, will be "seen" by the organisation simply as a perturbation from its environment which, if it does not serve its autopoiesis, should, and, if the organisation is viable, can be dissipated.

Explicit in much of the advice given about effecting intervention by change agents is that it should "begin at the highest levels and should involve individuals who have [the] power and autonomy ..." (Argyris, 1982, p.82). What has not been contemplated is that there may be no such levels, and no such powerful individuals. Power and individual autonomy, may have already been dissipated. The organisation itself may be the only autonomous and powerful entity around.

The distinction has been drawn between autopoietic [organism-like] and non-autopoietic [colony-like] organisations (Maturana, 1980; Robb, forthcoming). Humans and their collectivities are properly called "components" of autopoietic organisations and "members" of colony-like organisations. As components, their own unity is dependent on that of the system; as members their own unity remains unaffected. Beer's (1980) viable system exists only in so far as the relationships required are actually used. The continued existence of its components depends on their continued exercise of the autonomy prescribed for them. The entity generated by Pask's (1978) conversations, its consciousness, depends on the continued affirmation and reaffirmation of the agreements reached between its participants. However, particular components are not necessary; only some of the processes which they produce are necessary for the continuation of the entity. As Pask has pointed out, the real conversation stops when agreement is reached and when a consistent ideology is formed, and it is then that the potentially self-conscious system emerges.

Brunsson (1985) distinguishes actions taken in an organisational context from the ideology or set of ideas [the autopoietic process] which moulds and directs the action elements. Where organisational ideologies are "strong", consistent, complex and conclusive then planned organisational change which might entail an ideological shift is unlikely to be successful, although such an organisation may continue to be responsive to environmental change. They would thus be unable to influence the direction of organisational change because they would have reached one of Pask's (1980) "limits of togetherness".

Despite all this, the organisation might well be adaptive to changing conditions in its environment. It would exhibit what Smith (1982, p.318) has called "morphostasis", that is "making things look different while remaining basically as they always have been" or "making changes only within the instructions encoded into the system". Chester Barnard also anticipated the rise of the autopoietic organisation thus "... the most important single contribution required of the executive, certainly the most universal qualification, is loyalty, domination by the organisations personality" (see Burns, 1963). Burns and Stalker (1966) point to the pathologies of bureaucratic organisations and the problem of making interventions, ascribing the resistances encountered to the "strong forces" of interlocking systems of commitments, internal politics and career structure, which perpetuate out of date, and maladaptive organisations. Galbraith (1974) points in the same general direction when deploring the inherent contradictions between the priorities established by the powerful planning sector and the needs and desires of the public at large.

Foucault (1976, p.104) saw the genesis of this in the emergence, in the seventeenth and eighteenth centuries, of "... a new mechanism of power possessed of highly specific procedural techniques, completely novel instruments ... [which type of power] is exercised by means of [continuous and permanent] surveillance ... [and] presupposes a tightly knit grid of material coercions". A somewhat similar view is to be found in Habermas (1984), who recognises in our inability to maintain the differentiation of the technical system [the self-steering mechanisms of social organisation] from the social life-world, the invasion of the latter by the former.

So we uncover a "consilience of inductions" coupling the works of Barnard, Beer, Brunsson, Maturana, Pask, Smith, Foucault, Habermas and others, each of whom has converged on, but maybe not quite reached, the notion that some organisations are autopoietic and, though responsive to environmental changes, are highly resistant to human intervention. Laughlin posits the possibility that change of a type he calls "colonisation" could entail a shift of the nature of the organisation against the better judgements of the participants (Laughlin, 1988).

IMPLICATIONS FOR MANAGEMENT

If humans come to believe that, through the organisation, their perception of the world can be identified with that of their fellows in the organisation and that they can realise themselves within the organisation and only in that way, then they truly become "components" of it. If humans resist the temptation to relinquish their identities to the organisation, then if autopoiesis is being generated because the organisation is under stress, they risk alienation and rejection (Weick, 1979).

Contrary to the received wisdom of the day, we come to the conclusion, not unfamiliar to critical management theorists, that subservience to a particular viewpoint leads to suppression and exploitation. We differ from them in taking the view that it is not a conscious conspiracy of the managerial classes which engenders this undesirable condition, but the operation of forces to be seen elsewhere in nature. As Foucault (1976, p.98) puts it, "... individuals are the vehicles of power, not its application ... The individual is an effect of power ... The individual which power has constituted is at the same time its vehicle". It is argued that the managers are as much victims as actors. They are formed by the systems of which they are components and remain as individuals realised and constrained by those systems only in so far as they exercise power.

Human purposes can be served by organisations which respect human individuality and do not institutionalise the expression of that respect. Organisations, of the colonial type, can be effective despite the fact that they do not absorb the whole of a human's life time or require a unified conceptual reality. Total absorption is not a necessary condition for organisational success, only for autopoiesis. What is becoming ever more clear is that in order for humans to affect morphogenic processes, those which maintain the autopoiesis of organisationally closed systems, we must intervene into the strongly held ideologies, the well-structured concepts, which hold the organisation together as an entity. Just how this can be done without at the same time inducing the destruction of its human components remains an open question to be answered only in terms of a new more adequate language of description which can comprehend notions such as organisational closure and the relations between the processes which sustain it. At present the indications are that the emergence of autopoiesis is irreversible and that reversion to colony-type organisation is not possible without organisational decomposition.

CONCLUSION

The insights, so briefly outlined here, develop the foundations for a "second-order construct" already suggested by cybernetics. By means of this we may observe the behaviour of real-world organisations and be alerted to the very real possibility that some of these may have emerged from a period of crisis to become autopoietic systems of such an order that they defy description and cannot be controlled by human actors. This construct may enable us to consider what effects might arise from human intervention by way of design and control of managerial organisations in different states of development and what the limits to intervention may be. The received wisdom (e.g. Peters and Waterman, 1982) that we shall always be able to make interventions which will loosen up organisations and induce cultural changes so as to direct the organisations activities to serving human purposes is very much open to question.

REFERENCES

Argyris, C., 1982, How learning and reasoning processes affect change. In "Change in Organisations; New Perspectives on Theory, Research, and Practice," P.S. Goodman and Associates (Eds.), Jossey-Bass, San Francisco and London, 1982, 47-86.

Beer, S., 1980, Preface to Maturana, H.R. and Varela, F.J., Autopoiesis and cognition," D. Reidel, Dortrecht, 63-72.

Brunsson, N., 1985, "Irrational Organisation," Wiley, London.

Burns, T., 1963, Mechanistic and organismic structures. New Society,

January 1963, 17-20. In "Organisation Theory," D.S. Pugh (Ed.), Harmondsworth, Penguin, 40-50.

Burns, T. and Stalker, G.M., 1966, "The Management of Innovation," Tavistock, London.

Bittner, E., 1965, The concept of organisation, Social Research, 32:239-255.

Calder, B.J., 1977, An attribution theory of leadership. In B.M. Staw and G.R. Salanic (Eds.), "New Directions in Organisational Behaviour," St Clair Press, Chicago.

Foucault, M., 1976, Lecture Two. In "Michel Foucault, Power/Knowledge; Selected interviews and other writings 1972-1977," (Ed. C. Gordon), The Harvester Press, Brighton (1980), 92-108.

Galbraith, J.K., 1974, "Economics and the Public Purpose," Deutsch, London.

Laughlin, R.C., 1988, Accounting and organisational change. Paper to the Second Interdisciplinary Perspectives on Accounting Conference, Manchester.

Habermas, J., 1984, "The Theory of Communicative Action," (Trans. T. McCarthy), Heinemann, London.

Maturana, H.R., 1980, Autopoiesis: reproduction, heredity and evolution. In M. Zeleny (Ed.), "Autopoiesis, dissipative structures and spontaneous social orders," Westview, Boulder, Colorado, 45-79.

Maturana, H.R., 1981, Autopoiesis. In M. Zeleny (Ed.), "Autopoiesis: A theory of living organisation," North Holland, Oxford and New York, 21-33.

Pask, G., 1978, A conversation theoretic approach to social systems. In "Sociocybernetics,", Vol.1, R.F. Geyer and J. van der Zouwen (Eds.), Martinus Nijhoff, Leiden, Boston and London, 15-26.

Pask, G., 1980, The limits of togetherness. In "Information Processing," S.H. Lavington (Ed.), North-Holland, Amsterdam, 999-1012.

Pask, G., 1981, Organisational closure of potentially conscious systems. In M. Zeleny (Ed.), "Autopoiesis: A theory of living organisation," North-Holland, Oxford and New York, 265-308.

Peters, T.J. and Waterman, R.H., 1982, "In search of excellence," Harper and Row, New York.

Prigogine, I., 1980, "From being to becoming: Time and complexity in the physical sciences," Freeman, San Francisco.

Robb, F.F. [forthcoming], Cybernetic and supra-human autopoietic systems, Systems Practice.

Smith, K.K., 1982, Philosophical problems in thinking about organisational change. In "Change in Organisations; New Perspectives on Theory, Research and Practice," P.S. Goodman and Associates (Eds.), Jossey-Bass, San Francisco and London, 316-374.

Tritton, D.J., 1986, Ordered and chaotic motion of a forced spherical pendulum, European Journal of Physics, 7:162.

Weick, K.E., 1979, "The social psychology of organizing," Random House, New York.

ACKNOWLEDGEMENTS

Your author is in debt to Dr R.L. Flood of the City University and Richard Laughlin of the University of Sheffield for their introductions to Critical Theory.

FROM MEASUREMENT AND CONTROL TO LEARNING AND REDIRECTION:

USING 'SOFT' SYSTEMS THINKING IN THE EVALUATION OF HUMAN ACTION

P.J. Lewis

Department of Systems and Information Management
The University of Lancaster

ABSTRACT

Social action programmes remain an important area of social concern and national expenditure, but their evaluation remains problematical. It is argued that the predominant, systematic approaches to evaluation are based upon a rational-technical view of evaluation and its relationship to organisational decision making, which is not well suited to the complex, social processes of human activity systems such as social action programmes. Action research within the Department of Systems and Information Management at the University of Lancaster has led to a systemic view of evaluation, based upon 'soft' systems thinking, which emphasises organisational decision making as a social process, within which evaluation acts as a mechanism for organisational self-examination and learning, rather than for mechanistic control in relation to pre-defined programme objectives. This view recognises alternative stakeholders' perceptions of purpose as an important feature of such programmes and distinguishes between the processes of judgement making and enabling judgement making through the provision of evaluative data. This leads to the suggestion that the design of an appropriate and effective information system, through which evaluative data may aid those involved in the situation to form evaluative judgements, is an important, though often neglected, component of evaluation.

INTRODUCTION

A great deal has been published upon the subject of evaluation since the 1950's and descriptions of the evaluation of educational schemes, health care delivery systems, management training or defense programmes are easily found. The majority of the published studies are concerned with the evaluation of social, educational and other social action programmes of a type which Checkland (1971,1981) has shown may be usefully regarded as human activity systems, but the methods employed have not always recognised the special nature of such systems.

In discussing the different approaches to evaluation Easterby-Smith (1986) has described many of the approaches as being rooted in the systems model but it would be more correct however to say that most of the approaches described in the literature are based upon what Checkland (1978) has characterised as 'hard' systems thinking with its twin strands of development within the systems engineering and systems analysis traditions Despite discussion of alternative, social science based, approaches to the evaluation of social action programmes there remains in practice an over-riding emphasis upon the use of scientific/experimental approaches, upon the use of quantitative data and upon evaluations based upon the programme owners' view of the situation. Underlying such approaches are the assumptions that a programme, its goals and its effects have a real, objective existence and that knowledge about that programme is to be found by discovering the 'real', (though presently unknown), relationships between these three thing things. The acquisition of such knowledge will (it is assumed) allow the programme to be engineered so as to maximise the achievement of the programme goals in a resource efficient way.

Vickers (1970) has pointed out the danger of this view that human affairs may be managed in the same way as designed technical systems:

> "..the dominance of technology has infected policy making with three bogus simplifications, just admissible in the workshop but lethal in the council chamber. One of these is the habit of accepting goals - states to be attained once for all - rather than norms to be held through time, as the typical object of policy. The second is the further reduction of multiple objectives to a single goal, yielding a single criterion of success. The third is the acceptance of effectiveness as the sole criterion by which to choose between alternative operations which can be regarded as means to one desired end. The combined effect of these three has been to dehumanize and distort beyond measure the high human function of government - that is regulation - at all levels." (p 116)

House (1984) has discussed the way in which the adoption of a particular metaphor will inevitably direct and distort both our appreciation of a problem situation and the language with which we describe it. His identification of the 'social programmes are like machines' metaphor as having a particularly strong effect upon the conduct of evaluation echoes the view of Steinmetz (1975) that the 'evaluation for decision making' view of evaluation, which he finds a dominant in educational evaluation, may have serious and undesirable consequences. Steinmetz expresses concern over much the same point as Vickers by use of Habermas' (1970) distinction between purposive-rational social systems and social systems based upon symbolic interaction, arguing that the difficulties experienced in doing educational evaluation have led to the adoption of evaluation approaches which

> "..conceptualize organized educational efforts as though these were or should become social systems in which purposive-rational action predominates entirely. They do, after all, provide a model for the conduct of an activity according to which educational practice is to be improved. The pressure exerted by these evaluation systems is toward the rationalization of the conduct and organization of the educational effort into a particular means-ends logic."   (p. 55)

The adoption of such a conceptualisation of a programme implies a regulative role for evaluation which is very closely related to the management and control of the programme and leads to evaluation concentrating upon the collection of data about the extent to which, and the efficiency with which, the planned programme activities are carried out. Consideration of the essential 'humanness' of social action programmes (political factors, the diversity of values, the possibility of those involved 'changing their minds') is precluded.

Although there has been much discussion of the weaknesses of such approaches to evaluation and calls for more recognition of the special nature of social situations there have been few suggestions as to how alternative approaches may be practically applied. The difficulties with applying the 'hard' systems approach to human affairs have however been widely investigated by action research and described within the literature relating to 'soft' systems methodology and recent experience in action research has shown that the SSM may be combined with social science techniques to make a particularly powerful approach to the evaluation of social action programmes.

A 'SOFT' SYSTEMS VIEW OF EVALUATION

In the transformation from 'hard' to 'soft' systems thinking the notion of evaluation was carried across from the 'hard' tradition with little change; the inclusion of a 'measure of performance' as a requirement within the formal systems model and the inclusion of an activity labelled 'monitor and control' within every conceptual model show how this was reflected in the language of the 'soft' systems methodology (SSM).

Even though a root definition was seen as an epistemological device which was open to debate it was assumed that for any given root definition there was a logically correct conceptual model and that from the activities within this model measures of performance could be logically derived. Thus if we examine early applications of SSM to problem situations concerned with evaluation we find that though the purposes of any programme or business are not taken as given, and are open to debate, the concept of evaluation which is used is still one of the feedback loop within the goal-seeking-with-feedback model of the 'hard' systems paradigm. Such debate as took place in this early studies tended to be dominated by those with most power in the situation, and by considering the actors and owners of a particular relevant system rather than the stakeholders in the situation, and examining the situation via a single agreed model the criticism could be made that the identification with the needs of a particular group (those that agreed the model) makes the approach inherently unlikely to lead to radical change. The result is that this use of SSM led to what MacDonald (1976) has characterised as an autocratic evaluation strategy, serving those managing the activities under review, rather than a democratic evaluation, providing information for the world at large. This should not be seen as evidence of any inherent conservatism (Jackson (1982, 1983)) of the 'soft' systems methodology however but as the result of the way it was used and the requirements of the particular situation in which it was used. Using SSM in this way has been useful in situations where there has been agreement over the purposes underlying a given set of real world events, for the methodology may be usefully applied in such situations to break down high level purposes and generate acceptable measures of the efficiency, effectiveness and efficacy which may be used to examine these activities' implementation in the real world situation.

Such situations are however not typical of social action programmes and more recent action research has shown that in such situations SSM may be more usefully applied, in conjunction with various social science techniques, in a way which utilises an alternative concept of evaluation, as the enabling mechanism for organisational learning and development.The recognition that

> "The peculiarity of human and social systems lies not so much in their ability to store and handle information as in their ability to evolve and change the manifold and competing standards by which they live and in the uncertainties involved in the resolution of these conflicts."   (Vickers (1970) p 26)

has led to a view of evaluation which is not the simple feedback of information within the goal-seeking-with-feedback model of cybernetics but instead the enabling mechanism within a learning process; the role of evaluation becomes more than the simple aid to decision making accepted elsewhere, it becomes the mechanism which sustains through time the relationship between the ever changing (either through external events or through changes in perceptions) problem situation and a programme of planned activities which satisfies the needs of the stakeholders.

Whereas the 'hard' systems approaches to evaluation, for example the goal based methods, look for the achievement of certain predefined goals, this 'soft' systems view argues that some set of measures of performance may be used to measure progress towards achieving a given purpose but the decision as to whether the progress made is sufficient is something to be debated, and so too is the validity and relative importance of the given purpose itself. The evaluation activities may not only provide information about progress in respect to given perceptions of purpose but may themselves amend or change those perceptions. This 'double learning' is shown diagrammatically in Figure 1.

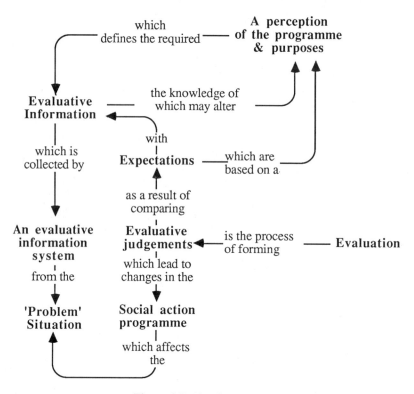

Figure 1.Evaluation as Learning

The central process here is not measurement but the social process of forming evaluative judgements. The process of making such judgements is complex, involving 'irrational' political factors as well as considerations of the cost and effectiveness of alternatives, requiring evaluative information rather than simply control information and the involvement of stakeholders at all stages of the evaluation process.
Thus whilst the end result of the 'hard' systems approach to evaluation is a measurement of the achievement of some particular perceived purpose this 'soft' systems approach aims towards discussion and opinion formulation amongst the stakeholders in the situation.

The separation of the processes of collecting relevant evaluative information, defining the criteria for what is 'relevant', and making evaluative judgements implies a changed role for the analyst, for it is morally correct that control of the latter two processes should belong to the stakeholders in the situation. In one application of this approach a representative committee was formed from all of the stakeholders, including some who had no representation in any of the decision making processes; this committee regularly examined progress towards each of seven separate areas and debated whether this did or did not constitute 'success' of the programme. Here the role of 'evaluator', the person who placed value upon and judged the worth of an activity, was taken not by the analyst, nor by the programme manager or owner but by this representative committee. The role of the analyst consisted of providing that committee with an understanding of the issues and relevant information and facilitating the process by which evaluative judgements were made, rather than making such judgements.

The problems of designing a suitable 'evaluative information system' are rarely considered within the evaluation literature but adopting this view of evaluation as a continuing (rather than a 'one time only') process makes this an important task. It has been

found that whilst SSM may be used as a means of defining the evaluative infc
requirements the work of others in information systems design can be used to
advantage; the work of Mumford and others (Mumford, Land & Hawgood (1
Mumford & Weir (1979), Mumford & Henshall (1979)) concerning participati
particularly relevant in view of the moral correctness of involving stakeholders
stage of the evaluation process.

RELATIONSHIP WITH OTHER APPROACHES TO EVALUATION

This view of evaluation may, when considered as part of the larger evaluation field, be
most accurately located with those approaches which may be grouped together under the
general heading of illuminative evaluation (Lewis (1988)), the small but important body of
evaluation literature which declares its roots to be in phenomenology rather than the
functionalism of systems engineering. It is related to other approaches within that grouping
both in terms of how it regards the process of evaluation and in its rejection of 'hard'
systems thinking in relation to what may be perceived as human activity systems. The
similarities can be seen if Guba's (1978) comparison of naturalistic and conventional
methods of inquiry is modified, as in Table 1, in order to include 'soft' systems thinking.

Table 1. Basic differences between naturalistic, conventional methods of inquiry and 'soft'
systems thinking.

Philosophical base	Logical positivism	Phenomenology	Phenomenology, hermeneutics.
Inquiry paradigm	Experimental physics	Ethnography; investigative journalism	Singerian
Purpose	Verification	Discovery	Learning
Stance	Reductionist	Expansionist	Holistic
Framework/design	Preordinate/fixed	Emergent/variable	Emergent, negotiated,adaptive
Style	Intervention	Selection	Facilitating learning
Reality manifold	Singular	Multiple	Multiple
Value structure	Singular	Pluralistic	Pluralistic. Expressed via concept of weltanschauungen and appreciative systems
Setting	Laboratory	Nature	Real-World
Context	Unrelated	Relevant	Relevant (specifically investigated via analysis 1, 2 and 3)
Conditions	Controlled	Invited interference	Naturally occurring. Method of inquiry adapts to changes in problem situation
Treatment	Stable	Variable	Variable (implicit in iterative nature of methodology)
Scope	Molecular	Molar	Molar in early stages, focussing later through definition of 'relevant' systems
Methods	Objective - in sense of inter-subjective agreement	Objective - in sense of factual / confirmable	Negotiated agreement of perceptions through debate and learning

Such approaches have in the past been criticised for their non-practicality in real world
problem situations and inherent conservatism; Eraut (1982) has pointed out too that
illuminative evaluation's concentration upon issues identified by those in the situation may
have disadvantages when those individuals are not aware of alternatives or motivated
towards their achievement. The use of SSM, however, allows the exploration of alternative
issues and concerns, and the explicit expression and debate of these by means of root
definitions and conceptual models; it tackles the problems of values and multiple perceptions

great rigour and in a way which is conducive to the formulation of useful and socially acceptable evaluative judgements. In particular the richness of concepts such as 'appreciative system', 'relevant system' and the methodology itself, provide the operational support which is lacking in so many of the proposed illuminative approaches to evaluation.

In many other situations 'soft' systems thinking has proved to be a most powerful way of achieving illuminative evaluation's intentions of

"...challenging conventional assumptions, disentangling complexities usually left in a muddle, isolating the most significant processes, and generally by raising consciousness and the level of discussion." (Parlett et al (1977) p.34)

Recent work suggests that in evaluation also the 'soft' systems methodology is a particularly flexible and useful approach to the needs of evaluation in complex social settings.

REFERENCES

Checkland, P.B., 1971, A Systems Map of the Universe, Journal of Systems Engineering, 2, 2, 107-114.
Checkland, P.B., 1978, The Origins and Nature of 'Hard' Systems Thinking, Journal of Applied Systems Analysis, 5, 2, 99-110.
Checkland, P.B., 1981, "Systems Thinking, Systems Practice." Wiley, Chichester.
Easterby-Smith, M., 1986, "Evaluation of Management Education, Training & Development", Gower, Aldershot.
Eraut, M., 1982, Handling Value Issues. in: "Evaluation Studies Review Annual. Volume 7.", E.R. House, S. Mathison, J.A. Pearsol and H. Preskill, eds., Sage, Beverly Hills.
Guba, E.G., 1978, "Toward a Methodology of Naturalistic Inquiry in Educational Evaluation.", CSE Monograph Series in Evaluation. 8. Center for the Study of Evaluation, University of California. Los Angeles.
Habermas, J., 1970, "Toward a Rational Society", Beacon Press, Boston.
House, E.R., 1982, Introduction: Scientific and Humanistic Evaluations. in: "Evaluation Studies Review Annual. Volume 7.", E.R. House, S. Mathison, J.A. Pearsol and H. Preskill, eds., Sage, Beverly Hills.
House. E.R., 1984, How We Think About Evaluation in: "Evaluation Studies Review Annual. Volume 9.", R.F. Conner, D.G. Altman and C. Jackson, eds., Sage, Beverly Hills.
Jackson, M., 1982, The Nature of 'Soft' Systems Thinking: The Work of Churchman, Ackoff and Checkland., Journal of Applied Systems Analysis, 9, 17-29.
Jackson, M., 1983, The Nature of Soft Systems Thinking: Comment on the Three Replies., Journal of Applied Systems Analysis., 10, 109-113.
Lewis, P.J.,1988, "The Evaluation of Social Action Programmes: A 'Soft' Systems Approach.", Ph.D. thesis, University of Lancaster, Lancaster.
MacDonald, B., 1976, Evaluation and the Control of Education in: "Curriculum Evaluation Today. Trends and implications.",D. Fawney, ed., MacMillan Schools Council, London.
Mumford, E., Land F. and Hawgood J., 1978, A Participative Approach to the Design of Computer Systems, Impact of Science on Society 28, 3,.235-253.
Mumford E. and Weir M., 1979, "Computer Systems in Work Design, the ETHICS method.", Associated Business Press, London.
Mumford, E. and Henshall, D., 1979, "A Participative Approach to Computer Systems Design.", Associated Business Press, London.
Parlett, M., Jamieson, M. and Pocklington, K. ,1977, Concentrating on the Audience in: "Introduction to Illuminative Evaluation: Studies in Higher Education.", M.Parlett, and G. Dearden, eds., Pacific Soundings Press.
Steinmetz, A., 1975 The Ideology of Educational Evaluation., Educational Technology, May, 51-58.
Vickers, G., 1970, "Freedom in a Rocking Boat.", Penguin, Harmondsworth.

IS SPEAKING MAN AN ETHICAL CREATURE?

Otto Hansen

Departments of Systematic Theology and General and Applied Linguistics
University of Copenhagen

INTRODUCTION

More than a century and a half ago Wilhelm von Humboldt wrote a paragraph the first sentence of which reads: "The connexion between the form of sound and inherent laws of language constitutes its completion." That Humboldt's thoughts about inherent laws of language were ahead of his time is almost pathetically indicated by Steinthal (1884, pp. 362-367), Humboldt's learned commentator from whose critical edition I have translated. He states immediately that he has little or nothing to contribute for elucidation of the paragraph. Ordinarily quite level-headed, he gets rather desperate and declares that while expressly resigning his "office as interpreter", he would not dare just to reject a thought by Humboldt. Steinthal (1884, pp. 1-5) had also in his preface explained that through his many years of work on Humboldt's linguistic texts, he often had the feeling that there was still more behind the words - something he could not fully grasp. Probably the difficulty had been summarized by Humboldt (1820) himself: "Man is only human through language, but to invent language he had to be human."

Then after the narrow and generally trivial period of nineteenth century historical language study, Noam Chomsky catalyzed renewal of the philosophical approach to human language with the theory of universal grammar (1976 a, pp. 207-209) as a hypothesis concerning the aquisition of the language faculty, with language being (1976 b) "a fixed function, characteristic of the species, one component of the human mind, a function which maps experience into grammar". The many misunderstandings concerning Chomsky's "universal grammar" and transformational technique generally have largely been due to not fully grasping that the value of transformational linguistics must be sought in its methodological scientific perspective, not in any practical pedagogical application.

THE BIOLOGY OF THE LANGUAGE FACULTY

If there are any inherent laws of language in Humboldt's meaning of that which made the human species human, or as a species characteristic mapping experience into grammar, as Chomsky put it, then the language capacity cannot be understood as the construction of a highly developed intelligence generally. Therefore appropriately Chomsky (1976 b) found it "not unreasonable to approach the study of language as we would some organ of the body". There would be good biological reasons to take this suggestion seriously. What clearly appears the special characteristic of the mammalian brain as such is an increasing development of the mid-dorsal upper part of the cerebral hemispheres, the neocortex. Deeper down in the brain we have the older limbic system. The latter does show increased elaboration with ascent towards man, and it does indicate evolutionary relation to the developing neocortex. Furthermore there are many similarities between the human limbic system and that of lower forms. Nevertheless animal vocalization is generated by tissue within the limbic system, while human speech is generated by tissue within the neocortex. This distiction is firmly evidenced on both the animal and the human side. Planned experiments have been carried out with animals, and although one cannot operate experimentally on the human brain, already the accumulated clinical material made it largely impossible to question the conclusion which eventually was confirmed by noninvasive isotope techniques. It was the philosophical conflict about the descent of man that produced the old Darwinian notion that the differences between human language and animal cries were of degree and not of kind. It is among non-biologists the misconception is kept alive that the neural apparatus supporting animal vocalization is related to that producing human speech. I have extensively dealt with this unfortunate psychologists' archaism (Hansen, 1978, pp. 4-5, 32, 42-43, 45; 1979; 1980, pp. 24-25, 29-30; 1981 a; 1981 b; 1985, pp. 35-37).

Human Language As Genetically Determined

Reformulation in neurological terms of the observations of transformational grammar as it dealt with the rules necessarily followed in the generation of one syntactic structure from another, with the necessity of the rules taking them back to the mode of conversion of thought into extended speech, implied the hypothesis of human language as genetically determined (Hansen, 1981 a: 1981 b; 1985, pp. xi-xii, 51, 80-81; 1986 a), and when made explicit this hypothesis was considerably strengthened by clinical investigation of communication impairments (Hansen, 1985, pp. 81-86; 1986 b; Hansen et al., 1986; Hansen et al, 1987).

THE ETHICAL IMPLICATIONS

With biological confirmation of the uniqueness of human language, a modern agreement as it were with the claim that only through language became man human, the corollary followed that ethics entered evolution with language (Hansen, 1979). Evolution has linked all other elements in human behaviour with language, and if language caused man's dominance of the earth, then ethics entered in any case with language. Without that dominance there would be no moral choices to make; with it they had to be made (Hansen, 1978, pp. 11-12, 50; 1979; 1980, pp. x, 66; 1981 a; 1981 b). If then language was genetically determined, we might also find that there were biologic rights and wrongs (Hansen, 1985, pp. x-xii, 23, 61, 66). Chomsky (1976 a, p. 133) found it "reasonable to suppose that just as intrinsic structures of mind underlie the development of cognitive structures, so a "species character" provides the framework for the growth of moral consciousness, cultural achievement, and even participation in a free and just community". And Young (1974, p. 623) suggested fundamental religious tenets to be inherent in neural structures closely associated with the areas of the brain involved in the production of language.

A Double Systems Approach

It seems pertinent to examine if a double systems approach including both our biological knowledge concerning language and the Medieval European theological views on the endowement of speech might throw light on the deontological canons we have developed and the ethical perspective of human society of today. First two false systems must be invalidated, however. I have maintained (Hansen, 1985, pp. 43-44, 67-68) that if we did not accept human language to have arisen de novo, i.e. in evolutionary terms without any gradual development from any form of animal communication to the superior coordination of human behaviour that transformed humanity's place in the natural order (McNeill, 1980, pp. 10-11), then we had accepted a behaviourist view on human nature and behaviour. Actually the behaviourist view has wide acceptance, and superficially it might seem to offer a theoretical possibility of opposition to the inseparability of human language and ethics. The behaviourists distinguish between two learning processes (Hansen, 1978, p. 36). In rational learning, in which essentially the central nervous system is involved, competence is acquired through reinforcement. The process of conditioning works by contiguity. There is no reward in any sense of the term. According to the behaviourists, moral behaviour is conditioned behaviour. As far as the individual is concerned, it is accidental if something is moral or immoral. Pavlov noted (Salter, 1965) that for the human species, speech provides conditioned stimuli just as real as any other stimuli. Apparently language acquisition and language use could be separated. Behaviourism has not, however, produced any evolutionarily applicable theory. In Chomsky's (1976 a, p. 20) words there is no reason to believe that nonexistent learning theory has produced any theory of behaviour. It may be noted that it is a behaviourist prerequisite that language is something a child learns by imitation. That this is not true is easily seen. If, as Crystal (1976, pp. 34-35) observed, imitation were the governing principle in childrens' language-acquisition, then we would expect them to produce rather different patterns in their language than in fact they do, and not to produce some of the patterns they do.

It is a different matter that the neural basis of conditioning does exist, and the clinical behaviourists can show not a few successful cures of certain sexual and compulsive neuroses which are generally intractable to dynamic psychotherapy. But the behaviourists have produced no theory in the scientific meaning, and a successful technique can therefore validate nothing. Although I mean no disrespect to the nuclear physicists, to their toil, or to the importance of their work, there is some similarity to the behaviourists' situation in the transfer to philosophy of the uncertainty relations and complementarity concept of quantum mechanics.

Heisenberg worked out his uncertainty principle as a methodological solution to the problem that while in classical dynamics independent numerical values could be ascribed to coordinates, in quantum mechanics position and momentum cannot be simultaneously determined. It is creating a false system when for example Lucas (1970, pp. 112-113) argues quantum mechanics as philosophically complete. Quantum mechanics cannot, he claims, be a partial expression of a more complete system which after addition of further parameters would prove deterministic. Indeterminacies could not logically be removed, he felt. His formulation is that whichever hidden variable were added, non-commuting operators would continue to have different sets of eigenvalues. The eigenvalue as used by the physicists is nothing more, however, than the numerical value of the operator under consideration. Heisenberg's problem was randomness at the atomic level. Bohr's extension of the uncertainty relations into his principle of complementarity did not change anything philosophically but probably carry much responsibility for the meaningless association with the question of randomness at the macroscopic level. The error, whether made by physicists drawing philosophical consequences from the uncertainty relations, or by philosophers misinterpreting the physicists, is at the evolutionarily conceptual level. It is because of the lack of a common eigenfunction that the operators do not commute. This is logical enough on the conditions of the language the physicists have chosen. This language expresses a reality of measurement possibilities. But it does not allow the philosophical conclusion that God did throw dice, to paraphrase Einstein's famous wording of protest. The uncertainty relations of quantum mechanics do not include time. Time would remain a number, Prigogine and Stengers (1985, p. 226) said, not be an operator, and only operators can appear in Heisenberg's uncertainty relations. Probably the whole affair cannot be better summarized than they did (1985, p. 231): "In quantum mechanics, the wave function evolves in a deterministic fashion, except in the measurement process." Actually the philosophical misuse of the uncertainty relations is not only phenomenology and therefore scientifically invalid through lack of any theory to refute or confirm. It is also basically immoral as it denies reality.

The Medieval European View

Speaking at the University of Copenhagen in 1986, Umberto Eco said that without a deep understanding of Medieval Europe, we cannot hope to understand the human situation of today. And the importance of our Medieval European background is by now noted by several workers. Prigogine and Stengers (1985, p. 46) refer to the role of theology. "At the origin of modern science," they say, "a "resonance" appears to have been set up between theological discourse and theoretical and experimental activity." They further (p. 47) cite Whitehead (1967, p. 12) for the research secret of the European mind being the Medieval insistence on the rationality of God. Thomas Aquinas (<u>Summa contra Gentiles</u>, III, 104 (3)) maintains that "the very act of speaking is a characteristic of rational nature". He has two arguments. Firstly (III, 147 (2)) human beings have been given comprehension and reason enabling them to discern and investigate truth, and they were therefore also given the use of speech so that the truth absorbed could be made clear to others. That human beings should help each other to know the truth would follow, since man according to Aristotle (in <u>Politics</u>) was a naturally social animal. Secondly (III, 154 (7)) the knowledge obtained could not possibly be conveyed without dialogue, and the grace of speech was necessary for fulfilment of the obligation under divine order to propagate revelation received. If a modern reader might feel a lack of distinction between types of argument, the cause is not confusion in that mind whose clarity of thought may be rivalled, in the whole of European history of ideas, only by the philosopher Thomas held in such high regard: Aristotle. What spurred Thomas Aquinas was precisely his wish to demonstrate the coherence of human knowledge, this knowledge being both the truths open to inquiry by human reason and that truth which surpasses reason.

The evolutionary mixture of necessity and chance is to be understood through the concept of dissipative structures (Prigogine and Stengers, 1985, pp. 12-14, 142-143, 148, 170, 189, 300). Dissipative structures are those that arise from nonlinear processes in nonequilibrium systems. The theory of dissipative structures, his work on which was the citation for awarding Prigogine the Nobel Prize, must, Prigogine and Stengers (1985, p. 171) claim, take us back to Aristotelian conceptions with their inspiration in the organization and solidarity of biological functions. Historically Aristotle was brought back to European influence when retranslation into Latin directly from the Greek removed Arabic misrepresentations so that he would be used to solve what was called the "philosophical problems", i.e. the construed enmity between reason and faith. The point of connexion is very precisely put by Thomas (<u>Summa contra Gentiles</u>, III, 9 (1-2)) claiming the standard in moral matters to be reason, with the extended teaching that evil is the privation of good. I earlier (Hansen, 1980, p. 93) said that he is even specific to the point of agreement with modern systems researchers in admitting that something is more or less evil in reference to the good it lacks (III, 9 (5)). This formulation is of theological interest as it appears centrally in the Scholastic definition of original sin (<u>Summa contra Gentiles</u>, IV, 52), which, contrary to popular belief, was close to verbatim taken over by the Lutheran symbolic texts. As to the theological reality of original sin, I must here confine myself to Thomas' observation that certain signs "make it probable that original sin is present in mankind". Concerning the consquences Søren Kierkegaard in <u>The Concept of Dread</u> (1923, IV, p. 315) notes that while the reality of thought was the presupposition in the whole of classical and Medieval philosophy, then this presupposition became doubtful with Kant. As for the emergence of language, Kierkegaard is no less specific or biologically oriented than was Thomas. Kierkegaard (1923, IV, pp. 350-353) stresses that the major mistake is the mythical interpretation of the Fall. The Biblical account is the description of the reality of humanity's biological origin and condition. Adam and Eve are numerical repetition inasmuch as every individual is both this individual and the species, through sexuality becoming a synthesis of sould and body and getting history. "Adam had been created, and he had given names to the animals (language thus in being, even if in a way incomplete as when children from the hornbook learn to know an animal). Yet he had found no partner. Eve had been created; built up from his rib. She was as close to him as she could be. But still they were not intimate." Kierkegaard has a footnote (p. 352) saying that the question of the acquisition of language does not belong to the investigation at hand. But he shall not evade the absolute premise that "it is not permissible to have man having invented language himself".

IS THEN SPEAKING MAN AN ETHICAL CREATURE?

The historical Kantian view of the human being as a goal per se is after all only a well established but unfounded teleological interpretation of nature, or in other words just a form of the naturalistic fallacy of concluding from an "is" to an "ought", i.e. deriving moral claims from the factual. Looking closer at evolution as natural history, we shall soon se that nature is not "good". The biological history of the human species is the history of continuous war with natural selection. Without that war we would of course have no problems. We would not be here. It is easy to demonstrate our ethical mistakes, but to argue the very presence of the human species to be unethical is ridiculous. Fundamentally the change we must invoke is a new acceptance of the presence of evil defined as privation of good. One of the important biochemists of our time (Chargaff, 1976) asked (discussing recombinant DNA) if we "had the right to counteract, irreversibly, the evolutionary wisdom of millions of years?", and Michael Ruse (1981, pp. 184-185), one of the important philosophers of science of our time, answered that it is absolute nonsense to talk about the evolutionary wisdom of millions of years. One of the examples he gives of the naturalistic fallacy is the conclusion that the efforts of the WHO in eliminating smallpox were immoral, because they made a certain species go extinct.

The answer to the question if speaking man is an ethical creature is clearly in the affirmative. Human beings do decide between right and wrong, and the deliberation taking them to their decision is based in the logic of the inherent grammar of their language. In other words, the standard in moral matters is reason. The theological comment to the many unhappy decisions is that the privation of original justice as incurred by the Fall has the consequence that "the inferior powers are not perfectly subjected reason and the body not the soul" (<u>Summa contra Gentiles</u>, IV, 52 (6)). Irrespective of one's immediate attitude to the Medieval European belief behind the choice of words, it must seem quite reasonable that we thoroughly acquaint ourselves with that formative part of our past from which we were so strangely disconnected.

REFERENCES

Aquinatis, S. Thomæ Doctoris Angelici <u>Opera Omnia</u> (Sixteen Volumes). Iussu impensaque Leonis XIII P.M. edita ex Typographia Polyglotta, Rome 1882-1948.

Chargaff, Erwin (1976) On the dangers of genetic meddling. Science 192, 938-940.

Chomsky, Noam (1976 a) Reflections on Language. Pantheon Books, New York.

Chomsky, Noam (1976 b) On the nature of language. Ann. N.Y. Acad. Sci. 280, 46-57).

Crystal, David (1976) Child Language, Learning and Linguistics. Edward Arnold, London.

Hansen, Otto (1978) The Genes of Universal Grammar. An Attempt in Biologicized Philosophy. University of Lund, Lund (ISBN 91-970270-0-6).

Hansen, Otto (1979) "Human Language as a Biological Behaviour Determinant" in Improving the Human Condition: Quality and Stability in Social Systems, Richard F. Ericson, ed., Springer-Verlag, Berlin, Heidelberg, New York, pp. 961-970.

Hansen, Otto (1980) Speech and Language. A Compendium in Empirical Sociobiology. University of Lund, Lund (ISBN 91-970270-2-2).

Hansen, Otto (1981 a) Are the genes of universal grammar more than structural? Hereditas 95, 213-218.

Hansen, Otto (1981 b) "Biological Study of Language as a Key in General Systems Research" in General Systems Research and Design: Precursors and Futures, William J. Reckmeyer, ed., SGSR, Louisville, pp. 547-555.

Hansen, Otto (1985) Biological Linguistics. Akademisk Forlag, Copenhagen. (ISBN 87-500-2573-2).

Hansen, Otto (1986 a) "Sociobiology and Biological Linguistics" in Essays in Human Sociobiology Vol. 2, Jan Wind and Vernon Reynolds, eds., V.U.B. Study Series No. 26, Brussels, pp. 129-139.

Hansen, Otto (1986 b) "Sign Language of the Deaf as Natural Language" in Signs of Life, Bernard T. Tervoort, ed., Institute of General Linguistics, University of Amsterdam, Amsterdam, pp. 57-61.

Hansen, Otto, Nerup, Jørn, and Holbek, Bertha (1986) A common genetic origin of specific dyslexia and insulin-dependent diabetes mellitus? Hereditas 105, 165-167.

Hansen, Otto, Nerup, Jørn, and Holbek, Bertha (1987) Further indication of a possible common genetic origin of specific dyslexia and insulin-dependent diabetes mellitus. Hereditas 107, 257-258.

Humboldt, Wilhelm von (1820) Ueber das vergleichende Sprachstudium in Beziehung auf die verschiedenen Epochen der Sprachentwicklung. Abhandlungen der Akademie (Berlin) as printed in H. Steinthal's critical edition Die sprachphilosophischen Werke Wilhelm's von Humboldt. Ferd. Dümmlers Verlagsbuchhandlung, Berlin 1884.

Kierkegaard, Søren (1923) Samlede Værker Vol. IV, A.B. Drachmann, J.L. Heiberg, and H.O. Lange, eds., Gyldendal, Copenhagen.

Lucas, J.R. (1970) The Freedom of the Will. Clarendon Press, Oxford.

McNeill, William H. (1980) The Human Condition. An Ecological and Historical View. Princeton University Press, Princeton.

Prigogine, Ilya and Stengers, Isabelle (1985) Order Out of Chaos. Man's New Dialogue with Nature. Fontana Paperbacks, London.

Ruse, Michael (1981) Is Science Sexist? And Other Problems in the Biomedical Sciences. D. Reidel Publishing Company, Dordrecht, Boston, London.

Salter, Andrew (1965) "The Theory and Practice of Conditioned Reflex Therapy" in The Conditioning Therapies, J. Wolpe, A. Salter, and L.J. Reyna, eds., Holt, Rinehart and Winston, New York, Chicago, San Francisco, Toronto, London.

Steinthal, H. (1884) Die sprachphilosophischen Werke Wilhelm's von Humboldt. Ferd. Dümmlers Verlagsbuchhandlung, Berlin.

Whitehead, Alfred North (1967) Science and the Modern World. The Free Press, New York.

Young, J.Z. (1974) An Introduction to the Study of Man. Oxford University Press, London, Oxford, New York (Cited from the Paperback Edition, first published 1971).

INVENTION OF "MASTER SYSTEM" AS DISTINCT FROM EXPERT SYSTEM -

ITS SUPERIOR ROLE IN THE PERSPECTIVE OF THE COMPUTER EVOLUTION

Asim K. Datta

Operational Research Society of India
7/3 Mandeville Gardens
Calcutta - 700 019
India

PROLOGUE

"Can machines rule ?" is the question which forms the fundamental theme of this paper. There is no pretension to compare this with the epoch making question "Can machines think ?" as asked by Turing (1950), the legendary British Mathematician. That was an unparalleled event in the history of computer marking the very beginning of the man´s concept and endeavour towards Artificial Intelligence. The present question rather arises from the simple extrapolation of the future trend in the perspective of the evolution of computer generations viewed from a different angle.

EVOLUTION OF COMPUTER GENERATIONS - A DIFFERENT VIEWPOINT

Conventional Approach

The computer generations by the conventional approach based on the the types of hardware and software used are listed in Table 1.

A Different Viewpoint

An insight into the underlying role played by the computer reveals that a generation change characterised in terms of hardware and software is also marked by a radical improvement in the quality of humanlike intelligence discharged by it in performing its vastly advanced role. As a matter of fact there exists a strong correlation between the upgradation of computer generation and the level of intelligence discharged by the computer.

Table 1. Conventional Approach for Classifying
Computer Generations

Generation	Hardware	Software
I (1940s/50s)	Valve, Card Puncher and Reader(Mark-I,ENIAC)	Binary Machine Code
II (Late 1950s/ early 1960s)	Transistor,Magnetic Tape (IBM 1401)	Assembly Language
III (Late 1960s/ 1970s)	IC, Hard Disk (IBM 360/370)	High Level Languages, Compilers, MIS and OR/MS Packages
IV (1980s)	Microchip,Terminals,PCs, Networking	4GL, Time Sharing,User Friendly,Communication
V (years ahead)	Parallel Processor,Optical Disk, Neurocomputers, Super Global Networking	Artificial Intelligence (Expert System, Natural Language,Speech,Vision, Robotics),Global Family

Correlation. The underlying correlation is shown in Fig. 1.

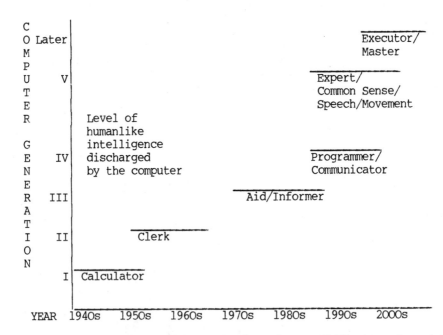

Fig. 1. Evolution of computer generations from a different viewpoint. The correlation showing the upgradation of computer generation as a function of the humanlike intelligence discharged by it.

CONCEPT OF "MASTER SYSTEM"

Extrapolating in the perspective of the evolution viewed from the above angle it is quite logical to expect that the "advisory" role of the expert systems will in course of time be evolving into a superior role of an "executor" and a system which can do so is baptised as "Master System" (MS). Hence the distiction between ES and MS is quite obvious. Whereas an ES plays the role of an "advisor" in recommending a course of action, the MS administers as a "ruler" in enforcing the proper action to be taken. Thus MS assumes a superior level of humanlike responsibility than that of an ES.

The major advantages of an MS are derived from the fact that it has no flesh and blood nor any sentiment and emotion and hence its administration is absolutely unbiased and impartial and more significantly completely immune from the human weaknesses like greed, favouritism, fear and succumbing to persecution or coercion.

INTRODUCTION OF EDFA LOOP STRUCTURE

In this paper an EDFA loop structure has been defined and introduced formally although in some applications this principle is already being followed.

EDFA Loop

EVENT-DATABASE-INFO-ACTION (EDFA) loop works in the manner of a feedback control loop system similar to that of a typical servocontrol mechanism used for mechanical control as shown in Fig. 2.

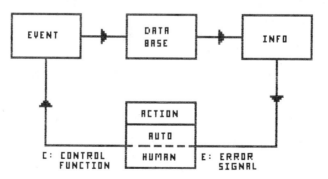

Fig. 2. EDFA loop structure working in the pattern of a Servocontrol Mechanism where the corrective control action C is a function of the error signal E, where E = D (Desired Level)- A (Actual Level), both D and A being available in the database. The corrective action C is aimed at controlling the next relevant event thus closing the loop.

<u>Measures</u>. The two important measures of an EDFA loop are:

$$SL = \frac{\text{Number of events properly controlled}}{\text{Total number of events}} \times 100 \qquad \text{(i)}$$

which is a measure of the level of success of a particular system in percentage of the events properly controlled by the action taken in the ACTION module whether automatic or human.

$$DA = \frac{\text{Total weighted measure of auto actions}}{\text{Weighted sum of all actions}} \times 100 \qquad \text{(ii)}$$

which is a measure of the degree of automation achieved in percentage.

<u>INFO Module</u>. The function of this module is to evaluate and generate error signal E by accessing D and A from the Database and feed this to the ´Action´ module of the on-line system.

<u>ACTION Module</u>. As shown in Fig. 2 the control action module has two distinct components - one is "auto" and the other is "human". Auto type of action is automatically initiated under normal fluctuation of circumstances according to some deterministic procedures within the specified tolerance limits. For this purpose built-in MS/OR models and other software packages may be used as necessary. However, there are some actions that can only be taken by the people having the right experience and expertise (unless replaced by competent Expert Systems).

<u>EVENT Module</u>. In this module the particulars relating to relevant events get entered into the main on-line data base. The events may be classified into two major types - internal and external. Internal events are initiated internally via the ´Action´ module (human or auto) of the EDFA loop such as purchase ordering. External events on the other hand originate from the external world outside the boundaries of the EDFA loop such as arrival of goods purchased.

<u>DATABASE Module</u>. This is the main database of the EDFA loop excluding the knowledge bases used exclusively by the corresponding Expert Systems, if any.

THE MASTER SYSTEM

<u>Definition</u>

The "Master System" (MS) applied in a specific area is defined as a computer based fully automated supreme head on all aspects of the execution, management and control of the operation in that area.

<u>Basic Structure</u>

The MS uses an EDFA loop working like a feedback control loop system. Its structure and the basic components with interaction between them are shown in Fig. 3.

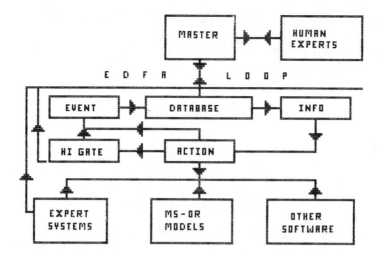

Fig. 3. The basic structure of a Master System with the "Master" module
sitting on the top of the EDFA loop performing two main
functions. (a) To control the total operation of the EDFA loop
which includes resolving of any conflict with the "Human
Interface" (HI) gate. (b) To seek the advice and knowledge of
the external human experts as necessary like an expert system
and then enforce the same in the area of operation discharging
the responsibility of that of the topmost "executive".

Essential Features and Major Characteristics

(a) The DA (Degree of Automation) as measured according to the
formula (ii) must always be 100 %. (b) However, there is a "Human
Interface" (HI) gate in the branch of the loop between the ACTION and
EVENT modules. The basic function of this gate is to either "agree" or
"differ" from the action initiated by the auto ACTION module. In case
of any conflict, the matter gets referred to the "Master" module giving
reasons of disagreement by the HI. (c) In case of such a conflict the
Master module will seek the services of two independent and best suited
human experts from its panel of experts in the knowledge base. If they
both agree either on the auto action or on a different action, their
decision will be binding. Otherwise the opinion of a third human expert
will be sought and his verdict will settle the issue.

Execution Engineering

We are already familiar with the Knowledge Engineering (KE) in the
development of an Expert System. Execution Engineering (EXE) is a new
role required for the designing of a Master System. The main difference
between the KE and the EXE is whereas KE works in isolation for the
development of an ES which is in itself a finished prduct and may be
used in any manner even as a part of a larger system requiring such
expert advice, EXE has to design the entire structure of an EDFA loop
which may contain a number of ES to work in the on-line mode mainly
concentrating on embodying the eventual "kill" attributes of a human
executor in a specific functional area and the major criteria that are
considered before a recommended action is put to final execution.

Backward Evaluation

"Backward Evaluation" is a convenient technique for designing an EDFA loop by listing the external events and then evaluating backwards to assess the control actions that can make those happen properly and the error signals required to initiate those corrective actions.

APPLICATION POTENTIAL

The potential area of the application of the MS is very wide with exciting possibility of even working towards a model for unbiased auto surveillance and maintenance of world peace. However, one application of the MS is highly feasible even today with tremendous social benefits in some of the developing countries where a substantial portion of fund allocated to development projects like construction of roads, bridges, hospitals and schools gets "eaten" away through shameful racket that exists involving some unscrupulous administrators, politicians and contractors. In a paper by Datta (1987) a prototype MS has been demonstrated with a model on road construction project.

EPILOGUE

The "Master System" is an absolutely original concept in the computer discipline. The evolution of the humanlike role discharged by the computer also provides an entirely new direction to foresee what is in store for the posterity. The invention of the MS only marks the beginning of the future era with totally automated operation universally and a super MS controlling and executing all global operations automatically. The human beings practically have to do nothing but to occasionally interact with the system staying at the comfort of the home to see whether all is going smoothly and very rarely put some effort to give response to the master module in case of extremely abnormal situation. Yet the mankind in general will enjoy the rich dividends of the wealth and affluence generated by the super MS without any exertion. Is this merely a day-dream ? Well, only time can tell.

All said and done, we began by asking the question "Can machines rule ?" succinctly. If we are somewhat hopeful about that why not look ahead at its heart and conclude by asking can machines love ?

REFERENCES

Datta, A. K., 1987, A Computerised Model to Combat Unethical Practices for Proper Utilisation of Development Funds in Developing Countries, Official proceedings of the SEARC Academy of International Business, Kuala Lumpur.

Turing, A. M., 1950, Computing Machinery and Intelligence, MindLIX.

THE SOCIAL SYSTEM OF NATIONAL HEALTH CARE

Ferenc Kun

Department of Social Medicine
University Medical School Debrecen
Debrecen, Hungary

Scientists who have a certain feeling of responsibility for the fate and future of mankind draw our attention much more often and more definitely to the fact that the development of society is impossible in the present way /1/.

Among the factors eliciting this new situation the results achieved in the research of natural sciences and their application in praxis have a major role /automatization, computer technique, nuclear energy, space research, chemistry, biology, genetics, etc/. The application of these results of research in praxis exert not only a positive /advantageous/ but also a negative influence with disadvantageous consequences in nature and the environment surrounding man and thus, in an indirect way in human biology and in several instances in a direct way in the biological nature, body of man.

This process which is gradually gaining more and more importance has become extremely dinamic recently. The new expressions indicating"man's taking possession of nature" show its dimensions. Thus, the terms "anthropogenous nature" and "noosphere" have been introduced in ecology and in philosophy, respectively.

Let us here remark that these changes and their possible consequences are even more radical and important than it was suspected earlier. It has turned out that not only the increase itself or the development are at stake but mankind has come to such a crisis regarding its prospects that the alternative of "survival" and "non-survival" has to be weighed. It is a good thing and a welcome condition improving our prospects that more and more politicians who feel themselves responsible for the fate of mankind recognize this situation.

The changes, however, are so expansive as regards their contents that everybody has to look for a solution in his special field of activity. The representatives of sciences have

---

1. The most important and well-known work of warning is "The Limits to Growth. A Report for the Club of Rome's Project on the Predicament of Mankind. N.Y.1972, Problems of World Modeling. Political and Social Implications." Ed. K.Deutsch et al. Cambridge /Mass./ 1977, etc.

an outstanding responsibility in this work. As far as we are concerned the question has arisen what the scientists dealing with system research, operation research can do in this respect.

Another feature originating from the change is: the world wide change in the conditions of conscious human activity due to the development in technique and conveyance and communication. A further characteristic of the new situation is that the various spheres of social life and social activity which are divided into a par excellance and a vulgar /i.e. exercised by the inhabitants/ part interweave with each other and they can often be separated on the basis of their function only. Health care which means taking care of people's health at a social level can be considered one of these social spheres /which is in this respect a social subsystem/ gaining gradually more and more importance.

This intervolvement and overlapping and the condition resulting from it that these processes can be assessed adequately globally and as a complexicity make their theoretical examination and the conscious interference, and especially the prognosis more difficult. The fact that the assessment has not been elaborated in theory and praxis yet, our work ends in failure day by day. Most authors think that this must result from fragmentation i.e. the connections between the parts, and the parts and the whole, respectively, in the theory and praxis of social life and in sciences dealing with their development are not taken into consideration as a totality. Our viewpoint in the case of health care is: the relevant manner of solution is to organise a global, institutionally established social subsystem which -beyond healing- validates in its all-social globality and complexity the preventive medicine and the health promotion, that is its activity extends all spheres /technical, economical, cultural, etc./ of life of society. The way of solution: establishing Health Care based on systems- and operational research and its enforcement adequately applied to the given sphere of society. The first step of solution: creation of such a philosophically-logically-semantically established and systematized social theory which allows the establishment of a system of conditions for conversation and promotion of the quality of life by forming the way of thinking, activities and relations of people. We call this system of conditions THE SOCIAL SYSTEM OF NATIONAL HEALTH CARE.

The most important conclusions drawn in the literature recently have come to the idea that the most effective way out is the view and praxis of human life and health culture as a central social factor, as a central reference system. Thus, the new way of assessment of the term "development" and of the criteria of social values has been raised taking these points of view into consideration /1/./see the bottom of the next page/.

The book by John Kenneth Galbraith "The New Industrial State" is of special importance as the author deals with the limits of development of the industrial system and the dangers resulting from the exclusive predominance of the aims of development /2/. It is suggested that the man of future will not be the "homo economicus" but the "homo sanus". However, the way of development bringing the conditions for changes in health service and the systems of activity on the one hand and on the level of the total social system on the other hand has not been discovered yet. Galbraith is very realistic when stating why the quality of life as the theory of the main human values and that of the basic criteria of development are thrust into the background. Besides several of his arguments this is what he writes in the Addendum /point 3/: "There is a further advantage in economic goals. The quality of life is subjective and disputable. Cultural and aesthetic progress cannot easily be measured. Who can say for sure what arrangements best allow for the development of individual personality? Who can be certain what advances the total of human happiness? Who can guess how much clean air or uncluttered highways are enjoyed? Gross National Product and the level of unemployment, on the other hand and objective and measurable. To many it will always seem better to have measurable progress toward the wrong goals than unmeasurable and hence uncertain progress toward the right ones. But this would hardly have served the purposes of this book."

The conclusion we have drawn are the following:

The development of economy is an important condition of the development of society, however, it is the means and not the goal when it serves the foundation of development. The only trouble is if economy is held to be an end in itself. It is industry, production, economy which allows us to change the ideas the great humanist thinkers of humanity dreamt of more hundred years ago into reality.[x] The situation is,however, paradoxical. These forces and possibilities involve besides facilitation of development also the possibility of destruction. On behalf of the solution we have two theoretical tasks /which can be realized in practice/ to face:

1.Interdisciplinary and intersectoral coordination of professional and social activities /where also environmental health, health education, etc. are involved/ performed on behalf of health protection.

According to the classification of the System of National Health Care 3 levels can be distinguished
- the level of medical and curative activity where also the professional, human relations originating from the mutual task are implied.
- the level of curative-preventive service providing social frames for the curative activities.
- social activity and relations /institutions, etc./, the level of systems outside them where we can find health relations in greater or less measure.

These three levels must complete each other but to coordinate them is still ahead.

Also Galbraith misses the surmounting of subjectivity, creation of measurability. The epidemiological data of mortality and diseases allowed in our case measurablity. These data have to be connected with social processes and transfer factors which serve as risk factors and which play a role in the development of certain diseases, consequently, also the social "background" factors have to be taken into consideration to be able to exert a beneficial influence on these diseases. The following figure shows the mortality rate in the case of the most frequent diseases during the past 60 years /from 1921 to 1980/ in Hungary.

---

[x]Thus e.g. Thomas Moore cites Horace and Plato in his "Utopia". They consider health to be of the greatest value as compared to gold, etc. Francis Bacon held the prolongation of life to be the most important task of medicine.
/1/ L.Ervin et al.: Goals for Mankind, p.322;M.Mesarovic and E. Pestel:Mankind at the Turning Point, p.154./2/ J.K.Galbraith: The New Industrial State.p.414.

MORTALITY OF INHABITANTS WITH TUBERCULOSIS, CANCER AND HEART DISEASE IN HUNGARY (/10 000 INHABITANTS)✻

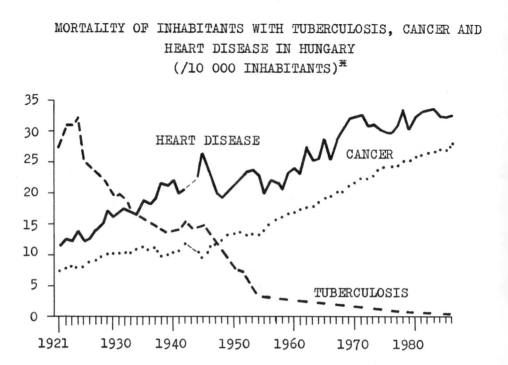

✻On the basis of the data of the central bureau of statistics

On the basis of the figure it is obvious that the structure of the diseases leading to death has changed between 1921 and 198 in Hungary radically. TBC which was held to be "morbus hunga- ricus" earlier has been driven into the background, the rate o mortality due to cancer and heart diseases has proved to be much higher than it was in the case of TBC. It would be diffi- cult to find a more significant alternative for the quality of life than disease and death. If it is so it would be difficult to find a more elevated goal of the creation of the future so- ciety than the fight against them. Research in the field of medicine provides more and more complete and exact informa- tion on the social risk factors eliciting these diseases or on those playing a role in them. The social causal system which i in their background can be influenced beneficially by social decision making supported by social sciences.

The scope of this paper allows only the introduction of the main trends of the work having been started in Hungary to find a solution to this problem. We have relied on the first place on the material issued by WHO, systems research, social medicine and the European Society for Philosophy of Medicine and Health Care[x].

---

[x]I gave a lecture on the same topic on the 31st Annual Meeting of the Society for General Systems Research held in Budapest, 1987, vol. II., p. 648.

OPTIMAL LOCATION OF SCHOOL FACILITIES

J. D. Coelho

Faculdade de Economia, UNL

Tv. Estevão Pinto
1000 Lisboa

1. THE PROBLEM

The location of educational facilities may be formulated in many different ways depending how the main components are considered and the nature of the decision process whose optimal solution is pursued.

A most typical approach (see, for example, Balinski, 1961; Hakimi, 1964; Revelle and Swain, 1970; Leonardi, 1978; Laporte, 1987) consists in assuming the study area divided into zones with the demand in each zone concentrated in one of its central points and a number of potential facility location sites that may coincide or note with the demand points. The optimization is accomplished by finding out which potential location sites offer greatest benefit according to the criteria assumed and the constraints placed to bind the feasible region. In this way, no consideration is given to existing facilities and an optimal set of new facilities would be provided.

In real life the scenario above is, however, very seldom found in the educational facility location context. It is uncommon that no facility of the type for which the optimization procedure is set up will exist in the study area. The location of educational facilities is often confined to marginal improvements to an existing situation, that may consist in opening some new facilities, closing a few others, increasing or decreasing capacities by transforming available spaces or adapting the facilities to different uses.

Another aspect that has started receiving attention (Laporte and Nobert, 1981; Branco and Coelho, 1986) is the joint location of facilities and routing of users (or goods, in different settings). The routing/location problem has a special relevance in low density regions where the attraction area for a facility can be quite extensive. Important savings may be obtained by having fewer facilities and school buses collecting facility users if a joint optimization is accomplished.

In the educational facility location modelling, we note that four main objectives are pursued:

1. Providing education services at full population coverage, this means ascribing a school to every student having in account some maximum distance or travel time constraints;

2. Minimization of overall costs - these include set up costs for new facilities, location dependant operation and transportation costs for open facilities and costs assigned to closing downs;

3. Maximization of users benefit - a measure of user benefit usually related to accessibility and therefore dependant on the location of facilities.

4. Political objectives - these are often ill-defined but, for example, may take the form of maximization of the voting population in the communities where new school are built or upgraded.

In addition, the constraints binding the optimization search are the balance between demand and supply for educational services, maximum distance or travel time imputed to users, budget and eventually some extra planning constraints.

2. MODELS

We carry on a brief review of optimization combinatorial models that fit to the location of educational facilities and point out some adjustments that make them more adequate to the peculiarities of the real world problems in the educational field that have been mentioned previously. The review will cover the following models:

A. Simple Facility Location Model
B. P-Median
C. Capacitated Facility Location Model
D. Surplus Maximization Capacitated Facility Location Model
E. Covering Location Model
F. Maximal Covering Location Model
G. Hamiltonean P-Median Model
H. Routing Location Models

The notation adopted in the formulation of the models is as follows:

I   - set of indices of zones
J   - set of indices of potential facility sites
     J_0 - set of indices of facility location sites fixed 'closed'
     J_1 - set of indices of facility location sites fixed 'open'
     J_d - set of indices of facility location sites subject to the decision
          process of open/close
P   - total number of users in the study area
p_i - number of users in zone i
y_j - boolean variable assigned to facility j
     y_j - is equal to one if site j is used to locate a facility and zero otherwise
y   - vector for location variables y_j, $j \in J$
     y_0 - vector of location variables y_j, $j \in J_0$
     y_1 - vector of location variables y_j, $j \in J_1$
     y_d - vector of location variables y_j, $j \in J_d$
ℓ_j - minimum size of the facility at site j
u_j - maximum capacity of the facility at site j
t_{ij} - flow of users from zone i to facility j
x_{ij} - proportion of users from zone i assigned to facility j
c_{ij} - 'generalized cost of assigning a user of zone i to a facility j
f_j - set up cost for facility j
f_{0j} - cost of closing down facility j
f_{1j} - cost of opening or maintaining open facility j
A,B - matrices of coefficients of planning constraints
b   - vector rhs of planning constraints
t_{ij} - apriori probability that users from zone i select facility j

A - Simple Facility Location Model

The simple (uncapacitated) facility location model was put forward by Balinski (1961) and received many contributions until a very efficient dual based solution method was proposed by Bilde and Krarup (1977) and Erlenkotter (1978). The simple facility location model is formulated as follows:

$$(A.1) \quad \underset{(t,y)}{\text{Min}} \ Z = \sum_{ij} c_{ij} \, t_{ij} + \sum_{j} f_j \, y_j$$

subject to

$$(A.2) \quad \sum_{j} t_{ij} = p_i \qquad\qquad\qquad\qquad\qquad i \in I$$

$$(A.3) \quad t_{ij} \leq p_i \, y_j \qquad\qquad\qquad\qquad\qquad i \in I, j \in J$$

$$(A.4) \quad t_{ij} \geq 0 \qquad\qquad\qquad\qquad\qquad\quad i \in I, j \in J$$

$$(A.5) \quad y_j = 0,1 \qquad\qquad\qquad\qquad\qquad\quad j \in J$$

In this formulation, constraint (A.2) indicates that all users are assigned to a facility and constraint (A.3) prevents users to be allocated to a closed facility. The objective is the minimization of set up costs for "open facilities" and the overall location dependant costs of serving users at facilities. An alternative formulation involving variables x_{ij} defined as

$$x_{ij} = \frac{t_{ij}}{p_i}$$

and cost parameters $c'_{ij} = c_{ij} \, p_i$ consists of:

$$(A.6) \quad \underset{(x,y)}{\text{Min}} \ Z = \sum_{ij} c'_{ij} \, x_{ij} + \sum_{j} f_j \, y_j$$

subject to

$$(A.7) \quad \sum_{j} x_{ij} = 1 \qquad\qquad\qquad\qquad\qquad i \in I$$

$$(A.8) \quad x_{ij} \leq y_j \qquad\qquad\qquad\qquad\qquad\quad i \in I, j \in J$$

$$(A.9) \quad x_{ij} \geq 0 \qquad\qquad\qquad\qquad\qquad\quad i \in I, j \in J$$

$$(A.10) \quad y_j = 0,1 \qquad\qquad\qquad\qquad\qquad\quad j \in J$$

Since no capacity constraint are considered in the formulations above it is easy to conclude that variables x_{ij} are at optimality either zero or one, and therefore users in the same zone are assigned to a single facility.

A step forward is achieved considering the vector $y = (y_1, y_0, y_d)$ and introducing the costs for closing down facilities. The model is than formulated as follows:

$$(A.11) \quad \underset{(x,y_d)}{\text{Min}} \ Z = \sum_{ij} c'_{ij} \, x_{ij} + \sum_{j} f_{1j} \, y_j + \sum_{j} f_{0j} \, (1-y_j)$$

subject to constraints (A.7) - (A.9) and

$$(A.12) \quad y_j = 0,1 \qquad\qquad\qquad\qquad\qquad\quad j \in J_d$$

The first term in the objective function denotes the location dependant operation and transportation costs, the second term the set up costs for new facilities and the last one the costs assigned to closing down facilities.

B - P-Median

The concept of p-median on a graph was introduced by Hakimi (1964) and since then the p-median model for facility location has attracted considerable attention. In this model, the number of facilities is fixed to p and total overall costs are minimized. A concise formulation is obtained if it is considered $J \subseteq I$ and assigned $x_{jj} = 1$ if a facility is located at site j. The p-median model is formulated as follows:

$$(B.1) \quad \underset{(x)}{\text{Min}} \ Z = \sum_{ij} c'_{ij} \, x_{ij} + \sum_{j} f_j \, x_{jj}$$

subject to

$$(B.2) \quad \sum_{j} x_{ij} = 1 \qquad\qquad\qquad\qquad\qquad i \in I$$

$$(B.3) \quad x_{ij} \leq x_{jj} \qquad\qquad\qquad\qquad\qquad\quad i \in I, j \in J$$

(B.4) $\qquad \sum_j x_{jj} = p$

(B.5) $\qquad x_{jj} = 0,1 \qquad\qquad\qquad\qquad\qquad\qquad\qquad j \in J$

     As above, constraint (B.2) ensures that all users cells are allocated to a facility and (B.3) prevents allocation to closed facilities, while (B.4) fixes the number of facilities to p. If it is assumed that facilities may also be shut down, then (B.1) is replaced by

(B.1) $\qquad \underset{(x)}{\text{Min}}\ Z = \sum_{ij} c'_{ij}\, x_{ij} + \sum_j f_{1j}\, x_{jj} + \sum_j f_{0j}\, (1-x_{jj})$

     A recent study of heuristic and exact methods to solve the p-median problem is provided by Captivo (1988).

C - Capacited Facility Location Model

     A natural extension of the simple facility location model is achieved by considering constraints regarding the minimum size and maximum capacity of facilities. Adopting again the flow variables (t_{ij}) extensively used in the geographical and transportation fields, the model is written down as:

(C.1) $\qquad \underset{(t,y)}{\text{Min}}\ Z = \sum_{ij} c_{ij}\, t_{ij} + \sum_j f_{1j}\, y_j + \sum_j f_{0j}\, (1-y_j)$

subject to

(C.2) $\qquad \sum_j t_{ij} = p_i \qquad\qquad\qquad\qquad\qquad\qquad\qquad i \in I$

(C.3) $\qquad \ell_j\, y_j \le \sum_i t_{ij} \le u_j\, y_j \qquad\qquad\qquad\qquad\qquad j \in J$

(C.4) $\qquad A t + B y \ge b$

(C.5) $\qquad t_{ij} \ge 0\ ,\ y_j = 0,1 \qquad\qquad\qquad\qquad\qquad i \in I, j \in J$

where (C.2) defines the demand for the facilities, (C.3) provides minimum size and maximum capacity bounds and (C.4) stands for any additional planning constraints.

     This model was put forward by Balinski (1961) and received contributions, among others, by Kuehn and Hamburger (1963), Feldman, Lehrer and Ray (1966), Sá (1969), Akinc and Khumawala (1977), Geoffrion and Macbride (1978) and Van Roy (1986). This last author has proposed a very efficient algorithm based on a cross decomposition technique.

     At this stage, we note that embedded in the simple facility location, p--median and capacitated facility location models is the linear transportation model. The behavioral assumption implicit there is that users select facility according to the minimum cost criterion. This is adequate when decision are taken centrally as in the transportation of goods. However, when individual users take their own decision on the facility where to go, some level of dispersion must be considered, since users may have many different reasons for deviating from the minimum cost alternative (Fig. 1).

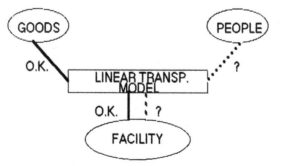

Fig. 1 - Facility location models with embedded linear
transportation sub-models

D - Surplus Maximization Capacitated Facility Location Model

In this model proposed by Coelho (1980) the users' dispersion of preferences is incorporated by a gravity type transportation sub-model. The model is, as follows:

(D.1)
$$\underset{(t,y)}{\text{Max}} \; S = -\frac{1}{\beta} \sum_{ij} t_{ij} \left(\log \frac{t_{ij}}{\bar{t}_{ij}} - 1\right) - \sum_{ij} c_{ij} t_{ij} - \sum_{j} f_{1j} y_j - \sum_{j} f_{0j} (1-y_j)$$

subject to

(D.2)
$$\sum_j t_{ij} = p_i \qquad\qquad i \in I$$

(D.3)
$$\ell_j \, y_j \leq \sum_i t_{ij} \leq u_j \, y_j \qquad\qquad j \in J$$

(D.4)
$$A t + B y \geq b$$

(D.5)
$$y_j = 0,1 \qquad\qquad j \in J$$

It may be shown that the first two terms in the objective function are a measure of consumers' surplus associated to the gravity spatial demand model (Wilson et al (1981), Coelho (1983)). The objective function (D.1) is therefore an aggregate measure of locational surplus. Numerical experience is provided by Erlenkotter and Leonardi (1981), Leonardi (1981) and Coelho (1988).

E - Covering Location Model

The aim in this model is to locate facilities in such a way that every demand zone is 'covered', for example, at a maximum distance, time or generalized transportation cost. The model is written as follows:

(E.1)
$$\underset{(y)}{\text{Min}} \; Z = \sum_j f_{1j} \, y_j + \sum_j f_{0j} \, (1-y_j)$$

subject to

(E.2)
$$\sum_j a_{ij} \, y_j \geq 1 \qquad\qquad i \in I$$

(E.3)
$$y_j = 0 \text{ or } 1 \qquad\qquad j \in J$$

where

$$a_{ij} = \begin{cases} 1 & \text{if zone } i \text{ is 'covered' by facility } j \\ 0 & \text{otherwise} \end{cases}$$

The covering location model was put forward by Toregas et al (1971). It is a particular application of the set-covering problem, which has received very substantial attention in the literature (see, for example, Pierce (1968), Balas and Padberg (1972), Christofides and Hey (1978), Almeida, Paixão and Coelho (1982). A very efficient algorithm for large scale set-covering problems is given by Paixão (1983).

F - Maximal Covering Location Model

Assume that from budgetary constraints or others, it is not possible to have more than p facilities which will not provide full population coverage. Then, the location of facilities that ensures maximal population covering is given by the maximal covering location model, proposed by Church and Revelle (1974), which is formulated, as follows:

(F.1)
$$\underset{(y,z)}{\text{Max}} \; Z = \sum_i p_i z_i$$

subject to

(F.2)
$$\sum_j a_{ij} \, y_j \geq z_i \qquad\qquad i \in I$$

(F.3)
$$\sum_j y_j = p$$

(F.4)
$$y_j = 0,1 \text{ and } z_i = 0,1 \qquad\qquad i \in I, j \in J$$

where z_i is a boolean variable that is equal to one if zone i is 'covered' by at least one facility. The objective function corresponds to the maximization of the population covered by the facility set. Constraints (F.2) ensures that z_i is equal to zero if no facility covers zone i and constraints (F.3) fixes at p the number of facilities.

G - Hamiltonean P-Median Model

The Hamiltonean p-Median problem (HPMP) is a mixed location and routing approach proposed by Branco and Coelho (1984 and 1986) which embeds a joint routing and location optimization.

It is clear that in school facility location it is essential to take into account simultaneously the locational aspects of the potential facility sites and the school buses routing.

Let N denote a set of potential facility sites and demand points. The Hamiltonean p-median problem consists in finding out p-Hamiltonean circuits serving all users in such a way that every user is assigned to a facility in the same circuit. This model is therefore an extension of the p-median and travelling salesman problems.

Several formulations for the HPMP may be considered (see Branco and Coelho (1984)) which may be explored for different solution methods. In particular, the HPMP can be formulated as follows:

(G.1)
$$\operatorname*{Min}_{(x,y)} Z = \sum_k \sum_{ij} c_{ij}\, x_{ijk}$$

subject to

(G.2)
$$\sum_k y_{ik} = 1 \qquad\qquad i \in I$$

(G.3)
$$\sum_i y_{ik} \geq 1 \qquad\qquad k \in K$$

(G.4)
$$\sum_i x_{ijk} = y_{jk} \qquad\qquad i \in I, k \in K$$

(G.5)
$$\sum_j x_{ijk} = y_{ik} \qquad\qquad i \in I, k \in K$$

(G.6)
$$\sum_{i,j \in S} x_{ijk} \leq |S|-1 \qquad\qquad \forall S \subseteq R_k : |S| \geq 1$$
$$k \in K$$

where

$$y_{ik} = \begin{cases} 1 & \text{if demand point i belongs to circuit k} \\ 0 & \text{otherwise} \end{cases}$$

$$x_{ijk} = \begin{cases} 1 & \text{if demand point i precedes j in circuit k} \\ 0 & \text{otherwise} \end{cases}$$

the cardinality of set K is p and $R_k = (i : y_{ik} = 1)$ is the set of vertices in circuit k.

Constraints (G.2) ensures that each demand point is assigned to a circuit, (G.3) defines p circuits, (G.4) and (G.5) together with (G.2) establishes that every demand points belongs to just one circuit and (G.6) prevents the existence of subcircuits in R_k.

H - Routing - Location Models

A more general setting regarding joint routing and location modelling than the one considered in the Hamiltonean p-median problem, would include multiple routes serving the facilities. The optimization process must obviously have into account, in this context, the location costs assigned to facilities and the transport cost associated with the routes that are designed to collect users. This framework is depicted in figure 2 diagrammatically.

280

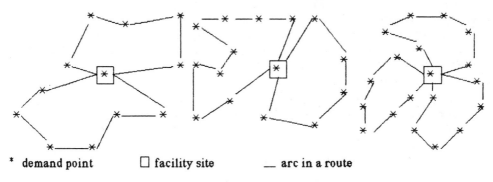

* demand point   ☐ facility site   __ arc in a route

Fig. 2 - Routing-Location Diagram

A recent survey of routing location problems is given by Laporte (1987). A particular routing location model for just one facility and m routes is studied by Laporte and Norbert (1981). It corresponds to the following formulation:

(H.1)
$$\min_{(x,y)} Z = \sum_{i<j} c_{ij} u_{ij} + \sum_{j} f_j y_j$$

subject to

(H.2)
$$\sum_{j} y_j = 1$$

(H.3)
$$\sum_{i<j} u_{ij} + \sum_{k>j} u_{jk} = 2 + 2(m-1) y_j \qquad\qquad j \in J$$

(H.4)
$$\sum_{\substack{i<j \\ i,j \in S}} u_{ij} \le |S| - 1 + (m-1) \sum_{j \in S} y_j, \qquad\qquad \forall S \subseteq I : |S| \ge 2$$

where $u_{ij} = 0,1,2$ is the number of times an arc (undirected) is used in a route. Constraint (H.2) fixes the number of facilities to one, (H.3) states that the degree of a mode is 2m if it corresponds to a open facility and 2 otherwise, while (H.4) prevents illegal subtours.

3. MODEL IMPLEMENTATION STRATEGY

The implementation of the modelling approach depends on the nature of the educational facilities being located, availability of school transport, whether schools shut downs are considered or not, type of constraints to be taken into consideration and objectives assumed to drive the search of optimal policies.

A common strategy may however be identified in several applications consisting of:

1. Definition of the study area - this may be at a national, regional or local scale depending on the facilities being located, attraction areas and degree of interdependence between alternative facilities;

2. Retrieving from database the information required to run the model or models according to the constraints and objectives set up;

3. Optimization phase consisting in exploring several algorithms and optimization procedures in relation with the criteria and constraints considered;

4. Analysis by a planning team of the results generated in the optimization phase to ensure adequacy to reality and add considerations that the optimization procedure has been unable to embrace;

5. Preparing and documenting consistent proposals to submit to political and financial approval;

6. Follow up the implementation of the plan.

281

4. DATA COLLECTING

The data collecting in our particular experience at the planning bureau of the Ministry of Education is based upon a relational data base management system which allows storing retrieving, querying and protecting data in a users friendly environment.

Clusters of information regarding educational facilities, students, teachers and other technical administrative staff and educational statistics are considered.

The cluster of educational facilities includes information for every school, such as name, address, postal code, telephone, school type and level, availability of special facilities and a unique index found on the geographical location of the school. Detailed information on buildings and grounds is also stored. The cluster on students includes information on registration, by age groups, classes and success/failure performances from kindergarden to high school. The information is registered by school unit and since this is referred to a geographical based index, a very fine spatial location grid of students by age groups is produced. The cluster on teachers and other staff includes detailed information on academic degrees, previous school related experience and other relevant information for imputation of costs. Finally, the cluster on educational statistics contains information on rates of approval and failure by grade, rates of premature school leaving, average size of classes, average number of students by teacher, etc., and many demographic and social indicators needed for school population forecasting.

Loading information into the database is a huge task demanding careful planning and long term persistance. A project involving the survey of schools, buildings, spaces and equipments assigned to educational activities at a national scale has been set up. Routine surveys of school population have been improved to feed the database. Data from the National Institute of Statitistics was transfered and substantial in house information previously dispersed was gathered and copied or inputed into the database.

A menu driven query system has been implemented, as well as procedures for validating and listing data and defining users protections. This overall data management system intends to support planning in the Ministry of Education and provide data consistency to the modelling efforts previously described.

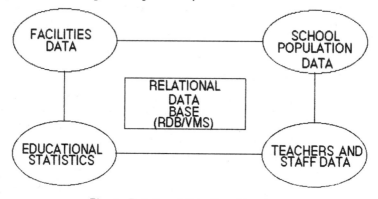

Fig. 3 - Relational Data Base Structure

5. OVERCOMING "TRAPS"

The size of the tasks underpinning the optimization of school facilities at a national scale, and the large number of bureaucrats that all big organizations tend to attract, for which the Portuguese Ministry of Education is no exception, creates an environment propitious to resistance to changes and growing fears of transfer of power.

In order to overcome this difficulty it is essential providing careful planning, conveying a substantial amount of effort explaining the usefulness of the approach and to show results fast.

A set of recommendations derived from our experience may be put forward:

1. Try to keep as many tasks as possible with those previously acquainted with them;

2. Whenever possible to decentralize do not hesitate in doing it;

3. Engage in making database users feel part of the project;

4. Stimulate users from other departments to benefit from the stock of information that the database has made available;

5. Provide facilities for querying the database in a users friendly environment;

6. Keep the computer system dedicated exclusively to database management in order to ensure short response time;

7. Search optimality but do not become slave of it - the uncertainty in data items such as costs and demographic forecasts and the existence of qualitative and political components that the models are unable to absorb, advocates precaution in proclaiming optimality;

8. Make all optimal model solutions pass through the sieve of common sense and the expertise of planners of different disciplines such as architects, urban planners, geographers and teachers;

9. Stand the project as a global integrated one and not as a project of the small team that has eventually had the opportunity of starting it.

6. REFERENCES

AKINC, V. and KHUMAWALA, B.M. (1974), "An Efficient Branch and Bound Algorithm for the Capacitated Warehouse Location Problem", Graduate School of Business Administration, University of North Carolina.

ALMEIDA, M.T., PAIXÃO, J.P. and COELHO, J.D. (1982), "Aplicações dos Problemas de Cobertura e Partição de um Conjunto", Economia, vol. 6, pp. 22-54.

BALAS, E. and PADBERG, M.W. (1972), "On the Set Covering Problem II: An Algorithm", Management Sciences Research Report nº 295, Carnegie - Mellon University.

BALINSKI, M.L. (1961), "Fixed Cost Transportation Problems", Naval Research Logistics Quarterly, vol. 8, pp. 41-54.

BILDE, O. and KRARUP, J. (1977), "Sharp Lower Bounds and Efficient Algorithms for the Simple Plant Location Problem", Annals of Discrete Mathematics, vol. 1, pp. 78-87.

BRANCO, I.M. and COELHO, J.D. (1984), "Formulações Matemáticas da p-Mediana Hamiltoneana", Centro de Estatística e Aplicações da Universidade de Lisboa, Nota nº 10/84.

BRANCO, I.M. and COELHO, J.D. (1986), "The Hamiltonean P-Median Problem", Faculdade de Economia, Universidade Nova de Lisboa, WP 59, Paper presented at EURO VII, Lisbon. Submitted for publication.

CAPTIVO, M.E. (1988), "Algoritmos para o Problema da P-Mediana", Ph.D. Thesis, Universidade de Lisboa.

CHRISTOFIDES, N. and HEY, A.M. (1978), "Lower Bounds for the Set Covering Problem from Network Flow Relaxation", Imperial College of Science and Technology, Dept. of Management Science.

CHURCH, R. and REVELLE, C. (1974), "The Maximal Covering Location Problem", Papers of the Regional Science Association, vol. 32, pp. 101-18.

COELHO, J.D. (1980), "A Locational Surplus Maximization Approach to Public Facility Location", Methods of Operations Research, 40, pp. 265-269.

COELHO, J.D. (1983), "Formulação em Programação Matemática do Modelo Gravitacional e sua Interpretação Económica", Economia, 7, pp. 471-517.

COELHO, J.D. (1988), "Surplus Maximization Capacitated Facility Location Model - A Report", Forthcoming.

ERLENKOTTER, D. (1978), "A Dual-Based Procedure for Uncapacitated Facility Location", Operations Research, vol. 26 (6), pp. 992-1009.

ERLENKOTTER, D. and LEONARDI, G. (1981), "Facility Location with Spatially--Interactive Travel Behavior", Western Management Science Institute, University of California, Los Angeles.

FELDMAN, E., LEHRER, F.A. and RAY, T.L. (1966), "A Warehouse Location under Continuous Economies of Scale", Management Science, vol. 12, pp. 620-684.

GEOFFRION, A. and MCBRIDE, R. (1978), "Lagrangean Relaxation Applied to Capacitated Facility Location Problems", AIIE Transactions, vol. 10(1), pp. 40-47.

HAKIMI, S. (1964), "Optimum Locations of Switching Centers and the Absolute Centers and Median of a Graph", Operations Research, vol. 12, pp. 450-461.

KUEHN, A.A. and HAMBURGER, M.J. (1963), "A Heuristic Program for Locating Warehouses", Management Science, vol. 9(4), pp. 643-666.

LAPORTE, G. (1987), "Location-Routing Problems", Université de Montréal, Centre de Recherche sur les Transports, Publication #506.

LAPORTE, G. and NOBERT, Y. (1981), "An Exact Algorithm for Minimizing Routing and Operating Costs in Depot Location", European Journal of Operational Research, vol. 6, pp. 224-226.

LEONARDI, G. (1978), "Optimum Facility Location by Accessibility Maximizing", Environment and Planning A, 10, pp. 1287-1305.

LEONARDI, G. (1981), "The Use of Random-Utility Theory in Building Location -Allocation Models", International Institute for Applied Systems Analysis, Laxenburg, Austria, WP-81-28.

PAIXÃO, J.P. (1983), "Algorithms for Large Scale Set Covering Problems", Ph.D. Thesis, Imperial College, University of London.

PIERCE, J.B. (1968), "Application of Combinatorial Programming to a Class of All--zero-one Integer Programming Problems", Management Science, vol. 15, pp. 191.

REVELLE, C. and SWAIN, R. (1970), "Central Facilities Location", Geographical Analysis, vol. 2, pp. 30-42.

SÁ, G. (1969), "Branch-and-Bound and Approximate Solutions to the Capacitated Plan-Location Problem", Operations Research, vol. 17, pp. 1005-1016.

TOREGAS, C., SWAIN, R., REVELLE, C. and BERGMAN, L. (1971), "The Location of Emergency Service Facilities", Operations Research, vol. 19, pp. 1363-1367.

VAN ROY, T.J. (1986), "A Cross Decomposition Algorithm for Capacitated Facility Location", Operations Research, vol. 34, pp. 145-163.

WILSON, A.G., COELHO, J.D., MACGILL, S.M. and WILLIAMS, H.C.W.L. (1981), "Optimization in Locational and Transport Analysis", John Wiley & Sons, New York.

INTERVENTION AND CHANGE

INTERVENTION AND CHANGE

David Knights

School of Management
University of Manchester Institute of Science and Technology
Manchester, U.K.

Almost by definition Operational Research (O.R.) involves intervening in the routines of conventional goal-oriented human practices for purposes of promoting change in the direction of a more rational and effective use and co-ordination of means towards specific ends. In this sense it is and cannot help being engaged in strategies of power. Now depending on whether one looks at power as wholly negative and repressive or as also positive and productive of subjective well-being, this may or may not imply direct criticism of O.R.. In this brief introduction to intervention and change, I want to suggest that the critics of positivist or conventional O.R. practice are right to draw attention to the neglect of the role of power and interpretation in the 'object' to which O.R. applies its expertise but wrong to assume that this - and especially their own - power can be eradicated. After a discussion of the epistemological and methodological dilemmas facing O.R., the paper assesses critically the critique of positivism and its relevance for the discipline. Drawing upon the work of Foucault (1970, 1980, 1982), it is concluded that as a major strategy of intervention designed to create change, O.R. has power effects that constitute subjects in its image which, in turn, will always be resisted in varying degrees. My argument is that to reform O.R. is to struggle with this resistance in ways that render the discipline more stable in its power relations and thereby productive of subjectivity.

Methodologically O.R. has adopted 'the paradigm of experimental natural science, but with attention not on the natural world but on human (initially military) operations' (Checkland, 1983:663). From its earliest roots of improving the effectiveness of the 1939 war effort to more recent applications in manufacturing industry, apart from a rigorous and technical concern with efficient and appropriate procedures for meeting given ends, O.R. has not normally subjected its interventions to critical examination. As a result of associations with the social sciences in general and perhaps the complexity of intervention in particular, that situation has begun to change considerably. As Jackson (1987:455) has argued, alternative approaches - 'soft systems thinking, organisational cybernetics and critical management science - are capable of guiding rational intervention in human affairs in areas where success frequently eluded the traditional approach'.

Weaned upon the application of mathematics and statistics to practical human (especially miltary) problems, O.R. could not have expected to be deeply reflexive about either its methodology or its epistemology. When national security is at risk, as in war, even the most philosophical of citizens may be willing to suspend disbelief in order to concentrate on the practical tasks of defending the country. Indeed, if that defence is focussed on protecting people from the totalising impact of a fascist dictatorship then even radicals and socialists would join forces with liberals and conservatives in support of the war effort. In this context of its birth and development O.R. can readily be excused its displacement of the analysis of ends or objectives in favour of a concern with the technical means to their achievement.

Perhaps as a result of the difficulty of evaluating the success or otherwise of intervention against non-intervention in human affairs given that rarely two situations are so identical as to perform an adequate experiment, O.R. went from strength to strength once its techniques were extended to peacetime production processes. Intervention necessarily creates change and, in the absence of knowing what might have happened had the intervention not occurred, there is a tendency to generate a halo effect around the change such that the techniques of O.R. may be defined as successful regardless of any critical assessment. Arguably, it was this practical success which rendered O.R. both more and less positivist and technicist in approach. For by emulating the positive methods of the natural sciences, it could guarantee status and respect for its expertise as a result of the production of quantifiable results. However, as a result of this success, O.R. extended its area of expertise to fields where not only the means to the achievement of a goal are problematic but so also are the ends themselves. To remain pragmatic in these circumstances would clearly demand a modification if not abandonment of 'traditional management science, with its emphasis upon optimisation and objectivity' (Ackoff, 1977, quoted in Jackson, 1987:457).

One question that begins to be asked at this stage is whether or not O.R. is scientific? As a comparatively late entrant among other of the management sciences (e.g. marketing, finance, production control and industrial relations) to the field of social science, O.R. could be seen to suffer a disadvantage. On the other hand, as a mathematically or statistically based discipline it could be viewed as enjoying an enviable position. If likened to engineering O.R. could claim scientific justification in its mathematical roots. However, the engineering of inanimate objects such as machines is one thing; the engineering of human beings quite another. Moreover, mathematics might be the 'purest' of sciences but it is so only when it remains 'pure'. Once it is applied or adapted to the world outside of its own internally consistent language, the appeal of science cannot be made unquestioningly.

The major appeal O.R. had the opportunity to make was to the philosophy of social science and especially the work of Popper (1963, 1966) who developed a rationally justified conception of social engineering. However, to do so was no easy option ·for even if the O.R. specialist was not drawn into the contentious and destabilising philosophical debates which surrounded this thesis (Kuhn, 1962, Lakatos and Musgrave, 1970, Dando and Bennett,1981), Popper demanded a more rigorous scientific approach than the mere application of mathematical techniques to practical problems. In other words, it was necessary for O.R. to abandon its detachment from - or avoid some of the more contentious philosophical and epistemological problems to resort to deductive model building on the basis of fundamental assumptions about human technical rationality in the pursuit of uncontested, politically neutral objectives. Indeed, this has continued in what might be seen as an 'isolationist' (Jackson, 1987: 460) fashion by many O.R. academics whose model building theory has become intellectually fascinating but completely divorced from reality (Checkland, 1983:662) much like the work of many econometricians. However, unlike some of the management sciences (e.g. accounting, marketing) that could readily draw upon positive economics with its claim to reflect individual economic rationality, O.R. is locked into the area of human interpretation and decision-making concerning problems that cannot easily be reduced to an economic calculus. That is to say, in Foucault's (1970) language it is caught in the space between the representations of life, language and labour - as studied by the positive sciences of biology, linguistics and economics - and the human subject whose existence cannot be reduced to that which is positively known about them. Accordingly, O.R. is equally as precarious and unstable a discipline as the other social sciences and cannot simply hide behind the scientific cloak of mathematics.

This epistemological and methodological uncertainty has reflected and reinforced a significant challenge to the objectivist conceptions of problems and practices adopted by mainstream positivist O.R. specialists. It has generated a debate covering much of the same ground as those in social theory concerning the epistemological, methodological and political dilemmas of the social sciences. This debate is already being rehearsed and developed in many of the other papers in this volume so all that is necessary here is to indicate its relevance for this stream on intervention and change. One issue, however, that may not have secured the attention it deserves in this debate is the role of 'truth' in the management/social sciences. Although many researchers within these sciences are prepared to challenge and reject positivist methods in their investigations, most cling on to the positivist legacy in regard to the production of 'truth'. In the following section this will also come under the microscope of critical analysis.

THE CHALLENGE TO POSITIVIST O.R.

In bare outline the criticism of 'hard' methodologists has been pursued from positions either of an hermeneutic or a critical tradition (Habermas, 1972). Thus, from an hermeneutic position, positive methods have been attacked for failing to capture the problematic nature of meaning and the incorrigibly shifting and unstable interpretive ground of human intercourse. Critical theorists, on the other hand, have been more concerned to question not only the methodological acceptance of an unproblematic status to meaning but also to ground what both O.R. specialists and the practitioners whose procedures they aim to improve take for granted. They do this by disclosing the social and economic structures that are seen to determine the material and symbolic universes that 'hard' and sometimes 'soft' O.R. reflects and reproduces. But, in doing so, these critical theorists return by default to an assertion of positive knowledge about social structures and their effects (Jackson, 1982). In this respect they are a sitting target for the 'soft' systems methodologists (Checkland, 1982) who can accuse them of imposing a closure upon the problematic nature of meaning through projecting their certain and absolute knowledge of the distortions of structural inequality and thereby placing a utopian barrier around unconstrained dialogue (Willmott, 1988). In response, of course, the critical theorists can equally appeal to Habermas (1979) in arguing that since the ideal speech situation is only a counterfactual, the dialogue which soft system methodologists wish to protect from utopian restrictions, is far from being unconstrained as it is already systematically distorted by power and inequality. To recognise this is not to descend into the positivist traps of objectivist interpretations of social reality.

In relation to this stream of intervention and change all three of these methodological perspectives: positive, hermeneutic and critical are of relevance since each will promote distinct interventions and each create different kinds of changes. But in one sense their battles (or indifference, Dando and Bennett (1981:101-2), in the case of the positivists) are misguided in that, though seeking to gain supremacy, no single one of these methodological approaches can supplant the other. They are, in effect, locked in an inescapable interdependence with one another for the social or management sciences must necessarily oscillate between <u>positive</u> knowledge of what defines human existence and an <u>hermeneutic</u> questioning of its determining power through a <u>critical</u> examination of the historical and philosophical conditions which make it possible (Foucault, 1970).

What from the perspective of intervention and change it is more important to focus upon are the 'truth' effects of knowledge as it is an inseparable element in the exercise of power (Foucault, 1980, 1982) and of self-legitimating, technical systems of adaptation to the goals of self-preservation (Habermas, 1976). Few clearer examples of this phenomenon within the management sciences can be found than in this area where, regardless of interpretive problematics, O.R. is about the application of a set of mechanisms and techniques precisely to a field (e.g. the military, private business) in which power-knowledge relations have the effect of producing the 'truth' of their own practices. In other words, the conditions of possibility of positive knowledge (e.g. O.R. means-ends logical chains) actually determining the behaviour of the human subjects concerning the 'truth' or accuracy of which that knowledge is dependent is extremely high.

Habermas (1971) described this as the domination of sub-systems of purposive -rational - instrumental or strategic - action (empirical-analytic science) over the institutional framework of symbolic or communicative interaction (culture) which is a condition of their existence. Interaction is governed by '<u>consensual norms</u>, which define reciprocal expectations' whose validity is 'grounded only in the intersubjectivity of the mutual understanding of intentions and secured by the general recognition of obligations' (ibid:92). By contrast, instrumental action reflects the application of technical rules that are developed through the production of empirical knowledge and strategic action is the result of analytical deductions based on a rational choice of alternative means to the achievement of valued goals.

It is a major plank of Habermas's thesis that as a result of its success in actively rendering nature adaptable to the culturally derived needs of human, collective self-preservation, the system of purposive-rational action is continually being extended to the institutional framework of interaction so as to bring society and social development 'under control in the same way as nature' (ibid: 117). Moreover, purposive-rational action is self-legitimating and self-regulating in terms of its efficient and effective production of quantifiable results.

Systems analysis and O.R. is 'the model according to which the planned reconstruction of society is to proceed' (ibid:106). No wonder, then, that the positivist O.R. practitioners refuse to debate with hermeneutic and critical theorists. For such debates would seem regressive to those who routinely transform practical social problems into matters of appropriate analytical design subordinated to strategic means-ends chains that are capable of technical solution. Indeed, notwithstanding the paradox of reducing practical problems exclusively to technical resolution, conventional O.R. theorists regard as impractical any participation in an open dialogue and discourse concerning the development of consensually agreed norms since it delays - or denies the possibility of - the application of a depoliticised, technical expertise to the continuous production of self-legitimating results.

Habermas is largely concerned about the ideological effects of knowledge which create insuperable difficulties for subjecting technical achievements to rational debate and consensual validation through communicative dialogue and interaction unconstrained by political domination. In this respect, he remains wedded to an understanding of technocratic consciousness as an ideology that, through a preoccupation with the technical control of material and socio-cultural environments, displaces any practical interests in communicative interaction free of domination.

In broad terms the work of Habermas and Foucault seems entirely compatible and mutually complementary and the latter has said as much (Foucault, 1982) although the compliment has not been returned (Habermas, 1988). But in subscribing to quite distinctive conceptions of power, they diverge considerably. Foucault does not actually deny the existence of ideologies; he is simply unwilling to concede that knowledge can only be free of ideology and thereby true once it is extricated from power relations. This is because he perceives knowledge as both productive of power and always created wherever power is exercised. Moreover, power is not the property of persons, a dominant class, or the sovereign state but is dispersed throughout a diverse range of social networks and exercised in a multiplicity of spheres often only loosely connected with the higher echelons of state and capitalist institutions. Nonetheless, relationships of power have as their limits, at one extreme, confrontation through which the free play of forces and resistances may displace or undermine power as an active yet stable mechanism that defines the actions of others and, at the other, complete domination where the subject is without the freedom to resist (Foucault, 1982). In the former situation, power is in transition and, in the latter, it no longer exists.

Finally, Foucault contrasts the negative and repressive conception of power held by Habermas with one that is positive and productive. Although wherever power resides so also one will find resistance, subjects often subordinate their actions to the demands - or in support of the actions - of others. In this sense, through the social practices that it embraces, power constitutes subjects in its image but insofar as it is also 'synonymous indeed with sociality' (Smart, 1985:132), there are no social occasions where it is absent. Consequently, the conditions of possibility of the 'ideal speech situation' upon which Habermas (1979) constructs his rational grounding of the emancipatory project are non-existent since they rely upon the eradication of power and its tendency to distort communication.

To this extent, both soft systems methodology as advocated by Checkland and critical theory proposed by Jackson are equally incapable of accomplishing their respective tasks of transforming O.R. into an emancipatory tool of social reconstruction. For whatever reforms are performed on the discipline it cannot but operate as a power over human subjects. What these critics seem to neglect is how in taking account of human interpretation and power, they are actually producing knowledge that creates and sustains mechanisms of discipline, networks of surveillance and processes of individualisation which tie subjects to the practices advanced by the revisions to O.R. For these practices not only attract the technical-rationality of subjects but they also appeal to the very sense (i.e. identity) of what it is to be a competent and valued human being. In short, subjective identity becomes an all-absorbing preoccupation of - and a major self-disciplining force upon - individualised subjects seeking to perform adequately within the practices defined by the science of O.R. Indeed, it could be argued that these reforms are precisely what were necessary to the further development of the discipline and a condition of increasing its power over the actions of practising managers and their employees.

Is this not what informs Churchman's (1979, quoted in Checkland, 1983: 662) description of 'The Dreary 60's' when "O.R. academically became 'modelling'; not really modelling at all, but a study of the delights of algorithms; nuances of game theory; fascinating but irrelevant things that can happen in queues"? It is equally the concern of Jackson (1987: 465) when, wearing his pluralist hat, he argues that a variety of different

291

approaches 'allow the management scientist to work, with a good chance of success, in the full range of problem situations found'. Insofar as O.R. knowledge is inevitably designed and utilised for purposes of intervention in the social world, practitioners should be aware not just of their power to change practices but perhaps, more importantly, of the effects these have upon subjectivity. The question remains as to what kind of subjects O.R. would feel morally, politically and socially justified in producing?

REFERENCES

Ackoff R.L., 1977, Optimisation + objectivity = opt out, Eur.Jnl.Op.Res., 1:1.
Checkland P.B., 1982, Soft systems methodology as a process: reply to M.C. Jackson, Jnl.Appl.Sys.Anal.,9:37.
Checkland P.B., 1983, OR and the systems movement: mappings and conflicts, Jnl.Op.Res.Soc., 34:661.
Dando M.R. and Bennett P.G., 1981, A Kuhnian crisis in management science?, Jnl.Op.Res.Soc., 32:91.
Foucault M., 1970, "The Order of Things", Tavistock, London.
Foucault M., 1980, "Power/Knowledge: Selected Interviews and Other Writings 1972-77", C.Gordon, ed., Harvester, Brighton.
Foucault M., 1982, The subject and power, in; "Michel Foucault: Beyond Structuralism and Hermeneutics", H.L.Dreyfus and P.Rabinow, eds., Harvester, Brighton.
Habermas J., 1971, "Toward a Rational Society", Heinemann, London.
Habermas J., 1976, "Legitimation Crisis", Heinemann, London.
Habermas J., 1979, "Communication and the Evolution of Society", Beacon Press, Boston.
Habermas J., 1988, "The Philosophical Discourse on Modernity", Polity Press, Oxford.
Jackson M.C., 1987, Present positions and future prospects in management science, Omega, 15:455.
Jackson M.C., 1982, The nature of soft systems thinking : the work of Churchman, Ackoff and Checkland, Jnl.Appl.Sys.Anal., 9:17.
Kuhn T., 1962, "The Structure of Scientific Revolutions", Chicago University Press, Chicago.
Lakatos I. and Musgrave A., eds., 1970, "Criticism and the Growth of Knowledge", Cambridge University Press, Cambridge.
Popper K., 1963, "Conjectures and Refutations", Routledge and Kegan Paul, London.
Popper K., 1966, "The Open Society and its Enemies", Routledge and Kegan Paul, London.
Smart K., 1985, "Foucault", Tavistock, London.
Willmott H., 1988, OR as a problem situation, (this volume).

FIRM   SPECIFIC   KNOWLEDGES: THEIR   CRITICAL   SCRUTINY

Peter Clark

Reader in Organization Behaviour
Aston Business School
Birmingham, England

INTRODUCTION

The paper is concerned with the capability of organizations to resolve the innovation/efficiency dilemma through expert intervention (Hall, 1984) and treats the wider societal diffusion of knowledges, including their definition as the context (Clark and Starkey, 1988; Clark and Staunton, 1989). The paper reviews the concept of social science intervention presented at the First Symposium and suggests a focus for this anniversary reflection.

FIRST SYMPOSIUM: CLAIMS AND QUESTIONS

During the early sixties leading social scientists proclaimed usability (e.g. Lupton, Woodward, Cherns, Mumford) in a manner which contrasted with an earlier opposition to the directions and efficacy of work study. It was implied that social science knowledge was a 'radical' alternative to existing firm specific knowledges. It had been postulated that an applied social science would require the re-cognitising of firms and that this possibility would be shaped by political forces (as Burns and Stalker, 1961). At the First Symposium practitioners in operations research reported themselves "...puzzled about how to deal with human reactions to change when they come to try to implement change...have read of descriptive research by social scientists...are sympathetic...but concerned about the lack of validated models and measurement...wonder...how it can be applied in solving real organizational problems...have heard of occasions when social scientists have been invited to study and improve organizations ...wonder how this kind of study compares with a broad operational research study...can the two be regarded as complementary?...is there a risk that the two would come to contradictory conclusions?" (Lawrence, 1967:5).

SOCIAL SCIENCE: A DIFFERENT EXPERTISE

Two key contributors to the First Symposium emphasised the unlikeness of OR and the social sciences, but did so in somewhat different ways. First, in a key note paper, Bennis (1967) emphasised that OR was and largely remains an institutionalised and intellectually homogeneous framework of expertise "owned" by management science professionals located inside organizations and with a specialised supply channel from higher education. OR possessed street credability from the sagas of wartime mystique and the association with computer based analysis. Social science in Britain (cf. USA: Clark, 1987) was and is a university centred disciplinary association of diverse orientations largely oriented to indirect intervention through changing the constructs used by members of society about their economic, individual and social situations. Bennis offered a searching exposure of the inapplicability of much of social science in its direct problem solving format. Existing approaches to usage were depicted as rationalistic, technocratic, individualistic (Bennis, 1967: Figure 1), especially social scientists as scholarly consultants. Bennis promoted the notion that social science was usable to implement change. Second, Burns (1967) emphasised the disjuncture with the notion of "system" in the singular which was attributed to OR and the plurality of social systems which are inherently political. Hence there are problems of using social sciences because of their value laden and interest driven attributes: whose social science and on whose behalf (cf. Hall, 1984)?

IMPLEMENTATION   ONLY - OR DESIGN FIRST?

The postwar era had been dominated by the image of social science as a source of "best practice" about how to implement decisions through heightening acceptability and through the various experientially based approaches of "planned change" (see Bennis, Benne and Chin, 1961). Bennis accepted that social science was about implementation and offered a future in which the existing problems of direct intervention were resolved by adopting the experiential approaches which he claimed "validated knowledge", created "collaborative power relationships" and were "systemic". Bennis was energetic about processes but neglected the choice of directions (Cook, 1967:27) and focused attention away from the vision of a design role at the organization level for social scientists.

The design claim was formulated in the Carnegie School of cognitive psychology (Cooper, Levitt and Shelly, 1964), in the open systems approach of Miller/Rice (1967) and in the British contributions to contingency theory. Moreover, the design claim was emphasised at the smaller, Second Symposium (Clark and Cherns, 1970) and became widely accepted in the business schools, especially in North America. In principle the notion of design should have provided many encounters between OR and social science, so why did these not occur? The vision of designing organizations became the mainstream in the American business schools where the abstract theory was shaped into a design format by J. R. Galbraith (1977), but there were few examples of practice (Clark, 1975a; Lewin, 1986). Even the most cogent contemporary approaches to design do not explain how usage might occur (e.g. Daft, 1986).

PROBLEMS OF INTERVENTION

It might be argued that cognitive organization design should have been fused with the experiential planned change. The First Symposium was the major trigger to the SSRC's six year programme on the use of the social sciences under the direction of the late Albert Cherns which aimed to construct a description of organizational intervention and to devise a descriptive quasi theory (Mohr, 1982). The programme, which was deeply shaped by the Bennis/Burns debate, engaged directly in organizational design to report the forms of social science being introduced, the degree of usage, the interactions between suppliers and users and the factors explaining the outcomes (Clark and Ford, 1970; Clark and Cherns, 1970). First, the types of social science which were supplied attempted to fuse the styles

of social structural analysis of the factory typical of industrial sociology
(Whipp and Clark, 1986) with contingency theory and with the Miller/Rice
(1967) variant of open systems analysis (Clark, 1972a: Ch. 7). The users
included systems experts, yet it was the social scientists who applied
elementary OR to an analysis of the changing system interdependencies. The
social scientists supplied "maps" of the existing and implied social
structures including the changing power bases for labour and for management.
Second, the degree of usage varied between firms and there were wide variations in the types of mapping which the users claimed were useful and how
they actually used them. In every case the trade unions declined to use
the studies whilst also noting the implications. Interestingly, one group
of users - the internal designers - made extensive use of the maps of the
power structure yet failed to appreciate the implications of the contingency perspective for the design of a new cigarette factory. Third,
contrary to the model of planned change the processes of supplying involved
several different types of role: the social scientist as expert, as creator
of DIY models, as technician and as partner (Clark, 1975b) whilst the power
relationships followed the standardization of knowledge curve: gradual
routinization. The claims of Bennis about experiential methods of collaboration and knowledge-validation were only partly sustained and required
extensive revision (Clark, 1975a). Third, career, political and power
factors as well the cognitive models of the users significantly influenced
the interactions and the users interpretations of the supply of social
science (Clark, 1975a). Explaining the levels of utilization requires the
rejection of the Bennis model and great revisions to the orthodox diffusion
perspective (e.g. Rogers, 1983). Finally, the programme failed to create
a vision of practice. In part this failure arose from the directions
taken by the social sciences in the 70s towards post-marxism, critical
theory, abstract organization design theory construction as well as the
separation between the cognitive and experiential branches of the organizational sciences. In part there was a failure to address the problems of
internal practice and the capabilities of practitioners to survive by
obtaining ownership over a problem portfolio (Larsen, 1977).

ALTERING FIRM SPECIFIC KNOWLEDGES

    Social scientists have learned to map the organizational dimension
to reveal the actors frames of reference and the power dynamics of existing
systems within an open multi-system perspective as well as revealing how
existing organization would be altered by proposed changes (e.g. Miller/
Rice, 1967; Clark, 1972a). So, the appearance from the OR side of a multi-
systems prospectus from Checkland (1981) including principles (e.g. CATWOE)
and a practice which was potentially relevant. Since then soft modelling
has been criticised by academic social scientists for: its notion of systems
theory (rather fixed); for its curious usage of Habermas (self-destructive);
for the absence of a theory of social structure and of man; for the neglect
of normative organization theory. But Checkland is concerned with changing
firm specific knowledges through taking account of the different cognitive
structures adopted within the managerial division of labour for different
types of problem (1981: 138-143) and the implications for integrative
problem working. The relevance of this prospectus can be explored with
reference to the generic problems of altering the firm specific knowledges
of British enterprises with respect to the diffusion, adoption and usage
of computer based management in the components industry. Clark and
Staunton (1988) contend that these issues should be explored through an
examination of the structures and processes affecting the British situation
which go beyond the firm (cf. Hall, 1984). It is necessary to combine two
kinds of analysis: (a) the sectoral/institutional (Scott, 1987; Clark,
1987); (b) the organization/sector (Clark and Starkey, 1988) and to examine

the structure and processes affecting the British situation. The innovation perspective emphasises that these "technologies" were developed in the USA and diffused widely and with different effects in Japan and in Britain so that the current British scene is littered with American and Japanese variants. Root definitions of the British situation suggest "fragmented diffusion" whilst the USA has been typified by "crusading homogeneity" (Clark, 1987). The implications are multi-level and many interests are involved. The only significant diagnostic framework for orchestrating intervention in such a situation is that of the centre periphery model (Rogers, 1983) yet because there is no single centre of diffusion an alternative diagnostic framework is required. The development of such a framework ought to be of immediate relevance to this conference.

REFERENCES

Bennis, W. G., 1967, Theory and method in applying behavioural science to planned organizational change, in Lawrence, J. R., ed.
Bennis, W. G., Benne, K. D. and Chin, R., 1961, "Planning of change" Holt, Rinehart & Winston, New York.
Burns, T., 1967, On the plurality of social systems, in Lawrence, J. R., ed.
Checkland, P., 1981, "Systems Thinking and Systems Practice", Wiley, New York.
Clark, P. A., and Ford, J. R., 1970, Methodological and theoretical problems in the investigation of planned organizational change, Sociological Review, 18.1. 29-52.
Clark, P. A., and Cherns, A. B., 1970, A role for social scientists in organizational design, in: "Approaches to Organizational Behaviour" G. Heald, ed.
Clark, P. A., 1972a, "Organizational design", Tavistock, London.
Clark, P. A., 1972b, "Action Research and Organization Change" Harper Row, London.
Clark, P. A., 1975a, Organization design: A review of key problems, Administration and Society, 7.2. 213-256.
Clark, P. A., 1975b, Intervention theory: matching role, focus and context, in: A. Cherns, and L. Davis, QWL: Vol. 1, Free Press.
Clark, P. A., 1987, "Anglo-American", de Gruyter, New York.
Clark, P. A., and Starkey, K., 1988, "Organization Transitions and Innovation Design", Pinter, London.
Clark, P. A., and Staunton, N. 1989, "Innovation in Technology and Organization", Routledge, London.
Cook, S. 1967, in: J. R. Lawrence, ed.
Cooper, W. W., Leavitt, H. J. and Shelley, M. W. 1964, "New Perspectives in Organization Research", Wiley, Chichester.
Daft, R. 1986, "Organization Theory and Design" (2nd Edn.) Western Pub.
Galbraith, J. R., 1977 "Organization Design", Addison Wesley, Reading.
Hall, R., 1984, The natural logic of management policy making, Management Science 30.4. 905-927.
Larsen, M. F., 1977, "The Rise of Professionalism", Campus.
Lawrence, J. R., 1967, ed., "Operational Research and the Social Sciences", Tavistock, London.
Lewin, A., 1986, "Report on the uses of organization design", NSF.
Miller, E. J. and Rice, A. K., 1967, "Systems of Organization", Tavistock, London.
Mohr, L. B., 1982, "Explaining Organization Behaviour", Jossey Bass, San Francisco.
Rogers, E., 1983, "Diffusion of Innovations", Free Press, New York.
Scott, W. R., 1987, The adolescence of institutional theory, Administrative Science Quarterly, 32. 493-511.
Whipp, R. and Clark, P. A., 1986, "Innovation and the Auto Industry: Product, Process and Work Organization", Pinter, London.

INTERVENTION AND CHANGE PROCESS:

INTERACTION BETWEEN COMMUNICATION AND FORMALIZATION

Luigi Mucci and Maria Franca Norese

Politecnico di Torino
Dipartimento di Automatica e Informatica
Torino - Italy

INTRODUCTION

Change is a complex situation in which communication plays a predominant role.

Recognizing elements of complexity and originating relevant forms of interaction are the basis for activating a change process, that is an individual and collective learning process by which the actors invent and fix new models of integration for the organized action.

The intervention of involved actors and analysts in a change process implies participation firstly through both cognitive and operational interactions with the representations that spring from the process. Its validity and efficacy are related to the possibility of both identifying the specific level of complexity and supporting the action of communication by relevant structures of formalization.

Very few valid tools are available today to help the analyst in the intervention and the actors in the change process. There is a clear tendency to develop new instruments or to integrate the valid old ones in a new reading of the actorial behaviour. The attention applied to the situations of change and to the forms of action and interaction can supply methodological indications able to produce valid and useful results.

CHANGE PROCESS AND INTERVENTION

Several interpretations of change are supplied in literature. At least two are very prevalent and the first, the most consolidated, explains change as a necessary and unilateral adaptation of the organization to the task environment.

The second defines change as <u>the transformation of a system of action</u> by the innovation of nature, rules and ways of controlling the actorial game and the modification of the freedom of action, the sources of power and the paradigms (cf. Perrow, 1967; Crozier, 1977; Pfeffer and Salancik, 1978). These conceptions of the change problem induce different definitions of perception of change situations and action.

In the first case, change perception is a task of strategic management that must be able to elaborate and implement a controlled change project. Operations Research classical techniques are often suggested as a support to this action.

In the second view the perception of a "problem situation" (cf. Landry et.al., 1985) on the part of one or more organization actors activates a process. Change derives from this process which is seen as an action of collective learning (cf. Crozier, 1977), a dialectical process in which old and new ways of understanding interact (cf. Bartunek, 1984). This second approach makes the relationships between intervention and change more evident (cf. Hatchuel, 1986).

The often misunderstood concept of intervention (cf. Boothroyd, 1984; Moscarola, 1984) implies an approach, on the part of an analyst or whichever actor is involved, applied to learning through the interaction with the other actors and the different logics of action and representation. Understanding, formalization and explication of elements of the system of action are some of the main results of this interaction, but also essential conditions of the transformation of a system of action. Every technical support to the change process, seen as a collective exchange that transforms a system of action, has to pertain to the specific level of complexity of the process and to the purposes and margins of action of the intervention.

In the following paragraphs we will propose a "reading" of some actual situations to identify forms of intervention in which communications and formalization are integrated in different ways.

LEARNING PROCESS AND CHANGE COMPLEXITY

The actor who "moves" in the organization in order to understand the nature of the "boundaries" of his action (cf. Crozier, 1977) activates learning processes "by which he modifies his knowledge about event contingencies, and adjusts and integrates behavior in terms of this contextual understanding" (cf. Hunt and Sanders, 1986). His intervention in the system of action is in relationship with his a priori freedom and the specific courses of action that he must discover, create or choose and make evident. Only with a real knowledge of the operating system can he discover his actual freedom of action, control it and if necessary, expand it and make the produced alternatives evident, connecting them with the changing system, the old and new logics of action and the facts that generated the situation.

From the point of view of the system, collective learning is a process by which some actors, in a system of action, learn - i.e. invent and fix - new models of integration for the organized action. This is both the discovering or, in certain cases, the creation and acceptance of new relational models, new reasoning modalities and a new collective capacity of solving problems of the collective action.

This reading of the change process helps us to recognize different levels of complexity of the change situations. They are not only characterized by the nature of the actorial game (number of actors, roles ...) and the specific object of deliberation, but also by the nature of the actorial representations, that is (cf. Morin, 1981; Le Moigne, 1985) of the representation that the observing system (the observer with his model) elaborates the observed system. In this perspective, an organizational change situation can be defined as complex in the presence of many representations from different points of view (multiplicity), related to partially common and interdependent topics (interdependence) and subject to multiple and conflicting interpretations (equivocality).

In a change process different levels of complexity can be recognized in the different associations and interactions of multiple, interdependent and equivocal representations.

It does not imply that a defined complex situation is a change situation. Really multiple, interdependent and equivocal representations can produce, in certain cases, a stabilization, rather than a change, of the pre-existing system of action, for instance in the presence of a dominant representation that conserves the situation control.

CHANGE AND COMMUNICATION

The action of communication is a particular type of action directed towards agreement between actors, mutual arrangement of courses of action and understanding and achievement of consensus (cf. Habermas, 1981). Communication tends to assume a predominant role in the global action when the complexity of the situation is growing. Generally it occurs when the habitual systems of action - where the actorial game develops - appear to be no more valid nor acceptable because they do not guarantee each actor the possibility of defending, consolidating or widening his margins of action. The need to define new rules of the actorial game requires a continuous revision of both the formal and informal system, by opportune forms of interaction and negotiation in which communication has a central role.

The action of communication becomes a means for agreement between norms, rules, constraints, degrees of freedom, prizes and sanctions to realize the integration of different logics and behaviour regulations (cf. Ostanello et.al., 1987).

The essential role of communication in the change process obliges one to take this action into account more than any other and to identify and activate forms (formalized representations, inquiry procedures ...) able to support communication before any other activity.

INTERACTION BETWEEN FORMALIZATION AND COMMUNICATION

An actor or an analyst intervenant in an evolving system of action can perceive the change complexity only through the actorial represent-ations. The high interdependence (i) of the representations - as also the high equivocality (e) or the high multiplicity (m) and the associations (e and i), (e and m), (i and m), (e, i and m) - require an action on the representations applied to acquiring, organizing, presenting and reorgan-izing knowledge on the system of action, at both an individual and collective level.

The specific level of complexity of the change situation has to be recognized and faced by an integrated action of formalization and communication. In this action the intervenent develops representations - i.e. elaborates and makes them formal - and proposes them in communi-cation contexts, so as to both understand and make an action operational.

Really among the different purposes of an intervention the more frequent and significant are the cognitive (acquiring and organizing knowledge of the problem, the decisional process, the involvement of actors, the state of information ... ; clarifying and making explicit the terms of reference; identifying a structure of preference ...) and the operational, applied to structuring the collective action that will develop in the new system of action.

The development of the intervention is relatively simpler when only one of these purposes is clearly defined or only one among the many others prevails. When both of them are present with the same significance much more time and resources are generally necessary.

Some of our interventions in public organizations and the analysis of decision processes in change situations (cf. Ostanello et.a.l., 1987; Norese, 1988) allowed us to recognize forms of action on the representations applied to facing complexity in the change process. The need to activate communication and control it combine with the more general requirement of supporting communication by formalized representations. Really proposing representations in communication contexts - where they can be explained, construed, validated, reinforced, made more precise or general, reconstructed, refined and attributed to the reality (cf. Ramaprasad, 1987) in a shared meaning - is an action on the representations that should be technically supported.

Forms of integrations between activities of communication and formalization, in the presence of complex representations, are suggested in literature (see for example Eden et.al., 1984; Hatchuel, 1986; Panayotopolos and Assimakopulos, 1987; Bowen, 1988; Norese and Ostanello, 1988).

The more evident and frequent recognized forms are positioned in Fig. 1 in relation to the two axes of reference - the level of complexity of the change situation and the cognitive and/or operational nature of the intervention - and will be described by the analysed cases of actorial and/or technical intervention.

Case X - Support to coordination

An Actor, intervenent in a process of industrial localization, in presence of (e, i and m) representations does not reduce, rather exploits the recognized high equivocality and draws out - and formalizes - elements in favour of his purpose from each basic representation. Thus he elaborates a new representation, a project, to support the interaction with the Public Actor and this representation develops so as to make evident the high coherence and the potential coordination of the project with the pre-existing system of representations.

Case Y - Support to meaning explanation

The existence of multiple and conflicting interpretations about an organizational situation imposes a shared definition of the problem situation. This is the main object - exclusively cognitive - of a technical intervention in which the Analyst is asked to "acquire" representations from the non intercommunicant involved actors and to formalize them in a model of planning (cf. Norese and Ostanello, 1985). This process of formalization includes activities (cf. Norese, 1988) to reduce ambiguity and integrate multiple points of view in a descriptive model whose possible use has to be "making communication easy".

Case Z - Support to cooperation

The presence of both cognitive and operational purposes induces a more articulate intervention. Really it aims to reduce or control the representation complexity and to structure the action by formalized representations. In these cases the action is, before everything else,

Levels of complexity / Intervention purposes	(e and i)	(e and m)	(i and m)	(e, i and m)
operational				X
cognitive		Y		
operat. and cognitive	Z1		Z2	Z3

Fig.1. Intervention forms

communication. Most of the analyzed interventions are of this kind and are mainly distinct in terms of complexity of the recognized representations.

Case Z1. An Actor elaborates a study where he explicates his point of view and defines aims and criteria of action in relation to the more consolidated and formalized representation. The two representations, the first and the originated from the interaction with the first, become the object of an interactive action, a negotiation, initial purpose of the actorial formalization. A result of the negotiation is the emerging of a new and shared representation, that is a compromise between the two precedents.

Case Z2. The Analyst-Actor of this case develops a global model in which all the identified actorial proposals (different ways of organizing a public service) are described and evaluated by the same dimensions (cf. Roy, 1985). He wants to demonstrate to other actors his technical competence to deal with the problem and to create an easy and immediate tool to support the decision process. Starting from the descriptive model he finishes by elaborating a new proposal that is a result of the learning process that characterizes an intervention. The action (of interaction with information sources and multicriteria formalization) and the nature of the formalized representations - in terms of reduced equivocality, wholeness and operational, immediate and accepted by all language - are specifically applied to activating communications and cooperation.

Case Z3. The same purpose characterizes the technical intervention of the last case where the presence of a few very interdependent and equivocal representations and a reduced disposition of the actors to interact operationally on the problem induce a different form of integration between formalization and communication. In this case the Analyst uses (i.e. combines and integrates) the few confused pieces of knowledge of the representations to formalize new terms of reference and stimulate communication and possibly cooperation.

CONCLUSIONS

In order to act on the representations and then to support communication, especially in change situations, we need schemes that are sufficiently elaborate to deal with complex situations and free enough from structural limits to become a vocabulary suitable for suggesting ideas, associating concepts, making points of view clear and measuring compatibility and consistency.

A representation structure suitable for every situation cannot exist, likely structures adequate in different contexts of action and logically and operationally related to a common methodological approach may be recognized.

Developing representations and presenting them in an interaction context are radically different activities that should be supported by representation tools such as:
- schemes suitable for directing the exploration and which can easily be modified by the exploration results and
- structured representations of a language, that is frequently graphic or textual, adapted for stating global ideas and then fragmenting them and for declaring the meaning of every included element and controlling its ambiguity level.

A global approach, common to each phase of the action on the representations, might guarantee specific pertinence and global consistency; in order to become an actual support to the individual and collective action it should integrate validation activities with conceptualization and communication ones.

From a theoretical point of view there is the open problem of "orienting the supporting tools" towards specific forms of communication. Then there is a need to identify and characterize these forms clearly.

REFERENCES

Bartunek, J.M., 1984, Changing Interpretive Schemes and Organizational Restructuring: The Example of a Religious Order, Admin.Sci.Quart., 29:355.
Boothroyd, H., 1984, The Deliberative Context of Systems Analysis, in "Rethinking the process of Operational Research and Systems Analysis", R. Tomlinson and I. Kiss, eds., Pergamon, Oxford.
Bowen, K., 1988, Research on Decision-Aiding - How to be a DSS, Paper presented at the VII International Conference on MCDM, Manchester.
Crozier, M., and Friedberg, E., 1977, "L'acteur et le systeme", Editions du Seuil, Paris.
Eden, C., Jones, S., and Sims, D., 1983, "Messing About in Problems", Pergamon, Oxford.
Habermas, J., 1981, "Theorie des Kommunikativen Handelns, Vol. I, Handlungzrationalitat und Gesellschaftliche Rationalisierung", Suhrkamp, Frankfurt am Main.
Hatchuel, A., 1986, L'entreprise sur longue periode: Incoherence et intelligibilite, in "Methodologies fondamentales en gestion", ISEOR, ed., Paris.
Hunt, R.G., Sanders, G.L., 1986, Propaedeutics of Decision-making: Supporting Managerial Learning and Innovation. Decision Support Systems, 2:125.
Landry, M., Pascot, D., and Briolat, D., 1985, Can DSS Evolve Without Changing Our View of the Concept of 'Problem'?, Decision Support Systems, 1:25.
Le Moigne, J.L., 1985, Progettazione della complessita e complessita della progettazione, in "La sfida della complessita", G. Bocchi and M. Ceruti, eds., Feltrinelli, Milano.
Morin, E., 1977, "La methode 1", Editions du Seuil, Paris.
Moscarola, J., 1984, Organizational Decision Process and ORASA Intervention, in "Rethinking the process of Operational Research and Systems Analysis", R. Tomlinson and I. Kiss, eds., Pergamon, Oxford.
Norese, M.F., 1988, A Multidimensional Model by a Multiactor System, in "Compromise, Negotiation and Group Decision", B.R. Munier and M.F. Shakun, eds., Reidel, Dordrecht.

Norese, M.F., and Ostanello, A., 1985, A model for an evaluation of supply/
demand of sport facilities to simulate intervention politics, in
"Atti AIRO 1985", Italian Association of Operations Research, ed.,
Venezia.

Norese, M.F., and Ostanello, A., 1988, Identification and Development of
Alternatives: Introduction to the recognition of process typologies,
Paper presented at the VII International Conference on MCDM,
Manchester, August.

Ostanello, A., Mucci, L., Tsoukias, A., 1987, Processus 'publics' et
notion d'espace d'interaction, Sistemi Urbani, 2/3:forthcoming.

Panayotopolos, A., and Assimakopoulos, N., 1987, Problem structuring in
a hospital, Europ.J.Operational Re., 29:135.

Perrow, C., 1967, A Framework for the Comparative Analysis of Organizations,
Amer. Sociological Rev., 32:194.

Pfeffer, J., and Salancik, G.R., 1978, "The External Control Organizations",
Harper & Row Pub., New York.

Ramaprasad, A., 1987, Cognitive Process as a Basis for MIS and DSS Design,
Management Sci., 33:139.

Roy, B., 1985, "Methodologie Multicritere d'Aide a la Decision", Economica,
Paris.

THE CHALLENGE OF THE NEW ORGANIZATION

George A. Tingley

SWISSAIR
Systems Engineering/CTX
CH-8058 Zurich Airport, Switzerland

VISIONS OF THE NEW ORGANIZATION

Drucker on the Effect of IS on Middle Management

Peter Drucker, in his Harvard Business Review article "The coming of the new organization" (1988) says that large businesses 20 years from now will have fewer than half the levels of management and no more than a third of the numbers of managers as their counterparts today. Rather than following the military command and control model, the new organizations will be more like hospitals or symphony orchestras, with their self-directed and disciplined groups of specialists. This change is dictated by the need for flexibility to respond to market forces and is made possible by developments in information technology.

Drucker illustrates the transformation of data to information and the change from opinion to analysis with the example of the impact of computer technology on capital-investment decisions. With an electronic spreadsheet one can, in a few hours, analyse the several aspects of the problem which would have taken man-years previously. Information technology makes this possible but it takes people with the specialist knowledge to exploit it.

These specialists will not be led according to the military command and control model, but will rather be organized in smaller, more responsible groups. These individuals will not be instructed in detail what to do, but for the organization to function, the members, like the players in the symphony orchestra, all have to be playing to the same score.

Drucker gives some hints on how business will be organized in the future, but lists some unresolved problems, such as, with a reduced number of middle managers and the increase in specialization, what will be the source and training ground of future general management? He leaves his HBR readers with the managerial challenge of the future: building the information-based organization, and leaves us computer types with the job of figuring out how to design and provide the information systems to support it.

Naisbitt on Networking

John Naisbitt in his book Megatrends (1982) describes the pyramid hierarchy as the traditional organization of human enterprise, from the Roman army to IBM. However, from the 1960's and 70's rigid hierachies were seen to be slowing down the information flow in an information economy. The USA and Europe were challenged by Japanese competition, which was clustered into small, decentralized work groups. As more technology was being introduced people were being irritated even more by bureaucratic hierarchy. Better educated and more rights-conscious workers did not accept as self-evident that the pyramid hierarchy was the natural order of things.

Networks are people communicating with each other, sharing information and recources. Networks based on the telephone, and its descendant the PC network, and relatively inexpensive airplane travel are the contemporary replacement of the networks previously provided by the family, church, and neighborhood, which have been weakened by social and physical mobility. Networks come into being when bureacracies fail, in particular, to provide information. Networks provide the human touch which can help to find the information we need in the flood of data. Networks provide what bureaucracies can't-the horizontal link.

This does not mean that companies will abandon formal controls, allowing employees to spend all day just talking to each other. Rather the new management style will be inspired by and based on networking. The communication style of the new organization will be lateral, diagonal, and bottom up and its structure will be cross-disciplinary. Naisbitt gives illustrations of trends in this direction at Intel, Hewlett-Packard, Tandem, and Honeywell.

Carlzon on Service Organizations

SAS has gone through the change from being product-oriented to customer- and market-driven and from loss to profitability. Their President, Jan Carlzon, has written a book about it. Its title Moments of Truth (1987) refers to the effect of the quality of each and every customer contact on the future business and survival of the company.

Carlzon describes the turnarounds at a tour operator, then Linjeflyg, and then SAS as the marketing response to costs: cut the costs which have nothing or little to do with service and spend money when it will improve service and consequently income. Such goals require a motivated, initiative, and informed work force, supportive rather than restrictive supervision and organization, and a management that clearly communicates overall objectives and encourages initiative.

Carlzon treats, in turn, the components of his recipe for success in a deregulated business environment: the leader as a definer and promoter of objectives, by words and example, setting of strategy, flattening the organizational pyramid to the three levels of service providers, support management, and general management, taking risks, communicating, getting the cooperation of boards and unions, measuring results, rewarding employees, and being ready for the next challenge.

Chapters 8 on communicating and 10 on measuring results have consequences for the specifiers, users, and customers of information systems. The service providers are organized in decentralized groups which are expected to exercise initiative and take responsibility. To take responsibility a group or

·individual must understand the overall situation. Also the group has to know if their efforts are contributing to the companies goals; feedback measurement is necessary, moreover measurement of the quantities which relate to the service to the customer. Do not promise one thing but measure another, for instance promising prompt and precise cargo delivery but measuring volume and whether paperwork and packages get separated en route. For me, his message for IS people is "A decentralized company is much more in need of good measurement methods than is a hierarchial, centralized organization."

Quantitative Models of Organizational Structures

Malone and Smith (1988) have developed models for comparing the performance of organizational structures, such as human organizations and computer systems. The four generic structures of product hierarchies, functional hierarchies, centralized markets, and decentralized markets are ranked in terms of the three performance measures of production costs, coordination costs, and vulnerability costs. Queueing models are needed to cope with the stochastic aspect of the tasks and messages.

Malone and Smith's models show how changes in the values of the performance measures affect the desirabilities, i. e. effectiveness or efficiencies, of the different structures. They show how the relationships between performance measures and the desirabilities of structures are consistent with (a) major shifts in the organizational forms used by American businesses since 1850 and (b) the evolution of computing system architectures. For me such models offer a systematic and quantifiable description for the possible future evolution of business systems described by Drucker, Naisbitt, and Carlzon and the information systems supporting them.

Reorganization at Swissair

The European Community plans to sweep away all barriers to trade among member states by 31 December 1992, creating an integrated internal market of about 320 million consumers. Helmut Schmidt and others have said that this integration should include a European central bank and common currency. Switzerland, although not a member, is in the middle of the EC territory, like the hole in the doughnut. The Swiss economics minister has been cultivating contacts with the EC members to avoid the isolation of Switzerland from access to this market. For the transporation industry in Europe, including Switzerland, 1992 means "liberalisation", the European version of deregulation. From consideration of basic principles and observation of experience in the USA, in a deregulated environment companies have to increase their flexibility and reduce their costs if they are to survive. Already now in Switzerland we have examples of companies which have to let people go because these companies did not adjust to the market fast enough.

Organizational changes in Swissair for the 1990's were announced at the annual general meeting of the shareholders on 28 April 1988 and then released to the press. The measures are (1) delegating authority and responsibility to the people on the front, especially those in customer service, (2) reducing the number of management levels, and (3) organizing marketing efforts according to route management. The objects of these measures are (a) to improve our response to the market, to the needs of the customer, (b) to shorten the communication and decision paths, reducing the effort for coordination and control, and (c) to manage our product in the units in which it is consumed by the customer, transport from A to B. This is to be achieved without letting people go. If we start now, while we are profitable, we have a much better chance, and with less pain, of being ready for 1992.

METHODS AND TOOLS FOR THE TECHNICAL CHALLENGE OF THE NEW ORGANIZATION

Enumerated here are the methods and tools available to attack the technical aspects of integration, adaptability, and standardization described in the original, expanded version of this paper (Tingley 1988). IBM's Business Systems Planning (IBM 1984) is available for top-down integrated business information system planning. The French counterpart of BSP is based on the modular analysis of systems (Mélèse 1984) which in turn is influenced by industrial dynamics (Forrester 1961). Hein (1985) describes computer support for BSP studies. Data dictionary support for BSP studies (Estep 1981, Sakamoto and Ball 1982) anticipated some features of today's CASE tools. Prototyping, a tool for interactive, evolutionary information system development, is a familiar, but unsung OR approach (Tingley 1983). "Structured Modeling" (Geoffrion 1987) is cited as an example of an effective response to the need for standardization, illustrated in the OR arena.

PEOPLE PROBLEMS

Although the computer, when applied to routine accounting tasks, billing, and order taking, was not primarily considered a job-killer it was sometimes considered a job-hassler, keeping up the pressure by monitoring performance (Terkel 1972). In manufacturing, on the shop floor, robots do replace human workers, although a rationale is that the robots take over the dangerous or monotonous work that humans should not be doing, or do not want to do. This time around we are talking about the reduction of the number of managers' jobs made possible by information technology.

The perceived threat of the computer as a job-killer has also been haunting office workers, including some managers, before Drucker's article appeared. We in OR have met this already in attempts to introduce planning models. A composite of this experience is illustrated in the "delegation" story (Tingley 1983).

The delegation story begins when the fast-track manager, inspired by his recent executive refresher course at business school, wants to modernize his planning process. In his shop, he has a crew of old-timers who have been producing budgets or schedules manually the same way for the last 20 years. In the modern, participative management style, the manager calls in his chief old-timer, delegating him to work together with the OR analyst to define and develop the new planning support system.

The old-timer has been through this before; every summer there are trainees from business school around with fancy ideas that wind up buried in the files. Because the budgeting or scheduling problem is really quite complicated, either because of combinatorial aspects or because of the rules some of them contradictory and unwritten, the old-timer is confident that th systems analyst will be unsuccessful, so there is no risk in giving him some information on the problem and good data to work with. The systems analyst produces a model, which if it works is a prototype.

Of course there are things which were left out in the original description which have to be put in now. In a collegial spirit these are included in the first modification. The results this time look better, but o course other aspects, which were not so obvious at the beginning have to be considered. Ideally, these iterations continue, say 4 to 5 times, until the model approximates sufficiently the features of the desired new planning system.

The better the results the model delivers, the more it is a credible alternative to the old manual method in which the old-timer has made a 20 year investment. His suggestions for improvements become more extreme and capricious the better the prototype performs. Finally he is able to pose a test case that the prototype cannot cope with. He reports back to his boss that the old tried and true manual methods are still the best. Since the fast-track manager did not plan to do the details of the budgeting or scheduling himself, and since he will not or can not replace the old-timer, unless he has done something really awful or the company is losing money, the project fizzles out.

Interactive, evolutionary methods of information system specification require a basis of cooperation and trust if they are to work. If the old-timer knows or suspects that he is going to be replaced, no real help will be coming from him in specifying the new information or planning system. His know-how will be lost to the organization. The company should (1) tell the old-timer that he will still have a job in spite of the computer, and (2) demonstrate this policy by example.

CONCLUSIONS

Here Is An Opportunity For Operational Research

Tools and methods for top-down integrated analysis and design of information systems for the new organization are already available or they soon will be. Top-down integrated analysis of companies and prototyping are particulary appropriate approaches for this job, and they are familiar approaches of operational research; operational researchers are thereby eligible to play a significant role in the analysis and design of information systems for the new organization.

Integration And Flexibility Require Standardization

Standardization facilitates not only integration but also flexibility and fast development. With standardized components we can more readily connect things together, to try things out, as in prototyping. We shall be able to devote more time to the application, rather than to the bits and bytes, when we can connect modules of reusable code which are documented in a familiar, predictable (standardized) format. Geoffrion's "Structured Modeling" illustrates this idea in the OR context. The OR societies may well follow the example of the engineering societies in formulating and promoting standards.

Human Factors Will Be Even More Important Than Before

In the past, workers in agriculture and manufacuting have been replaced by machines and computers to process raw materials and goods, objects that could not talk back or protest. Now we shall have the job of designing systems for managers to help them manage people, and for people to manage themselves. The survival and effectiveness of the new organization will be determined by the motivation, attitude, and self direction and discipline of the workers who provide the services to the customer. Managers, less in number in relation to the workers than before, will be playing a more supporting role. To design systems which will be considered as job aids rather than job killers, we must be, even more than before, aware of the concerns and motivations of the users. In this spirit these comments from one from industry are submitted to the participants, particularly the social scientists, of this Conference.

REFERENCES

Carlzon, Jan 1987, Moments of Truth, Harper & Row, Sydney and London

Day, J. D. and Zimmerman, H. 1983, "The OSI reference model," Proceedings of the IEEE, Vol. 71, No. 12 (December), 1334-1340

Drucker, Peter F. 1988, "The coming of the new organization," Harvard Business Review, Vol. 66, No. 1 (January-February), 45-53

Estep, C. 1981, "DB/DC Data Dictionary-A Business System Planning model," GUIDE 52, Session No. SD-33, Atlanta, GA, USA, (May)

Forrester, Jay W. 1961, Industrial Dynamics, MIT Press, Cambridge, Massachusetts

Geoffrion, Arthur M. 1987, "An introduction to structured modeling," Management Science, Vol. 33, No. 5 (May), 547-588

Hein, K. P. 1985, "Information System Model and Architecture Generator," IBM Systems Journal, Vol 24, No. 3/4, 213-235

IBM Corporation 1984, Business Systems Planning-Information Systems Planning Guide, 4th edition, GE20-0527-4,

Malone, Thomas W. and Smith, Stephen A. 1988, "Modeling the performance of organizational structures," Operations Research, Vol. 36, No. 3 (May-June), 421-436

Mélèse, Jacques 1984, L'analyse modulaire des systèmes de gestion, 3rd edition, Editions Hommes et Techniques, Paris

Naisbitt, John 1982, "From hierarchies to networking," Megatrends, 1984 Paperback Printing, Warner Books, New York, 211-229

Sakamoto, J. G. and Ball, F. W. 1982, "Supporting Business Systems Planning Studies with the DB/DC Data Dictionary," IBM Systems Journal, Vol. 21, No. 1 54-80

Shore, John 1988, "Viewpoint: why I never met a programmer I could trust," Communications of the ACM, Vol. 31, No. 4 (April), 372-375.

Terkel, Studs 1972, Working, 1975 British Edition, Wildwood House, London, 76-77

Tingley, G. A. 1983, "Prototyping: the new paradigm for EDP systems development but an old story for operations research," 23rd AGIFORS Symposium Proceedings, 449-472

Tingley, G. A. 1988, "The challenge of the new organization," 28th AGIFORS Symposium Proceedings

PROJECT BASED WORK GROUPS

AND THE ORGANISATION

Duncan A. Conway

Division of Statistics, OR and Economics
Hatfield Polytechnic, P.O. Box 109
Hatfield, Herts.   AL10 9AB,   England

PROJECT BASED WORK GROUPS

Much traditional thinking, teaching, research and scholarship about
organisations and their management concentrates upon groups, departments
and functions, which have a naturally continuing activity, once the
organisation has decided to include them within its structure. In a
manufacturing company, production is an inevitable and obviously
continuous activity, as is policy administration in a life assurance
company. Activities like selling and personnel may be less entwined
with the identity of the organisation, but once they are agreed as part
of the organisation's pattern of activities they continue into the
future in a naturally continuous flow of activity. These activities
will be described as 'routine based'. A number of other activities, of
which OR is an excellent example, do not possess this natural on-going
flow of work. They are in essence project based. At any point in time
there is no guaranteed future work programme beyond the currently
committed projects. This creates at least two major differences for
such groups as compared with routine based work groups, which have not
been widely discussed.

Firstly, the relationship between the organisation and the work
group is very different for a project based group from that for a
routine based group. For a routine based group the decision to include
the activity within the organisation rather than to sub-contract it, or
to do without altogether, may be intermittently (or occasionally, or
never) reviewed. For a project based group there is in effect a
frequently recurring review by many individuals within the organisation
who have to decide whether to pose the next problem to the group, or
whether to sponsor the next project. A continuing flow of new projects
is essential to the future of project based work groups. However, the
composition as well as the size of the flow needs active management in a
way which is outside the experience of routine based work groups.

The second major difference arises from this fundamentally
different form of relationship. Every member of a project based work
group who has contacts outside his group is immediately an ambassador
for the group and his performance and attitudes can influence the size
and composition of the flow of new projects. Equally the future destiny
of the group is determined wholly by the size and composition of that

flow, and so virtually all members of project based work groups have a direct influence on the future of the group. Obviously staff at the project leader, project manager, and group manager levels have more effect, and carry a correspondingly greater responsibility for shaping that future, and for helping the group to respond appropriately to changes in their organisational environment.

This paper describes two models which have been built of aspects of project based work groups. The first is a model of the relationship of such a group to its organisation. The second is a life cycle model of the states through which such groups can pass in response to organisational pressures, and in response to their own skill (or lack of it) in managing the flow of projects. The field research on which these two models are based was primarily concerned with the Process of OR, and with in-house OR groups in particular. However, it is clear that the concepts have much wider applicability and the future development of the work and its application to other, seemingly very different, situations is described in the final section.

THE ORGANISATIONAL RELATIONSHIP

As already discussed, every contact with someone outside the project based work group is an opportunity to influence the flow of future work. However, the major influences on the future flow of work are the current and past projects undertaken by the group. The processes by which this influence affects the future flow of work are varied. The obvious source of influence which is immediately looked to in informal discussions of this problem is the degree of success. However, the fieldwork (Conway 1985) showed virtually all OR groups investigated to be running at between 80-100% of projects being regarded by the client as producing enough satisfaction to be classified as successful. There were far too few unsuccessful projects to permit any serious investigation. Where there is a high perceived level of continuing success, it is possible and desirable to investigate other forms of influence. In particular it proved particularly valuable to investigate how a project came to be initiated in relation to previous projects. Six different sources for projects were identified, only one (or at most two) of which had had any previous discussion in the literature on the Process of OR/MS.

Type 0 projects result from an independent initiative by a potential client and are the type universally assumed in the literature (e.g. Taha 1987, Davis, et al 1986). Type 1 projects represent the request for some form of upgrading/extension of the resolution provided by a previous project and result from the increase in aspirations of the same manager inherent in satisficing behaviour (March and Simon 1957), or from some similar recurring cause. Type 2 projects represent the desire by clients for assistance which is sparked off by discovering the value of a specific project carried out for a similar manager, often within the same department or division. Type 3 projects represent projects requested as a result of the much more diffuse general reputation built up by a successful group over a period of time, without any direct knowledge of any specific projects. Type 4 projects are the other type occasionally referred to in the literature (e.g. Stewart 1985) which is the deliberate selling of a project idea by member(s) of the project based group. The final type, Type 5, represents a different form of selling, where the decision is made by the project based group to undertake the project without outside sponsorship and then to sell the results. Thus Types 0-3 are client initiated, and Types 4 and 5 are project group initiated. The six types of project source are summarised in Table 1 below.

Table 1.   Sources of Projects

Type	Title
0	External Stimulus
1	Closed Linkage
2	Internal Linkage
3	External Linkage
4	Selling
5	Risk Investment

Obviously the processes by which the different types of new projects are influenced, are essentially communication and perception processes, and these processes are integrated in the Dynamic Model in Figure 1 below (Conway 1985). At the heart of the model are the individual projects, and the professional learning process (marked PL) which over time develops the professional competence of the project group to handle complex problems in a manner well adapted to the needs of the organisation. The feedback from the project results in various communications and in perception and aspiration changes resulting in clients requesting new projects (these types are marked 0-3). The other project group based projects (marked 4-5) must develop from the professional learning process which enables the project group to create the project in the appropriate form.

Thus the Dynamic Model describes the symbiotic form of the relationship between a project based group and its organisation. Twelve years of fieldwork (Conway 1985, Holland 1988) have shown that the above set of six project sources is adequate to describe all forms of project encountered in OR.

THE LIFE CYCLE OF A PROJECT BASED WORK GROUP

The original concept of consultancy groups having some form of life cycle is due to Pettigrew (1975). The key features of Pettigrew's model were that after the pre-birth developments, a period of pioneering led to a period of self-doubt during which the results of the pioneering were reviewed. This period of reflection normally led to a period of project activity where the work was either well adapted to the organisation (adaptive response) or was not (maladaptive response). This failure to adapt could be due to mistake, poor strategic understanding or even professional pride. For whatever reason it led to one of two end points (demise or absorption) in which the group unwittingly either lost its independent identity or ceased to exist. An adaptive response led to one of two end points (continued independence through a dynamic equilibrium or planned absorption) in which its identity was maintained. The fieldwork amongst the OR groups showed that whilst this structure was valid, it did not capture the full range of movement between states that was possible.

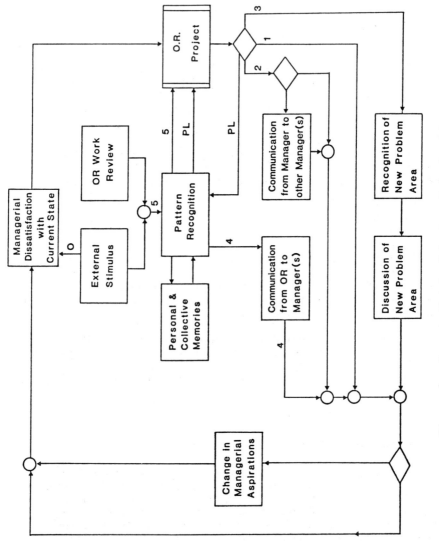

Figure 1. Dynamic Model of Project Relationships.

Two additional, transitional states have had to be added to the model. Specialisation within the professional field is possible in either form of response. The madadaptive form, complacent specialisation, can delay reaching the end points, but the disaster is always waiting round the corner. The adaptive form, strategic specialisation, is a transitional position which can be held for a very long time. However, it normally leads on to one of Pettigrew's end points. The second extension is much more significant. The end points of Pettigrew's model turn out not to be terminal. The addition of these return routes completes the Revised Life Cycle model in Figure 2, below. These return routes represent the changes that occur in the relationship with the organisation. Effective leadership by the group manager can correct a maladaptive response. Unfortunately poor management can move a group from adaptive to maladaptive response. In either case there are several re-entry points to the early part of the cycle. Indeed there is even life after death! The fieldwork has uncovered several groups which have died more than once, and which have been recreated at one of the pioneering stages of the life cycle. Virtually all the other groups have moved between the apparent end points and earlier stages of the cycle on several occasions. Indeed as the latest fieldwork enriches our picture (for some groups we now have three surveys of their work) it is clear that in the U.K. during the 1980's most groups are moving around the life cycle much faster than realised, even as fast as a complete cycle per year at times. Very few indeed are remaining in one state for a long period of time. Indeed the few that have achieved it are confined to the groups that have been able to manage the dynamic equilibrium successfully over a long period of time (10+ years).

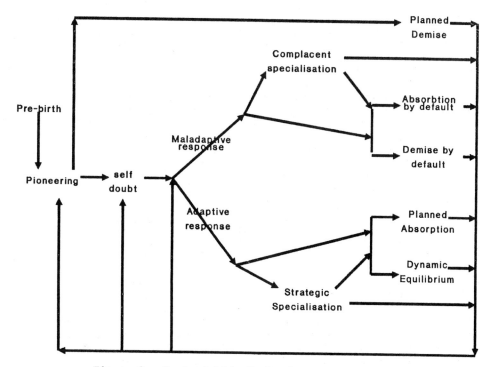

Figure 2. Revised Life Cycle for Project Groups

The fieldwork has also shown that the Dynamic Model classification gives significantly different project mixes at each stage of the life cycle. For instance, one group at the Pioneering stage had 40% Type 0 projects (a measure of its backing from its sponsor), 30% Type 4 projects where it had sold the idea in the first place, and 30% Type 2 projects, which indicated that it was being successful as other managers were taking problems to the group because they had heard about the completed projects (Conway 1985). On the other hand, the average pattern (% in each Type 0-5) for the two groups in the Dynamic Equilibrium state on the latest fieldwork is

$$0, \ 3, \ 27, \ 58, \ 12, \ 0.$$

From this Dynamic Model vector it can be seen that these groups have a widespread good reputation (58% of their new project work is based on this factor). This is backed up by a smaller but still substantial number of projects arising because of specific previous projects and by a small amount of work due to enhanced aspirations. The 12% of projects that results from selling is an indicator of their positive action to maintain the dynamic equilibrium.

Obviously this measure could be much enriched by looking at other characteristics of the project work load. Work is currently underway to investigate how some of the other characteristics of the projects might help to produce a more accurate measure of where the group is in the life cycle, and in turn produce a powerful predictive test of where the group is going in the current environment and under current policies. Another aspect of prediction is the development of a matrix of profile changes from one state to another. The choices available, as defined by Stewart (1982), to a manager of a project based work group must be considerably larger than for the manager of a routine based work group because of the ability to define the future work which is inherent in the nature of a project based group. The life cycle concept together with the developing quantitative measures of expected and required changes will enable those choices to be made more effectively.

APPLICATIONS TO NON-OR GROUPS

Pettigrew (1975) based his work on three groups, an OR group, an OD group and a Systems Analysis group. The Hatfield fieldwork (Conway 1985, Conway and Holland 1987, Holland 1988) has covered OR groups in the UK, both in-house and consultancy. Both types of OR conform totally to the structure of the Dynamic Model and the Revised Life Cycle Model, though the clients which form the environment for an external consultancy firm are clearly not within the same organisation. However the generalisation of the results depends on the acceptance of the concept that project based work groups are fundamentally different from routine based work groups, and pose different management problems requiring different approaches. The acceptance of this will have important consequences for the teaching of management. It is clear from the evidence already available that all the traditional management services activities (OR, O & M, Work Study, Systems Analysis, OD, etc.) are, and ought to be managed as, project based work groups. There are however a number of other activities which in some organisations, for all or part of the time, operate as project based work groups and must therefore be managed as such. Examples are research and development, management accounting and market research. Clearly all these areas need investigation, not only for model validation but also to provide the templates for the states of the life cycle and the consequent state change guidelines.

Potentially of much wider scope is the application of these ideas
to project based businesses and other similar organisations whether
profit making or not. Preliminary studies have shown that independent
research institutions operate in a similar manner, as do academic
research groups. Less obviously a preliminary study of the business
start up for a sculptor show a typical Dynamic Model pattern for a
pioneering phase to adaptive response development. The original
commission was unexpected and based on personal contacts not on the
quality of previous work. Subsequent commissions have clearly fallen
into Type 2 and later into Type 3 classes, with the volume of work
allowing the person concerned to consider Type 4 and Type 5 selling
activities. Again application of the ideas described in this paper will
allow a very different approach to be taken to understanding the growth
and decline of such businesses and to their management.

REFERENCES

Conway, D. A., 1985, "The Development and Application of a Dynamic Model
    of the Process of Operational Research", Ph.D. Thesis, Hatfield
    Polytechnic.

Conway, D. A. and Holland, J. C., 1987, This little piggy ...., OR
    Society Annual Conference (OR29)

Davis, K. R., McKeown, P. G. and Rakes, T. R., 1986, "Management
    Science: an Introduction", Kent, Boston, Mass.

Holland, J. C., 1988, This little piggy revisited, OR.Insight, 1, 4

March, J. G., and Simon, H. A., 1958, "Organisations", Wiley, New York.

Pettigrew, A. M., 1975, Strategic Aspects of the Management of
    Specialist Activity, Pers. Rev. 4, 5-13

Stewart, R., 1982, "Choices for the Manager: a Guide to Managerial Work
    and Behaviour", McGraw-Hill, London.

Stewart, R., 1985, "The Reality of Organisations", Macmillan, London.

Taha, H. A., 1987, "Operations Research: an Introduction", Collier
    Macmillan, New York.

NEW PARADIGM RESEARCH AS PART OF A HOLISTIC PRODUCTION PLANNING AND

CONTROL SYSTEM FOR A CARPET MANUFACTURING FIRM IN SCOTLAND

E A Buttery

Marketing Department
University of Strathclyde

INTRODUCTION

This paper deals with research carried out over a two-year period for a carpet manufacturing firm in Scotland. The objective of the research was to build a short term model and operating system which shared the same structure and data base, and to incorporate the decision maker as a necessary element with the model. Systems design was to be jointly undertaken by representatives of the firm and the researcher, and so this led to exploratory research into the use of new paradigm methodology as a basis for computer systems design.

The paper will explain how the model was completed particularly with reference to the use of new paradigm research, mixed scanning, and prototyping in the 'soft system' aspects of the work and in the spirit of new paradigm research the characteristics of the fieldwork together with the potential bias of the researcher will be considered. Mention will also be made of the 'hard system' elements to complete the description of the project and to demonstrate the 'holistic' nature of the approach.

THE HOLISTIC APPROACH

Central to systems theory is the concept of 'holism' which considers that all systems are composed of interrelated subsystems which together are greater than the sum of constituents, and therefore can only be explained in totality rather than a piecemeal way. With slight adaptation of the Kast & Rosenzweig (1972) systems model, any computerised production planning and control system can be reflected in systems terms as the following sub systems:-

Psycho/Social - The human content of the system.
Goals and Values - The aims and objectives of producing and operating the system.
Structure - Organisation of information capable of computerisation.

Technical   - The production conversion process, computer
hardware and associated programming language.
Management   - Effective guiding, co-ordinating, integrating, and
control activity for computerised production.

In enveloping the 'holistic' approach, the above subsystems had
to be reflected in the research. However, this immediately raised the
dichotomy which has been clearly recognised by others, i.e. that
certain aspects require a 'soft systems' approach, whilst others fair
better under a 'hard systems' approach. Checkland (1985) provides the
following explanation of these alternative approaches "If we ask what
is the single most marked change in moving from special (hard) case of
an unproblematual system which can be engineered, to the general
(soft) case of an issue-based situation in which accommodation must be
sought, it is then thought of as a shift from thinking in terms of
models of (parts of) the world to models relevant to arguing about the
world". Miles (1985) has stated that the dichotomy exists for system
developers who need "to bridge a gap between information needs of an
organisation and the technology which can be used to produce and
disseminate that information". He views the roots of computer systems
embedded in the hard systems tradition, but the need of the analyst to
contend with the political (or social) dimensions as being constrained
by the hard systems tradition, and best dealt with by a soft systems
approach. As the final computerised model reflects some aspects of
hard and soft systems, both have to be built into the approach. In
developing the soft systems aspect which deals with psycho/social,
goals and values and management subsystems, new paradigm research,
mixed scanning and prototyping were used. For the technical and
structural subsystems, which reflect the production process and the
computer hardware and software, the researcher drew on the literature
on M.I.S. and database design.

THE CHOICE OF METHODOLOGY FOR DEVELOPING THE SOFT SYSTEMS

There is ample evidence in the literature where researchers such
as Ackoff, Boland and Checkland, based on the principles of
hermaneutics, have successfully advised clients on the means of
structuring a rich debate. Mumford (1983) on the other hand, through
the ETHICS approach, has provided a more prescriptive methodology, but
at the expense, one might argue, of moving away from hermaneutic
ideals. In this research, it was necessary for the researcher to
become part of the development process rather than acting as mentor or
advisor, and this led to the utilisation of new paradigm research
methodology. New paradigm research argues that for a true analysis of
human action and experience there must be a firm collaboration between
'subjects' and 'researchers', and that this must be established in the
sphere of those actions, transactions and experience it is supposed to
be analysing. Its benefit according to Diesing (1972) is that the
approach "exhibits the most respect for human dignity and freedom
because it enables a person to work with, not on, his case, to treat
him (or them) as fellow human beings rather than as things".
Positivist research, and indeed much of the research carried out under
the Social Sciences, adopts a more detached stance in the interest of
maintaining objectivity. New paradigm researchers accept that
complete objectivity is difficult, indeed impossible, but make arduous
attempts to identify and define bias which may distort research
results. Supporters of the hermaneutic approach have been criticised
because people of differing status and power come to the 'rich

debate', and this inhibits the debating process. Habermas (1970) draws attention to what is required for the 'ideal speech situation' and details the distortions which arise from repressive domination in society. Taking on board these ideas, this research which culminated in a short term production planning and control model, utilised new paradigm research methodology. The researcher became a member of the development team and undertook as his input, the work of producing the computer program. He also tried to understand and evaluate his bias, described later in the paper.

In choosing the content of the debate, members of the firm were encouraged to select what should be included in the model and what should, for the time being, continue to be treated incrementally. For the successful implementation of this task, the mixed scanning approach was adopted which is used mainly in the public sector where objectives are often not quite as clear cut as the profit motive in the private sector, e.g. the dilemma of identifying objectives for the NHS. This research project drew on the literature available in this field, e.g. on the work of Etzioni (1967), and Smith and May (1980) who state "Each of the two elements in mixed scanning helps to reduce the effects of the particular shortcomings of the other; incrementalism reduces the unrealistic aspect of rationalism by limiting the details required for fundamental decisions, and contextuating rationalism helps to overcome the conservative slant of incrementalism by exploring longer-run alternatives". In order to debate and decide what must be treated rationally and what in a piecemeal way for this project, two groups were formed in line with on-going decision making levels, status and power. The group of senior managers concentrated on information requirements which would improve the carpet manufacturing process, whilst the middle managers and user group provided the details within the framework set by senior managers: Production manager and researcher were party to both groups. The outcome of various planning meetings was that many parts of the production planning and control system were in need of being included in the computerised system, but some aspects of the production planning process were left out of the project.

As both groups had to be satisfied of the viability of the system, prototyping was chosen as a means of producing the model. This meant that part of the system was produced, then sent for consideration, and approved or altered by the two development groups. This complements the spirit of new paradigm research as Jenkins (1985) remarks that "the prototyping methodology is based on one simple preposition: users can point to features they don't like about the existing system (or indicate when a feature is running well) more easily than they can describe what they think they would like in an imaginary system".

The soft system characteristic of the research and especially the use of new paradigm methodology, which recognises that bias exists, can only be truly understood in the context of the situation, and so the fieldwork, the bias, and the dynamics of the development process are now discussed.

THE FIELDWORK

The Jackson and Keys model (1984) suggests that the extreme forms of management are 'unitary' and 'pluralistic' considering problems which are perceived as 'mechanical' or 'systemic'. The situation within the production planning department tended towards pluralistic/systemic and the writer points to the proven usefulness of the methodology in these circumstances only. Applying Schein's (1987) perspectives of fieldwork, which focuses in part I on fundamental relationships and in part II on the process of inquiry and intervention, one can appreciate that as the writer was invited by a non-executive board member of the firm to investigate specifically the production system, the research was clinical as opposed to ethnographic. The production department favoured spoken contact, and so communication tended to be person to person. Throughout the research it was the firm's need to improve the production planning and control system which drove the process of enquiry and conditioned the scope of data gathered. The psychological contract represented the unspoken understanding of my principal duty to the University and my position as an academic researcher. For my part I conducted my work in keeping as far as possible with the culture of the firm. Although development of the model was exclusively concerned with the production of Axminster carpets, it was accepted that the process could be adopted and amended to suit the production of Wilton and Bonded carpets, indeed this was added at a later date but falls outside the scope of this paper.

Detailed data was gathered within the firm and tended to relate to carpet production. The scope of the study meant it was difficult to relate the work to the firm as a whole, and so the boundary of the study was the production department. Other departments of the firm were considered as major environmental influences. The validation process was on-going and undertaken jointly by the two development groups. The nature of the development did not call for past data being used to validate the system in a statistical manner, validation occurred partially by running the manual and computerised system concurrently for a short period, but more importantly was considered in terms of confidence expressed by the two teams which preambled the next part of the modular development. Confidence in the system in turn was related to how close the system operated when compared to the group specifications on structure, the output and input requirements, the scope of the task, the user friendliness etc. This type of validation is more akin to the methods used by Forrester (1961) and Coyle (1977) in systems dynamics than to the traditional statistical validation approach.

THE BIAS

An area for which the writer produced his own structure for analysis, was in retrospectively identifying the bias he may have brought to the research, so that readers may be able to evaluate any possible research result distortions, and other users of the new paradigm methodology may use the structure to identify their bias. Regarding this project, the main areas of bias were <u>initially in the research</u> - representing a concern for the writer to establish himself in the firm, and using tactics to secure the research project, also to 'win people over' in the early parts of the project, <u>because the writer's background</u> is marketing, computer studies, and academia, compared to other team members whose backgrounds were production systems and industry, <u>the writer's interests</u> which reflected his knowledge of concepts and techniques, and the <u>expediency of the research</u> which sometimes meant the writer was willing to forfeit further debate to meet pressing deadlines.

THE DYNAMICS OF THE DEVELOPMENT GROUP

When viewed as a cybernetic model one could see that, as the success of the project grew, the primary motivation for generating interest moved from 'quickly achieving interim goals' to 'people wanting to be part of a successful project'. This was fortunate as one can demonstrate, using cybernetics, that rapid development of a project initially is modified by a negative loop. This reflects the complexity of the project. In this project the negative loop related to the need by the researcher for time to understand the intricassies of the production process, and by the firm's members to become conversant with computers and to assimilate what has been achieved and what potential remains untapped.

THE HARD SYSTEMS ELEMENTS

As previously outlined, in this project the hard systems elements relate to the use and limitations of the computer hardware and programming language. Also the structure and content of the database and the information subsystems were perceived as fairly rigid, the latter comprising four subsystems, a data bank, a model bank, a statistics bank and a fairly standard input/output system. This aspect of the research was treated as a 'hard system'; the choice of hardware and programming language was left, subject to a financial constraint, entirely to the researcher, and it is necessary to point out, that sometimes, although rarely, the choice of hardware represented a real constraint on what the development teams wanted and what could be achieved on the computer.

The physical process of producing a carpet from the ordering of base yarn, which is followed by the dyeing of yarn, and the spooling and weaving processes, to the storage of finished carpet, reflects the characteristics of a hard system. However, the fact that throughout the various processes, decisions can be made about priorities in manufacture, together with the kind of intervention necessary to support decisions and how it should be presented introduces the soft systems approach to the model. In addition, views exist about how management can improve, its performance, underlining again the soft system approach.

SUMMARY AND CONCLUSION

The paper explains an approach to developing a production planning and control system for a specific firm. It recognises the 'soft' and 'hard' system elements in the design of the system. The soft system approach was developed using New Paradigm Research in conjunction with Mixed Scanning and Prototyping. In the spirit of New Paradigm Research it was recognised that bias entered into the research and so the fieldwork and individual bias of the researcher were considered wherever possible using, as a basis for analysis, models already available in the literature. The hard system context was identified in terms of the researcher developing the computer programme using a standard approach to data base design and Information System structure, and in the physical characteristics of the production process which have not changed considerably in the last century.

REFERENCES

Checkland, P, Achieving 'Desireable & Feasible' Change. An
     Application of Soft Systems Methodology, Journal of
     Operational Research SOC, Vol 36, No 9, pp 757-767, 1985.
Coyle, R, G, Management Systems Dynamics, London, Wiley, 1977.
Diesing, P, Patterns of Discovery in the Social Sciences, London,
     Routledge & Kegan Paul, 1972.
Etzioni, A, Mixed Scanning: A Third Approach To Decision Making,
     Public Admin Review, Vol 27, pp 385-92, 1967.
Forrester, J, W, Industrial Dynamics, Cambridge, Mit Press, 1961.
Habermas, J, On Systematically Distorted Communication and
     Towards a Theory of Communication Competence, Inquiry, Vol
     13, 1970.
Jackson, M, C, & Keys, P, Towards a System of System
     Methodologies Journal   of OPL Research Soc, Vol 35, No 6,
     pp 473-486, 1984.
Jenkins, A, M, Prototyping: A Methodology for the Design and
     Development of Application Systems, Spectrum, Vol 2, No 2,
     April 1985.
Kast, F, E, & Rosenzweig, J, E, Organisation & Management - A
     Systems Approach, London, McGraw-Hill, 1972.
Miles, R, K, Computer Systems Analysis: The Constraint of the
     Hard Systems   Paradigm, Journal of Applied Systems
     Analysis, Vol 12, 1985.
Schein, E, H, The Clinical Perspective in Fieldwork, Sage,
     Beverley Hills, 1987.
Smith, G, & May, D, The Artificial Debate Between Rationalist and
     Incrementalist Models of Decision Making, Policy & Politics,
     Vol 8, No 2, pp 147-161, 1980.

PRODUCTION PLANNING SYSTEMS FOR CORPORATIONS WITH SEPARATED AND COMPETING PRODUCTION UNITS

Sverre Inge Heimdal

Division of Economics
The Norwegian Institute of Technology
Trondheim, Norway

INTRODUCTION

Production of metals has traditionally been a dominating part of Norwegian industry. During the last decades, ferrous alloys and aluminium, next to oil and gas, have been the most important export articles. This paper will present experiences from an attempt to develop and implement a production planning system for the ferro-silicon division within a major Norwegian enterprise.

Ferrosilicon is a highly energy consuming product and is made in large electric furnaces. Quartzite and iron are melted together with coke and coal. About 50 % of all raw materials are imported, and almost every piece of produced metal is exported. Ferrosilicon is mainly used in production of steel and iron foundry.

PRODUCTION AND SALES

The company concerned in this paper is selling their products all over the world and has played a leading role in development of production technology for this kind of production. Within Scandinavia they own or have operating responsibility for 9 production plants. The products made at the different plants differ. Only 5 plants are capable of producing ferrosilicon.

Most sales activities are organized by a sales office located close to division management in Oslo, the capital of Norway. The various plants are located in small towns with little if any other industry. Unemployment problems in areas surrounding these small tows are usually above the average there. In addition, there are large geographical distances between the different plants and between the division management and the plants.

The prices of metals are not at all stable. Production equipment and raw materials are expensive. It is essential to adjust production to the demand for metals. A common strategy is to stop production on one or several furnaces when demand is decreasing and prices are low.

The different plants have few if any possibilities for deciding which prices to charge for their products. This is mainly because of the mechanisms in the market and, secondly due to the organization of sales activities within the company. The only way to compete is to be cost efficient and ship the demanded amount of metal with the proper quality and at the proper time.

When demand for ferrosilicon is decreasing, the company will decide to stop those furnaces which are most expensive to run. This fact makes it important to have competitive advantages not only compared to plants outside the company but also to other plants within the company.

All sales contracts have to be distributed to the plants to initiate production. During any one year there will be more than 1000 sales contracts and with more than 10 combinations of different products and qualities. Production planning is done both at the division level and the plant level. The most important decisions at each level are:

Plant:

* When do we produce for each sales contract?

* Which raw materials do we use?

* When should we change from one product quality to another?

* How do we utilize variations in power prices?

* What will the plant's cash flow be?

Division:

* Which plant will have to complete the different sales contracts?

* How do we get enough low priced energy to support our plants?

* How do we organize shipment of metals from plants to customers?

* What will the division's cash flow be?

THE HISTORY OF A PROPOSED PRODUCTION PLANNING SYSTEM

The large quantity of sales contracts implies an enormous number of possible production plans. Within the 5 different plants there are 14 different furnaces. The furnaces are of different sizes. For each kind of product quality, there will normally be possible to choose among several furnaces for completing any given sales contract.

This very complex planning situation, together with experiences with a production planning system for the ferromanganese plants, initiated the idea of making a new planning system for the ferrosilicon activities. The main goals for this new planning system should be:

* To assist in making the decisions as to which plants and furnaces should be used for fulfilling the respective sales contracts.

* To assist those who are planning shipment of ferrosilicon from the plants to the customers.

A consequence of this kind of planning will be to decide if and when to stop the different furnaces. There will also exist possibilities to change among different products and qualitylevels on most of the furnaces.

The problem to be solved will be to find those plans which can maximize the net operating income within a planning period of 6 months. Decisions for the first 3 months should be based on actual sales contracts and for the last 3 months on sales estimates.

It was decided to calculate the net operating income based on net present value techniques. The algorithm chosen to solve the problem was based on Monte Carlo simulation and heuristics.

The planning system was designed and developed by people employed at the division's own edp-department, and the project was managed by a senior manager within the ferrosilicon division. The different plants had few possibilities for correcting the system during development. Their main contribution were to release data describing productivity and costs for each of their furnaces.

During calculation the system will keep a list of the best solutions. In addition, there are built in some features which increase the possibility of choosing good solutions. A normal run will test several hundred solutions at an expense of 10-20 cpu minutes on a powerful mini.

After development of the most vital parts of the system, the solutions chosen by it were tested and compared with plans developed by traditional methods. The solutions suggested by the system improved the net present value of estimated income and expenses by 10 to 15 percent in comparison to the reference plans.

The investment made for developing the production planning system is estimated to be 3 man-years of design and programming work. The most important events during the period in which the system was developed are:

Apr 86   Design and development of the production
            system is initiated.

May 87   Version 1 of the production planning
            system is completed.

Nov 87   Version 2 of the production planning
          system is completed. The main difference
          between version 1 and version 2 is the
          ability to calculate solutions with co-
          shipment from two or more plants to the
          different markets.

          The planning system is presented to people
          representing the different plants.

Dec 87   The project manager stops further
          development. A decision is evidently made
          not to use the planning system.

Apr 88   Production planning previously done within
          the staff of the division management will
          from now on, be done by people employed at
          one of the plants.

EXPLANATIONS

It is not easy to understand what actually happend within the
company during the last two months of 1987 and the first months of
1988. It is also not easy to understand why the division is unwilling
to use a planning system which might improve it's net operating
income by 10 or maybe 15 percent. No official explanation exists, but
there are important pieces of information which might give some
ideas.

During the period of 1986 and 1987 it became more and more
evident that the company was heading for a financial disaster.
Unfortunate strategic decisions, combined with low prices on their
main products, gave an increasingly uneasy feeling among the
employees. This feeling was easy to recognize among people working
within the divison staff and those working at the plants.

Another aspect of interest is the requirements for decentralized
decisionmaking. During this period such requirements had been
articulated with an increasing strength. The great distances between
the plants and the division will always give situations in which it
is hard to understand decisions made by the division management. The
reason for this may be lack of information by those who are making
the decisions or the problem of recognizing global optimal solutions
when you are only really interested in the solution to a small part
of the problem.

The production planning system had, all the way been regarded as
a division management product. People from the different plants had
only contributed to a small extent during the design and development
period. The response after the 4-hour meeting late in November 1987
gave the planning system no chance. The most important arguments
given by people representing the plants are:

    * No computer program can ever be made to handle
       this kind of problem.

    * Production planning based on Monte Carlo
       simulation will never work.

* This production planning system does not
  contribute to or support planning at the plant
  level.

* Important decisions are based on a set of productivity parameters for each furnace. The values
  of these parameters can never be established by
  an agreement among the plants and would have to
  be decided by people not knowing the furnaces
  and the plants well enough to give the parameters
  the correct values.

* Development costs have been too high, and there
  should be some questioning about how this project
  has been managed.

Contrastinging the prediction of a 10 to 15 percent increase in
net profits with the arguments against the planning system, the
essential question is: Will the planning system benefit the company?
So far it is impossible to answer this question. The production
planning system never got the chance to prove its capabilities.

CONSEQUENCES OF NOT IMPLEMENTING

Given the fact that the system actually is designed and
programmed, what are the consequences of not even testing the system
properly and correcting possible errors?

* The planning system itself represents a lot of
  knowledge about production planning within the
  company. This knowledge is easier to transfer
  from one person to another when it is represented
  by a well documented computer program.

* The company has decided not to use a program
  which could increase its capability of doing
  sensitivity analysis.

* The company proves to be unwilling to explore
  new technology to increase profits if it has to
  change established ways of decision making.

* Stopping projects when the arguments are not
  understood or accepted may destroy that part of
  the organization which has developed the system.

A STRATEGY OF LOCAL PRODUCTION PLANNING

The last word is probably not spoken concerning production
planning within the company. It is hard to believe that the division
for some years will be without a computerized decision support system
dedicated to production planning. The complexity of production will
only increase due to an increasing number of specialized products.

The first step to take should probably be to introduce computer-
based production planning at each plant. Such systems would have to
be developed by people who know the plants and their production
capabilities. A lot of work would have to be done by those who later
are supposed to use the systems. We can hope they will regard this
effort as an investment which would later yield them high payoffs.

In addition to suggesting which raw materials to use and when to complete each sales contract, the system should be able to calculate the marginal cost function for each furnace. Some economies of scale exist in addition to stopping and starting the furnaces.

Sooner or later the division will implement a decision support system which assists in generating production plans. The different marginal cost functions will then be an important input.

The company should then also take a closer look at the heuristics which tell when to stop the furnaces. These decisions are difficult first of all because the different plants usually will be unwilling to exercise them and second because it is expensive to stop a furnace and later start it again.

WHY LOOK AT UNSUCCSESSFUL ATTEMPTS TO IMPLEMENT PLANNING SYSTEMS

A lot of people have realized that it is not simple to introduce decision support systems based on operational research techniques. This paper only confirms that fact. Possible explanations are the difficulties in understanding the techniques and their limitations, the great effort which has to be spent on using such techniques, and the way in which they might change decision making.

The planning system described in this paper was intended to coordinate production for plants which have an element of competition against eachother. Such situations are complicated because the plants will ask for incentives to accept the system. In one way or another there have to be mechanisms which gives all plants an increase in profits or a reduction of risk. If such incentives are missing even the very best planning system will be difficult to implement.

The repeated experience of how thumbs are turned down when someone tries to implement operational research or decision support raises important questions. In each case there are good reasons to ask:

* Will new systems or techniques benefit the shareholders?

* What makes a "good" system, it's ability to give good solutions or the way it is introduced?

* Is it reasonable to teach students optimization techniques noone or few are willing to use?

There are probably no simple answers to these questions except the recognition that a system has to be implemented before it is possible to prove its capabilities. The way you introduce it is essential.

OPERATIONAL RESEARCH AND PRODUCTION MANAGEMENT : THE CASE OF "OPT"

Ian Graham

Department Of Business Studies
University of Edinburgh
Scotland, U.K.

In 1979 E Goldratt, an Israeli physicist, founded Creative Output Inc. to market Optimized Production Timetable (OPT), a production control package. In 1980 Goldratt outlined the principles behind this package to the annual conference of the American Production & Inventory Control Society (Goldratt, 1980). He claimed, "this breakthrough encompasses a reformulation of certain of the fundamental bases used in production planning; a mathematical solution to the newly formulated and enormously difficult, multivariate mathematical problem; a computer program package called OPT which enables a user to obtain an optimized, realistic production schedule". The references to "mathematical solution", "mutivariate mathematical problem" and "optimized schedule" suggest that Goldratt's approach was based around a mathematical programming algorithm and could be regarded as being within the operational research discipline. In naming the algorithm "OPT", Goldratt gave it an easily remembered name which suggested its claimed "optimality", in contrast to most operational research algorithms.

By 1981 Creative Output had renamed the algorithm "Optimized Production Technology", and Goldratt presented a paper to that year's APICS conference further describing his approach to scheduling based on finite capacity loading of "bottleneck" resources (Goldratt, 1981). At the same conference a paper was presented describing a practical implementation of OPT in an area manufacturing investment casting moulds at Howmet Turbines (David & LeFevre, 1981). David and LeFevre redefined "optimization" as being "optimization ... in the sense that empirical tests using other approaches have not developed 'better' schedules", without describing the empirical tests upon which they based this assertion.

David and LeFevre claimed that OPT led to "output increased by 25% over previous 'maximum' levels", but their description was of how an "OPT simulation was initially run for one product to define the number of pattern assembly workers needed". So the benefits were at least in part due to the use of OPT as a simulation package for assessing the design of the system rather than from its use to produce "optimum" schedules. David and LeFevre concluded by stating that "most importantly, OPT had an educational impact". In all following descriptions of OPT implementations it is impossible to apportion the claimed benefits between OPT's roles as scheduling package, simulation program and educational aid.

Awareness of OPT was increased by four articles in Inventories and Production Magazine by Robert Fox between July 1982 and March 1983. Fox, previously a vice-president with management consultants Booz, Allen & Hamilton, was a vice-president of Creative Output. In the first article (Fox, 1982a) Fox defined OPT in relation to "materials requirement planning" (MRP), the dominant Western production control paradigm, and the Japanese kanban system. Fox anthropomorphised Goldratt's algorithm, calling it the "OPT Brain", but he still did not divulge the nature of the algorithm. In the second article (Fox, 1982b) Fox codified the "philosophy" underlying OPT as ten rules. OPT had mutated from being a model, which followed an operational research approach, to become a "philosophy" independent of any algorithm or model. Clearly increasing emphasis was being placed by Creative Output on the marketing of OPT's educational benefits. The codifying of previously disparate concepts and guidelines as a system of simple rules had been used earlier in 1982 by Schonberger in his book which defined "just-in-time" manufacturing, "Japanese Manufacturing Techniques: Nine Hidden Lessons in Simplicity" (Schonberger, 1982). By the third article (Fox, 1983), Fox had adopted a biblical metaphor to describe his rules, calling them the "ten commandments of scheduling".

OPT'S TEN COMMANDMENTS
1. The utilization of a non bottleneck is not determined by its own potential, but by some other constraint in the system.
2. Activating a resource is not synonymous with utilizing a resource.
3. An hour lost at a bottleneck is an hour lost for the total system.
4. An hour saved at a non bottleneck is a mirage.
5. The transfer batch may not and many times should not be equal to the process batch.
6. The process batch size should be variable and not fixed.
7. Capacity and priority need to be considered simultaneously and not sequentially.
8. Murphy is not an unknown and his damage can be isolated and minimized.
9. Plant capacity should not be balanced.
10. The sum of local optimums is not equal to the optimum of the whole.

The first critical coverage of OPT came in Fortune (Bylinsky, 1983). Bylinsky reiterated Fox's description of OPT and stated that 20 of the Fortune 500 companies had purchased the system, including GE, Westinghouse, GM, Ford and RCA. However Bylinsky noted that, "Oddly, given Goldratt's insistence on keeping his equations secret, it's not the OPT software or even his algorithm that he considers crucial-it is, rather, his 'philosophy'. Even without buying OPT software, Goldratt claims, a receptive manufacturer could achieve a large improvement in operations simply by following his ideas". This is a remarkable argument for the vice-president of a company selling computer systems. Bylinsky reported that not all OPT users were satisfied, with Black and Decker finding the system too slow and John Deere finding it less effective than their existing systems, but these negative cases were balanced by claimed inventory reductions of $30 million at GE, Wilmington. This was followed by a report in International Management (Whiteside and Arbose, 1984) which questioned whether OPT was significantly different to other finite capacity scheduling and simulation tools, and accused Creative Output of over-selling their product.

In 1984 Goldratt published a novel, "The Goal: Excellence In

Manufacturing" (Goldratt & Cox, 1984). The novel tells the story of how one manager came to accept the OPT principles and saved his marriage. As its cover says, "'The Goal' is about the OPT principles of manufacturing. It's about trying to understand what makes our world tick so that we can make it better. When we think logically and consistently about our problems we can determine the 'cause and effect' relationships between actions and results. Using the OPT principles we can compete with anyone. The Western world does not have to become a second or third rate manufacturing power if we just understand and apply these principles.". "The Goal" belongs to the fiction genre typified by "Zen and Art of Motorcycle Maintenance", implying that OPT's "philosophy" is not only of relevance to production management but also to life in general. The novel clearly owes some debt, alluded to in its sub-title, to "In Search Of Excellence" (Peters and Waterman, 1982), one of the most successful management books of the past decade. In "The Goal" OPT has split away from the scheduling algorithm, with no mention of OPT as a computer-based system. The success of this novel (selling 2000 copies per week in 1987, (Jayson, 1987)) has been instrumental in the wide dissemination of Goldratt's ideas.

Prior to 1985 OPT had not been discussed in academic journals: for production management academics there were few full implementations to write up as case-studies (unlike the wealth of just-in-time implementations studied), whilst for operational research academics it was hard to discuss or investigate OPT without a description of the OPT algorithm. Creative Output's reticence in discussing their algorithm was justified in that if they had described it there would have been a deluge of academic papers proving how its claims to being optimal were false. In 1985 Gelders and Van Wassenhove (Gelders & Van Wassenhove, 1985) compared the use of MRP, JIT and OPT for capacity planning. They accepted the "ten commandments", stating that "few people will disagree with them and it is clear that a good production-inventory system should try to implement these rules as well as possible". They stated this even though only four of the "commandments" were instructions, the other six being observations. Also the third rule, "an hour lost at a bottleneck is an hour lost to the total system", assumes a hard capacity constraint on a machine scheduled to work at 100% utilisation, in which case any uncertainty in the estimation of load or capacity would expose the system to the risk of a gross over-load. Whereas Fox defined OPT as a further advance on MRP and JIT, Gelders and Van Wassenhove proposed that a synthesis of the three elements would be the ideal solution to capacity planning problems. In arguing this they were treating OPT not as a specific software product but as a generic name for any finite capacity scheduling system which followed the OPT rules.

The relationship of OPT to JIT and MRP was also discussed in Autumn 1985 by Aggarwal in two papers (Aggarwal, 1985; Aggarwal & Aggarwal, 1985). In Aggarwal's article in the Harvard Business Review OPT is described as being an alternative production control system to MRP or kanban, but with a minimum purchase price of $2 million. (To gain some impression of the scarcity of full implementations of OPT which could be studied in case-studies, this cost should be compared to the total turnover of the U.K. agents for OPT of only £5.6 million for the five years up to March 1987).

In the second quarter of 1986 the APICS journal, Production and Inventory Control, devoted an issue to papers on OPT. Of the five papers, three (Plenert & Best, 1986; Swann, 1986; Vollman, 1986) used "OPT" as a generic term for finite capacity schedulers, and two

(Lundrigan, 1986; Meleton, 1986) focussed on the use of OPT as a simulation tool. Lundrigan of GE claimed that "OPT is particularly valuable as a plant start up tool...", while Meleton of Rockwell noted that "several companies have mentioned OPT's modeling and simulation capabilities as one (sic) of the most positive aspects of the system". All five papers saw OPT as a piece of computer software rather than as a philosophy.

In 1987 it was reported in Management Accounting (Jayson, 1987) that Fox and Goldratt had established the A. V. Goldratt Institute in Newhaven, Connecticut, "near the campus of Yale University", to "develop the know-how" to close the "dangerous imbalance of manufacturing capabilities between the West and East". OPT was described as being "a concept, a philosophy, and approach that happens to have as part of its offering a software tool". Goldratt and Fox have moved away from using a software company to market a tangible product to setting up a quasi-academic institute to sell a "philosophy", which they claim will be the salvation of Western industry. Again the development of OPT shadowed that of JIT, where Schonberger's second book (Schonberger, 1986) defined a vague ideal of "World Class Manufacturing" which each company should strive to attain in order to remain competitive with south-east Asian industry. Both Goldratt and Schonberger were following the path beaten in general management texts by Peters and Waterman in "In Search of Excellence: Lessons from America's Best-Run Companies" and Ouchi in "Theory Z: How American Business Can Meet the Japanese Challenge" (Ouchi, 1981).

Goldratt's quasi-academic phase has given birth to "constraint theory". In May 1988 an article written by a manager at Valmont/ALS (Koziol, 1988) described how he had attended a seminar at the Goldratt Institute on "Logistics of Constraint and Finance" and then implemented the "theory of constraints" in his own company. The theory embodies the concepts covered in Goldratt's paper of 1980, but they have now become totally divorced from any specific software product. Koziol explains that it was necessary for Valmont/ALS to develop in-house the software required to implement the theory.

Meanwhile OPT lives on, at least as a term for Goldratt's rules (Feather & Cross, 1988) and as a generic term for a finite capacity scheduling system (Eloranta, 1988).

The development of OPT demonstrates the marginalising of production operational research from the mainstream of manufacturing management writing and practice. For manufacturing problems, production operational research practitioners have traditionally sought to develop optimal solutions to specific problems, whereas management as a discipline seeks general solutions to generic problems. Whilst they are ostensibly tackling the same problems, operational research and production management are discourses in conflict.

Management discourses, whether by Peters, Schonberger, or Fox, depend upon the argument: "These are the companies that are successful, these are the common factors, and these are the "golden rules" to achieve success". Management discourses do not need a model to validate the lessons drawn - they are anecdotal and anti-analytical. Peters and Waterman in "In Search of Excellence" explicitly rejected model-based approaches to management, stating "But if America is to regain its competitive position in the world, or even hold what it has, we have to stop overdoing things on the rational side.". The history of OPT has shown how the rejection of rationalism has removed the need even for claims of "optimum" policies to be validated by models.

OPT was launched without the user-base to link itself to successful companies, so it had to be marketed to its early users as an operational research based solution to their scheduling and modelling problems, its strong selling point being the system's claimed optimality when compared to more conventional heuristic scheduling algorithms. At this stage an operational research perspective could have been used to question Creative Output's claims, pointing out the technical and theoretical difficulties in developing an implementable generalised optimal scheduling system in complex non-deterministic environments. Once a bridgehead of prestigious users had been established, OPT entered the mainstream of production management discourse, shadowing closely the lead set by just-in-time inventory techniques. The names of users were stressed in descriptions, but explaining how or why benefits had been gained was elided, and the view of OPT as a computer-based model or algorithm was progressively displaced by descriptions of OPT as a "philosophy" or set of "commandments".

The influence of the ideological needs of Western industry on the lessons drawn by Western observers seeking to explain the success of Japanese industry was discussed in an earlier paper on the development of "just-in-time" manufacturing (Graham, 1988). Similarly, in the case of OPT, it was the needs and values of the audience of potential users that shaped the development of OPT from algorithm to philosophy. The set of OPT "commandments" contained valid insights into the management of manufacturing systems, particularly as a critique of orthodox management accounting practices and performance measures. The coalescing of disparate guidelines and observations into the single entity, OPT, created a framework which allows the inter-relationships between load, capacity, inventory and lead times to be viewed and explored. It is the provision of a framework which allows managers to better understand their environment. This is the major value of management texts and training. To be "successful" the framework should appear to be a radical break with past ideas, but it must not mount a serious challenge to the present management ideology. The history of OPT has been shaped by the needs and ideological perspectives of its users. Released from any need to justify OPT analytically, it could be claimed that OPT was a radical breakthrough whilst it was being progressively shaped into a concept readily marketable to manufacturing managers. In contrast, genuine operational research model-based approaches are inherently less malleable.

REFERENCES

Aggarwal, S. C., and Aggarwal, S., 1985, The Management of Manufacturing Operations: An Appraisal of Recent Developments, Int. Jour. of Ops. & Prod. Mang., 5:3.
Aggarwal, S. C., 1985, MRP, JIT, OPT, FMS? Making Sense of Production Operations Systems, Harvard Business Review, 63:5.
Bylinsky, G., 1983, An Efficiency Guru with a Brown Box, Fortune, 108:5.
David, H. I. and LeFevre, D. C., 1982, Finite Capacity Scheduling With OPT in "APICS Conference Proceedings", APICS, Falls Church, VA.
Eloranta, E., 1988, Managing Production for Customer Service and Plant Efficiency, Industrial Engineering, 20:3.
Feather, J. J. and Cross, K. F., 1988, Workflow Analysis, Just-In-Time Techniques Simplify Administrative Process in Paperwork Operation, Industrial Engineering, 20:1.

Fox, R., 1982a, OPT - An Answer for America, I, <u>Inventories & Production Magazine</u>, 2:4.

Fox, R., 1982b, OPT - An Answer for America, II, <u>Inventories & Production Magazine</u>, 2:6.

Fox, R., 1983, OPT - An Answer for America, III, <u>Inventories & Production Magazine</u>, 3:1.

Gelders, L. F., and Van Wassenhove, L. N., 1985, Capacity Planning in MRP, JIT and OPT: A Critique, <u>Engineering Costs & Production Economics</u>, 9:1.

Goldratt, E. M., 1980, Optimized Production Timetable: A Revolutionary Program For Industry <u>in</u> "APICS Conference Proceedings", APICS, Falls Church, VA.

Goldratt, E. M., 1982, The Unbalanced Plant <u>in</u> "APICS Conference Proceedings", APICS, Falls Church, VA.

Goldratt, E. and Cox, J, 1984, "The Goal: Excellence in Manufacturing", North River Press, Croton-on-Hudson, NY.

Graham, I., 1988, Japanisation as Mythology, <u>Industrial Relations Journal</u>,19:1.

Jayson, S., 1987, Goldratt & Fox: Revolutionizing the Factory Floor, <u>Management Accounting</u>, 68:11.

Koziol, D. S., 1988, How The Constraint Theory Improved a Job-Shop Operation, <u>Management Accounting</u>, 69:11.

Lundrigan, R., 1986, What is this Thing Called OPT?, <u>Production & Inventory Management</u>, 27:2.

Meleton, M. P., 1986, OPT - Fantasy or Breakthrough?, <u>Production & Inventory Management</u>, 27:2.

Ouchi, W. G., 1981, "Theory Z: How American Business Can Meet the Japanese Challenge", Addison-Wesley, Reading, Mass.

Peters, T. J., and Waterman, R. H., 1982, "In Search of Excellence: Lessons from America's Best-Run Companies", Harper & Row, New York, NY.

Plenert, G. and Best, T. D., 1986, MRP, JIT and OPT: What's "Best"?, <u>Production & Inventory Management</u>, 27:2.

Schonberger, R. J., 1986, "World Class Manufacturing: The Lessons of Simplicity Applied", The Free Press, New York, NY.

Schonberger, R. J., 1982, "Japanese Manufacturing Techniques: Nine Hidden Lessons in Simplicity", The Free Press, New York, NY

Swann, D., 1986, Using MRP for Optimized Schedules, <u>Production & Inventory Management</u>, 27:2.

Vollman, T. E., 1986, OPT as an Enhancement to MRPII, <u>Production & Inventory Management</u>, 27:2.

A FRAMEWORK FOR THE DESIGN OF EFFECTIVE STRUCTURES FOR ORGANISATIONS FACING RAPID TECHNOLOGICAL CHANGE

R. Keith Ellis

Department of Systems Science
The City University
Northampton Square, London. EC1V 0HB

INTRODUCTION

Technological Change (TC), and the complexity associated with it, cannot be viewed as a single issue topic. It is more appropriate to consider TC as a set of factors emanating from the 'push-pull' interactions contained within an organisational setting. These factors include Markets; Process, Product and Knowledge Innovation; Managerial Style; Corporate Culture; Human Values, Attitudes and Beliefs; Organisational Structure; etc. (Ellis and Flood, 1987; Ellis, 1988). The list of factors, although not exhaustive, can be a daunting prospect for managers faced with the task of designing organisational structures capable of coping with the complexity and variety associated with rapid TC.

This paper will propose a framework within which practising managers can address the problems of maintaining organisational and environmental congruency, commensurate with varying rates of TC.

TECHNOLOGY AND CHANGE AND ORGANISATIONS

Technological Change (TC) is endemic in modern organisational life. The pace of TC has been accelerating since the 1940's and that increasing rate of change shows no sign of slackening. Indeed, it can be argued that the reverse is true, as evidenced by ever shortening product life cycles. The impact of the ubiquitous silicon chip on organisations - and society in general - has been significant, and this impact is by no means complete or fully understood.

The impact of TC on organisations can be disastrous if it is inadequately understood and managed. Conversely it can, if well understood and managed, be beneficial. For the latter to be the (preferred) case, managers must adopt a proactive boundary scanning approach taking into account contingency factors such as age, size and technology used by the organisation. (Mintzberg, 1979). Such an approach can facilitate flexibility which is likely to encourage an organisational culture receptive to change, and capable of recognising that its structure need not be like a rigid steel framework which will snap if the winds of change become too strong.

TC can emerge in three distinct forms in organisations, namely :

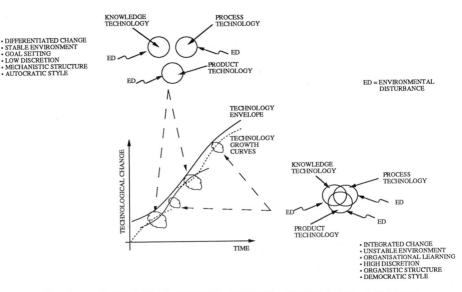

Fig I. RELATIONSHIP BETWEEN PRODUCT, PROCESS AND KNOWLEDGE
TECHNOLOGY AND THE RATE OF TECHNOLOGICAL CHANGE

innovation in products; innovation in the process of production; knowledge innovation. It has been argued that changes in 'product' technology can affect the structure or architecture of an organisation. (Broekstrar, 1986). Changes in 'process' technology can bring about a requirement for change in both the structure of the organisation and in the skill capabilities of those occupying positions within that structure. Changes in 'knowledge' technology can result in a need for changes in individual behaviour leading to changes in the operating characteristics of segments of the organisational structure, or indeed, of the whole organisational system. (Ellis, 1988; Dimond and Ellis, 1989).

For organisations operating in a stable environment and a relatively slow rate of change, changes in product, process or knowledge technology tend to occur with little or no interaction with each other. A variation in one form of TC has little effect upon the other two. That is, they are differentiated from each other, and the resulting effects on organisational structure and working practices are readily absorbed. Such conditions existed up to 1940, and the engine for change was largely fuelled by concern for improved productivity, resulting in a Tayloristic approach to organisational management which considered organisations as closed systems.

However, the emergence of dynamic and highly competitive business conditions from the 1970's onward has meant that organisations need to innovate over shortening time cycles. This has led to situations in which organisational managers have to integrate changes in technology which takes account of the exponential characteristics of TC. With rapidly increasing rates of TC and emergence of new technologies replacing older technologies, changes in product, process and knowledge technology have significant consequences for each other. Under such conditions the changes must be considered to be integrated, and the resulting effects on organisational structure and working practices are less easily absorbed. Thus is, in part, due to the fact that managers are faced with change situations which are significantly more complex than was evident during times of less rapid

338

change. FIGURE 1 is an illustration of the 'differentiated'-integrated' forms of TC. (Ellis, 1988: Dimond and Ellis, 1989).

A DESIGN FRAMEWORK FOR ORGANISATIONAL STRUCTURES

The ontological and epistemological underpinnings for the framework are rooted in the functionalist paradigm, in that it adopts a realist, positivist stance. An organisation is considered to be a rational goal seeking entity, but the framework permits inclusion of the human dimension.

Within the functionalist paradigm a Contingency Model for organisational analysis has been developed. (Burrell and Morgan, 1979). Fundamental to the model is the notion of a differentiated environment which ranges from stable to unstable. Contained within the model is the view of an organisational system made up of four sub-systems: Strategic Control; Operational; Human; and Managerial. Organisational system effectiveness is predicated upon sub-system functional imperatives, and congruence between sub-system response to environmental differentiated conditions. FIGURE 2 refers.

The Contingency Model lacks a temporal dimension which impairs its use in a predictive sense. This may be a reason why the model has received little attention by practising managers. It is clearly necessary to bring time into the picture. This can be achieved by combining the 'differentiated' - 'integrated' forms of TC with the congruency of organisational Sub-System response to the environment in a framework for the design of appropriate organisational structures.

It is now possible to visualise 'differentiated' TC during times of environmental stability and 'integrated' TC during the unstable environment. That is the stable environment will be apparent at the lower and upper portions of a technology growth curve, and the unstable environment at the point of inflexion (maximum rate of change) of the curve.

Under conditions of 'differentiated' TC (slow change and stable environment) it would be appropriate for an organisation to adopt goal setting approach which would demand routine low discretion tasks. That would require a hierarchical, mechanistic organisational structure based on bureaucratic rule based operating procedures under managerial control. Creativity is not required and management style would be autocratic.

However, under conditions of 'integrated' TC (rapid change and unstable environment) it would be necessary for the organisation to develop a strategy which established systems for organisational learning requiring complex high discretion roles. The appropriate organisational structure would be open with a preponderance for lateralism. Creativity would become a paramount feature, with expertise not necessarily restricted by organisation status. Management style would need to be democratic.

It should be recognised that the framework proposed is likely to be problematic. Environmental changes may occur independently of TC. The rate of TC may be so rapid as to make it difficult to adjust the organisational response quickly enough to be appropriate for the conditions. Differing organisational functions may be subject to differing rates of TC and variable environmental conditions; production may be operating with routine technology in a relatively stable environment, whereas R&D may operate with complex changing technology in an unstable environment. Such problems will need to be addressed by ongoing research work. There may be other difficulties not readily apparent at this stage of development of the framework.

ENVIRONMENTAL DIFFERENTIATION
Stable ◀──────────────────────▶ Unstable
STRATEGIC CONTROL SUB-SYSTEM
Goal Setting ◀──────────────────────▶ Organisational Learning
OPERATIONAL SUB-SYSTEM
Routine, Low ◀──────────────────────▶ Complex, High
Discretion   Discretion
HUMAN SUB-SYSTEM
Economic Man ◀──────────────────────▶ Self-Actualising Man
MANAGERIAL SUB-SYSTEM
Bureaucratic   Democratic
Mechanistic ◀──────────────────────▶ Organistic
Theory X   Theory Y

Fig 2.   CONGRUENCY OF ORGANISATIONAL SUB-SYSTEM RESPONSE TO
DIFFERENTIATED ENVIRONMENTAL CONDITIONS

It must also be pointed out that it is at times necessary for organisations to create structures and managerial control systems which vary from mechanistic to organistic and from autocratic to democratic respectively.

However it is considered that the proposed framework can provide practising managers with an approach, which will assist in tackling the complex multi-faceted problems associated with designing organisation structures capable of coping with technological change.

REFERENCES

1. Broekstrar, G., 1986, Organisational Humanity and Architecture: Duality and Complementarity of PAPA logic and MAMA logic. in: "Managerial Conceptualisations of Change". Cybernetics and Systems an International Journal., 17:13-41.

2. Burrell,G., and Morgan, G., 1979, "Sociological Paradigms and Organisational Analysis." Heinemann, London.

3. Dimond, A.J., and Ellis, R.K., 1988 The Development of a Prescriptive Approach to Modelling Technological Change. Unpublished Research Memorandum (DSS/AJD/RKE/154)., City University, London.

4. Dimond, A.J., and Ellis, R.K., 1989, Technological Change and the Human Aspect, in: "Systems prospects : The Next Ten Years of Systems Research" R.L. Flood, M.C. Jackson, and P. Keys, eds. Plenum, London. (In press)

5. Ellis, R.K., 1988, Modelling Technological Change: A Socio-Technical Approach, in: "Proceedings of the Second International Industrial Engineering Conference". Nancy.

6. Ellis, R.K., and Flood R.L., 1987, Managing Technological Change : Systems Theory or Systems Technology? in : "Problems of Constancy and Change. Proceedings of 31st Annual Meeting of the International Society for General Systems Research". Vol 1 : 270-277. Budapest.

7. Flood, R.L., and Ellis, R.K., 1989 A Critical Analysis of Systems Theory and Systems Technology in the Management of Technological Change, in: "Systems Redesign". B. Banathy. ed., ISGSR. New York. (In press)

8. Mintzberg, H., 1979, "The Structuring of Organizations". Prentice-Hall Inc, Englewood Cliffs, NJ

HOW INTERVENTION PROCEDURES MAY BE RELATED TO DIFFERENT KINDS OF CHANGE

Marion Le Louarn

LAMSADE, Université de Paris-Dauphine
Place de Lattre de Tassigny
F - 75775 Paris Cedex 16

INTRODUCTION

Change is either "descriptive" (perceived and assessed) or "normative" (desired and planned). Only normative change is considered here, as we analyse intervention procedures in organizations. Intervention procedures are of three types : power-driven (an authoritarian process aimed at inciting employees to meet organization objectives), relationship-driven (a communication-oriented process aimed at developing positive attitudes, participation and creativity) or cognition-driven (an individualistically-oriented approach, encouraging simultaneous optimization of personal and organization goals). We discuss to what extent each of these procedures may be appropriate to the different types of change considered.

WHY CHANGE ? A CRITICAL OUTLOOK

According to the "Longman Webster English College" dictionary, "change" usually describes a complete substitution or an essential difference (that we shall call "drastic"), while terms like "alter", "vary" and "modify" indicate less important differences (that we shall call "incremental"). For organized bodies, incremental modifications may appear safer than drastic changes, as the latter require more energy, preparation and guidance, and are less likely to be reversible processes in case of dissatisfaction. Why then do organizational studies particularly favour the concept of change ? One reason might be that change appeals not only because of its direct effects but also because of its psychological implications. Indeed change may be desired for at least three kinds of reasons : (i) some dissatisfaction (whether specified or not) with a perceived situation - usually in relation to defined goals ; (ii) the hope to modify some balance of power (whether economical, political or affective) in favour of oneself or some group (see Brunsson, 1985) ; (iii) the belief that the production process can be enhanced by uniting the energies of those involved in it (without regard for the nature of the goals) - see Arendt (1972). Change can also be desired as an end in itself (versatility being viewed as a quality and a sign of intelligent behaviour in western cultures - in contrast with the conservatism of some other cultures*), and defined in terms of personal "life tempo" (product of the number of recorded experiences of change by the importance of these changes - see Sim-

---

*See for example the references to the Book of Changes in de Bary (1981.)

mel, 1977). Considering the intricateness of these psychological, political and ideological factors, change should not be mistaken for progress, especially in complex organizations. Similarly, change should not be thought of as the natural outcome of reasoning and willpower (see Sfez, 1984 for a critical view of decision-making as a linear, rational and voluntary process). One could even be tempted to reduce the reality of change to a poetical emotion, well rendered by Byron in one of his poems : "A change came o'er the spirit of my dream" (The Dream, St. 3) ; we would then prefer the use of words like "innovation" and "alteration", as they are better defined, and more appropriate to Business Administration. However, the problems of "change" have to be addressed as such, if only because at present intervention requests are often expressed in those terms. So, before reviewing the different kinds of change, let us learn more about the types of intervention procedures available.

INTERVENTION PROCEDURES : THREE APPROACHES

Intervention is the job of an analyst called upon by people in an organization to solve problems that they have not been able to deal with at some level of the hierarchy. We assume that when the intervention addresses problems across different levels of the hierarchy, specific procedures are to be defined for each level. We can thus categorize intervention procedures according to the kind of relationship that is being developed at the hierarchical level dealt with. This relationship is either authority-oriented (i.e. between a manager and his subordinates), communication-oriented (i.e. between employees of equal status) or individualistically-oriented (i.e. between an employee and himself, as is the case in self-appraisal).

Authority-oriented Intervention Procedures

Their salient feature is their use of pre-defined evaluation grids, established on the ground of formal authority or expertise, to describe and assess the activities and attributes of employees (see Schwebig, 1988). They often aim at helping a manager incite his subordinates to meet the objectives set forth for the period by higher-level staff ; in that case, the analysis is akin to an aerial photograph, in that it is a brief survey, representative only of the period under review, aimed at recording relevant synthetical information about each of the employees ; and the recommendations, drawn from comparisons between actual performances and set objectives, are of limited scope. Less frequently, another kind of authority-oriented procedure is used, presented as a "research intervention" ; in that case, what justifies the intervention is essentially the authority of Science (of the Social Sciences in particular - see Dreyfus and Rabinow, 1982 ; Astley, 1985), to which the personal authority of the analyst contributes ; the analysis involved is then more akin to an "underground" search for recurring unconscious patterns, hidden motives or structural dependencies among sets of individuals interacting in an organization ; in turn, the outcome is often believed to stem from the intervention process itself (frequently long and intricate) as much as from particular recommendations ; this is because research procedures oriented towards depicting systematic behaviour, with the help of pre-defined interpretative tools, design their objects of analysis as closed systems ; such closed systems are supposed to maintain their course of action by resisting or integrating adverse forces (cf. de Gaulejac and Pagès, 1979), unless drastic change intervenes from outside, in which case reorganization is difficult to control, and to scientifically justify (see Le Louarn, 1986). The driving force of these various intervention procedures being authority, we also call them "power-driven". In contrast, intervention procedures of the second category, defined as "communication-oriented", obey more democratic rules, and will be called "relationship-driven".

Communication-oriented Intervention Procedures

In relationship-driven procedures, attention is focussed on the Communication process itself, which usually takes place between employees of comparable status and is aimed at : (i) sharing and dispatching information to make up for non-existing or malfunctioning channels (i.e. between non-communicating departments of a hierarchical structure) ; (ii) improving the work atmosphere and social climate, by favouring greater freedom in team work and by developing social activities ; (iii) promoting engagement in a common project, by encouraging participation to decision-making and to problem-solving ; (iv) fostering creativity, by instituting and maintaining a favourable psychological climate, through "democratically-oriented" rules (i.e. non-evaluative rules, and competition-enhancing procedures - see Abric, 1984) ; and (v) enabling people on unfriendly terms to develop a more positive relationship, by giving them opportunities of making better acquaintance. In all these procedures, the analyst's role is rather that of an organizer and arbitrator, and his intervention is of less systematic nature, more like a blind trial and error process, than in authority-driven procedures. The ideology prevailing in relationship-driven procedures is what Dumont (1966) attributes to modernity, namely the "egalitarian ideology", which on the contemporary scene takes the form of the "Communication" epic described by Sfez (1988) ; when communication is considered as an end in itself, meaning tends to disappear, while the interaction and noise brought about by communication-oriented procedures tend to hinder the thinking activities (d'Unrug, 1976). In this respect, such procedures contrast much with individualistically-oriented ones.

Individualistically-oriented Intervention Procedures

"Thinking is sheer activity" (...) "and is a solitary but not a lonely business" (Arendt, 1971) ; that is why we call cognition-driven procedures "individualistically-oriented" even though they originate in specific social conditions. Usually they aim at helping one decision-maker (whether real or fictional) elicit and solve a problem to the best of his perceived interests, by making appropriate use of the available resources of Reason. The driving force here is Rationality (whether the search be for optimality, or only for "satisficing solutions" - as in the theory of "Bounded Rationality" of Simon, 1983) ... and much resembles an ideological figure, unless one interprets rationality in a constructivist way, i.e. ex post instead of ex ante : one is then led to admit that the decision-maker attributes actual meaning to the problem-situation, constraints and solutions only in the course of the problem-solving process, so that the "rationality" implied does not pre-exist the elicitation phase (and may be discarded before the decision implementation) while its main purpose may be of self-justification rather than of enlightenment. Such a view is compatible with recent decision-making research in the line of Montgomery's Dominance Search Theory, that defines the decision-making process as a "search for a cognitive structure in which one alternative can be seen as dominant over the others" and that is aimed at "building a stable enough intention to act in a certain way" (Montgomery, 1989). Applying these ideas to decision-aid should lead to the creation of methods for helping the decision-maker be more aware of his own patterns of behaviour, style of thinking, motivations and ethical stands, so as to endow him with more stability and consistency in his positioning of the problem in question (on value-system representations see Le Louarn, 1988). Additionally, implication in work practices might be enhanced and decision-making facilitated if it were recognized as natural for an executive in his work-related decisions to openly take into account his own interests and standards alongside those of the firm (see March, 1978 on the idea of "optimal sin"); this is suggested by some psychological studies on the importance of implication for action (see Beauvois and Joule, 1988). One could then describe work-related decision-making (at any level in the organization hierarchy) in terms of "doubly-bounded" rationality, bounded both in H.A. Simon's sense of limited cognition abilities, and in the sense of a compromise between personal will and will imposed upon by the organization. The advantages of cognition-driven intervention procedures may turn out to be drawbacks : as thinking produces creative ideas, it also tends to destabilize beliefs, and this might not be appropriate in certain kinds of situations.

CHANGE PATTERNS AND THEIR RELATION TO INTERVENTION PROCEDURES

We already saw the need to distinguish between incremental and drastic change. We define incremental change as composed of minor alterations, or of major modifications that are progressively implemented ; drastic change in turn is defined as a set of major modifications implemented at a given time. The other useful distinction seems to be between "strategical" and "peripheral" change. Strategical change concerns all modifications related to the work itself (i.e. technical activities, evaluation modes, decision-making procedures...), whereas peripheral change concerns the modifications related to the environmental setting of work (geographical location, working hours, interpersonal relationships, composition of teams...). This enables us to distinguish between four change patterns : incremental strategical change, incremental peripheral change, drastic strategical change and drastic peripheral change. We shall examine, for each of these change patterns, how appropriate the three kinds of intervention procedures are.

Incremental Strategical Change (I.S.C.)

One is here concerned with somehow changing for the better current procedures of work, evaluation and decision-making. One quite obvious way of tackling this is through critical analysis (of the situation, of the relevant improvement factors, of the related risks...) ; therefore cognition-driven intervention procedures are most appropriate to I.S.C., at least at the planning stage. Power-driven procedures may in turn be useful at the implementation stage, as they enable to enforce unpopular decisions (through injunctions, incitations, penalties...). In contrast, relationship-driven procedures are inappropriate for I.S.C., as they tend to hinder critical reasoning (through reification and "fetishism", described by d'Unrug, 1976 as "premature neutralization" of speech acts) ; but they can be marginally useful when applied alongside cognition-driven or authority-driven procedures (for instance to facilitate brainstorming, or to enable strategic communication to take place). Let us finally note how important it is for strategic innovations to be introduced in an appropriate way : according to Doms and Moscovici (1984), innovations have little chance of getting accepted unless they are presented by people who appear to hold very consistent views ; this confirms our idea that methods for supporting reflection on personal value systems are useful tools in the context of cognition-driven procedures.

Incremental Peripheral Change (I.P.C.)

I.P.C. includes modifications in the work atmosphere and interpersonal relationships, in office disposition, timetable arrangements, social activities, cultural habits... The most efficient way of bringing about I.P.C. is through the use of communication-oriented procedures, which are specifically designed for that purpose. But cognition-driven procedures may also be relevant, because : (i) genuine thinking is an invitation to authenticity and to truthful relationships (Sartre, 1983 : "La réflexion pure est bonne foi et comme telle appel à la bonne foi de l'autre") ; (ii) the use of cognitive-driven self-assessment tools may strengthen confidence in oneself and foster trust in other people. In contrast, authority-oriented procedures are irrelevant to I.P.C., and have even hindering effects : according to the theory of "double-bind", ordering someone to act naturally is a contradictory injunction that calls for a schizophrenic response (see Watzlawick's work) ; similarly, managing Quality Circles in an authoritative manner leads to disenchantment and disengagement on the part of the employees ; another instance of this can be found in the recent fashion for "Management Ethics" (see Lattman, 1988), that may be interpreted (because Ethics are individually-oriented rather than collectively-oriented - see Arendt, 1981) as a cover-up for the resurgence of an authoritarian stand. Power-driven procedures should therefore be avoided when communication-oriented procedures have been adopted as the main source of change.

Drastic Strategical Change (D.S.C.)

D.S.C. is the case of any major change in policy, staff or technology abruptly taking place in an organization. It implies a thorough shake-up of the strategical positions of people, who need some time to adapt to the new situation. Intervention procedures that are authority-oriented are therefore the most effective way of managing the change, and also the safest one for all parties. Cognition-driven procedures in turn may be useful for directing efforts in appropriate directions, but their destabilizing effects restrict their usefulness in D.S.C. situations. Relationship-driven procedures are, this time, of reduced utility, and would rather tend to be counterproductive when applied together with another kind of procedure : promoting participative management methods after a large number of employees have been dismissed would for instance be resented as a very frustrating prospect. Relationship-driven procedures should therefore be banned in the case of D.S.C.

Drastic Peripheral Change (D.P.C.)

D.P.C. takes the form of important modifications in the environmental setting of work, like geographical move, team shift, merger into a firm with a very different culture, major change in working hours... As for any drastic change, authority-driven procedures are usually safest. But as the adaptation required is mostly in the domain of interpersonal relationships, communication-driven procedures are successful as well. In contrast, cognition-driven procedures do not yield many results, because they tend to be interpreted in terms of egotism as opposed to group solidarity, and because they originally destabilize beliefs.

CONCLUDING REMARKS

Taking a critical view of normative change and of different manners in which it can be brought about has led us to distinguish between three kinds of intervention procedures more or less appropriate to the various types of change patterns. In Table 1, we have summed up the degree of appropriateness of Power-driven (P), Relationship-driven (R) and Cognition-driven (C) interaction procedures to Incremental Strategical Change, Incremental Peripheral Change, Drastic Strategical Change and Drastic Peripheral Change.

Looking at the table, one sees that, were an organization faced with implementing the four patterns of change at the same time, the kinds of interaction procedures most beneficial to change would be Power-driven and Cognition-driven procedures ; this gives us a general idea of the overall efficiency of the three kinds of procedures, although they cannot quite be compared. One also notes that applying any kind of procedures without regard for the change pattern addressed is more risky in the case of Power-driven and Relationship-driven procedures than in the case of Cognition-driven procedures. Conversely, the change patterns for which it is most important to apply appropriate procedures are I.P.C. and D.S.C., since in both cases change may be hindered by the use of one of the three procedures. This goes to show that normative change is not a simple matter to handle.

Table 1. Appropriateness of Intervention Procedures
to the different Patterns of Change

Change	Incremental	Drastical
Strategical	C, (P), ⌀	P, (C), ⚠
Peripheral	R, (C), ⚠	P, (R), ⌀

P : Power-driven Intervention Procedure ; R : Relationship-driven I.P. ; C : Cognitive-driven I.P.
T : very appropriate ; (T) : appropriate ; ⌀ : not
appropriate or indifferent ; ⚠ : very inappropriate

REFERENCES

Abric, J.-C., 1984, La créativité des groupes, in: "Psychologie Socia-
    le," S. Moscovici, ed., Presses Universitaires de France, Paris.
Arendt, H., 1971, "The Life of the Mind," H. B. Jovanovich, New York.
Arendt, H., 1972, "Le Système Totalitaire," Seuil, Paris.
Astley, W. G., 1985, Administrative Science as Socially Constructed
    Truth, Administrative Science Quarterly, 30:497-513.
Beauvois, J.-L., and Joule, R.-V., 1988, La psychologie de la soumis-
    sion, La Recherche, Sept., 19, 202:1050-1057.
Brunsson, N., 1985, "The irrational organization," Wiley, Bath.
de Bary, W. T., 1981, "Neo-confucian Orthodoxy and the Learning of the
    Mind-and-Heart," Columbia University Press, New York.
de Gaulejac, V., and Pagès, M., 1979, "L'emprise de l'organisation,"
    Presses Universitaires de France, Paris.
Dumont, L., 1966, "Homo-Hierarchicus," Gallimard, Paris.
Dreyfus, H. L., and Rabinow, P., 1982, "Beyond structuralism and her-
    meneutics, with an afterword by Foucault," University of Chica-
    go Press, Chicago.
d'Unrug, M.-C., 1976, "Les techniques psycho-sociologiques dans la for-
    mation, Usages et Abus," Ed. ESF & Entreprise Moderne d'Edition,
    Paris.
Lattman, C., 1988, "Ethik und Unternehmensführung," Physica Verlag,
    Heidelberg.
Le Louarn, M., 1986, "Epistémologie et Décision," D.E.A. n°103, Uni-
    versité Dauphine, Paris.
Le Louarn, M., 1988, Introducing the values and value-systems of actors
    in the decision process, in VIIIth Int. Conf. on M.C.D.M., 22-
    26 August, Manchester.
March, J., 1978, Bounded Rationality, ambiguity and the engineering of
    choice, Bell Journal of Economics, Autumn, Vol. 9-2:587-608.
Montgomery, H., 1989, From cognition to action : the search for domi-
    nance in decision making, in: "Process and Structure in Human
    Decision Making," H. Montgomery and O. Svenson, ed, Wiley.
Moscovici S., and Doms, M., 1984, Innovation et influence des minorités,
    in: "Psychologie Sociale," S. Moscovici, ed, P.U.F., Paris.
Sartre, J.-P., 1983, "Cahiers pour une morale," Gallimard, Paris.
Schwebig, P., 1988, "Les Communications de l'entreprise," Mc Graw Hill.
Sfez, L., 1984, "La Décision," P.U.F., Paris (QSJ 2181, 2nd ed. 1988).
Sfez, L., 1988, "Critique de la Communication," Seuil, Paris.
Simmel, G., 1977, "Philosophie des Geldes," Duncker & Humblot, Berlin.
Simon, H. A., 1983, "Reason in Human Affairs," Blackwell, Oxford.

CHANGING SOCIAL CHANGE

Slawomir Magala

Erasmus University, Rotterdam

P.O.Box 1738, 3000 DR R'dam, NL

INTRODUCTION

The most interesting transformations of organized forms of social interaction resulting in social change emerged from new social movements
which appeared in mid-sixties and continue to influence our societies
until the present day.

The appearance of movements in question changed not only the patterns
of social interaction but also the way in which social sciences tackle
the explanation of social change. Among the changes in social sciences
one notes the revival of interest in cultural determinants of social action
(communication, symbols, religion, etc.), a stress on a relative autonomy
of the political processes vis a vis the economic ones and an attempt to
use social psychology in order to provide the missing link between an
individual decision and macrochanges. It is therefore quite legitimate
to speak of changing the very social change, i.e. the conceptualization
of it.

The emerging paradigms and subparadigms of research on new social
movements are best characterized as being strongly interdisciplinary,
historical and focusing on a subtle interplay of intersubjective agreements and subjective constructions of social reality. In order to construct
a sound explanation of social action the researcher is often encouraged
to participate in analyzed actions and to negotiate his or her findings
and conclusions with the very people he is trying to scrutinize (e.g.
A. Touraine in his studies of the Polish "Solidarity" and of the French
working class movements).

Recent history provides an ample threshold which marks off the period
of long term stability of the paradigms of social research from the
period of turmoil and attempts to achieve social mobilization and social
change as perceived in modern social sciences. This threshold can be found
in the wave of student and youth unrest which started with an unconventio-
nal civil rights movement in the southern states of the USA. It gained
momentum with the anti-war rebellion of students whose numbers grew within
the still predominantly traditional, conservative educational institutions.
The peak has been reached with the mass protests against the war in
Vietnam. Later on the movement dissolved into a broad array of peace
initiatives, citizens' groups, feminist movements and countercultural
experiments in the post-1968 culture and politics.

Another interesting feature of the 1968 threshold of student and youth protest movements and their counterculture is that they appear not to have respected the system borders of state socialism and modern capitalism emerging, as it were, across all possible frontiers of nation states and supranational alliances. It is plausible that intersystemic differences between the three worlds are therefore less vital for the daily functioning of modern societies than those features which the three have in common, i.e. bureaucratization and dynamic development of mass media, technological transformations and global interconnections.

1. SOCIAL PSYCHOLOGY OF ORGANIZING

There has been a considerable shift in explaining social change over the past ten years. Social scientists seem to have shifted away from the macro- to mezo- and micro-level of theoretical analysis. We have ceased to think of a social change according to a paradigm which has evolved as a historical alliance of sociology and socialism. This paradigm emerged in the period of struggles of the working class movements against the capitalist states. The workers have been armed with ideological explanations of their respective national societies provided according to the rules of the class model of Marx. It has been assumed within this paradigm that the screenplay of political struggles has to be based on an analysis of an economic division of labour. Moreover, one assumed that a mass political party with a professional core of organizers should be transmitting the economic identity of actors – by means of a political mobilization – into an overall, wholesale, "total" revolution whose criterion of success lied in the takeover of the national state.

The abovementioned shift has been mainly due to two factors. First, there has been a remarkable lack of correspondence between a political success in taking national states over on the one hand and the construction of a more egalitarian social order on the other. None of the communist parties which came to state power in the present century could either radically overtake rival economic system with respect to the living standards or introduce a definitely more egalitarian social order. The inequalities in the USSR – albeit organized differently – are comparable to the inequalities in the USA, the inequalities in Poland to those in Spain or in the Netherlands. Standards of living which the state socialist economy can secure lag behind those which the capitalist one furnishes at least in comparable European capitalist countries.

Needless to say, the political democracy and cultural creativity have not been well served either by the communist parties in power. Compared to the political and cultural patterns of interaction observable in the capitalist states, the ones in state socialist societies appear to be busy "catching up" with the degree of freedom of expression worked out in the former and not with setting a qualitatively different, presumably superior, example.

Second, the new social movements failed to respect the iron laws of history as formulated in the descending paradigm. They are not brought about by the most numerous groups of individuals in any given society. They are not based on a predominantly economic positioning within the social structure. They do not follow the principle of a mass mobilization organized by a professional political party which aims at the state takeover.

Moreover, it is not assumed that any single theoretical diagnosis of social reality should be considered superior to the others and become a message beamed down towards the masses. The complexity of styling of the radical will, negotiating between ends and means and using modern mass media did not favour the emergence of all-round professionals of change.

If the Bolshevik revolution conveyed an image of armed men storming the public buildings, the new social movements stress local initiatives informed by global concerns, piecemeal engineering based on a holistic worldview (environmentalists) and a mass mobilization through the mass media rather than through an armed insurrection. Social psychologists stress the importance of the "basic pockets of trust" between human individuals which enable small social groups to design and implement alternative patterns of interaction and alter the political decision-making process. Their work (cf. Weick,1979, Axelrod,1984) has not yet reached the stage of a direct "in situ" testability, but some studies of the minimal social situation indicate that complex matters, as, for instance, sophisticated patterns of interaction can be established with a relatively little effort and energy input if some ethical principles of an individual conduct are respected thus standarizing behaviour and expectations of the others in new and uncertain situations. Moreover, they can also be established to a large extent unintentionally, even unconsciously and tacitly (cf.Weick, 1979). The discovery of the role played by tacit assumptions of trustworthiness has revised and reshaped many research programs directing them towards social psychology rather than towards economic sociology in search for the regularities underlying human conduct.

2.THE POLISH CASE

It has often been said (Negt,1987) that organizational success of the working class parties in the XIXth century has to a certain extent been due to a free ride effect of the concentration of the workers with their families around large industrial centers. It is quite possible that the new technologies of mass communication also offer a similar chance of a partly free ride to new social movements by connecting their potential participants. Moreover, the emergence of the Polish "Solidarity" can be traced back to a negatively egalitarian exclusion of all social groups from political decision making in a state socialist society. This exclusion, however, did not mean a total deprivation of all alternative channels of communication and of symbolic communities independent of the state (i.e. of the Catholic Church, independent intellectuals, professionals guarded by the autonomous associations, and a limited but considerable private ownership of some means of production). The emergence of ecological movements, in turn, can be traced back to the global reach of pollution, to the relative speed with which the dissemination of information about environmental threats is accomplished and to a relative uselessness of national states as regulating agencies. Similar observations can be made, mutatis mutandis, about the student and feminist movements.

It is interesting to note that new social movements mobilize individuals not as representatives of some "deep structures" underlying social reality, but qua individuals whose rights are perceived as independent of their position in social division of labour. "Think globally, act locally" acknowledges the double bind of a human individual in global communication networks and in the network of local struggles against inequalities and against regional exclusion from decision-making processes.

On the other hand to label new social movements as clearly expressive and cultural rather than instrumental and contingent obscures the issue and leads to a theoretical polarization (or even inflation). One notices all theoretical explanations at once and concludes that new social movements tend to be overdetermined (cf. Bunce,1988).
The decisive factors which influenced the making of the Polish "Solidarity" were: a discreet transformation of social space, changes in the timing of socialization and a growing significance of collective imagery (the access to which is facilitated by promptly responding mass media and locally available communication networks, or, vice versa, by the blocking of access to the mass media and the reinforcement of local communication networks).

The discreet transformation of social space works through the "opening" of free spaces outside of the structured interactions (institutions) subjected to the existing forms of control (e.g. cenzorship). For instance, the appearance of satellite television opened up a free space which has not been a priori allotted to the nation state as a legitimate controller of broadcasted products for an enclosed population.

Changes in the timing of socialization correspond mainly to the emergence of "underdetermined" periods in the life of socialized individuals. They have to be socialized by specific institutions for increasingly long periods of time. The very nature of the future professional socialization, usually involving the university level education, leaves whole "chunks" of individual life-time free from a direct control of an overall social structure.

Third, future patterns of socialization can be modelled, negotiated and tested for their popularity before and during the attempts to implement utopian ideas. If social utopias are "option markets" for social imagination their stock ratings can be checked much more quickly than ever before and without having to wait until an utopia degenerates into an ideology according to a well-known Mannheimian pattern. Current ratings of social utopias can be checked much more quickly because utopias have nowadays an immediate impact upon the alternative forms of "organic solidarity" (Collins, 1988). The latter arises with the new identities of social actors in the course of a rapid mobilization around issues rather than institutions.

This reverses the causal link between "deep structure" of socialization and its surface manifestations. Political mobilization and cultural identity formation are not simple consequences of a manifestation of "core" divisions, but can influence some basic divisions within a society which emerge in the course of social conflicts.

All three factors seem to have been at work in the case of the Polish "Solidarity". The creation of free spaces has been a legacy of a destalinization campaign waged by many groups of civil society from the mim-fifties onwards. Though never completely successful, the campaign left a model of action for intellectuals combatting cenzorship and state monopoly of cultural communication. It also left a model of action for workers who persisted in calling for an increased participation in the economic decision making locally and in the political decision making nationally.

Thus in 1956 a loose coalition of "young angry intellectuals" and desperate workers dismantled the stalinist monopoly in culture and politics. To be sure, the coalition has been gradually suppressed, the most active bi-weekly "Po prostu" closed down and workers' self-governments reduced to the role of an approval mechanism after a party and management decision has been announced. But after 1968 and 1970 the coalition re-emerged leading to a mass social movement and to the emergence of "Solidarity" in the summer of 1980. The re-directing of the causal links can also be observed. Waves of destalinization campaigns: 1956, 1968, 1970, 1976, 1980-1981, 1988 demonstrate a growing tendency to organize the movement according to the principle of supragenerational and supraoccupational solidarity. The "organic solidarity" of the participants plays a vital role in redefining social reality, creating pockets of trust and designing new political and economic structures ("the round table" discussions of the representatives of "Solidarity" and of the ruling class, the emerging consensus on the economic reform, etc.).

There have been two types of free spaces which have been opened through and around "Solidarity" as a new social movement. The interstate area in central Europe is commonly perceived as less threatened by the Soviet "core" military intervention now that the "perestrojka" is accompanied by an intense lobbying of a political elite for a mass support (the Baltic nations, for instance, regained their right to use national flags and anthems in return for voicing support of Gorbatchev's "perestrojka"). The armies of local states have to remain in the Warsaw pact, and local communist parties have to continue playing a significant role in domestic political systems. But in a recent round of talks with the "Solidarity's" leader, Lech Walesa,the Polish Minister of the Interior, general Kiszczak, could not evoke the threat of an armed Soviet intervention. His arguments were that the "dogmatic" and "hawkish" elements within the armed forces and secret police might be conspiring if he showed too much willingness to compromise. Using domestic "hard-liners" instead of the Soviet tanks as examples of the enemies of national free spaces is a very significant indication in the change of perception of the limits of political and economic transformations and reforms in central Europe.

The second type of free spaces is being established domestically. Individuals and groups are founding and running associations and networks, some of them legalized and some informal (or even underground). Thus they are filling in the "blanks" left by state institutions. Among these free spaces in the Polish society in 1988 there are, for instance, such economic areas of activity as vegetable, fruit and - to a large extent - meat markets, private and cooperative housing construction, educational and political organizing (the latter on a subparty level), some social services, publishing and the press, video and computer markets,etc. Political organizing includes local ecological initiatives, alternatives to draft and military service or military training at a university, gradual removal of the Russian language from a privileged position in school curriculae, and some countercultural manifestations - for instance "an orange alternative".

The time factor played crucial role in the reemergence of "Solidarity" from the strikes and negotiations of April and August 1988. First, a new generation of industrial workers grew up within the economic stagnation and political repression. The reviving of a harassed trade union became a convenient focus of the next generational struggle for access to the political and economic decision making. Taking the estrangement of the intellectuals into account we may notice that the period 1982-1988 has been defined by broad social groups as the waiting and struggling period for a gradual opening of renegotiations with respect to the political participation of those groups in national and local decision making.

Finally, the imposition of martial law has left a symbolic mark on a social imagery of the Poles. It has been widely perceived as an attempt to stop the process which has originated (in 1980-1981) a possibility of articulating one's interests and finding the roots of maintaining solidarity and "Solidarity". Political socialization is now initiated at the high school level and collective learning processes include organizing, associating and communicating outside of the areas controlled by the state. Political actors emerging in the nineties will not be bound by the matrix of the reproduction of state socialist political order.

The overall strategy of the ruling class is presently linked to the radical distinction between the democratization of the "visible structure" of state authorities and the preservation of the "core" network including the top police, party and army officers forming the "invisible" element of the state power. In September 1988 the regime trade unions which are still struggling to win some legitimacy were given a chance to ask the parliament to revoke the prime minister Messner and his cabinet. The "core" power elite hoped to add some social prestige to the regime trade unions showing that they "work" and hoping they might replace the reemerging "Solidarity".

The crucial issue of the current reversal of a theoretical perception

of the causal links (which are supposed to lead from cultural identity
formation and the political articulation towards the changes in "deep
structure" of social order) can be settled when we have more empirical
data on a relative success or failure of the post-1988 changes in Poland.
At present one may only cautiously state that the price of a political
defeat of "Solidarity" on December 13, 1981 (when the military started
recapturing free spaces won by civil society and lost to the monopolists
of state power) has been not paid in vain. In the long run, after seven
years of clandestine political existence, the legitimate nature of the
Polish civil society's claims to local truth, autonomous initiative and
solidarity symbolized by "Solidarity" have been recognized and became
the issue of negotiations and conflicts. This would appear to confirm
a hypothesis that new social movements act as vehicles of social change
but that they are not designing their own criteria of success in terms
of an institutional survival and continuity. Even politically defeated
they nevertheless continue to originate collective learning processes
and to dismantle the existing social order through new nuclei of change.

3.CONCLUSIONS

Modern social movements reveal features which barely fit the theoretical
explanations offered by the theories which had emerged during the old
alliance of "sociology and socialism". Among those features one should
mention individualism (think globally, but act locally and involve individuals as autonomous and responsible agents), non-institutionalized
participation and anti-authoritarian mobilization tactics. The movements
are best examplified by the West German "Greens", the Polish "Solidarity"
and an array of the feminist and citizens' initiatives.

Changes of social change can be best analyzed in terms of changes in
social space (the emergence of free social spaces outside of the state or
market control), social time (redefinition of free time due to the exclusion from the market in case of students and youth, marginalization in
case of women, or insufficiency of social inclusion in case of the unemployed),
and the direction of social causality. The re-emergence of the utopian
ideals in media-linked "global village" testifies also to changes in
social imagery. The processes of cultural identity formation and of the
political mobilization result in an articulation and self-creation of
agents. These agents emerge in the course of social conflicts and their
emergence is not simply a secondary manifestation of "deep structures"
of social order. The making of social agents contributes to the establishment and change of a social order.

The Polish case appears to confirm the validity of these factors in
the process of changing social change be new social movements. Being defeated
in terms of political struggle for power the movement has nevertheless
managed to re-emerge and to claim free spaces lost to armed intervention.
The ongoing process of political changes in Poland in 1988 seems to sonfirm
the thesis that new social movements prompt their participants to think
globally and act locally and that local truth in a global context and
global setup does carry a transformational potential even if the original
instruments and tactics are being defeated.

One should also note a very interesting role played by the Catholic
Church and organized religion in present political struggles in Poland.
It is clear that the Church provides an alternative custody for the national traditions, symbols and imagery. It also runs a semi-independent
network of media and educational institutions thus securing the "moral
community" outside of state control. However, its direct political role
remains very limited. It could be even argued that it provides a formal

background for the emergence of a new civic religion of civil society. One is tempted to assume that in the triangle of powerful forces in Poland "Solidarity" emerged as a carrier of this new civic religion partaking both of the socialist state's egalitarian claims and of the Catholic Church's ideological individualism and practical rituals. The arguments Tiryakian mentions comparing Iran, Poland and Nicaragua, while inconclusive, appear to support this assumption (Tiryakian, 1988).

REFERENCE

Axelrod, Robert, The Evolution of Cooperation, Basic Books, New York,1984
Bunce, Valerie, "Why some rebel and others comply: the Polish crisis of
            1980-1981. Eastern Europe and theories of revolution"
            (unpublished paper, August,1988)
Collins, Randall, "The Durkheimian tradition in conflict sociology",in:
            Durkheimian Sociology: Cultural Studies, Alexander, Jeffrey,
            ed.,Cambridge University Press, 1988
Hegedus, Zsuzsa, "Social movements and social change in self-creative so-
            ciety: new civil alternatives in the international arena"
            (a paper presented at the Goteborg workshop on·new social
            movements in Europe, Goteborg, September 19-22,1987)
Negt, Oscar, a personal communication at the international symposium on
            socialist theory, Groningen, March 27-28,1987
Tiryakian, Edward A., "From Durkheim to Managua: revolutions as religious
            revivals", in: Durkheimian Sociology...,op.cit.
Weick, Karl, The Social Psychology of Organizing, Random House, New York,
            1979

DESIGN AND PLANNING

DESIGN AND PLANNING

Christopher J.L. Yewlett

Department of Town Planning
University of Wales, College of Cardiff
PO Box 906, Cardiff CF1 3YN

Generally, this conference is an occasion both for taking stock and for looking forward. Specifically, this introduction will be a very personal overview, but one from a perhaps uniquely privileged position for discussing OR in the context of Planning and Design. The aspirations abroad in parts of the OR Community in the early 1960s, which led to the previous (1964) Operational Research and the Social Sciences Conference, (Lawrence, 1966), also led to the establishment of the Institute for Operational Research (IOR), with a remit to extend the scope of OR in terms of both applications and methods (Friend, Norris, and Stringer, 1988). This mission led to work in a number of 'multi-organizational' settings (Stringer, 1967), including a programme on urban and regional planning proceses, where the author, following initial studies, cut his OR teeth. (See Friend and Jessop, 1969; Friend and Yewlett, 1974; Friend, Power, and Yewlett, 1974; and Friend and Hickling, 1987, for details.) Following a period in Local Government on first returning to Wales, the author took up his present position, in Wales' only Town Planning School, with particular teaching responsibilities in planning methods, planning practice, and management. This background thus offers a possibly unique personal perspective on many issues.

However, there is also a less immediately personal justification for this stance. Like OR, Town Planning has also been through the experience of an encounter with the social sciences. Many of the controversies surfacing in the OR world over the last decade (see e.g. Dando and Bennet, 1981)) have been presaged by similar controversies in Town Planning. Although much of what will be said here is rooted in Town Planning experience, 'Planning' is of course a much broader field; a lot of concepts are drawn from (and applicable to) a far wider field of endeavour. Moreover, when we come to discuss relevant experiences in the areas of planning and design, these parallel experiences converge, not least because first systems and then OR approaches and concepts have had a major influence on Town Planning. Indeed, the field of planning and design embraces most of the technical, philosophical, and ethical issues exercising OR today.

First, the all-embracing nature of the task in hand, and thus of the approaches towards resolving it. Planning and Design, almost by definition, involve a total process, from initial brief through to some kind of product, including significant issues about the nature of the initial brief itself, and the form and use of the product. In contrast, much (but by no means all) OR has concerned the improvement, often attempted optimisation, of only part of a process.

Historically, Town Planning was regarded as an intuitive, artistically creative, discipline in education and training terms. Town Planning education essentially involved apprenticeship with a 'Master'. The neophyte would produce many attempts, usually in the form of drawn plans, for critical discussion by tutors. In many ways, Architecture education still reflects this approach, but planning has switched to a more expressly 'multi-disciplinary' stance, involving substantial social science inputs, as well as many 'techniques' familiar to the OR world. Nevertheless, all planning still involves a major element of design. Now design is essentially the activity of synthesis. As such, it is a key professional activity; indeed, intrinsic in all professions and activities which seek to intervene in the world. Moreover, the concept is not, popular usage notwithstanding, limited to 2- or 3-dimensional graphic work - design embraces verbal policy design as well.

However, as Simon (1969) points out, the expansion of formal professional education has been accompanied by a marked tendency to de-emphasise the traditional 'design' skills, in favour of more emphasis on 'science'. The reasons suggested, incidentally, include the wish to introduce more readily teachable (or, to be honest, examinable!) material, coupled with a derogatory view in education towards the apparent lack of rigour in much design teaching. The result, he argues, has been an increase in substantive knowledge in many professional fields, but a tendency to distance education (and academics!) from practice. The author detects more than a hint of this in practitioners' criticisms of both Planning and OR education; and, indeed, in the initiatives which led to the launch of 'Interfaces' (USA) and 'Insight' (UK). Speculatively, a reversal of this trend might just be beginning, through a resurgence of interest in design in planning (Rodriguez-Bachiller, 1988), and the increasing interest in process and methodology in OR.

At a more fundamental philosophical level, though, this emphasis on science leads to the second key point: 'What is science?'. This is clearly of importance to Town Planning, but, given OR's classic definition, which opens by stressing that 'Operational Research is the application of the methods of science to complex problems ...(1)', it is of crucial importance to the perennial question: 'What is OR?'.

This is not just semantics. Questions of what constitutes science are the starting point of many philosophical debates (see e.g. Chalmers, 1982, or Johnston, 1986, for some recent discussions). They are also directly germane to our present concerns. A particular kind of 'mechanistic' and 'reductionist' scientific world view undoubtedly led to major advances in scientific knowledge in the 19th century. There has been a continuing tendency for social scientists to cast envious glances at these advances, and thus to seek to import some of the perceived key elements into social science. However, if we look at what is actually going on in the physical sciences currently, we see some fundamental changes. Ideas rooted in early 20th Century developments, which themselves undermined the mechanistic view, are increasingly supplemented by more emphasis on a wholistic (or holistic) approach, and a growing acceptance of the importance of process (see e.g. Capra, 1975; Bohm, 1980). Increasingly, social scientists (and OR workers!) are being encouraged to look in this direction (Capra, 1982; Robertson, 1983, 1985). Co-incident with this, there has been, in both planning and OR, an increasing interest in process, leading specifically in OR to an increasing interest in developing and combining problem formulation methodologies (Bennett, 1985).

For OR, these developments have internal political, technical, and external political, implications, the last two shared with planning amongst other fields. The internal political implication refers to the perennial debate as to what is OR. It is perhaps significant that, although no agreement was ever achieved on a new formal definition by the (U.K.) OR Society, the old one was quietly dropped from the Journal in April 1984! Although in practice, in parallel to many other fields, we are moving towards the position 'OR is what OR workers do', some people are uncomfortable with this, or perhaps especially, with the uncertainty entailed. Interestingly, the same point is often made in Town Planning: that the continuing debates as to what planning is, (or, rather, should be - see Reade, 1987, for a social scientist's view) are somehow peculiar, and not to be found in other, more established, fields.

Given the self-confidence to pass over this, the next two implications have direct relevance to OR in Planning and Design. At the technical level, the emphasis on process leads into a consideration of the overall decision process, and ways of enhancing it. Moreover, design entails creativity. Now creativity cannot be 'programmed'. All the research indicates that conditions conducive to creativity can be arranged, but creativity itself cannot be forced (Stein, 1974; Rickards, 1974). However, various methods for imposing structure on a problem can assist the creative process, partly by alleviating mental overload induced by complexity, and partly by pointing to possible new insights. Design studies has been an important area for interaction between Operational Research and related conceptual fields. Of particular interest is Jones' 'Design Methods' catalogued review (Jones, 1970). Here, Jones hit on the valuable tactic of not only summarising (and ordering) many other workers' contributions, but actually inviting them to comment on his reviews, thus overcoming some of the problems caused by idiosyncratic terminology, and allowing a focus on the 'real' problems. Interestingly, the major change in the second edition (1980) is a new foreword, in which Jones disowns attempts at programmed or systematic design in favour of a more open, almost 'mystical', approach. He replaces creativity, for instance, with 'inventiveness' or 'imagination', as 'Creativity seems to imply too much control, too little sensitivity. It is often thought of as an ability to think of alternatives... A more profound notion of creativity is that of being able to change one's view of things, and of oneself, to the point of attempting something you thought was impossible, beyond you. Creativity in design methods shows itself in the originality of one's questions, aims, classifications, processes, etc.' (Jones, 1980). His more recent work seeks to stress the significance of discontinuities and disorder as well. Interestingly, he also now stressses the fact that 'making the future is a collective work', (Jones, 1988) a concept which emerges, slightly less explicitly, in the earlier work as 'collaborative design'.

This notion of collective responsibility is also crucial in Town Planning, with the shift away fom 'the expert planner' towards 'collective participation' by many actors and interests with, as Jones puts it, the 'leading designer' needing to know 'how to switch from being the person responsible for the result to being the one who ensures that 'the process is right''. This then leads us into both technical and political implications; indeed, it is often hard to separate the two. Technically, we are into a multi-actor process, which rules out the 'traditional' stance of 'advising the decision-maker'; rather, OR workers become participants in a political process (Stringer, 1967; Friend, Power, and Yewlett, 1974). We also come face-to-face with the existence of conflict, and the ethical question of just who are we trying to help? One consequence of this has been the development of methods which can best be characteriised as 'Consequence Display' rather than 'Optimising', with an emphasis on structuring problems to make key decisions explicit (see Rosenhead, 1981; Friend and Hickling, 1987), rather than subsuming everything into a single 'optimisable' metric.

Whilst technically this leads into issues of multi-criteria analysis (Zeleny, 1982), in Town Planning these kinds of consideration have lead into extensive debate of issues that have hardly surfaced in OR. There is a continuing debate in (Town) Planning Theory circles about the appropriateness, or even desirability, of developing a procedural theoretical framework for planning decisions, within which many of the techniques developed by OR could be applied. Faludi is the best known exponent of this approach (1973;1986a); significant collections edited by Healey, McDougall, and Thomas (1982), and Paris (1982) offer an introduction to the critiques, and the underpinning social science concepts.

At a more practical level, the U.K. Town Planning profession came to feel that many individuals did not command access to the necessary advice to put their case through the the planning system, leading to the establishment of a 'Planning Aid' movement, largely staffed by professionals on an 'extra-mural' voluntary basis (Curtis and Edwards, 1980). We have recently seen a very similar parallel development in the U.K. OR world, leading to the Community OR initiative (Rosenhead, 1986; Jackson, 1988).

The recognition of conflict also leads beyond consequence display, to consideration of issues of control over decisions, and thus of bargaining and negotiation. Here again, there have been parallel developments in Town Planning (see Bruton, 1983). Some of these issues have of course been extensively investigated, notably through developments in Game Theory.

Most real-life situations are not, however, zero-sum games. Indeed, recent work in social science (Mant, 1983) and negotiation theory (Fisher and Ury, 1981), supplemented by Axelrod's classic (1984) simulation of the notorious 'Prisoner's Dilemma', all point in the direction of constructive, co-operative resolution of conflict, rather than seeking outright 'victory' (or capitulation!). This also brings us back to a direct reflection on ethical issues (Faludi, 1986b).

These developments offer perhaps one of the most exciting arenas for the further extension of OR. As outlined above, successful design involves creativity; for OR, this indicates methods for the integration of rationality with intuition (Yewlett, 1985). Whilst the literature indicates that successful initiatives in constructive bargaining will need a certain attitude of mind on behalf of the parties (which alas is not always present), the effective resolution of the problems will clearly also need appropriate 'Solution Building' tools and skills. It is here that developments in Planning and Design over the past 25 years offer a solid base for further work, which may confidently be expected both to be further refined in its parent field, and indeed to filter into other application areas of OR, over the next 25 years.

NOTE

Note 1: OR Society 'Official Definition' (printed in J. Opl Res. Soc. until April 1984)

Operational Research is the application of the methods of science to complex problems arising in the direction and management of large systems of men, machines, materials, and money in industry, business, government, and defence. The distinctive approach is to develop a scientific model of the system, incorporating measurements of factors such as chance and risk, with which to predict and compare the outcomes of alternative decisions, strategies, or controls. The purpose is to help management determine its policy and actions scientifically.

REFERENCES

Axelrod, R., 1984, "The Evolution of Co-operation," Basic Books, New York

Bennett, P.G., 1985, On Linking approaches to Decision aiding: Issues and Prospects, J. Opl Res. Soc., 36:659-669

Bohm, D., 1980, "Wholeness and the Implicate Order," Routledge and Kegan Paul, London

Breheny, M. and A. Hooper (Eds), 1985, "Rationality in Planning", Pion, London

Bruton, M.J., 1983, "Bargaining and the Development Control

Process," Papers in Planning Research 60, Department of Town Planning, UWIST, Cardiff

Capra, F., 1979, "The Tao of Physics," Shambhala, Berkeley, CA.

Capra, F., 1983, "The Turning Point: Science, Society, and the Rising Culture," Wildwood House, London

Chalmers, A.F., 1982, 2nd Edn, "What is this thing called Science?," Open University, Milton Keynes

Curtis, B. and Edwards, D., 1980, "Planning Aid - An analysis based on the Planning Aid service of the TCPA," Occasional Papers OP1, School of Planning Studies, Reading University

Dando, M.R., and Bennett, P.G., 1981, A Kuhnian crisis in management science?, J. Opl Res. Soc., 32:94-103; see also comments by Gault (1982) and rejoinder, J. Opl Res. Soc., 33:91-101

Faludi, A., 1973, (2nd Edn 1984), "Planning Theory," Pergamon, Oxford

Faludi, A., 1986a, "Critical Rationalism and Planning Methodology," Pion, London

Faludi, A., 1986b, "Procedural Rationality and Ethical Theory," Planologisch et Demografisch Instituut, University of Amsterdam

Faludi, A., 1987, "A Decision-centred View of Environmental Planing," Pergamon, Oxford

Fisher, R. and W. Ury, 1981, "Getting to Yes," Hutchinson, London

Friend, J.K., and W.N. Jessop, 1969, "Local Government and Strategic Choice: an Operational Research Approach to the Processes of Public Planning," Tavistock Publications, London (2nd Edn, 1977, Pergamon Press, Oxford)

Friend, J.K. and C.J.L. Yewlett, 1974, A resume of work in the Institute for Operational Research 1963-74, Planning Theory Group, Technische Hogeschool, Delft, Netherlands

Friend, J.K., J.M. Power, and C.J.L. Yewlett, 1974, "Public Planning: The Intercorporate Dimension," Tavistock Publications, London

Friend, J.K., and A. Hickling, 1987, "Planning Under Pressure," Pergamon Press, Oxford

Friend, J.K., M.E. Norris, and J. Stringer, 1988, The Institute for Operational Research: An Initiative to extend the Scope of OR, J. Opl Res. Soc., 39:705-713

Healey, P., McDougall, G., and Thomas, M.J., Eds, 1982, "Planning Theory: Prospects for the 1980s," Pergamon, Oxford

Jackson, M.C., 1988, Some methodologies for Community OR, J. Opl Res. Soc., 39:715-24

Johnston, R.J., 1986, 2nd Edn, "Philosophy and Human Geography," Edward Arnold, London

Jones, J.C., 1970, & 1980 (2nd Edn) "Design Methods: Seeds of Human Futres," Wiley, Chichester

Jones, J.C., 1988, Writing the Future, Futures, 20:317-321 (June)

Lawrence, J.R., Ed., 1966, "Operational Research and the Social Sciences," Tavistock Publications, London

Paris,C., 1982, Ed., "Critical Readings In Planning Theory," Pergamon, Oxford

Rickards, T., 1974, "Problem Solving Through Creative Analysis", Gower, Epping

Robertson, J., 1983, 2nd Edn, "The Sane Alternative," Robertson, Wallingford, Oxon

Robertson, J., 1985, "Future Work," Gower, Aldershot

Rodriguez-Bachiller, A., 1988, "Town Planning Education: An International Survey", Avebury, Aldershot, Hants, UK

Rosenhead, J., 1981, Operational research in urban planning, Omega, 9:345:64

Rosenhead, J., 1986, Custom and Practice, J. Opl Res. Soc., 37:335-43

Simon, H.A., 1969, (1981 2nd Edn), "The Sciences of the Artificial", MIT Press, Boston, MA

Stein, M.I., 1974, "Stimulating Creativity", (2 Vols), Academic Press, New York

Stringer, J., 1967, Operational research for "multi-organisations", Opl Res. Q., 18:105-120

Yewlett, C.J.L., 1984, Polishing Practice _ The Reconciliation of Scientists and Practitioners, J. Opl Res. Soc., 35:487-498

Yewlett, C.J.L., 1985, Rationality in planmaking: a profesional perspective, in Breheny, M. and A. Hooper (Eds), 1985, "Rationality in Planning", Pion, London 209-228

Zeleny, M., 1982, "Multiple Criteria Decision Making," McGraw Hill, London

PLANNING AS RESPONSIBLE SCHEMING

Some Reflections on the Role of the Management Scientist in Public,
Semi-Public and Private Domains

John Friend

IOP Consulting

Bamford, near Sheffield, BR30 2BR, UK

Words and Meanings

There are many contexts in which the two English nouns "plan" and "scheme",
or their counterparts in other languages, can be used more or less
interchangeably.   An architect may submit a scheme or a plan for
modernising an office block, or a traffic engineer for changing the
circulation system in a town. A company may design a pension plan, or more
often a pension scheme, for its employees. Most dictionaries attempt to
define each of the nouns with reference to the other. And the saying which
is most often applied to the failure of ambitious plans comes from a well-
known poem written by Rabbie Burns two centuries ago:

The best laid schemes o' mice and men gang aft a'gley"

It is when the two words become used as verbs or as adjectives that some
interesting distinctions of emphasis begin to emerge. To plan is usually
regarded as idealistic, rational, far-sighted, noble; to scheme, on the
other hand, is regarded as devious, irresponsible, subversive, underhand.
Large organisations usually contain planning departments, managers,
officers. Yet if the adjective scheming is applied to an individual, the
it is almost always with derogatory intent - and it may be no accident that
it is usually coupled with a noun implying illegitimacy.

Planning as Responsible Scheming

My thesis in this paper is that people responsible for planning and
management support in complex organisations - and, even more so, those who
work between organisations - cannot be effective, in terms of influencing
decisions which matter, unless they are prepared to cultivate and deploy
many of the skills associated with the ignoble business of scheming.

Yet I suggest that this need not imply that they should abandon all sense
of responsibility in their work. I believe that there is such a thing as
the *art of responsible scheming*; and that many who have made their mark in
strategic roles in both the private and public domains have done so by
becoming skilled practitioners in this art. To "scheme responsibly" may
seem to involve an element of paradox; to be almost a contradiction in
terms. Yet I believe that we can go at least some way towards articulating

what the skills of responsible scheming might be; and that, once these have been more clearly articulated, two kinds of benefit can flow.

First, those who have cultivated skills of responsible scheming will begin to understand better the nature of their own intuitive judgements, and thus adapt them to new situations in a more conscious way. Furthermore, it may help them in coming to terms with the sense of guilt which seasoned managers and planners often experience when they find themselves making departures from those taught principles of rationality and objectivity which may have served as benchmarks during their earlier careers.

Secondly, if the skills of responsible scheming can be more clearly articulated, then it should be possible to make them more accessible, through educational and training programmes, to new generations of planners, managers and advisers who may find themselves thrust into positions of responsibility in complex organisational domains at an early stage of their careers. Herein lies a major challenge both to the OR worker and the social scientist; and I believe it is one in which neither is likely to make much progress without being prepared to learn something from the insights of the other.

The thesis of this paper is one which I would not have been able to develop were it not for my involvement over more than two decades in a particular programme of work concerned with the understanding and enrichment of planning processes, especially though not exclusively in the public domain. At this point, I should say a little more about the context of this programme, and its relationship to the theme of the conference. Then I will be able to elaborate on my theme, and offer some suggestions as to how operational research and the social sciences may be able to contribute in future to the theory and practice of responsible scheming in public, private and intermediate domains.

A Formative Experience

It is fitting that I should begin my story in September 1964, when the first conference on OR and the Social Sciences was held in Cambridge. I was then an operational researcher in my early thirties, with ten years of conventional industrial OR experience behind me. But I now found myself on more unfamiliar ground; for six months earlier I had joined the newly-formed Institute for Operational Research (IOR) - a joint venture of the UK Operational Research Society and the Tavistock Institute of Human Relations, dedicated to the application of OR and social science insights to major policy issues, especially in the public domain (Friend, Norris and Stringer, 1988).

As my first assignment in IOR, I had found myself immersed in a four-year research project, funded by the Nuffield Foundation, concerned with the processes of policy-making and planning in city government - seen, in terms of the project proposal, as a "microcosm of government as a whole". The location of this project was the city of Coventry, where the City Council had offered the researchers privileged access to its decision processes at all levels - formal and informal, technical, administrative and political. I was one of two full-time members of an interdisciplinary team, the other being a social anthropologist at a similar career stage to my own. The two more senior part-time members of the team reflected the same deliberate balance between OR and the social sciences; and we could call on the support of a small but distinguished band of advisers including Russell

Ackoff and Eric Trist - both of them key actors in the formation of an operational research institute within the Tavistock setting.

So I had only six months of experience in working with social scientists behind me when I found myself coming to Cambridge - indeed, to the very College to which I had come to read mathematics fifteen years earlier - to participate in an international conference on OR and the social sciences. At that stage of the project, I still had little idea as to how a synthesis between the two traditions might be achieved, even though my full-time team colleague and I were by then deeply immersed in the same field experiences. Between us, we were spending most of our time attending committee meetings as observers; talking with senior officials and politicians; and keeping track of as many as possible of the many interrelated issues - capital budgeting, traffic management, school reorganisation, land-use planning - with which the decision-makers were wrestling at the time.

Searching for Relevant Theory

Yet it was at that time far from apparent how we could bring the contributions of OR and the social sciences together in such a way as to offer something useful either to our hosts in Coventry, or to the understanding of policy-making and planning in the public domain. Both of us, in our own ways, struggled for understanding at a holistic systems level. Indeed, the original research proposal had suggested that the synthesis between the OR and social science contributions might come through bringing together an OR capability to model the interrelationships of a city, viewed as a system, with the social scientist's skills in understanding the intricate system of relationships among those responsible for its governance. This argument drew on the Tavistock concept of the "socio-technical system" - a system in which the human and non-human elements were matched, by design, so as to form a harmonious whole.

However, to model the complexity around us in socio-technical terms was easier said than done. In retrospect, however, our ideas only began to converge when we turned aside from this macro level of systems analysis, and turned to the microscopic level of trying to represent the complexity of the *processes* by which specific and often urgent issues were steered through the Council's elaborate machinery of collective decision-making.

For it became more and more apparent to us that our real opportunity lay in the priveleged access we had been granted to the Council's committees; to its departmental and interdepartmental meetings; and to the private caucus meetings at which the opposing party political groups agreed their tactics for the monthly Council meetings at which major decisions were publicly endorsed. Furthermore, we had access to many more *ad hoc* interfaces between those with formal responsibilities for decision-making and representatives of the rich variety of local and wider interest groups with a stake in particular issues that arose.

What we found significant was not only the processes of collective decision-making *within* these various group settings; but the subtle judgements being made, and the agreements being negotiated, as to how particular issues should be steered *between* one setting and another; and, In the process, choices arose as to how far interrelated issues should be handled together; and what kind of balance should be struck between urgency and uncertainty, between flexibility and commitment, between technical and political concerns.

Progress in Understanding Processes

It was not until 1966 – just over midway through the project, and two years after the first Cambridge conference – that we could begin to see signs of a breakthrough in bringing together the OR and social science dimensions of the project. The breakthrough came in interpreting what we were witnessing around us as a *process of strategic choice* – a phrase we chose to avoid what we saw as conventional yet artificial distinctions between management and planning, between politics and administration. In this view of the wider process, we adopted an orientation not so much towards *systems* as towards *decisions*. In particular, we found ourselves beginning to understand the ways in which decisions on substantive issues – often complex enough in themselves – related to those *process decisions* by which individuals and groups chose how to route those issues through complex and sometimes turbulent organisational and inter-organisational domains.

The key to this understanding lay in a tripartite classification of basic types of *uncertainty* in deciding how to tackle difficult decision problems. The three categories were:

Uncertainties in the *operating environment*, calling for an *investigative* response, for example through costing, forecasting, technical analyses;

Uncertainties about *guiding values*, calling for a *political* response, whether through policy debate, objective-setting or informal soundings;

Uncertainties about *related decisions* to that currently being addressed, calling a *co-ordinative* response through planning or liaison activities.

It was easy enough, we observed, for people to say that it was important to invest in all three types of response if better decisions were to be made. But time was usually pressing; resources were always limited; and the decision-makers faced some highly significant process choices in agreeing how to move forward at any moment. In effect, they had to judge which kinds of uncertainty were significant enough for them to invest time and other resources in appropriate forms of exploratory action, whether investigative, political or coordinative. At the same time, they had to judge which substantive decisions to take now and which could be deferred until these explorations had been carried through. And all these choices had potentially significant implications for the routeing of future decision processes – affecting who should be involved, and how and when.

Project Outcomes

This decision-centred view of planning as a process of strategic choice emerged gradually, and rather painfully, through attempts to piece together various insights from the Coventry project – insights not only from observing group meetings and talking with individuals, but also from analysis of the structures of assumptions underlying written policy documents, and observation of the challenges to which they were subjected in both internal and more public debates.

The resulting view of a dynamic, evolutionary process of strategic choice, driven by judgements about how best to manage uncertainties, was presented, somewhat speculatively, in the book *Local Government and Strategic Choice* (Friend and Jessop 1969; second edition 1977). The OR/social science collaboration was reflected not only in the underlying view of process, but also in some tentative suggestions as to the nature of an "appropriate technology of strategic choice", with ideas about the creative management of uncertainty at its core - and with some even more speculative ideas about what its organisational implications might be.

The more practical outcome of the project, however, was a programme of further research, consulting and training activities which, during the seventies, expanded to involve IOR staff with decision-makers in many other organisations - most but not all of them in the public sector. There was a series of government-financed "action research" projects in which the proposed technology of strategic choice was applied on-line to real-time urban development problems, proving its value as an aid to communication in inter-disciplinary groups.   There was a series of research projects concerned with inter-organisational decision-making, at levels from the local community to the regional strategy team.   There were also numerous training courses in Britain and overseas at which practising managers and planners were introduced to the new strategic choice approach.

Scheming and Responsibility in the Public Domain

The ideas on responsible scheming which I should now like to develop a little further are a product of these subsequent interactions with decision-makers as much as of the seminal project experience in Coventry. My IOR colleagues and I learnt a great deal more about the process judgements facing not only managers, planners and politicians who were influential within organisations; but also those more exposed individuals whose influence depended on operating effectively across the boundaries *between* organisations at all levels. The nature of these network-managing or "reticulist" skills was explored in a further book (Friend, Power and Yewlett, 1974), and subsequently in an occasional journal called *Linkage* to which practising managers and planners contributed insights of their own.

It was in an issue of *Linkage* that I first put forward the notion of responsible scheming (Friend, 1983), in an attempt to describe in realistic terms the challenges faced by planners and coordinators operating in complex organisational domains who, unlike the classical entrepreneur, had to reconcile effective influencing over decisions - often though informal channels - with formal accountability to complex constituencies of public interest.   Those who had become effective in such roles seemed to have become skilled in appreciating both the ever-changing connections between the substantive decisions to be made, within and beyond their own spheres of direct influence, and also the intricate webs of formal and informal relationships among the organisations, parties and individuals involved.

On the basis of this dual understanding, such people had to exercise what influence they could over the routeing of decision processes, operating in an incremental way.   In discussion, people seemed to recognise clearly the elements of scheming in their process decisions, as contrasted with more conventional and rationalistic forms of management and planning skill. Also, most of them are uncomfortably aware of the problem of reconciling this with the elements of public accountability in their working roles.   In professions such as town and country planning or social work - usually attracting entrants with an idealistic and non-commercial orientation - the dilemma can be especially acute.   At one level, such people can be held publicly accountable for their visible contributions to decision-making; at another, however, they are thrown back on a more private sense of personal responsibility in the exercise of their all-important scheming skills.

Scheming in Public, Private and Semi-Public Domains

Recent applications of strategic choice methods have taken myself and my associates (Friend and Hickling, 1987) into a steadily expanding range of public, private and intermediate organisational domains. In Allen Hickling's case, these include government-industry collaboration in tackling threats to the global environment; in my own, they range from support to a small housing cooperative to strategic decision support in an international high technology firm. The need for skills of responsible scheming seems to pervade all of these organisational domains - though the challenge of reconciling effective process influence with a sense of personal responsibility may take on a subtly different form in every case.

There is much still to be learnt about how the challenge of responsible scheming impinges on different organisational and inter-organisational domains. There are, I believe, significant barriers in some contexts - more perhaps in the private than the public sector - to the exchange of personal experiences in the exercise of informal scheming skills. For example there is a risk that, if some of the subtle skills of process influence are made more overt, their effectiveness will be diminished.

I believe that the challenges to OR and the social sciences arise at several levels. First, there is the challenge of surmounting disciplinary barriers - at the level of language but also at that of what Checkland (1981) would call *Weltanschauung*. This challenge demands, I believe, the intensive sharing of field experiences, coupled with the willingness to abandon cherished theories and beliefs. Then, there is the challenge of sustaining rich dialogues with practitioners in responsible scheming, through programmes of training, consultancy and research. In this way, more can be learnt about their intuitive skills, and advances at a broader scientific level - such as the ideas mentioned here about the creative management of different categories of uncertainty, can be tested and adapted to give a better match with the realities people face.

Lastly, my experience has convinced me that practitioners in management science - whether from an OR, social science or any other background - must learn to cultivate effective scheming skills themselves. For they can draw on their own working experiences, as much as those of others, in addressing the difficult business of reconciling scheming with responsibility in whatever management domains they may operate.

References

Checkland, P, 1981, "Systems Thinking, Systems Practice" Wiley, Chichester.

Friend, J.K.., 1983, Planning: the Art of Responsible Scheming?, in "Linkage Seven", Tavistock Institute of Human Relations, London.

Friend, J.K. and Hickling, A, 1987, "Planning under Pressure: the Strategic Choice Approach", Pergamon, Oxford.

Friend, J.K., and Jessop, W.N., 1977, "Local Government and Strategic Choice". Second edition Pergamon, Oxford.

Friend, J.K., Norris, M.E., and Stringer, J., 1988, The Institute for Operational Research: an Initiative to Extend the Scope of OR. In Journal of the Operational research Society 39:8.

Friend, J.K., Power, J.M. and Yewlett, C.J.L., 1974, "Public Planning: the Inter-Corporate Dimension". Tavistock Publications, London.

AN APPLICATION OF UTILITY FUNCTION THEORY

TO LARGE-SCALE PUBLIC DECISION PROBLEM

Seo Kyoo Ahn

Department of Business Administration
Kyung Hee University
Seoul, Korea

INTRODUCTION

This paper applies utility function theory to a large-scale public decision problems which often involve multiple objectives. One way of attac-
king this kind of problems is first to obtain an objective function involving multiple measures of effectiveness. Such an objective function describes
a ranking of preference, where we identify trade-offs between different
levels of various attributes. Since most real decision problems involve
uncertaintites, we need to specify an objective function with special characteristic of using the expected value. The objective function is a utility
function. This utility function generates the necessary information to rank
consequences and identify trade-offs between attributes.

Utility function theory by nature is very flexible and sound. Considering there is no standardized procedures for obtaining multiattributed
utility functions, one might think it is the most difficult part of both
using the utility function and making a model-validity check. Many people,
however, make assumptions about preferences and then modify the functional
forms of the utility function satisfying these assumptions. In most cases,
once the assumptions seem to be verified, the functional form is to be used
to assess the utility function with its simplicity.

In this paper, as a real problem of applying utility function to
large-scale public decision issue, we have considered the problem of
selecting a strategy for developing the major station facilities of the
Busan metropolitan area.

INFORMATION OF BACKGROUND

Busan, located south-east most corner of the korean peninsular, is the
second largest in population and first largest port in Korea. As the Pacific Basin Era is approaching, the city of Busan will play an important
role as a core of social, political and economic activities.

Recent figure shows that the population of Busan grows about 7% annually and the percentage of the growth in number of vehicles is even worse
as the automobile industry in Korea is quite bomming: number of vehicles
in Busan was 9.6 (thousands) in 1978 and 70.2 (thousands) in 1988. The
authorities feel that the existing train station facility is quite over

capacitated, so they face with the problem of whether expanding the existing station facility or moving the facity to the new place. The existing station facility is located quite a central part of Busan and it was built in close to a seashore. And the new selected candidate, Hwa-Myung Dong, is located 24Km (15 miles) away from north-west of the city in a less developed farming area. When the Busan train station was built in 1908, it was out in the country, but the station has now been surrounded on three sides due to rapid growth of population and economy. This has created problems of noise, pollution, traffic conjestion and safety. Even though major facilities already exist in Busan, expanding the existing station facilities may turn to disadvantages.

For example, the expansion of the Busan station built on a seashore not only makes construction expensive, but also disfigure the landscape. At the same time, the facilities in Busan do not meet the standards of accomodating the population. On the other land, the construction of new station at Hwa-Myung Dong is not expected to have the same kind of difficulties occurred in Busan. However, it will creat a new set of problems such as purchase of new land, construction of a new high way linked directly to the city, etc.. Such a problem is not simple. Many factors are interrelated to get into decision making process: how to measure the quality of service in terms of many defferent criteria.

We need a scheme to reconciliate the differences of judgement, and opinion of different interest groups concerned with station development. We also need to consider what strategies for the development of the station facilities are best in the light of social and budgetary restrictions.

MODEL DESCRIPTION

We are going to deliver sufficient conditions of a multiattribute utility function being either multiplicative or additive. The conditions described later will require the decision maker neither to consider trade-offs between attributes nor to consider lotteries over more than one attribute. Let us introduce some notations. X is a consequence space, denoted by a cartesian product of $X_1 x X_2 x$ --- $x X_n$, representing a subset of a finite-dimention Euclidean space. Then a specific consequence is to be designated by $\vec{x}$ whose element x_i is a particular amount of attribute X_i. Function u is a mapping of X into a scalar quantity ranged from zero to one.

It is assumed that $u(X)$ is continuous and preferences over X is bounded. We also designated $\bar{x}$ if it is the most desirable and $\underline{x}$ if it is the least desirable consequence. $\tilde{X}_{ij}$ describes $X_1 x$ --- $X_{i-1} x X_{i+1}$ --- $X_{j-1} x X_{j+1} x$ --- $x X_n$ and $\tilde{x}_{ij}$ is a subset of $\tilde{X}_{ij}$. Similarly, the notation $\tilde{X}_i$ is defined as $X_1 x$ --- $X_{i-1} x X_{i+1} x$ --- $x X_n$ and $\tilde{x}_i$ is a member of $\tilde{X}_i$.

Theorem. Let $X \triangleq X_1 x$ --- $x X_n$ be a consequence space. If $X_i x X_j$ is preferentially independent of $\tilde{X}_{ij}$ for all i,j, i$\neq$j and X_i is utility independent of $\tilde{X}_i$, then either utility function of multiplicative or additive form is valid, i.e.

$$u(\vec{x}) = \sum_i^n k_i u_i(x_i)$$
or
$$1 + ku(x) = \prod_{i=1}^n [1 + kk_i u_i(x_i)]$$

where u and u_i are utility functions and k_is are scaling constant.

Proof. see Keeney (1974)

To develop a model, we first need to specify the types of train operating at each of the two possible sites for the coming decades. We catagorize

374

trains as follows: Cargo train(C), General passenger train(G),Oil train(O) Express train(E) and Military train(M). Even though trains are classified into five, some of them have quite similar characteristics with respect to operating or traffic loads. To account for changes in operating level over the 30 year period, we set up the time horizon of 1990, 2000, 2010. The discreteness of three time-horizon contributes to keep the analysis tractable

To evaluate the alternatives, we need to specify criteria of effectiveness and the measuring scheme for the objective as a whole. As the objectives,

a) Minimize total construction costs
b) Minimize the access time to the station
c) Provide the Maximum safety of the system
d) Provide the adequate capacity to meet demands
e) Maximize the benefit from tourist industry by maintaining the landscape of the seashore
f) Minimize the noise and air pollution

Those objectives are overlap in the sense that access time to the station, for example, has something to do with the cost. Besides the overlap, objective a) mainly accounts for the government's stake; objective b), c) and d) for the user's; and c), e), f) for the nonuser's. Lets define the measure of effectiveness for these objectives as follows:

X1 = total cost in billion wons
X2 = access time to and from the station in minutes
X3 = # people injured or killed at the crossing
X4 = practical capacity in terms of the # of trains for accomodation of the demands
X5 = income per persion in thousand wons
X6 = # people subjected to either high noise level or polluted air

Decisions will be made at three different points (1990, 2000, 2010) and at each decision point, a consequence for each possible alternative will be quantified by a probability distribution. The probability density function was assessed by the use of the fractile method. For example, consider the maximum and minimum amount of access times to and from the station for all classes of trains. Let it be 60 minutes and 10 minutes in 1990. Next we evaluate the point of 50% fractile, i.e. the point from which cumulative probability is .5. If the point is at 45 minutes, chance of being either longer than or shorter than 45 minutes is equally likely. Then, access time of 45 minutes corresponds to .5 of cumulative probability. Similarly, the interval between 10 minutes and 45 minutes are divided into equally likely parts by choosing .25 fractile as 32 minutes as well as .75 fractile as 50 minutes. In succession, each quatile is subsequently divided into equally likely parts in a similar manner. Then, we made the smooth lines which represent the cumulative probability distribution.

As is shown above, there are three different measures at three different time spans. To aggregate the three measures, we take a weighted average of probability assessments to account for impact over the 30-year period to the year 2010. For access time, we define X_2 as

$X_2 = (X_2^1 + 1.5X_2^2 + 2X_2^3)/4.5$
where X_2^i denotes the access time of year i
(the year of 1990 for i = 1, 2000 for i = 2, 2010 i = 3)

Those weighted aggregation will eventually end up with the probability distribution for X_2 to account for the overall impact of a particular strategy in terms of access time.

Now, we need to assess a utility function $u(\vec{x})$. $u(\vec{x})$ will have a exact form of

$$u(\vec{x}) = \sum_{i=1}^{6} k_i u_i(x_i) + k \sum_{i=1}^{6} \sum_{j>i} k_{ij} u_i(x_i) u_j(x_j) + k^2 \sum \sum \sum k_{ijk} \text{ ----}$$

$$\text{--- } + k^5 k_1 k_2 k_3 k_4 k_5 k_6 u_1(x_1) u_2(x_2) u_3(x_3) \text{ ---- } u_6(x_6)$$

where the utility function $u(\vec{x})$, $u_i(x_i) \forall i$ are all scaled zero to one and $k_i \forall i$ are scaling constant.

The main theoretical results shows that if preferential independence and utility indenpendence hold, the utility function turns out to be either additive or multiplicative depending upon the value of sum of ki's. If $\sum k_i = 1$, then k = 0 and it reduces to the additive form of $u(x_1, \text{---}, x_6)$ $= \sum k_i u_i(x_i)$, if $\sum k_i \neq 1$, then k can be found by using the multiplicative form of $ku(x_1, \text{ --- }, x_6) + 1 = \prod_{i=1}^{6}(kk_i u_i(x_i) + 1)$. Here, the perferential independence only concerns ordinal preferences and no probabilistic elements are involved. To verify the preferential independence, we partition the set of attributes into x and y. If we confirm ourselves that the rankings of consequences at different levels of attribute x are the same regardless of the fixed level of attribute y, then x is preferentially independent of y. The cardinal preferences of the decision maker, on the other hand, involved in utility independence. To verify this, if the specific values of x attribute over the 50-50 lottery yielding either the most desirable or the least desirable are the same regardless of the fixed levels of the attribute y, then x is utility independent of y.

Let's assume preferential independence and utility independence in our case. To assess the k_1 scaling factors, let us take x_1 as an example. We asked the decision maker to find the value p_1 such that $u(\overline{x_1}, x_2, x_3, x_4, x_5, x_6) = p_1 * u(\overline{x_1}, \overline{x_2}, \text{ --- }, \overline{x_6}) + (1 - p_1) * u(x_4, x_2, \text{ --- }, x_6)$. Then k_1 is equal to p_1. By using the same procedure, we can estimate all k_is'. Then we have to make consistency checks by asking decision makers, "Would you prefer to have access time or cost changed to its most desirable level with the set of all attributes at their least desirable level?". If the answer was access time, the coefficient of access time utility has to be greater than k_1, coefficient of cost utility. Once the k_is' are all estimated, then k is evaluated by making use of either additive or multiplicative form of utility function.

CONCLUSION

Decision making is one of the most important issues in substance. If the problem has the form of unidimension, it could sometimes be quite straightforward, but in most decision problems, it has the form of multi-dimension where profit, ethics, traditional identification, aesthetics, personal values, corporate values, etc. are involved in decision making process. These considerations including other societal externalities create conflicts; what is good in one aspect turns out to be bad in the other. Those value judgement is looked upon subjective measures. Utility function theory helps us to develop frame work for assessing and quantifying these subjective values and systematically including them in the decision making process. After decision makers structure their problem and assign utility values to consequences associated with all possible events, they calculate their optimal strategy that maximizes expected utility. This strategy indicates what they should do in their courses of action. There are various techniques those strategies can be employed, i.e. a logical sequence of operation would perform the strategies either dynamically or statically.

The original purpose of this paper is to identify effective strategies for developing the station facilities in Busan, located south-east most corner of koeran peninsular. The analysis in this paper has only demonstrated master plans. Since those strategies only help to give master plans defining broad activities for a 30-year period, it is considered as static model. In this model, they don't indicate what actions should be taken by the government to meet its needs. As the second step in the decision making process, dynamic model is considered for giving us action plan. Initial plans, in dynamic system, can be revised depending upon how subsequent events progress. Revised strategies are again fed into the dynamic model as an input both to see how it affects and to make things fit in the system.

In decision making contexts, utility function theory plays very important role especially when problems require subjective value judgements. We could broadly extend it to those problems whose objective function is not crystal clear.

REFERENCES

Arrow, K. J., 1951, "Social Choice and Individual Values," Wiley, New York.

Carter, G. and Ignall, E. L., 1970, A simulation model of fire department operations: design and preliminary results, IEEE Transactions on Systems Science and Cybernetics, SSC - 6, 282-293.

Cochrane, J. L. and Zeleny, M., 1973, "Multiple Criteria Decision Making," University of South Carolina Press, Columbia, S.C.

Keeney, R. L., 1974, Multiplicative Utility Function, Operations Research, Vol. 22, 22-34.

Keeney, R. L. and Raiffa, H., 1976, "Decisions with Multiple Objectives: Preferences and Value Tradeoffs," Wiley, New York.

Mehrez, A. and Sinuany-Stern, Z., 1983, An interactive approach for project selection, J. Opl Res. Soc., Vol. 34, 621-626.

Megrez, A. and Sinuary-Stern, Z., 1983, A note on multiplicative utility in interactive project selection, J. Opl Res. Soc., Vol. 34, 1123-1124.

DECISION ANALYSIS MODELS AS A FRAMEWORK FOR POLICY DIALOGUE

Matthew Jones and Chris Hope

University of Cambridge, Management Studies Group
Department of Engineering, Mill Lane
Cambridge CB2 1RX

INTRODUCTION

There are three possible ways in which Operational Research techniques may be used to study social issues. The first, and historically the most common, has been to transfer techniques developed for the study of the traditional applications of OR, such as process control, directly to the study of social issues. The initial lack of success of this approach did not deter its development, since the failure of social actors to adopt the prescriptive solutions derived from these approaches was usually attributed to the restrictive assumptions upon which the techniques were based. Much effort was expended therefore in refining models and methodologies to account for the complexity encountered in studying social issues. Increasing the sophistication of the models, however, does not seem to have overcome the limitations of this approach, and its value is now widely questioned (Ackoff, 1979).

The second way, arising, at least in part, in response to the observed failures of the first approach, was to develop new methods, such as the soft systems methodologies (Checkland, 1981), explicitly designed to study social issues. Whilst these methods have undoubtedly contributed greatly to the more effective application of OR in the social sciences, there exists a third option - the adaptation and extension of OR techniques developed within the "hard" tradition, to make them more suitable for the study of social issues. This third way, building upon the established techniques of OR, but adapting them to cope with the very different character of the issues to be addressed within the social sciences was the one used in the study described in this paper.

The application area for the study was UK energy policy, and more specifically, the integration of environmental issues into energy policy, an area of increasing political concern in recent years. The characteristics of energy policy issues differ substantially from the customary areas of application of hard OR techniques. As a result, the experience of trying to transfer these into the energy policy area has not been a happy one (Keepin and Wynne, 1987). Rather than an individual decision-maker addressing a specific decision, there are a large number of organisations concerned (in varying degrees) with a process of policy formation.

In such circumstances, the primary contribution of a model may be the provision of a channel for communication between different organisations, rather than support for specific decisions. Thomas & Sansom (1986) describe this as supporting decision structuring and policy dialogue. The 'conversation' established through the model provides a means by which organisations can learn about each others viewpoints (Thomas & Harri-Augstein, 1986).

Decision Analysis (DA) appeared to us to be well suited to the study of energy policy, since it explicitly recognises the subjective aspects of human judgements. This led us to develop a simple multi-attribute value theory model to address this issue.

THE ENERGY AND ENVIRONMENT POLICY MODEL

The model has two main endogenous components. A list of 41 attributes, which represent the important issues which need to be considered in determining energy policy, and five different options for UK energy policy to the year 2010 drawn from a range of publications, intended to represent the spectrum of opinion on possible policy options.

Use of the model involves four stages: attribute selection, attribute rating, option scoring and experimentation. Users are invited to select up to 15 attributes from the 41 available, although users can add to or modify the list if they consider that the default attributes are inappropriate. The user then rates the relative importance of the selected attributes on a scale from 0 to 100. Following this they are asked to score the performance of each of the options on each of the attributes. The product of the scores and ratings for each attribute (weighted scores) and the sum of the products for each option are then calculated. These results are presented in tabular and graphical form (showing the weighted scores for each option on each attribute). Users are encouraged to modify ratings or scores to obtain a satisfactory set of results, such that the options with the highest weighted scores are indeed those that they feel to be most attractive.

The model is described in detail in Jones et al. (1988). It does not perform any complicated mathematical manipulation of the inputs, or evaluate the viewpoints, but simply provides a framework for representing them in a structured and coherent form. Users typically take a couple of hours to perform all four stages of model operation, and produce a set of results that they are willing to accept.

There are two ways in which this model may be used to aid policy dialogue. The first is through an analysis of the numerical results produced by representatives from different organisations concerned with energy policy. This might highlight, for instance, areas of agreement or particular conflicts, which could become the subject of further investigation. We report such an analysis elsewhere, which indicates that the model is effective as a means of characterising viewpoints, and in eliciting differences of both 'facts' and 'values' between the representatives (Hope et al., 1988).

The second way involves the establishment of a formal dialogue between invited representatives of organisations involved in the energy policy debate, through "Interfacing" workshops. These bring together participants from two stakeholder organisations (generally with a strongly contrasting perspective on energy policy) in the presence of a facilitator. The participants use the model twice, once to describe their own views and once to represent what they believe to be the viewpoint of the other participant (their 'partner'). They then meet face to face to explain the reasoning behind their representations and compare their representations with their partner's description of their true views.

The Interfacing approach is similar to role reversal (or bilateral focus) techniques used in negotiation exercises (Walcott et al., 1977). It differs from conventional role reversal, however, in that the computer model is used as the means of describing viewpoints, rather than relying solely upon verbal presentation of viewpoints. In addition, the focus of the Interfacing workshops is on communication and learning, rather than conflict resolution (the traditional aim of role reversal exercises), although this is not to say that Interfacing may not also contribute to conflict resolution.

THE POTENTIAL OF INTERFACING

Four inter-linked aims have so far been mentioned as reasonable goals for "hard" OR techniques in addressing social issues: decision structuring; policy dialogue; communication; and conflict resolution. Tape-recordings of the workshops held to date and participants' responses to questionnaires have been used to assess the potential contribution of Interfacing in each of these areas.

Decision structuring

We have already stated that in energy policy formation there is no single specific decision to be addressed. That is one reason why energy policy is so difficult to frame. So Interfacing cannot help to structure a particular decision, but it can help the participants to think systematically about their own viewpoints and those of their partner. It is the process of using the model that contributes most to this. Some participants discovered the need to structure their views to produce a coherent set of results was quite a challenge: "...it was quite useful for me personally as an exercise; what's really a priority, what's really important in your perspective, forcing you to actually think about that..."

In structuring the energy policy debate by its use of default attribute lists and pre-set options for action, however, the model also limits the debate. One participant commented that "There's a lot of attributes that I would have chosen which didn't feature here." This could undoubtedly be a problem if the other aspects of the Interfacing approach, the face-to-face explanation and the reversal of roles, were not also present. Some users did indeed include a number of additional attributes to represent missing concerns when representing their own viewpoints.

However, when role-playing, this restriction was generally not a problem, and no new attributes were added by the workshop participants. Thus, although the modelling approach may mask some of the subtleties of the debate, it appears to provide an adequate foundation for the Interfacing process to lead to dialogue and communication . For example one participant commented: "You wouldn't learn without, because in actually pressing a button that says 7 as opposed to 3 or something you're really putting your finger on the line. Whereas if you discuss it you hedge and fight...". Another observed that "if you get it [ie your partner's views] badly wrong you can actually see what it was that you misjudged or forgot about or whatever."

Policy dialogue

Although, in theory, policy dialogue could have arisen from each participant running the model in isolation and sending their results to their partner, we would not be sanguine about the quality of the inferences that would be drawn from such an exercise. In practice the commitment to attend an Interfacing workshop appeared to act as an incentive to the participants to undertake a more thorough analysis of their own views and those of their partner than would otherwise have been the case. Thus the establishment of a formal procedure for debate between the participants contributed to the creation of a more fruitful policy dialogue.

The formalisation of debate also has other consequences. By bringing together representatives from environmental groups and official bodies to discuss their views on an equal basis, the environmentalist contribution is given greater legitimacy. This may be of benefit to them in enhancing their perceived status, but also offers the official bodies benefits in terms of forcing the environmentalists to provide a formal description of their views.

Communication

Results from the questionnaires give some indication of the effectiveness of Interfacing in increasing understanding. The mean results are shown in Table 1 below.

Table 1. Participant understanding before and after the workshop[a].

	Before	After	Gain[b]
Understand partner's views on energy policy	3.7	5.1	1.4
Partner understands my views on energy policy	2.9	4.8	1.9

[a] Means of ten participants' responses. A score of 1 corresponds to 'Not at all', a score of 7 indicates 'completely'.
[b] 'After' minus 'Before'.

Ten of the eleven participants completed the questionnaire (an early workshop involved three participants). Both gains are significant at the 1% level on a Wilcoxon matched pairs signed-ranks test. So it appears that participants felt the workshops to be effective, particularly in helping others to understand their views. However, they do not tell us whether some other process, such as a simple discussion, would have been just as effective. Participants' comments on the importance of the role-reversal elements included: "You understand much more if you try to get inside the other person's skin." and "..the crucial element, otherwise you would just reinforce your prejudices." Clearly, for some participants at least, the reversal of roles was beneficial in increasing their understanding of their partner's views.

Thus all three aspects of Interfacing (the use of a model, the face-to-face explanation and the reversal of roles) seem to be important for communication, rather than just structuring or dialogue, to occur.

Conflict resolution

Although not primarily intended as a conflict resolution procedure, the shared experience of participation in the workshops may encourage the development of greater personal understanding amongst participants. Meeting their opponents on neutral ground, going through the same process with them and reviewing and learning about their views may encourage participants to take a more constructive attitude toward their partners. However, evidence from role-reversal experiments suggest that the technique may not be particularly effective in modifying viewpoints in areas of particularly entrenched disagreement (Muney and Deutsch, 1968). It is even possible that the workshops may enhance conflict by informing participants of their opponents' views and priorities.

We have insufficient evidence to settle this issue either way at present. However, given the depth of disagreement between some partners, providing a common framework for discussion, and encouraging reflection on each others viewpoints would appear to be a positive development on the out-of-hand dismissal, or narrow criticism which has characterised much of the interchange in the UK energy debate in recent years (de Man, 1987).

THE USE OF "HARD" TECHNIQUES TO ADDRESS "SOFT" PROBLEMS

What conclusions can be drawn, therefore, about the adaptation of "hard" techniques to address "soft" social problems. From the experience of the Interfacing workshops four principal conclusions may be identified. Firstly, the workshops indicated that it was possible to obtain useful results from the application of a "hard" technique to a very "soft" problem which was substantially different from its normal domain of application. Whether this finding can be generalised to techniques other than DA is not clear. Other techniques, such as cost benefit analysis, which have usually sought to determine 'objective' value data, would seem less likely to be so readily adaptable to tackling issues where there is so little agreement over the 'objective facts' of the situation as is the case in the energy policy debate.

Where the Interfacing approach diverges most widely from the traditional application of DA has been in the focus on description rather than prescription. The model used in the Interfacing workshops does not seek to evaluate the viewpoints of the participants, but simply to provide a common framework for describing them. The second conclusion would therefore be that we may need to accept that the normative assumptions underlying much "hard" OR may be inappropriate when addressing social issues. In part this arises from the legitimacy problem alluded to earlier. Even if the viewpoints expressed by participants can be taken to be representative of the organisations they come from (and this itself is questionable as many participants commented), the representativeness of the organisations themselves can be challenged. Who is to say what weight should be attributed to a particular organisation's views? In addition, in recognising that all viewpoints involve subjective judgements, the subjective nature of any evaluation procedure must also be accepted.

If the outcome from the Interfacing workshops is not to be used as a basis for decision-making, then what is the purpose of the exercise? The third conclusion from the workshops would be that the benefits should be seen in terms of process as well as product. The focus of effort in adapting "hard" OR techniques to study social issues, needs therefore to be transferred from obtaining a solution to the problem, to facilitating the participants in communicating their different perceptions of it. In traditional "hard" OR most of the effort is directed to obtaining the right answers, and relatively little attention is paid to how the answers are obtained. This imbalance needs to be remedied. Modelling effort would thus seem likely to be more productively spent on improving the ease of use of models rather than increasing the accuracy of their output. An increased emphasis on process also implies that the nature of the data to be collected needs to change. Analysis of recordings of the Interfacing workshops has indicated that they are a fruitful source of data which may provide insights on the processes involved. Suitable questionnaires and structured interviews may be another source of such data.

The final conclusion which the Interfacing workshops have highlighted, follows from the other three, and concerns the modelling process itself. A "hard" model used in a social context is not the end, but the means. A requisite approach (Phillips, 1984) needs to be taken toward the modelling process therefore, in which the form and content of the model is determined by whether it is considered adequate for the task for which it has been developed rather than whether it is the theoretically ideal form. As the description of the model used in this study has emphasised, its structure was extremely simple. All values used to obtain the results were derived solely from the users' inputs. Some users suggested that this meant that it was not a "proper" model as there was no internal mechanism. The evidence of the workshops however has shown that the process of using this "model" encouraged the users to think through their views in a systematic and critical way and to obtain insights on their own and their partner's viewpoints. If this 'proto-model' can achieve these results, then a requisite approach to model development would suggest that there should be strong grounds for further sophistication before rejecting such a simple model as inadequate.

REFERENCES

Ackoff, R.L., 1979, The future of operational research is past, J. Opl. Res. Soc., 30:93-104.
Checkland, P., 1981, "Systems Thinking, Systems Practise", John Wiley, Chichester.
deMan, R., 1987, United Kingdom energy policy and forecasting: technocratic conflict resolution, in: "The Politics of Energy Forecasting", Baumgartner, T. and Midttun, A., ed., Clarendon Press, Oxford.
Hope, C.W., Hughes, R.M. and Jones, M.R., 1988, "Examining Energy Options Using a Simple Computer Model", Management Studies Research Paper, University of Cambridge Management Studies Group: Cambridge.
Jones, M.R., 1989, UK energy policy: ceasefire or new consensus?, Energy Policy, (in press).
Jones, M.R., Hope, C.W. and Hughes, R.M., 1988, "A Multi-attribute Value Model for the Study of UK Energy Policy", Management Studies Research Paper, University of Cambridge Management Studies Group, Cambridge.
Keepin, B, and Wynne, B., 1987, The IIASA world energy model, in: "The Politics of Energy Forecasting", Baumgartner, T. and Midttun, A., ed., Clarendon Press, Oxford.
Muney, B.F. and Deutsch, M., 1968, The effects of role-reversal during the discussion of opposing viewpoints, Conflict Resolution, 12:345-356.
Phillips, L.D., 1984, A theory of requisite decision models, Acta Psychologica, 56:29-48.
Thomas, H. and Samson, D., 1986, Subjective aspects of the art of decision analysis: exploring the role of decision analysis in decision structuring, decision support and policy dialogue J. Opl. Res. Soc., 37:249-265.
Thomas, L.F. and Harri-Augstein, E.S., 1985, "Self-organised Learning: Foundations of a Conversational Science for Psychology", Routledge & Kegan Paul, London.
Walcott, C., Hopmann, P.T. and King, T.D., 1977, The role of debate in negotiation, in: "Negotiations: Social-Psychological Perspectives", Druckman, D., ed., Sage, Beverly Hills.

COMPUTER SUPPORTED DECISION MAKING SYSTEM

IN THE PRELIMINARY DESIGN

W. Tarnowski

Technical College
Koszalin, Poland

J. Misiakiewicz

Technical University
Gliwice, Poland

INTRODUCTION

A designer (or a manager) often faces a decision situation when he must choose one variant from among a finite (numerable) set of variants. The variants may be physical objects or actions to be undertaken. For example, the designer tries to find the best electric motor (from a catalogue) for a machine-tool he is designing, or to choose the best operating concept of the machine, or to determine the best way of marketing of the machine etc. Such decision task is typical for a preli-minary or a conceptual design, when the designer is to deter-mine an idea or an operation principle of an object which he is designing, or he must appoint a tender, a licensor etc.

The set of variants is usualy not very numerous: from a few to some scores, at most. In the same time, what is a rule, there are many aspects which should be taken into account: these are objectives (performance parameters) of the being designed object. Some of them are constraints, and majority are of the limit character (of an equality or non-equality type), but still a good number of aspects are such that we want to minimize, or respectively, to maximize their values. These are called criteria of optimization.

In respect to the accuracy of the analysis, the criteria often should be formalized in probabilistic terms as being de-pendent on random influencing variables (as the future price of energy, for example). The assessed variants themselves should be described in the stochastic way, as well, as being manu-factured in random conditions, and being dependent on random interferencing variables during operation , as the ambient temperature, for example. Besides, some criteria are to be adopted as the fuzzy variables (as the safety, robustness or comfort, for instance).

The pecularities of the decision problem on a finite set of variants in design are:

i) many conflicted criteria, and majority of them being of the probabilistic character,

ii) the performance data of variants given explicitly (not as the functions of decision variables), but with a limited accuracy, and often of a probabilistic or of a fuzzy character,

iii) the constraints are given explicitly, consequently the set of feasible variants is given explicitly.

To solve the discussed problem the MADM (Multiattribute Decision Making) concepts and methods are useful. Problems of a choice in this complex conditions are discussed in the book of Hwang and Yoon (1981), where a theory of the choice in design is proposed, as well as some methods and procedures that can aid the choice process. On this basis, a dialog computer system is devised, which is described below.

THEORETICAL BASIS

Assume that given are:

1) The set of variants (names or numbers)

$$\mathbf{A} = \{a_v : v = 1,..,V\} \qquad \qquad ... \ (1)$$

2) The set of criteria

$$\mathbf{K} = \{k_i : i = 1,..,I\} \qquad \qquad ... \ (2)$$

Some of them should be minimize (k_i=min!, i = 1,..,n), the rest should be maximize (k_j=max!, j = n+1,..,I)

3) Matrix of assessments

$$|k_{iv}| \ , \ i = 1,..,I, \quad v =1,..,V \qquad \qquad ... \ (3)$$

Assessments $|k_{iv}|$ may be real numbers (the deterministic form), density probability distributions $f_{iv}(k_i)$ (the stochastic form) or membership functions $\mu_v(k_i)$ (the fuzzy form); they describe performance characteristics of an object,

4) Constraints imposed on performance parameters:

$$g_j(k_{1v}, ..., k_{Iv}) \geq 0 \ : \ j = 1,..,J \qquad \qquad ... \ (4)$$

Now the task may be articulated as follows: it is to find such variant $a^* \in \mathbf{A}$ that satisfies all the constraints (4) and meets in the 'best' way all criteria $\mathbf{K}$ (2):

$$a^* = a_p \in \mathbf{A} : \{ \bigwedge_{i \in [1,I]} k_i(a_p)= min! \cup max! \ \bigcap$$

$$\bigcap \ \bigwedge_{j \in [1,J]} g_j(k_{ip}, ..., k_{Ip}) \geq 0\} \qquad ... \ (5)$$

The above problem is the polyoptimization task, and being such has not the unique solution but in general case rather a subset **P** of variants **A**, **P** ⊂ **A**, is the solution. In other words, there may be many variants a* satysfying definition (5) and we call them polyoptimal variants (there are used synonims, as well: effective, nondominated or Pareto-optimal variants). Thus, having this subset a human decision-maker is to determine one variant as the final solution.

POSSIBLE APPROACHES TO THE MULTICRITERIA DECISION TASK IN A CHOICE PROCESS

There are three common methodologies of handling of the subject.

1. The most general approach is to determine and explicitly demonstrate to the human decision-maker a complete set **P** of the polyoptimal variants. This set may be presented in a form of a diagram or a table. The latter is possible if the set is discrete, otherwise it must be segmented. If the set **P** is very numerous an adequate representation must be found which is to be produced to the user. There are some methods to cope with the problem.

On the other hand, the drawing may be produced if there are two criteria k_1 and k_2 (or three, at most, when the family of curves may be depicted). This approach is called the poly-optimization or the vector optimization. The designer is presented a set, and he must make a decision which element of a set is to be accepted as a unique, final solution.

2. To remove the ambiguity, or to avoid the subjective decision, it is necessary to have one and only one optimality criterion. So in the presence of the set of criteria, there has to be created a compound or aggregated criterion F:

$$F = F(k_1, ..., k_I) \qquad ... (6)$$

Often it is an artificial function of criteria, i.e. it has not necessarily have any physical or economical interpretation, except that it is a measure of a 'goodness' of an object or it is a degree of a 'fitness' or a compatibility to the requirements and functions the object must do. Many authors accept the linear additive form of the function:

$$F = \sum_{i=1}^{I} w_i \cdot u_i(k_i); \qquad ... (7)$$

where w_i are the normalized weighting factors and u_i are utility functions, as they were proposed by von Neumann and Morgenstern. All the values are within the [0,1] range. It may be proved that (7) is the very general form of (6), simple to handle and easy for comprehension by the designer. Later the Authors discuss what are the premisses to determine w_i and u_i.

Having established the overall criterion F we are to solve the classical optimization task with one criterion F. In the presented computer programme other techiques are incorporated, too.

3. The third possible approach is to identify the most impor-

tant criterion $k_o \in K$ and take it as the single criterion of optimality, and for the rest criteria to determine the upper and the lower limits and adopt them as constraints. This is a very common but not very reasonable technique. A version of this approach is the lexicographic method.

In the proposed computer system there are implemented some diverted techiques, in majority known in the literature.

COMPUTER SYSTEM FOR THE CHOICE MAKING

The basic role of a choice making system is to supply a user (a decision maker) a full range of information about a choice task and its possible solutions. Particulary, such system for a preliminary design should include:
- the choice strategies as described above;
- some choice methods for each strategy;
- support facilities for guiding the user through the choice process;
- a kind of postprocessor helping the user to interpretate the results obtained by different choice methods.

In addition, it should provide the communication between the system and the user in a dialog form and should have the possibility of repeating some stages of the choice process to allow the user to examine the choice task results obtained with changed process parameters. Besides, the system should be user-friendly and should be able for implementation on a mini/micro-computer just to stand-by at the designer's working place.

The programme CDSS-PD (Computer Decision Support System in Preliminary Design) has been worked-out to meet these requirements. The programme is written in Pascal and works under MS(PC)-DOS on the IBM microcomputers (IBM PC/XT/AT/PS 2).

The CDSS-PD system has a modular structure and consists of several functional moduls performing particular tasks during a choice in progress.

A short description of basic system functions is given below.

Inputting and Editing Data

This function allows the user to input and modify data as the formulas (1)..(4) show. The set of constraints imposed on performance parameters (4) plays the role of qualifier and delimits the set of variants to the set of feasible solutions.

Two structers of the criteria system can be used: one-level (simple) or multi-level (hierarchical).

Formalization of a Value System

The realization of a choice process requires formalization of a value system, it means: stating an overall optimization criterion, choosing suitable strategy and a choice method and performing some additional tasks - transforming criteria and setting weighting factors, if necessary.

Choice Methods

Many methods for choice making are known in the Multi Attribute Decision Making theory (MADM). Several of them are included into the CDSS-PD system. At the present state of the development of the system there are implemented the following methods: maxmin, weighted maxmin, weighted linear sum, level of satisfaction, TOPSIS (Technique for Order Preference by Similiarity to Ideal Solution), Electre, lexicographics, dominance (Pareto set). The description and discussion of their properties can be found, for example in the survey of Hwang and Yoon (1981).

The system has supporting facility for choosing a choice method for the particular task (it requires inputting some additional information during a dialogue between the system and the user).

Transformation of Assessments

Some of MADM methods require a uniform scale for criteria values. The scale of [0, 1] is usually used.

In the *CDSS-PD* system the user can apply several transformation methods, including the linear scale normalization (two different types), the vector normalization (required by TOPSIS and Electre), the point-to-point construction of the utility function by the user applying the lottery developed by von Neumann and Morgenstern or just can choose one from the "typical" shapes of utility functions.

In the case when the utility function depends on the states of an enviroment and when the probabilististic distributions of these states are known, the system allows constructing a suplementary utility function for such a criterion. The methodology is described by Tarnowski (1984).

Assessment of Weights

The information about the relative importance of each criterion is essential for many MADM methods. It is usually given by a set of normalized weights (weighting factors) w_i, i=1,..,I, where

$$\sum_{i=1}^{I} w_i = 1 \qquad \qquad ...(8)$$

In some cases, e.g. in the lexicographic method only a rank order of criteria is required.

In the *CDSS-PD* system the user can input weighting factors directly or can use one of supporting methods, for instance a decision table.

The Choice Phase

After inputting all necessary data which are required by the choosen strategy and the method, all calculations can be made. Then the results are presented to the user and they may be stored for the later use by the postprocessor (the postoptimalization analysys).

The choice process or only some of its stages can be repeated with changed process parameters (another method, another transformation, another set of weights).

Postoptimalization Analysis

This function is used for comparing results of the choices

obtained with various process parameters. The *CDSS-PD* system proposes some methods for comparison. A particular attention will be focused on this part of the system in its future developments.

Each of the choice process phases may be documented by special printable reports.

The system has been already used for solving some decision tasks, among others for choosing a measuring instrument, for assessing the quality of a being design truck crain or for aiding the choice of tenders (Tarnowski and Misiakiewicz, 1986).

Further Future Development of the System

The *CDSS-PD* system has an open architecture, so it can be easily modified to suit individual needs.

In the nearest future it is planned to extend the system with:
- adding new MADM methods, especialy methods based on the users preferences on the given alternatives (variants);
- introducing fuzzy criteria, fuzzy assessments and/or fuzzy weights;
- developing the more sophisticated postprocessor;
- perfectioning the user support procedures;
- including the interface to databases and some operations on data (the preprocessor).

CONCLUSIONS

Growing popularity of professional microcomputers opens a new perspectives for introducing methods of rational decision making in designers every-day practice. It will help to eliminate bad projects and to reduce expences, but elaborating special computer based systems are needed.

Decision supporting systems are closly related to expert systems. The cooperation of these two fields of knowlegde could yield interesting results, especially when supported by modern computers.

Bibliography

Hwang Ch., Yoon K., 1981, "Multiple Attribute Decision Making - Methods and Applications", Springer - Verlag, Berlin,

Tarnowski W., 1984, "Model procesu wyboru w projektowaniu technicznym" ("General Model of the Choice Process in Design"), Zeszyty Naukowe Politechniki Śląskiej, seria Automatyka, z. 72, Gliwice (in Polish),

Tarnowski W., Misiakiewicz J., 1986, "Heuristic algorithms for tenders choice and quality appraisal as examples of multiple criteria choice making in the design and practice", in: "Proceedings of the Seminar on Nonconventional Problems of Optimization", Publisher: Instytut Badan Systemowych PAN, no.134. Warsaw, pp. 95-114.

PRAGMATIC STRUCTURES FOR SOCIAL INTERACTION

Rosaria Conte, Cristiano Castelfranchi,
Amedeo Cesta, and Maria Miceli

Istituto di Psicologia
Consiglio Nazionale delle Ricerche
Viale Marx 15, 00137 Rome (Italy)

INTRODUCTION

This paper is concerned with the representation of deep knowledge about actions. Pragmatic Structures (PSs) are presented and discussed as structures of knowledge about actions. They are intended to unify the two separate notions of plan and script. On one hand, PSs attempt to unify current tendencies with memory based and hierarchical planning (Swartout, 1988); on the other, they include circumstantial knowledge, but unlike scripts do not limit themselves to describing typical action sequences. Somewhat in line with schemata (Dyer, 1983), they consist of causal and final relationships.

This work is part of a Project aimed at the Simulation of Social Behaviour (SBS Project). However, the issues addressed are of interest in different areas of Artificial Intelligence: among others, the developement of 2nd Generation Expert Systems with augmented capacity for explanation and deep knowledge representation (Steels, 1987); discourse understanding and management (Allen, 1987, Part III) in Natural Language Processing.

The paper is organized as follows:
-PSs are presented;
-claims about the underlying philosophy of social action are made somewhat in contrast with traditional script theories;
-Multi-Agent-Pragmatic-Structures are sketched;
-the SBS Project is outlined and its sequenced objectives mentioned.

PRAGMATIC STRUCTURES

The structure for representing knowledge we propose is quite general, and aims at describing any domain in terms of: (a) goals; (b) actions; (c) states of the world (both conditions for action and effects, be they intended or unintended); (d) pragmatic structures. Pragmatic Structures (PSs) imply both hierarchical and circumstantial relationships, since they may be linked also with circumstantially related goals, i.e. side-goals (see later on in the text).

Each PS is associated with one goal, and specifies how to achieve that goal (it is a sort of enriched planning operator (Sacerdoti 1977)). It consists of a set of

propositions, expressed as predicates with one or more arguments, and related to one another *via* the following kinds of links:
- *causal*, including both means-ends relationships (FOR-, AVOID- links) and cause-effect relationships (CAUSE-links);
- *temporal*, including sequentiality (AFTER-) and contemporaneity (AT-SAME-TIME-) relationships;
- *compositional*, including the possibility of specifying conditional or alternative solutions (IF-, COND-) and iteration (REPEAT-UNTIL-).

In fig.1 we show an example of very simplified PSs in a commonsense domain (taking a train); for the sake of clarity only some of its relevant links are showed.

The main feature of PSs is their immersion in the general knowledge representation formalism, namely, in a propositional (assertional) one. The various items can be connected by means of propositions predicating a particular property of an object. Moreover, the items can be connected hierarchically, and can be composed using Knowledge Representation language operators (such as AND, OR, etc.) to express more complex relations.

These characteristics are meant to provide a simulation system with a variety of capabilities (see Cesta et al., 1988; Conte et al., 1988):
1) *explicitness* of the various links, especially causal ones, providing the system with "deep knowledge";
2) *flexibility*: a system is able to use stored plans in a "flexible" way, that is to use them in all or in part, to combine their steps, and then to integrate separate plans for solving new problems or achieving new goals;
3) *generativeness*: a system is able to explore preexisting knowledge and draw new knowledge from it, to use it both for planning anew as well as for storing new pragmatic knowledge.

Explicit links among PSs and interrelationships with both general knowledge and side-goals allow PSs also to motivate circumstantiality. We define a side-goal as a goal achieved by means of a plan step, but not converging on the top level goal of that plan. Thus a side-goal is only circumstantially related to that PS. This gives rise to what might be called a principle of motivated circumstantiality: a connection-by-circumstance always implies a FOR-link from a PS step to a side-goal. (Look at the step "buy-newspaper" in figure 1: whilst it is temporarily and conditionally related to the rest of the PS "take-train" it is causally linked with the side-goal "not-get-bored".)

SCRIPTS OR PLANS: TWO PHILOSOPHIES OF SOCIAL ACTION.

What relationship do PSs bear to the modelling of social behaviour? Beside circumstantiality, do they have any other connection with scripts? Indeed, these are quite representative in the modelling of social behaviour.

However, scripts have a widespread sociological and social psychological bias: the basic role of knowledge, expectations, and socially shared rules is to facilitate interaction, social identity and a sense of belonging. In other words, knowledge allows both a frightening anomy and the impact with undefined situations to be avoided. In this view, proper social conduct turns into a sequence of situationally adequate actions ("what is to be done in this circumstance?"). Moreover, roles (a word with the same etymology as scripts) are seen as sets of expected behaviours.

To our mind, this view is both old-fashioned and precognitive. Rather, our claims are the following:
- memory of plans is not just a means for "ordering the world". Memorized plans are traces of (socially) already solved problems;

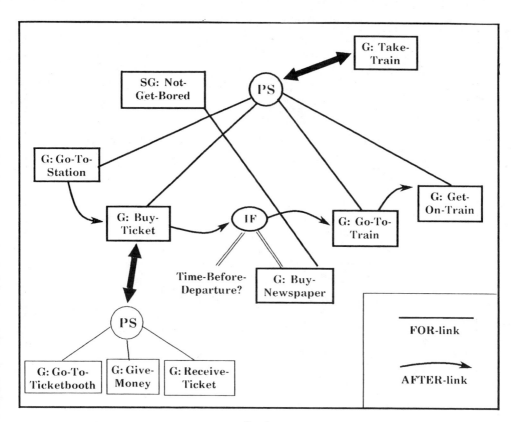

fig.1

- a long-term memory of plans provides ready-made macro-structures for planning, understanding others' intentions and plans, explaining, advising, evaluating, and monitoring one's own and others' behaviour;
- memorized plans are often socially shared, allowing plan-recognizing and influencing. Many shared plans are also multi-agent: they involve two or more agents with different tasks;
- social roles are plans prescribed by social functions and/or institutions, often included in multi-agent plans. Let us now turn briefly to describing how PSs may include multi-agent plans.

MULTI-AGENT PRAGMATIC STRUCTURES (MAPSs) AND MODELLING SOCIAL BEHAVIOUR.

Let us give a trivial example of a PS mentioning more than one agent. Consider the PS for "Buy-ticket", which is a subtask in the PS for "take-train" (fig.1). In fact, buy-ticket shows an elementary structure in which two agents, namely a customer and a seller, must coordinate their actions and goals (see fig.2). Such a structure allows two embryos of micro-roles in social interaction to be outlined (look at the norm impinging on Agents2's goal in fig. 2). Also, it shows two individual "points of view": one and the same action sequence and, at least in part, goal-hierarchy lead to two different ends depending on each agent's point of view (look at the difference between full and dashed FOR-links in fig.2). This is because the propositional approach allows plans to be represented as mental states (Pollack,in press) predicable of different agents. A MAPS may then include both overlapping and agent-specific subparts.

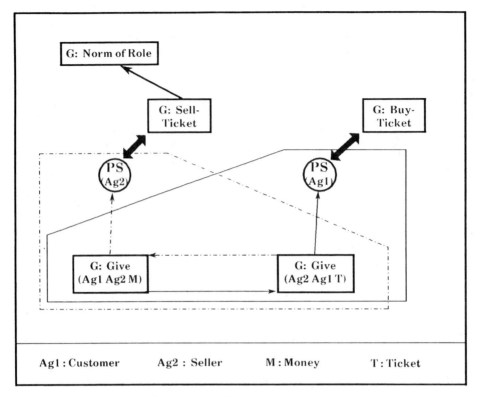

fig.2

MAPSs have two advantages over applying to social interaction a model from derived distributed problem-solving (Huhns, 1987):
-MAPSs are already-solved-problems: they do not entail reiterated problem-solving and negotiation. What remains to be done is to adapt well-known cooperative strategies to novel situations, and at most to establish who-is-to-do-what in the ready-made structure of roles and tasks, and to check for task-fulfillment;
- MAPSs allow for unintended cooperation: suppose Agent X performs a MAPS subpart because she received an order, or because previous performances have been reinforced: she has no idea that her actions form part of a complex plan mentioning other agents as well. In such a case, cooperation is either external to the participants' minds or even "functional": it occurred by chance and, proving successful, was socially selected. We believe that situations of this kind are quite common in social life.

THE SBS PROJECT

The main goal of our project is to simulate interaction among autonomous intelligent agents. Useful features of such an agent are: a) to be able to show social capability, that is, both to play a social and cooperative role, accepting the others' influence (for instance, adopting others' requests), and to perform actions not only in the physical world but also in a social context, so as to modify others' goals and beliefs (social influencing); b) to be cognitively modelled, that is to be able: to represent knowledge about both shared and Multiple Agent plans; to follow a principle of uniformity of knowledge, that is the capability of using the same item or chunk of knowledge to perform different functions; c) to be able to reason about one's own and others' knowledge and, if required, to modify one's knowledge according to external inputs.

The steps needed to reach these goals have different levels of generality. There are at least three levels of implementation:
- the low level steps, now being implemented, are: a) the design and implementation of a complex editor of our world's objects (Goals, Actions, Events and PSs); b) the implementation of the capability of question/answering about such a world (e.g., to use pragmatic knowledge to answer questions such as "What should one do in order to achieve X?", or "What is doing Y needed for?");
- the more complex goals of the middle level are: a) to get the system to perform planning activities both in recognition of intentions and in problem solving, by means of PSs; b) to get the system to work with incomplete knowledge, or in the absence of knowledge, broadly using the flexibility and generativeness of PSs;
- the higher level functionalities are those which allow such system to be used as an autonomous mind and embedded in an environment with other agents to simulate interaction.

Each of the low and intermediate levels has a demonstrative objective; as for the former, the objective is an advice-giving and helping expert system; as for the latter, fields of application will be both discourse analysis and understanding, and flexible planning.

CONCLUSIONS

Plan representation in long term memory and plan transmission and sharing are a basic means for coordinating social action.

This knowledge about socially "already solved problems" allows an agent to: a)understand others' actions; b) apply influencing strategies; c) play well-known roles either in face-to-face interaction or, more generally, in social life.

The problem of mental representation of knowledge about plans is indeed crucial for a theory of social regulation.

REFERENCES

Allen, J.F., 1987, "Natural Language Understanding", The Benjamin/Cummings Publ., Menlo Park, CA.
Cesta, A., Conte, R., and Castelfranchi, C., 1988, "New Knowledge from Previous Knowledge: Explicit and Flexible Pragmatic Structures", Tech. Rep. No.RT-IP-PSCS-6, Istituto di Psicologia del CNR, Rome (Italy).
Conte, R., Castelfranchi, C., and Cesta, A., 1988, "Natural Topic-Change in Plan-Recognition", Tech. Rep. No.RT-IP-PSCS-8, Istituto di Psicologia del CNR, Rome (Italy).
Dyer, M.G., 1983, "In-Depth Understanding", The MIT Press, Cambridge, MA.
Huhns, M.N., ed., 1987, "Distributed Artificial Intelligence", Morgan Kaufmann Publ., Los Altos, CA.
Pollack, M., in press, Plans as Complex Mental Attitudes, in: "The Role of Intentions and Plans in Communications and Discourse", P.Cohen, J.Morgan, and M.Pollack, eds., The MIT Press, Cambridge, MA.
Sacerdoti, E.D., 1977, "A Structure for Plans and Behavior", Elsevier, New York, NY.
Steels, L., 1987, The Deepening of Expert Systems, AICOM, 1:9.
Swartout, W, ed., 1988, DARPA Santa Cruz Workshop on Planning, AI Magazine, 2:115.

OPERATIONAL RESEARCH IN ORGANISATIONAL DESIGN WITH SPECIAL REFERENCE TO DESIGN OF PUBLIC SECTOR ORGANISATIONS IN DEVELOPING ECONOMIES

R. Bandyopadhyay

National Institute of Bank Management
Pune, India

INTRODUCTION

Historically organisational theory developed along with management theory. However, even to this date, we do not know exactly the principles and criteria which are applicable for designing various types of organisations under varying environmental conditions. In what follows we propose to briefly review the empirical/conceptual literature related to organisational design. We then try to identify the role that O.R. can play in such an exercise. We attempt an integrative framework for such design exercises, taking special care that the framework takes into consideration the special aspects related to public sector organisations in developing economies. We then discuss two case studies where such a framework was applied in practice. In concluding section we bring together our threads of discussion and identify areas where further research is called for.

OVERVIEW OF LITERATURE

Broadly speaking there are two approaches to organisational design. The first approach is concerned with viewing the organisation as a network of decision making nodes (March and Simon, 1957, Beckmann, 1983) and analysing the information and control system problems in relations to the network of nodes (Hess, 1983).

The approach is analytical and often mathematical models and tools have been used. (Marschak and Radner, 1972). Further O.R. methods are mostly concerned with this approach. Recently Takahashi has provided a mathematical framework for hierarchical task analysis. (Takahashi, 1988).

According to this approach, the structural design of an organisation is essentially determined by the information and communication needs. An organisatinal task or activity is divided into sub tasks or activities and designing is concerned with grouping of these activities in a certain manner (Bandyopadhyay, 1972). Manner of grouping of activities will determine the number of 'supervisory nodes' and will indirectly determine the 'tallness' or 'flat'ness of an organisation. Simple principles of graph theory and algebra have been applied to illustrate these concepts (Beckmann, 1983). This approach tries to link the span of control with 'fragmentation'. These two determine the degree of flatness and tallness of organisation and they in turn determine the information and communication needs of an organisation.

Large span makes an organisation flat, lower span increases the number of hierarchical tiers and the supervisory nodes through which information and communications must pass. Every such node introduces delay, distoration and bias in the information and communication channels. Lower span leads to distorted and delayed information, larger span creates information overload. One has to therefore (in any practical design exercise) strike a balance between span and fragmentation. There are certain aspects implicit in this approach.

(a) Non-hierarchical organisation is not feasible because information overload increases as square of number of activities. When the number of tasks increases from 3 to 6, the information overload becomes 4 times and not 2 times as it is normally assumed. Thus hierarchy is essential for performing large number of tasks in an organisation.

(b) For optimal design we should be able to measure objectively the distortion, delay and bias under different values of span.

Behavioural scientists point out that there are many other aspects which distorts information and decision making. Individual knowledge and awareness of information will determine decision making time and nature of delay and distortion. In any case our measurement tools are imperfect. Mathematical formulations fail to capture the entire reality. Thus optimal methods of O.R. may ignore a number of important variables which may make the organisation as designed (following this approach) relatively less effective. The approach is promising and must be pursued further but it has to be integrated with the behavioral aspects of the people system and work culture. This leads to the discussion of the second approach.

Behavioral scientists advocated a people-centred approach for organisational design. According to this approach an organisation is defined as an entity consisting of people (groups of people) acting together to achieve a common purpose.

Instead of concentrating our attention anywhere else, we should understand the people who provide the key to the success of the organisation. Thus inter-personal relations, group behaviour, resolution of conflict are to be stressed in our design. We should try to achieve maximal result out of the people system through co-ordinated and harmonious development of the organisation. Such an approach tries to mitigate the dysfunctional effects of hierarchy, encourages openness and participative style (Galbraith, 1977). Organisational design following this approach has been tackled by behavioural scientists mostly based on results of empirical research.

Decision theorists and O.R. scientists attempt now to introduce along with structural variables, certain behavioural variables like bounded rationality, organisational slack, adaptive search, power, conflict and organisational learning.

In both the above approaches certain assumptions are in-built - (a) organisations are meant to maximise the benefit of the people constituting the organisation and (b) the efficient resource allocation is ensured through price mechanism of a market system.

In a developing economy market system is not perfect and market imperfection can produce distortions if resource allocation is done on the basis of market prices. Most public sector organisations are not meant for maximising benefits to the participants constituting the organisation. Such organisations are basically meant to maximise benefits to the community. Further there are limits to price systems ability to allocate resources optimally even in developed market systems

(Arrow, 1974). Thus for public sector organisations in developing economies the above approaches may require suitable modifications, to ensure realistic assessment of (i) the purpose of the organisation and (ii) criteria of efficient allocation of resource in a given economy. Before we end this section we may also like to point out that in actual practice most of the design exercises are, really speaking, re-design exercises or renewal exercises. We are seldom asked to design a new organisation for a given set of purposes or tasks. More often we are constrained with an existing organisation requiring renewal because of its ineffectiveness to deal with environmental changes (Arrow, 1974). We should therefore be concerned with organisational redesign. Much that we have said earlier will be applicable with suitable modification. We have to appreciate the constraints and barriers to the process of change in an existing organisation.

We are really concerned with managing effectively two types of changes - (i) System structure may remain unchanged but properties of individual elements may change - thus education/skill level/technology adopted may change - though hierarchical relationships may remain unchanged. (ii) The structure also undergoes change.

The first type of change has been called a first order change and the second type of change has been called second order change. In case of organisational renewal we have to manage both types of changes in a functional manner, so that they produce synergy and help in co-ordinated achievement of organisational goals. O.R. approaches can contribute in this.

AN INTEGRATIVE FRAMEWORK

It is our view that O.R. can provide an integrative framework where all the aspects tackled in the two approaches can be put together. Specific issues relating to relevance of a concept in public sector organisation in a developing economy can also be tackled within this framework. In evaluating any organisation (new or renewed) we must follow certain criteria. The criteria to be followed in our model are as follows. (Ansoff and Brandenburg, 1969).

i)   Steady state efficiency criteria: Classical efficiency criteria for routine and normal operations.

ii)   Operative responsiveness criteria: Whether the structure can cope with sudden change of demand of existing goods and services. This calls for flexibility and local initiative. Criteria (i) calls for standardisation. Thus there is a conflict between (i) and (ii).

iii)   Strategic responsiveness criteria: Organisation should be effective even when new goods and services are demanded and new markets open up.

iv)   Structural responsiveness criteria: With the changes of environmental challenges, organisation should be able to initiate a process of self renewal.

       Above four criteria are needed for all organisations; in case public sector organisation we need to add further

v)   Conformity criteria: Various units of decision making in the organisation should conform to the same general norms dealing with similar situations.

vi)   Consistency criteria: In different periods of time in similar situations decision making behaviour should follow a constant pattern.

The above are mainly structural considerations. Here O.R. approaches and techniques like mathematical programming, networks, simulation and queueing can be applied successfully. In our next set of criteria namely feasibility criteria we explicitly take behavioural aspects into consideration.

vii)   Human resource feasibility criteria: This should evaluate effects of changes on motivation, morale and inter-personal relations .

viii) Economic feasibility criteria: This evaluates availability of resources and costs to the organisation including human cost through dislocation. These costs have to be compared with the proposed (likely) benefits in terms of increased returns, better performance and organisational effectiveness, improvement of employee morale and career progression.

In a developing economy like India public sector organisations like nationalised commercial banks have their branch network spread all over the country. For such organisations monitoring and control, coupled with delegation and decentralisation of decision making in response to local developmental needs become very important.

Based on our discussion so far, we now propose a framework for design (redesign or renewal) of an organisation. The framework consists of the following components and processes :-

i)   Study of organisation, (history and evolution) - geographical spread and market segments.

ii)   Study of environment - past, present and future.

iii)   Relating environmental changes with performance of the organisation - past and present - projected future.

iv)   Study of present structure - of decision system, various decision making nodes, roles, delegation, and decentralisation .

v)   Study of related information system .

vi)   Study of inter-action and inter-relatedness between nodes.

vii)   Study of the people system - recruitment, training, allocation, transfer, performance appraisal, promotion, career progression .

viii) In respect of changed environment and emerging challenges study of decision making needs - at various levels, degree of delegation and decentralisation needed.

ix)   Study of need for information/communication and new work technology.

x)   In view of (viii) and (ix) suggestions of restructuring and changes in respect of (iv), (v), (vi)and(vii) and study of the effects of proposed suggestions and changes.

xi)   Ascertaining resource needs and changes in allocation for (ix) above and checking of feasibility in respect of economic resources, human resources and infrastructural requirements.

xii)   Determination of needs for changes in people system through
         O.D. intervention.

         It may be seen that above framework though sequentially des-
         cribed may be interactive and we may have to go back and
         forth till we reach a feasible redesign which will improve
         overall effectiveness of the organisation.

     Many of the steps suggested above are also valid in case of
private organisations in developed economies and are therefore generally
applicable as a methodology. However, details of barriers and constraints and the nature of feasible modifications will vary substantially
in case of public sector organisations like commercial banks in India.

CASE STUDIES

     We mention two case studies to illustrate the application of the
framework. We were engaged in redesigning a relatively small size
nationalised bank, whose headquarter was in Calcutta. Following
certain developmental priorities, branch network had to be extended to
rural areas and it increased 4 fold in 10 years. New roles and functions demanded new information, new decision making skills and attitudinal changes in the staff. It was found that historically this bank
had been more conservative in outlook, quality of human material had
been relatively less skilled. The bank operated in a restricted well
charted market segment less disturbed by competition. With nationalisation and rapid expansion all these changed. From a local institution
the bank became a national institution. With the changes in recruitment though the bank was attracting good talent after nationalisation,
at control levels it lacked talent, leadership and initiative. Further,
decision and information systems were not suited for expanded roles
and functions of the bank.

     We therefore developed alternative environmental scenarios for the
bank for those areas, where the bank was likely to have major concentration of branches in the next 10 years. Mathematical models relating
such environment and market with business growth in terms of deposits,
credit manpower and infrastructure (including branches) were drawn up.
Decision and information systems (initially ignoring constraints due to
people system) ideally suited for the organisation were then designed.
This structure and related implications were then discussed with the
top management and representatives of other levels of executives.
Overall design as it emerged after discussion with the top and senior
level executives were discussed with union representatives and modifications suggested were evaluated by our criteria as suggested in the
framework. When it was pointed out to the concerned people that certain
modifications would be inefficient in respect of new roles and functions
(even though more desirable from steady state efficiency point of view)
a compromise level was reached by objective assessment of conflicting needs.

     In the second case the size of the bank was bigger, and its head-
quarter was in Bombay and it had better quality manpower at the senior
level. In this case a similar approach was followed. However, in the
second case it was found that people system was more geared to urban
banking whereas in the first case the bank was better suited to adjust
to rural banking. The first bank had interpersonal/intergroup conflicts
based on caste and language. The second bank had also intergroup/
internpersonal conflicts but not to the extent of the first bank. Purely
structurally the second bank had overmanning in urban, metropolitan and
controlling centres. Rural branches were suffering due to infrastructural
and manpower resources. The first bank had difficulties in providing

proper talent in controlling positions. They had also problems in rural areas but these were less serious compared to the second bank.

Adequate role specification and proper decision allocation coupled with role clarification seminars brought significant improvements in performance of both the banks. The problems of the banks were identified as behavioural problems, but many of the behavioural constraints became relatively tractable when structural modifications were carried out and some degree of role clarity emerged. The first bank needed strengthening of officers and given the present constraint of promotion, had to recruit top level executives from open market to make the modified structure work and to provide right kind of leadership. However, this infusion of new blood had to be kept to the minimal level so as not to demotivate staff down the line. The likely effects of these changes could be dispassionately assessed by means of our framework. In case of the second bank we had to determine norms of manning for controlling offices. This was done by mathematical modelling and comparing results with number of controlling nodes of banks of similar size.

CONCLUDING REMARKS

In summary following points may be noted :

(i)   Multi-disciplinary approach - O.R. providing the framework where conflicting views can be assessed and accommodated. We may have to resolve many conflicts subjectively through discussion.

Further research areas are listed below (illustrative not exhaustive).

(ii)   We have not yet any valid theory of public sector decision making. Research in this area is urgently called for.

(iii) Relationships of behavioural and structural variables under different environmental conditions should be better specified.

(iv)   Our measures of span, fragmentation, information overload and other related concepts need to be refined further.

REFERENCES

Ansoff,H I and Brandenburg, R G, 1969, " A Language for Organisation Design in Perspectives Planning, by Jantsch E (Ed), OECD, Paris.
Arrow, K J, 1974,"The Limits of Organisation", W W Norton and Company, New York.
Bandyopadhyay, R, 1972, "Planning in Commercial Banks, National Institute of Bank Management, Bombay
Beckmann, M J , 1983, "Tinbergen Lectures on Organisation Theory", Springer Verlag, Berlin.
Galbraith, J R, 1977, "Organisation Design",Addison Wedley, Reading,Mass.
Hess, J D, 1983, "The Economics of Organisation", North Holland,Amsterdam.
March, J G, 1988, "Decisions and Organisations", Basil Blackwell,Oxford.
March J G, and Simon, H A, 1958, "Organisations", Wiley, New York.
Marschak, J and Radner, R, 1972,"Economic Theory of Teams", Wiley, New York.
Takahashi, N, 1988, "Sequential Analysis of Organisation Design : A Model and a Case of Japanese Firms", EJOR, Vol. 36, No. 1.

INFORMATION SYSTEMS

INFORMATION SYSTEMS

Veronica Symons and Geoff Walsham

Cambridge University Management Studies Group
Mill Lane, Cambridge CB2 1RX
United Kingdom

As organisations become increasingly dependent on computer-based information systems for their routine operations there is a steady growth in awareness and understanding of information technology (IT) and its impacts. The importance of this is reinforced as many companies use IT for strategic advantage, as a "competitive weapon". Numerous examples have been cited in the literature in areas such as airline reservation systems, banking systems, and pharmaceutical supply systems. Changes in competitive positioning are generally achieved by information systems which enable better communications and interactions outside the organisation, or alter the balance of power between customer and supplier.

The impacts of information systems within organisations can be equally profound and wide-ranging. Changed routines and procedures and information flows may result in shifts in the distribution of power and control. Specialised resources and suppliers can give rise to new forms of dependency. Jobs may be deskilled or upgraded, lost or gained; the quality of working life may be radically altered. Such effects can be critical in user acceptance of the new technologies. Resistance to technological and organisational change is a very natural phenomenon, which necessitates the careful planning and management of the introduction of information technology.

Potential impacts do not follow the introduction of information systems in a deterministic fashion. Rather, they are dependent on human choices and judgements made during development and implementation, and are rarely explicable by reference to the technology alone. "Human factors" as well as technical ones are a common source of problems during implementation. Information systems, after all, affect the way people work, which means much more than how they perform certain business tasks. People engage in work partly for economic reasons but also for the social interaction and stimulation which accompanies it. The tendency has been for systems developers to pay little attention to social and organisational aspects, considering them only in so far as appears necessary for staff training and motivation. But the behavioural and social consequences which are frequently problematic will not be understood as long as information technology applications are designed in purely technical terms. Information systems need to be conceptualised as social systems, reliant on technology but also constrained by their organisational context. The network of social relationships, communication processes and information flows around the technology cannot be divorced from it without a corresponding loss in understanding. The complexities of the function of information systems within organisations can best be addressed by means of methodologies from the social sciences. These provide a rich source which has been underutilised in the area of information systems but is the theme of this stream at the conference.

We turn now to some of the specific topics which are discussed in the stream papers. Recent years have seen an increasing interest in the development of expert systems, at the "high-tech" end of the information systems spectrum. So far the claims that artificial intelligence will supplant human experts remain unsubstantiated. Expert systems applications are currently limited to very narrow fields of human expertise, but they will probably have an important role in the future in supporting human judgment. One area which is the subject of major research at present is artificial intelligence. For example, extensive work is being carried out on the development of artificial intelligence interfaces for future computer users who will have access to extensive networks and a proliferation of databases. The function of the interface would be to provide an intelligent screening function, in consultation with the user, and thus give rapid access to the most relevant information or people as required by the user. Another area of current research is the use of expert systems as teaching tools; expert knowledge is captured on a database for assimilation according to the student's interest, and at his or her own pace.

Decision support systems (DSS), computer-based systems which help decision makers to confront semi-structured problems through direct interaction with databases and related analytical models, overlap to some extent with expert systems and artificial intelligence. The growth in personal workstations, user-friendly interfaces, access to databases and networking opens up much wider possibilities for the application of decision support systems in the future. Such information systems are examples of the likely trend towards user-centred rather than machine-centred systems. A greater proportion of computing power in the future will be directed towards helping the end user; for example in end-user computing such as spreadsheets and desktop publishing and, more generally, to help people who are not information technology specialists to make effective use of IT. Issues concerning the interface between humans and computers will be of increasing importance. The development and implementation of decision support systems is a natural development of "old-style" operational research, organised intervention into human activity using computer-based tools.

Many of the newer, "soft" methods of OR have proved useful in work on information systems. Systems thinking does not study the technology in isolation, but in relationship with its organisational environment Checkland's soft systems methodology, in particular, has been successfully applied in information systems development. Its strength lies in it detailed, holistic analysis and the way it makes explicit the different often conflicting perspectives of the various interest groups involved Checkland's work is only one example of the social science methodologie that attempt to put some structure on the analysis of organisational life Conceptualising an organisation in a particular way, for example as a information system, provides a model with specific insights as well a specific limitations. Methodologies such as these are interpretive, no accepting the everyday world at face value but attempting to show how it i perceived by the actors within it. This involves a formal treatment o goals, beliefs and values, and of conflict and negotiation as intrinsic t decision making and action.

An example of decision making as a political process and the inadequacy of formal, "rational" models can be given in the area of the choice or evaluation of information technology applications. The formal-rational method which is most widely used is cost-benefit analysis. This approach purports to quantify all costs and benefits related to a project and to aggregate them into a single figure which represents the project's worth. However, many of the benefits of information systems are of a qualitative and intangible nature, and their quantification is highly subjective and uncertain. Thus cost-benefit analysis is often "used" but its purpose is as part of the socio-political process of decision making and the figures are designed to "persuade the accountants" or "prepare a case for the Board" where the key reasons for the proposal are to "back a judgment" or "support future strategy" or, more covertly, to change the balance of power in the organisation. The pros and cons of an information system may be perceived very differently by the various actors and interest groups involved.

Work on information systems strategy has often been prescriptive and concerned with the formulation of strategy through a conscious and explicit process designed to align information systems with business and organisational goals and, in the commercial context, to seek competitive advantage from information technology. Empirical evidence in organisations shows, however, that strategy often does not arise from such a process but is better conceptualised as a consistent pattern in the stream of decisions. Strategy forms gradually and sometimes unintentionally, and the term strategy formation is more appropriate than formulation for this emergent process. Information systems, being social systems, are often used as sources of power and influence. The actors involved will point to "technical benefits" such as labour reduction, internal efficiencies and increased managerial control in order to rationalise the adoption and use of information systems, but other social and political goals often underly these publicly expressed objectives. Where there are conflicts in the hidden goals, information systems become an arena for organisational politics.

The predicted benefits of computer-based information systems have always been harder to realise than conventional wisdom seems to suggest. Management information systems, database systems, decision support systems, expert systems, and fifth-generation computer systems have in turn captured the imagination of prospective users with extravagant claims about their widespread effects, but the reality has proved both less glamorous and less frightening than the reports. The direct links between computer use and some social outcome, whether desirable or troublesome, are weakened by contextual constraints and conditions. If new computing resources or organisations which adopt them depend upon the resources or cooperation of others, then it is less certain that participants will experience the most extreme capabilities of information systems. The interaction of key actors, their diversity of perceptions, ways in which sources of power and influence are mobilised and exploited, approaches to decision making and conflict resolution and the role of formal mechanisms are all fundamental to the understanding of how information systems function in organisations. The analysis of such factors requires the use of methodologies from the social sciences.

ROLE ANALYSIS IN INFORMATION SYSTEMS DEVELOPMENT

Brian Wilson

Department of Systems and Information Management
University of Lancaster

INTRODUCTION

Recent work associated with information systems development has been
concerned with two areas. The first area is related to the development of
a primary task model and, as this forms the basis of the subsequent
'information needs' derivation it is a crucial stage in the whole methodology (Wilson 1987).   It has been recognised for some time that the
multiple perceptions of the various managers in an organisation make a
considerable impact on the direction undertaken by that organisation and
hence, the activities that are managed within it. If these multiple
perceptions are taken seriously then they provide a major problem when
answering the question: 'What do we take the primary task of this
organisation to be, for this particular set of managers in their situation
having lived through a particular history?'

The second area is associated with information systems development
but arises as a result of the introduction of new technology. It is the
case that the use of new technology within an organisation, demands new
ways of working and frequently the introduction of new roles. Primary
task models have also been used to explore the nature and scope of such
roles. It is this latter development which will be discussed in the
paper. The development of primary task models has been described in detail
in the above reference.

ROLE INVESTIGATION

In a number of organisations service departments have grown up which
are essentially concerned with exploiting the developments of IT and the
particular boundary and responsibility of these departments is frequently
of concern to the organisation they are supporting. Models can be
developed which aim to represent particular interpretations of the roles
of these departments and can be used to explore the feasibility of
adopting any one role in terms of an allowable area of responsibility and
the necessary interactions with the rest of the organisation which will
be demanded by that role. It is usual for a role exploration such as this
to develop primary task models whose boundary, and hence area of authority,
is greater than that which is likely to be considered acceptable. Such
an investigation was undertaken in an international petrochemical company
in 1987.

Within the Engineering and Technical Centre of this company the use of PCs had proliferated to the extent that each Engineer had his own. This was used to produce, analyse and store data which effectively became 'owned' by the individual Engineer. It was the case, however, that the data was not the personal property of the originator, it belonged to the organisation and was shared by others. The individual ownership of data produced particular structures and interpretations which led to inconsistencies and misuse when the data was shared. The company concerned believed that a change in attitude was required which would result in a shift from individual to corporate ownership. Their response was to make a structural alteration and appoint a 'data administrator'. The intention was that the appointment of a central responsibility for data would bring about the desired change. The big question of course was what could the role and responsibility of this data administrator be within the current environment within the Company. It is current philosophy in many organisations, as it was here, that 'each manager should have a PC on his/her desk' and if that is the case then this problem of individual data ownership is one that must be faced generally.

Within the situation described above, the role of data administrator could lie somewhere on a spectrum which extended from a mere store-keeper (with responsibility for the capture, storage and availability of data) to the other extreme of 'information manager' (with a responsibility for the planning, progressing maintenance and control of an information network).

We investigated this problem by developing a model of the task of information management (see Fig. 1) and then used it to explore the implications of taking different responsibility boundaries for the data administrator, (see Fig. 2).

The model of Fig. 1 consists of a set of activities connected together on the basis of their logical relationships and the whole model could be looked at as a job specification for the role of data administrator (if it were to reside at the end of the spectrum represented by the task of information management in total). Each boundary which is drawn on the model represents a reduction in the area of responsibility. The set of activities within the boundary could be seen as a reduced job specification and the interactions between these activities and those outside the boundary represent the communication processes and procedures that would need to exist to link the data administrator with those other Managers who are undertaking these external activities (if, in fact, they are undertaken).

Each alternative was examined both in terms of its practical feasibility as an individual management task and also in terms of its feasibility within the existing culture of the Engineering and Technical Centre.

CONCLUSION

The approach briefly described above, together with developments in primary task model building, are illustrative of the kind of work which is being undertaken at present in relation to information systems development (Checkland, 1981; Wilson, 1984). The emphasis in the last couple of years has been in relation to role exploration rather than to the specific analysis of information requirements. Two approaches are possible in relation to the former problem. One is to adopt the consensus

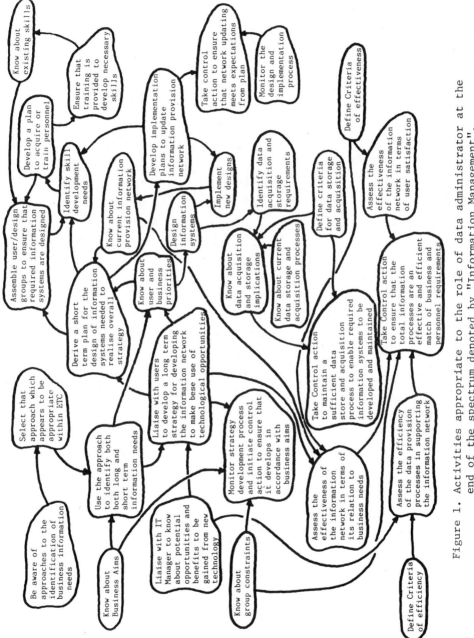

Figure 1. Activities appropriate to the role of data administrator at the
end of the spectrum denoted by "Information Management".

411

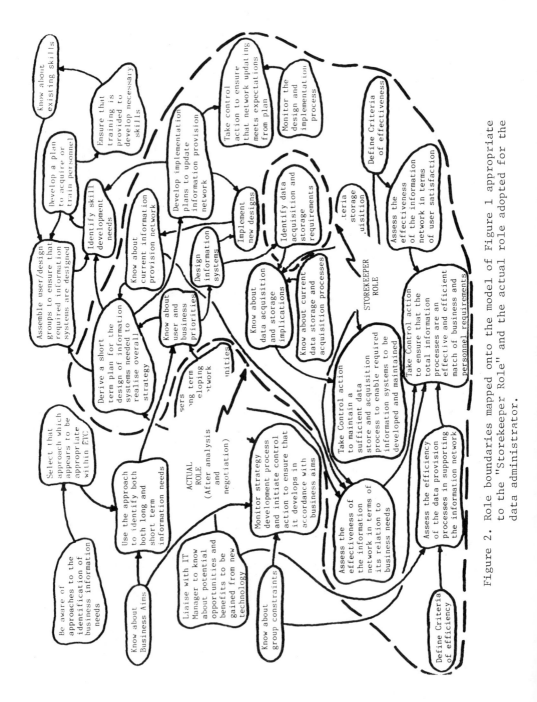

Figure 2. Role boundaries mapped onto the model of Figure 1 appropriate to the "Storekeeper Role" and the actual role adopted for the data administrator.

412

primary task approach. An alternative is to investigate a wide range of potential role descriptions and through a comparison with what exists to identify incremental change which could move an existing organisation towards any one or more of the potential roles explored. In this approach a single role is not defined but merely changes from the present state in terms of the activities undertaken by the particular organisation unit under review.

REFERENCES

Checkland, P.B., 1981, "Systems Thinking, Systems Practice", Wiley, Chichester.

Wilson, B., 1984, "Systems: Concepts, Methdologies and Applications", Wiley, Chichester.

Wilson, B., 1987, A Systems Methodology for Information Requirements Analysis, Proc. I.F.I.P. Conf. "Information Systems Development for Human Progress in Organisations", Atlanta.

INFORMATION SYSTEMS EVALUATION:

UNDERSTANDING SUBJECTIVE ASSESSMENTS

Steve Smithson

London School of Economics
Houghton Street
London WC2A 2AE

INTRODUCTION

The evaluation of computer-based information systems (CBISs) is an important yet difficult area. Its importance can be seen in the high investment that organizations are prepared to make in major new systems, in terms of both monetary costs and general disruption. Regularly, horror stories appear in the literature (e.g. Lucas, 1975), describing CBISs that fail to meet the requirements of users, as well as being delivered late or over-budget. Some form of evaluation of implemented systems is thus essential as feedback for organizational learning, to identify aspects of success and failure, and to help prevent the repetition of errors.

It is argued that CBISs should be treated as social systems (Hirschheim, 1985) rather than technical systems, although the evaluation of complex social systems, such as hospitals, is extremely difficult. In a recent survey of the CBIS evaluation literature, Hirschheim and Smithson (1988) argue that too much attention has been paid to the detail of particular evaluation methods and not enough consideration has been given to the function and substance of evaluation. They argue that the paradigms of the natural sciences have been inappropriately applied to information systems, where a more interpretive approach (Burrell and Morgan, 1979) should be adopted. Noting the subjectivity inherent in evaluation, they call for renewed efforts to understand the evaluation process.

Recently, this notion of attempting to understand evaluation has been applied in a number of case studies in the area of information retrieval systems (Smithson, 1988). These systems, also known as online bibliographic database systems, comprise large textual databases, mounted on centralized mainframes, and accessed via data communication networks.

The traditional functional model of the online search process (see for example, Salton and McGill, 1983) is too system-based for our purposes. In this model, a user, with an information need (and usually assisted by a skilled intermediary) inputs a query, made up of appropriate subject terms. The retrieval software matches these terms with references (citations) in the database and the output comprises a set of references to the literature.

A more comprehensive, user-oriented approach is to examine the online search in the context of the user's information seeking behaviour (Belkin and Vickery, 1985; Smithson, 1988). A user is assumed to be in a problem situation, requiring information to help resolve the problem. The user selects a particular search strategy, involving a certain source of information, executes the strategy, and evaluates the outcome with respect to the problem/information need. This process is repeated, with the problem and information need continually changing as new information is received, until either the problem is solved, or the user gives up. In this model, the online search is just one potential search strategy; others include browsing through libraries, and informally asking colleagues.

The objective of this research is to examine the evaluation of an IRS, from the perspective of the user, in the context of that user's information seeking behaviour. This paper briefly examines the approach in the light of the interim results available.

METHODOLOGY

A case study approach was chosen for three reasons. Firstly, in order to understand the evaluation process, a relatively deep analysis is required of the complex information seeking context. Secondly, the diversity of users and information needs suggests that any hasty aggregation, or averaging, is likely to lead to a loss of information. Finally, controlled experiments were rejected because of the difficulty in replicating 'real' information seeking behaviour in an artificial laboratory environment (Ellis, 1984).

IRS evaluation is dependent upon the judgements of users, regarding the usefulness, or relevance, of individual retrieved items. Relevance is an abstract, elusive concept, that has been much debated in the literature (see Saracevic, 1975); here we shall use the term synonymously with 'usefulness'. Users are often unable to make definitive judgements from citations; they can only make predictions of usefulness, which may change when/if the user finds the complete document. Therefore, the evaluation process is a longitudinal one, and our study must also be longitudinal.

Given that information seeking is a complex process, involving the interaction of the user with the IRS and individual documents, it would seem foolhardy to expect to evaluate a system using a single indicator, on a single scale. Thus we have adopted a multi-dimensional approach, drawing on some of the traditional tools of IRS evaluation (Belkin and Vickery, 1985): effectiveness, utility, satisfaction, and use.

Effectiveness measures rely on counting the number of useful (relevant) documents retrieved, to give such ratios as the proportion of relevant documents retrieved (known as recall) and the proportion of retrieved documents that are relevant (known as precision). The analysis may be extended by a failure analysis of the non-relevant items retrieved and the relevant items not retrieved, to identify the causes of failures. The problems of relevance judgements, mentioned above, plus the difficulties of estimating the number of relevant documents not retrieved, constrain the meaningfulness of this approach.

The utility-theoretic approach (or cost-benefit analysis) poses problems in identifying and quantifying both costs and benefits, e.g. it is not clear on what basis to value the information (or documents) retrieved. Measures of satisfaction, through questionnaires or interviews, give users a chance to air their opinions but, taken alone, they may be ambiguous, imprecise and uninformative - the "is everybody happy?" syndrome (Suchman, 1967, p.27). Usage measures would be ideal, but it is rarely clear how to measure the use of documents (or information).

The overall stance taken is an interpretive one, which seeks to describe and provide understanding, without attempting to discover generalizable laws. We feel that such laws are unlikely to exist in a complex social activity such as information seeking.

EXPERIMENTAL DESIGN

It was felt that the cases should be as comparable as possible and thus a group of fairly homogeneous users, with similar (and real) problems, information needs, and deadlines/timescales, was needed. Such a group presented itself in the shape of students following the M.Sc. course in the Analysis, Design and Management of Information Systems, at the London School of Economics. As part of the course, students have to produce a project report of up to 10,000 words, on a topic in the field of Information Systems. Students were offered a free online search (aided by a skilled intermediary) and altogether twenty-three complete cases were recorded.

It could be argued that this group were not ideal, being relatively inexperienced in both online searching and in their chosen topic areas. Also, the offer of a free search tends to render any cost figures somewhat artificial, but IRSs are expensive to use and would not have been chosen by students otherwise. It was felt reasonable to pay this methodological price for access to a relatively large number of comparable, manageable cases. Thus, the evaluation is based on a single search for each user but, given the price of searches, this is the normal pattern of use.

Data was collected at various points within the information seeking cycle, using a variety of methods:

pre-online search	questionnaire
online search	observation & search log
post-online search	structured interview
end of project	structured interview

In addition, the list of citations in each student's project was examined and, in many cases, the database was searched again for cited items not originally retrieved. Much of the information seeking activity takes place inside the minds of users and, whilst it is easy to fault individual data collection techniques, it is hoped that using a variety of techniques has compensated for the shortcomings of individual ones. Particular difficulties include the problem of what people 'say' that they do, compared to what they actually do.

To illustrate the research method, a summary of a typical case is presented below; the actual evaluation figures are less important, in this essay, than the approach taken.

RESULTS OF A SINGLE CASE

The figures shown below refer to the numbers of items (documents), where an item may be either a citation or a complete document. Satisfaction was measured on a four-point scale ('very satisfied', 'fairly satisfied', 'fairly dissatisfied', 'very dissatisfied'). The initial and final satisfaction measurements and usefulness judgements were given immediately after the online search and at the end of the project respectively. Similarly, the user's intentions to search for complete documents were recorded immediately after the online search, and those that the user actually searched for were recorded at the end of the project.

The number of 'new' retrieved documents cited refers to those contributed exclusively by the online search. In the failure analysis, as the number of relevant documents in the database was not available, retrievals are compared with documents cited (from other sources). The merits of employing 'documents cited' as a surrogate for 'use' is debatable.

Information Need

Information need (in the user's own words):
"Any literature on the role, responsibilities, skills of a systems analyst in developing and implementing computer-based information systems in organizations."

No. of useful items already held: 10-19
No. of useful items expected from online search 30-39

Effectiveness

No. of items retrieved 32
No. of retrieved items in correct topic area 32

Usefulness judgements

	Initial (items)	Final (items)
very useful	5	0
useful	10	1
background interest	3	1
cannot say	3	0
little use	2	1
not useful	9	5
Total	32	8

Behavioural Data: Searching for Complete Documents

Intentions to search	23
Actually searched	18
Found	8

Satisfaction

Initial	"fairly satisfied"
Final	"fairly dissatisfied"

Document Usage

no of documents cited	37
no of retrieved documents cited	1
no of new retrieved documents cited	0

Failure Analysis

Retrieval of non-relevant items		Non-retrieval of cited items (misses)	
foreign language	3	not in database	28
judged not useful	11	in database	8
not found	10		
found not useful	6		
Total	30	Total	36

DISCUSSION

The online search, in this case, was not particularly successful, as can be seen from the fact that the online search did not contribute any new items to the cited set. The only finally 'useful' item was already known to the user before the search. Whilst 32 items were retrieved, all of which were in the appropriate topic area, almost half were judged to be of doubtful value, by the user on the basis of the citation details only.

The user had found it "fairly easy" to explain her information need to the intermediary and had been "fairly satisfied" immediately after the online search. She had intended to look for most of the items retrieved, including some that were apparently not very useful, but were likely to be easy to find. In the event, the user searched for most of those originally intended, but was noticeably unsuccessful, finding less than half of those sought. The main complaint of the user, when giving her final satisfaction judgement was the difficulty in finding the complete items. These items were from relatively obscure journals and conference proceedings. At the end of the project the user stated that she had "excessively satisfied" her information need; i.e. she had too much information (from all sources). Thus, it seems that the online search did not perform very well, in comparison to other information sources.

Regarding the failure analysis, the retrieval of non-relevant items is described in the table above. The failures to retrieve cited items are mostly due to the user citing a large number of books, which are outside the coverage of the database. There were however eight items in the database, that 'should' have been retrieved; the reason for non-retrieval being the non-occurrence of query terms in the document records. This could be attributed to either query formulation, or indexing, shortcomings.

CONCLUSION

Using this longitudinal, multi-dimensional evaluation method, we have been able to paint a fairly clear picture of a single search. We can understand why the user evaluated the search at a relatively low level, compared to alternative sources. The results are of course not generalizable to other cases, but provide deeper information than a simple survey of user opinions. Should we have evaluated the search solely on the basis of the number of items retrieved in the appropriate topic area, we would have mistakenly concluded that the search had been successful and had matched the user's expectations.

REFERENCES

Belkin, N. J., and Vickery, A., 1985, "Interaction in Information Systems," British Library LIR Report, No. 35, British Library, London.

Burrell, G., and Morgan, G., 1979, "Sociological Paradigms and Organisational Analysis," Heinemann, London.

Ellis D., 1984, Theory and explanation in information retrieval research, Jour. Inf. Science, 8:25.

Hirschheim, R. A., 1985, Information systems epistemology: An historical perspective, in: "Research Methods in Information Systems," Proc. of the IFIP W.G. 8.2 Colloquium, Manchester Business School, 1-3 September 1984, Mumford, E., Hirschheim, R. A., Fitzgerald, G., and Wood-Harper, A. T., eds., North Holland, Amsterdam.

Hirschheim, R. A., and Smithson, S. C., 1988, A critical analysis of information systems evaluation, in: "Information Systems Assessment: Issues and Challenges," Proc. IFIP W.G. 8.2 Working Conf., Noordwijkerhout, The Netherlands, 27-29 August, 1986, Bjorn-Andersen, N., and Davis, G. B., eds., North Holland, Amsterdam.

Lucas, H. C., 1975, "Why Information Systems Fail," Columbia University Press, New York.

Salton, G., and McGill, M. J., 1983, "Introduction to Modern Information Retrieval," McGraw Hill, New York.

Saracevic, T., 1975, Relevance: A review of and a framework for the thinking on the notion in information science, Jour. Amer. Soc. Inf. Science, 26:321.

Smithson, S. C., 1988, "The evaluation of information retrieval systems: A user perspective," Ph.D thesis (in preparation), University of London, London.

Suchman, E. A., 1967, "Evaluative Research: Principles and Practices in Public Service and Social Action Programs," Russell Sage Foundation, New York.

EVALUATION OF INFORMATION SYSTEMS:

A CASE STUDY OF A MANUFACTURING ORGANISATION

Veronica Symons and Geoff Walsham

Cambridge University Management Studies Group
Mill Lane, Cambridge CB2 1RX
United Kingdom

INTRODUCTION

Computer-based information systems are increasingly used to enhance organisational performance without necessarily reducing headcount or operational costs. In consequence they can be extremely difficult to cost-justify. An evaluation of improved corporate performance must take into consideration a range of benefits which are rarely easy to quantify, and in many cases are intangible and uncertain. The costs to the organisation may be equally difficult to assess. Costs such as capital and operational expenses for hardware and software, physical changes to the work environment, and staff training are reasonably clear, but there may also be more insidious long-term effects, e.g. a deskilling of work or decline in job satisfaction. In addition, user resistance to technological and accompanying organisational change is a natural phenomenon which is likely to manifest itself in some form, and at some cost, during implementation. Evaluation of information systems, to be comprehensive, must therefore go beyond the initial selection of hardware and software to include the organisational and technical issues of development, implementation and operation.

Information systems (IS) can have profound impacts on the ways in which people work and the type of jobs they do; as a result they facilitate changes in organisational structure and the allocation of roles and responsibilities. Specialised resources and suppliers give rise to new dependencies; altered information flows can cause shifts in the distribution of power and control. Human factors such as these have tended to receive less emphasis than the technical aspects of IS, although they are a very common source of problems. In order to anticipate and plan for the behavioural and social consequences it is necessary to regard IS not as technical systems discrete from their organisational environment, but as social systems in which the technology is only one, although a very important one, of the components.

Contextual and social aspects are an integral part of the decisions and actions taken within organisations, as they are to an understanding of IS. Evaluation is itself a social process with formal and informal elements and both overt and covert functions. Although formal evaluation studies may be commissioned primarily to contribute to cost-effective, "rational" decision making, the considerable legitimacy they often have within and outside the organisation may make them an arena for organisational politics. Questions of who carries out the evaluation, when and where it is conducted and what criteria are used can become important tactical issues and mask a variety of

421

hidden goals. IS development and evaluation thus necessitates an interpretive methodology, one which does not accept the everyday world at face value but attempts to make explicit how it is perceived by the actors within it. Our approach is a descriptive one, aiming to understand IS as social systems and evaluation as a social process. In the following section we outline our methodology; we then briefly describe a case study we have carried out, from which we draw illustrations of some of the concepts.

THEORETICAL BASIS

Data collected over a period of several months from interviews and internal documents is collated to form a "rich picture" [Checkland, 1981] of the information system in its organisational context. The concepts of web models [Kling, 1987] provide the structure within which to do this. Web models of computing describe IS as complex social organisations constrained by their context, supporting infrastructure, and history. There is no specially separable "human factor": the development and routine operations of computer-based technologies hinge on many human judgments and actions, often shaped by political interests, structural constraints, and participants' definition of their situations. The network of resource dependencies and of social relationships between those who depend on the IS and those on whom its usage depends is traced through the organisation and sometimes extends outside it; this is the "web" from which the model derives its name.

From this holistic description are built conceptual models of evaluation. These are ideal types in the Weberian sense: their comparison with perceived reality offers a means of gaining insight into problematic situations in the real world [Checkland, 1981]. Each conceptual model embodies a particular perspective on information systems and a corresponding image of organisation [Morgan, 1986]. Conceptual models are compared with the actual behaviour of actors, highlighting similarities and differences; the resulting insights are used to explain problems, issues and conflicts, for example why outcomes do not match objectives.

We now briefly describe the history of information systems development and implementation in a small manufacturing organisation, giving some elements of the rich picture we have built up. The identity of the company has been disguised for purposes of confidentiality.

CASE DESCRIPTION

The *Processing Company* is a small wholly-owned subsidiary of a large international manufacturing organisation. Its original area of business was the manufacture of a product for which the market has shrunk gradually since the late 1970s with the entry of newer, technically superior products manufactured by large international firms. Increasing competition therefore forced the Processing Company to diversify into the area of derivative, "converted", products, for which the market was expanding.

Traditionally the Processing Company had had a small range of products with few customers and a limited number of large orders. Lead times were long and no stocks were held without an order. The business was controlled by means of manual card systems and programs for stocks of manufactured base material running on a Data General minicomputer. In addition an accounts and statistics package was run as a bureau system on the IBM mainframe at divisional headquarters. This software had been developed over the years by the Computer Services Manager. The situation in the 1980s is radically different: there are currently some 2000 converted products, of which approximately half are manufactured for stock and half to customer-specified orders. The number of customers has increased and orders are smaller and more numerous, with shorter lead times. As early as 1982 it had become clear that the existing systems were no longer adequate: customer service was slipping and delivery dates were being missed on some 70% of orders. Management determined that computer-based techniques were needed to increase their control over the business, and began to look within the holding company for a suitable system. When the search proved unsuccessful they decided to employ outside consultants, and in mid 1984 the management consultancy branch of their auditors was brought in.

The consultants drew up an Invitation To Tender (ITT) in conjunction with the Project Leader seconded from business control. It was proposed to introduce systems in the areas of sales order processing, production planning, shop floor production control, finished goods stock control, packaging stock control, purchasing, and production statistics. The ITT was sent to five systems houses including, at the insistence of the Processing Company, IBM. The reason for this was that IBM and DEC are the preferred suppliers of the holding company, and that to get agreement from the Divisional Board for the required level of funding the Processing Company would have either to choose IBM equipment or to demonstrate a good reason for not doing so. IBM in fact concluded that their own package software was not appropriate for the company's requirements; they were then urged to nominate an IBM systems house, and the ITT was passed on to this firm.

Four proposals were received, which were evaluated according to equipment, applications programs, and costs, together with the supplier's experience and support available. One of the proposals was rejected on the grounds of excessively high cost, and one because it did not demonstrate a sufficiently deep understanding of the requirements detailed in the ITT. The remaining two were a proposal from *Systems House* for *Manufacturing Systems* software and Data General hardware, and a slightly cheaper proposal for a package to run on a relatively small IBM configuration with limited expansion capability, which would require significant bespoke development. The former was felt to be a less risky option, having a standard package already developed and a wider user base. In addition it would make use of the Computer Services Manager's extensive experience with Data General equipment, and enable existing programs to be transferred without significant conversion costs. This was viewed as important, and an addendum had been added to the ITT to that effect. The consultants thus recommended the Systems House proposal, subject to software modifications which were to be agreed.

By March 1985 the contract was awaiting clearance for funding. When the Divisional Board realised that the chosen system was not an IBM one they expressed concern. The Processing Company was directed to demonstrate the feasibility of data transmission between Data General and the IBM mainframe at divisional headquarters; this was done satisfactorily. In addition they were required to send the ITT to a different IBM systems house which was implementing a similar package at another of the subsidiary companies. The Project Leader and one of the consultants then compared Manufacturing Systems with the alternative package, point by point against the ITT; the Manufacturing Systems software was again felt to be superior. Two employees from the central Management Services Department of the holding company acted as referees, and at a meeting in August 1985 it was reluctantly agreed that the contract be awarded to Systems House.

Once the contract was signed, work started on drawing up specification documents for modifications to the standard Manufacturing Systems modules. The departments involved were identified, and a "key user" appointed in each to represent the users in discussions on modifications, train users, and coordinate implementation when the time came. The sales order processing module was to be implemented first, followed by purchasing and manufacturing, with the aim of completing the project by the end of 1986. The work on specification continued over many months, as it was quickly realised that many more modifications were required than had originally been envisaged. Software began to come in from Systems House in May 1986, and over the summer the Project Team was simultaneously testing software, setting up the database and revising procedures. A major change was the replacement of the old 4-digit part numbering system with a precise 13-digit one. Attempts to explain the importance of this met with enormous resistance.

Pressure from management to implement the system was building up, and in October 1986 the sales order processing department started dual running. This was not very successful: it was difficult to use old and new procedures in parallel, and the Project Team had not completed their software testing and were not satisfied that the system was robust enough. Management decided, however, that cutover could not be delayed any further; the switch was made in December 1986 for home orders, and January 1987 for exports. During these months there was "total chaos" in order processing and despatch. By Christmas hundreds of orders were late, and a lot of business and several customers were lost. Repetitive training and continual stock takes to improve the accuracy of the records gradually increased staff familiarity and confidence with the system, until by April 1987 delivery promises were being kept on 60-70% of orders. At this stage work was resumed on specification documents for the purchasing and manufacturing modules, and the Project Team and senior management attended a three-day workshop run by Systems House personnel, on the whole area of materials requirements planning. It was then that they began to understand the system not simply as a piece of software, but as a whole new philosophy of working.

The implementation schedule had been extended by 12 months, but management was now concerned that the company did not have the experience, skills and resources available to undertake all the system implementation and organisation tasks even within the revised timescale. The consultants were brought in again to address these issues. In October 1987 they produced a review of Manufacturing Systems implementation and operation, followed in early 1988 by three more detailed reports suggesting a revised organisational structure, implementation plan, and education and training programme. Management accepted their findings but took no immediate action to implement the recommendations, one of which was the appointment of an operations manager. It was agreed that while the restarting of the implementation project was not dependent on this appointment, it should not go ahead without adequate senior management resource. In particular the chairman of the revised Project Team would need to spend at least one day per week on it. Since the relevant person was unable to spare the time required, it was decided to postpone the start of the project until a new operations manager could be appointed.

Soon after this the holding company announced that the Processing Company was to be merged with another subsidiary, and would from then on be the main manufacturing technical research site in the division. A series of new products was to be launched. The organisation was completely restructured, and seven times as many senior managers brought in. This sudden infusion of investment was accompanied by a complete reevaluation of the IS required by the new company, for finance as well as for manufacturing.

CASE ANALYSIS

In this section we use examples from the case study to illustrate some methodological ideas. One of the fundamental concepts of web models is context. In the case of the Processing Company the Divisional Board was an aspect of the organisational context which acted as a powerful constraint on the otherwise technical evaluation procedure by which the hardware and software was selected. The preference of the holding company for IBM equipment determined the inclusion among the list of suppliers invited to

424

tender of first IBM and then an IBM-recommended systems house. Not satisfied that IBM had been given a fair chance, the Divisional Board then insisted on the Manufacturing Systems proposal being compared with another IBM package. The objective of the project, as far as Processing Company management was concerned, was to improve control and management information within the business. To the more senior divisional managers, however, data communications within the division and the holding company were a strategic issue.

The importance of infrastructure can be seen in the specification phase, where the functionality of the system was being evaluated. The essential infrastructural resources supporting an information system include not just physical aspects, but experience and skills and the authority to produce these more concrete resources as required. In drawing up the specification documents the necessary resources in terms of skilled personnel and commitment to organisational change were lacking. The Project Leader had significant experience of computers but his team, comprising an implementation leader, an ex-machine operator who had been injured in a works accident, a temporary employee from purchasing, and two sandwich students, did not. Top management in the Processing Company were keen to have the system up and running but were not themselves experienced with computers and did not appreciate the extent of organisational change involved. Consequently they did not support the Project Team in its attempts to gain user participation, and the Project Leader was not sufficiently senior to enforce this.

A pilot system based on a few home orders ran through the summer of 1986 but the orders raised were not picked up in the warehouse so the system was never fully tested. The Project Leader pointed out to management that the users were not relying on the system, but nothing was done about it. At cutover when the old systems were removed it was immediately discovered that the lot-number control of the system was incompatible with the way products were stored with one lot of material split between several pallets. A major modification to the software was required, which took several months to implement. This example highlights one of the dysfunctional effects of inadequate infrastructure. The lack of managerial and user commitment to the project resulted in an emphasis, in specification, on the functionality of the system, so that evaluation was overly technical and excluded any consideration of the way staff actually carried out their tasks.

Much work on information systems treats the technology as separable from its organisational context and supporting infrastructure. Often allied to this view of IS is the image of an organisation as a relatively well-defined machine, best described by objectives, procedures and administrative arrangements over which there is assumed to be consensus. These approaches to information systems and organisation constitute a conceptual model with particular directives for IS development. The perspective underlying the evaluation work of selection and specification in the Processing Company matched many aspects of this model.

The initial selection procedure, contrasting suppliers' proposals with the ITT, focused almost exclusively on the technology. "Requirements" were defined in terms of the procedures and functions over which management wished to tighten control. The evaluation was done not by employees of the Processing Company but by the consultants, and no attempt was made to understand how staff regarded their lines of work. The old manual systems were frequently over-ridden by "informal procedures", and it was expected that the proposed systems should minimise the extent to which this was possible without official authorisation. The increased regulation was not successful, however. For example, while management viewed the information system as a tool by which to increase their overall control of the business, to the warehousemen it appeared as an institution which removed much of their discretion over their work. They no longer had responsibility for home transport arrangements and for making optimal use of the scarce warehousing space, but because they could see where improvements in these would save money for the company they found ways of working around the system.

As a consequence of the initial technical evaluation and the lack of user participation in drawing up the specification documents, the project was systems-led. It corresponds to the view of systems development as instrumental reasoning [Hirschheim and Klein, 1988], which suggests that all IS are designed to contribute to specific ends. The role of management is to specify ends which are then translated into systems goals. The primary role of the systems developer is to be the expert in technology, tools and methods of system design, and project management. Systems development is formal and rational, with little reliance on human intuition, judgment and politics; it is primarily a technical process.

In the Processing Company, however, the ends were not articulated and shared, nor were they objective. The perspective of the systems developers was markedly different from that of the users. The Project Team had so many problems with technical aspects of the system that when their initial attempts to get the key users to attend project meetings met with resistance they did not persevere. They viewed the users as having no experience of computers, being rather apprehensive about them, and not being prepared to invest time and effort in learning how to use the system before they were forced to do so. From the users' perspective, on the other hand, consultation and communication about the project was non-existent. There was no attempt to improve basic disciplines and procedures before implementation. Training was minimal, consisting of a single seminar which was too long and too technical, and documentation was inadequate. The system itself was very inflexible and did not provide all the facilities needed to deal with exports, which comprised 75% of orders.

The assumptions associated with the instrumental reasoning view of IS development can have potentially dysfunctional consequences, as can be seen in the case of the Processing Company. In reality ends are not agreed: they are controversial and the subject of considerable disagreement and debate. The prespecified ends benefit certain system stakeholders at the expense of others, and as a result the development process meets with resistance.

CONCLUSION

Information systems development in the Processing Company has been shown to be a social process in which the actors involved have diverse and conflicting perspectives. Succeeding stages of formal evaluation run up against actors' own informal evaluations of the situation, often with dysfunctional effects. Conceptual models are a useful vehicle for analysis of IS and the contextual and infrastructural elements, history and social relationships with which they are inextricably interlinked.

REFERENCES

Checkland P. (1981), "Systems Thinking, Systems Practice", Wiley: Chichester.
Hirschheim R. and Klein H. (1986), "The Emergence of Pluralism in Information
    Systems Development: Stories, Consequences and Implications for the
    Legitimation of Systems Objectives", OXIIM RDP 86/15, Templeton College:
    Oxford.
Kling R. (1987), Defining the Boundaries of Computing Across Complex
    Organizations, in "Critical Issues in Information Systems Research",
    Boland R. and Hirschheim R. (eds.), Wiley: Chichester.
Morgan G. (1986), "Images of Organization", Sage: Beverly Hills.

THE ORGANISATION AS AN INFORMATION SYSTEM

David Stanley

Organisation and System Innovations Limited
Tectonic Place, Holyport Road, Maidenhead, UK

INTRODUCTION

What are the issues for management in an era increasingly dominated by information? The key value-added contribution from management has always been in decision-making from an integrated business viewpoint, both for operations and planning, both tactical and strategic and spanning the complete spectrum of the organisation's functions. In an age where the timescale available for making decisions has shrunk dramatically, this role is even more crucial in running excellent businesses. Strategic marketing decisions on new product developments may have to match a seasonal opportunity window of only a few weeks, while some tactical trading decisions may be made in seconds with computer support. Non-optimum performance is punished cruelly when there is no time for a second chance. The days have gone when management could muddle through, patch up mistakes later and still expect to succeed.

It is into this scenario, where a human headful of information and processing power is becoming increasingly stretched by any key organisational function from salesman to Managing Director, that the technologies which support information processes now have so much to offer. The potential benefits are no longer to be measured in terms of saving in administrative costs, but in increased profits, market share, market capitalisation and, not least, quality of life for employees. The goal is to provide accurate, timely and comprehensive information to the people who can add most value to the business working in the optimum organisational context.

IT is, however, not the panacea it is often claimed to be since report after report highlights the dissatisfaction of management with IT effectiveness and demonstrates their failure to derive personal benefit from it. Automation has achieved major benefits when applied to well understood operations where costs could be saved, but comparatively little success has resulted from applying systems to "softer" organisational processes. How is it possible that an enterprise which has undoubted corporate strength in its approach to business strategy can get its technology planning so wrong?

The reason for this is not hard to find. IT has often been seen by senior management as an end in itself rather than as a means to an end and so computer applications have been conceived in isolation from the overall strategic thrust of the business. Furthermore, the term IT itself has led planners to believe that the real problem is getting the technology right, forgetting that any successful system must include the people who interact with it and the information in their heads. Too many really innovative strategies have foundered on the organisation's inability to implement them effectively. This is a particular problem in large organisations where the IT Department has evolved into a large autonomous function and is rarely drawn into non-technical strategic business planning.

Because of this, in many medium to large businesses improvements in organisational capability are the most promising remaining area for substantial performance gains. But managements are understandably cynical about major consulting exercises which simply review current practice and recommend a change. They are now looking for a more systematic and benefits-oriented approach.

THE OASiS CONCEPT

What if the organisation were one in which all functions were totally efficient, all staff were perfectly matched to their jobs, the right (and only the right) information were available exactly when and where it was needed and the response to changes in the external environment were immediate and optimum? It would be unbeatable! Although the real world may not be like that, organisations *do* behave systematically, albeit very imperfectly. Managers should not dismiss a systematic approach prematurely just because it is difficult. Systems thinking can make these complex patterns explicit and susceptible to rational critique, helping us to get away from an obsession with *computers* and focused on *management processes*. Our thesis is that, by understanding organisational dynamics, a much greater proportion of the enormous scope for performance improvement through using information systems can be realised.

Many organisational arrangements have evolved in the context of a different set of internal pressures and external forces which now no longer apply. The environment has changed beyond recognition: old power structures, control systems and motivational techniques are often totally inappropriate to the new economics. Confusion and conflict often result from attempts to achieve the necessary radical change and this further saps the resources of the enterprise and blunts its cutting edge.

In most successful organisations, the whole is more than the sum of the parts and there is an identifiable corporate added-value. But that alone is not sufficient to generate competitive advantage. In highly innovative organisations, the organisation *itself* adds value by stimulating and nurturing processes which harness the innate creative potential of each of its members. No Chief Executive can *cause* innovation to happen. Rather, in a carefully managed networked organisation, many individual contributions, inspired by a shared vision of the corporate goals, are orchestrated to create an integrated product or service. Staff are aligned in the sense that their personal success factors are in harmony with each other's and the organisation's. The key role of the CEO is to get his top team to reappraise the very nature of organisation and its interaction with systems and the innovation process. They then have to mobilise around a new corporate information infrastructure that recognises that the organisation *is* an information system, managers *are* information processors and that new systems represent changes in behaviour.

Achieving such harmonisation also requires excellent internal communications, just as those who were building the Tower of Babel were thwarted when their common language was undermined. Perhaps they underestimated the ability of the competition to penetrate their corporate strategy! The most challenging issue for management is, therefore, how do we generate the necessary common understanding, effective communications and, above all, shared commitment? No longer should we be worrying about how to get more out of our systems, but about whether our systems are helping or hindering the process of getting more out of our people.

With this in mind, we have chosen a deliberately provocative analogy in thinking of organisations from an Operational Research viewpoint as information processing "factories" that take in materials (data and facts from a range of sources) and process them using a variety of procedures at different people-based processors, arranged (or at least behaving) in a network of activity linked to databases. Each processor is triggered by the arrival of Work in Progress (WIP), or an originating event, plus an associated set of instructions and information specifying the "algorithm" for the process required. The output of each processor becomes the WIP for another further down the value chain, or a result to be output by the organisation.

Whatever state of automation exists in an organisation's "transaction" processing (and it is unusual for more than 10% of a company's key information to be resident on corporate databases), there will always be areas of the business that defy economic analysis because little is understood about the business dynamics and problem solving that take place. This is especially true when there are major changes of activity, either arising from a new business enterprise or external business pressures.

People are often used to plug the gap and fulfil difficult and unpredictable organisational functions, yet are rarely the subject of disciplined optimisation in the way that capital investment would be. This is because of their ability to design "on the fly" solutions to problems as they arise. These networks of people account for most of the costs in the service and manufacturing industries in many sectors, yet little is done to improve their productivity, management effectiveness or learning curve.

This problem is compounded by the pressures for responsive handling of tactical issues such as quotations, price cuts, complaints, design repairs, questions from the Minister, etc. These pressures place time dependencies on the organisation which may span many "processing nodes" in the people network. Progress management and dependency management, which may be deemed essential for the well understood and systematised areas of the business, are often ignored as a way out of this deadly embrace.

Because of the nature of the network, the transactions are characteristically complex, poorly analysed, high variety/low volume, compared with the high volume transactions which will already have been computerised. This makes the network doubly difficult to *manage* which, as we have seen, is a key requirement for extracting the optimum performance. Notwithstanding these problematic features, the people network inevitably functions as an information system. Each of the nodes sometimes behaves in a quasi-autonomous mode, giving rise to the phenomenon of creativity, and sometimes in a way which is strongly correlated with its neighbours.

Such systematic behaviour is susceptible to techniques such as OR, systems analysis, modelling and even Computer-Aided Design, but these disciplines have rarely been consciously applied by those concerned with developing IT strategies or organisational design. The OASiS contribution has been to bridge these two worlds and weld them together through the development of a pragmatic methodology.

A detailed model of the organisation as an information processing system is both surprisingly rigorous and useful for deriving insights and methodologies for improving its capability. Figure 1 illustrates the key elements, encompassing:

- the input of information, relating both to the control function exercised by management and "raw material" for processing, from the external world and the output of results

- networks of people processors who control other functions and processes or add value to the "work in progress" through internal transactions based on well understood algorithms

- access to data and knowledge stored in memory or a mix of on-line or archival media

- adaptability through feedback loops which enables the "system" to improve its own performance.

As with any system, a change in the external inputs will stimulate a change in the outputs. The effectiveness of the organisation, whether public or private, can very largely be measured by how optimum this response is, together with the "capital" resources (ie both human and material) required to generate a given output. For instance, an imitative "me too" response to the launch of a new competitive product is often the knee-jerk reaction of a company with only limited ability to analyse the full implications from a marketing or value-chain perspective. If the cumulative corporate wisdom based on past experience were represented in a more immediately accessible way, perhaps through a knowledge-based system, a more considered response would be possible without sacrificing speed. Again, the focus in much of the public sector over the last few years has been to reassess the value of the functions and services performed by a given number of civil servants and to optimise the output per head.

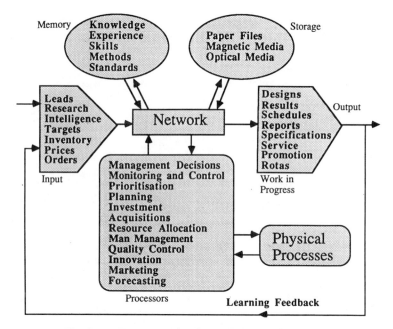

Fig 1.   The organisational information system

The aim is to maximise the value added to the WIP by each people processor, while minimising the time taken to do it, and minimise redundant or irrelevant processes. This is inevitably a rather mechanistic view of complex human systems, but it nevertheless generates some valuable ideas. It certainly is not limited to trivial deterministic processes. For instance, knowledge-based systems techniques even allow us to model self-learning, adaptation and experience-building behaviour in organisations.

THE BENEFITS

Such a concept will merely have academic elegance unless it is also useful in providing a route towards more effective and flexible enterprises. We have demonstrated that substantial benefits to the business can be achieved by creating harmony between people- and technology-oriented information processes, rather than leaving it to chance, by using system design methodologies (and IT itself) for both aspects of the overall system and then actively managing the associated change. This follows directly from our observations (Figure 2) that the cost of organisational introduction can easily be 2 to 3 times that of the technology in a strategic application. The origins of this are not difficult to identify: training, education, opportunity cost, project management, parallel running, productivity loss, demotivation, etc.

IT, representing the automation of what would otherwise be manual processes, must not therefore be planned and executed in isolation from the rest of the system. The approach must be to "informate" rather than automate; to see technology as a means, not the end. IT makes a good friend but a despotic master and many organisations are waking up to the fact that it is often easier to lose than to win through using it.

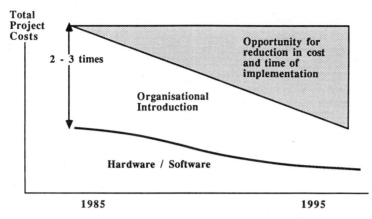

Fig 2. The real cost of system introduction

Nor must IT simply automate the old manual system, but must optimise the execution of the relevant function. The failure of early attempts to introduce robots into manufacturing compared with the best examples of the far more effective CIM approach is a good example. The difference in a CIM plant is not just the flexible routing of material to processes which in themselves are flexible, but the massive amount of manufacturing information that is created on almost every aspect of the process or product. These databases have wide access since the plant manager no longer has a monopoly over information, and the environment is such that as learning occurs, small changes can be initiated and information feedback confirms that improvement has been achieved.

Requirements for strategic change should now be derived directly from the corporate objectives and Critical Success Factors and will be expressed in implementable terms such as the initiation of new internal information transactions or tracking management WIP. The latter cannot be measured by the thickness of papers in the in-tray. Instead, relevant measures must be derived for the specific business which specify the task, the entities required for it, the origin of the input, the destination of the output and the desirable time to undertake the task, all relating to the overall business goals. Where IT systems form part of this plan, they can be conceived and specified from the perspective of their ultimate organisational function, thereby easing the often difficult and expensive introduction process. Organisational structure can also be designed from a systems perspective to avoid duplication (except where desirable), optimise functional effectiveness, maximise value-added and, above all, serve the business needs.

Since nearly all systems, particularly those with strategic impact, require changes in the way the organisation functions, it is desirable to identify how benefits can be systematically predicted and subsequently measured. To do this, it is important to build up a detailed picture of the people transactions, interactions and information used, in much the same way as the now unfashionable O & M techniques used to do. But unless this discipline is accepted, it will be impossible to design, scope, cost and introduce change systematically.

Most of all, an effective organisation will improve its performance by learning: changing priorities, better informed responses, eliminating poor processes, leveraging good ones. Peters and Waterman make the case that the real advantages of IBM and Procter and Gamble are the decades of investment in creating "the truly insuperable barriers to entry based on *people capital* tied up in ironclad traditions of service, reliability and quality" [italics mine]. An organisation cannot learn if it does not monitor its behaviour. Monitoring activity can be the focus for pinpointing what needs further automation or organisational development.

Like many fertile ideas, however, these concepts cannot be applied in a random or mechanical way. OASiS has had the opportunity of developing this approach into a rigorous and pragmatic methodology for application to several real life situations. Such a process delivers many more benefits than just an effective organisation and set of systems, including:

- establishment of consensus and commitment within both the user community and the management team

- surfacing and prioritisation of critical information-based applications

- new insights for management through the business-oriented functional and entity models

- dynamic understanding of the organisation's reporting structures and information flows

- clarity of management objectives related to information needs and data generation and ownership

- better fit of people to job functions and therefore better return on the "human capital"

- a productivity improvement achieved by reducing staff needed by as much as 40% as a direct result of enhanced organisational performance.

In other situations the business models have been implemented as games for the executives to play as part of their education and as a simulation of the prototyping method we used. This can be a very powerful technique for refining the organisational processes in the light of experience with pressures and problems that might not surface during the prototyping exercise.

Perhaps the most important spin-off was that the managers appreciated how much was not going to end up in computer applications and why. This generated a focus on the need to examine ways in which the many people transactions and paper-based interactions could be managed in order to create a business learning curve. In turn this led to an approach to standard forms management as a step towards structuring the really important people interactions with a view to automating them later.

CONCLUSION

It is perhaps inevitable that, like most useful ideas, the systems model of the organisation has its limitations. A change or update of a computer system can be achieved, in theory at least, through installation of a new software release or connection of an additional processor or peripheral. Such autocratic imposition of change is doomed to failure if attempted with a human system, which is not composed of uncomplaining processors. Many good demonstrators have failed to reach operational status because of the failure to convince line management of the soundness of the business case. Not surprisingly, considering the pace at which the technology changes, there has been confusion, fear, cynicism or just plain obstinacy: all "emotional" reactions that a purely logical system would not display. Performance improvements only become visible when managers see through the information system the real cost of quality, of doing things twice, of doing unnecessary things because that's the tradition or of failing to harness someone else's expertise because of the not-invented-here syndrome.

Perhaps real organisations are still more complex than we can model in detail, but the discipline that systems thinking can bring to what is often a hand-waving exercise can be immensely beneficial. It not only provides a way of linking the qualitative strategic objectives of the business to the hard decisions on details of structure and systems, but also is a major investment in building that lasting quality into the business which will improve competitive performance in the future, namely that harmonious integration of human expertise and technology which we call organisational capability.

THE FORMATION OF IS STRATEGY : A CASE STUDY OF A BUILDING SOCIETY

T.M. Waema and G. Walsham

Management Studies Group
Cambridge University

This paper describes a longitudinal case study of the formation
of information systems strategy in a U.K. building society.
A summarised description is given of the changing external
context of the building societies, the internal contexts of the
society studied and the process of information systems strategy
formation in a particular application area. The case is
used to illustrate a number of aspects of information systems
strategy formation which emerge from the interpretivist
approach adopted in the study.

INTRODUCTION

Information technology (IT) is an important strategic resource in an
increasing number of organisations. Managerial concern in these
organisations has shifted from routine automation to the overall
management of information systems (IS). One of the important ways to
manage IS in an organisation is by forming implementable IS strategies
that are aimed at achieving business objectives. Organisations that have
attempted to do this have reported a mixture of successes and failures and
existing IS research has not offered adequate explanations for this. Most
IS research has taken insufficient account of historical, processual or
contextual aspects of information systems strategy formation (ISSF). There
is a need for more detailed explanatory research and the study outlined in
this article departs from the normative approach of prescribing ISSF
activities and the factors that affect them to exploring how IS strategies
actually form in organisations and the consequences of such a process of
formation.

This paper describes part of an on-going research project on ISSF in
U.K. financial services sector organisations. These organisations were
chosen because of the high strategic importance of IT in coping with the

severe competitive pressure and the turbulent environment within which these organisations operate. Only a brief account of the project is possible in this article but further reference on our approach can be found in an earlier working paper by the authors (Waema and Walsham, 1988). In the next section we will briefly outline the approach to ISSF that has been adopted in our research. This will be followed by a description of a particular case study. The paper ends by illustrating some aspects of ISSF that emerge from our approach and by drawing some conclusions.

IS STRATEGY FORMATION

The concept of strategy formation is used in this paper rather than the more traditional concept of strategy formulation in recognition of the fact that a 'consistent sequence of decisions over time' may emerge rather than be planned for (Mintzberg, 1978). Strategy formulation has a purposive connotation which is not truly reflective of organisational reality. We are of the view that, although most organisations will consciously attempt to formulate strategy, strategies will indeed form out of a mixture of planned and emergent processes. In fact it can further be argued that the process of strategy formulation helps shape the context in which strategy formation takes place. There is a constant interplay between the two and they will interact in organisations to differing degrees depending on organisational contexts.

In our work on strategy formation, we use an approach which is essentially interpretivist in that it is historical, contextual and acknowledges the existence of multiple perspectives. It is also holistic in that the ISSF process and the contexts in which it is embedded are analysed and the relationships between them explored. Our approach attempts to integrate a social systems view of IS in organisations and the allied models of organisational decision-making with the processual study of ISSF in its context. Some important aspects of ISSF emerging from our approach are summarised as follows:

* Information systems are social systems which form part of the internal context of an organisation (Land and Hirschheim, 1983; Kling and Scacchi, 1982). As with any other context, information systems are not neutral and will be mobilised and/or exploited by powerful and influential actors to shift the balance of power (Keen, 1981; Kling and Scacchi, 1982; Markus, 1983).

* ISSF should be viewed as a socio-political process in which formal mechanisms and rational concepts form only one part. These mechanisms and concepts can also be used to legitimate actions and decisions of those who sponsor, support or use them and to pursue political motives which are often covert (Kling, 1980; Keen, 1981; Waema and Walsham, 1988). A combination of the formal-rational and power-behavioural process decision-making paradigms can provide valuable perspectives on ISSF (Quinn, 1980; Hax and Majluf, 1988).

* Over time, ISSF processes and organisational contexts have a mutual and causal influence over each other. Context has a dual purpose in this relationship; it is both a medium and an outcome of the ISSF process (Willmott, 1987; Giddens, 1984). Aspects of context are drawn upon by actors to produce and reproduce the ISSF process while at the same time this process reproduces some aspects of context.

A CASE STUDY OF THE "REGIONAL SOCIETY"

The case study outlined here has been written in such a way as to conceal the identity of the organisation and the name "Regional Society" is fictitious. The data for the case study is mainly based on interviews with a total of ten key managers in the Regional Society. These managers included senior executives, IS and user department personnel. The interviews were conducted over a period of more than a year from early 1987. Most of the managers were interviewed more than once and the data was triangulated by cross-interviews and analysis of data from both primary and secondary sources. Before describing the Regional Society itself, we now give a brief description of the changing external context of the building societies over the last decade.

The traditional operation of buiding societies has involved borrowing money from the retail savings market and lending it for house loans, using interest rate differentials to cover overhead costs and provide profits. This operation, however, has been under continuous transformation in recent years mainly as a result of changes in the two primary building society markets. The Conservative Government introduced competition in the personal savings market when it decided, in 1979, to fund part of the public sector borrowing requirement from personal savings. Further competition ensued when the Government systematically, between 1983 and 1984, abolished the tax advantages building societies enjoyed over the banks. These tax advantages, together with the Building Societies Association (BSA) interest rate cartel, which was also abolished in 1983, protected building societies' markets from competition at a time when there was excessive demand for house loans. Thus the protective regulations were removed which, coupled with other factors such as the privatisation programme and cheaper wholesale money markets to which building societies had little access, caused severe competition in both personal savings and mortgage markets.

The banks made a serious entry into the mortgage market after the Conservative Government's decision to remove the "corset" that constrained bank lending. The Government's decision was in line with a policy to encourage competition in the financial services sector and came at a time when the clearing banks faced difficulties in the more turbulent corporate and national lending markets. In comparison to these, the house loans market was more stable and secure and this attracted other financial entrants into the mortgage business. These entrants were encouraged by the reduced entry barriers to the sector largely due to advances in IT (Waema and Walsham, 1988). A combination of all the competitive forces led to a significant erosion of building societies' market shares in both the savings and mortgage markets which in turn forced many large societies to demand changes in the regulatory framework. This culminated in the 1986 Building Societies Act which enabled societies to diversify into hitherto prohibited markets.

The above description has outlined the turbulant external context within which all the building societies, including the Regional Society, have been operating in recent years. The Regional Society is a medium-sized building society with over one billion sterling pounds worth of assets and over 250 branches and agencies. Over most of the 1980s, the Regional Society's financial performance was very good. Between 1981 and 1987, its assets more than quadrupled, operating costs went down from over 1.7% to 1.0% of mean total assets, and advances on mortgages increased by about eight times, while the total number of employees only went up by 50 people.

A former general manager (GM) of one of the largest building societies was appointed as the chief executive (CE) in 1981. At this time, the building societies were facing serious competition from the banks and other financial institutions, as explained earlier. Before the appointment of the new CE, the Regional Society had stagnated in its use of IS and was performing very poorly. The new CE instilled an enterprise culture, changed the management structure to a functional basis, reduced the Board size and the number of administrative centres from three to one, and used IT to decentralise both investment and mortgage processing. He centralised decision-making to himself, pursued business strategies that were not made explicit to others in the Society and dominated decisions on IS planning and implementation.

The biggest IS project during the first half of the 1980s was the decentralisation of processing to the branches. The start of this project had been given impetus by the CE's knowledge of how other societies were using IT in their business and also his previous micro-computer experience. The project started with investment processing decentralisation. After this was completed, the same systems were slightly modified to decentralise mortgage applications which had previously formed a backlog in the administration centre. After this had been achieved in early 1985, a further need was identified to decentralise mortgage administration and to combine both the mortgage and the investments databases into a customer-based database. A database system based on customer profiles was perceived by top management, particularly by the deputy CE, to be of strategic potential in anticipation of the enactment of the 1986 Building Societies Bill. But the project was cut short at this stage because the CE thought that it would take up most of the Society's IS resources at a time when there were more profitable short-term projects for which these resources could be better utilised. As the Society's business expanded, the CE continued directing the modification of the database system so that it could cope with increased mortgage processing requirements. For the two years up to early 1987, there was no major IS project and systems development staff were kept busy making systems enhancements aimed at achieving short-term top management objectives. The computer systems, including those concerned with databases, became more complex and difficult to maintain as further enhancements were added. This, according to the systems development manager, was a result of the CE insisting on the use of what the manager termed "inflexible software" and "non-standard systems development techniques" in attempting to implement objectives that were largely oriented to profit and growth.

When one of the largest building societies advertised for a chief executive, the CE of the Regional Society applied and got the job. He then led merger discussions between the two societies but these failed in late 1986 for a variety of reasons ranging from differences in culture and business focus, to the personality and approach of the CE of the Regional Society. He resigned in early 1987 to become the managing director of a major financial institution in the U.K.

A former operations GM of one of the largest building societies was appointed as the new CE from the middle of 1987. He immediately introduced a formal strategy formulation process and fostered a participative management culture. The deputy CE re-asserted himself in the direction of the IS function, and IS and user department personnel made demands to formalise IS planning and implementation. These demands were approved. The new culture also made it possible to pick up the decentralisation project from where it had stopped and to undertake the re-development of the Society's databases into a customer-based system. These changes were partly facilitated by the opportunities the 1986 Building Societies Act had presented to the societies, the fact that the societies' markets were becoming even more competitive, and the difficulties IS staff experienced in maintaining the inflexible systems that had been designed under the supervision of the previous CE. The DP manager, whose role had previously been to meet time scheduled demands of the former CE, was promoted to assistant computing GM. His role focus was transformed from day-to-day activities to managing the IS function. This increased autonomy in IS decision-making enabled the database re-development project, as the new project was now called, to be directly controlled by the computing function and user departments.

CASE ILLUSTRATIONS

The earlier section on ISSF noted three important aspects emerging from our approach. These aspects can now be illustrated by the case study:

* **The use of formal-rational concepts and mechanisms to legitimate decisions and to pursue political motives.** When top management leadership changed, the new management culture encouraged more formalisation of the ISSF process. A formal and comprehensive strategy formulation process was started in which wide participation was encouraged. This formal strategy formulation concept and its participative nature was partly a means through which senior management legitimated strategic decisions. A notable example was how the deputy CE used the the formalised process to justify the priority resourcing of a customer-based system project. His previous personal experience of retail banking lead him to favour this project but, to justify the project, he linked the development of such a system to the Society's profitability objective which had been set by the new strategy formulation team. He argued that a customer-based system would achieve this by enabling the Society to take advantage of business opportunities which had been opened up by the 1986 Act, such as the offering of unsecured personal loans and the re-packaging of loans for sale in secondary markets. A further example of the use of formal mechanisms concerned the demand by IS and user department personnel to change to structured IS design methods and user-driven steering committees. The expressed rationales for the use of formal methodologies in the IS area, such as removing inflexibilities in systems development and allowing for increased user participation, whereas true, were also a means to pursue political motives of the IS and user department personnel. These motives centred on shifting the control of IS planning and implementation from top management to IS managers and users, particularly after the experience of tight control of these activities by the previous CE.

* **The mobilisation of IS as a source of power.** The use of information systems as a source of power was evident in the processing decentralisation project. The decentralisation of processing had been accompanied by decentralisation of decision-making power to the branches. The previous CE, however, installed a micro-computer in his London office which had on-line access to the mainframe database. Using this computer, he could access the low level management information that branch managers used to make autonomous decisions. In this way, he monitored the activities of branches from the London head office. The CE had thus used IS to shift the balance of power, first from the centralised administration office to the branches, and then onto himself.

* **The mutual and causal influence between ISSF process and contexts.** The previous CE created an ISSF process that had an underlying culture of informality, scant communication, centralised and autocratic decision-making, and incremental technology-driven business strategies. This process was at the same time an agent in creating a context of little shared understanding of the business strategies underlying IS work and little control by others over IS planning and implementation. When top leadership changed however, both top management and IS and user department staff worked in concert to help create a process of ISSF that favoured their different interests. This new ISSF process enabled a view of the relationship between IS strategic planning and business strategy to be widely shared, IS and user department personnel to have more power in IS planning and implementation and more communication between various levels of management. This illustrates how the previous informal and autocratic ISSF process created a context which, with the advent of a new management culture, helped change the process of ISSF to be more formal-rational and participative. At the same time, the changed ISSF process was creating an emerging context of less centralised control of IS decisions, more communication, and more shared views of the strategic nature of IS in the organisation.

CONCLUSIONS

In this paper, we have given a brief outline of a case study of IS strategy formation in a building society. We have used the case study to illustrate some important aspects of ISSF emerging from our approach : the use of formal-rational concepts and methods for legitimation, the mobilisation of IS as a source of power, and the duality of context in relating to the ISSF process. A more comprehensive description of the ISSF process and our underlying methodology was beyond the scope of this short paper but, in line with the focus of the conference, the paper has illustrated the value of social science methodologies in the study of the use of information technology and information systems in organisations.

References

Giddens, A. (1984), "The Constitution of Society", Polity Press.

Hax, A.C. and Majluf, N.S. (1988), "The concept of strategy and the strategy formation process", Interfaces 18(3) : 99–109.

Keen, P.G.W. (1981), "Information Systems and Organisational Change", Communications of the ACM 24(1) : 24–32.

Kling, R. (1980), "Social analysis of computing : Theoretical perspectives in recent empirical research", Computing Surveys 12(1) : 61–110.

Kling, R. and Scacchi, W. (1982), "The web of computing : Computer technology and social organisation", Advances in Computers 21 : 1–90.

Land, R. and Hirschheim, R. (1983), "Participative systems design : Rationale, tools and techniques", Journal of Applied Systems Analysis 10.

Markus, M.L. (1983), "Power, Politics, and MIS Implementation", Communications of the ACM 26(6) : 430–444.

Mintzberg, H. (1978), "Patterns in strategy formation", Management Science 24(9) : 934–948.

Quinn, J.B. (1980), "Managing strategic change", Sloan Management Review 21(4) : 3–17.

Waema, T.M. and Walsham, G. (1988), "Information Systems Strategy Formulation", Management Studies Group WP number 5/88, Cambridge University Engineering Department.

Willmott, H. (1987), "Studying managerial work : A critique and a proposal", Journal of Management Studies 24(3) : 249–270.

AN ARTIFICIAL INTELLIGENCE APPROACH TO BUILDING

A PROCESS MODEL OF MANAGEMENT POLICY MAKING

Roger I. Hall*

Faculty of Management
University of Manitoba
Winnipeg, Canada R3T 2N2

INTRODUCTION

A process model is a device borrowed from basic science where it is often used to describe the form, matter and motion of phenomena in a flow-of-events style (e.g., Bohr's model of the atom, Crick and Watson's model of the DNA molecule). In an organizational context, a process model would describe how things are decided and done by people and groups. But as Mohr (1982) has pointed out:

> Process models are used little in organization theory and even less in many other social science subfields. When they are used, they are often underdeveloped. There is a tendency to present and conceptualize the stages in the process but to omit the forces that drive the movement from one stage to another. The latter, however are essential (p. 14).

In order to capitalize on the strength of process modeling in providing a rich picture of the workings of an organization and yet avoid the weakness in omitting the driving forces, a two part modeling methodology has been devised (see Figure 1). The first part models how the organization works as an integrated flow of resources (the CORPORATE SYSTEM MODEL), and draws on the author's previous work in modeling corporate systems (Hall,1976; Hall and Menzies,1983). The second part attempts to provide an Artificial Intelligence that models the collective decision making behavior of managers in an organizational setting (the POLICY MAKING MODEL). The focus of this paper will be on the design of this latter POLICY MAKING MODEL.

THE POLICY MAKING MODEL

The Policy Making model provides the driving forces for the Corporate model. It is based on a process modeling approach developed by the author using 'cause mapping' and other decision-behavioral theories within an 'evolutionary learning' framework. The process model of organizational decision making is shown in outline form in Figure 2. The model is, in effect, a skeletal Artificial Intelligence (in the Psychological and Philosophical tradition of Artificial Intelligence of building models that think like humans) that converts inputs, such as the periodic reports and

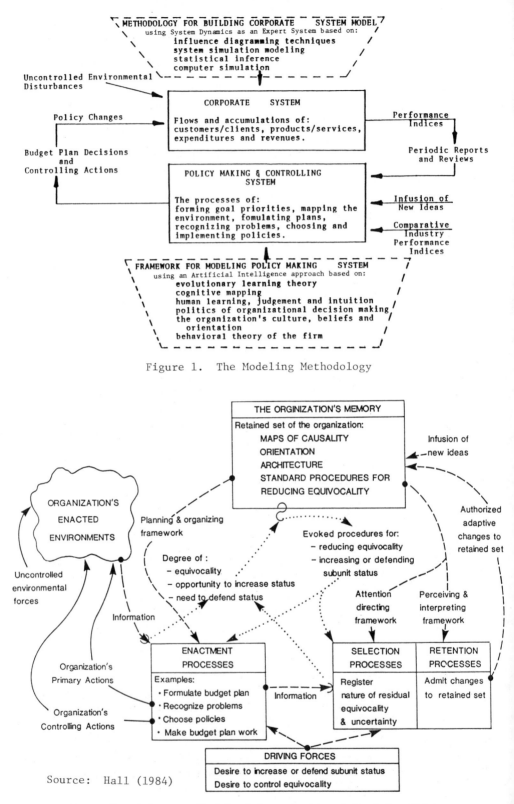

METHODOLOGY FOR BUILDING CORPORATE SYSTEM MODEL
using System Dynamics as an Expert System based on:
influence diagramming techniques
system simulation modeling
statistical inference
computer simulation

Uncontrolled Environmental
Disturbances

Policy Changes

Budget Plan Decisions
and
Controlling Actions

CORPORATE SYSTEM

Flows and accumulations of:
customers/clients, products/services,
expenditures and revenues.

Performance
Indices

Periodic Reports
and Reviews

POLICY MAKING & CONTROLLING
SYSTEM

The processes of:
forming goal priorities, mapping the
environment, fomulating plans,
recognizing problems, choosing and
implementing policies.

Infusion of
New Ideas

Comparative
Industry
Performance
Indices

FRAMEWORK FOR MODELING POLICY MAKING SYSTEM
using an Artificial Intelligence approach based on:
evolutionary learning theory
cognitive mapping
human learning, judgement and intuition
politics of organizational decision making
the organization's culture, beliefs and
orientation
behavioral theory of the firm

Figure 1. The Modeling Methodology

THE ORGINIZATION'S MEMORY

Retained set of the organization:
MAPS OF CAUSALITY
ORIENTATION
ARCHITECTURE
STANDARD PROCEDURES FOR
REDUCING EQUIVOCALITY

Infusion of
new ideas

Authorized
adaptive
changes to
retained set

ORGANIZATION'S
ENACTED
ENVIRONMENTS

Planning & organizing
framework

Evoked procedures for:
– reducing equivocality
– increasing or defending
subunit status

Uncontrolled
environmental
forces

Degree of :
– equivocality
– opportunity to increase status
– need to defend status

Attention
directing
framework

Perceiving &
interpreting
framework

Information

Organization's
Primary Actions

Organization's
Controlling Actions

ENACTMENT
PROCESSES

Examples:
• Formulate budget plan
• Recognize problems
• Choose policies
• Make budget plan work

Information

SELECTION
PROCESSES

Register
nature of residual
equivocality
& uncertainty

RETENTION
PROCESSES

Admit changes
to retained set

DRIVING FORCES

Desire to increase or defend subunit status
Desire to control equivocality

Source: Hall (1984)

Figure 2. A Process Model of Organizational Decision Making

reviews of the current performance from a Corporate System Simulation Model, into simulated budget plans and controlling actions, that, in turn impact on the Corporate System Simulation Model to drive it to new states. The model, if properly designed, should make policy decisions in a way that closely follows the social-psychological and socio-political processes of organizational decision making that have been systematically observed. The model has been used, albeit in a qualitative fashion, to explain the "natural logic" used by the senior management of a magazine company that resulted in some, seemingly, strange decisions; particularly as they led to the demise of the company (Hall, 1984). Similarly, the cognitive 'cause map' of a hospital unit head was put together from interviews and observations to provide a credible explanation for his decision making behavior (Roos and Hall, 1980).

It is not the purpose here to present the theory underlying Figure 2 that has already been described in detail elsewhere (Hall,1984). Rather the purpose is to attempt to demonstrate that generic protocols used by managerial groups for organizational policy making are now well enough understood to be eligible for programming as an interlocking set of routines; where each routine represents a particular behavioral process. The author is in the process of programming these protocols as a set of interlocking routines to make the policy decisions that will drive the corporate model to provide its own history and, in turn, to learn from it.

POLICY MAKING PROTOCOLS

The feasibility of synthesising a Corporate System Simulation Model has already been demonstrated (Hall, 1976; Hall and Menzies, 1983). It is assumed as a starting point, therefore, that such a model exists or can be so constructed to realistically simulate (for the purpose in hand) the processing of policy variables (the inputs) into performance variables (the outputs).

What has not been attempted before (to the author's knowledge) is the synthesis of a generic Policy Making Simulation Model that would realistically simulate (also for the purpose in hand) the derivation of the policies (the inputs to the Corporate System Simulation Model) from the values of the performance variables (the outputs of the Corporate System Simulation Model). What follows is a description of the possible design for such a Policy Making Simulation Model. At the moment of writing this is not yet completed.

The protocols (routines) with which it is proposed to simulate the management policy making (the Artificial Intelligence referred to above) are described in outline form as follows:

REPORT will process the output from a Corporate Simulation System Model to provide an annual Operating Statement complete with financial operating ratios, growth percentages and other performance indices generally used by management for appraisal purposes in the routine REVIEW. It will also generate Progress Reports periodically (say, monthly or quarterly) showing performance against budget to aid in the routine CONTROL.

REVIEW will simulate the routine end-of-year appraisal of the operating results by the management. For example, the current level of achieving goals will be registered. The values of 'factual relations' (e.g.,the latest estimate of the promotional cost of obtaining an extra customer) will be computed from the simulated observations . Policy results will be registered in the form of 'positive hits' (when the intended policies worked) and 'false-positive hits' (the policies clearly did not work). The notion of 'positive' and 'false-positive hits' is borrowed from the work on the process by which managers gain confidence in their judgements by Einhorn and Hogarth (1978).

BUDGET will provide a proforma operating statement based on typical assumptions used in budgeting and forecasting. Budget lines will be computed either from 'factual relations' supplied from the organization's MEMORY (where they are stored), or from forecasts based on the behavioral processes suggested by Kahneman et al.(1982) and Sterman (1987). The budget will compute typical 'bottom line' projections used by managers to identify problems likely to arise in the coming year. By comparing the 'bottom line' projections with the aspiration levels of the organization (supplied from MEMORY), shortfalls and surpluses can be identified. These are passed to DECIDE as problems that need resolving.

CONTROL will scan periodically (monthly or quarterly) the Progress Report (variances against budget) generated in REPORT. It will invoke 'slack reduction' or 'slack absorption' programs in an attempt to make the budget plan work. Authorized changes to the decision variables will be passed to the Corporate System Model.

DECIDE attempts to choose which policy variables to change and by how much using behavioral protocols (see Hall, 1981 for a list of these protocols based on the Behavioral Theory of the Firm of Cyert and March, 1963). Preferred policies to solve each problem are supplied by CAUSE MAP. Each time a policy variable is chosen, the direction and amount of change to close a particular performance gap is calculated using BUDGET. Passes are made through BUDGET with the proposed changes until all problems are solved or no solution can be found. The new values of the policy variables are used to drive the Corporate System Model for the next simulated year of operations.

CAUSE MAP will simulate the social-psychological processes by which a group a managers try to make a complex system of cause-and-effect tractable so that decisions can be made with it. The process of generating group 'cause maps' is based on the seminal work of Axelrod (1976) and Weick (1969). Examples of cause mapping can be found in Bougon et al.(1977), Roos and Hall (1980) and Hall (1984). In the first step, a list is generated from the Corporate System Model of all causal paths (chains of arguments) from policy variables to performance indices (goals). This list is simplified in accordance with behavioral theory by removing all confounding feedback loops of causality (after Axelrod, 1976). Also, where 'indeterminancy' is found (i.e., two or more parallel paths from a policy to a goal with opposing path correlations), the path with the fewest links is adopted as the 'official' line of reasoning, until disconfirmed by experience. At such time, it is replaced by the path of opposite sign with the fewest links. The list is searched for 'acceptable' policies that will simultaneously satisfy 'bottom line' goals and departmental goals. The behavioral protocols for simplifying group cause maps are described in greater detail in Hall (1984).

REORIENT is intended to simulate the kind of internal cultural readjustment that takes place in an organization following a 'crisis'. The socio-political processes involved in a reorientation of goals and strategy were distilled from published sociological studies found in Pettigrew (1973 and 1985), Mumford and Pettigrew (1975) and Turner (1976). According to the nature of the 'crisis', as diagnosed in the routine REVIEW, a new 'dominant' department is selected. Its goals and preferred policies will now be given priority over the other departments in policy making. So, for example, if the diagnosis is 'loss of customers', the Marketing Department and its goals will be given prominence. The routine will have several options representing various degrees of rigidity in accepting a new orientation (new dominant department). Option #1 (default option) will trigger a change in orientation only after a survival-threatening crisis (e.g., financial loss). Option #2 will cause a transfer of power to a new department after disappointing results for a predetermined number of years (say, two). Option #3 will not accept any change whatever the severity of the crisis (i.e., the dominant department is well entrenched and can resist the transfer of power). The goal priorities and preferred policies associated with the new dominant coalition will be stored in the organization's MEMORY.

The MEMORY of the organization will process and store the key information representing the current state of the management's knowledge, beliefs and goals. There will be a number of options for simulating a challenge to established beliefs that are intended to reflect the degree of willingness of the management to undertake such a major review, (1) when a survival crisis develops with a 'bottom line' goal, (2) after disappointing results for a number of years (say two), (3) after accumulating a number of pieces (say 2) of disconfirming information ('false-positive hits'), and (4) no challenge to established beliefs allowed. Similarly, the effects can be simulated on long-run performance of (1) changes to the organizational structure (e.g., divisional vs. functional), (2) different organizational cultures (i.e., emphasis on different 'bottom line' goals), and (3) different levels of perceptions of senior managers of the complex corporate system they seek to control, following, for example, a Critical Success Factor analysis; Hall and Munro, 1987).

POSSIBLE USES OF THE MODELING METHOD

The method has possible uses as: (1) an alternative policy tool for companies in crisis (where the effect of the natural policy making process on the corporate system is not producing the desired results and the reason is not obvious), (2) a vehicle for uncovering "Critical Success Factors" (a subject of particular interest to MIS specialists--see Hall and Munro, 1986) and (3) as a training tool for new executives (see, for example, Hall and Lai, 1984; Hall,1988). The feasibility of building a gaming 'shell' (programmed in "C") is currently being tested. Such a 'shell' would facilitate the quick assembly of a Corporate System Model with the necessary input and output interface (e.g. report generator) to make it operational. These, together with various built in analysis and debriefing aides developed by the author, could be used as a training game with decisions being inputted by teams of management trainees. Alternatively, (4) the Corporate System could be driven by a Policy Making model (the Artificial Intelligence described above) and its behavior explored as a self-regulating entity creating its own history and, in turn, learning from it.

The author is under no illusion that these modelled protocols will trap all influences on policy making and corporate system performance; it is after all but a beginning. It is hoped, though, that, in a macro sense, at least the major endogenous influences can be modelled with enough reality to enable the technique to generate some useful results that might cast some further light on the interactions between policy making and corporate system performance.

REFERENCES

Axelrod, R., The Structure of Decision: The Cognitive Maps of Policy Elites. Princeton University Press, Princeton, N.J., 1976.

Bougon, M., Weick, K. and Binkhorst, D., "Cognition in Organizations: an analysis of the Utrecht Jazz Orchestra", Admin. Sci. Qtrly., Vol. 22, 1977, pp. 606-639.

Cyert, R.M. and March, J.G., A Behavioral Theory of the Firm, Prentice-Hall, Englewood Cliffs, N.J., 1963.

Einhorn, H.J.and Hogarth, R.M., "Confidence in Judgement: persistence of the illusion of validity", Psychological Review, Vol. 85, 1978, pp. 395-416.

Hall, Roger I.,"A training Game and Behavioral Decision Making Research Tool: an alternative use of System Dynamics simulation', Working Paper, Faculty of Management, University of Manitoba, 1988, available from the author.

_____, "The Natural Logic of Management Policy Making: Its Implications for the Survival of an Organization", Management Science, vol 30, 1984, pp. 905-927.

_____, "A System Pathology of an Organization: the rise and fall of the old Saturday Evening Post", Admin. Sci. Qtrly., Vol. 21, 1976, pp. 185-211.

_____' "Decision Making in a Complex Organization"' in England, G.W., A. Neghandi and B. Wilpert (Eds.), The Functioning of Complex Organizations, Oelgeschlager, Gunn and Hain, Cambridge, MA, 1981, Ch. 5, pp. 111-144.

_____, and Menzies, William. "A Corporate System Model of a Sports Club: Using Simulation as an Aid to Policy Making in a Crisis", Management Science, Vol. 29, 1983, pp. 52-64.

_____, and Lai, Daniel. Magazine Publishing Game and Research Tool Operation Manual, Working Document, Faculty of Management, University of Manitoba, Winnipeg, Canada, 1984.

_____, and Malcolm Munro, "Corporate System Modeling as an Aid to Defining Critical Success Factors." Working Paper, Faculty of Management, University of Manitoba, 1986.

Kahneman, D., Slovic, P. and Tversky, A., Judgment under Uncertainty: Heuristics and Biases. Cambridge University Press, Cambridge (U.K.), 1982.

Lindblom, C.E., The Policy Making Process. Prentice-Hall, Englewood Cliffs, N.J.,1968.

Mohr, Lawrence B., Explaining Organizational Behavior. Jossey-Bass, San Francisco, 1982.

Mumford, Enid and Pettigrew, Andrew M., Implementing Strategic Decisions. Longman, London, 1975.

Pettigrew, Andrew M.,The Politics of Organizational Decision Making. Tavistock, London, 1973.

Roos, Leslie, L. and Hall, Roger I. "Influence Diagrams and Organizational Power", Admin. Sci. Quart., Vol. 25, 1980, pp. 57-71.

Sterman, J. D., "Expected Formation in Behavioral Simulation Models", Behavioral Sci., Vol. 32, 1987, pp.190-211.

Turner, Barry A., "The Organizational and Interorganizational Development of Disasters." Admin. Sci. Quart., Vol. 21, 1976, pp. 378-397.

Weick,Karl E., The Social Psychology of Organizing. Addison-Wesley, Reading, MA, 1969.

Tel: (204) 474-9709[*]
Fax: (204) 474-8851
El-Mail: RHall@ccm.UManitoba.CA

INFORMATION SYSTEMS AND OPERATIONAL RESEARCH

Geoff Lockett

Professor of Management Science and Information
Management
Manchester Business School, Booth Street West
Manchester, M15 6PB

INTRODUCTION

The history of computers and operational research have occurred over
roughly the same time, and there have been many interactions. It has
been a period of great change, where fortunes have fluctuated. In this
paper we will discuss the major developments that took place and point
out their respective roles, and differing emphases. Information
Technology is now all pervasive whereas OR is probably not. Are there
any clear reasons behind this and lessons to be learned, and what should
OR do in the future? In the next sections we will discuss the practice
and philosophy of OR, followed by a look at todays Information Systems
environment. We will then discuss where OR fits, and illustrate with
some cases. The future role of OR within the IS field will be
developed.

OR METHODS

Much has been written about Operational Research and its methodologies
(e.g. see Churchman et al (1957)). Such texts were written in the very
early days when the operational research was seen as a process which
required the use of model building (Ackoff (1962)). Typically the
phases would be:

    Formulating the problem
    Constructing the model
    Testing the model
    Deriving a solution
    Testing and controlling the solution
    Implementation of the solution

There are variations from this, but in essence they are all the same.
This is how the OR world would look like to see itself, and this was
probably the way things did happen. The technical tool kit did not
exist, as it does today, and therefore more time was spent on the non-
mathematical aspects of OR. In fact the best OR that was done mainly
involved "looking and understanding". Often taking a different view of
the operations under scrutiny helped bring out the problems clearly and
seemingly simple solutions presented themselves.

If we look at OR today it however is seen more to be controlled by mathematics than by the other disciplines. Many techniques have been developed which are freely available to many managers. Computer packages for linear programming, simulation, forecasting and critical path analysis are cheap, easy to use and do not need as many experts to operate (eg. see Lockett (1984)). We will return to this later on. If we look at OR from an Information Systems perspective we get a somewhat narrower view e.g.

"Management science emphasizes the development of normative models of decision making and management practice. Operations research focuses on mathematical techniques for optimizing selected parameters of organisations such as transport costs, inventory control, and transaction costs". (Laudon and Laudon (1983)).

"Operations research is important relative to management information systems because of the developed procedures for the analysis and computer-based solutions of many types of decision problems. The systematic approach to problem solving, use of models, and computer-based solution algorithms are generally incorporated in the decision support system component of MIS". (Davis and Olson (1985)).

These are typical of the writings of IS specialists, where OR is seen to be important but not central to the future developments in IS. In order to gain a better understanding of this position let us consider a potted history of OR and computers.

COMPUTERS AND OR

When operational research started computers were really just beginning. After the war when OR "took off" the developments in computers were tremendous, and OR made great use of the new tools - more than most other disciplines. The techniques developed could not have been applied without computers. For example linear programming and simulation in practice cannot be done without modern machines. In the forties and fifties OR departments tended to buy their own computers which were different from the ones used for basic accounting. OR lead in the use of modern technology and often had the best computer people within the organisation. However OR had little to do with the accounting machines. An excellent account of the history of one Operational Researcher's involvement with computers can be found in Ranyard (1988), and will not be repeated here. OR models tended to be stand alone, intermittent and not needed every day. There are some exceptions e.g. stock control. Here OR promised a lot with EOQ type models, but the facts are that in general it has been a failure (e.g. see Lockett and Muhlemann (1978)) and the models are rarely used. So OR has been used for important one-off problems rather than the day-to-day computing, Examples can be shown where Operational Researchers have been heavily involved in developing new computer systems. But normally this has not been by using the OR approach.

When minis and time-sharing came along, again OR took part in their introduction, but was not really in the driving seat. Most of the applications were the old fashioned commercial ones i.e. accounting, purchasing, etc. The production of computerised invoices was not seen as an interesting innovation and OR spent more of its time on the new ideas, rather than making a series of modest innovations.

Hence when the concepts Management Information Systems (MIS) and Decision Support Systems (DSS) were developed, they came from the

446

accounting disciplines not OR. Huge computers controlling large data bases were set up which were managed by Data Processing (DP) Departments, and again OR did not really take part. OR was struggling to solve large simulation and LP type models of ever greater complexity. Gradually OR did not become involved in strategic problems - and helped more on tactical activities. That is not to say that they were not important, but OR became less involved in the main day to day computational operations. In fact sometimes tensions arose, because of the large and different computational needs of OR compared to the transaction processing of day to day commercial activities. The two did not always mix. OR was seen as a large computer user rather than a computer developer.

This trend has continued for a long time. With the advent of the ubiquitous micro it has reached its limit - OR no longer needs the mainframe - it can stand alone. We all know of the tremendous achievements that have been possible because of the PC. Many OR departments have helped introduce them to organisations. But the PC has been seen as the logical development of miniaturization - a mainframe at ones fingertips. A recent article by James (1988) delights in its introduction and states "Today the mainframe is moving towards the periphery in the OR teaching context - long live the micro!". In essence he is talking about using computers to help OR, rather than the broader use in companies. Again the present narrow focus of OR is apparent and the developments in Information Systems seem of little interest.

However, one result of the PC revolution is that in many cases what was once OR territory has now been overtaken by other disciplines. Linear Programming packages make application easy and most managers have been on a course showing them what it is about. The era of the self starter has arrived (Lockett (1984)) and OR is having to move to other fields. Therefore OR may not totally benefit from the introduction of the PC, and in one sense it may be a cul-de-sac. Let us now concentrate on the general development in Information Systems.

INFORMATION SYSTEMS ENVIRONMENT

There is no doubt that organisations are undergoing a period of tremendous change and information systems are an integral part of that process. It is not the place to discuss the real cause which is probably the new communication technologies eg. see Keen (1986) but just to look at what is happening. One of the first fruits of the new technologies including information technology, was the gradual automation of many jobs. This led to a loss of some jobs which were mainly of the manual type, and a period of work force reduction took place. Computers were a major contribution to that process. Lately we have seen similar changes in the office where word processing has gradually increased the efficiency of secretaries. These trends are still continuing, but at a much slower pace. In parallel to these two effects, business have been applying the information and communication technologies to the main functional areas e.g. accounting, marketing, production, etc. Using the power of PC's, the new networking systems, and the ever increasing mainframe capacities, a complex interlocking net of computer systems can now be put in place. There are three driving forces behind these; they are

1. Shorter Time Horizons
2. Move to Globalisation
3. Changing Managerial Roles

Today most organisations have less time to spare, lead times for delivery are shorter, and communication easier. It is easy to hop on to a plane in the UK, visit the USA for a meeting, get a decision and action and be back at work in a day or two. The speed of action is increasing. In parallel, many organisations (even small ones) see themselves as international and have plants and associates in other countries. We are all about to witness the changes in 1992. Most of the multi-nationals will be unaffected, it is the medium and small companies that will have to face the new forms of competition.

The third and maybe more important effect is one of changing managerial roles. Information systems technology now allows us to design organisations which require fewer managers. The automation of parts of management is occurring. This is slowly taking effect but with much more difficulty than the other developments. IS gives us great potential, but it is not easy to profit from that advantage. We are now faced with a choice of many ways of exploiting information opportunities, and the best course of action is not easy to isolate. The very success of IT gives most organisations tremendous future IS problems. They are not insurmountable, but will require much more detailed analysis than has been applied to computer systems previously.

Unfortunately OR has not had a seat at this particular table. It took little or no interest and is only just beginning to realise the effect information systems are having on organisations. Computers are at last coming into their own - not because of accounting, not because of OR, but because they are being given back to the users. Where can OR fit in this process? Before considering this question, let us discuss two case sketches to bring to light the points hinted at above.

SOME CASES

The first case concerns an international chemical company. A full description is given in Lockett and Palmer (1989). The organisation had grown rapidly for over a decade and had bought and sold a number of companies, reuslting in a complex set of plants and procedures. Over 70% of the output crossed at least an international frontier. The computers varied with over four different manufacturers, and each country had its own systems and ways of working. It was decided to put in one commercial system for the whole of the European division. This would enable the orders to be processed by one system and common stocks to be utilised. It would mean the company acting as a single entity instead of separate organisations and the operations of the company would be transformed. The problem of deciding which way forward is reminiscent of the original ones tackled by OR i.e. a systematic and logical approach etc. In this instance there is also the added problem of differing cultures and styles. How do you get agreement to replace three other systems - buy one new one? It is the sort of problem that should appeal to OR, an overall systems problem in practice. Unfortunately OR in the company is still playing with its micros and does not see the real interesting problems that are facing information systems. In this instance a series of 13 minis have been connected on-line across Europe. A large mainframe is the central point keeping the overall data base, and each mini deals with local customers. The system has the potential to radically alter the way the business is done.

Market data is now readily available so that advantage can be taken of the volatile price changes that take place. There isn't the space here to discuss the case in detail, and all we can highlight is the complexity of the problem - which classically would have been OR's. It, of all the disciplines should have the conceptual tool kit. At present it doesn't seem to be part of the leading edge changes. Micros are not the only future, systems are - and they are getting more complex.

The second case is much smaller and involves a company supplying the airlines with fuel. It only has a few standard products, but has to deliver on a world wide basis. At present it has a series of computer systems which do not interact and it takes a long time to get the overall management information together. The systems are sometimes run in house, in a bureau, or by other parts of the organisation. A new managing director decided that he would like to have one common system which would give him accurate and timely management information. Which is the best way to provide for his needs? Again this cries out for the OR approach. There may even be some mathematical modelling to be done, but in the event the OR profession was not called in. However, a system was chosen, developed and implemented. The list of options and associated pluses and minuses was large. Some small scale decision diagrams greatly assisted the decision making process, and in the end a system of regional machines has been introduced which use package switching systems to interact. It is cheap, easy to use and efficient. They have developed a world wide computer system which is impressive, and has changed the organisation.

The two case sketches are an attempt to illustrate some of the exciting IS possibilities. Many organisations face difficult choices and need all the help they can obtain if the systems are to be successful. As well as those above, there are many failures. Changing organisations and the relationship to IS is a complex and evolving problem.

BACK TO THE FUTURE

The two illustrations above clearly show some of the complexity that is becoming inherent in IS. With the new technologies the list of possible choices is growing and it is becoming harder to predict eventual success. The way organisations are behaving is being greatly influenced by IS - back to operation. Originally that is what OR did - go look at the operation, and from that build a model describing what was happening. Some researchers in the early days actually looked at these types of problems eg. Amiry (1967) and Ranyard (1968) and were ahead of their time. Mathematics only came later, and as we have seen a lot of the methods are no longer owned by OR. We have been through that phase and maybe the time is for OR to get back to its roots (Ackoff (1979)). There is still a crying need for a group of people to be able to look coldly at the operation of the organisation, and predict the outcome of certain proposed changes. Information Systems fit squarely into OR methodology.

Data Processing did not really need much OR to help it along. MIS began to use bits of OR, but the interaction was not large. We then gradually became aware of the value of information rather than data, and the IT age was born. Companies are still grappling with the concepts, and how to gain advantage from their information. Much of the effort has gone into designing systems that are "user-friendly", which managers, who take decisions, will want to use. However, very little work has gone into the design of such systems. They are still developed using old fashioned methods, with very fancy presentation e.g. graphics. There are some new methods which enable us to get at the information quicker e.g. data bases. The use of prototypes is gaining favour at last eg. Kraushaal (1985), whereas OR has always built models and tested them - in essence made prototypes. But, in general, the hidden assumption is that all you have to do is provide the information. What has not been thought about is <u>communication</u>. How do you get managers to act on the information? In the next decade we will be given tools of enormous power, but will we be able to make the most of them. IS is crying out for some new ideas in this area, and companies want systems implemented. They are becoming dependent on their introduction. OR could be of tremendous help if it decided to take the subject seriously, because the new problems are organisational, technical and logistical conundrums. They will become increasingly complex and hard to disentangle. The original methodological approach of OR would seem to hold out great hope if it were to be applied to this area. Using our phases of problem solution will be of great benefit in deciding what are the real problems and possible solutions in designing and implementing information systems in the future.

REFERENCES

Ackoff, R.L., 1962, Scientific Method: Optimizing Applied Research Decisions, Wiley, New York.

Ackoff, R.L., 1979, The Future of Operational Research is Past, JORS, Vol. 30, No. 2, pp. 189-200.

Aimiry, P.A., 1967, The Simulation of Information Flow in a Steel Making Plant, Proceedings of the NATO Conference, 'Digital Simulation in Operations Research', p. 157-165.

Churchman, C.W., Ackoff, R.L., and Arnoff, E.L., 1957, Introduction to Operations Research, Wiley, New York.

James, J., 1988, Teaching Operational Research Using IBM-PC Compatibles, JORS, Vol. 39, No. 8, August, pp. 787.

Keen, P., 1986, Competing in Time, Ballinger Publishing Co., Cambridge, Massachusetts.

Kraushaal, J.M., and Swirland, L.E., 1985, A Prototyping Method for Applications Development by End Users and Information Systems Specialists, MIS Quarterly, September.

Lockett, A.G., 1984, Self Starters in MS/OR - A Growth Area, Interfaces 14, May-June, pp. 59-61.

Lockett, A.G., and Muhlemann, A.P., 1978, The Use of Formal Inventory Control Models: A Preliminary Survey, OMEGA, Vol. 6, No. 3, pp. 227-230.

Lockett, A.G., and Palmer, B.M., 1989, Organising For Chance: Managing an IS Project Across Countries, in "Competitive Advantage and IT", Ed. H. Thomas, A. Huff, Wiley, New York.

Ranyard, J.C., 1968, The Value of Different Methods of Providing Information in Colliery Control Centres, ORQ 19, pp. 61-62.

Ranyard, J.C., 1988, A History of OR and Computing, JORS, Vol. 39, No. 12, December.

OPERATIONAL RESEARCH APPLICATIONS

IN INFORMATION SYSTEM DESIGN

R Bandyopadhyay

National Institute of Bank Management
Pune 22
India

INTRODUCTION

Inspite of significant growth in literature on information system
a well structured methodology for design incorporating both behavioral
and structural aspects is still not available. An attempt here is made
to provide a framework for such design methodology. In what follows
we shall make a quick review of the salient aspects dealt with in
literature in respect of information systems for organisational decision
making. On the basis of the review we shall attempt to develop a
framework for design of information system. Such a framework we hope
will also be useful in designing MIS or DSS in an organisation. We
shall also discuss briefly two case studies where the framework was
successfully applied.

REVIEW OF LITERATURE

One can broadly classify the literature on information and information system into three schools of thought (Bandyopadhyay, 1977).
a)   the classical school of thought
b)   the management science school of thought
c)   the behavioral science school of thought

Classical school - Proponents of the classical school tend to apply the
concepts and measures of statistical communication theory to the information problems of the organisation (Cherry, 1966). The amount of information in a message or observation is defined as the unexpectedness
or surprise value. This is measured by negative entropy. These
concepts and ideas of classical school are relevant only in respect of
storage and channel capacity. However, the basic weakness of classical
theory from our point of view is its inability to link the concepts to
the needs of organisational decision making.

Management science school - The management science school of thought
deals with formal mathematical models (predictive and normative) of infor-
mation-decision systems of an organisation. This school of thought can
be subdivided into four sub schools of thought - (a) system science and
cybernetics sub school (b) operational research sub school (c) economics
sub school and (d) accountancy sub school. Scholars of system science
and cybernetics sub school study the information problems in an

organisation in total systems terms. Basically an information-decis-
ion system within an organisation is a multi level system. Apart from
stressing flow aspects a few authors in this sub school discuss the
problems of information system design (Rappaport, 1970). Value and
cost aspects of information are not considered by this sub school.

In O.R. sub school Ackoff (1967) makes significant contribution
towards a better design. He suggests classification of decision and
prescribing information system from such classification. He divides the
decisions as (a) modellable and solution obtainable, (b) modellable but
solutions not known and (c) non-modellable. In cases (a) and (b)
the models prescribe the information needs. Case (c) becomes the case
of unstructured or illstructured problems. Such problems have to be
resolved through partial formalisation and use of judgement and intuition
(Bandyopadhyay, 1975). Various useful concepts like timeliness, accuracy and relevance required for adequate analysis of informational
problems have been discussed by scholars of this school (for details
see Bandyopadhyay 1977). White (1969) discusses information value.
Issues of information structure and its relationship with hierarchical
decision structure are ignored by scholars of this sub school.

Authors (Marschak and Radner 1972) of the economics sub school
consider information as a commodity. Principles of market mechanism
are then applied to the problems of information economics. Monetary
cost and value aspects are brought into this type of analysis. It is
debatable whether marginal analysis of the economists can be applicable
in human decision making in an organisation in view of the large
number of interacting variables and uncertainties involved. The problems
of spatial decentralisation and cost of information flow to the centre,
have been discussed by regional economists and planners.

Scholars of the accountancy sub school are basically concerned
with following types of information problems :- (Anthony, 1965).
(a)   problems of designing financial information system, and
(b)   problems of structuring accounting data to provide information
       to various levels of the organisation for planning, control
       and operational purposes.

Some general comments regarding the contribution of the management
science school of thought may be pertinent here. In general, greater
recognition of the systemic nature of the problem of design exists among
the scholars of this group. However, aspects like decentralisation,
delegation and provision of information at various levels of organisational hierarchy have not been adequately dealt with. Behavioral
variables though recognised are not explicitly considered in the model
structure and interaction problems are often ignored.

Behavioral science school - Behavioral science school gives adequate
recognition to such behavioral variables like motivation, organisational
stress, uncertainty absorption, information ambiguity and organisational
power sharing through information. Ideas of this sub school can again
be studied under three sub schools of thought - (a) administrative
science sub school (b) behavioral theory of firm sub school (c) psychology sub school. Administrative science sub school explicitly recognises the importance of information and communication in organisational
decision making.

Scholars of behavioral theory of firm sub school direct their
analysis of information and decision systems by concentrating on such
behavioral factors as quasi-resolution of conflict, problematic search
and organisation learning. Large scale simulation studies have been
conducted. (Bandyopadhyay, 1977, March, 1988).

452

Studies of the psychology sub school have been concerned with observing the decision making processes of individual decision makers in complex problem solving situations under varying conditions of information availabilities, and communication net arrangements (Schroder et al 1967).

Taking the contribution of behavioral science school of thought as a whole following general remarks are in order :-

(a) Behavioral variables are explicitly considered
(b) Structural issues are inadequately dealt with
(c) Information value aspects and inter actions of decision nodes within an organisation are not considered.

Our review suggests absence of an integrative framework incorporating behavioral and structural aspects. Similar conclusion was derived in Padwal (1978).

In the next section we shall therefore attempt to develop an integrative framework.

AN INTEGRATIVE FRAMEWORK

O.R. can provide an integrative framework for information system design where both behavioral and structural aspects of information can be considered. Validity of any such framework will be assessed from the answer to the following questions.

(a) Does the framework incorporate concepts and definitions?
(b) Does the framework provide guidelines for information for better decision making in an organisation?
(c) Does it try to identify information needed at various levels of the organisation?
(d) Does the framework provide any insight or guidelines into nature of desirable channels of timely flow of adequate information to various parts of the organisation?
(e) Does the framework recognise explicitly the problem of information structure?
(f) Does it specifically concern itself with defining and measuring the informational attributes like accuracy, relevance etc. and relate these attributes to various attributes of decision making like importance, complexity, criticalilty?
(g) Does it consider various costs and value aspects of organisational decision making?

Information can be classified as state type and relation type information (White, 1969). State type information describes status of the system (past, present or future). Information that relates different states gives us more decidability. Generation of relation type information is achieved through appropriate modelling of the problem situations. Further a good taxonomic scheme will help in structuring the information and in converting potentially relevant information into relevant information. Computers as information processing tools are useful by providing relevant and timely information and by providing relation type information at decision making nodes through appropriate modelling exercises in various decision areas. A broad framework (avoiding details for economy of space) which integrates various aspects of information system design is shown in Figure 1.

CASE STUDIES

The framework developed in Figure 1 was applied in designing information system for a nationalised bank.

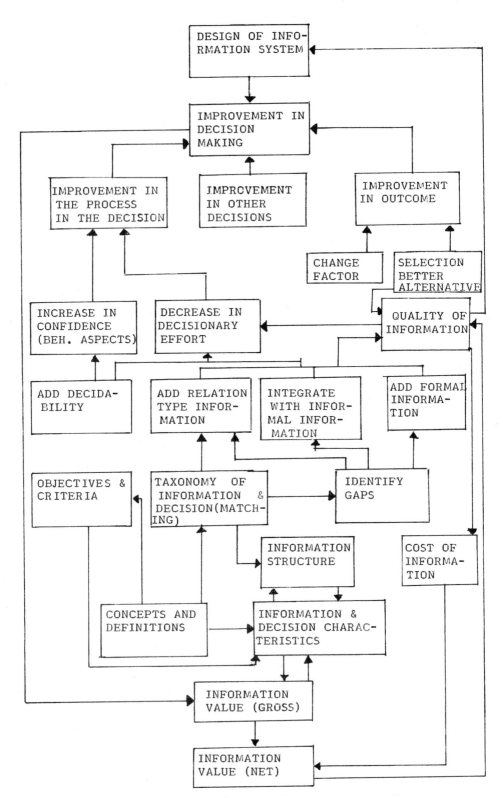

FIGURE 1 : MODEL OF INFORMATION SYSTEM DESIGN IN A ORGANISATION

454

We studied hierarchy of decision making in the organisation and classified the decisions into various homogeneous groups (functional groups like credit, personnel). Normative requirements of information for each decision node of the hierarchy were then identified in consultation with the executives concerned. The existing information system was then thoroughly studied and catalogued and classified in the same way as decision areas. Information available in a particular area and normative requirements were checked and compared and gaps if any were identified. This resulted in identification of redundant and duplicate information. Frequency of provision of information was matched with the frequency of corresponding decisions and mismatches were identified. These mismatches, redundancies and duplications were discussed with decision makers. Cost, availability and structure of new information provision were also discussed with them.

In carefully identified areas needing relational information regression, linear programming, scording and simulation models were used for providing (relation type) information for improving decision making. The organisational decision making improved as a whole when the system was implemented. The system when implemented resulted in saving in information provision as it eliminated 50% of the existing information as redundant, duplicate or irrelevant. However, the project did face certain behavioral impediments. Information in organisational context is a source of power. Redistribution of information and formalisation of informal source may create loss of power to executives. Such conflict areas became evident after our first round of discussion and was resolved through discussion based on the framework. Many informal information sources were identified and attempts were made to integrate these into the overall system.

In the second case, we are concerned with designing information system for decentralised planning. Administratively India is divided into a number of states. Each state has a number of district. Each district is divided into a number of administrative/development units called blocks. Each block consists of a number of villages (varying from 50 to 100). It is considered necessary to design planning systems at block and district level. Computers are installed for local level planning decisions. Large amount of natural resource data has been made available through (satellite) remote sensing.

Computerised Natural Resource Data Management System at the district level has been created. Socio-economic and demographic and infrastructural data for the district have been added to the data system. Thus the information system has been designed purely on the basis of state type data. Decision makers at district and block level have started questioning the utility of the system. So we have been asked to design the information system. We have identified the following inconsistencies -

(a) Decision system is unspecified.
(b) Understanding of future environment in relation to the planning requirements is lacking.
(c) No relation type information is available in the system.

Under different environmental and alternative resources availability conditions, models connecting inputs and outputs are constructed and results are discussed with the decision makers. We are involved in extensive modelling exercises to create alternatives and consequence of alternatives in respect of spatial, temporal and sectoral dimensions for allocation of water and fertilizer under different cropping patterns within a block. These are "if-then" type of models where the decision makers can take decisions after examining various consequences which have sociological, behavioral, technical and economic dimensions.

CONCLUDING REMARKS

We have made a broad overview of literature and discussed two case studies of different types. It may be seen computerised information system in itself will not improve the quality of information system in an organisation. We require a framework integrating various aspects of information and decision system and we must explicitly try to model decision situation and provide relation type information through application of O.R. approaches, such O.R. modelling exercises must try to capture the real problem situations involving behavioral, structural and economic and technological variables.

REFERENCES

Ackoff, R L, 1967, Management Misinformation System Management Science B 14/4

Anthony, R N, 1965, "Planning and Control Systems - A Framework for Analysis Graduate School of Business Administration", Boston.

Bandyopadhyay R, 1975, "Use of Taxonomies of Information and Decision - A Case Study", in the Role and Effectiveness of Theories in Decision in Practice by White, D J and Bowen, K C (Eds), Hodder and Stoughton, London.

Bandyopadhyay R, 1977, "Information for Organisational Decision Making " A Literature Review, IEEE Transactions, SMC 7/1.

Cherry, C, 1966, "On Human Communications", MIT Press, Cambridge, MB.

March, J G, 1988, "Decisions and Organisations", Basil Blackwell, Oxford.

Marschak J and Radner, R, 1972, "Economic Theory of Teams", Wiley, New York.

Padwal, S M, 1978, "Information System Design - A Few Cases", Unpublished Research Report, National Institute of Bank Management, Bombay.

Rappaport, A, 1970, "Information for Decision Making", Prentice Hall Engelwood Cliffs, N J.

Schroder, H M, Driver, M J and Strentfert, 1967, "Human Information Processing", Holt Rinehart and Winston, NewYork.

White, D J, 1969, "Decision Theory", Allen and Union, London.

A STRUCTURING APPROACH IN DESIGNING INFORMATION SYSTEMS ENVIRONMENTS

Nikitas Assimakopoulos

Agricultural Bank of Greece
54, Eratosthenous Street
GR-173 43 Athens
Greece

ABSTRACT. The Problem Structuring Methodology is presented in this paper as a Systems Approach methodology to design the operation and the management functions of an enterprise. It includes the main interfaces between the enterprise as a System and its environment. This methodology is a new way to identify in a general manner the data needed to both operate and manage the enterprise. It goes further than the usual diagrammatical presentations of subsystems, elements and individuals of a System in order to determine their positioning and role. It gives the opportunity to create system models which can be used to represent the subsystems of an integrated Ω system.

1.   INTRODUCTION

The computer information systems are the main constructs for an enterprise in order to gain a competitive edge in the market. Initially, each manual information system was simply automated usually beginning with the financial information systems. As more and more of the manual information systems were individually automated, the well defined problems of data redundancy, data dependency and data inflexibility [5] arose. The solution was the development of the database management system (DBMS) and the concept of data resourse management [6] (DRM). This was followed by such innovations as structured programming, structured design and fourth generation languages along with concepts such as decision support systems (DSS), information resourse management (IRM) and information systems planning (ISP) [4]. All of these have been classified and validated in [1] in order to improve the design and the development of information systems.
        The enterprise can also be considered as a system and the need for information to operate and manage the system made more apparent through the use of an appropriate system model. General systems models are not widely used in texts on information systems [3], [4]. The text [4] does cover general systems concepts use it primarily to show that information systems are also systems. This paper outlines a general system model using the concepts of the Problem Structuring Methodology (PSM) [2], [7] that represents the enterprise in a way that assists in identifying the needed information systems. It does not replace other design techniques but augments them by providing the designer with a new design language,

457

useful to communicate with the individuals of the enterprise, as a standard way of modelling and identifying the information needs.

2. THE MODEL OF THE SYSTEM

In this model the inputs to the system from the environment have been separated into the resources used in the system operation and the request for the output referred to as the "order". The "input" in this control model is really the "order". Two special groups in the environment are also identified. These are the "suppliers" of the input resources and the "customers/clients" who are the cause of the orders to be placed and who receive the output.

The system outputs to the environment have also been divided into the desirable ones that are used to fill the orders and the undesirable ones that cannot be avoided in the process. The former are called the "outputs" to maintain agreement with the common use of the term and the latter are called the "wastes". The wastes can have an economic and/or social cost and data about them may be needed.

The negative feedback loop of the control system is assumed to be included within the system block. The model does allow the negative feedback loop in the environment that is represented by the customers/clients placing orders and receiving output.

The operation of the control system is determined by its physical structure and the various laws or algorithms involved. The operation of the enterprise is not only defined by the physical processes involved but also by the rules, procedures, directives, etc, that establish how employees carry out their tasks. It is also determined by the operational plan (budget) and by the standards of performance and quality. It is the function of "operating" management to ensure that the actual operation conforms to the desired operation as defined by the information listed above.

Once a system is operating in a controlled manner, consideration can be given to making adjustments to the operation so as to provide the required outputs, according to the defined standards, for the minimum amount of the input resourses. That is, maximizing the efficiency of the operation which will be referred to as "optimizing" management.

Optimizing is done by investigating possible changes in the operations that would result in improved efficiency and implementing the appropriate changes.

It is conceivable that the control theory [8] and the PSM [7] could be used to design the operation to be optimal assuming there is sufficient information available about the system and its environment. This is not normally done in business systems. Therefore, this management function is included to make it easier to identify the information needs.

Optimizing management is concerned with improving the system in a static environment. As the environment is normally dynamic, the "adapting" management function is added. This function is concerned with identifying changes in the environment and then determining the changes that must be made to the system to ensure its continued operation. Ideally, the changes should be identified before they occur so that the system changes can be implemented at the best time. With this consideration the steps are :
1. Gather data about the expected environment from outside sources or by developing a forecast based on current environmental data.
2. Estimate the expected performance of the current system operation using the forecast data.
3. Prepare proposed changes to the system operation based on the results of 2 above.
4. Determine the expected performance based on implementing the proposed changes.

5. When a set of satisfactory changes in the system operation have been defined, implement them at the appropriate time.

It may be possible to make use of control theory to provide a more analytical method of selecting the new values for system parameters. However, the purpose is to identify the adapting function and then to use the resulting model to assist in identifying the required information systems.

3.   THE Ω SYSTEM OF SYSTEMS

The next stage is to consider this basic model as an element in an integrated Ω system of systems. Such an Ω system can be considered from a management or an operations point of view as it has been described in [2] and [7]. The operations view will be considered first.

From an operations point of view the systems are connected by the output of one system acting as a source of orders or input resources to another. If the rate at which the various systems operate is variable and not common, then there must be a storage element between the systems to absorb this variation. Such storage can take several forms :

1. Inventories of supplies or products.
2. Extra capacity within the system (including man-hours).
3. Storage of orders (e.g appointment schedules).

Fig. 1 illustrates a system's structure and operation using the PSM that is made up of several subsystems each with storage shown between them. The connection at the operating level may be information or products/services keeping a certain rate of flow. Information about such storage elements is required in the management of such Ω systems.

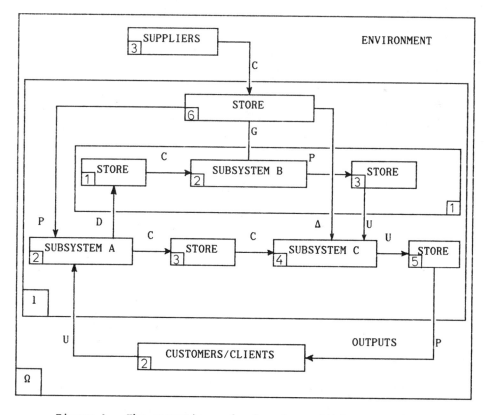

Figure 1 . The operations of subsystems within an Ω system

459

It may be needed to develop dynamic models to assist the management functions.

The management view of the Ω system will be explained through the use of an example. The organization division is one of the 20 main divisions of the Agricultural Bank of Greece. This division is consists of several departments such as Organization and Operation, Development of branch-network, Information Systems etc. The manager of each department is concerned with the operation of the department, its efficiency and its ability to adapt to the changing environment of the division. The director of the division is concenred with ensuring that each manager does his job with the global optimization and adaptation of the whole division within the changing environment of the Bank.

The Ω system is modelled from a management perspective by adding the optimizing and adapting loops around all the subsystems that make up the Ω system. Regardless of the number of levels of management, the actual operation only takes place at the lowest level.

Management responsible for optimizing or adapting, especially at global levels, can modify the system operation by one or more of the following means :
1. Changing the components used in the operation.
2. Change the rules, procedures, etc, that define the operation.
3. Modify the operation plan.
4. Modify the constraints imposed on the subsystem.

The first three result in direct action at the operating level and can be done by any level in the management hierarchy authorized to do so. The fourth means alters the environment to pressure the subsystems to modify themselves so as to achieve the desired result. Though this does not provide the direct control of the first three means, it does allow for a greater measure of freedom on the part of the subsystem management.

4. THE INFORMATION REQUIREMENTS FOR THE SYSTEM

The system operation can be defined by the set of functions that are performed in converting the input resources into the outputs required to fill the orders. While the actual data required to support the operation depends upon the details of functions performed, the following approach can be used to identify that data :
1. Determine the functions that are carried out to fill the orders.
2. Identify the data required to carry out each function.
3. Identify any additional data that results from the completion of any function.
4. Determine what portion of the required data can be captured when the order is placed (typically about the output required and about the customer).
5. Determine what portion of the remaining data is provided in completing some function and what must be provided by supporting files.
6. Determine the magnitude of the order transaction rate and its variation.
7. Determine the volume, as well as the source of modifications, for each file of supporting data.

With the above knowledge of the system operation and its data requirements, the conceptual data base can be designed. Then using this design, the necessary input forms/screens and report form/screens developed. The use of a fourth generation language, prototyping methodology and the PSM can be used effectively at this point.

The information required to manage the operation can be summarized as follows :
1. Information about deviations from the rules, procedures, etc.

2. Information about deviations from standards of function performance and output quality.
3. Information about deviations from the operating plan.

Now, the information for optimizing management function must be taken into account in order to maximise the efficiency of the system operation for a given environment. That is, using the minimum quantity of input resources to provide a unit of output. This requires a model of the operation that relates the use of the resources in each function employed in producing each type of output. Such a model is developed from data about resource consumption relative to a unit of output for each function involved.

Automated information systems that support operating management provide data about the total quantities of resources used and output provided over some period of time but seldom in detail by function. The detailed data must be obtained through special studies conducted from time to time. Data about the quantity of items in each storage element is also used in efficiency determination.

Data about the actual orders is used as input to the model to test possible changes in the operation that would improve efficiency. Information used to suggest possible changes comes mainly from the manager's knowledge of the system and its constraints.

The management function which is concerned with modifying the system to ensure its continued operation in a changing environment is the information for adapting management. It requires information about the system itself plus information about the customers/clients, the suppliers, the constraints imposed by higher levels of control, emerging technology and competitors.

Assuming the system is functioning satisfactorally in the present environment, then information is required to provide a forecast of the future environment. This, along with the data required to develop the system level and the information needed to propose system changes to test, is all that is required.

Information about the future environment can be obtained from those who specialize in providing such data or the system can prepare such information itself using its own techniques. Information used to suggest possible changes comes mainly from the manager's knowledge of the system and its environment. The model is developed through the use of operating data plus that required to establish relationships with the environmental factors.

Management responsible for optimizing or adapting, especially at global levels, can modify the system operation by one or more of the following :
1. Change the components used in the operation.
2. Change the rules, procedures, etc, that define the operation.
3. Modify the operating plan.
4. Modify the constraints imposed on the subsystem.

The first three result in direct action at the operating level and can be done by any level in the management hierarchy authorized to do so. The fourth approach alters the environment to pressure the subsystems to modify themselves so as to achieve the desired result. Though this data does not provide the direct control of the first three means, it does allow for a greater measure of freedom on the part of the subsystem management.

The functions identified so far have only implied innovation. At any management level it is possible to be innovative or "original" in identifying ways in which to optimize or adapt. It is also possible for the system as a whole to be innovative by modifying the goals or objectives of the system so as to better meet the changing needs of the customers/clients.

It is difficult to define even in a general way the information required to support innovation. The information required to adapt the system is a beginning, but the inspiration may be triggered by some characteristic of a totally unrelated system. Perhaps when innovation is better understood, then the information needs can be better defined.

5.   CONCLUSIONS

The literature includes many ways of looking at the operation, control and evolution of enterprises. In fact the list continues to grow. The approach of this paper was to present a simple model that would assist in identifying the information needs of the enterprise with the design of PSM. It does not deal with many important management needs as people skills, leadership, etc.

One area that has an impact on the information system needs of the enterprise has not been included in this paper. This is the information systems required to support the enterprise as a customer. Much of the data required is available as it supports other functions and the remainder can be clearly identified. However, to explain how the model is expanded to include this function and show how it is integrated into a system operation as well as the optimizing and adapting functions would cause the paper to be too long. Accordingly, this topic will be covered in another paper.

The model and approach outlined has been used successfully in a limited number of applications in the Agricultural Bank of Greece. It is offered as a good tool to the system analyst/designer that may help him and his client in understanding a new system in a dynamic environment.

REFERENCES

1. Assimakopoulos, N., 1988, Management Methodologies for Computer Information Systems. Resource Management and Optimization. To appear.
2. Assimakopoulos, N., 1988, The Routing and Cost of the Information Flow in a System. Systems Practice, 1(3): 297-303.
3. Burch, J., Strater, F. and Grudnitski, G., 1983, "Information Systems : Theory and Practice," John Wiley & Sons, New York.
4. Davis, G.B. and Olsen, M.H., 1985, "Management Information Systems," McGraw-Hill, New York.
5. McFadden, F.R. and Hoffer, J.A., 1985, "Data Base Management," Benjamin, Menlo Park.
6. Nolan, R.L., 1973, Computer Data Bases : the future is now. Harvard Business Review, Sept-Oct., 98-111.
7. Panayotopoulos, A. and Assimakopoulos, N., 1987, Problem Structuring in a hospital, Eur. Jour. of O.R., 29(2): 135-143.
8. Sage, A. and White, C., 1977, "Optimum System Control," 2nd. Ed., Prentice-Hall, Englewood Cliffs.

OR AND RULE-BASED EXPERT SYSTEMS

S.R.K. Prasad and R. Prabhakar

Kasbah Systems Software
India

INTRODUCTION

Although since early 1950's many theoretical OR models and techniques have been developed, only a few of these are continuously used in actual practice. Till the 1970's OR analysts and practitioners were still concerned mainly with general methods (to improve representation and to search efficiently) and to use them to create specialized computer programs. The basic idea and aim of improved improved representation was to find useful formulations of problems that are feasible to solve. The aim of the efficient search process was to minimize computer time and capacity in the quest for obtaining useful and meaningful solutions to complex real-world problems, by controlling the sifting of feasible solutions.

The experiences of OR practitioners over the years, brought about the realization that specialized techniques would be used, with high quality, specialized and specific knowledge to create specialized computer programs. These programs were expert in some narrow problem area and were called expert systems. At first, although designing and building an expert system was considered more an artistic endeavor than a scientific enterprise, now, however the process is better understood and more clearly designed.

CONCEPT OF EXPERT SYSTEMS

The process of building an expert system is often called knowledge engineering. It typically involves a special form of interaction between the expert system builder, called the knowledge engineer, and one or more human experts in some problem area. Knowledge engineering relies heavily on the study of human experts in order to develop intelligent, skilled programs as pointed out Hayes-Roth (1983) in the book Building Expert Systems.

Features of an Expert System

The Heart of an expert system is the powerful corpus of knowledge that accumulates during system building. The most useful feature of an expert system is high level expertise it provides to aid in problem solving. Another useful feature of an expert system is its Predictive Modeling Power. The corpus of knowledge that defines the proficiency of an expert system can also provide an additional feature, an institutional memory.

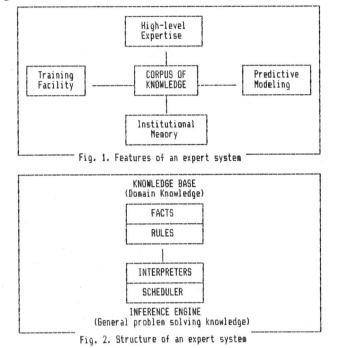

Fig. 1. Features of an expert system

Fig. 2. Structure of an expert system

A final feature of an expert system is its activity to provide a training facility for key personnel and important staff members. Fig.1 depicts the general features of an expert system.

Organization and Structure of Expert Systems

AI scientists use the term knowledge to mean the information a computer program needs before it can behave intelligently. The knowledge in an expert system is organized in a way that separates the knowledge about the problem domain from the system's other knowledge, such as general knowledge about how to solve problems or knowledge about how to interact with the user. This collection of domain knowledge is called the knowledge base, while the general problem solving knowledge is called the inference engine. A program with knowledge organized this way is called a knowledge based system. Fig. 2 depicts the structure of an expert system.

Rule-Based Expert Systems

The three basic components of a rule-based expert system are (a) a rule interpreter, (b) a set of production rules, and (c) a database. In expert systems jargon, the term rule refers to the most popular type of knowledge representation technique, the rule based representation.

When a rule is used, its actions make changes to the internal database, which contain the system's decisions or deductions. The process of trying rules and taking actions can be compared to reasoning, and explanations require displays of how the rules use the information provided by the user to make various intermediate deductions and finally to arrive at the answer.

Rules are expressed as IF-THEN statements. An example is

i)   If a flammable liquid is spilled, call the fire department.
ii) If the pH of the spill is less than 6, the spill material is an acid.
iii) If the spill material is an acid, and the spill smells like vinegar, the spill material is acetic acid.

These are rules that might exist in a crisis management expert system for containing oil and chemical spills as explained in Building Expert Systems by Hayes-Roth, Waterman and Lenat (1983).

Fig.3 shows a rule based consultation system with explanation capability.

The purpose of an explanation capability is to give the user access to as much of the system's knowledge as possible.

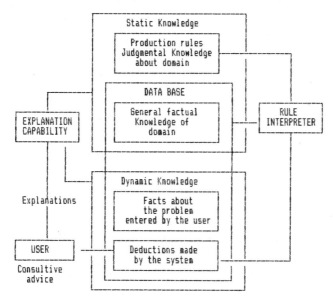

Fig.3. Rule-based consultation system

The database is made up of general facts, about the system's domain of expertise, facts that the user enters about a specific problem, and deductions made about the problem by the system's rules. These deductions from the basis of the system's consultative advice. The explanation capability makes the use of the system's knowledge base to give the users explanations. This knowledge base is made up of static domain specific knowledge (both factual and judgmental) and dynamic knowledge specific to a particular problem.

AN EXPERT SYSTEM AS AN OR TEACHING TOOL

Each expert system is designed to solve problems from some area. In the past, universities all over the world have required students to write run and interpret the output of programs for micro computers which implement various OR models and techniques. In recent times these courses are now requiring students to work in teams to develop full fledged expert systems. The team must weave together programes written to implement various OR models to form, eventually an expert system. This final system is to have the capability of dealing with a variety of problems from one area of OR.

Some examples of the common prototype expert systems which are used in actual OR laboratory courses are SCHEDULE for machine scheduling developed by Lamatsch, et.al (to appear) and Rubach in his report WIOR-241 (1985), PROJECT PLAN for planning, scheduling and controlling of large projects which comprise numerous inter related activities as explained by Elmaghraby (1977). The aim of these courses commonly include the provision of:

(i)   a working knowledge of micro computer systems.
(ii)   Specialist skills in handling data structures and other computer programming skills.
(iii)   experience in implementing certain OR models and techniques.
(iv)   the ability to work as part of a team and
(v)   familiarity with the basic concepts of expert systems.

These prototype expert systems which are used in actual OR laboratory courses attempt to achieve these objectives.

CATALOGUE OF EXPERT SYSTEMS

A catalogue containing summaries of selected expert systems from application areas like Computer systems, Electronics, Information Management, Manufacturing, Military Science, Process Control is presented. This catalogue is only a small portion of the many expert systems that have been developed and in each application area one example has been cited.

(i) ISA schedules customer computer system orders against the current and planned material allocations. ISA takes customer orders, including changes and cancellations, and produces a schedule date for each order. It displays additional information about a problematic order, including the schedule proposed, the problems uncovered and alternative proposals for scheduling the order. ISA contains significant scheduling expertise, such as knowledge about order cancellation probabilities, material availability, and strategies for relaxing scheduling constraints. ISA is a forward chaining rule-based system implemented in OPS5. It was developed by Digital Equipment Corporation and is in use at DEC's manufacturing plants. It reached the stage of a commercial system. (Intelligent Scheduling Assistant) (O'Connor (1984) and Orcinch and Frost (1984)).

(ii)   CADHELP simulates an expert demonstrating the operation of the graphical features of a computer-aided design (CAD) subsystem for designing digital logic circuits. It explains to the user how to use the CAD subsystem, tailoring its explanations to fit the needs and desires of the user. It provides explanations in English when the user makes an error or asks for help and generates the test of the explanation from its knowledge base in a dynamic way, relying on scripts associated with the different features of the CAD sub system. As the user becomes more experienced, the explanations created by CADHELP become more terse. Knowledge in the system is organized as a set of cooperating subsystems or "experts" controlled by a higher level task manager program. The system is implemented in FRANZ LISP and runs on a DEC VAX 11/780 under UNIX, CADHELP was developed at the University of Connecticut and reached the stage of a research prototype. (Computer-Aided Design HELP) Cullingford et. al (1982).

(iii) FOLIO helps portfolio managers determine client investment goals and select portfolios that best meet these goals. The system determines the client's needs during an interview and then recommends percentages of each fund that provide an optimum fit to the client's goals. FOLIO recognizes a small number of classes of securities (e.g. dividend-oriented, lower-risk stocks and commodity-sensitive, higher-risk stocks) and maintains aggregate knowledge about the properties (e.g., rate of return) of the securities in each class. The system uses a forward chaining, rule-based representation scheme to infer client goals and a linear programming scheme to maximize the fit between the goals and the portfolio. FOLIO is implemented in MRS. It was developed at Stanford University and reached the stage of a demonstration prototype.   Cohen and Lieberman (1983).

(iv)   ISIS constructs factory job shop schedules. The system selects a sequence of operations needed to complete an order, determines start and end times, and assigns resources to each operation. It can also act as an intelligent assistant, using its expertise to help plant schedulers maintain schedule consistency and identify decisions that result in unsatisfied constraints. Knowledge in the system includes organizational goals such as due dates and costs, physical constraints such as limitations of particular machines, and casual constraints such as the order in which operations must be performed. ISIS uses a frame-based knowledge representation scheme together with rules for resolving conflicting constraints. The system is implemented in SRL. It was developed at Cornegie-Mellon University and tested in the context of Westinghouse Electric Corporation turbine component plant. ISIS reached the stage of a research prototype. (Intelligent Scheduling and Information System). Fox and Smith (1984).
(v)   AMUID assists military commanders with land battlefield analysis. The system integrates information from intelligence reports, infrared and visual imaging sensors, and MTI(moving target indicator) radar. AMUID classifies targets and organize them into higher level units (e.g.battalions and regiments). It provides real-time analysis and situation updating as data arrive continuously over time. AMUID'S expertise is encoded as rules which operate on domain knowledge (e.g. types of military equipment, deployment patterns for military units, and tactics) maintained in a semantic net. Certainty factors are employed to handle the uncertainty typically involved in the analysis of

sensor data. The control structure is event driven, where
events are new sensor reports, major decisions made by the
system itself, or user queries. AMUID was developed at Advanced
Information & Decision Systems and reached the stage of
demonstration prototype. (Automated Multisensor Unit
Identification system) DraZovich (1983).

(vi) FALCON identifies probable causes of process disturbances
in a chemical process plant by interpreting data consisting of
numerical values from gauges and the status of alarms and
switches. The system interprets the data by using knowledge of
the effects induced by a fault in a given component and how
disturbances in the input of a component will lead to
disturbances in the output. Knowledge is represented in two
ways as a set of rules controlled by forward chaining and as a
casual model in network form. The system is implemented in LISP
and was developed at the University of Delaware. It reached the
stage of a demonstration prototype. Chestu et. al (1984).

REFERENCES

Cohen, P. and Lieberman, M.D., 1983, A report on FOLIO : An
    expert assistant for portfolio managers. Proceedings
    IJCAI-83, pp.212-214.
Chester, D., Lamb, D., and Dhujati,P, March 1984 Rule-based
    computer alarm analysis in chemical process plants.
    Proceedings of the seventh Annual Conference on
    Computer Technology , MICRO-DELCON 84, IEEE , pp.22-29.
Cullingford, R.E., Krueger, M.W.,Selfridge, M. and
    Bienkowski, M.A., April 1982, Automated explanations as a
    component of a computer-aided design system. IEEE
    Transactions on Systems, Man, and Cybernetics,
    vol.SMC-12. no.2, pp.168-181.
Drazovich, R.J, October, 1983, Sensor fusion in tactical
    warfare. AIAA Computers in Aerospace IV Conference.
S.E.Elamaghraby, 1977, Activity Networks :" Project Planning
    and Control by Network Models ", John Wiley, New York.
Fox, M.S. and Smith,S.F., 1984, ISIS : a knowledge-based
    system for factory scheduling. Expert Systems,
    vol.1, no.1.
Hayes-Roth, F.Waterman, D.A., and Lenat, D.(eds), 1983
    "Building Expert Systems", Addison-Wesley.
Hayes-Roth, F.Waterman, D.A., and Lenat, D., 1983, An
    overview of expert systems. In Hayes-Roth, Waterman
    and Lenat (eds), "Building Expert Systems", Addison-
    Wesley.
H.Hoffmann and A.Lamatsch, 1986, ProgrammSystem
    Projektplanung Vorstufe eines Expertsystems, Report
    WIOR-277, Institutfur Wirtschaftstheorie and
    Operations Research, Universitat of Karlsruhe.
O'Connor, D.E., 1984, "Using Expert systems to manage change
    and complexity in manufacturing". In W.Reitman(ed)
    Artificial Intelligence Applications for Business,
    Norwood,N.J. : Ablex.
Orciuch, D. and Frost, J., December 1984, ISA : an intelligent
    scheduling assistant. Proceedings of the First
    Conference on Artificial Intelligent Applications for
    Business, IEEE Computer Society.
T.Rubach, 1985, Programmabiliothek and Expertensystem zur
    Mashinenbelegungsplanung, Report WIOR-241, Institufur Wir-
    tschaftstheorieund Operations Research, Univ of Karlsruhe.

PROBLEMS OF MEASUREMENT

PROBLEMS OF MEASUREMENT

John Mingers

School of Industrial and Business Studies
University of Warwick
Coventry CV4 7AL
England
0203 523523
bsrcd@sky.warwick.ac.uk

INTRODUCTION

Operational Research can be seen as an approach towards intervening in human organizations in a purposeful and rational manner in order to improve the workings of such organizations. Other disciplines would also lay claim to such a broad definition, e.g,. organizational behaviour, marketing and strategic management, systems, organizational development, management consultancy.

The distinctive nature of OR lies, I would suggest, in its historical development from science, in particular hard sciences, and its concomitant emphasis on an empirical base. Its classical modus operandi has been to observe and measure various aspects of the world and then use mathematical and statistical techniques on the resultant data. Of course, most Operational Researchers would argue that OR was much more than quantification and mathematical manipulation, and increasingly it has been moving towards the realization that the softer aspects of the human world must be more explicitly recognised. Yet empirical measurement and the collection of data is still the fuel of OR, both in practice and, if one is to judge the subject by its writings, in terms of textbooks and journals.

One might expect therefore that such a vital input to a practical subject would have a large literature in its own right, and its problems and pitfalls would be the subject of chapters in textbooks and lectures on courses. In fact however data turns out to be a rather ghostly substance as far as the academic and educational spheres of OR are concerned.

In a small survey of introductory OR, business statistics and quantitative methods textbooks in the university library,

I found much data and many techniques to use on it. However, almost without exception, such data was given and assumed to be correct, and there was not even any discussion of possible inaccuracy or error, let alone whole chapters on the problems of measurement or the philosophy of empiricism. In fact, out of 47 such textbooks only five (Cuming, 1984; Churchman et al, 1957; White, 1985; Davies and Yoder, 1941; Neter and Wasserman, 1966) had significant chapters, or sections of chapters, concerning the problems of measurement (as opposed to merely describing sources of data), and a further eleven made a passing reference. Thus nearly 70% made *no* mention of such problems at all. The International Abstracts in OR has a category entitled "Measurement" which recorded a total of 80 papers from 1981-1986 inclusive, around 11 per year. Hardly an impressive world-wide total for such an important subject. My personal experience in education, both as student and lecturer, is that generally OR and Business Studies courses remain equally silent on such matters.

In contrast, my experience as a practitioner was similar to Graham (1982), that data was perennially a problem - what should one measure? Was it possible to measure it? How reliable was the data already available? What to do about the unquantifiable aspects? What about the vital data that was missing? In short, data and measurement do involve a lot of problems.

The rest of this paper will provide an overview of the different levels at which problems can occur - both philosophical and methodological - and the pragmatics of response to these problems. Generally it will only be able to ask questions and prompt further work, rather than provide solutions.

PHILOSOPHY - THE OBJECTIVITY OF DATA MEASUREMENT

To begin, a definition of measurement in the abstract (Phanzagl, 1959):

"The general aim of measurement is to map a set M on the set of real numbers in such a way that - to the greatest possible extent - conclusions concerning relations between elements of M can be drawn from corresponding relations between their assigned numbers."

Within OR, the philosophical underpinnings of empirical research are rarely discussed and yet within social science frequent debates and disputes indicate major differences about the very possibility of valid empirical work. See for example Hughes (1980), Irvine (1979), Pratt (1978), Morison (1986) and McNeill (1985) for general surveys.

I shall begin this summary of a very complex area by outlining two alternative paradigms, the objectivist (positivist) and the subjectivist (interpretive) (Burrell and Morgan, 1979). These differ both in terms of ontology - what is assumed to exist in the world, and epistemology - how we gain valid knowledge of the world.

472

For the positivist, the world (both physical and social) consist of objects and entities which can and should be observed and wherever possible measured. These objects are causally related to each other. Scientific knowledge rests on such observation. In principle everything observable can be measured, and the results used in models which mimic reality. Data is ready and waiting to be "collected", and provided that this process is done correctly, the results will be *objective* – they will reflect the world as it actually is, and be independent of the observer collecting the data.

A major critique of this viewpoint has been mounted by interpretivists. They maintain that the social world in particular is not objectively given to us, but is a product – an active creation – of our interactions with each other. Human actions are intrinsically meaningful. Our actions have meaning for ourselves, and we have to attribute meaning and interpretation to the actions of others. From this viewpoint, there is no objective world waiting to be measured and portrayed in collected data. Rather, observers bring with them a structure of meanings, beliefs, theories and methods which they then impose on the world in a *process of production* of data. The results reflect this process as much as, or even more than, they do the real world. Many studies, particularly about official statistics concerning, for example, crime, health, wealth and unemployment figures, show the extent to which the results depend on the perceptions, definitions, practices and resources put into their production.

Within this paradigm there are significantly different standpoints (Mingers, 1984). Phenomenologists, such as Schutz (1967), whose focus of analysis is the individual consciousness; ethnomethodologists, such as Cicourel (1964), who examine the structured practices of groups of actors; and Marxists, for example Hindess (1975), who would claim that practices of measurement are not purely arbitrary, but are determined by the nature of (capitalist) society.

I would argue that, by default if nothing else, OR assumes the objectivist standpoint. The very lack of attention outlined in the introduction is itself testimony to a belief, ay a philosophical level, in the unproblematic nature of data *collection*. In contrast, both experience and logic suggest to me that the interpretive view of data *production* is much more realistic. Its implications need to be seriously considered by Operational Researchers.

METHODOLOGY – PROBLEMS IN THE PRODUCTION OF DATA

This section moves from the philosophical level to consider the types of practical problems encountered in the production of reliable and valid empirical measurements. Such problems are encountered firstly in the very *possibility* of measuring certain factors, secondly in assessing the *validity* of such measurements and finally in terms of their *reliability*.

Possibility of Measurement

There are difficulties which inhibit the very possibility of measurement. In particular, concerning what to measure and whether a factor could ever be measured.

The first, and most fundamental, problem that may be encountered is in knowing *what* to measure. This is not a measurement problem per se, but relates to the whole of the problem analysis. What are to be considered as the main measures of performance, and what are the varied factors which influence them? For instance, what should be measured in comparing the performance of a number of operating subsiduaries? On what basis should funds be allocated in a national development plan? What factors account for the high variability of average consumption in similar retail outlets? All to often, analysts take the easy options such as assuming profit maximization, or ignoring factors that cannot easily be quantified.

The second problem occurs when it is virtually impossible to measure what is required. A classic example is stock control where one should measure *demand* for products, but can often only measure actual *sales*, which will have been affected by the availability of stock. Similarly, in trying to forecast sales of a highly promoted product, major influences are competitor's promotions and the total stock in the trade. Neither of these can be measured in any systematic way. Even if historical figures could be produced, how could they be forecast? In fact, many forecasts fit into this category as there are seldom strong grounds for believing that they accurately map what will actually happen.

Validity of Measurement

Assuming that one knows what to measure and that it is possible to measure it, the next level is to ensure that the results are valid. That is, that they do accurately reflect the real world attributes of which they are a measure, rather than simply their process of production. In the case of primary data production this means determining *how* to convert a concept into some appropriate type of measure, whereas in the case of already existing data it involves determining whether it actually does measure what is required.

Many concepts which can be stated clearly are difficult, if not impossible, to quantify into an interval or even ordinal measurement. This must be one of the most common problems, and examples abound. How does one measure managerial performance, staff quality, the real cost of being out of stock, the effects of a particular type of advertising, the effect of particular activities in deterring crime? In such cases there are three options – try to use a proxy or surrogate variable which is believed to be closely correlated and which can be measured; be content with subjective and qualitative measures; ignore the factors.

As for secondary data, particularly official statistics, the key question is what is *actually* being measured as a result of the process of measurement. For example, data on production costs are often required in OR studies and are usually produced routinely by accounting and management information systems. Such figures, maintained for recording purposes, will often be quite misleading if used for decision making. As for official statistics such as crime, health, suicide and unemployment, many studies have shown the extent to which these reflect the practices of their production rather than any underlying reality (Irvine et al, 1979; Atkinson, 1978; Slattery, 1986).

Reliability of Measurement

Finally, assuming data is available and that it is believed to be valid, there remains the problem of its reliability. That is to what extent is it complete and free from error? If a further set of data were produced in a similar manner would it replicate the first? Here all the common problems arise such as missing data, mis-measurement, transcription errors and the recalculation of old estimates as new information is available.

PRAGMATICS - RESPONSES TO THE PROBLEMS

Given the many problems and inadequacies outlined above, what is to be the practical response?. I would like to mention just two areas worthy of attention - the incorporation of qualitative variables in an analysis and the choice between rectification and robustness.

Incorporating Qualitative Variables

Almost inevitably, an OR project ends up with various factors quantified (to varying degrees of validity and reliability) and incorporated into a model, but with others, often of major importance, left unable to be measured. How can these be dealt with? Some possibilities are:

- Use the quantified results of the analysis to provide a background and general framework.

- Use the quantitative model to answer 'what if' questions that are based on possibilities and scenarios suggested for the qualitative aspects, as in financial modelling.

- Have management specify required levels and use the model to find ways of achieving them, as in setting protection levels in stock control.

- Accept the more general results available from a purely qualitative analysis as in Dyson (1983).

Rectification versus Robustness

There would seem to be two possible responses the problems of uncertain data. These are either to spend time and effort trying to make it better - to rectify its inadequacies; or to accept that it will always be inadequate and uncertain, and concentrate on developing methods which are robust

(Rosenhead, 1980) and will provide reasonable answers despite the problems. These are not of course mutually exclusive, but I feel that the intractability of the problems of measurement and the relative under-development of robust methods make this a particularly attractive research area.

CONCLUSION

This paper has outlined some of the many problems inherent, but often unmentioned, in undertaking measurement, and has suggested that it is better to view this as a process of data *production* rather than data *collection*. These problems arise at a number of levels:

- Determining *what* factors to measure.

- Discovering if measurement is *possible*.

- Establishing *how* they should be measured.

- Testing whether existing data is appropriate.

- Assessing and allowing for the *reliability* of data.

The final step in the chain, which takes us back full circle to the importance of interpretation and meaning, is to recognise that numbers on a piece of paper have no meaning *in themselves*. *Data* only becomes *information* when it is interpreted and invested with meaning(s) by people.

Suggestions have been made for research in a number of these areas, and it is hoped that the papers at this conference will contribute to that process.

REFERENCES

Atkinson, J. M., 1978, "Discovering suicide: Studies in the social organization of sudden death," Macmillan, London.
Burrell, G. and Morgan, G., 1979, "Sociological Paradigms and Organisational Analysis," Heinemann, London.
Churchman, C. W., Ackoff, R. L., and Arnoff, E. L., 1957, "Introduction to Operations Research," Wiley, New York.
Cicourel, A. V., 1964, "Method and Measurement in Sociology," Free Press, Glencoe.
Cuming, M., 1984, "A Manager's Guide to Quantitative Methods," Elm Publications, Cambridge.
Davies, G. and Yoder, D., 1941, "Business Statistics," Wiley, New York.
Dyson, R. G., 1983, Operational research on the peat bog: A case for qualitative modelling, J. Opl. Res. Soc., 34:127.
Graham, R. J., 1982, "Give the kid a number": An essay on the folly and consequences of trusting your data, Interfaces, 12:40.

Hindess, B., 1975, "The Use of Official Statistics in
     Sociology: A Critique of Positivism and
     Ethnomethodology," Macmillan, London.
Hughes, J., 1980, "The Philosophy of Social Research,"
     Longman, London.
Irvine, J., Miles, I., and Evans, J., "Demystifying Social
     Statistics," Pluto Press, London.
McNeill, P., "Research Methods," Tavistock, London.
Morison, M., "Methods in Sociology," Longman, London.
Mingers, J. C., 1984, Subjectivism and soft systems
     methodology - a critique, J. Appl. Sys. Anal., 11:85.
Neter, J. and Wasserman, W., 1966, "Fundamental Statistics for
     Business and Economics," Allyn and Bacon, New York.
Phanzagl, J., 1959, A general theory of measurement:
     Applications to utility, Naval Research Logistics
     Quarterly, 6:283
Pratt, V., 1978, "The Philosophy of the Social Sciences,"
     Methuen, London.
Rosenhead, J., 1980, Planning under uncertainty I: The
     inflexibility of methodologies, J Opl. Res. Soc.
     31:209.
Rosenhead, J., 1980, Planning under uncertainty II: A
     methodology for robustness analysis, J Opl. Res. Soc.
     31:331.
Schutz, A., 1967, "The Phenomenology of the Social World,"
     Northwestern University Press, Evanston.
Slattery, M., 1986, "Official Statistics," Tavistock, London.
White, D. J., 1985, "Operational Research," Wiley, Chichester.

PRACTICAL ISSUES IN PLANNING:

A CASE OF METHODOLOGY

Dick Martin

Department of Management
Ealing College of Higher Education, London

INTRODUCTION

All practical project work requires decisions concerning measurement
and methodology. When one is working in a familiar area it is possible to
make such decisions implicitly and even without a recognition of the
assumptions underlying the decisions. However, when working in an
unfamiliar area of work, the serious practitioner is confronted with major
issues relating to measurement and methodology. This paper discusses how
some of these issues were raised and coped with in a practical project on
industrial planning in Nigeria. From the outset I, as Project Manager, was
confronted with a number of questions. How does one gain an understanding
of the present situation regarding industry and 'development' and possible
options for the future. What are the relevant sources of information,
qualitative and quantitative? How does one judge the reliability of
information and uncover the assumptions on which it is based? How reliable
are eyes used to making these judgments in a European setting or other parts
of the Third World? On a practical project these are not issues for
academic debate but have to be considered concurrently with the development
of recommendations for implementation in practice. But, are they compatible
with an OR approach?

THE PROJECT

The project was concerned with assessing existing industry, examining
industrial potential and planning industrial development in Borno State in
the north of Nigeria. The clients were the State Ministry of Economic
Affairs and the Ministry of Industry.

Since doing the project I have met many people with long experience of
working in Nigeria; however, none except my direct contacts on the project,
had as much experience as I now have of working in that rural, northern
State and very few had even visited it. Viewing the project with the
confidence of urban Lagos eyes could have be even more dangerous than
viewing it through London eyes yet realising they would have a lot of
learning to do. I soon saw physical examples of failed projects caused by
implementing a technological system designed in London, modified in Lagos
and still inappropriate to rural Nigeria, and the major errors made by an
expert who applied his twenty-plus years of experience designing housing in

Malaysia as if the same solutions were directly applicable to this part of Nigeria.

Here, we immediately come up against one of the fundamental problems not just of operation research but of people in general. On the one hand my 'nakedness' was so great that there was no damage that I would believe that I was wearing a complete and totally adequate set of clothes. My lack of knowledge was so great that I knew that I would have a lot of learning to do. On the other hand, if I had had a great deal of experience, there would be the danger of my not opening my eyes and therefore of my believing that my experience was directly relevant to this situation. There is then the great temptation to apply the slot-in technique or give the slot-in solution in a totally inappropriate setting. Thus my awareness of my nakedness was an advantage; but there was also a danger that my fast learning process would leave me with a lot of half-truths, and misconceptions and misperceptions.

But overriding these matters was the fact that the problem that I was tackling was in Rittel & Webber's term a 'wicked problem' as are all problems of social policy. However good my knowledge, however deep my insight I could not guarantee that I would come up with the right answer:

"The search for scientific bases for confronting problems of social policy is bound to fail because of the nature of these problems. They are 'wicked' problems whereas science has developed to deal with 'tame problems'. Policy problems cannot be definitively described ... there is nothing like the indisputable public good ... no objective definition of equity ... (policies) cannot be meaningfully correct or false ... it makes no sense to talk about 'optional solutions' ... there are no "solutions" in the sense of definitive and objective answers." Rittel & Webber (1973).

Commenting on Rittel's earlier adoption of the term 'wicked problem' Churchman points out the danger of the OR person "carving off" a piece of the problem and finding a rational and feasible solution to this piece a process which he describes as 'morally wrong' Churchman (1967).

In my case I had no toolbag of technology, no formal methodology, no panacea solutions and a really wicked problem.

GAINING UNDERSTANDING AND CREDIBILITY

Once I had agreed to go, my first and most pressing objective was to get a base knowledge of Nigeria. It was knowledge of Nigeria rather than just Nigerian industry that I considered important. Further than that, I attempted to get information of a comparative nature; Nigeria in the African context, Nigeria vis-a-vis the UK. Maps, books and the Development Plan gave factual information; anecdotes gave partial insights; novels gave another 'feel' for the situation. Gradually I came to get sufficient knowledge that would enable me not only to hold my own in conversation but also to ask the pertinent questions. As usual at this stage of a project the data raised more questions than it answered. However, it achieved its aim of putting some of the factual evidence into perspective so that one began to get a picture of the State and Industry in the State in the context of the country as a whole. What it could not give, of course, was an understanding of the aspirations and motivations of the population, local people, the ultimate 'beneficiaries' of development schemes.

To gain credibility it was necessary not only to know the locations of each industry and each State. I had to learn about Borno State and to know where Damaturu was and not to confuse it with Damagum or Damboa, nor to confuse Geidam, Gwoza, Gashua and Gujba nor Ngala and Nguru.

It can be argued that this process was only necessary because of the unfamiliarity of the area. But, if I had been on seemingly more familiar ground, there would have been a pressure to omit much of this stage of the work or consider it unnecessary. The major danger here is that my preconceptions (which could be quite erroneous) would never have come under scrutiny, the implicit assumptions would never have been made explicit and perhaps not even recognised. This could apply to values as much as to facts. So, for example, somebody working in in their own company in the UK in familiar surroundings is not faced with the nakedness I was faced with; how much easier is it then to fall into the trap of not questioning one's assumptions? The important point is that the process is essential and, if omitted, can lead to 'erroneous' conclusions. The 'right' method for exploration will depend on individual circumstances and individuals' preferences.

PERCEPTIONS

Before leaving England I had avidly read anything about Nigeria that I could find and that time would allow. Although there were problems, the country did seem to be developing rapidly. These views were strongly reinforced in the first week of my visit and by the end of the week I was very optimistic about Nigeria's future.

By the end of my second week my views had changed drastically. I had got to know people better, better work and social contacts. Maybe it was this that led them to temper my optimism. I heard horror stories from all sides; massive agricultural schemes that had failed, developments just under way that were bound to fail, the over-ambition and wrong direction of development plans, industrial projects that would have no market, that the subsistence farmers were getting poorer, corruption, waste... I had a very depressed view of Nigeria and any report I had sent back to the UK would have made very depressing reading. By the end of the third week my views had changed again. The picture didn't seem entirely good or entirely bad, yet I wasn't sure which was which. It would have been a confused report that I sent back.

OPERATIONAL RESEARCH

It will be argued by some readers that most of what I have written above has little to do with Operational Research and is actually the antithesis of 'the OR Approach'. This they will contend is to act as an objective observer and to act scientifically by building a model of the situation which can be independently validated; optimism and pessimism portray an unhealthy involvement rather than the required detachment.

I could not disagree more. In this situation there is little that could be done in the name of objective detachment. The objectives to which the Development Plan was working, the extent to which particular projects contributed to these objectives, the measures by which these projects were to be judged, all held within them intrinsic value judgments; and my Western eyes were likely to make judgments according to criteria I recognised.

A common cop-out is to set up and hide behind an 'objective scientific method.' As one of my tasks I had to review a number of feasibility studies of projects proposed for the next Development Plan; including industries such as a corn mill and a tomato canning factory; it was not uncommon for such a feasibility study to use discounted Cash Flow and the argument to be:

If the land in area A is all turned over to Product B then there will be C hectares available. With proper irrigation and fertilisation and good management it should be possible to achieve a yield of D/hectare. With the factory working at E% efficiency and F% conversion rate we should be able to produce G tons/yr at a cost H/ton. If the farmers enter into the contact at the price we have proposed and if import controls keep the retail price at J and if labour costs can be kept to K and if the original machinery costs L and has a life of M years and we use a discount rate rate of N%, ...and, ...and... then the project will be profitable and it is recommended that the project proceed.

I have written [Martin (1986)] of the myriad of assumptions and uncertainties that lie behind these purportedly value-free calculations. Few of these are amendable to 'what - if'? sensitivity analysis. Many of the inputs involve assumptions about the results of actions by other parts of the government and those may not take place; they also assume particular actions on behalf of the farmer without any assessment of their likelihood. Failure to take these issues into account is likely to lead to the failure of the project. More fundamental still, is the problem of measurement of the success of the project. Even if the scheme is a financial success (as measured above) it may be a disaster as measured against the aims of the development plan including those of reducing poverty and spreading wealth. It can be argued that if I, the consultant tried to take these issues into account I was, wrongly, attempting to act as if omniscient. This is to get the emphasis wrong. I was working in a situation where, at the end of the project I, in conjunction with the client, had to make recommendations for action; these were to be on the basis of contribution to 'development'. Judgements of relative contributions would influence recommendations on the extent of industrial development relative, say, to agriculture and the relative importance of large and small-scale schemes. There is no simple metric to aid these judgements. There is no objective and value-free way of making them. Yet these considerations underlay all the work on the project; if OR is to operate in the real world, these considerations must come within the realm of OR.

IT DOES NOT HAVE THE NOSE

At the opposite extreme to the problem of the contribution to development was that of reliability of the hard data that we were given. Of the three basic sources of current industry in the State, one was the State Directory of Business Establishments. In one section, two thirds of the total entries, the Directory listed 112 establishments, 60 of which were in two smallish towns, Gashua and Geidam. Seeing this I decided that, in the terms of a French colleague 'it does not have the nose', or as we would say 'it doesn't smell right.' Further investigation showed that the 60 businesses were market stalls in these towns, presumably supplied by conscientious clerks the situation became more ludicrous when a separate (Town Planning) survey showed 4000 market stalls and 13000 employees in the main Maidugari market and these were not in the Directory.

The important point here is not that the data was wrong but that somehow I had developed some 'feel' so that I could say 'it doesn't smell right.' Here it was fairly simple; I had some knowledge of the size of the towns relative to others and the conclusion was obvious. But the more fundamental points are that checks on data as to whether it smells right are often substituted by technical checks such as do the row and column totals each total 100% thus missing the point; that we do not make efforts to understand how we get the sense of 'it doesn't smell right' nor of how to transmit this in education; that even if we subject data to the 'smelling' routine we may well accept 'wrong' data because it is not in discord with our wrong sense of smell.

A further example came from the conflict between the published statistics and what I had seen simply driving around the streets of the city and its outskirts - on many streets people were involved in a variety of industrial activities, hammering at water tanks, working with wood, making bricks and concrete blocks. It would have been beyond our resources to do a full survey of such activity so we limited ourselves to given major streets. In one afternoon we counted over 1300 people working in such industries; our number of people working in manufacturing industry in Maiduguri had suddenly doubled. This finding had a major bearing on the balance and scale of industry in the city. We were already aware that many of the larger industries were in the construction and furniture businesses but now we had found a further 700 employees in this area. All of these jobs rested on increasing construction and that rested on continuing investment by the Federal Government; there were suggestions that it would actually decline. The situation was further compounded by the fact that many of the applications for small industry loans were also for the construction industry. This led us to examine the vulnerability of the construction industry. Fairly simple facts, prompted through casual observation had thrown up questions of strategy.

AND THE POLITICS ARISE

However 'thorough' an investigation may be, it is not possible, nor advisable to question all information. In many areas we worked in we had to take things on trust. One area we decided to look at more deeply was the Small Industries Credit Scheme. This had been set up to provide loans for small industries repayable over a number of years. Repayments would provide the basis of a revolving fund and thus would be topped up to allow for expansion. We and our clients took the view that this scheme deserved extremely high priority as it seemed a potentially viable way of spreading small-scale industry throughout the state and providing an infant industrial base away from the urban cities. We were also told that it was very popular and that there were a large number of outstanding applications for grants.

This in itself warranted deep consideration. However there was an even more important consideration; it was time that this European consultant 'got his hands dirty.' We needed to be seen as serious and hard working and able to do a time-consuming, nitty-gritty job. This led to a potential problem. I had been told by the senior civil servant that there were 3445 applications for loans, of which 300 had been granted and that 50% of the remainder were good valid applications which deserved a loan. The types of industries in the outstanding applications were similar to those granted; for example about a quarter were for cement block factories and even more than this were for bread bakeries. So, this implied, there were 2000 good applications. He further told us that he and his staff personally sorted all the applications every six months. Our problems started when we went to the filing cabinets containing the application files. They started at number 3000. Further investigation showed that few files less then number 2500 existed and they were six years old. So instead of 4000 outstanding applications there were less than 1500 and many of these might not be viable. The problem now was how to confront the official in charge of the applications with the fact that his description of the situation was wrong. We told him what we had found; we didn't mention his previous statement. "If that was the situation we had found, then that was the actual situation", he told us. The political problem avoided we now delved more deeply into the files and a new picture emerged. Over 90% of loan capital approved had been in the last two years. In the last two years the applications had grown from less than 100/yr to 1100/yr. And, possibly more importantly, the figures that we had on the rate and poor success of loan repayment were based only on the early approvals before the great increase in approvals; there was now ten times as much loan outstanding. What had not come through to us in conversation was this massive growth. This threw up more problems. The revolving fund did not have the capital to cope with further expansion, the scheme did not have the administrative capacity to collect the repayments from the expanded list of projects and many applications were in a potentially contracting area, construction. The reality of the information instead of the data we were initially given led the project off in a new direction.

CONCLUSION

I have attempted to show, through the medium of one case study that the problems of measurement in the social arena are not capable of simple resolution. I am not alone in my view that OR has a major role to play in working on these problems but, if it is to be 'useful' it must not hide behind the myth of scientific objectivity. Rather, what is required is much more research into effective ways of operating in such an environment. Sagasti (1976) has forcefully pointed out many of the pitfalls of abusing OR in developing countries. Chambers (1983) takes the issue of 'measurement' and what is seen in the Third World much further than I can here and provides a good starting point for discussion by OR workers. Moreover these discussions whilst voiced here through issues of Third World Planning are of universal significance to all practitioners of OR.

(This paper is a modified version of 'Planning in Nigeria: A Case of Methodology' which discusses the particular case at greater length).

REFERENCES

Chambers R   1983 'Rural Poverty': Putting the Last First', Longman, London.

Churchman C W   1967 'Wicked Problems', Mgmt. Sci. 14, 4.

Martin R C   1986 'Uncertainty and Planning in the Third World,' paper to O.R. Society Conference.

Rittel H W J &   1973 'Dilemmas in a General Theory of Planning', Policy
Webber M W   Sciences 4.

Sagasti F R   1976 'Thoughts on the Use (and Abuse) of OR/MS in the Planning and Management of Development, Opl. Res.Q, 27, 4 ii.

TOWARD A STAKEHOLDER PARADIGM IN CORPORATE PRODUCTIVITY

MEASUREMENT AND ANALYSIS

Foo Check Teck

School of Accountancy
Nanyang Technological Institute
Nanyang Avenue, Singapore 2263

"...a theory is just a model of the universe or a restricted part of it...." Stephen W. Hawkins, 1988.

ABSTRACT

Twenty-five years ago, the idea of reporting corporate value-added as a measure of performance would probably be seen by the financial community, at best as something of a controversy : return (profit) on investment is the only meaningful indicator. However in very recent years, countries like United Kingdom and Singapore are beginning to realise the potential viability of the value added concept as a meaningful measure of corporate performance. In 1984, for instance the Singapore Society of Accountants promulgated the Recommended Accounting Practice No.3 - Reporting Value-added Information - as a guide to publicly-listed companies. In U.K. attempts were also made, as in the manner of the Corporate Report, to encourage companies to provide relevant value-added data. Japanese corporations had been known to emphasize corporate value-added productivity as an overall measure of performance. Impetus towards value added measure can be speculated as centrally-planned economies like China and the Soviet Union are beginning to realise that the practical route to economic progress is via the capitalistic vehicle of `profit-making´ corporation - perhaps re-cast in an ideologically more palatable social entity - value-added creating and sharing corporation. This paper explores the use of the measure of value-added as a the primary stakeholder measure of corporate performance : using Singapore´s RAP as a methodological starting point for discussion.

VALUE ADDED

Various measures of productivity performance (Norman and Bahiri, 1972) had been advocated in the literature. However those centering on value-added as the primary means of measuring performance are still scarce as compared to those that are profit-based. Various authors had dealt with the different uses and aspects of the value-added concept (see for example: Cox B. 1979, Morley 1979, Smith, 1980, Silver, 1979, Wood 1979, Denton 1980, Harper 1984, Kang-Hong 1984, 1985, Check Teck, 1988). The objective of this paper is to explore the development of an integrative paradigm of corporate value-added performance - creating and sharing - from the perspective of the employee as a corporate stakeholder. Such a framework could meet the need for a conceptual as well as analytical framework for assessing the structural configuration of corporate systems in improving value-added productivity performance.

Using notations defined in Check-Teck (1989), (T turnover, V value-added, En number of employees employed, Bm bought-in materials, Bsu bought-in services and utilities, @ ratio of quantities sold Qs to quantities manufactured Qm, A total capital, At technology-related capital, Ab the balance (A less At), P mean price of product sold); value added is the net measurable result of continuous cycles of production flow activities and from the creation side, it is measurable by turnover less bought-ins { T - (Bm + Bsu) } and sharing side by the addition of stakeholders' stakes as reflected in employment cost, dividends, interest, taxes, and of depreciation and re-invested profits. Corporate performance is reflected by value-added productivity measured by value-added per employee. Although indicators like descriptive averages do embed performance information content, analytical needs may not be wholly met by such data (Goldman, 1984). There is also a need, in productivity measurement and analysis, to distinguish between the requirements of purists and those managing the corporations (Chew, 1988). In the even more competitive environment of the 1990's, there will be greater need by the practitioners for conceptual and analytical frameworks capturing in the main, key factors that may influence corporate productivity performance. To meet such analytical needs, changes in the primary ratio of VA/En had being related to key performance factors at each stage of value-added creation cycle as captured in the measures of Bsu/T, Bm/T, At/En, @, Q, and P (Check Teck, 1989). The sharing side of value-added need however to be considered in order to configure a more complete (and also dynamic, if not causal) paradigm. Clearly the motivational forces of the employee as corporate stakeholder should also be built into such a scheme.

STAKEHOLDER

Different definitions of `stakeholder´ are found in the literature for example Einshoff and Freeman (1978) as : `..any group where collective behaviour can directly affect the organisation´s future, but which is not under the organisation´s direct control´. The approach here, is to view from a far more restrictive perspective of directness with value-added creating and sharing. From the the sharing side of the published value-added statement these stakeholders are those of employees, shareholders and lenders, and the government. If one however, looks at the selling and buying-in (of materials and services) side, other groups of stakeholder can be identified as the customer and supplier. The approach here however is to develop the discussion by focusing primarily on the employees(labour) and secondarily shareholder (capital); since the former are directly involved in value-creation by application of skills whilst the latter provided the risk capital (as distinguishable from lenders in giving secured loans) for the micro-structure of space, plant and machinery, technology and cash.

CAPITAL AND LABOUR

Marx (1957) abstracted society to one fundamental relation of capital to labour and on assumption of the dominance of the relationship developed a dynamic model of societal change. (Unlike the more static Weberian model) . The central thesis being that the rate of capitalist profit is a function of the rate of labour exploitation. Although a more in-depth discussion is, outside the scope of this paper, it is obvious that current developments in the real world do not accord with the outcome being predicated by the model. Given the current trends including that of greater use of complex technology in businesses, increasing market-driven competitiveness in a global (China and Russia) free-market, a more educated and skilled workforce; the next quarter´s changes is more likely to be one of more co-operativeness with labour in the maximisation of corporate value added productivity. Precisely because K/L (K being a measure of the technology-related means of production being utilised) had increased (implying a lower proportionate share of labour in the value-add) that co-operativeness with labour (in contrast to exploitativeness) become more critical in determining the rate of `capitalist profit growth´. Goal congruency among labour (skills provider) and capital (risk money provider) may emerge to be a predictor of corporate performance as well as survival. Thus the need for a non-conflictive, stakeholding orientation by management towards labour. Both labour and capital stakeholders derive income streams from the value creation activities. With labour however there may be other intrinsic, non-money benefits that a job provides : satisfaction from social interaction and other intrinsic benefits. Assuming however that financial returns in terms of size and security of income stream are primary, then clearly stakeholders have vested interests in the continuation and the maximisation of gains from the flow of value-added creation activities. In the general case representing any absolute value of shares of any stakeholder i as Si, then

> Proportionately, Si/VA or Ri
> Where Sum of Stakeholders´ Absolute Shares Ss+Sl+V = VA
>   (assuming only two stakeholder groups of `s´ for
>   shareholders and `l´ for labour and V being re-investment)
> Mean absolute value Si/Ni (N being the number of stakeholders)
> Thus, an identity, Si/Ni = Si/VA X VA/Ni
> Differentiating with respect to time,

$$\overset{*}{Si/Ni} = \overset{*}{Si/VA} + \overset{*}{VA/Ni}$$

Assuming that persons within a stakeholder group hold equal stakes in the corporation then increases in the annual rate of mean return can flow either from productivity increases and/or that of an increased share of the pie by the stakeholder group. In the case of employees (given their most direct human intervention in the value creation process),

$$\overset{*}{Sl/(En)} = \overset{*}{Sl/VA} + \overset{*}{VA/(En)}$$

Assuming also constant total productive factor (and factor-mix) endowment (eg. plant and machinery, technology) per employee, then changes in VA/En (though a partial measure) will mirror changes in (total factor) corporate value-added productivity performance. In a conflictive setting, labour demand for increases in wage rates will be made independent of corporate performance. If productivity declines, then the meeting of increases can only be from increasing labour proportionate share. Such increases are unlikely to be motivators since such increases are secured without regard to corporate performance but due to the `demand´ of labour. In an ´a day´s pay for a day´s work´ setting, improvement in corporate productivity performance is also not

translated into additional gains by labour, but a larger proportionate
share is distributed as dividends to shareholders as capital providers.
In a ´cooperative´ setting, labour is motivated into contributing to
corporate productivity performance because of a direct nexus (say
through an institutionalisation of gainsharing or flexible wage system)
between their wage level and corporate productivity performance. Or
more precisely,

$$\overset{(+)}{\underset{*}{Sl/En}} \quad \text{as a result of} \quad \overset{(+)}{\underset{*}{VA/En}}$$

In promoting goal congruence between the shareholder and
jobholder, value-added is conceptually a more ´powerful´ integrating
concept than profitability since profits is but a share of value-added.
Assuming that profits takes a constant proportionate share of value-
added, increases in the latter will imply higher profits available for
shareholders. The corporation´s need for greater re-investment
(implying higher V/VA) into research, development and technology to
cope with competitive pressures will mean declining proportionate
shares of the value-added flow to both labour (Rl) and capital
providers (Rs). This can be viewed positively only for so long as both
see a continuing stake in the longer term viability of the corporation.
Thus the issue of optimality of a specific value-added distributional
structure.

OPTIMAL VALUE-ADDED DISTRIBUTIONAL STRUCTURE

Clearly, any corporation inspite of poor financial performance
will be driven by competitive forces to pay a minimum level necessary
to attract persons with the requisite skills but will not be able to
sustain a pay level beyond a certain maximum limit if it desires to
stay in business (Whittney, 1988). Likewise a minimum dividend
payout ratio may be necessary for the shareholders to retain the shares
as long-term income stream assets; too high a payout could mean
inadequate re-investment. Within these constraints, management of
corporations need to re-configure the value-added distributional
structure (as reflected ex-post in the aggregate by reported ratios)
to achieve optimality. The definition of what is an optimal
distributional structure in relation to employees is a difficult one.
One element is however clear : it should motivate the employees to
maximise their stakes in the corporation. In the case of capital
stakeholder, the distribution of dividends to each or any particular
shareholder is determined by the quantity of shares held. For labour
stakeholder however, the distribution to employees is far more complex
and given the motivational impacts need to be considered from various
levels/types of disaggregation e.g. at stakeholder level : Sl/VA or
Rl (1 being labour); of skill-type category – Sli/VA or Rli (i for
skill type); for each skill-type, the individual employee´s share –
Slie/VA or Rlie (e for specific employee, accounting for other
differentials). The weights given to each skill-type can vary from
company to company. For example, a software company being more
dependent on software specialists for quality computer programs may
place a greater weight on this particular skill category than others.

Also as argued earlier market forces constraints will set an upper
and lower limit within which an optimal ratio must lie : Rmi (min) <
Rop (optimal) < Rma (maxi). Further implicit in any design of a
value-added distributional structure is the need to distinguish between
the corporate performance-independent (fixed) and dependent (variable)

component. For instance, assuming a corporation had designed a
sharing structure such that any value-added in excess of the moving
mean average value-added for the last five years will be shared with
employee stakeholder at a constant proportion without an upper limit,
then the reported employee stakeholder ratio for the year Rt will
really be comprised of an independent (Ri) and dependent component (Rd)
i.e. Rt = Ri + Rd. Thus configuring a structure of value-added
distributional ratios to achieve optimality is more complex than what
is apparent at first glance.

A STAKEHOLDER VALUE-CREATING AND SHARING PARADIGM

    A paradigm of corporate value-added creating, sharing and the
motivating (of the employee stakeholder) is proposed here. Key
performance factors related to corporate value-added productivity
performance (as measured by VA/En) are presented in the manner of
production flow (buys-in, adds value, sells). Distribution ratios at
various levels of disaggregation reflect the sharing dimension. The
paradigm implicitly assumes a system of linking the rate of wage
increase to that of value-added productivity. The configuration of
distribution ratios is assumed to have motivational impacts on the
employee stakeholders' efforts in improving productivity. There is
also the need to account, in designing any particular form of value-
added sharing for (Ronen and Livingstone, 1975), the different
extrinsic (E) and intrinsic (I) valences. The aim is to maximise the
motivational impact. The paradigm is represented below:

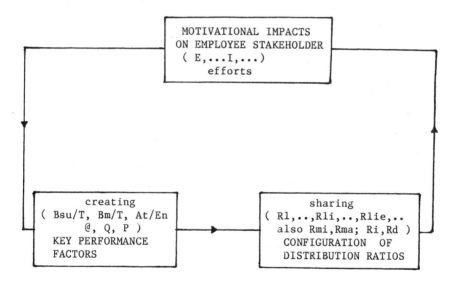

    Management, in order to improve corporate performance as measured
by value-added per employee, will have to consider the optimality of
the existing structure of value-added distribution. The needed changes
to a key performance factor has a higher chance of being successfully
effected if this is so. For example, if the ratio of At/En is low, and
funds had to be raised, the raising of funds through further issues of
shares will be more easily facilitated if the corporation had
maintained an optimal distribution of value added that results in
dividend policies that motivate the shareholder to extend his stake in
the company. In the same vein, where there is a need to cope with rush

orders or reduce reject rate, employees are more likely to cooperate if there exist a system for sharing of gains. Although the likelihood of corporations being viewed as collectives is seen to be unlikely (Simon, 1986), the perception of labour as stakeholder will grow as capital takes on a more technologically-oriented character.

CONCLUDING REMARK

Finance literature treatment of a firm in terms of maximisation of shareholders´ wealth should perhaps be broadened to that of maximisation of stakeholders´ utility of ´stakes´ : shareholders being seen as only one subset (albeit an important one). The paradigm here is a restrictive one – only of employee as stakeholder – there remains for the need for further development and refinement of a more total stakeholder framework in the measurement and analysis of value added corporate performance.

REFERENCES

Check-Teck F., 1988, How To Achieve Value-added Goals, Productivity Digest, June:13.

Check-Teck F., (forthcoming) 1989, Managing Performance in Technolgical Industries, Singapore Management Review, 11(1).

Chew B.W., 1988, No Nonsense Guide To Measuring Productivity, Harvard Business Review, 1:110.

Cox B., 1979, Value Added: An Appreciation for the Accountant Concerned with the Industry, Heinemann, London.

Denton N., 1980, Value added : The Accountant´s Key to Sane Economics, The Accountant, 182:957.

Einshoff J.R. and R.E. Freeman, 1978, Stakeholder Management, A Working Paper from the Wharton Applied Research Center.

Gold B., 1984, Productivity Measures : Descriptive Averages Versus Analytical Needs, in ´Managerial Issues In Productivity Analysis´, Dogramaci A. and N.A. Adam, eds., Kluwer-Nijhof Publishing, Amsterdam.

Kang-Hong H., 1984, Problems of Productivity Measurement, Stock Exchange of Singapore Journal, 12(5):4.

Kang-Hong H., 1985, The Value Added Statement, Singapore Accountant, 1(14):11.

Marx K., 1957, 1958, 1962, Capital, Lawrence & Wishart, London.

Morley M.F., 1979, Value-added: The Fashionable Choice for Annual Reports and Incentive Schemes, The Accountant´s Magazine, LXXXIII:234.

Norman R.G. and S. Bahiri, 1972, Productivity Measurement and Incentives, Butterworth, London.

Ronen J. and Livingstone, 1975, An Expectancy Approach to the Motivational Impacts of Budgets, The Accounting Review, 50(4):671.

Simon M.A., 1986, Do Machines Make Humans More Productive ?, Some Preliminary Considerations, in ´Technology and Human Productivity´, Murphy J.W. and J.T. Pardeck, eds., Quorum Books, New York.

Singapore Society of Accountants, 1984, No.3, Recommended Accounting Practice – Reporting Value Added Information.

Whitney J.L., 1988, Pay Concepts for the 1990s, Part 2, Compensation and Benefits Review, 20(3):45.

Wood E.G., 1979, Setting Objectives In Terms Of Value-added, Long Range Planning, 12(4):2.

ASSESSING ECONOMIC INEQUALITY[1]

Bernard Sinclair-Desgagné and Mihkel M. Tombak

The European Institute of Business Administration (INSEAD)
Blvd. de Constance, 77305 Fontainebleau Cedex, France

Abstract

Important contributions to the assessment of economic inequality have recently been made using concepts such as stochastic dominance, that are well known in Operational Research. We reformulate and extend this approach in a way that allows for identifying the range of welfare improving income redistributions in a given society.

1 Introduction

The bicentennial year of the French Revolution seems to be an appropriate time to reflect on one of the causes of this upheaval - economic inequality. During the past two centuries little progress has been made in characterizing optimal, or even desireable, income distributions. Should individual incomes be absolutely equal, or proportional to individual contributions, or adapted to personal needs, or, as Rousseau said, should income inequalities reflect "Physical" disparities while preserving a balance of power? [2] Clearly, none of these suggestions can ultimately be accepted without making a value judgement.

(**Keywords: Poverty orderings, stochastic dominance, risk aversion, Farkas's lemma, linear programming**)

---

[1] Acknowledgements: The authors would like to thank Prof. Ludo Van der Heyden and the participants of the INSEAD business economics seminar for their helpful comments.

[2] In the Discours sur l'origine et les fondements de l'inégalité parmi les hommes, Jean-Jacques Rousseau argues that: *"... l'inégalité morale, autorisée par le seul droit positif, est contraire au Droit Naturel, toutes les fois qu'elle ne concourt pas en même proportion avec l'inégalité Physique ... ".* With regards to preserving a balance of power he stipulates in Du contrat social that *"... à l'égard de l'inégalité, il ne faut pas entendre par ce mot que les degréde puissance et de richesse soient exactement les même, mais que,... quant à la richesse, que nul citoyen ne soit assez opulent pour en pouvoir acheter un autre, et nul assez pauvre pour être contraint de se vendre."*

As argued by Sen (1979), this does not, however, make the assessment of economic inequality in a society a value judgement. One can simply try to elicit the conventions of a given society, take them as data, and derive their consequences for the distribution of income. This constitutes an excercise in positive science, for which Operational Research (OR) can be relevant.

Indeed, some authors have recently shown the usefulness of stochastic dominance and risk aversion, both well known OR concepts, for the positive assessment of economic inequality.[3] In this paper, we extend these contributions and propose a new procedure for identifying welfare improving income redistributions in a given society. Our procedure directly considers empirical income distributions and is based on standard OR tools: Farkas's lemma and linear programming. The procedure is described in §3 below. It first requires a new framework that is presented in the next section, along with a reformulation and interpretation of stochastic dominance.

2 Ranking Income Distributions

The distribution of income in a given society can be described by a non-decreasing function $F:\mathcal{R} \to [0,1]$, $F(y) = p$ meaning that a proportion p of the population in the country earns at most y.[4] This is an abstraction from empirical data, where individual incomes are grouped into a finite number of income classes. Adopting the empirical framework we work directly with the given income distributions and assume that there are K income classes indexed by k = 1, 2, ..., K. A lower k indicates a lower income.

Each income class k yields a certain amount of social welfare w_k. These welfare weights constitute a row vector $\bar{w}$ in the K-dimensional Euclidean space $\mathcal{R}^K$. Higher incomes are agreed to be better for society than lower incomes, leading to a requirement on the $\bar{w}$ vectors that they be component-wise nondecreasing: $0 \leq w_1 \leq w_2 \leq \ldots \leq w_K$. Put in matrix form, this assumption becomes:

(1) $$A\bar{w}^T \leq 0$$

where

$$A = \begin{bmatrix} -1 & 0 & 0 & \ldots & 0 & 0 \\ 1 & -1 & 0 & \ldots & 0 & 0 \\ 0 & 1 & -1 & \ldots & 0 & 0 \\ \vdots & & & & & \\ 0 & 0 & 0 & \ldots & 1 & -1 \end{bmatrix}$$

---

[3]Classical references on risk aversion are Arrow (1965) and Pratt (1964). Bawa (1975) establishes relations between stochastic dominance and risk aversion. The relevance of these concepts for comparing income distributions is then stressed in Foster and Shorrocks (1988).

[4]See the article by Foster and Shorrocks (1988) and the references therein.

and the superscript T refers to a transposed vector. We shall call any vector $\bar{w}$ satisfying (1) a <u>welfare vector</u>.

Now let $p_{i,k}$ be the percentage of people in society i, who are within the kth income class. Suppose that the income distribution in society j is unambiguously preferred to that of society i, that is, for all welfare functions the following holds:

$$(2) \qquad\qquad (p_{i.} - p_{j.})\bar{w}^T \leq 0$$

By the well known Farkas's lemma [5] of linear programming, (2) holds if and only if there exist nonnegative real numbers $\alpha_1, \alpha_2, \ldots, \alpha_k$ such that:

$$p_{i.} - p_{j.} = \sum_{k=1}^{K} \alpha_k a_{k.},$$

or equivalently,

$$(3) \qquad\qquad p_{j.} = p_{i.} - \sum_{k=1}^{K} \alpha_k a_{k.}.$$

In the above, $a_{k.}$ represents the kth row of the matrix A. Geometrically, (3) means that $p_{j.}$ is the convex cone pointed at $p_{i.}$ and generated by the directions $-a_{k.}$. Writing equation (3) component by component we get,

$$p_{j1} = p_{i1} + \alpha_1 - \alpha_2,$$

$$p_{jk} = p_{ik} + \alpha_k - \alpha_{k+1} \quad for \quad 1 < k < K,$$

$$p_{jK} = p_{iK} + \alpha_K.$$

The α_k's clearly indicate the modifications in society i's income distribution that are applied to achieve the income distribution of society j - a proportion α_k of people would have to be transferred from income class k-1 into the kth class, and a proportion α_{k+1} of people would have to move from the kth class into the next higher income class. Since α_K is nonnegative $p_{jK} \geq p_{iK}$, that is, the proportion of people in the highest income class must not be lower in society j than it is in society i. Moreover, the relative size of the lowest income class cannot be greater in society j than it is in society i; this is a consequence of the following proposition.

Proposition 1   *If (3) holds, then $\alpha_1 = 0$.*

Proof:

Let $\bar{e}$ denote the K- component vector $(1, 1, \ldots, 1)$. Multiplying (3) by $\bar{e}^T$ we get:

$$(4) \qquad\qquad 1 = 1 - \sum_{k=1}^{K} \alpha_k a_{k.} \bar{e}^T$$

---

[5] A proof of Farkas's lemma can be found in Rockafellar (1970), pg. 200.

Noting that $a_1.\bar{e}^T = -1$ and that $a_k.\bar{e}^T = 0$ for $2 \leq k \leq K$, equation (3) then becomes $\alpha_1 = 0$.

$$Q.E.D.$$

That is, society j is unambiguously preferred to society i if and only if the relative size of the lowest income class is smaller in j than in i. The following proposition generalizes this statement.

Proposition 2 *If (3) holds, then $\sum_{r=1}^{S} p_{jr} \leq \sum_{r=1}^{S} p_{ir}$ for all $1 \leq S \leq K$.*

Proof:

When $S = 1$, since $\alpha_2 \geq 0$ we already have $p_{j1} \leq p_{i1}$ by Proposition 1. Let us consider $S > 1$. Summing the terms of (3) we get:

$$\sum_{r=1}^{S} p_{jr} = \sum_{r=1}^{S} p_{ir} - \sum_{k=2}^{K} \alpha_k \sum_{r=1}^{S} a_{kr}$$

by Proposition 1. Note that for any S and for any k such that $2 \leq k \leq K$:

$$\sum_{r=1}^{S} a_{kr} = 0 \quad or \quad 1.$$

Hence, since the α_k's are nonnegative we have $\sum_{r=1}^{S} p_{jr} \leq \sum_{r=1}^{S} p_{ir}$ for all S such that $2 \leq S \leq K$.

$$Q.E.D.$$

Proposition 2 essentially says that the income distribution of society j must stochastically dominate the income distribution of society i, if it is to be unambiguously better. This result has already been proven by many authors. [6] However, we provide here a constructive definition of stochastic dominance. The utility of this method for policy makers will be shown in §3.

3 Welfare Improving Income Redistributions

Suppose that a given society i reveals a preference for income distribution $p_{1.}$ over income distribution $q_{1.}$, for income distribution $p_{2.}$ over income distribution $q_{2.}$, ..., and for income distribution $p_{n.}$ over income distribution $q_{n.}$. This society's welfare function must then agree with the following set of conditions:

---

[6]See, for instance, the papers by Atkinson (1970), and Foster and Shorrocks (1988).

$$\begin{bmatrix} A \\ \cdots \\ q_{1.} - p_{1.} \\ \vdots \\ q_{n.} - p_{n.} \end{bmatrix} \overline{w}^T \leq 0$$

Note that some of these conditions might be,

$$w_{k-1} - 2w_k + w_{k+1} \leq 0 \quad for \ all \ 2 \leq k \leq K - 1$$

that is

$$w_{k+1} - w_k \leq w_k - w_{k-1} \quad for \ all \ 2 \leq k \leq K - 1.$$

This would imply that *ceteris paribus* society's welfare improves to a greater extent when a proportion of poor people are shifted to a higher income class than when an identical proportion of rich people are shifted. Formally, the welfare functions must be concave, that is, exhibit risk aversion.

Let $q_{i.}$ be a description of society's current distribution of income. Again, applying Farkas's lemma society prefers another income distribution $p_{i.}$ if and only if there exist nonnegative real numbers $\alpha_1, \alpha_2, \ldots, \alpha_K, \beta_1, \beta_2, \ldots, \beta_n$ such that:

$$(5) \qquad p_{i.} = q_{i.} - \sum_{k=1}^{K} \alpha_k a_{k.} - \sum_{l=1}^{n} \beta_l (q_{l.} - p_{l.})$$

Clearly, the α_k's and the β_l's indicate the income class modifications that are welfare improving for this society. If a precise income distribution $p_{i.}^*$ is sought, the α_k's and β_l's of equation (5) define the adequate redistribution policy, and they can be computed by obtaining the dual variables of a simple linear program.[7]

4 Conclusions

Attempts to modify the distribution of income are often criticized for being subjective. As Okun (1975) pointed out: "The concept of economic equality is hard to define or to measure. It might be impossible to recognize complete equality if it existed; but inequality is very easy to recognize." So, in this paper we suggest that, in a given society, acceptable policies to redistribute income can be deduced from previous revelations of this society's norms with respect to inequality. Performing such an excercise in positive science requires working directly with empirical income distributions and using standard Operational Research tools - Farkas's lemma and linear programming.

The procedure proposed in this paper can be refined in many ways. A method for eliciting society's opinion about some income distributions needs to be designed.

---

[7]Generic problems are formulated and solution methods, eg. the simplex method, are explained in, for instance, Chvátal (1983).

Sensitivity of the procedure with respect to ambiguities at the elicitation stage or changes in the number and definition of income classes must be examined.

References

Arrow, K.J., (1965): "Aspects of the Theory of Risk-Bearing", *Yrjö Jahnsson Lectures*, Yrjö Jahnssonin Säätiö, Helsinki, Finland.

Atkinson, A. B., (1970): "On the Measurement of Inequality", *Journal of Economic Theory*, 2, pp.244-263.

Bawa, V.S., (1975): "Optimal Rules for Ordering Uncertain Prospects", *Journal of Financial Economics*, 2, pp. 95-121.

Chvátal, V., (1983): Linear Programming, W.H. Freeman and Co., New York.

Foster, J. E., and A. F. Shorrocks (1988): "Poverty Ordering", *Econometrica*, 56, pp. 173-177.

Okun, A.M. (1975): Equality and Efficiency: the Big Trade-Off, Washington: The Brookings Institution.

Pratt, J.W., (1964): "Risk aversion in the small and large", *Econometrica*, 32, pp. 122-136.

Rockafellar, R.T.,(1970): Convex Analysis, Princeton University Press, Princeton, NJ.

Sen, A., (1979): "Issues in the Measurement of Poverty", *Scandinavian Journal of Economics*, 81, pp.285-307.

OPTIMAL SCALING FOR ORDERED CATEGORIES

Werner Pölz

Department of Applied Systems Sciences
Johannes-Kepler-University
Linz, Austria

INTRODUCTION

When measuring the responses of individuals on scoring scales, it is usually necessary to assign numerical values to points on these scales and to proceed with statistical analyses based on these numerical scores. Generally, we can assume that the scale points are ordered. In common practice we consider points on scoring scales as evenly spaced, and we thus imply an arithmetic scale. But such an assignment of scale values may not be optimal. In order to find a solution to this problem we can use the method of correspondence analysis (optimal scaling).

A precise basis for this procedure is given in Bock's statement (Bock, 1960, p. 1): The approach of optimal scaling "is to assign numerical values to alternatives, or categories, so as to discriminate optimally among the objects ... in some sense. Usually it is the least-squares sense, and the values are chosen so that the variance between objects after scaling is maximum with respect to that within objects".

The application of optimal scaling to ordered categories, however, frequently involves problems concerning the mapping of the existing relation of order upon optimal weights.

The existing (traditional) solutions to this problem are not satisfactory, as the number of scale points (categories of variables) have to be reduced until the given relation of order in the optimal weights is reached (Bradley and Katti, 1962, Nishisato, 1980, 1986). Uniting categories, however, means a loss of information.

In this paper a short introduction to a new approach avoiding the loss of information resulting from the uniting of categories will be given.

METHOD

Suppose that a scoring scale has n points (a variable X has n categories) designated by $x_1 < x_2 < ... < x_n$. In an experiment the corresponding relative frequencies of occurrences are p_1, p_2, ... p_n.

We use "statistical scaling" ("Markierungsskala", Pölz, 1988, p. 107 ff) in order to get starting points (x_i^*) for our procedure according to the following formula:

$$x_i^* = \sum_{k=1}^{i-1} p_k - \sum_{k=i+1}^{n} p_k \quad . \tag{1}$$

The starting points are only based on the relative frequencies of the categories and are standardized between -1 and +1.

If we assign a constant frequency to the intervals $x_i^* \pm p_i$, we can show (Pölz, 1988) that the mean of these starting points is zero and that the variance is given by

$$s^2 = \frac{1}{3} \left(1 - \sum_{i=1}^{n} p_i^3\right) \quad . \tag{2}$$

These starting points are excellently suited for setting up a system of discrete Legendre polynominals.

Let L be the matrix with the Legendre polynominals in rising order as rows and the categories as columns.

If we think of a crosstable we can easily proceed with the other variable - let's call it Y with m categories - in the same way: We begin with starting points y_i^* and compute the Legendre polynominals forming matrix S.

Let P be the matrix of two dimensional relative frequencies of the crosstable. We calculate:

$$Q_{yx} = S \ P \ L' \quad . \tag{3}$$

Because of the orthogonality of the Legendre polynominals the expressions

$$Q_{xx} = L \ R \ L' \quad \text{and} \quad Q_{yy} = S \ C \ S' \tag{4}$$

are equal the identity matrix (I_m and I_n respectively).

Indices m and n respectively show the dimension of the identity matrices, where C is the matrix of the marginal relative frequencies of Y and R the matrix of the marginal relative frequencies of X in the principal diagonal and zero in the non-principal elements.

We get the familiar quadratic form:

$$\begin{bmatrix} v' & w' \end{bmatrix} \cdot \begin{bmatrix} I_m & Q_{yx} \\ Q'_{yx} & I_n \end{bmatrix} \cdot \begin{bmatrix} v \\ w \end{bmatrix} \quad , \tag{5}$$

v, w are the vectors of the optimal weights corresponding to the variables Y and X respectively,

$$\text{with } Q = \begin{bmatrix} I_m & Q_{yx} \\ Q'_{yx} & I_n \end{bmatrix} \quad . \tag{6}$$

We are looking for these **v** and **w**, which maximize the canonical correlation.

If we solve the characteristic equation, the resulting squared canonical correlation coefficients are identical and, of course, equal to the biggest non-trivial eigenvalue (λ).

If we postulate

$$\mathbf{w'}\ \mathbf{w} = 1 \text{ and } \mathbf{v'}\ \mathbf{v} = 1\ , \tag{7}$$

the optimal weights corresponding to x_1, x_2, ... x_n, forming vector **x**, can be easily obtained by calculating

$$\mathbf{x} = \mathbf{L'}\ \mathbf{w}\ . \tag{8}$$

Analogously, we calculate for Y:

$$\mathbf{y} = \mathbf{S'}\ \mathbf{v}\ . \tag{9}$$

If the optimal weights (**x**), let's say for X, do not reflect the existing relation of order we can produce any given order by reducing the degree of orthogonal polynominals step by step. So we reduce matrix **L** by cutting off one row after the other, beginning with the last row (this is the row with the Legendre polynominals of the highest degree) and going on with (3). We do this until the weights reflect the existing relation of order.

If there is an ordered constraint for variable Y, too, we proceed simultaneously in the way shown above.

The scope of this paper does not allow a more detailed discussion of this procedure (for further information confer Pölz, 1988).

A brief example will contribute to a better understanding of the new approach:

EXAMPLE

Let's take example number 3 from Bradley and Katti (1962, p. 366). The following table shows the data for a three-treatment experiment with a fivepoint scoring scale:

Scale values (X)	Treatments (Y)			Totals
	I y_1	II y_2	III y_3	
x_1 Excellent	9	4	0	13
x_2 Good	9	14	1	24
x_3 Fair	16	17	3	36
x_4 Poor	5	4	27	36
x_5 Terrible	2	5	12	19
Totals	41	44	43	128

The starting points can be calculated by means of (1). We get:

$$\mathbf{y^*} = \begin{bmatrix} -0{,}898 & -0{,}609 & -0{,}141 & 0{,}422 & 0{,}852 \end{bmatrix}$$
$$\mathbf{x^*} = \begin{bmatrix} -0{,}680 & -0{,}016 & 0{,}664 \end{bmatrix}\ .$$

The matrices S and L built up on y* and x* are:

$$S = \begin{bmatrix} 1 & 1 & 1 & 1 & 1 \\ -1,601 & -1,086 & -0,251 & 0,752 & 1,518 \\ 1,700 & 0,172 & -1,050 & -0,465 & 1,490 \\ -1,453 & 0,902 & 0,548 & -1,174 & 1,041 \\ 1,132 & -1,520 & 1,044 & -0,629 & 0,360 \end{bmatrix}$$

$$L = \begin{bmatrix} 1 & 1 & 1 \\ -1,249 & -0,029 & 1,220 \\ 0,750 & -1,381 & 0,699 \end{bmatrix} .$$

To calculate Q, we need the following matrices:

$$P = \frac{1}{128} \begin{bmatrix} 9 & 4 & 0 \\ 9 & 14 & 1 \\ 16 & 17 & 3 \\ 5 & 4 & 27 \\ 2 & 5 & 12 \end{bmatrix} \qquad R = \frac{1}{128} \begin{bmatrix} 13 & 0 & 0 & 0 & 0 \\ 0 & 24 & 0 & 0 & 0 \\ 0 & 0 & 36 & 0 & 0 \\ 0 & 0 & 0 & 36 & 0 \\ 0 & 0 & 0 & 0 & 19 \end{bmatrix}$$

$$C = \frac{1}{128} \begin{bmatrix} 41 & 0 & 0 \\ 0 & 44 & 0 \\ 0 & 0 & 43 \end{bmatrix} .$$

In order to avoid the trivial solution $\lambda = 1$, we cut off the first row of L and S; because of (3) and (4) we get (6):

$$Q = \left[\begin{array}{cccc|cc} 1 & 0 & 0 & 0 & 0,562 & 0,240 \\ 0 & 1 & 0 & 0 & 0,016 & 0,050 \\ 0 & 0 & 1 & 0 & -0,163 & -0,270 \\ 0 & 0 & 0 & 1 & -0,211 & -0,000 \\ \hline 0,562 & 0,016 & -0,163 & -0,211 & 1 & 0 \\ 0,240 & 0,050 & -0,270 & -0,000 & 0 & 1 \end{array} \right] .$$

Starting from (5) and solving the eigenequation we get $\lambda = 0,4794$, the biggest non-trivial eigenvalue, and through (8) and (9), because of (7), the optimal weights result as follows:

$$y' = \begin{bmatrix} -1,057 & -0,886 & -0,772 & 1,262 & 0,914 \end{bmatrix}$$
$$x' = \begin{bmatrix} 0,764 & 0,661 & -1,405 \end{bmatrix} .$$

These weights are optimal on the one hand because they maximize λ, but they are not optimal on the other hand because of the condition of ordered constraints for variable Y $(y_4 > y_5)$. For this reason we reduce the degree of orthogonal polynominals for Y by cutting off the last row.

Doing this and calculating (6), (8) and (9) we get

$$\lambda = 0,4455,$$

and $y' = \begin{bmatrix} -0,755 & -1,350 & -0,516 & 1,142 & 1,036 \end{bmatrix} .$

We see $y_1 > y_2$ and $y_4 > y_5$, so we continue reducing S and get Q_2 (Q after two iterations):

$$Q_2 = \left[\begin{array}{cc|cc} 1 & 0 & 0,562 & 0,240 \\ 0 & 1 & 0,016 & 0,050 \\ \hline 0,562 & 0,016 & 1 & 0 \\ 0,240 & 0,050 & 0 & 1 \end{array} \right] ,$$

$\lambda = 0,374$ and

$$\mathbf{y}' = \begin{bmatrix} -1,502 & -1,075 & -0,301 & 0,724 & 1,600 \end{bmatrix}$$
$$\mathbf{x}' = \begin{bmatrix} 0,850 & 0,573 & -1,400 \end{bmatrix}$$

Now the ordered constraints for variable Y are fulfilled. If we compare this result ($\lambda = 0,374$) to the result of Bradley and Katti (1962, p. 367)- ($\lambda = 0,124$)- we see how much better our result is.

FINAL REMARKS

This approach is not only suited for finding optimal weights for ordered categories. Because of the matrix notation it can as well be used for calculating the maximum correlation between several variables, which is the square of the multiple correlation.

REFERENCES

Bock, R. D., 1960, Methods and applications of optimal scaling, Psychometric Laboratory, Report No. 25, University of North Carolina.
Bradley, R. A., Katti, S. K., 1962, Optimal scaling for ordered categories, Psychometrika, 4, p. 355-374.
Nishisato, S., 1973, "Optimal scaling of partially ordered categories," Paper presented at the Spring Meeting of the Psychometric Society, Chicago.
Nishisato, S., 1980, "Analysis of categorical data: dual scaling and its applications," University of Toronto Press, Toronto.
Nishisato, S., and Arri, P. S., 1975, Nonlinear programming approach to optimal scaling of partially ordered categories, Psychometrika, 40, 525-548.
Pölz, W., 1988, "Ein dispositionsstatistisches Verfahren zur optimalen Informationsausschöpfung aus Datensystemen mit unterschiedlichem Ausprägungsniveau der Merkmale," Peter Lang, Frankfurt.

MODELLING SOCIAL BEHAVIOUR

MODELLING SOCIAL BEHAVIOUR

Joyce Tait and Geoff Peters

The Open University
Milton Keynes, U.K.

For some people, operational research will always be treated as a branch
of mathematics involving the use of modelling techniques such as dynamic
programming, linear programming and simulation. However, when such models
are deployed in a decision making context, the analyst must also be aware
of social influences that often run contrary to strict mathematical or
economic rationality. These can determine how, or even whether, the results
of an OR analysis are implemented and they can exert their effects at any
stage. Such problems are particularly relevant when more than one organisation is involved in decision making.

There are several approaches to furthering our understanding of such
issues. One is to attempt to integrate 'soft' considerations throughout
the traditionally 'hard' process of mathematical modelling. Another is to
look at the social aspects of the framing of problems that precede a mathematical analysis, or alternatively to concentrate on individual or organisational influences affecting implementation. All of these can involve the
modelling of social behaviour to some extent. However, it is also possible
to apply operational research and modelling techniques to the analysis of
social behaviour for its own sake, and not in relation to the use of other
models as aids to decision making.

In the public arena, for example, organisational interactions become
a major source of constraints on decision making. Modelling such interactions can help in understanding the roles adopted by clients towards a
problem and its analysis. The modelling of apparently irrational behaviour
in complex situations, or of behaviour based on moral considerations can help
us to understand some of the more perplexing aspects of decision making.
Other modelling techniques can be applied to aspects such as the improved
understanding of social conflict or the influence of social norms on our
behaviour. Where working models have been developed they can be used to
expand the range of experience of managers or to improve their skills.

Such research is set apart from that carried out and reported as
social science, pure and simple, by either its emphasis on improving our
understanding of how decisions are made or its contribution to improved
methods of decision making.

COGNITIVE MAPPING AS CONCEPTUAL MODELLING

Sylvia M. Brown

Dept of Agricultural Sciences
University of Bristol, BS18 9AF
On secondment from
Oxford School of Business
Wheatley, Oxford

ABSTRACT

   In recent years have emerged needs to clarify organizational problems
and their ramifications before embarking on elaborate OR techniques to
solve them.

   Soft Systems Methodology (SSM) and Cognitive Mapping have both addressed these needs. Some users report problems with SSM at the Conceptual
Modelling Stage. Cognitive Mapping and Repertory Grids can also cause
problems because of their insistence on polarity.

   During a major study in the Crop Protection Industry using Cognitive
Mapping, disjunction between poles suggested cognitive structures in some
respondents analogous to the "real world/ideal world" dichotomy favoured
by SSM. Use of this polarity is discussed as a possible means to overcome
problems in building Conceptual Models, drawing upon experiences in 82
organisations.

INTRODUCTION

   A pilot study aims to illuminate strategic decisions that are affecting availability in UK of Crop Protection technology. Choice of method
was based on many theoretical and practical criteria and at least one
assumption. (All are discussed in more detail elsewhere).

   Theoretical criteria included the need for "emic" analysis (working
within the conceptual framework of those studied), if what was really
important to them was to emerge.

   Practical criteria included Face-Validity, gaining willing participation of decision-makers at the right organizational level (i.e. Directors,
General Managers, etc.) and obtaining good data with only one interview
per person.

507

A crucial assumption was that the majority of strategic decision-makers sampled would believe their own strategy to be logical and systematic, whether or not they used some formalized Rational Decision procedure. They were likely, therefore, to recognize or wish to appear to recognize "single form" problem-solving methodologies within the convergent traditions of systems analysis.

Thus evolved a Problem Analysis approach to the study, which is described below.

1. CHOICE OF METHODOLOGY

Single Form Problem-Solving Methodologies

These may be stated, crudely, as addressing the questions;

"In respect of problem X:
A - where are we now - (i.e. what are our problems and
    opportunities?)
B - where would we like to be and
C - what are the feasible ways of getting from A to B?"

Management Science in general and OR in particular has a long history of strength in approaches that started at B but allowed the decision-maker to ignore A, at least until C was in progress and possibly altogether. For example, definition of the business mission is translated into operational objectives; thereafter criteria for feasibility are set and alternative strategies for achieving objectives are evaluated against these. Criteria for feasibility may/may not include A. However, as Checkland (1981) has so ably demonstrated, problems and opportunities cannot be taken as given and are often not easy to define.

More recently has arisen the body of work focussed upon A, i.e. on problem clarification and definition. Much of this work may entail other implicit assumptions, e.g. that definition of the business mission is independent of analysis of the problem-field (i.e. falls within the same domain as Conceptual Models?).

Soft Systems Methodology (SSM)

In its requirement to generate a logically necessary ideal system by decomposing a rigorously defined Root Definition, the logical structure of Conceptual Modelling is exactly similar to single-form problem-solving methodologies that begin with a precise objective and decompose it into logical requirements for its achievement. This relationship might have made it attractive for our purpose. Moreover, SSM is prominent in problem-clarification methodologies.

However SSM was not chosen for the present study. Among the reasons were my unquantified and unsystematic observations that a number of respondents find conceptual modelling very difficult at their first attempt.

It is crucial to SSM that the Conceptual Model is logically derived from the ideal state as defined by the Root Definitions(s) and that it does not attempt to describe any actual system.

This is problematic, since many planners do not know what their new sytem should be like; they just know that their present one is unsatisfactory and intuit that causes of and remedies for the unsatisfactory aspects are not all under their control.

508

CATWOE takes care of this aspect satisfactorily enough, but as Checkland himself notes (Systems Thinking, Systems Practice 1981) "... once conceptual model building starts there is a noticeable tendency for it to slide into becoming a description of actual activity systems known to exist in the real world."

That there would be difficulties is predictable from the work of Johnson-Laird & Steadman (1978) on mental models and Short Term Memory (STM) capacity, from which they were able to derive the following "law": "The difficulty of syllogistic inference depends on the number of mental models of the premises", from which he predicted; "The major source of difficulty should be the construction and validation of alternative models ... the task should be harder when it is necessary to construct two models ... harder still when ... three ..." which is just what their experiments showed. There are, however, cognitive and motivational factors to be explained in addition to SSM incapacity.

Anecdotally, some respondents rebel against ideal descriptions as "airy-fairy". This may be particularly likely if they also hate producing Rich Pictures and may be because they find such descriptions inconsistent with their self-image as logical and practical people, or both may be Face-Validity problems.

Some possible explanations for cognitive difficulties

Pessimistic explanations would draw on evidence of the irreversibility of subjective hypotheses under certain conditions (e.g. Bruner & Potter, 1964; Peterson & du Charme, 1967; Valins, 1974; Ross, Lepper & Hubbard, 1977). Alternatively or in addition, the problem may be something to do with inability to "break set".

A further possibility is incapacity to make explicit all the logical steps in the Conceptual Model. Cognitive styles vary as between people and as between problems (see e.g. Pask, G. & Scott, B.C.E., 1972; Bruner, J.S., Goodnow, J.J. & Austin, G.A., 1956; Hudson, L., 1966).

More optimistically, the ability to work purely in the realm of the conceptual may be involved. A Piagetian analysis would explain this; few attain wide-ranging ability in Formal Operations; most remain locked in Concrete Operations (i.e. in the realm of the empirical) for most purposes (Piaget, J., 1950). Where training is possible, ability in Formal Operations can be improved. However, this project only collects data from respondents once. It was also important that respondents succeeded in operating the chosen method and thus felt good. (As a practical aside, even where SSM is being employed conventionally and for its more usual purposes, consideration is due to the demotivational effects that may follow from inability of some respondents to succeed early).

Personal Construct Theory and Cognitive Mapping

Cognitive Mapping was chosen for the following reasons:

Firstly, respondents seem to grasp it very quickly, find it involving, interesting, illuminating and non-threatening. There is room for research as to why this should be so, but a few hypotheses, mostly related to Face-Validity, might be:
- Cognitive Mapping is visually related to spray diagrams, with which many managers are already familiar.
- It deals with causal and consequential relationships and thereby may have a quasi-scientific aura.

- It treats problem-situations as non-simplistic but clarifiable, which may appeal both to "easy answer" and "yes but" managers.
- it externalises "what I've always suspected" half-formed thoughts: (this can inspire no end of self congratulation!).

Secondly, it does not have to proceed linearly; backtracking and filling in are possible.

Thirdly, bringing about organizational change was not our purpose; if respondents were irreversibly committed to their scenarios, that was fascinating evidence, not a problem.

Fourthly, the method avoids many of the problems of traditional survey research, whether this be Marketing Research or Ethnography. For example, since what is being recorded is at all stages visible to and "owned" by the respondent, worries about misrepresentation are allayed.

Fifthly, the need for an emic approach whilst still keeping some structure and analysability is met.

The technique is rooted in Kelly's (1955) Personal Construct Theory (PCT). It is an important claim of this theory that it avoids "ethic" analysis (imposition of the cognitive or attitudinal structures of the researcher). Kelly's Grid Technique requires respondents to describe their world in polarized constructs that can then be dimensionalized and rated or ranked. This requirement to generate the "psychological oppo-site" of each initial construct can create considerable frustration in novice Grid-respondents and presents PCT theoretical problems, as Eden (1984) has noted. The requirement can impede a free-flowing thought-process or push the respondent into logical negatives that reveal little. Many hate doing ratings. Ratings can be highly ambiguous, for many reasons. For instance, whilst some constructs are obviously scalar, others appear to represent discrete either/or oppositions.

"Laddering", as developed by Hinkle (1965) was used to elicit super-ordinate constructs (which may indicate values) by doing roughly what the persistently curious infant does - repeatedly asking "Why?".

Interactive modelling using Cognitive Mappings retains some elements of Grid Technique, modifies or abandons others. It retains bi-polar psychological opposites, uses a version of laddering to explore causes and consequences, allows expression of these and other linkages overtly as you go along (instead of these emerging from the rating analysis later) and does away with ratings.

The Cognitive Mapping software COPE partially retains bipolarity. Simple assertions are treated as bipolar but both poles similar. Mono-tonic concepts (i.e. Constructs) are represented analogously to interval scales.

What are termed "Bipolar" constructs are non-dimensional. COPE works with complete constructs, as do Grid analyses. Linkages between constructs, (called "concepts" by Eden) may be made positively or nega-tively. Links may be causal, consequential or connotational. Making links is not mandatory.

New Uses for Cognitive Mapping?

Some respondents object to linking whole concepts. For example, they may wish to link positive poles only. This suggested that some Mappings may represent not unitary but pluralistic conceptualizations, i.e. the web of "Psychological opposites" describes at least two maps held in parallel with tangential links between them.

It should be emphasized that what is being referred to here is not subgroups of whole concepts, however linked.

There seemed at least a possibility that this duality could aid conceptual modelling. Respondents' freedom to generate just any category of "Psychological Opposite" was withdrawn; they were required to generate concepts of the form "The way it is now ... The way I would like it to be".

2.   COGNITIVE MAPPING AND CONCEPTUAL MODELLING?

At the time of writing, too few results were available for firm conclusions to be drawn but some comparisons are possible, nonetheless, as follows:

Conceptual Model	Split Cognitive Map
Starts with Root Definition	Starts with set of concepts
Models ideal activity-system	Models ideal system-state (in activity/other terms)
Represents a view of sytem	Represents a view of sytem
Uses list of activity-verbs	Uses list of ideal states (in activity/other terms)
Uses influence diagramming	Uses influence diagramming
Includes entities and activities	Includes entities and activities
Requires activity verbs to be connected logically	Items already connected cognitively
May lack connectivity	May lack connectivity
Is formal system-model	Enables formal system-model
Is flawed formal model, requiring improvement by iteration	Is set of conditions (i.e. "wants") for ideal system
Overloads STM store	Well within capacity of STM
Enables actual/ideal comparison	Enables actual/ideal comparison
Enables revelation of inadequacies in model or Root Definition	Enables revelation of inconsistencies, logical gaps and incompatible wants
May "fight" client's normal thought processses	Works with client's normal thought processes
"Appreciative system" largely implicit	"Appreciative system" made more explicit
Client confined in conceptual	Client may move back and forth empirical/conceptual
May be difficulty describing ideal state	"Futures" emerge easily
Pre-evaluated options included	Pre-evaluated options included
Can cause frustration	Causes enjoyment

CONCLUSIONS

Empirically, SSM presents difficulties for some respondents at the Conceptual Modelling Stage.

Reasons for these are related to STM capacity and may also be related to cognitive and motivational factors.

Observations during a large study suggest that use of an adaptation of Cognitive Mapping may help to overcome many of these difficulties.

Comparative methodological studies are necessary before it will be possible objectively to assess the potential of the approach outlined. Proposals for such research are already in train.

REFERENCES

Bruner, J.S., Goodnow, J.J. and Austin, G.A., 1956, "A Study of Thinking". Science Editions Inc., New York.
Checkland, P., 1981, "Systems Thinking, Systems Practice". John Wiley.
Eden, C. and Jones, S., 1984, Using Repertory Grids for Problem Construction. Jnl of Operations Research Society 35(9):779-790.
Hinkle, D., 1965, The Change of Personal Constructs from the Viewpoint of a Theory of Construct Implications. Unpublished Ph.D. thesis in Fransella F. and Bannister, D., 1977, A Manual for Repertory Grid Technique. Academic Press Inc., London.
Hudson, L., 1966, Contrary Imaginations. Penguin.
Johnson-Laird, P.N. and Steedman, M.J., 1978, The Psychology of Syllogisms. Cognitive Psychology 10:64-99.
Kelly, G.A., 1955, The Psychology of Personal Constructs. Norton, New York.
Pask, G. and Scott, B.C.E., 1972, Learning Strategies and Individual Competence. International Journal of Man-Machine Studies 1972 4:217-253.
Peterson, C.R. and Du Charme, W.M. A primacy effect in subjective probability, revision 1967. Journal of Experimental Psychology 73:61-65.
Piaget, J., 1950, "The Psychology of Intelligence" trans. Piercy M. and Berlyne, D.E. Routledge Kegan Paul, London.
Ross, L., Lepper, M.R., Strack, F. and Steinmetz, J., 1977, Social explanation and social expectation: effects of real and hypothetical explanations on subjective likelihood. Journal of Personality and Social Psychology 35:817-829.
Valins, S., 1974, Persistent effects of information about internal reactions: ineffectiveness of de-briefing, in: "Thought and Feeling: Cognitive Modification of Feeling States", H. London and R.E. Nisbett, eds. Aldine, Chicago.

MODELLING SOCIAL BEHAVIOUR IN MANAGEMENT EDUCATION

Hugh Gunz

Manchester Business School
Manchester, U.K.

INTRODUCTION

Managing is a social activity in which the tasks are highly
interdependent, context-dependent and systemic (e.g. Hales, 1986; Whitley,
1987). The work is characterised by brevity, variety and fragmentation, and
is about getting action by pursuing many different agendas through extensive
networks of contacts. It involves balancing the interests of the many
stakeholders in the organization, and being able to discern patterns in a
mass of variety. Doing it successfully depends on non-rational elements such
as vision, creativity, leadership and the ability to operate in a social and
political environment (Luthans et al., 1985) as much as it does on so-called
"rational" analytical skills.

It is the task of the management educator to find ways of preparing
students for this kind of work. If its realities are not faced there is a
danger that management education programmes will be designed which perpetuate
what Anthony (1986: 140) calls the "general banality" of courses which
operate within "an intellectual framework that is non-discursive, closed,
dependent upon acceptance of axioms about human behaviour rather than
critical examination". The techniques of management science, while obviously
a vital part of the manager's armoury, are not by themselves enough
(Mulligan, 1987).

The more that management teachers are able to draw on real experience,
the more useful the learning potentially might be (Mumford, 1987).
Management courses commonly try to capture some of the complexity and
ambiguity of managerial work by means of case studies and projects involving,
as often as possible, real-life people and organizations. But this has a
cost: the penalties for making mistakes become much higher, so students are
discouraged from taking the kinds of risk from which they might learn. For
this reason it could be argued that it is important that education should
insulate people from the real world to some extent. Flight simulators, for
example, form an essential part of pilot training because they allow trainees
to experience situations which would never consciously be risked in a real
aircraft. In addition, real-life projects have the limitation that in
general they are about producing recommendations for someone else to carry
out, so that in many respects they teach the skills of consulting more than
those of managing.

The development of computer-based simulations in management education can be seen, at least in part, as a response to these challenges (Larréché, 1987). From the point of view of the management educator a particularly useful property of computers is their ability to simulate an environment which reacts realistically to students' actions (Robinson, 1985). This makes them ideal for teaching people how to manage highly complex systems which, in real life, are unforgiving of failure. To some extent at least they are attempts to simulate social systems, sometimes whole organizations and sometimes just parts of them. Simulations have been designed to illustrate problems in most branches of management education. They do not necessarily need computer models (e.g. Kaplan et al., 1985); "manual" simulations can also be used to investigate executive decision-making processes (Moynihan, 1987).

Modelling a social system is a substantial challenge. In many cases the simulations used in business games do little more than work from the assumptions of microeconomists about the behaviour of, for instance, markets and production functions. Although computer networks are beginning to be used to simulate communication behaviour (e.g. Noel et al., 1987; Rickards, 1987), the models underlying most computer-based management games are algorithms which, with few exceptions (e.g. Wooliams and Moul, 1982), have to be supplied with data (the "decision") at the end of each "period" so that the game can move forward.

It proves far from easy to move beyond simulations based on conventional microeconomic assumptions towards modelling social systems in their entirety. The state of organization theory is such that the behaviour of any such model is likely to be controversial. Furthermore even apparently simple, formal situations involving just two people, such as a selection interview, produce an enormous range of possible outcomes which the modeller must anticipate (Rushby and Schofield, 1988). A common way of handling this difficulty is to provide students, or experimental subjects in a research context, with the description of a situation and a predetermined range of actions they can choose from (Holden, 1988). Some writers argue that it is now becoming possible to get computer models to simulate human decision-making in an increasingly realistic way (Sylvan, 1987). Others disagree: Chapman (1987), for instance, uses examples of the use of gaming in systems analysis to demonstrate that, in his view, human intervention is vital for the realistic simulation of complex systems such as human communities.

The difficulties of moving towards full social system simulations probably mean that, in management education at least, emphasis will remain for some time on what Martin (1988) calls "out-screen" simulations. These, which he contrasts with what he labels "in-screen" simulations which focus on the behaviour of the model itself, concentrate on the behaviour of the people interacting with the model. For out-screen simulations the computer is seen as a "tool supporting an essentially real-world activity. This real-world activity is the simulation, seen as a microcosm of a wider slice of reality" (p. 23). Nevertheless it is possible, using more recent developments in information technology, to develop the in-screen aspects of what is essentially an out-screen simulation.

If a trend can be discerned in the development of computer-based simulations it is to do with moving them on to smaller and more portable stand-alone machines (Wolfe and Teach, 1987). Certainly, micro-based simulations have the great advantage of making the simulations more available. But it is proving increasingly possible to exploit newer technologies such as wide area networks (Jones, Scanlon and O'Shea, 1987), videodisc (e.g. Rushby and Schofield, 1988) and mainframe computers to design more comprehensive and realistic simulations. In the next part of this paper I shall describe a mainframe simulation, Proteus, which represents one such attempt. Then I shall review some experience with this simulation which highlights the logistical and pedagogical problems such products seem to give rise to.

THE PROTEUS PROGRAMME

The Proteus programme (Gunz, 1988) is an attempt to simulate a real-life managerial environment in the classroom, designed to allow students to experience complexity and learn how to manage it. A company is simulated on a computer and students take the roles of members of its management team. There are, however, a number of features which differentiate Proteus from other such products. First, it operates in simulated real time so that, as in real life, the "managers" of the simulated company can implement decisions by doing something or by leaving things alone: either way, the production, finance and marketing trends to which the firm is subject continue to operate and its business carries on. Second, the members of the management team operate from information appropriate to their functional roles so that, for instance, marketing managers are not supplied with detailed production figures unless the teacher wants them to be. Third, Proteus can be operated from anywhere with a link to the mainframe; the managers communicate with each other using a commercial electronic office system. This gives the opportunity of simulating an operation in which the management team are geographically remote from each other (such as in a multi-site company) and also to use it to teach many forms of business administration in distance mode. Fourth, not only is Proteus a stochastic discrete event simulation, but it is non-deterministic in other respects. The teacher can adjust the corporate environment in many different ways as the simulation progresses, and introduce ambiguous and confusing information which the managers have to evaluate and filter. This makes it difficult for students to "beat the model", a frequent goal of participants in management games. Finally, Proteus is an architecture for organizational simulation, a framework capable of being used in many different ways. Teachers can use the architecture to design simulations for their particular educational purposes, rather than be confined to just one scenario or kind of organization. So Proteus is the name of a family of simulations rather than being one single simulation.

In Martin's (1988) terms Proteus is an out-screen simulation. The social system model is of the "manual" kind characteristic of role-plays in the sense that, other than the algorithm-driven responses of the quantitative model which simulates the company's production, marketing and financial operations, the behaviour of the social system is determined by the actions of participants and teacher. But a great deal of this social behaviour impacts directly on the behaviour of the quantitative model. For instance, if the teacher is dissatisfied with the way in which the managers are treating a customer, he can make the customer stop ordering. He may also tell the sales manager that this has happened, or he may simply wait to see whether anyone notices. So it can seem to the players that a social system is being simulated, even though not all of the responses are machine-driven.

There are many ways in which the Proteus architecture might be developed, ranging from simple demonstrations of how information technology can be used in a business environment to full industry simulations of many kinds (Gunz, 1988). The most complex application to date, Protocom, was designed to teach organizational communication by simulating the way managers must assemble information from jigsaws of data, some of it the so-called "hard" quantitative data produced by management information systems and some the "soft" qualitative data that come from notes, letters, memos, minutes of meetings, face-to-face contact and telephone calls. The simulation combines the features of a complex computer-based management game and an in-basket exercise using a modern commercial electronic office environment. The students' task is to manage a medium-sized rainwear company facing a number of problems which can only successfully be defined and solved if they learn to locate information buried in a mass of unimportant detail, share it effectively and ensure that the right people are told what they need to know to take corrective action. They must make commercial and operational decisions which affect the firm's operations, and deal with a large number of people and groups both inside and outside the company including subordinates, unions, colleagues, customers, suppliers, local government officers and so on. A more detailed description is given in Gunz (1988).

USING PROTOCOM: THE PROBLEMS OF REALISM IN SIMULATIONS

Protocom has been used on an experimental basis in a number of environments, and experience suggests that in principle it is a valuable teaching tool but that the major advance in realism it brings with it poses significant problems. In terms of its original teaching aim the product succeeds well in demonstrating that managers must use both hard and soft data to manage effectively, and that both under- and over-communication can lead to trouble. It has also become evident that Protocom raises many more managerial issues than just communication, in areas ranging from strategic to functional, operational management, and as such is capable of a much broader teaching application than originally intended.

Some of the problems which are encountered running Protocom come about because the behaviours that are observed are highly realistic, rich and complex, and analysing them can prove difficult simply because of their richness and complexity. These difficulties are being tackled by a combination of teacher training and product modification, are probably not insuperable, and space does not allow a more complete discussion here.

The technology itself poses interesting problems which are clearly a reflection of real-world issues. For instance British managers on executive courses typically lack keyboard skills and so are unable to make much use of the electronic office. Others who surmount that barrier find some of the standard commercial software used for Protocom less than user-friendly, and of course the kind of hardware and software problems to be expected with such a new product exacerbate this. But of greatest interest is students' reactions to the realism of the simulation.

A great deal of time can be taken up during the exercise and subsequent debriefing dealing with complaints about "unrealistic" aspects of the exercise. Some of these complaints are quite reasonable, such as the point that one would not, in real life, use electronic mail alone to discipline someone. Others reflect strongly the kind of problems managers have in real-life companies, although they are not seen in this way by the complainers. Older executives, for instance, remark that they would not expect to use an electronic office themselves. Younger, but nevertheless practising, managers have remarked caustically that they cannot imagine an organization in which the chief executive would not have full and free access to all of the finance director's raw data. Participants often complain of the unfairness of one manager having a lot more routine work to do than his colleagues. Groups will criticise the difficulties and unsuitability of communicating through the electronic office with colleagues literally sitting at a terminal at the next desk.

Protocom, in other words, seems to be paradoxical in its impact. No simulation can ever faithfully duplicate reality, but the Protocom environment comes closer than many classroom exercises in this respect. This is done not just for its own sake but because it is not possible otherwise to expose students to organizational phenomena which simpler simulations conceal by virtue of their simplicity. Everyone who has run some form of educational simulation is familiar with the response: "we wouldn't have behaved like that in real life" even though it is obvious that the respondent has been behaving in a highly realistic way which, in all probability, does reflect their real-life behaviour. This is, of course, one of the great benefits of simulations as teaching tools. A consequence of Protocom's greater realism and departure from being "just a game" seems to be that it encourages participants to compare the simulated experience even more closely with reality.

Although many of the criticisms which this process of comparison gives rise to are justified, it is evident that they also result from a form of displacement activity. Protocom presents participants with a difficult set of tasks which they are unlikely to perform well. It is not easy to confront one's inabilities, and it is much easier to find explanations outside oneself which excuse mistakes. The more complex the task, the more explanations there are potentially available to distract attention. This, in turn, can lead to the simulation designer searching for ways of blocking off the excuses, increasing the complexity of the situation yet further. For instance, by analogy with officer training in the armed services, "real" managers can be brought in to take roles in the simulation and to debrief the students, to give the experience the stamp of authority. But this makes the simulation yet harder to manage and requires managers who have the kind of "authentic" experience which will impress the students, the time available to become closely involved in teaching, and the necessary teaching skills to help the students learn. It could also be seen as a form of displacement activity on the part of the teacher, moving the issue of learning to that of authority in the sense that each side avoids examining the learning taking place, instead trying to convince the other side of the authoritativeness of their viewpoint.

One is therefore drawn to wonder whether there is some kind of practical limit to the level of realism and detail with which social systems can usefully be simulated in the classroom. Is there an asymptotic relationship between the complexity of simulations and learning, mediated by the resistance to learning which comes from students' problems in confronting their learning difficulties? This question raises important issues because developing simulations such as Proteus (which was only made possible by a major grant from IBM) is very costly and probably beyond the means of most educational institutions on their own. These demands on resources mean that, increasingly, institutions are either being drawn into collaborative work to develop teaching material of this kind, or they are buying material developed by larger, richer institutions. Both outcomes have important implications. Collaboration increases interdependencies which are hard to manage and lead to difficult situations if one or more collaborating institutions should withdraw. The increasing use of common teaching material may lead to an unplanned, de facto standardization of teaching across schools which is not necessarily in the best long-run interests of management education (Gunz, 1988).

To summarise, there are good educational reasons for expecting management teachers to continue their search for greater realism in modelling social systems. Developments in the use of information technology, of which Protocom is an example, are opening new possibilities for generating realistic behaviour in the classroom, but these developments raise a number of important issues. Although it is often said that information technology is becoming cheaper the actual cost to institutions of developing and running such simulations is high, and this has major implications for the development of management education. But perhaps the most important conclusion that can be drawn from experience with using Protocom is that the problem with such simulations does not lie in the difficulty of generating realistic behaviour, but in helping students to learn from their behaviour. The greater the realism, paradoxically, the harder this becomes. Simulations of complex social systems, it seems, certainly have the potential to enhance educational experiences, but they do not automate the teacher's job away. In many ways they make the teacher's job more, rather than less, demanding.

REFERENCES

Anthony, P. D., 1986, "The Foundation of Management", London, Tavistock.
Chapman, G. P., 1987, Gaming Simulations and Systems Analysis: Two Factions of the Truth, J. Applied Systems Analysis, 14, 3-15.
Gunz, H. P., 1988, Information technology in management education: Myths and potentialities, Personnel Review, 17(5).
Hales, C. P., 1986, What Do Managers Do? A Critical Review of the Evidence, Journal of Management Studies, 23(1), January, 88-115.
Holden, G. W., 1988, Using computers for research into social relations, Simulation/Games for Learning, 18(1), March, 61-68.
Jones, A., Scanlon, E. and O'Shea, T. (eds.), 1987, "The Computer Revolution in Education: New Technologies for Distance Teaching", Sussex, The Harvester Press.
Kaplan, R. E., Lombardo, M. M. and Mazique, M. S., 1985, A Mirror for Managers: Using Simulation to Develop Management Teams, J. Applied Behavioural Science, 21(3), 241-253.
King, W. R., Raghunathan, T. S. and Teng, J., 1986, Personal Computers in Business Education: An Experimental Assessement, Omega, 14(4), 317-323.
Larréché, J.-C., 1987, On Simulations in Business Education and Research, J. Business Research, 15, 559-571.
Luthans, F., Rosenkrantz, S. A. and Hennessey, H. W., 1985, What Do Successful Managers Really Do? An Observation Study of Managerial Activities, J. Behavioural Science, 21(3), 255-270.
Martin, A., 1988, Out of the screen: computers and simulation, Simulation/Games for Learning, 18(1), March, 21-29.
Moynihan, P., 1987, Expert Gaming: A Means to Investigate the Executive Decision-Process, J. Operational Research Society, 38(3), 215-231.
Mulligan, T. M., 1987, The two cultures in business education, Academy of Management Review, 12(4), 593-599.
Mumford, A., 1987, Using Reality in Management Development, Management Education and Development, 18(3), Autumn, 223-243.
Noel, R. C., Crookall, D., Wilkenfeld, J. and Schapira, L., 1987, Network gaming: a vehicle for international communication, in "Simulation-Gaming in the late 1980s", Crookall, D et al., eds., Oxford, Pergamon.
Rickards, T., 1987, Can Computers Stimulate Creativity? Training Implications from a Postgraduate MBA, Management Education and Development, 18(2), Summer, 129-139.
Robinson, J. N., 1985, Using Games to Present Economics to Managers: A Survey of the Literature, Management Education and Development, 16(1), Spring, 17-30.
Rushby, N. and Schofield, A., 1988, Conversations with a Simalcrum, Simulation/Games for Learning, 18(1), March, 30-39.
Sylvan, D. A., 1987, Supplementing Global Models with Computational Models: An Assessment and an Energy Example, Behav. Sci., 32(3), 212-231.
Whitley, R. D., 1987, On the nature of managerial tasks and skills, their distinguishing characteristics and organisation, Manchester Business School Working Paper 153.
Wolfe, J. and Teach, R., 1987, Three Down-loaded Mainframe Business Games: A Review, Academy of Management Review, 12(1), 181-192.
Wooliams, P. and Moul, D., 1982, Real time company simulation for management development, in "Simulation in Management and Business Education", Gray, L. and Waitt, I., eds., London, Kogan Page.

QUALITATIVE AND QUANTITATIVE MODELS OF SOCIAL BEHAVIOUR

WITH RESPECT TO HEALTH PROMOTION AND WASTE DISPOSAL

Alan Jones[1], Mike Luck[2], Bob Pocock[1], and
Andrea Rivlin[3]

[1]Midland Environment Ltd, Aston Science Park, Love Lane
Birmingham B7 4BJ

[2]Aston Business School, Aston University, Aston Triangle
Birmingham B4 7ET

[3]West Midlands Regional Health Authority, 146 Hagley
Road, Birmingham B16 9PA

INTRODUCTION

At the 1964 Operational Research and Social Sciences Conference all the papers dealt with decision making in formal organisations (Lawrence, 1966). There were no papers concerned with health promotion or waste disposal nor were there any papers which dealt with the social behaviour of individuals and households. In the intervening 25 years there has been only limited progress.

The maintenance of good health is an important aim for most people. OR has made a significant contribution in the U.K. to the National Health Service but has been almost entirely concerned with services for curing ill-health (Luck et al, 1971, for example). Very little has been done for health promotion and the preventive approach to health care. This is despite the Black Report (Townsend and Davidson, 1982) and the follow-up by Whitehead (1987) which have shown that preventable ill-health in the U.K. lags behind most other industrialised countries and that social class inequalities have not reduced.

Waste disposal is an essential service in an urbanised society. The process of collection and disposal of waste is large both in cost and volume. Here OR has made a contribution to issues such as the efficient routing of collection and transport to waste disposal sites (Clark and Gillean, 1977; Riccio, 1984). However, little work has been done on individual and household behaviour and decision making.

The authors have worked together on a number of projects for Health and Local Authorities. This experience has convinced us that there is the danger of an over-emphasis on collecting more data whereas what is needed now is the development of models which provide a greater understanding of social behaviour. This will help to design, plan and manage health promotion and waste disposal services which are more effective and responsive to consumer needs and more efficient.

519

HOUSEHOLD WASTE

Household waste is a by-product of a miscellany of household activities. The process starts with decisions to purchase goods, these are then consumed and, finally, decisions are made about disposal of the waste. This is defined as 'post-consumption' waste. There are other household activities such as gardening which produce 'non-consumption' waste. It is surprising that these aspects of household behaviour have gone largely unresearched and, in consequence, the processes by which waste is generated are poorly understood. This is difficult to justify in view of the large quantities of household waste disposed of each year, currently 19.5 million tonnes in the U.K. costing £500 million, and the environmental problems this poses.

Two basic approaches have been used in the modelling household waste generation: input-output models and explanatory models. The former approach bases predictions of outputs (waste) on a knowledge of inputs (consumption). Whilst such models are reliable in their prediction of the generation of certain items of waste, e.g. glass, where the delay between purchase and disposal is short, they are less successful in predicting other types of waste, obsolete consumer durables for example, where the delay between purchase and disposal can be years rather than days (Rufford, 1984). Input-output models assume a 'mass balance' flow of materials which, although a reasonable assumption for post-consumption waste, is inappropriate in explaining the generation of non-consumption waste such as garden refuse.

Explanatory models have been used successfully in predicting the quantities and composition of dustbin waste (Rufford, 1984). Since the primary constituent is post-consumption waste it is not surprising that the factors which correlate with dustbin waste generation are also those which we most often associate with food purchasing (Wenlock, 1980; MAFF, 1981). The multiplicity of routes for the disposal of non-consumption waste has hindered the development of a model of non-consumption waste generation, since two sets of behaviour must be considered, one relating to its generation and another to its disposal. Household size appears to be unrelated to the disposal of non-consumption waste at civic amenity sites. Factors which appear to be more important are home ownership and, not surprisingly, car ownership (MEL, 1988).

HEALTH PROMOTION

Health Promotion has been defined in many ways. One definition which encompasses the essence of the subject is:

"Health Promotion is the process of enabling people to increase control over, and to improve their health...it involves the population as a whole in the context of their everyday life rather than focusing on people at risk for specific diseases."
(World Health Organisation, 1984)

This definition includes the ideas that health promotion is positive and involves everyone. One dimension that is not included is that of community participation and responsibility of society. The emphasis of the definition is on the individual. However, much that affects our health is not controlled by the individual e.g. housing conditions, smoke free atmosphere, provision of healthy foods. Health promotion should include both of these ideas of community and individual responsibility.

Within the NHS, health promotion has concentrated on providing individuals with the information on which to make choices. This had been followed in the belief that by providing information (knowledge) then people will change their behaviour accordingly. This link has not been proven so health promotion has had to develop strategies for helping people understand why the desired behaviour is the correct thing to do. One of the more innovative approaches is to use a community development project. This involves enabling a community to look at its own health needs and providing only the professional assistance asked for by the community.

In order to make effective plans for health promotion it is essential to have quantified targets and then progress can be monitored against these targets. Most health promotion activities are aimed at the whole population or sub-groups defined by factors such as gender, age, social class and geographical area of residence. A relatively simple population model of this type can be set up with a microcomputer spreadsheet and can be 'rolled forward' with various sets of assumptions about trends and effects of interventions. Smoking is a major risk factor which can be modelled in this way. The prevalence of smokers in the population is well established and transitions between states due to starting and giving up have been estimated and can be applied in the model.

The most widely accepted qualitative model used in health promotion is that of Fishbein and Ajzen (1975). The antecedents of behavioural intentions are attitudes which are based on an individual's assessment of the likely outcomes of behaviour and the value placed on those outcomes. This model, however, does not include the social and economic barriers such as lack of facilities which may prevent a behavioural intention being acted upon.

SOCIAL SCIENCE

Measurement in social science has developed in a different way to that in physical science. The tools that are used are based on asking people for information. There are many survey methods of which the most familiar are: postal questionnaires, household interviews, street interviews, telephone interviews and discussion groups. These can be divided into the more quantitative and the more qualitative methods and they have different purposes. Details can be found in a textbook such as Moser and Kalton (1971).

At the present time postal questionnaires are not so popular because of their traditionally poor response rate. However, in the health promotion field they have been used with considerable success (Luck et al, 1988). The growth area is in the use of qualitative methods such as in-depth interviews and group discussions. These methods are used to gain a better understanding of a subject area rather than collecting representative views. A recent book by Gordon and Langmaid (1988) looks at the development of group methods.

MODELLING

In this section we analyse the various approaches to modelling that have been described above. The purpose is to demonstrate the roles and functions of different types of model and also to highlight areas where there appears to be scope and potential for improving the quality of the models. The framework for our analysis of models is based on two dimensions: function and data quality. We have categorised the functional nature of models into four types: (a) descriptive, (b) predictive, (c) forecasting and (d) explanatory and diagnostic.

Descriptive models help to illuminate and classify some phenomena of interest. The apportioning of household waste into 'post-consumption' and 'non-consumption' represents a simple descriptive model for classifying waste according to its origins. Another common descriptive model of waste is its apportionment into 'combustible' and 'non-combustible'. This is done by waste management engineers when investigating the potential energy value of waste which may be incinerated, as in the district heating schemes found more often on the Continent of Europe, or turned into fuel pellets.

Predictive models attempt to produce estimates of the state of the dependent variables of some phenomena of interest from knowledge of other (independent) variables. One example is where the weight of weekly wastes arising in a town is predicted from the population of the town multiplied by a 'mean weight per capita' coefficient. More advanced versions exist where various census data parameters act as the independent variables.

Forecasting models are distinguished from predictive models by the explicit use of time-variants in the independent variables. Predictive models can be turned into forecasting models so long as the relationship between independent and dependent variables is stable over time and forecasts of the independent variables, e.g. population size in the example above, are available.

Finally, explanatory or diagnostic models may be defined as models that attempt to articulate a theory. Evidently the theories of behaviour applied in the health promotion context are examples of this approach. The 'health beliefs' model proposed by Fishbein and Ajzen (1975) is an attempt to explain behaviour on the basis of knowledge and attitudes. Explanatory models of this sort are often effective in predictive or forecasting applications, but predictive or forecasting models only acquire explanatory power when causality is established in the relationship upon which the model is based.

Having classified models in this way we can stand back and classify again on the basis of data quality. Crudely, there is a dichotomy between models that are: (a) qualitative, and (b) quantitative. Qualitative models are not expressed algebraically because their parameters are usually not measurable beyond ordinal scales. They should not, however, be regarded as intrinsically inferior to quantitative models. For example, to assert that "smoking stems from social isolation" is to pose a qualitative model. If valid, it does not particularly matter that 'social isolation' cannot be measured on interval or ratio scales so long as the model helps to target health promotion initiatives more effectively.

Quantitative models are generally put into mathematical expressions and the variables are measured on interval or ratio scales. Waste collection rounds, for example, can be formulated in a mathematical model which can then be computerised and an optimal, or at least very good, solution found.

Issues which present difficulties in measurement such as 'stress' and 'bulky household waste' are also those where the state of theory is poor and therefore where the models are qualitative and descriptive. Thus, the two axes for appraising models are fundamentally linked.

Having established the above framework for classifying models, it is now pertinent to consider the OR context. Here there are two levels of need: first, the practical need for models to aid decision makers; second, the need for the state of the art of modelling and underlying theory and measurement to be developed. The highest degree of need is where the practical demand is high and yet models are poorly developed. Clearly we are now entering a judgemental domain where national and local circumstances, the policy agenda, and changing scientific and cultural perspectives dictate the perceived need. These judgements contrast with the more rigorous basis of classifying models by functional type and data quality set out earlier.

From our brief appraisal it is evident that within the waste context there are opportunities for applying quantitative computer waste management models. While some progress has been made, these models lack a sound behavioural basis and this is where more research is needed. This is likely to require qualitative research before behaviourally robust quantitative models can be developed.

For health promotion it is less easy to adopt a systems management perspective and the research interest has been more closely applied to understanding behaviour change and, in particular, the social and environmental factors mediating between intended behaviour and actual behaviour. The relative importance of these contrasting factors is poorly understood and the models are weakened by this gap in knowledge.

DISCUSSION

In the field of household waste, surveys have focused on inputs and outputs. There has been little study of people, behaviour and motives. Policies to encourage recycling or reduce littering are hampered by lack of understanding of attitudes and behaviour. For health promotion, by contrast, studies of knowledge, attitudes and behaviour have been more common, but the study of inputs and outputs is less well developed, perhaps because of the difficulty of observation and measurement. Models of health-related behaviour exist but so far research has failed to study and incorporate the role of economic and social factors such as access to facilities, low income and other constraints on behaviour. For waste disposal, economic and social factors, such as access to civic amenity sites and recycling facilities or lack of a car, also intervene between intention and behaviour. Here the basic behavioural models have not yet been developed.

Narrow academic and professional boundaries around waste disposal have meant that OR has tended to focus on limited technocratic applications such as refuse collection rounds. In contrast, health promotion has been more multi-disciplinary and this has facilitated cross-fertilisation between disciplines. However, OR has been little involved in health promotion to-date. Both waste disposal and health promotion are 'buoyant' items on the political and professional agendas. There is likely to be sustained growth in them and hence there are considerable opportunities for OR to contribute to both development of theory and practical applications. Both subjects can learn more from consumer market research. This is an area of rich and largely untapped potential. Both subjects are concerned with improving social provision and enhancing individual choice. Neither is sufficient in itself and both require the further development of behavioural models.

REFERENCES

Clark, R. M., and Gillean, J. I., 1977, Solid waste collection. A case study, Operational Research Quarterly, vol 28, pp. 795-806.

Fishbein, M., and Ajzen, I., 1975, Belief, Attitude, Intention and Behaviour, Addison-Wesley, Boston.

Gordon, W., and Langmaid, R., 1988, Qualitative Market Research: A Practitioners and Buyers Guide, Gower, Aldershot.

Lawrence, J. R., (ed) 1966, Operational Research and the Social Sciences, Tavistock Publications, London.

Luck, G. M., Luckman, J., Smith, B. W., and Stringer, J., 1971, Patients, Hospitals and Operational Research, Tavistock Publications, London.

Luck, M., Lawrence, B., Pocock, R., and Reilly, K., 1988, Consumer and Market Research Methods in Health Care, Chapman and Hall, London.

MAFF, 1981, Household Food Consumption and Expenditure, HMSO, London.

Midland Environment Ltd., 1988, Factors Affecting Civic Amenity Site Catchment Areas, Midland Environment Ltd., Aston Science Park, Birmingham.

Moser, C., and Kalton, G., 1971, Survey Methods in Social Investigation, Heinemann, London.

Riccio, L. J., 1984, Management Science in New York's Department of Sanitation, Interfaces, vol 14, pp. 1-13.

Rufford, N. M., 1984, The Analysis and Prediction of the Quantity and Composition of Household Refuse, PhD thesis, Aston University, Birmingham.

Townsend, P., and Davidson, N., 1982, Inequalities in Health. The Black Report, Penguin, Harmondsworth.

Wenlock, R. W., 1980, Household food wastage in Britain. British Journal of Nutrition, vol 43, pp. 53-70.

Whitehead, M., 1987, The Health Divide: Inequalities in Health in the 1980s, The Health Education Council, London.

World Health Organisation, 1984, Health Promotion: A Discussion Document on the Concept and Principles, World Health Organisation, Copenhagen.

A FORMAL REPRESENTATION OF INTERORGANIZATIONAL INTERACTIONS

FOR PUBLIC DECISIONS

Luigi Mucci, Anna Ostanello, and Alexis Tsoukias

Politecnico di Torino
Dipartimento Automatica e Informatica
Torino, Italy

INTRODUCTION

The interorganizational processes developing about "public decisions" (cfr. Ostanello et al., 1987) show the configuration of a generally non organized interaction of a multiplicity of actors, both individuals and organizations. The structuring, the evolutive dynamics and the effects of these processes are not well known. Descriptive/interpretative models of the actors' behaviour in actual interorganizational processes are lacking. OR interventions are especially problematical within these contexts. An empirical research has been conducted about these kinds of processes based on an approach integrating a political-organizational paradigm with the classical-rational one. That permits factors traditionally neglected by the classical OR approach to be taken into consideration and Social Science concepts to be used to develop both descriptive and operational softer tools.

The concept of *interaction space* (IS) has been introduced (Ostanello et al., 1987) on the basis of a long standing observation of situations where the simultaneous presence of "public" and "private" actors appeared as a reciprocal necessity. An IS is conceived to represent an informal "meeting domain" of subjects from different organizations. It is an informal, abstract structure, defined and recognized by the intervening actors themselves, allowing exchange and communication conditions by a public confrontation (see for instance Habermas, 1981). It is identifiable by the observer, mainly due to the public character that some of the intervening subjects give to it.

An interaction space is generally "activated" following an action that we call "primer". Such an action is promoted by one or a few subjects (organizational or individual actors) that may be in a more or less competing reciprocal position. A primer action may be generated by some combination and accumulation in the time of different factors as, for instance: chance, an intersection of programmed or foreseen actions by actors of different organizations, or a strategy. A primer action can be conceived as the result of an activity of abstract association of objects (both concrete and abstract) that an acting subject (promoter) estimates in relation with a particular object. Normally, the promoter has a privileged relation, for instance of property or of formal pertinence, with that object.

This association can produce a new object, a *meta-object* (m.o.), permitting the attribution of a "sense" to every considered object. The m.o. generally has a double function, projective (from the objects to the m.o.) and evocative (from the m.o. to the objects); that allows on the one hand, an integration of the representations of different processes developing about objects which do not apparently have any other relation, and, on the other hand, a legitimate participation of other actors and then their "contribution" of other objects.

The notion of meta-object is then substitutive or a generalization of that of "problem definition", usual in the OR literature.

The meta-object - as it has been conceived (or identified)- is used by the primer actor to develop an interaction space, with the aim of making a "problem situation" (in the French sense of "problématique" (Roy, 1985)) about some object public. That "problématique" implies the interaction of some other actors, belonging to different organizations, and thus generally the introduction of other abstract and concrete objects, in the space. Within such situations, the interaction space appears as a "tool structure" (in the sense of Organization Theory (see for instance Mackenzie, 1986)): on the one hand, the actors need or use it to "solve" some problems, to take opportunities of action and more frequently to act on the environment; on the other hand, this structure progressively defines a tight network of action constraints (cf Boudon, 1977; Schwarz and Thompson, 1985).

An IS is not static; its evolution is perceivable by other individuals or organizations that may decide to "enter it" directly or to use it to pursue some individual purposes, interests (indirect "entrance"). The consensual sense, attributed by the intervening actors to the meta-object, can also evolve with time.

Some elements of an interaction space are visible and recognizable by the observer/researcher which may correlate them, so as to produce a reference frame for an integrated reading of a variety of observed episodic phenomena. If the implicitly consensual "sense" of the meta-object is somehow stable, for a given interval time (stability phase), it can allow the observer to develop a formal representation of the interorganizational interactions.

FORMAL REPRESENTATION OF AN INTERACTION SPACE

A formal representation of an IS has been outlined, by assuming the following principal structural elements: three sets of elements, respectively named "acting subjects" or "participants", "objects" and "resources"; a relation structure on these sets; time as an independent variable. Sets and relations may evolve with time. The acting subjects may have different perceptions and evaluations of this variable and use it as a resource.

Sets of elements

Acting subjects or participants. Depending on the observer's point of view, the participants can be conceived both as individuals and organizations. From an empirical basis, two main typologies of subjects are distinguished: public and private. The presence of both subjects shows a reciprocal necessity to control different uncertainty dimensions about the objects (Ostanello et al., ,1987); Depending on their "position" in the space, the participants can be differently characterized. A characterization can be essentially based on relationships that the subject can have both with resources and with other actors.

Objects. Concrete and abstract elements are both denoted by this denomination. They may not even be explicitly defined, conceived by the intervening actors at individual or collective level. Generally they are complex objects (Landry, 1987) presenting different and interdependent dimensions of uncertainty (economical, technical, legal, social...) (see Mackenzie, 1986). Referring to the actors, uncertainty is here intended as a reduced actor's capacity of controlling some kinds of resources estimated necessary to pursue the action or to define a behaviour. An acting subject may have relations of different nature or intensity, with one or some of the objects, for instance: property, institutional pertinence, interest, attention. Some kind relationship can also exist between objects; it can be direct or indirect through some intervening subjects. As it will be shown in the following, different object typologies can be recognized. An object can shift from one class to another depending on the acting subjects, time and of course the observer.

Resources. Three principal kinds of resources are distinguished:
- *quantifiable* (physical, of production ..), normally considered in the classic OR models;
- *non quantifiable*: qualitative or abstract they may be of different nature, as for instance, information, consensus, legitimacy, relational resources;
- *of behaviour*, respectively, passive (authorization, rules, norms, laws,...) and active: each participant shows a capacity of mobilizing his own resources and of activating new ones in different ways (cf the concept of "habitual domain" in Yu, 1979). Following Yu (1979), these resources could be considered as "operators" that may be activated by the actor to develop and structure the interaction space. Examples of operators may be identification, association, projection, structuring... By association and structuring, a promoter can develop a meta-object which may induce other actors, by identification and association, to interact.

Relations structure

The structuring and evolution of an interaction space, as empirically observed, may be based essentially on the following relations: two binary relations, Sa and So, on the sets respectively of participants, A, and of objects, O; a relation between A and O, Sao, and a ternary relation, Saor, on the sets A,O and of resources, R. Identification and qualification of the relations depend on the criteria adopted by the observer.

The expected functions of an interaction space, are not necessarily developed. That depends on the dynamics of the space and on the evolution of its states, the control of which may be difficult.

POSSIBLE STATES OF AN INTERACTION SPACE

In the evolution of an interaction space, different possible states may be recognized. Some states have been distinguished on an empirical basis. They have been named: "controlled expansion", "non controlled expansion", "stalemate", "controlled contraction", "dissolution", "institutionalization". The state identification has been conducted using some typologies of objects and of participants, summarized in the Tables 1, 2 and 3 (for a full description see Ostanello et al., 1987)

Table 1 . Typologies of objects by actor and resource relations

TYPOLOGY	RELATION	FREQUENCY
"Strong concurrential"	Saor	Several participants and high level of involved resources
"Weak concurrential"	Saor	Several participants and low level of involved resources
"Oligopolistic"	Sao	Few participants
"Monopolistic"	Sao	One participant

At a given moment, a *profile* of a participant may be outlined by a multicriteria model (Roy, 1985), referring to the resources relevant to a given object and to the actor's behaviour (see Table 2).

Table 2 . A multicriteria model for the actor profile

CRITERIA	EVALUATION SCALES
"Endowment of resources"	Weak, Sufficient, Solid
"Need of resources"	Limited, High, Critical
"Mobilization capability"	Null, Limited, High
"Present behaviour"	Passive, Active

A "dominant" or "non dominant" actor property may be deduced by confronting the profiles of different actors (see Ostanello et al., 1987).

Table 3 . Some typologies of participant

TYPOLOGY	CHARACTERIZATION
"Principal"	Not dominated about any strong concurrential object such as the one Sao holds
"Strong"	Not dominated about any weak concurrential object such as the one Sao holds
"Key"	Not dominated about any oligopolistic object such as the one Sao holds
"Opportunist"	Not belonging to any of the previous classes not dominated about any elementary monopolistic object such as the one Sao holds

The IS states are synthetically illustrated as follows.

"Controlled expansion". Situation characterized by a relatively limited number of participants who have a well characterized profile and of elementary objects all pertinent to the m.o. (control at the "entrance").

One principal actor or some strong participants are identified which can reach: i) either some answers to "uncertainty problems" (for instance, obtaining public legitimacy), or ii)negotiated solutions to problems that could find a clear definition by IS, or iii) some control mechanisms (Ouchi, 1979) on identified uncertainty sources.

Time is also controlled by the principal and strong participants and can be differently evaluated.

"Non controlled expansion". Situation characterized by a great number of different type participants, each anchored to one or some objects that have found a sense in or by the IS.

One or several strong concurrential objects and a variety of concrete and abstract objects are present.

The great number of objects make either the structuring of collective congruent representations (a meta-object) or the attainment of some solution of individual problems by principal and strong actors impossible.

This kind of situation can develop in a "crisis" concerning some principal actor (a public administration, for instance). The lack of control at the "entrance" allows public alliances suitable for weakening the position of some participants. This situation may also correspond to cases where none of the participants wants or has the capability of assuming a dominant position. The IS thus fulfills a function of "communication space".

"Stalemate". This kind of situation is normally encountered following states of non controlled expansion. Generally a limited number of participants is found, "survivor" by choice or necessity.

It corresponds to a paralysis of the participants to the action, without however any risk of a crisis.

There are no problems of control; time seems to have no meaning.

The activated IS can allow some strong profile actor to enter it, thus unlocking the situation with a solution of a own problem (new object, ideas) which has been conceived on the elements of IS. IS can evolve to a state of controlled expansion or contraction.

"*Controlled contraction*". This kind of situation can be recognized following other different states of IS. It is characterized by a "selective expulsion" of elements of any kind.
Few principal participants are present. At the aim of reducing disturbing factors, they succeed in imposing a given representation of relationships.

"*Dissolution*". It is a possible finale state of the evolution of an IS.
Actors have no more reason or possibility to participate since they do not identify a m.o. evoking a sense for their own objects.

"*Institutionalization*". It is a possible final state of the evolution of an IS, normally following a "controlled contraction" after a legal regulation of some relations is reached.
The IS enters the official endowment of resources of the participants.

Table 4. A short description of the process. The client is a regional local authority. The end of the process, t_4, corresponds to the approval of a new law to finance a new highway and to the dissolution of the IS.

Time	t_0	t_1	t_2	t_3
Main Events	1975: Law 492 stops the funds for new highways. 1976: A committee promoted by the client is set up about the Susa Valley road network; all interested actors are invited to enter	February 1978: Conference promoted by the client, about the Susa Valley road network. A solution, proposed by the client, is generally accepted	April 1979: Conference promoted by some private actors, about a new road. End 1979: The PW Ministry presents a first proposal of law to finance the new road	June 1980: A new highway tunnell opens to traffic. The Susa Valley municipalities organize demonstrations
Client uncert. factors	Relationships with: central government, local municipalities; intervention behaviour; kind of planning policy	To find ordinary funds for the solution; consensus on the new road project; to meet private actor demands	Relationships with private actors, central government to find any kind of funds; identify the new road proposal	Financing a new highway; relationships with Ministry; controlling the funds; internal stability
Princ. actors	Client	Client	No principal actors; some strong actors	PW Ministry
Meta-object	Non existing; activation of an association process	"Communication policies in the Susa Valley"	"The new road"	"Financing the new highway"
States of IS	Some identifiable relations have a sense for the present client's situation	Controlled expansion	Non controlled expansion	Controlled contraction

AN APPLICATION TO A REAL CASE

The model has been used to analyze the behaviour of a public actor (p.a.) in a complex interorganizational process concerning the road network in a North Italy area. A short description of the process is presented in Table 4.

The dissolution of the IS, activated by p.a., generated for this promoter a "problem situation" for which an ex-post study was required. An initial "controlled expansion" enabled p.a. to reduce some uncertainty factors temporarily. A subsequent "non controlled expansion", induced by the behaviour of p.a., caused a shifting of the m.o. sense and a restructuring of the IS in a "controlled contraction", where p.a. lost his principal actor position.

Referring to the client (p.a.), the model could have a function of representation, of support to an understanding of the context situation; it allowed the identification of an adopted behaviour unsuitable for the actual dynamics of the space. Results of such a nature could not have been possible by a classic OR approach.

CONCLUDING REMARKS

The model which has been presented is one of the first results of an ongoing empirically based methodological research. It has been conceived as a soft tool to support representation and conceptualization activities of an analyst/researcher in complex cases of intervention.

Many research problems remain open, mainly concerning the dynamics of the interaction space. As the application shows, the evolution of the states of IS may induce a transformation of the "sense" that the participants may attribute to the meta- object. To reach a formal representation of these transformations is one of the aims of our present research. Results in that direction could be operationally useful, for instance, to outline possible behaviour to exert or to reach some control on the IS, in order to meet some expected result.

REFERENCES

Boudon, R., 1977, "Effets pervers et ordre social", Presses Universitaires de France, Paris.

Habermas, J., 1981, "Handlungrazionalität und Gesellschaftliche Rationalisierung", Suhrkamp, Frankfurt am Main.

Landry, M., 1987, Les rapports entre la complexité et la dimension cognitive de la formulation des problèmes, in "L'aide à la décision dans l'organisation", AFCET ed., Paris.

Mackenzie, K.D., 1986, Virtual Positions and Power, Management Science, 32:642.

Ostanello, A., Mucci, L. and Tsoukias, A., 1987, Processus "publics" et notion d'espace d'interaction, Sistemi Urbani, 2/3:forthcoming.

Ouchi, W.G., 1979, A conceptual framework for the design of Organizational Control Mechanisms, Management Science, 25:833.

Roy, B., 1985, "Méthodologie multicritère d'aide à la décision", Economica, Paris.

Schwarz, M. and Thompson, M., 1985, Beyond the politics of interests, in "Plural Rationality and Interactive Decision Processes", M. Grauer, M. Thompson and A. Wierzbicki eds., Springer-Verlag, Berlin.

Yu, P.L., 1979, Behaviour basis and Habitual Domains of Human Decision Behaviour. Concepts and Applications, in "Multiple Criteria Decision Making Theory and Applications", G. Fandel and T. Gal eds., Springer-Verlag, Berlin.

THANK YOU FOR NOT SMOKING:

SIMULATING THE GROWTH OF A NEW SOCIAL NORM[*]

Chanoch Jacobsen and Richard Bronson

Technion, Haifa, Israel

Fairley Dickinson University, Teaneck, NJ, U.S.A.

BACKGROUND

Social norms about smoking in company have changed greatly in the last few decades. Forty years ago it was still customary for hosts to offer cigarettes or cigars to their guests, and people smoked freely and companionably. The occasional discomfitted non-smoker would ask to be excused, or at most request that a window be opened. Today, fewer people smoke and many object to others' smoking in their presence, claiming a legitimate right to clean air. The major reason for this change is that research evidence has associated tobacco smoke with heart disease and lung cancer, both in smokers and in those exposed to smoke, the so-called "passive smokers". These findings have been well publicized and given official sanction in the reports of the U.S. Surgeon General. Voluntary associations, such as the American Cancer Society, A.S.H., and Non-Smokers' Rights, have alerted the public to the dangers of smoking and are lobbying for formal legislation to restrict smoking in public places.

In light of our theory of norm systems in contemporary societies (Jacobsen and Bronson, 1985), this development carries particular interest. The theory explains the dynamics of changing norms, and our simulation model of the theory has successfully replicated five different data sets of such changes. Most informal norms, however, have changed towards greater leniency, being part of the growing permissiveness in modern liberal societies. Smoking in company is different in that the change is towards more restriction. In fact, a new informal norm has emerged that prohibits smoking in the presence of others without their consent. If our theory can account also for this change, it will be an important extension of its applicability.

THEORY AND MODEL

The following propositions summarize the theoretical argument.

---

[*]The financial support of the United States-Israel Binational Science Foundation under grant 85-0057 is gratefully acknowledged.

531

1. Contemporary industrialized societies have structural impediments (e.g., anonymity, transience, cultural heterogeneity) to effective autonomous and informal social control. Thus a greater part of normative regulation falls to the formal mechanisms of social control, which are themselves limited by technical, political, and economic constraints (Wirth, 1938; Simmel, 1950). Therefore, the more structural impediments there are, the more norm violators there will be.

2. Rapid technological and demographic changes create structural strain in the norm system, which lends partial legitimacy to many norm violations (Williams, 1970:420). Therefore the more rapid and far-reaching the changes, the greater will be the legitimacy of violations.

3. As a result of (1) and (2), isolated and sporadic norm evasions coalesce into institutionalized evasions and create norm ambiguity (Merton, 1957:318). Therefore, the more norm violators there are, and the greater the legitimacy of violations, the more institutionalized evasions and norm ambiguity there will be.

4. For informal norms, increasing legitimacy of violations and norm ambiguity are normal transitory phases in their crescive institutionalization. Therefore, the greater the legitimacy of violations of the old informal norm, the more institutionalized the new informal norm will be.

5. For formal norms, increasing ambiguity increases the number of violators, who then trigger more formal control. If that fails to reduce the level of violation, the norm ambiguity will lower predictability in social interaction to unacceptable levels and trigger the enactment of new formal norms. Therefore, the greater the norm ambiguity of formal norms, the more violators there will be, and the greater is the probability for the enactment of new formal norms.

6. A permissive social climate constrains social audiences and formal control agents from reacting to over nonconformity, and limits the range and severity of social sanctions. Therefore, the more permissive the social climate, the fewer and the more lenient will be the social sanctions that are applied.

7. Exogenous crises (e.g., war, depression) temporarily affect norm violations, permissiveness, and the legitimacy of norm violations (Fritz, 1961:682-5). Therefore, the greater the perceived threat to the whole social system, the greater will be the crisis effect on norm violations, permissiveness, and legitimacy of violations.

This theory contains three positive and three negative feedback loops. Structural impediments and strain in contemporary industrialized societies drive the positive loops towards an increase in violators, to which the system reacts in the negative loops by changing the norms (see Figure 1). Our System Dynamics model, representing this theory, contains 24 variables, 9 time delays, and 19 fixed relationship functions. Six constants and three exogenous variables initialize the variety of social conditions in which the theory should apply. For the complete and documented model see Jacobsen and Bronson (1985).

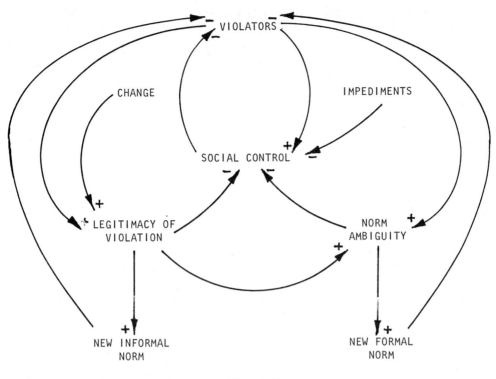

Figure 1
SIMPLIFIED CAUSAL-LOOP DIAGRAM

DATA AND SIMULATION

To test the empirical adequacy of this theory, we need a reliable and sufficiently extended time series of data which would be valid evidence for the change in the informal norm. The 1986 Surgeon General's Report contains an index for the "average restrictiveness of laws in effect", based on the number of states with regulations about smoking in public places, and the range of locations covered by these regulations (see Table 1, cols. 1-4)). As these enactments reflect the degree to which the no-smoking norm is gaining public legitimacy, the inverse of the yearly average restrictiveness should fit the model variable LEGIT, i.e. the percentage of the population who consider it reasonable to violate this emerging social norm.

A second data series was found in the archives of the American Cancer Society, containing the Society's annual expenditures on public education from 1949 to 1985. Since such education is aimed primarily against smoking, these statistics indicate the effort that was being invested to socialize the public towards the new norm. We converted these data into cents per person spent annually, adjusted them for inflation, and entered them into the model to initialize the variable SOCIAL, that is, the propor-portion of the population who are socialized into compliance with the norm (Table 1, cols. 5-7).

Table 1. Average Restrictiveness of Anti-Smoking Laws,
and Educational Expenses of the A.C.S. (1960-1985)

YEAR	NUMBER OF STATES	AVERAGE RESTRIC- TIVENESS	INVERSE IN PCT. (DATA)	EDUCATION EXPENSES (MILL. $)	INFLATION ADJUSTED cts. p/P	INDEX FOR SOCIAL
1960	14	.250	75.0	9.62	5.98	BASE
1961	14	.250	75.0	9.83	5.92	0.99
1962	14	.250	75.0	10.73	6.40	1.07
1963	14	.250	75.0	10.88	6.32	1.06
1964	14	.250	75.0	9.76	5.49	0.92
1965	14	.250	75.0	9.82	5.40	0.90
1966	14	.250	75.0	10.86	5.66	0.95
1967	14	.250	75.0	11.95	6.00	1.00
1968	14	.250	75.0	15.25	7.29	1.22
1969	14	.250	75.0	16.68	7.47	1.25
1970	14	.250	75.0	17.48	7.31	1.22
1971	16	.250	75.0	18.44	7.34	1.23
1972	17	.250	75.0	20.14	7.66	1.28
1973	20	.263	73.7	22.23	7.89	1.32
1974	22	.295	70.5	24.94	7.92	1.32
1975	27	.389	61.1	27.71	7.94	1.32
1976	30	.425	57.5	30.41	8.21	1.37
1977	33	.462	53.8	32.85	8.04	1.34
1978	34	.478	52.2	35.85	8.24	1.38
1979	37	.507	49.3	37.97	7.77	1.30
1980	37	.507	49.3	42.66	7.58	1.27
1981	39	.513	48.7	46.78	7.45	1.25
1982	39	.513	48.7	50.51	7.51	1.26
1983	40	.538	46.2	55.52	7.94	1.33
1984	41	.549	45.1	60.03	8.16	1.36
1985	42	.619	38.1	70.16	9.12	1.53

Sources: 1986 Surgeon General's Report, pp. 266-7.
American Cancer Society, Annual Reports, 1961-1986.

The simulation results are shown in Figure 2. Clearly, the data were reproduced quite accurately. In fact, 91.38% of the variance in the data was reproduced by the corresponding model variable. Also shown in Figure 2 is the rising percentage of the population who are deterred by peer pressure from smoking in company without prior consent (I), and the declining percentage who still violate the new norm (O). This suggests that the structural impediments to informal social control in modern industrialized society can be overcome by intensive and consistent socialization.

Following our standard procedure, we ran a series of experiments to establish the limits of the most important initialized variables within which reproducibility remains acceptable. Table 2 shows the results.

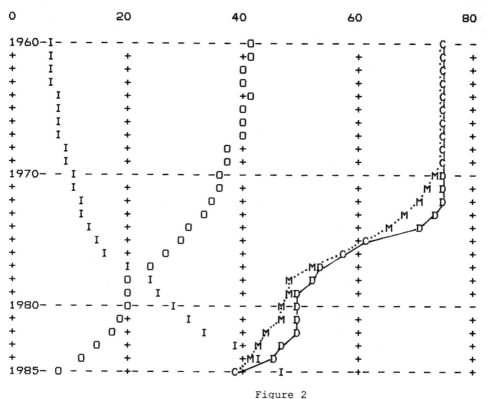

Figure 2
SIMULATION RESULTS

D = DATA (inverse pct. of average restrictiveness)
M = MODEL variable LEGIT (legitimacy of violation)
C = COMMON (identical values)
O = OVERT violators (smokers in company)
I = INFORMAL social control (peer pressure)

This brings the number of successfully replicated data sets to six:
three formal norms (illegal immigration to the U.S., violation of factory
safety regulations in Gt. Britain and in Israel), and three informal norms
(unmarried cohabitation in the U.S., emigration from Israel, and smoking in
company). The theory and model therefore apply not only to cases of
growing permissiveness, but also to the less commmon case of increasing
restriction in contemporary norm systems.

Table 2. Limits of Initialization Values

VARIABLE	LOWER LIMIT	ASSUMED VALUE	UPPER LIMIT	RANGE
LEGITIMACY OF VIOLATION	73.5	75	79.0	5.5
INITIAL PERMISSIVENESS	43.0	55	78.0	35
ULTIMATE PERMISSIVENESS	0.0	2	10.0	10
INITIAL SOCIALIZATION	59.0	70	82.0	23
IMPEDIMENTS TO INFORMAL CONTROL	0.0	2	37.0	37
OVERT VIOLATORS (SMOKERS)	38.0	41	44.0	6
PRESSURE TO VIOLATE THE NORM	47.0	50	54.0	7

REFERENCES

Fritz, C.E., 1961, Disaster, pp. 651-694, in "Contemporary Social Problems", R.K. Merton and R.A. Nisbet, eds., Harcourt, New York.

Jacobsen, C. and Bronson, R., 1985, "Simulating Violators", O.R.S.A., Baltimore.

Merton, R.K., 1957, "Social Theory and Social Structure", Free Press, Glencoe.

Simmel, G, 1950, The Metropolis and Mental Life, pp. 409-424, in: "The Sociology of Georg Simmel", K.H. Wolff, ed., Free Press, New York.

Williams, R.M., 1970, "American Society", 3rd Ed. Knopf, New York.

Wirth, L., 1938, Urbanism as a Way of Life, American J. of Sociology, 44, 3-24.

ON LEADERSHIP AND EFFECTIVE

CONTROL IN POLICE FORCES

M A P Willmer and K Gaston

Manchester Business School
Booth Street West
Manchester M15 6PB
Great Britain

INTRODUCTION

In recent times there has been an increased emphasis on the efficient and effective use of police resources with chief officers having to show that their existing resources are being used to the optimum. For instance, central government will only grant funds for additional manpower or equipment if a specific need together with anticipated benefits has been demonstrated. Indeed, in relation to force establishments Home Office circular 114/83 indicates that the Home Secretary"... will not normally be prepared to authorise additional posts unless he is satisfied that the force's existing resources are used to best advantage....it will not be sufficient for applications to be cast in general terms; a specific case for additional posts will need to be made." This necessity to justify closely requests for additional resources has led to an increased emphasis on the measurement of performance. HOC 114/83 indicates that the Home Secretary "...will look to H.M. Inspectors for their professional assessment of whether, for example, resources are directed in accordance with properly determined objectives and priorities..."

Over the years individual forces have developed a variety of mechanisms for target setting and performance monitoring mainly based on concepts borrowed from industrial management. However, the cybernetic philosophy on which many of these ideas are based has been criticised by Hofstede (1978). In general the cybernetic control paradigm is most valid when the process under consideration is routine and when there is a high level of objectivity associated with the output. When these conditions do not hold, the methods of control can still be effective but their application requires considerably more sensitive use of leadership skills, see Willmer (1983). In policing much of the work is routine and, as new technology enables forces to cope with ever increasing amounts of statistical material, the quantification of many aspects of policing has created an atmosphere of objectivity associated with its output: e.g. computerised crime statistics, crime pattern analysis, etc.

This enhanced ability to quantify has been used not only to justify the performance of a force to outside agencies but also to assess the performance of individual parts of the organisation itself. These recent trends are illustrated graphically in fig.1 in which the horizontal axis

represents the assessment of individual officers and the vertical
dimension the assessment of the force. The past is characterised by
informal procedures for assessing individuals and the effectiveness of
the force is evaluated against local or internal criteria. These
criteria were often made explicit and resulted in the use of notions such
as "reliable", "sound fellow" and "can be trusted to do the right thing".
The future is characterised by effectiveness of the force being made by
external authority and with individuals being assessed against formal
criteria. At the present time we are somewhere in between.

As procedures become more formal, as more reliance is placed on
complex internal information systems, senior officers may over-estimate
the ability of their control systems to monitor the performance of their
subordinates. For example, the usefulness of crime information for
control purposes depends on the willingness of police officers at all
levels to report and transmit data as faithfully as possible following
accepted procedures. Thus, if statistical criteria are made the dominant
factor in the way the worth and promotion potential of officers are
judged, the temptation to fudge the figures is increased, e.g. "New York
Police cheat on crime figures", Times (1972), "Faked confessions vanish
in probe at police station", Daily Mail (1987).

The basic methods of using numbers to appear to be more successful
than one's efforts truly merit - strategic numeracy - have been discussed
by Willmer (1984). However, as far as police work is concerned, the most
common ways of fudging are: 1) not recording crimes where the chances
of catching the culprits are small, 2) not looking too closely into
those cases where a ready culprit is to hand to see whether a crime has
actually been committed, and 3) persuading offenders to ask for other
offences to be taken into consideration and then not looking too closely
into whether the confessions are justified.

In view of the need for accurate data about the crime scene, this
paper explores the consequences of increasing the emphasis given to crime
statistics for evaluating the effectiveness of the police and assessing
the contributions of individual officers. Data has been obtained by
means of an especially designed computer simulation which reflects a
possible simplified future scenario when advancement prospects of
individual officers is heavily influenced by recorded crime statistics.
Its objective is to highlight the issues involved in the use of crime
statistics in leading, motivating and controlling subordinates.

THE SIMULATION

Each participant is asked to take the role of a Chief of Police and
to achieve as accurate an understanding as he can of the crime scene in
his force area. A handout is issued which describes organisational
changes which are occurring in forces, sketches a possible future
scenario in which greater emphasis is placed on the quantifiable aspects
of police work, and invites respondents to participate in such a future.
These organisational changes centre around the increased pressure from
both central and local government for police forces to explain closely
their deployment of human and other resources and to justify requests for
additional resources. As Chief Officer the participant is asked to
obtain the highest possible appreciation of the true nature of the crime
scene in the force under his command. Information is received from three
subordinates in control of divisions within the force concerning the
number of crimes committed, the number of criminals they have caught, and
the number of crimes they have cleared-up.

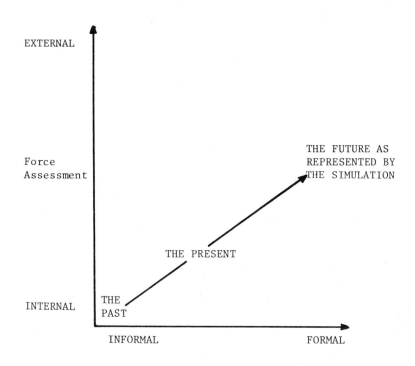

Method of Individual Assessment

Fig.1.   Illustration of Future Trends in Police Management Assessment

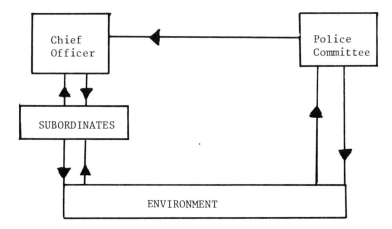

Fig.2.   Diagrammatic Representation of Major Information Flows

The effectiveness of the Chief Officer is measured by how accurately he can estimate the true picture of events in his area, specifically: 1. The actual number of crimes committed, 2. the actual number of active criminals, 3. the number of crimes due to the criminals who are caught, and 4. the number of crimes due to the criminals who are not caught.

The simulation challenges the participant to deal with both the interpretation of statistical data and the human behavioural problems associated with target setting and rewarding the work of subordinates. Background information is supplied about the number of residences in the force area: the number of potential criminals, and the percentage who are thought to be active at any one time. Brief descriptions are given of the personalities of the three subordinates in charge of the territorial divisions. These officers supply details of the crime scene in their area. These details need to be interpreted in the light of the officers being differentially rewarded for their performance.

The simulation is based on a series of 6 stage decision cycles:
1. Data is received from each subordinate about the number of criminals caught, the number of crimes reported, the number "cleared-up", the number classified as "no-crimes", and the "clear-up" rate.
2. The participant is then asked to assess the performance of each subordinate on a scale of 1-100: the higher the score, the better the performance. Participants are warned that subordinates may not agree with the way they are assessed and may react differently. For instance, the unscrupulous may fudge the statistics so that they appear to be better policemen than their performance actually warrants.
3. The participant is asked to assess the true crime scene for each area under the headings listed earlier, i.e. total crime, number of active criminals, crime due caught and uncaught criminals.
4. The performance of the participant as the chief officer is evaluated in terms of how his view accords with what actually occurred. The evaluation is given for each area separately followed by an overall performance score. The scale ranges from 0-100, the higher the score the more accurate the assessment.
5. The next step is the target negotiation process with the subordinates for the following period. The participant is asked to nominate an initial objective and set an initial target value to be achieved. The subordinates respond by offering their view as to what the target should be. The participant is then asked the degree to which he is willing to compromise on the difference between his initial target and that suggested by his subordinates. In this way a final target is determined.
6. The above final targets and the participant's appreciation of their subordinates performance is used to generate the crime data for the following period.

The aim of the exercise is for the participant to demonstrate that he can handle his subordinates in such a way that he can obtain an accurate picture of the crime scene in his force area.

DISCUSSION AND RESULTS

The information flows of the situation are represented diagrammatically in fig.2. Data about the crime in the force area and those responsible for committing it reach the chief officer via two channels. First, there is the link with the people on police committees and other worthies. For the purpose of the simulation it is assumed that they know the true crime picture and are thus able to assess the performance of the chief officer. Secondly, there is the information that reaches the chief officer via his subordinates. Although he sets them targets and has the power to reward their efforts by increasing

540

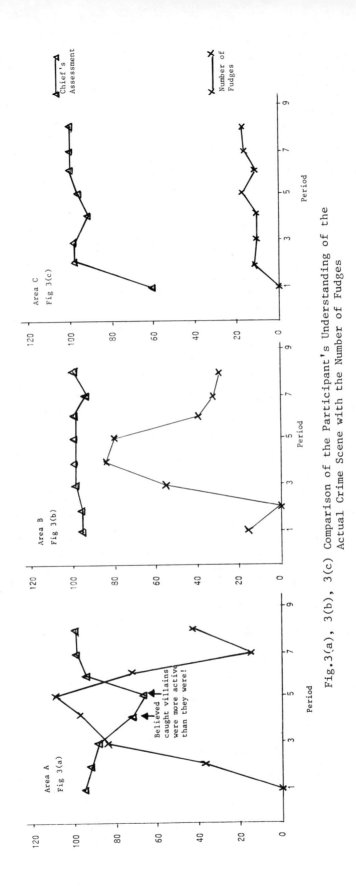

Fig.3(a), 3(b), 3(c) Comparison of the Participant's Understanding of the Actual Crime Scene with the Number of Fudges

their assessment scores, they may, should they be so inclined, manipulate to a considerable extent the data that he receives. To succeed the chief officer must be able to motivate and control his subordinates under conditions where they may well decide to use their influence over information flows to ensure that they obtain a greater reward than their efforts genuinely warrant. The exercise is concerned, therefore, not so much about the chief officer's ability to control but with his ability to lead. To be successful participants have to make assumptions about human nature. For example: i) how to people behave if they are set an impossible target? ii) how do they react when they find that others less competent are rewarded more highly? iii) do they take advantage of you when you show compassion and understanding?

The subordinates have many opportunities when they can manipulate their superior's perceptions of their activities. They can suggest a target that is higher than their superior's and then achieve it on paper by fudging the crime statistics. Or, they can persuade their superior to set a target that is far too low so that when they achieve it he rewards them highly. By skilfully manipulating the target setting procedure and fudging the statistics they may well start to control their superior! Those who seek power and advancement will naturally tend towards fudging the figures. Others may be pressured into adopting such tactics by the need to survive in a highly competitive environment. This pressure is particularly intense in cases where the superior fails to heed the warning of Wensley (1931), that great London detective of some fifty years ago. Commenting on the consequences of rewarding subordinates unfairly he notes: "there are men...with a streak of vanity that impels them to adopt a pose at the expense of those who have really done the work. They like to bask in the limelight...As Chief Constable I was always on my guard against this sort of thing and made it my business to see that credit went in the right direction. Unless kept in check it breeds resentment and checks energy."

The simulation is concerned with leadership rather than control. To receive reliable information, the participant must learn to appreciate how his subordinates view their jobs, their personal set of values and objectives, and the strengths/weaknesses of their characters. He must appreciate the ways in which his subordinates can fudge the statistics.

The vulnerability of the clear-up rate to the machinations of police officers has long been recognised. To overcome some of the inherent problems, a new method of evaluating achievement based on the concept of reducing uncertainty was proposed by Willmer (1973). As some preliminary results using this method were favourable, Willmer (1979), participants were allowed to select whether they wished to assess a subordinate by means of the clear-up rate or in terms of unresolved uncertainty.

To illustrate, a senior police officer from a major city force was asked to lead three subordinates: A: intellectually gifted, experienced admin. officer, wife keen that he should gain promotion, B: blunt old-fashioned copper, experienced detective. C: elderly, solid and dependable.

Figs.3(a), (b) and (c) show the comparison between the participant's understanding of the actual crime scene and the number of fudges generated by the subordinates. In each area the participant was able to establish a high degree of success despite changes in the fudging levels. His only failures occur in Area A in periods 4 and 5 when he believed that the caught villains had been more active than was the case. However, he was able to quickly adjust his perceptions and return to a high level of success. It would seem that this officer appreciates that

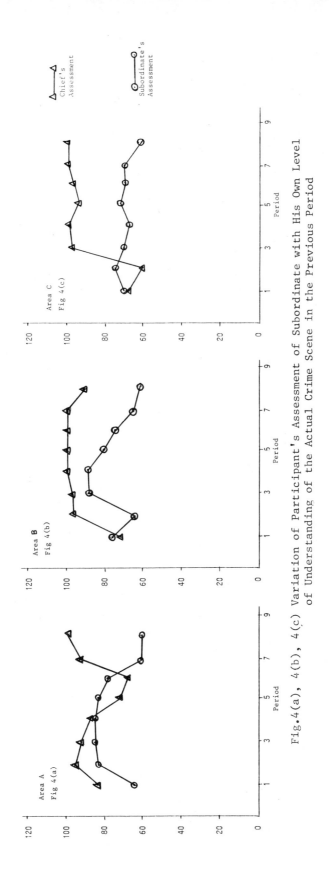

Fig. 4(a), 4(b), 4(c) Variation of Participant's Assessment of Subordinate with His Own Level of Understanding of the Actual Crime Scene in the Previous Period

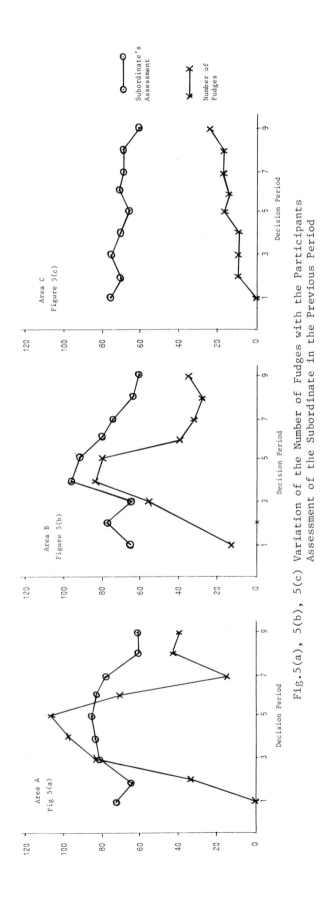

Fig.5(a), 5(b), 5(c) Variation of the Number of Fudges with the Participants
Assessment of the Subordinate in the Previous Period

544

his understanding is due to his won ability to interpret crime data rather than the accuracy of the data itself.

Figs.4(a), (b) and (C) show how the participant's assessment of his subordinate varies with his own level of understanding in the previous period. In Area A it can be seen that the participant did not immediately react to a loss in his own rating by penalising his subordinate. Instead he waited for a couple of periods before reducing his subordinate's assessment even though his own was beginning to improve. In Area B it can be seen that, even though his own assessment is high, his assessment of his subordinate continuously falls. This figure is consistent with a situation in which the subordinate initially gains by fudging but is subsequently penalised.

Figs. 5(a), (b) and (c) show how the number of fudges varies with the assessment given to the subordinate in the previous period. The first two figures show that the subordinates initially tried to gain a higher rating by fudging but were later penalised. Despite falling assessments from the participant neither subordinate reduces his fudging level to zero. In contrast fig 5(c) shows the situation for a rather dull but straightforward copper.

By asking people to participate in the simulation a whole range of issues connected with the accountability and assessment of subordinates, their reactions and motivations, is raised. This paper shows how one person struggled with the problems involved. Significantly he demonstrated that it is possible to interpret fudged data correctly even when the degree of fudging changes rapidly. However, if he had put greater emphasis on improving the accuracy of information, would he have reduced his ability to interpret the data that he received?

The simulation offers a useful tool for probing into many areas of the management and leadership of the police forces. Some areas under consideration include: 1) Development of methods of determining from statistical data the true crime picture, see for instance Willmer (1977). 2) Explore the use of alternative measures of subordinate efficiency. 3) Comparing the differences in management style in the simulation including reactions to subordinate behaviour with other measures of leadership capabilities. 4) Examine the relationship between putting pressure on subordinates to report more accurately and the skill of the chief to interpret the data when it is received. 5) When combined with post event interviews with participants can an effective diagnostic tool for selecting senior officers be developed?

References

P Burden(1987), The Daily Mail, 4 March.
G H Hofstede (1978), "The Poverty of the Management Control Philosophy", Academy of Management Review, July.
Home Office Circular to Chief Officers of Police No 114/83.
The Times, 16th May 1972.
F P Wensley (1931), "Detective Days", Cassell & Co., London.
M A P Willmer (1973), "Information Theory and the Measurement of Detective Performance", Kybernetes, Vol.2, 1973.
M A P Willmer (1979), "Information from Statistics", The Police Journal, Vol.L, No.4.
M A P Willmer (1983), "The Contribution of the Cybernetic Approach to Management Control", New Perspectives in Management Control, Ed: Tony Lowe and John L J Machin, Macmillan, London.
M A P Willmer (1984), "Freedom in a Numerate Society", Paper presented at the Orwellian Symposium on System Research, Information and Cybernetics, Baden-Baden, West Germany, August, Man. Bus. Sch. Working Paper No.11.

MORALITY AND MORAL BEHAVIOR

A STUDY ON OPERATIONALITY

Manfred J. Holler*

Institute of Economics and Statistics
University of Aarhus, DK-8000 Aarhus C
Denmark

ABSTRACT

In this note basic features of modelling morality and moral behavior are reconsidered and strategic problems of moral behavior are illustrated by examples. The discussion shows that, in game situations, the problem of moral behavior cannot always be solved by assigning positive, intrinsic or extrinsic, utilities to the payoffs related with moral behavior strategies.

BASIC FEATURES OF MORAL BEHAVIOR

The definitions of <u>moral</u> <u>behavior</u> can be assigned to two classes of attributes. Type A, the moral behavior of an individual is defined as behavior which concurs to moral norms, i.e., morality, in spite of conflicting self-interest. Type B, the individual rational behavior of an individual, conforms to morality and thus individual rational behavior concurs with moral behavior. It should be obvious that moral behavior which is contained in the class A is not open to analysis which derives from individual rationality, while moral behavior which falls in the class of type B is, in principle, accessible to be modelled and analysed with respect to some concepts of individual rationality.

It is standard to model the problem of moral behavior with reference to the well-known (two-person) Prisoner's Dilemma. A general description of this game is given in Figure 1. C_i represents player i's strategy "cooperation" (where i = 1,2) while D_i ("defection") represents non-cooperative behavior; a, b, c, and d are the payoffs which accrue to the players from the various combinations of strategies - more precise, from the events which evolve from the strategy combinations.

The matrix in Figure 1 is a Prisoner's Dilemma if (a) the payoffs of each player are ordered as a > b > c > d and (b) the game is of one period ("one shot"), only. The non-cooperative strategies, D_i, of the two players are dominating and (c,c) is the unique equilibrium outcome. Obviously, (c,c) is Pareto inferior to (b,b). The pair (b,b) is an element of the set of Pareto optimal payoff vectors which is the union of the two sets (τ(a,d) +

* The author would like to thank Hartmut Kliemt for helpful comments on an earlier version of this paper.

Figure 1.   Prisoner's Dilemma: $a > b > c > d$

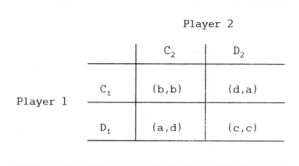

Player 2

		C_2	D_2
Player 1	C_1	(b,b)	(d,a)
	D_1	(a,d)	(c,c)

$(1-\tau)(b,b))$ and $(\sigma(d,a) + (1-\sigma)(b,b))$ where $0 \leq \tau, \sigma \leq 1$ if players can randomize on their pure strategies C_i and D_i.

In a Prisoner's Dilemma player i's moral behavior of type A coincides with choosing the cooperative strategy C and thus satisfying one of the conditions so that the Pareto optimal, cooperative outcome (b,b) prevails. However, it is profitable for the second palyer to exploit this potential by choosing the non-cooperative strategy D and the moral behavior strategy is honored by the minimum payoff of the game. As a consequence, following the self-interest perspective, none of the players will select the cooperative strategy and non-cooperative outcome (c,c) results.

There is a considerable literature modelling variations of the Prisoner's Dilemma game which allow for deriving the cooperative outcome from individual rational behavior. The best-known variation is the iterated Prisoner's Dilemma game. Following an idea noted in Luce and Raiffa (1957, pp. 94-102), it is assumed that (a) the game in Figure 1 is repeated an infinite number of times and (b) the players maximize the sum of discounted payoffs over time resulting from this infinite sequence.

We will not go in further detail of this approach here because there is an extremely rich literature on this (see, e.g., Taylor, 1976; Schotter, 1981). It is obvious that inasmuch as the cooperative outcome is considered as moral, the underlying normative concept and the evaluation derived from it are exogenous to the game even when the cooperative solution derives endogenously. For example, the two suspects of the original Prisoner's Dilemma story, who were submitted to a decision situation of the type described in Figure 1 where to "confess" is the dominating strategy, do not behave morally if they do not confess their crime though this induces the cooperative outcome if followed by both. (See Luce and Raiffa (1957, p. 95) for the original story of the Prisoner's Dilemma.)

We have to distinguish between morality, i.e., the norm of moral behavior, and moral behavior. The latter is costly for the individual if it is of type A. The production of the norm can be a by-product of producing a private good, and thus be without costs, or it might be profitable if professionalized. The latter applies, e.g., to teaching morality in various forms (at schools, in churches, etc.). There might be a public good problem involved in the production of the norm and thus the Prisoner's Dilemma model also applies as a proxy. However, this is not necessarily the case as many historical examples show. Often the creating of norms goes along, or is a means of, accruing individual power or wealth.

STANDARD MODELLING OF MORAL BEHAVIOR

A large share of the literature on morality, however, is dedicated to exercises which aim for deriving moral behavior of type A from individual rational behavior. In general, these exercises model the individual rational decisionmaking in such a way that the link of moral behavior to rational behavior is concealed by smoke so that one gets the impression that the analysed moral behavior does (a) not conform to self-interest, but (b) derives from individual rational behavior. The iterated Prisoner's Dilemma approach, outlined in the preceding section, is an example if interpreted correspondingly. In Kurz (1976) a variation of it is applied to model altruism. In this model, altruistic behavior becomes a consequence of self-interest - however, with respect to the standard connotation of altruism one is tempted to ask what is altruistic of altruistic behavior which derives from self-interest.

The cover-up strategy of deriving moral (or altruistic) behavior from self-interest is supported by the fact that there are many ways of how to model moral behavior of type B. The intrinsic or extrinsic utility from moral behavior, at present or in the future, can either be modelled by an enlarged set of feasible alternative (e.g., budget) or, directly, by an increase of utility which an individual derives from choices if they conform to morality. The latter case can be straightforwardly captured by an individual utility function which shows moral behavior as one of the arguments. This type of direct moral utility is, e.g., contained in Margolis (1982).

A less straightforward way is to assume a hierarchy of utility functions (or preference orders) in which the domain of highest ranking function consists of lower ranking utility functions (e.g., Sen, 1977). The individual chooses his or her lower ranking utility function from which, at the end, his or her behavior results. The lower ranking utility function can prescribe moral behavior in the sense of Type A, nevertheless it might be individually rational to choose this function in accordance with the higher ranking utility function. In general, strategic problems in the selection of the lower ranking utility function, especially related with multi-equilibria, leave enough space for rationalizing the choice of a utility function of action which conforms to type A and thus the choice of type A behavior does not conflict with individual rationality. (An example of rationalizable strategic behavior is discussed in the following section; see also Bernheim, 1984, and Pearce, 1984, on rationalizable strategic behavior.)

In principle, models which assume a long-run utility function controlling behavior which derives from a short-run utility function, either by choosing adequate institutional settings which restrict the domain of the short-run utility function in the interest of long-run utilities (see Thaler and Shefrin, 1981), or by influencing the short-run utility functions directly, also belong to the category of utility function hierachies. A related approach is characterized by the fact that higher ranking utility functions get created by lower ranking utilities. An alternative conceptualization assumes that a utility function u_1 willingly creates a utility function u_2 which substitutes u_1 (see Bolle, 1987). That is, in each point of time, an individual is characterized by one, and only one, utility function which, however, possesses the power to create a successor that derives its existence and values from it.

A STRAIGHTFORWARD MODEL OF MORAL BEHAVIOR

In this section we will discuss a simple approach to model moral behavior which sets out from the assumptions that (a) morality as norm exists independent of the action of the considered individuals and (b) behavior

Figure 2.   The ε-Biased Prisoner's Dilemma: a > b > c > d

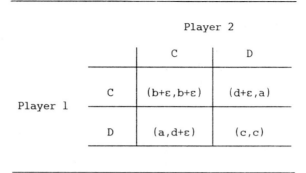

Player 2

		C	D
Player 1	C	(b+ε,b+ε)	(d+ε,a)
	D	(a,d+ε)	(c,c)

which concurs with the morality is directly beneficial to the agents. This concurs with the direct moral utility approach of the Margolis type. The novelty here, if there is any novelty at all, is to apply this concept in a strategic context. We will see that moral behavior is not straightforward in strategic situations even when it is honored by "large", intrinsic or extrinsic, benefits. First, however, we will discuss whether the Prisoner's Dilemma resolves in moral behavior if the selection of a cooperative strategy is honored by an extra utility ε, given C conforms with morality.

Again, in Figure 2 the payoffs are a > b > c > d. If ε > a-b or ε > c-d, then the non-cooperative strategy, D, is no longer dominant. Let us assume that there is a ε so that ε > c-d and a > b+ε. Then the Matrix in Figure 2 contains a Chicken Game; the strategy combinations (C,D) and D,C) are equilibria implying Pareto optimal outcomes. (There is a Nash equilibrium in mixed strategies, too. It will not be discussed here, since it implies substantial incentive and coordination problems as pointed out in Holler (1988).)

A third Pareto optimal outcome in pure strategies derives from the strategy combination (C,C) which corresponds to the implicit norm of morality. Note that (C,C) does not constitute a (Nash) equilibrium for this game. Thus, in addition to augmenting the payoffs related to the cooperative strategy we have to make additional assumptions in order to guarantee the moral outcome (b+ε,b+ε) if the (non-cooperative) equilibrium is the applied solution concept. In other words, it is not sufficient to introduce benefits from moral behavior into the payoff function, large enough to overcome the Prisoner's Dilemma situation, to guarantee moral behavior and thus inducing a moral outcome. There are, in general, other strategic problems luring. Only if ε > a-b and ε > c-d, so that cooperative strategies become dominant, the moral outcome is guaranteed.

There is, however, a rather substantial recent discussion of the monopoly of the Nash equilibrium (and its refinements) as solution concept for non-cooperative games. Bernheim (1984) and Pearce (1984) are in favor of the alternative concept of rationalizable strategic behavior. In short, since each of the strategies in a Chicken Game is related with an equilibrium, the choice of any of these strategies cannot be objected as unreasonable or irrational, i.e., each strategy is rationalizable. Thus, the moral outcome can be justified as the outcome of the game in correspondence with individual rational behavior. However, there is no unique outcome which can be justified on these grounds. In fact, all payoff vectors in the matrix of Figure 2 are supported by rationalizable strategic behavior including the Pareto dominated

Figure 3. Congestion Game: a > b = c > d

___

Player 2

		C	D
		C	D
Player 1	C	(b,b)	(a,c)
	D	(c,a)	(d,d)

___

outcome (c,c) if ε is such that Figure 2 represents a Chicken. (In the unmodified Prisoner's Dilemma of Figure 2, however, the choice of the dominated cooperative strategy cannot be rationalized.)

For the discussion of rationalizable strategic behavior it is crucial to remember that in game situations an individual player has no full control on the selection of a specific strategy combination and thus the outcome, but only on the selection of his/her strategy. The Nash equilibrium, however, applies to strategy combinations.

At least potentially, there are strategic situations in which (a) the outcome of a joint selection of cooperative strategies leads to a (weakly) Pareto dominated outcome and (b) the introduction of a general ε premium on playing the cooperative strategy does not change this "defect". An example is given in Figure 3 where the payoffs rank a > b = c > d. (This game is analyzed in Holler, 1988).

The Congestion Game in Figure 3 has three equilibria in pure strategies: (C,D), (C,C), and (D,C). Only the outcomes related to the asymmetric equilibria, (C,D) and (D,C), are Pareto optimal. (C,C) implies that from the point of view of social efficiency there is "too much" cooperation. That is, to some extent moral behavior is in conflict with social efficiency, however, the prevalence of moral behavior is a precondition of social efficiency as contained in (C,D) and (D,C).

Note that adding direct moral utility ε of any positive size to the payoffs related to the cooperative strategy, C, does not resolve the conflict between social efficiency and moral behavior. The dilemma is that morality is, in general, understood as a norm which applies to all individuals in a population and discriminating in this generalization is seen as a violation of the norm. However, as history shows, often societies were characterized by applying different moral norms to different subgroups. With reference to the Congestion Game in Figure 3 one may conclude that in many situations of social decisionmaking this differentiation was a precondition for achieving socially efficient outcomes. We may even conjecture that the degree of moral differentiation was positively correlated to achieving efficient social outcomes. It would be interesting to see whether empirical data are available to support this conjecture.

REFERENCES

Bernheim, D., 1984, Rationalizable strategic behavior, Econometrica, 52:1007-1028

Bolle, F., 1987, Level of aspiration behaviour, promises, and the possibility of revaluation, Understanding Economic Behaviour, 12th Annual Colloquium of IAREP, II:613-627.

Etzioni, A., 1986, The case for a multiple-utility conception, Economics and Philosophy, 2, 159-183.

Holler, M. J., 1988, An indifference trap of voting, Quality and Quantity, 22:279-292.

Kurz, M., 1976, Altruistic equilibrium, in "Economic Progress, Private Values and Public Policy," Bela Balassa and Richard Nelson, eds., North Holland, Amsterdam.

Luce, R. D. and Raiffa, H., 1957, "Games and Decisions," Wiley, New York.

Margolis, H., 1982, "Selfishness, Altruism and Rationality: A Theory of Social Choice," Cambridge University Press, Cambridge.

Pearce, D. G., 1984, Rationalizable strategic behavior and the problem of perfection, Econometrica, 52:1029-1050.

Schotter, A., 1981, "The Economic Theory of Social Institution," Cambridge University Press, Cambridge.

Sen, A., 1977, Rational fools: a critique of the behavioral foundations of economic theory, Philosophy and Public Affairs, 6:317-344.

Taylor, M., 1976, "Anarchy and Cooperation," Wiley, London.

Thaler, R.H. and Shefrin, H. M., 1981, An economic theory of self-control, Journal of Political Economy, 89:392-406.

AN EXPERIMENTAL INVESTIGATION OF SUBJECTIVE DISCOUNT RATES

Uri Ben-Zion*, Joseph Yagil**, and Alon Granot*

*Technion, Israel Institute of Technology
**Haifa University and New York University

ABSTRACT

This study extends a recent work by Ben-Zion, Rapoport and Yagil (1988). Two hundred and fifty MBA students at New York University have participated in an intertemporal choice experiment which manipulated three dimensions in a 3 × 3 factorial design (scenario, time delay and size of cashflow). Individual discount rates were inferred from the response and then used to test four hypotheses regarding the behavior of discount rates. The discount rates inferred from the experiment are quite high, particularly for short periods of time and small sums. However, they decline with both the time and sum dimensions. The findings in this study support the added compensation hypothesis which asserts that individuals require a compensation for a change in financial position.

1. INTRODUCTION

This study extends an earlier work by Ben-Zion, Rapoport and Yagil (1988) hereafter referred to as BRY. The study is based on an intertemporal-choice experiment designed to estimate the discount rates of individuals from their decisions. The subjects in this study are MBA students at New York University (NYU). The time period underlying the experiment is the summer of 1988 – a period which seems to be characterized by relatively stable prices.

The subjective discount rate is a basic concept in Economics and Finance and also in Psychology (see, e.g., Loewenstein (1986) and 1987), Stevenson (1986), Thaler (1981), and Thaler and Shefrin (1981)). Individual discount rates may differ, and they may also depend on the scenario in which the decision is made (e.g., postponing a payment or a receipt, and expediting a payment or a receipt). For a more detailed discussion of earlier works see BRY (1988). We extend BRY's study by using a different set of subjects in a different country and under different economic conditions. In addition, some of the statistical tests are refined, and an attempt is made to investigate the impact on the discount rates of the personal characteristics of the subjects as well as their required rate of interest and degree of risk aversion.

The plan of this short paper is as follows: Section 2 discusses the experimental design; Section 3 presents briefly the alternative hypotheses; Section 4 presents the results, and the last section presents a short summary and concluding comments.

2. THE EXPERIMENT

The experiment involved a questionnaire which was given to 250 MBA students at NYU. The students recorded their responses to postponing or expediting a payment or receipt of a certain sum of money ($40, $200, $1,000 and $5,000). The length of the time period (t) is 0.5, 1, 2 and 4 years, and the four scenarios are: postponing a receipt (A); postponing a payment (B); expediting a receipt (C); and expediting a payment (D). For any given sum in a question (Q), the individual has to give a subjective answer (A). We assume that $A = Qe^{Rt}$, where R is the continuous discount rate and is given by $R = Ln(A/Q)/t$, and where t is positive for postponing a receipt or a payment (Scenarios A and B), and negative for expediting a receipt or a payment (Scenario C and D). Equivalently, we use the notation F and P to indicate whether the relevant sum is in terms of future value or present value. In this case t will always be positive, and the discount rate will simply be $R = Ln(F/P)/t$.

In addition to making specific choice decisions, subjects were asked to provide information on various personal characteristics such as age, sex, ownership of financial instruments (such as savings plans, bonds and stocks) and their attitude toward risk. In addition, some information has been gathered about their knowledge of the prevailing rates of interest.

3. HYPOTHESES AND TESTS

Following BRY, four hypotheses are tested. The specific statistical tests which we have performed are outlined below.

I. The Classical Approach and the Segmented Markets Hypothesis

In our interpretation, the classical approach asserts that the discount rate is constant over the three dimensions of scenario, time and sum.

In contrast to the classical approach, the segmented markets hypothesis asserts that the discount rate will vary across scenarios. A joint test of these two competing theories is the following regression form:

$$R = \beta_1 C + \beta_2 t + \gamma_1 D_A + \gamma_2 D_B + \gamma_3 D_C + \gamma_4 D_D \qquad (1)$$

where R is the computed discount rate, t and c denote time and sum of cashflow respectively, and D_A, D_B, D_C and D_D are dummy variables for scenarios A, B, C and D respectively. The segmented markets hypothesis allows for differences between scenarios (i.e., different γ_j), but imply that $\beta_1 = \beta_2 = 0$. The classical hypothesis suggests in addition that $\gamma_1 = \gamma_2 = \gamma_3 = \gamma_4$.

II. The Implicit Risk Approach

According to the implicit risk approach, possible risk is involved in delaying for a cashflow. Two alternative hypotheses regarding the time-pattern of risk are the multi-period realization (MPR) hypothesis and the one-period realization (OPR) hypothesis. The first hypothesis assumes that risk increases exponentially with time while the second hypothesis assumes that only the delay itself involves risk. Formally stated the MPR and the OPR lead to equations 2A and 2B respectively.

$$F = Pe^{dt}e^{it} = Pe^{(d+i)t} = Pe^{Rt}, \tag{2A}$$

$$F = Pe^{d}e^{it}, \tag{2b}$$

where d is the "one-period" risk premium, i is the risk-free rate of interest (reflecting the pure time value of money) and R is the combined discount rate which incorporates both the effects of time and risk. As demonstrated by Eqs. (2A) and (2B) while the MPR assumes that the risk premium (d) applies to every future time period, the OPR assumes that d applies only once, and the time discount procedure uses the risk free rate. We test these alternative implicit-risk hypotheses by estimating the following regression

$$Ln(F/P) = \beta_0 + \beta_1 t \tag{3}$$

According to the MPR $\beta_0 = 0$, and $\beta_1 = R = i + d$; whereas according to the OPR $\beta_0 = d > 0$, and $\beta_1 = i$.

III. The Added Compensation Approach

The added compensation (AC) approach asserts that in addition to the discounting procedure an added compensation will be required for an undesired position change. For simplicity, we assume, initially, that the AC is a fixed component (independent of time and sum). Formally stated

$$F = \alpha e^{Rt} + Pe^{Rt} = (\alpha + P)e^{Rt}, \tag{4}$$

or, in a regression form

$$Ln(F|P) = \beta_1(1/P) + \beta_2 t \tag{5}$$

where $\beta_1 = \alpha = $ the AC component, and $\beta_2 = R$. The AC approach predicts higher β_1 for the more difficult position changes (scenarios A and D) and lower β_1, for the less difficult position changes (Scenarios B and C). The no-compensation alternative suggests that $\beta_1 = 0$.

If we assume that the added compensation also depends on the sum involved, we can replace α by $(\alpha_1 + \alpha_2 P)$, and the F - P relationship becomes

$$(F/P) = (\alpha_2 + \alpha_1/P + 1)e^{Rt} \tag{6}$$

or

$$Ln(F/P) = \beta_0 + \beta_1(1/P) + \beta_2 t \tag{7}$$

where $\beta_0 = \alpha_2$; $\beta_1 = \alpha_1$; and $\beta_2 = R$. The added compensation approach can explain the decline of the discount factor with sum and time.

4. RESULTS

The average discount rates across scenario and time are given in Tables 1.1, 1.2 and 1.3. The results indicate discount rates which are similar in magnitude and patterns to those of BRY. However, only the discount rates for $5,000 are in the magnitude of the interest rates in the U.S. (borrowing rates for Scenarios A and D, lending rates for Scenarios B and C). While Scenarios A and D indicate similar discount rates, Scenarios B and C appear to have quite different discount rates.

The regression results of Eq. (1) are given in Table 2. As demonstrated by Table 2, the discount rate varies across scenarios (lending support to the segmented market hypotheses). However, the discount rate also varies significantly with time and sum. Thus, we cannot accept the simple form of the classical approach and the segmented markets approach.

The results of Eq. (3) are given in Table 3, and they indicate that the intercept term is positive for all scenarios implying a rejection of the multi-period realization of risk approach (MPR). The results appear more consistent with the one-period realization of risk approach (OPR). However, the OPR cannot explain why the discount rate declines with the sums involved.

Table 4 reports the regression results for Eq. 5, and it demonstrates that the (required) added compensation component is of a magnitude of $5, with some differences between scenarios (e.g., $6.7 in Scenario A, and $3.9 in Scenario B). Taking into account the added compensation, the corrected discount rates are lower and appear to be more in line with the existing interest rates for borrowing and lending in the U.S. for the time-period underlying the experiment.

The regression results of Eq. (7) are contained in Table 5, and they indicate that there is a variable element in the added compensation (AC) component in a magnitude of 5% in Scenarios A and D, and 2% in Scenarios B and C. The fixed element of the AC component remains in the magnitude of $3-$5.

An initial investigation of the impact of personal characteristics on the discount rate indicates some differences across the subjects in the experiment. As demonstrated in the following examples, the average discount rate is associated with personal characteristics, such as: Male vs. Female: 16.6% and 21.9% respectively; Risk-lover vs. Risk-averter: 18.7% and 17.4% respectively; Married vs. Single: 16.7% and 19.2% respectively. We shall be making a more detailed analysis of differences in a later study.

5. SUMMARY

The findings of this experiment indicate a similarity between subjects in different countries under different economic circumstances (High inflation in Israel and lower inflation in the USA). The findings appear to support the added-compensation hypothesis which is the only approach that explains why discount rates decline with sum and with time. Further research is needed to integrate the individual subjective discount rates with the market behavior of the macro variables.

Table 1. Mean Discount Rates across Each of the Three
Different Design Factors

1.1 Mean Discount Rates across Sum by Time Period and Scenario

Scenario\Time(years)	0.5	1	2	4	Mean Across Time
A	0.476	0.236	0.200	0.156	0.267
B	0.220	0.137	0.114	0.093	0.141
C	0.310	0.184	0.136	0.112	0.185
D	0.371	0.232	0.177	0.146	0.232
Mean across Scenario	0.344	0.197	0.157	0.127	0.206

1.2 Mean Discount Rates across Time by Sum and Scenario

Sum ($) \ Scenario	A	B	C	D	Mean across Scenario
40	0.370	0.214	0.297	0.312	0.298
200	0.314	0.151	0.171	0.223	0.215
1,000	0.253	0.124	0.160	0.228	0.191
5,000	0.131	0.074	0.114	0.163	0.120
Mean across Sum	0.267	0.141	0.185	0.232	0.206

1.3 Mean Discount Rates across Scenario by Sum and Time

Sum ($)\Time(years)	0.5	1	2	4	Mean across Time
40	0.518	0.296	0.224	0.155	0.298
200	0.366	0.199	0.158	0.136	0.215
1,000	0.335	0.174	0.131	0.125	0.191
5,000	0.157	0.120	0.114	0.092	0.120
Mean across Sum	0.344	0.197	0.157	0.127	0.206

Table 2. Regression Test for the Classical Approach and the
Segmented Markets Hypothesis*

$$R_i = \beta_1 C + \beta_2 t + \gamma_1 D_A + \gamma_2 D_B + \gamma_3 D_C + \gamma_4 D_D$$

	β_1	β_2	γ_1	γ_2	γ_3	γ_4	R^2
			Dummy variable scenario				
X Coefficient(s)	-0.031	-0.092	0.484	0.362	0.401	0.444	0.07
	(0.003)	(0.006)	(0.018)	(0.018)	(0.018)	(0.018)	

Table 3. Regression Test for the Implicit Risk Approach*

$$LN(\frac{F}{P}) = \beta_0 + \beta_1 T$$

Variable\Scenario	A	B	C	D
Constant (β_0)	0.092	0.052	0.056	0.069
	(0.012)	(0.007)	(0.010)	(0.012)
Time (β_1)	0.016	0.076	0.087	0.113
	(0.005)	(0.003)	(0.004)	(0.005)

R^2 and F values for the regression are on the magnitude of .25 and 500
respectively and N = 1,488.

Table 4. Regression Test for the Added Compensation Approach*

$$\ln(\frac{F}{P}) = \beta_1(\frac{1}{P}) + \beta_2 t$$

Variable\Scenario	A	B	C	D
Sum (β_1)	6.669	3.859	5.571	4.446
	(0.628)	(0.387)	(0.505)	(0.611)
Time (β_2)	0.130	0.083	0.092	0.125
	(0.003)	(0.002)	(0.003)	(0.003)
R^2	0.274	0.295	0.258	0.259

Table 5. Regression Test for the Added Compensation Approach*

$$\ln(\frac{F}{P}) = \beta_0 + \beta_1(\frac{1}{P}) + \beta_2 t$$

Variable\Scenario	A	B	C	D
Constant (β_0)	0.047	0.025	0.016	0.041
	(0.013)	(0.008)	(0.011)	(0.013)
Sum (β_1)	5.670	3.330	5.235	3.558
	(0.686)	(0.423)	(0.553)	(0.668)
Time (β_2)	0.116	0.076	0.087	0.113
	(0.005)	(0.003)	(0.004)	(0.005)
R^2	0.278	0.299	0.262	0.264

* In all regression the standard error of the coefficients are given in parentheses.

REFERENCES

Ben-Zion, U., A. Rapoport & J. Yagil, "Discount Rates Inferred from Decisions: An Experimental Study in Management Science", (forthcoming).

Loewenstein, G., 1986, "Frames of Mind in Intertemporal Choice", Center for Decision Research, The University of Chicago.

Loewenstein, G., 1987, "Anticipation and the Valuation of Delayed Consumption", Economic Journal, (forthcoming).

Stevenson, M.K., 1986, "A Discounting Model for Decisions with Delayed Positive or Negative Outcomes," Journal of Experimental Psychology, 115: 131-154.

Thaler, R., 1981, "Some Empirical Evidence on Dynamic Inconsistency", Economic Letters, 8: 201-207.

Yates, F. & R.A. Watts, 1975, "Preferences for Delayed Losses", Organizational Behavior and Human Performance, 13: 294-306.

POWER, CONFLICT AND CONTROL

POWER, CONFLICT AND CONTROL

Mats Alvesson

Dept of Business Administration
University of Stockholm
Sweden

INTRODUCTION

The purpose of this paper is to give an introduction to some contemporary positions on the topic power, conflict and control. Of these three concepts, power is the most basic and also the trickiest to handle. It will therefore be treated more extensively than conflict and control. In the text, the distinction between the deeper aspects of power, conflict and control, and the surface manifestations of these, will be emphasized.

POWER

Despite its obvious significance, power was for a very long time not paid much attention by writers on management and organization. Zey-Ferrell and Aiken (1981) criticized dominating perspectives in organization theory for neglecting power relations and argued that

"If power is analyzed at all, it is in terms of influence or authority, the legitimized position-based power of management or organizational elites. Almost never are domination, subjugation, coercion, manipulation, or extortion of one group or class or organizational members by a more powerful group or class analyzed." (P 16)

Pfeffer (1981) explains the neglect of studies of power in business and management writings (particularly in USA) by referring to the complexity of the concept, the persuasiveness of other perspective on decision-making, organizational design, etc, and the interests and preferences of managers and business students for downplaying the role of power in explaining decisions, structures, careers, managerial work and so on, while it does not harmonize with the ideal picture of how things are.

During the 1980's an increase in interest in power, politics, control and related topics can be found in management and organization science (e.g. Daudi, 1986; Frost, 1987; Mintzberg, 1983a; Morgan, 1986; Pfeffer, 1981, etc). Even textbooks nowadays often include a chapter or two on the topic. Most of what is written is informed by a pluralist perspective, but also critical and radical frameworks inform a growing number of studies.

Like most concepts commonly used and believed relevant in explaining all sorts of social phenomena, power is a complex and elusive one. Many modern writers on the topic appear to be reluctant to give power a clear definition. It is possible that the complexity of the concept makes it impossible or at least unfruitful to define it sharply. Galbraith (1983) points at a tendency among authors to let the subject lead them to an unneccessary degree of complexity and subjectivity, which functions as a protection from criticism. Considering the criticism against authors who have defined power in a brief and clear way, such as the capacity of actor A to get actor B to do what actor B would not otherwise do (Dahl), it is understandable that many authors refrain from formal definitions of the concept.

The understanding of various positions on power demands that the deeper, paradigmatic underpinnings of various definitions and views are considered. As Lukes (1978) writes:

"The history of political theory and of sociology is in part a history of unending disagreement as to how power and authority are to be conceptualized and how they relate to one another. Moreover, that disagreement is endemic, and it is so for deep reasons. These concepts are not labels for discrete phenomena: they have distinct roles in social and political theorizing and in social and political life. Different and contending theories and world views yield different ways of conceiving power and authority and the relations between them." (P 633)

Lukes puts forward a number of important questions of value for the student of the subject to consider carefully: Is power a property or a relationship? Is it potential or actual, a capacity or the exercise of a capacity? Is it agents (individual or collective) or structures (systems) that possess or exercise power? Is it, by definition, intentional, or might power exist beyond intention and will? Must it be effective? What types of outcome does it produce: does it affect interests, options, preferences, policies or behaviour? Is it a zero-sum concept? Must it rest on or employ force, coercion or manipulation, or might it build upon consent?

It is fruitful to distinguish three frames of reference for approaching power, interests and conflict in organization theory (and social science in general) often referred to, the *unitary*, *pluralist* and *radical* perspectives (Burrell and Morgan, 1979. The unitary frame of reference, which has traditionally dominated management/organization research, largely ignores the role of power in organizational life. Concepts like authority and leadership are preferred in describing top management's prerogative of guiding the organization towards common goals. The pluralist perspective places emphasis on the diversity of individual and group interests. The organization is seen as a loose coalition of people. The internal relations are partly governed by power relations. It is assumed that a plurality of power holders exist, and that these are empowered by various power sources. The pluralist framework appears to be on its way to be the conventional wisdom of management/organization theory, but in a rather weak version, perhaps best to be described as a combination of the unitary and the pluralist frames of references.

Like proponents of pluralism, radicals believe that power relations are crucial for understanding organizations. Radicals emphasize class contradictions, which are seen as the basis of social conflicts and power struggles, partly being manifested on the organizational level. The radical framework assumes that these conflicts are of a much more basic nature than pluralists believe. It is not sufficient to focus only on the local level, while this partly is a reflection of and strongly influenced by wider processes of social control. In opposition to pluralists, who often emphasize horizontal relations and conflicts, the radical frame of reference views vertical conflicts (between management and workers) as primary. While pluralists view conflict and somewhat restricted power struggles as part of social life and thus eternal, radicals assume that the present (capitalistic) society is especially conflict-ridden and that a more positive form of power (authority), less coercive, and grounded in consensus is historically possible. (See Burrell and Morgan, 1979; Morgan, 1986.)

Leaving aside, for a moment, these overall frameworks for the study of power, I shall indicate some complexities involved in the concept of power proceeding from the critique raised against traditional conceptualizations. These proposed that power is the ability of an actor to determine the behaviour of another actor. As formulated by, for example, Emerson:

"The power of actor A over actor B is the amount of resistance on the part of B which can be potentially overcome by A." (Quoted by Pfeffer, 1981.)

As Lukes (1978) says, the notion of the bringing about of consequences is the common core to all conceptions of power. But the ability to affect visible outcomes, captures only some aspects and might give avisible only some aspects and might give a misleading picture of power.

An important critique was put forward by Bachrach and Baratz (1962), who talk about "the two faces of power". The first is the one indicated above, i. e. power manifested in concrete decisions and outcomes produced by these. The second face of power concerns the extent to which a person or a group can create or reinforce barriers to other people challenging the values or interest of the former, i. e. the power of "non-decisions", of maintaining status quo.

Lukes (1978), among others, has argued for an even more multifaceted view, and suggests a three-dimensional theory, in which not only observable conflicts and overt behaviour are involved, but also "the subtler and less visible ways in which potential issues are kept out of politics through the behaviour of groups and practices of institutions". The supreme exercise of power is a matter of "influencing, shaping, and determinating the perceptions and preferences of others" (p 669). While the first two dimensions or faces of power might be seen as surface manifestations of power, the third dimension concerns its deep structures (Frost, 1987). These are hard to detect and difficult to investigate. Most students of managerial and organizational power have concentrated on the surface aspects. The development of management/organization studies informed by critical theory and the organization culture perspective brings along a more extensive interest in the depth level of power (Deetz and Kersten, 1983; Frost, 1987; Riley, 1983; etc). Power is then seen as anchored in the minds, the language, the discourse and the interpretive scheme of people:

"The power holders have constituted and institutionalized their provinces of meaning in the very structuring of organizational interactions so that assumptions, interpretations and relevances become the interpretive frame, the cognitive map, of organizational members." (Ranson, et al, 1980 p 8)

The following observation in an empirical study clearly indicates the importance of going beyond the traditional view on power ("the first face"):

"The most important kinds of power were already constituted as being those occasions when A's didn't have to get B's to do things because B's would do those sorts of things anyway." (Clegg, 1987 p 65)

Congruent with the depth view on power — although it does not necessarily go that far — is the criticism of older, and still quite popular and perhaps also dominant views of conceptualizing power as an attribute. Power is not something one can possess, but a social relation, critiques say (Daudi, 1986; Knights and Roberts, 1982; etc). Power is then seen as a quality of the relationship between people, which manifests autonomy and dependence in both directions. Thus, power is viewed as dyadic. It invokes the self-understandings of all involved.

"These self-understandings are conceived as present in those relations (of leadership, manipulation, coercion, or force) in which one group commands or even dominates another, and as crucial in those relations in which members of a group enlist one another to become active in their own regard. ...Power, like all social interactions of active beings, is rooted in part in the reflections and will of those interacting, both the powerless as well as the powerful." (Fay, 1987 p 130)

This view is, of course, quite different from the conception of power as a dependent/independent variable, possible to isolate and measure (e. g. Hickson et al, 1971; Pfeffer, 1981; etc.)

564

Another criticism of the conventional and pluralist views on power from a radical perspective focuses upon the narrow range of phenomena which are discussed in terms of power and politics. The majority of studies are accused of paying attention to power and politics only when it comes to departures or deviations from an officially sanctioned order. Hierarchy and the "formal" and "technical" spheres of management and organization are, in practice, seen as on the whole apolitical. Pluralists start from the assumptions of a balance of power and a respect for the "rules of the game" that maintain the system in equilibrium. Much of what could be investigated in terms of power is therefore taken for granted (Knights and Roberts, 1982; Willmott, 1987; etc).

Most writers on power devote a lot of attention to the sources of power. Hickson et al (1971) suggested that the power of an actor (an organizational subunit) is a function of its ability to cope with uncertainty, its (non-)substitutability, and its centrality (the degree to which it affects a wide range of the work of other units, and the speed and severity with which its workflow affects the final output of the organization). This view is called the resource dependency model of power. Several authors have added that the willingness and ability of actors to transform resource dependency based power into exercised power is crucial. Resource dependencies are sometimes seen rather as a background to power relations than as something that mechanically brings about a certain distribution of power. History and culture are important here. The resource dependency relationship between actors at an earlier point in time is the origin of future deep structures of power:

"Successful efforts by these actors to reproduce and perpetuate that power, but in a disguised form (so that its sectional nature, benefiting powerful actors at the expense of other actors, is hidden from view), create deep structure power that operates beyond the time of the initiating dependency relationship." (Frost, 1987 p 512)

Morgan (1986) lists a number of other important sources of power: formal authority, interpersonal alliances, networks and control of "informal organizations", symbolism, gender and the power one already has. Ownership and property are, of course, also crucial.

Related to the criticism of traditional and narrow conceptions of power, referred above, is the suggestion of proponents of a radical frame of reference to take the social and cultural context of local manifestations of power into account. This context, involving meaning patterns, and taken for granted assumptions on rationality, authority, superior-subordinate relations, etc, provides important sources of power. Local actors draw upon these. This is seldom acknowledged in mainstream studies within the unitary or pluralist perspectives. In these the institutional grounds of managerial work as an expression of politico-economic relations of power are overlooked or obscured (Willmott, 1987).

In principle, it is possible for proponents of each of the three frames of reference discussed — the unitary, the pluralist and the radical — to investigate both the surface and the deep aspects of power. There is, however, a clear tendency that radicals take the deep structures most seriously. Those pluralists take the cultural dimension of management/organization carefully into consideration also who treat the depth dimension. While radical authors often also relate organizations to a wider totality and sometimes investigate how local actors are empowered by resources provided by the history, culture and socio-economic conditions of society at large, often end with the most complex view on power. This might, of course, be difficult to use empirically. It might also bring about a sense of unease to the individual eager to avoid areas in which power and politics put strong imprints. The more carefully one reflects upon power, the more difficult it becomes to find areas or issues "free" from power and politics.

CONFLICT

The three frames of reference for the study of power are also relevant for capturing different positions to conflict. They might be grouped along the consensus-conflict continuum. The unitary perspective assumes that conflict is a rare and transient phenomenon. It deviates from the normal order of things and might be avoided or removed through appropriate managerial action. This view has become increasingly unpopular among writers on conflict. More people adhere to the pluralist view, which regards conflict as an inherent and ineradicable characteristic of organizational affairs. Often the positive and functional aspects of conflict are stressed. Robbins (1983), for example, assumes that

"...a harmonious, peaceful, tranquil, and fully cooperative organization is prone to becoming static, apathetic, and nonresponsive to needs for change and innovation. Conflict can stimulate new ideas, improve intragroup cohesiveness, result in better decisions through the input of divergent opinions, allow for venting of frustrations, and offer other similar benefits to the organization." (P 290)

It is suggested that the "right" level of conflict increases organizational effectiveness. Of course, this demands that conflicts do take place against a background of converging basic goals and shared values. If people do not stick to the basic rules of the game, then conflicts hardly pay off. This version of pluralist thought, like most other which do not draw the ideas of the power conflicts and diverging interests so far that the existence and reproduction of the organization is seen as being in danger, can be said to focus upon restricted types of conflicts taken place against a background of consensus. Within the unitary/pluralist frameworks, authors take an interest in "the management of conflict", which might not only involve resolution techniques, but also methods for the stimulation of conflict, such as creating heterogeneous units or competition between units.

Conflict is typically seen within these frameworks in a way which corresponds to the traditional view on power. Conflict might, for example, be defined as "a process in which an effort is purposely made by A to off-set the efforts of B by some form of blocking that will result in frustrating B in attaining his goals or furthering his interests" (Robbins, 1983, p 289). This means that only overt conflicts are taken into consideration. Proponents of the radical frame of reference favour a broader view on conflict. Social conflicts are assumed to take place in society at large and these are an important driving force behind structural change. The local, organizational level is heavily affected by these macro-conditions and pro- cesses.

The radical framework does not concentrate only on manifest conflicts but presumes that conflict is often suppressed and thus often exists as a latent rather than a manifest characteristic of both organizations and society (Burrell and Morgan, 1979). For this reason, the concepts of disintegration and, in particular, contradiction are central to radical analysis. Contradictions are seen as a key property of social relations. They may characterize the relationship between existing structure and processes, between different groups etc. Ruptures, breaks and inconsistencies are unavoidable parts of social life, and the contradictions that follow sometimes become suppressed (for a shorter or longer time), sometimes they are manifested in social conflicts (Benson, 1977).

CONTROL

Control is a difficult concept in much the same way as power. It also appears to refer to the same type of phenomenon, making power and control synonymous concepts. Like power, control restricts discretion and produces a certain type of outcome, more or less governed by somebody's intentions and wishes. However, control normally has a slightly different connotation. It is less closely connected to particular actors (agents). Often it is perceived as more functional than power, thus more neutral in terms of social interests. Control is often also viewed as being of a static and stable character, a property of relatively fixed social relations and conditions, while power has to do with bringing about/changing outcomes.

The concept of control is ambiguous in terms of political connotations, which partly has to do with the wide range of phenomena it can refer to. Sometimes proponents of the unitary frame of reference talk about control instead of power because of its relatively "neutral" character (Morgan, 1986). On the other hand it is common among Marxists and other proponents of radical thinking to emphasize social control as a crucial dimension in corporations, being what is at stake in the relationship between capital and labour. The crucial task for management is to transform "formal" labour power into "real" labour, i. e. to make a more or less reluctant labour force produce surplus value. Control then refers to the political and contested nature of the enterprise. In a situation in which the direct producers themselves have a lot to say about the work-place, and production is grounded on a real consent, management and administration become less a matter of control and rather a question of "coordination" (Edwards, 1979). The unpleasant connotation of the control concept probably accounts for the fact that when Mintzberg (1983b) describes the functioning of "machine bureaucracies" he talks a lot of control: "the Machine bureaucracy is a structure with an obsession — namely, control. A control mentality pervades it from top to bottom" (p 167). His other four configurations are primarily described in terms of coordination.

An overview of studies on managerial/organizational control indicates a transfer in focus from an emphasis on the harsher, coercive, behaviour-restricting aspects of control to also considering consensual aspects. The debate in labour process theory for a long time dealt only with Taylorism and other forms of technical control. Other types of structural and behavioural control have also received attention (Burris, 1988, Edwards, 1979; Simpson, 1985; etc). Recently, also the "subjective", ideational, i. e. ideological and cultural forms of control have been studied (Alvesson, 1988; Czarniawska-Joerges, 1988; etc). A parallell can here be drawn to the development of studies of power in management/organization. An interest in its objectivistic and behaviouristic features and surface manifestations has been complemented with approaches focusing upon its less visible features.

Many authors argue that a particular type of control is dominating in a particular industry at a particular time (e. g. Edwards, 1979). Others are of the opinion that a complex mixture of means of control are in operation at the same time (Clegg, 1981; Storey, 1985). Clegg (1981) proposes the metaphor of sedimentation to conceptualize this:

"Within the organization, different control rules evolve at different times and at different stages of functional complexity. Earlier rules may persist at specific levels in the organization, despite the latter development of more complex rules. These rules may be represented as a superimposed series. As in the nature of social reality, the articulation of relations among the different levels may produce not only unanticipated consequences but also contradictions." (P 551)

A case study of a professional service company illustrates how a multiplicity of control forms might be in operation. Besides the structural types of control, such as economic control, decentralization and autonomy under tight financial control and legal contracts tying up the employees to the employer, also "softer" means such as ideological control, efforts from management to affect social perception and to support social bonds between managers and employees, were effective in the corporation (Alvesson, 1988).

Many authors writing about power now treat it in a much more complex way than before. This might be the result of theoretical development. The shifting view on power follows the general development of management/organization theory and its break with the domination of functionalist thought and increased paradigmatic diversity, involving a total picture of the objects of study which is quite multifaceted and complex (cf Reed, 1985). In addition to this, the theoretical development of power studies can also be connected to changes in the material basis for the exercize of power in society and organizations. As Galbraith (1983) proposes, the relative impact of earlier dominating forms of power, coercive and compensatory power (punishments and rewards), has been weakened in modern, affluent, permissive society and the importance of conditioned or persuasive power has increased heavily. This makes the problem of power more difficult and calls for theoretical responses in which the subtler forms of power are paid more attention to. Linking power closer to ideology, culture, discourse, etc becomes appropriate.

This broader and deeper understanding of power has clear implications for the role of research workers in relationship to practioners. Management science in general and also its more "technical" versions (O R etc) are rather often accused of subordinating themselves to present power conditions, of serving the powerful, strengthening their interests and thus reinforcing the status quo (Jackson, 1987; Zey-Ferrell and Aiken, 1981; etc). Realizing the complexities of power makes it even more important and difficult to take the critique seriously for those research workers eager to avoid being involved not only in surface manifestations, but also the deep plays of politics. As management scientists

"We tend to reinforce the play of many games that disproportionately empower some players over other players. ...(We should) recognize the political act we perform whenever we communicate knowledge to one interest group rather than another." (Frost, 1987 p 538)

Acknowledging the power dimension involved in (almost) all social action in combination with the reasonable assumption that many (most?, all?) complex problems to some degree are accompanied by divergent and opposing interests puts the political role of the expert dealing with "real world problems" into focus. Ironically, the expert eager to avoid being involved in power and politics by seeing him-/herself as a "technician" reporting to a CEO or another appointed official, responds in a political way to what is really an intellectual problem, Checkland (1983) says:

"Intellectually there is no reason at all why the expertise should not be equally at the disposal of the likely victims of the system in question, those who decree its creation, its designers or, indeed, any other group with any interest in the outcome of the study." (P 667)

One of the lessons of the study of power for management researchers and other applied social scientists concerns the multiplicity of levels in which dominating coalitions (or other groups) might be empowered. One level concerns the issues at stake in a particular situation, to what extent different actors determine a particular outcome. (This is the "first face" of power). Another level concerns to what extent a (dominating) group succeeds in blocking other groups' resistance to a particular issue. This "second face" of power, might lead to an expert not meeting groups opposing the view of the (dominating, recognized) client in his/her work. This can be a result of the limits in terms of interaction/communication channels as a consequence of an unreflective view on who the client(s) is (or are) (an effect of the defining out stakeholders other than the dominating one). The "third face", perhaps the most interesting one, involves the whole range of intended and unintended communications affecting the construction and reproducting of social reality. Such processes are not politically neutral. Especially not those versions which aim at the "scientisation" and "depoliticisation" of managerial problems and social reality. Increased awareness of the deep structures of power make technocracy as a bastion against the intrusion of the problematics of power a very unsafe position.

REFERENCES

Alvesson, M., 1988, Management, Corporate Culture and Labour Process in a Professional Service Company, Paper presented at 6th Labour Process Conference, Aston University, Birmingham, March 1988.
Bachrach, P., and Baratz, M., 1962, Two Faces of Power, *Am. Pol. Sc. Rev.*, 56:947-952.
Benson, J. K., 1977, Organizations: A Dialectical View, *Adm. Sc. Quart.*, 22:1-21
Burrell, G. and Morgan, G., 1979, "Sociological Paradigms and Organizational analysis", Heinemann, London
Burris, B., 1988, The Transformation of Organizational Control and the Emergence of Technocracy, paper, Dept. of Sociology, University of New Mexico
Checkland, P., 1983, O. R. and the Systems Movement: Mappings and Conflicts, *J. Opl.Res. Soc.*, 34:661-675
Clegg, S., 1981, Organization and Control, *Adm. Sc. Quart.*, 26:545-562
Clegg, S., 1987, The Language of Power and the Power of Language, *Org. Stud.*, 8:61-70
Czarniawska-Joerges, B., 1988, "Ideological Control of Nonideological Organizations", Praeger, New York
Daudi, Ph., 1986. "Power in Organization", Blackwell, Oxford
Deetz, S. and Kersten, A., 1983, Critical Models of Interpretive Research, in: "Communication and Organizations", Putnam, L. and Pacanowsky, M., eds., Sage, Beverly Hills
Edwards, R., 1979, "Contested Terrain", Heinemann, London
Fay, B., 1987, "Critical Social Science", Polity Press, Cambridge

Frost, P., 1987, Power, Politics, and Influence, in: "Handbook of Organizational Communication", Jablin, F.et al., eds., Sage, Beverly Hills

Galbraith, J. K., 1983, "The Anatomy of Power", Simon and Schuster, New York

Hickson, D., Hinings, C. R., Less, C., Schneck, R. and Pennings, J., 1971. A Strategic Contingencies Theory of Intraorganizational Power, *Adm. Sc. Quart.,16: 216-229*

Jackson, M., 1987, Present Positions and Future Prospects in Management Science, *Omega*, 15:455-466

Knights, D. and Roberts, J.,1982, The Power of Organization or the Organization of Power, *Org. Stud.*, 3:47-63

Lukes, S., 1978, Power and Authority, in: "A History of Sociological Analysis", Bottomore, T. and Nisbet, R., eds., Heinemann, London

Mintzberg, H., 1983a, "Power in and around Organizations", Prentice-Hall, Englewood Cliffs

Mintzberg, H., 1983b, "Structure in Fives", Prentice-Hall, Englewood Cliffs

Morgan, G., 1986, "Images of Organization", Sage, Beverly Hills

Pfeffer, J., 1981, "Power in Organizations", Pitman, Boston

Ranson, S., Hinings, B. and Greenwood, R., 1980, The Structuring of Organizational Structures, *Adm. Sc. Quart.,25:1-17*

Reed, M., 1985, "Redirections in Organizational Analysis", Tavistock, London

Riley, P., 1983, A Structurationist Account of Political Culture, *Adm. Sc. Quart.,28:414-437*

Robbins, S., 1983, "Organization Theory: The Structure and Design of Organizations", Prentice-Hall, Englewood Cliffs

Simpson, R., 1985, Social Control of Occupations and Work, *Ann. Rev. Sociol.,11:415-436*

Storey, J., 1985, The Means of Management Control,*Sociology*, 19:193-211

Willmott, H., 1987, Studying Managerial Work: A Critique and a Proposal, *J. Mgmt. Stud.*, 24:249-270

Zey-Ferrell, M. and Aiken, M., 1981, "Complex Organizations: Critical Perpectives", Scott, Foresman & Co., Glenview

DOMINATION: STRATEGIC PREMISE SETTING AS IDEOLOGICAL INSTRUMENT

W. P. Hetrick

Department of Marketing
Loyola Marymount University
Los Angeles, CA

A. J. Grimes

Department of Management
University of Kentucky
Lexington, KY

INTRODUCTION

"If you allow me to determine the constraints, I don't care who selects the optimization criterion" (Simon, 1964: 3).

Human nature is made from the materials of bureaucracy (Perrow, 1986: 138).

Premise setting, a form of unobtrusive control, is instituted by top management to further induce what they deem as necessary and desirable participation from organizational members. Reed (1985) contends that both Barnard (1938) and Simon (1976) modified their original view of organizations as cooperative systems to include "elements of ideological indoctrination and structural control to deal with an irredeemable human recalcitrance on the part of 'lower participants'" (p. 6). In their modified view, Barnard (1938) and Simon (1976) both recognized that material inducements must be supplemented by other means of social control (e.g., ideology) to assure and maintain compliance of organizational members (Reed, 1985).

Premise setting is an ideological device used not only to control the labor force, but also to perpetuate the dominance of some management coalitions over others. Specifically, those coalitions that control strategic premise setting can enhance their power. According to strategic contingencies theory (Hinings et al., 1974), "certain departments will benefit more than others the extent to which they are able to shape strategies to meet their needs" (p. 5). Provan (1989) defines departmental power "as the ability of a department to influence organization-level strategy decisions and their implementation so that top management is more supportive of the needs and perspective of that department than other departments within the organization" (p. 5). Premise controls are especially important when "work" consists of decision making and strategy-related activity. Perrow (1986) suggests that while the control of premises can be attributed to the effects of organizational culture, it is just as plausible to attribute controls of this nature to "indoctrination, brainwashing, manipulation, or false consciousness" (p. 130).

The following questions will guide our examination of premise setting as a form of ideology: How are "unobtrusive controls" a form of ideology? Do premise controls dominate managers? How does "bounded rationality" relate to unobtrusive controls and domination? How do departmental interest and organizational means/goal ambiguity influence these processes?

Bringing together a variety of critical theory (Geuss, 1981; Horkheimer, 1972; Marcuse, 1966), organization theory (Burrell and Morgan, 1979; Cohen, March, and Olsen, 1972; Nord, 1978; Perrow, 1986; Pfeffer, 1982), and critical organization theory perspectives (Alvesson, 1987; Reed, 1985; Steffy and Grimes, 1986), we offer an approach and insights that may answer the above questions and illuminate premise setting as a "third face" of power (Lukes, 1974) in the exercise of departmental power and in the perpetuation of domination in bureaucracy. Because of brevity, we will not be able to make many theoretical distinctions. While we paint with a wide brush on a short canvas, we hope our view is both plausible and provocative.

First, we consider ideology and domination. Next, we add the complications of bounded rationality, departmental interests, and decision making ambiguity. Third, we explore the advantage of unobtrusive controls to the management elite. Finally, we offer a brief summary of the paper.

IDEOLOGY AND DOMINATION

The theses of the Frankfurt School allow us to evaluate the crucial relationship between ideology and domination (Thompson, 1983). Our position: there is an inherent connection between the concept of ideology and the critique of domination.

Geuss (1981) views pejorative ideology as exhibiting negative characteristics: "so the term 'ideology' is used in a pejorative sense to criticize a form of consciousness because it incorporates beliefs which are false, or because it functions in a reprehensible way, or because it has a tainted origin. I will call these kinds of criticism: criticism along the epistemic, the functional, and the genetic dimensions respectively" (p. 21). Descriptive ideology for Geuss is simply a "form of consciousness." The interesting question is 'What makes a form of consciousness a false consciousness?'. For Geuss (1981), finding epistemic, functional, or genetic properties within a form of consciousness indicates it is ideologically false.

The bureaucratic ethos is a form of functionally false consciousness because it supports reprehensible practices (e.g., hierarchy and minute division of labor) that result in domination (Habermas, 1970). The bureaucratic ethos is a form of genetically false consciousness because it promotes calculability and efficiency, values particular to sectional interests, as values of the whole.

Formal rationality is also a form of ideology. Formal rationality reproduces the "distribution of scarcity": "no matter how useful this rationality was for the progress of the whole, it remained the rationality of domination, and the gradual conquest of scarcity was inextricably bound up with and shaped by the interest of domination (Marcuse, 1966: 36). Marcuse (1966) has in mind an engineered scarcity which has led to the introduction of controls that exceed those necessary for societal advancement. These unnecessary controls ("surplus repression") are required by those interested in perpetuating social domination. Marcuse (1966) distinguishes between authority and domination. Authority is derived from knowledge and exercised for the advancement of the whole. Authority provides the "basic repression" necessary for the perpetuation of civilization (Marcuse, 1966). Domination, for Marcuse (1966) as well as Thompson (1984), exists when power relations are systematically asymmetrical, and when some "agents are institutionally endowed with power in a way which excludes, and to some significant degree remains inaccessible to, other agents" (p. 45).

Kellner (1984) contends that domination has both external and internal dimensions. The internal dimension has particular relevance here in that it "involves the internalization of prohibitions, values and social demands through which the individual (manager) disciplines him- or herself, acting out internalized social roles and behavior" (Kellner, 1984: 165). When strategic premise setting is ideological, it is an example of this internal dimension of domination. Strategic premise setting is ideological when it facilitates control of strategic processes by elites in order to enhance their position.

CONTROL COMPLICATIONS

Several factors both facilitate and complicate the operation of strategic decision making processes in organizations. We briefly consider bounded rationality (March and Simon, 1958; Perrow, 1986), departmental power/interests (Pfeffer, 1982; Provan, 1989), and means/goal ambiguity (Cohen et al., 1972; Cohen and March, 1974), before we discuss premise control.

Bounded Rationality

Both Barnard (1938) and Simon (1976) argue that individuals are not rational decision makers until placed in a well-designed organization, after which they may still be only intendedly rational "tools" (Simon, 1976) of a rational organization. While bounded rationality creates limits, it also permits organization elites to shape individuals as indicated in the Perrow (1986) cite at the beginning of this paper. "With bounded rationality, even a small and temporary advantage in information, resources, or goal clarity can be transformed into a large and stable one. With a small advantage, information is controlled and selectively used; premises alter the unstable goal structure and stabilize it in the interests of the elites; ideologies justify the differences in power. Bounded rationality is, to a small degree, an impediment to organizational effectiveness, but to a larger degree it makes hierarchical structures possible" (Perrow, 1986: 123).

Departmental Power

In addition to bounded rationality, the existence of individual and departmental interests (Provan, 1989) at variance with organizational goals increases the need to control decision makers. Departments in organizations come to articulate local values which lead to a bifurcation of departmental and organizational interests. Departmental interests may clash with one another and with global, organizational-level interests. The former may lead to the domination of one department over another - an example of lateral management domination.

Ambiguity of Organizational Decision Making

Boundedly rational decision makers, confronted with their own department interests, are also hampered in their quest for rational choices by the ambiguity of organizational goals and the relationship of means to goals. Organizational goals are contradictory and not "operational" (teaching, research, and service is a familiar U.S. university example). Individuals may not engage in the behavior we call "decision making". Frequently, preference ordering is not possible. Solutions are in search of problems (Cohen et al., 1972). Rational organizational decision making is often a myth (Pfeffer, 1977).

ILLUSION OF SHARED INTEREST

While we have discussed organizational demands for control, we have not
discussed the fact that the control of managers is undertaken at some risk.
Contrary to most organization and management literature, Reed (1985) claims
that, since "a moral basis" exists for hierarchical control, top management
is limited in the use of inducement and constraint controls. A dialectic
between manager (lower level) commitment and control (by upper management) is
ignored at the latter's peril (Reed, 1985: 130). Such a dialectic leads to
control situations which are "unobtrusive." These are symbolized by Simon's
(1964) claim in our introduction. Control structures and mechanisms not
viewed as "legitimate" by managers undermine the ability of elites to achieve
control.

We have examined the necessity of controls that do not reduce managers'
commitment to organizational goals. These controls contribute to an
"illusion of manager's shared interest". We next briefly consider premise
setting as a typical example of unobtrusive controls.

PREMISE CONTROL

March and Simon (who, according to Perrow (1986), "know the devil best")
argue that to change decision makers' behavior you only need to change the
premises of decision makers (1958). Perrow (1986) elaborates: "These
premises are to be found in the 'vocabulary' of the organization, the
structure of communication, rules and regulations and standard programs,
selection criteria for personnel, and so on ---- in short, in the structural
aspects" (p. 127).

Such mechanisms affect organizational behavior in the following ways:
they limit information content and flow, thus controlling the premises
available for decisions; they set up expectations so as to highlight some
aspects of the situation and play down others; they limit the search for
alternatives when problems are confronted, thus ensuring more predictable
and consistent solutions; they indicate the threshold levels as to when a
danger signal is being emitted ... (Perrow, 1986: 128).

We can be more specific about premise setting and control by examining a
"strategic issue" (Ansoff, 1980; King, 1982). Strategic issues result from
"strategic issue diagnosis" (SID) which Dutton and Duncan (1987) view as a
"general interpretive process where data confronting decision-makers are
given meaning" (p. 280). Attaching meaning to strategic data is a highly
subjective process allowing those in control to "create" interpretations
that foster their interests. Dutton and Jackson (1987) identify two popular
strategic issue categories, "opportunity" and "threat". These terms for
strategic issues are linguistic labels that once applied, "initiate a
categorization process that affects the subsequent cognitions and motivations
of key decision makers; these, in turn, systematically affect the process and
content of organizational action" (Dutton and Jackson, 1987: 77). The
ability to label a strategic issue consolidates the control of organizational
premises. Subsequent acceptance of the label strengthens dominant coalitions
because premises are internalized by other participants. "Here the
subordinate voluntarily restricts the range of stimuli that will be attended
to ('Those sorts of things are irrelevant,' or 'What has that got to do with
the matter?') and the range of alternatives that would be considered ('It
would never occur to me to do that')" (Perrow, 1986: 126).

Dutton, Fahey, and Narayanan (1983) suggest that once the consequences
of issue diagnosis are understood in terms of departmental interest,
political interests of participants will be aroused, and these participants

will then attempt to influence the diagnostic process to fulfill their self-interests. When the process is ideological (Pfeffer, 1977), sectional interests become disguised as benefits for all. While we agree with Perrow (1986) that dominant coalitions of managers have the power "to create technologies that will support a hierarchical, bureaucratic structure where such a structure is not necessary" (p. 265), we wish to stress that these coalitions also have the power to enforce the unobtrusive, lateral domination of other managers.

SUMMARY

Bounded rationality, individual interest (recalcitrance), departmental interest (departmental recalcitrance), and ambiguity of means/ends relations and organizational goals complicate the exercise of control over organizational members, whether the specific exercise of control is directed toward achieving the common good (authority or basic repression) or some sectional interest (surplus repression). At the inter-management level, the control commitment-dialectic maintains an "illusion of shared interest" which necessitates the use of unobtrusive controls like premise setting. To the extent that surplus repression is exercised, premise setting is ideological. The control of cognitive premises, achieved in part through control of the strategic issue process (i.e., strategic issue formulation and diagnosis), is one ideological instrument used by dominant organizational members in support of the ideology of bureaucracy and departmental interest.

A critical analysis such as this allows one to evaluate the ulterior motives possessed by elite and/or dominant coalitions. It is in the best interest of these coalitions to support the formally rational bureaucracy with its emphasis on rational decision making, the division of labor, and organizational hierarchy. The tasks of the critical organizational researcher are to critique the prevailing ideology present in modern organizations and to reveal instruments and devices used in the perpetuation of this ideology. Demystification of the ethos of bureaucracy and its corresponding support mechanisms will lead individuals to be aware of ideology and its true function.

REFERENCES

Alvesson, M., 1987, "Organization Theory and Technocratic Consciousness," de Gruyter, New York

Ansoff, I., 1980, Strategic Issue Management, Strategic Mgt. Journal, 1:131-148.

Barnard, C.I., 1938, "The Functions of the Executive," Harvard University, Cambridge, MA.

Burrell, G. and Morgan, G., 1979, "Sociological Paradigms and Organizational Analysis," Heinemann, Portsmouth, NH.

Cohen, M.D., March, J.G., and Olsen, J.P., 1972, A Garbage Can Model of Organizational Choice, Admin. Sci. Quart., 17:1-25.

Cohen, M.D. and March, J.G., 1974, "Leadership and Ambiguity: The American College President," McGraw-Hill, New York.

Dutton, J.E. and Duncan, R.B., 1987, The Creation of Momentum for Change Through the Process of Strategic Issue Diagnosis, Strategic Mgt. Journal, 8:279-295.

Dutton, J.E., Fahey, L., and Narayanan, V.K., 1983, Toward Understanding Strategic Issue Diagnosis, Strategic Mgt. Journal, 4:307-323.

Dutton, J.E. and Jackson, S.E., 1987, Categorizing Strategic Issues: Links to Organizational Action, Academy of Mgt. Review, 12:76-90.

Geuss, R., 1981, "The Idea of a Critical Theory," Cambridge University, New York.

Habermas, J., 1970, "Toward a Rational Society," Beacon, Boston.

Hinings, C.R., Hickman, D.J., Pennings, J.M., and Schneck, R.E., 1974, Structural Conditions of Intraorganizational Power, Admin. Sci. Quart., 19:22-44.

Horkheimer, M., 1972, "Critical Theory," Herder and Herder, New York.

Kellner, D., 1984, "Herbert Marcuse and the Crisis of Marxism," University of California, Berkeley.

King, W.R., 1982, Using Strategic Issue Analysis, Long Range Planning, 15(4):45-49.

Lukes, S., 1974, "Power: A Radical View," MacMillan, London.

March, J.G. and Simon, H.A. 1958, "Organizations," John Wiley & Sons, New York.

Marcuse, H., 1966, "Eros and Civilization," Beacon, Boston.

Nord, W., 1978, Dreams of Humanization and the Realities of Power, Academy of Mgt. Review, 3:674-679.

Perrow, C., 1986, "Complex Organizations: A Critical Essay," Random House, New York.

Pfeffer, J., 1977, Power and Resource Allocation in Organizations," in "New Directions in Organizational Behavior," B.M. Staw and G.R.Salancik, eds. St. Clair, Chicago.

Pfeffer, J., 1982, "Organizations and Organization Theory," Pitman, Marshfield, MA.

Provan, K.G., 1989, Environment, Power, and Strategic Decision Making in Organizations: An Integrative Approach, Forthcoming Journal of Mgt..

Reed, M., 1985, Redirections in Organizational Analysis, Tavistock, New York.

Simon, H.A., 1964, On the Concept of Organizational Goal, Admin. Sci. Quart., 9:1-22.

Simon, H.A., 1976, "Administrative Behavior," Free Press, New York.

Steffy, B.D. and A.J. Grimes, 1986, A Critical Theory of Organization Science, Academy of Mgt. Review, 11:322-336.

Thompson, J.B., 1983, Ideology and the Critique of Domination I, Canadian Journal of Political and Social Theory, 7(1-2):163-183.

Thompson, J.B., 1984, Ideology and the Critique of Domination II, Canadian Journal of Political and Social Theory, 8(1-2):179-196.

CONFLICT & CONSENSUS: ETHICAL OPERATIONAL RESEARCH

Peter Pruzan and Ole Thyssen

Institute of Computer & Systems Sciences
Copenhagen Business School
Copenhagen, Denmark

INTRODUCTION

Since its inception as a discipline in the 1950's, the societal relevance of Operational Research has been restricted due to two major self-deceptions: a. Operational Researchers are *researchers/analysts, not decision-makers,* and b. Operational Research is a *scientific* methodology for formulating and solving *complex decision problems.* We contend that these two self-imposed restrictions can be loosened, and that OR can become both more realistic and relevant by explicitly introducing *ethics* as a guide to rational OR behavior. Our arguments are as follows:

1. OR analysts are also decision-makers. Their choice of terminology, problem representation and solution procedures predetermine the kind of results they obtain, results which may affect many parties. Therefore, Operational Researchers have social responsibility.

2. In its attempt to live up to the notion of "scientific", OR methodology is rooted in traditions from the physical sciences and neo-classical economics. Decision-makers are portrayed as utility maximizing individuals and therefore OR analysts should expurgate intuition, emotion and moral values from their notions of "problem" and "solution". It follows that "complexity" in OR interpretation refers to the algorithmic rather than the substantive characteristics of the models employed to extract "solutions" from "problems".

3. An ethical frame of reference is a necessary supplement to the prevailing OR perspective if it is to live up to its proclaimed systemic approach. In so doing it can significantly improve both the quality and the chances for success of the implementation of OR projects performed in social systems. Therefore ethics should explicitly be included in OR curricula and actively promoted as a legitimate - and therefore effective - systemic approach to good management.

ANETHICAL OR

We commence our reflections with the observation that the systemic aspirations of OR's progenitors are still rephrased in the introduction to virtually all OR texts in the Western world. Nevertheless, "solving complex decision problems" still appears to be synonomous with assigning numerical values to a set of variables so that some simple measure of performance for a model of a "mechanical" system is optimized. Consideration is only given to the objectives of a decision-maker - and not to collectivities of decision-makers nor to those affected by the decisions. And the system modelled is depicted in purely physical and economic terms. Human social behaviour, if considered at all, is represented by technical coefficients while the possibility that humans may have other values than self-interest and economic optimization is suppressed and implicitly considered to be irrelevant - or even immoral. Within this frame of reference, problems objectively exist - like stones. And solving such problems is the primary task of the Operational Researcher. The prevailing OR paradigm excludes notions of conflict, collective rationality and ethics.

This reflects the apparently widely held, but anachronistic, tenet in OR that the only social goal of the firm is to maximize its effectiveness. According to this line of thought, even though managers as private persons can experience social responsibility, their responsibility as economic functionaries is tightly delimited to serving the economic interests of the organization; activities to improve product quality, worker satisfaction, supplier relations, the integration of the organization within the local community, the enviroment and the situation of other parties affected by managerial decisions, should only be considered as investments. They are to be evaluated with the same measuring rod as is used to evaluate financial transactions. Paradoxically speaking, however, only if social responsibility is attributed intrinsic value can it be an instrument to further economic goals.

Since the last conference on OR & the Social Sciences 25 years ago, developments in employee expectations, organizational design, management practice, information technology and societal demands have led to new focii for the decision and behavioural sciences. Plans are no longer just the products of a planning process but are signals, communication to the organization's many stakeholders. And therefore their quality is not only being evaluated internally in the organization by top management but also by all those parties who are affected by the planning. A result is that self-organisation and social responsibility are gradually supplementing hierarchy and authority as prototypal organisational concepts. And search and insight are replacing optimization and solution as prototypal planning concepts. In such a world there is an increasing mismatch between OR's conceptual underpinnings and real-world needs.

In its enthusiasm to be "scientific", OR has remained a field of analysis. And the OR practitioner is still depicted as an objective analyst, an observer of - but not a participant in - the political and social processes in the organization, who works with complicated, but well-defined "technical" problems. When he ignores the vital concepts of conflict and concensus and relegates them to the realm of "art" he effectively removes them from intellectual consideration and discourse. This decision by the OR analyst effectively - and self-deceptively - relieves him of an inherent ethical responsibility.

CONFLICT AND CONSENSUS

With this point of departure, we will make our terminology explicit. We consider *ethics* to consist of basic principles and procedures to guide the rational dissolution of conflict and to create consensus. The context for this approach to operationalizing what hitherto has been considered to be a personal matter is the increasingly pluralistic character of modern, autopoietic organisations. Both society and the firm provide a framework for many ways of life, each with its own set of values. It is therefore important to differentiate between morals and ethics. In an organisation with many subcultures no single moral codex enjoys universal approval or dominates the organisational identity; morality has become a matter to be decided upon by the individual subculture. At the same time, it is vital for the organisation's stability that the many subcultures tolerate each other and cohere via their acceptance of membership in the organisation. The principles which underlie the coherence cannot be derived from the individual moralities; logically, they must be found at a higher level. Therefore, we refer to ethics as second order morality.

Ethical principles have no objective basis. They create their own autopoietic order. Even if they are universal and abstract, their function as a normative guide to behaviour is context-dependent. In transcending the moral perspectives of the subcultures, ethic's only tools are formal values and rationality. Therefore, a decision is ethically acceptable if it can be rationally approved by all parties affected. This is determined in an open, self-reflexive discussion. The ethical discourse relates to that whole which is constituted by the parties affected by a decision, the stakeholders. It follows that the stakeholders have an interest in participating in the making of decisions even though they may lack the formal power or authority to do so. And when no single part can prove that its standpoint has universal applicability, a universally applicable principle must be that all the stakeholders must approve of a decision if it is to be ethical. This is what we refer to as consensus via conflict. Just as there is a need for disagreement which creates diversity and development, there is also a need for common principles which can reconcile conflict between stakeholders and transform it into constructive, systemic consensus.

Conflict arises when *decision-makers*, who set out to solve a *problem*, that is, to select a course of action, cannot agree on a *solution*. If a decision is arrived at procedurally, for example by voting, tossing a coin or hierarchical referral, then the conflict has not been *dissolved* even though the problem has been solved. If on the other hand, the parties involved have openly discussed their values and preferences and arrived at a joint decision, they have both *solved the problem* and *dissolved the conflict*; they have achieved *consensus*.

Of course not all conflicts are ethical in nature. Many are technical and can be resolved by technical compromises. A conflict becomes ethical when it over and above considerations of functional effectiveness involves individuals' or groups' basic values. Thus, what determines whether a conflict is "ethical" is context-dependent as values vary with time and place.

Within this framework, we will consider three major classes of conflict which should become central themes in the OR curricula: 1) *intra*-personal conflict, where the individual decision-maker has difficulty in ariving at a decision when no single decision is optimal with respect to all his values, 2) *inter*-personal conflict, where the members of a group of decision-makers have difficulty in balancing their own values and aspirations with those of the other members of the group, and 3) *systemic* conflict, where the decision-makers not only have to consider their own and the group's values but also those of all the stakeholders who are affected by the potential decisions.

Intra-personal conflict

Before we can consider inter-personal and systemic conflict, we must first reflect upon an assumption which implicitly underlies most OR models. Namely that the values of a decision-maker can be reflected by a simple, single-criterion, preference function to be maximized or minimized. Focusing upon optimization of a single criterion may, at a superficial level, appeal to our desires for simple, quantifiable, accepted and communicable measures of performance. "Economic man", endowed with neoclassical economic rationality and whose only interests are in maximizing profits or minimizing costs (or other such surrogate measures of effectiveness), still appears to be the template used to structure OR problem formulations.

This is contrary to experience. When faced with complex problems in social systems, no matter how much information they may have at their disposal, decision-makers experience doubt and often seek security in non-decisions. Such behaviour is typically rooted in the decision-maker's multiplicity of unclear and unstable values. It is rare that any decision can simultaneously be best according to all the these values. In addition, the ability to rationally identify a "best" decision may be severely delimited due to a difficulty in expressing fundamental values in terms of measurable criteria, and therefore in being able to determine the best balance between them. A result may be intra-personal conflict, the inability of the individual to determine what he feels is the best course of action to follow. This difficulty becomes amplified when the decision-making environment is expanded to include other decision-makers (the subject of inter-personal conflict) and reaches its ethical expression when consensus is to be obtained not only between decision-makers, but when consideration is also actively given to those parties who will be affected by the decision (the subject of systemic-conflict).

It should be noted that in recent years OR has accepted multiple-criteria-decision-making techniques as members of its tool kit. However this acceptance is not one of principle and has not affected the mainstream approach to the OR interpretation of social reality. Its major affect has been to supplement the notion of "optimal solution" with one of "efficient solution" and the concept of an optimization algorithm with that of search algorithms for a "best" efficient solution. The emphasis is upon techniques, rather than on more realistic structuring and interpretating of an individual's decision-making processes.

Inter-personal conflict

By implicitly assuming that there is one decision-maker (or, equivalently, a monolithic group of decision-makers), OR has neglected the inter-personal conflicts which typify significant planning situations. Although some OR texts do give lip-service to concepts eminating from game theory and social choice theory, inter-personal conflict and consensus are essentially ignored.

Dissolution of such conflict presumes that all the participants achieve "consensus with one's self" and "consensus with the others". It is a process which requires the participants to determine what they want as well as what they can obtain. Since in general it is not possible to aggregate individual measures of value to a group measure of value, analysis of such processes requires the consideration of collective - rather than individual - rationality and therefore of different types of models than hitherto have characterized the OR profession.

Organisational theorists are concerned with notions of "corporate culture and values" originally developed in anthropology, systems scientists utilize notions of self-governance eminating from studies of autopoietic systems in biology, and the fields of decision-support and artificial intelligence have borrowed strongly upon the cognitive and computer sciences to develop frames of reference for group decision-making. There is a wealth of methodological inspiration to be found in these fields as well as in the "soft systems methodology" developed as a reaction to the "hard systems" methodology of OR.

Systemic conflict

As indicated earlier, traditional OR implicitly focuses upon the decision-*maker* and ignores the fact that the success of his behaviour depends upon the decision-*receivers* affected by the decisions. In other words, OR neglects the influence of the organization's stakeholders and therefore how these parties may exert significant influence upon the implementation of an OR project.

This focus upon those who decide and the concommitant neglect of those who are affected by the decisions, i.e. a neglect of ethics, is itself unacceptable from an ethical point of view. As noted earlier, operational researchers are also decision-makers and therefore have social responsibility. The lack of attention to this responsibility supports an attitude that what matters is not what is ethically acceptable or what is legitimate, but only what is legal; that which isn't forbidden is acceptable.

The neglect of ethics is also inefficient. It disregards what is obvious for modern management culture: appreciation of organisational needs and motivations as well as of societal acceptance is a prerequisite for managerial success. OR which does not respect such attitudes and which does not attempt to design and employ its tools accordingly will be relegated to the operating rather than the managerial level it aspires to.

AN ETHICAL OR: CONSENSUS & ASSENT

We have presented ethics as principles and procedures for dissolving conflicts between all the parties affected by a decision. It will therefore primarily manifest itself in OR methodology in the form of consensus-motivating procedures. Two warnings are called for before we proceed to a discussion of such procedures.

First of all, ethics is a special kind of "tool". Normally a tool can be employed without its user becoming personally involved. This is not the case here. If ethics is cynically employed as a tool to control others, i.e if the manager considers himself above the ethical perspective, it will quickly lose its effect. It is therefore risky to employ such an expression as "planning tool", which smells of ice-cold management philosophy, where what counts is not whether ethics is valid but if it is effective. Our attitude is the opposite; ethics is only effective if it is valid. And whether it is valid becomes apparent when ethical considerations clash with other considerations, especially costs and profits, the proxy measures traditionally employed as criteria in OR models.

Secondly, ethics focuses upon wholeness and on assent among stake-holders. "Whole" is those parties which are affected by the organiza-tion's decisions. These can include its employees and owners as well as its customers, suppliers, financers, competitors, local society and the environment. If ethics only deals with what happens inside the organiza-tion, while relations to the external stakeholders are ruthless, as in Mafia families, then it is difficult to speak of ethics. Ideally speak-ing, *all* the parties, who are affected by the organization's operations are to be involved in its planning. This is a demanding expansion of the prototype planning concept, which focuses only upon those who plan, not on the planned-for. It requires at a minimum that the stakeholders and their values and aspirations are identified by the organization's lead-ership and that the necessary "conversations" are established. Here is a huge and challenging area for OR: to partake in the development of this planning of planning and in the techniques, which will be useful in the analysis of conflict and the promotion of consensus.

By promoting consensus, ethics can contribute to the successfull *implementation* of OR projects:
    1. the need for control is reduced since all parties have respon-sibility for a decision.
    2. all stakeholders are obliged to transcend their narrow, private- and group interests.
    3. a decision which is an agreement between interested parties is, ceteris paribus, preferable to a decision process which is blocked by conflict and where either hierarchical referral or formal procedures lead to the same decision.
    4. enthusiasm, and the "goals of the organization" receive a ra-tionale.

COLLECTIVE RATIONALITY AND THE ETHICAL OPERATING STATEMENT

In conclusion we will suggest two new areas for methodological de-velopment. First of all we recommend that *collective rationality* - and not just the traditional individual, means-ends, rationality - becomes the focus of methodological research. Without operational concepts of collective rationality, it will be impossible to extend the OR horizon to include inter-personal and systemic conflicts, extensions which are necessary conditions for the fruitful analysis of social systems.

Secondly, we suggest that the concept of an operating statement be expanded to include the *"ethical operating statement"*. This will illumi-nate the conflicts between the organization and its stakeholders. It can include information as to how conflicts arise and how they are dissolved and can reflect such matters as employee satisfaction and pride, custo-mer evaluation of product quality, the local community's views on the organization's contributions to its development, the environment's "re-actions" to its production processes etc. Such information will enable the organization to openly discuss existing and potential conflicts and to thereby save resources which would otherwise be allocated to treating symptoms and to placating stakeholders.

In both cases, operational researchers and social scientists can extend the "scientific" basis for solving problems in complex systems by introducing more appropriate concepts and measures; unless OR methodolo-gy tackles the realities of collective rationality and systemic-conflict and introduces ethics as a perspective for conflict dissolution and con-sensus, the systemic relevance of OR will be limited to technical, rath-er than social systems.

584

THREATS TO INNOVATION; ROADBLOCKS TO IMPLEMENTATION:

THE POLITICS OF THE PRODUCTIVE PROCESS

Carolyn P. Egri and Peter J. Frost[*]

University of British Columbia
Vancouver, British Columbia Canada

Why is it that the initial man-machine interface in the modern world of computer technology relies on a keyboard designed over 100 years ago? An arrangement of symbols purposefully "anti-engineered" in order to slow down typists on a machine which relied solely on the forces of gravity to return the typewriter keys to their initial resting places. Even after the invention of spring-loaded keys and electric typewriters which negated the necessity of such an arrangement, we are still using the same basic keyboard designed to alleviate the constraints imposed by an machine invented in 1873. Has the innovative spirit for increased efficiency bypassed this most fundamental feature of office machinery? Is it because no better alternative has been designed?

For those who have struggled to master the awkward arrangement of the universal "QWERTY" keyboard, it may be a particular source of frustration to know that such an innovation has been in existence since 1932! Based on extensive time and motion studies, Dr. August Dvorak invented a simplified keyboard (known as the DSK) which overcame many of the physical limitations of the initial keyboard (Dvorak, et al., 1932; Parkinson, 1972; Rogers, 1983). Experimental test results have proven the DSK to be immensely user-friendly. Typists can increase their productivity by 35% to 100% on a keyboard which takes one-third of the time to learn than the universal keyboard. Through scientific study, Dvorak and his associates have been able to overcome the problems of excessive finger and hand motion, awkward key strokes, and left-hand overload thereby resulting in increased typewriting efficiency and accuracy with less physical stress and fatigue. Given its demonstrated technological superiority, why hasn't the Dvorak Simplified Keyboard been widely adopted?

To answer this question, one needs to go beyond the rational scientific world view into the sometimes irrational realm of power and politics operating on the individual, organizational and societal levels. We have found in our review of a literature replete with case studies of innovation that a recurrent theme has been the integral role played by power and

---

[*]The authors wish to acknowledge the thoughtful contributions of Craig C. Pinder in preparation of this manuscript.

politics in the emergence and/or rejection of numerous technological innovations.

Adopting this perspective highlights the proposition that having a better product, system or idea is not enough to ensure the adoption and diffusion of an innovation. In fact, it quickly becomes apparent that without the political astuteness of the innovator(s), many technologically superior products and concepts have been the victims of competitive political struggles. Innovation and change are often the focal point of organizational politics, for as Nicolo Machiavelli (1513) asserted in The Prince, "The innovator makes enemies of all those who prospered under the old order, and only lukewarm support is forthcoming from those who would prosper under the new...because men are generally incredulous, never really trusting new things unless they have tested them by experience."

In this framework, power is defined as potential, as the capacity to get others to do things they might otherwise not want to do and/or to resist others' efforts to get one to do what they want to do. Politics is defined as enacted power, as goal directed action that is first of all self-interested and that would be resisted if detected by others with different, competing self interests (Frost, 1987; Porter, Allen and Angle, 1981).

Power and politics operate on two levels--on the surface and in the deep structure of organizations and society. On the more readily accessible and observable surface level, power is often exhibited in political games played out by individual actors, groups, and organizations. On the deep structural level, power is embedded in the structure of power relationships and interactions among parties. As the outcome of past surface political games, the deep structure game serves to systematically distort reality in favour of a dominant order. At this level, the political game involves keeping individuals focussed on the agenda of the dominant order in society and in organizations in ways which preempt individual actors from recognizing, questioning or challenging the political underpinnings of that order.

This game is played through vehicles such as: (1) *socialization* in which learning schemes selectively reinforce only those attitudes and behaviours which are supportive of the dominant world view; (2) *legitimation* in which a higher order (for example, loyalty to country, to God, to an ideal) is invoked to sanction the dominant order's agenda; (3) *neutralization* in which the value bases of the dominant order are hidden (for example, "economics is value free" therefore its bases cannot be challenged); and (4) *naturalization* in which the agenda of the dominant order is presented as a historically determined natural evolution of ideas and systems when that development has actually stemmed from political actions now buried (for example, management has the natural "right" to determine worker outcomes).

INNNOVATION AS A POLITICIZED PROCESS

The focus of this investigation was to trace the power and politics surrounding technological product and system innovation. An analysis of the published accounts of numerous product and systems innovations (successful and failed) was conducted to examine a number of propositions regarding the role that power and politics play in the innovation process. We will present the political trials of Dr. Dvorak and other innovators as exemplars to test the veracity of these propositions. Drawing from these experiences, we will then explore a number of possible avenues by which organizations can enhance the emergence and adoption of technological innovation in the interests of scientific and economic progress.

Proposition 1. A good idea, product or system is very often not enough. The acceptance and diffusion of a technological innovation is highly dependent on the interplay of power and politics in all arenas: individual, group, organizational, and societal.

We have found that the course of innovation has been one fraught with political intrigue. At the organizational level, the most common roadblocks to innovative initiatives have been the political (self-interested) use of the competitive intra-organizational games of *restrictive budgeting* (as found in the cases of the NASA Moonlander Monitor in Pinchot, 1985; the ulcer drug, Tagamet in Nayak and Ketteringham, 1986); *line versus staff territoriality* (3M Post-It Notes in Nayak and Ketteringham, 1986, and in Pinchot, 1985; the Fiero sports car in Pinchot, 1985); *expertise* where only formal qualifications count (Geoffrey Hounsfield and the CAT scanner in Nayak and Ketteringham, 1986).

At the inter-organizational and societal levels, we found that *coalition building* and *networking*, *gatekeeping*, *covering up* or *secrecy*, *managing committee agendas*, and *isolating opposition* and *terminating deviants* have been particularly effective political tactics (Dvorak Simplified Keyboard; numerical control versus record playback machine tool automation technologies in Noble, 1984).

For advocates of an innovation the non-political strategies of *reasoning*, *assertiveness* and *rational discourse* have proven to be ineffective as solitary strategies (Dvorak Simplified Keyboard; record playback machine tool automation). To combat anti-innovation forces and parties, successful innovators must also be particularly skillful in *developing sponsors and champions* among senior level management (Fiero sports car, 3M Post-It Notes); *covering up* their innovation during the development and testing phases (Tagamet, Fiero sports car, Moonlander Monitor); *leaking information* (Moonlander Monitor, 3M Post-It Notes); and *insurgency* or guerilla warfare (Moonlander Monitor).

Proposition 2. Successful organizational innovators exhibit both the will and skill to effectively utilize a wide range of political games in order to overcome opposition to the change inherent in technological innovation.

To overcome the organizational impediments to innovation, Hulki Aldikacti of General Motors--Pontiac became a defensive political player. The resulting Fiero sports car proved to be a manufacturing breakthrough in that Aldikacti's team invented a machine that "would simultaneously trim all the places where body parts mount on the frame to insure a panel's perfect alignment" (Pinchot, 1985, p. 76). To neutralize the dominant corporate political games of *rule citing* (playing by the book), *line versus staff territoriality*, and *budget control*, Aldikacti *isolated* his project team 10 miles away from GM operations. He practiced *insurgency* by disregarding corporate orders to kill the project on three separate occasions, and by refusing to file the required corporate paperwork on project development. This was facilitated by the *sponsorship* of two top level Pontiac executives (*developing champions*).

Hewlett-Packard's Chuck House, inventor of the NASA Moonlander Monitor was another who resisted similar organizational anti-innovation processes Pinchot, 1985). Despite orders from David Packard himself to "get the project out of the lab in one year" after it had failed to meet the technical requirements of the initial FAA contract, House "chose" to interpret Packard's order to mean the project was to be on the market within one year (*selective use of objective criteria*). He continued development of the product in secrecy (*covering up*) and broke Hewlett-Packard procedures by conducting his own market research (*insurgency* and negating *line versus staff rivalry*). Fortunately, for House and NASA, his monitor technology proved to be immensely successful.

Proposition 3. When existing power relationships are threatened by technological change, the contest over adoption and diffusion of an innovation operates at both the surface and deep structural levels of organizations and of society. In those cases of deep structural opposition, the ability of individual innovators to effect change is often constrained and/or neutralized.

In the case of the Dvorak Simplified Keyboard, Dr. Dvorak fought the organized resistance of change by the "vested interests in hewing to the old design: manufacturers, sales outlets, typing teachers, and typists themselves" (Rogers, 1983, p. 10). Typewriter manufacturers opposed a new invention which conceivably could translate into fewer sales and additional royalty costs--a not inconsequential concern given the economic depression of the 1930's.

Dvorak's strategy to gain acceptance of his innovation was based on the rational model of decision making. He conducted scientific experiments on the efficiency of his invention; he and others published the results of these studies; he sent DSK-trained typists to the "world typing contests" of the 1930's where they consistently won the top prizes for typing speed and accuracy (Parkinson, 1972). However these contests were sponsored by the typewriter manufacturers who first tried to ban the DSK typists from future competition (*terminating deviants*), an action rescinded when Dvorak threatened to go public. The contest sponsors were successful in denying Dvorak favourable publicity by refusing to identify (*covering up*) the type of keyboard used by the winning typists. Contest officials asserted that it was the manufacturer of the typewriter and not the type of keyboard which was relevant (*selective use of objective criteria*).

In an effort to create user demand for his invention, Dvorak tried to obtain federal government contracts. Although experimental tests in the U.S. Navy Department and General Systems Administration proved the superior productivity of the Dvorak keyboard, procurement of the new system was denied for reasons of "economy". At the time, the U.S. Navy performed the ultimate political game of *covering up* the Dvorak test results by assigning it a security classification. Dvorak also encountered resistance from the American National Standards Institute (ANSI). Typewriter manufacturers were heavily represented on the ANSI Keyboard Committee (evidence of *networking* and *managing committees*) which had resisted inclusion of the DSK in their standards for office equipment. The *gatekeeping* role of the ANSI has proven to be instrumental in preventing the widespread adoption of the new keyboard.

Additional evidence of the power of the deep structure to resist change is provided by Noble in his account of the development of machine tool automation in post WWII United States. Noble illustrates how the dominant industrial, scientific, and government interests proved to be particularly adept at all the deep structure political games of *legitimation, neutralization, naturalization,* and *socialization* to forestall the acceptance and diffusion of a worker controlled record playback technology in favour of an ideologically compatible, but technologically inferior, numerical control technology.

Wilkinson (1983) also studied the introduction of numerical control technology in the production process. Using a class dialectical approach, Wilkinson details how managerial attempts to unilaterally wrest control of the manufacturing production process away from the workers never achieved their full productive potential. In these cases, workers effectively undermined ownership initiatives to deskill production work by engaging in political games of *sabotage* and *insurgency*.

Proposition 4. When a proposed technological innovation is consistent or supportive of existing power relationships, the deep structural context permits and encourages the emergence of that change. Contests over resource allocation and utilization remain at the more manageable surface level thereby resulting in a higher probability of innovation adoption and diffusion.

For example, even through the account of 3M Post-It Notes details the political gamesmanship of 3M chemist Arthur Fry, the eventual emergence of this innovation was assisted by an organizational culture which legitimated independent action. 3M has a corporate policy which provides for scientists to "bootleg" 15% of their time on unapproved projects (a policy which in effect sanctions such endeavors). They also hold regular in-house seminars at which inventors can present their "unapproved" research findings. At the deep structural level, "maverick" political behaviour is not only condoned but encouraged through the emulation of these mavericks as corporate heros with the end result being that others are *socialized* to follow their lead.

Another case is that of the NASA Moonlander Monitor. Hewlett-Packard prides itself as being an innovative company in a highly competitive industry. Although Chuck House was forced to go underground with his invention for one year, he had the implicit support that he could experiment and if successful, he would be rewarded rather than fired for violating corporate procedures. What these corporate cultures communicate to their employees is that the ends can justify the means and rule-breaking is permissable if the end result is successful and beneficial for the corporation.

MEANS TO ENHANCE INNOVATION

Working within the context of political processes, how then can organizations promote and enhance innovation? What are the necessary activities to be played out through individuals, organizations and society?

In the innovation process, the role of the individual innovator is to create a new vision of a product or system. In addition to the innovator, key players are often the senior executives who set the stage for innovative inquiry. Through the processes of creative awareness and intuition, executives are a critical resource in their ability and capacity to create a vision of organizational functioning which regards innovative endeavors as opportunities rather than as threats. Executives serve as role models for the individual dedication and political astuteness needed to overcome the roadblocks to innovation implementation and to facilitate, rather than to sabotage, promising innovations. (Frost and Egri, 1989)

As a collective, the role of the organization is one of providing an organizational culture with norms supportive of innovation. As Burns and Stalker (1961) and Kanter (1988) detail, organic organizational structures and participative organizational cultures are particularly beneficial to the emergence, development and diffusion of innovation. However, in a world of limited resources, the role of the organization is also one of providing a system of checks and balances which direct the course of organizational innovation. As evidenced in our innovation case studies, it is a fragile balance--one which is vulnerable to the destructive rather than constructive forces of political activity.

Finally, there is the role of the wider society in providing the deep structure norms and values supportive of productive innovation. Numerous organization theorists and observers regard industrial innovation as the necessary ingredient for individual organization survival and growth in their prescriptions for macro-economic revitalization. To effect a societal transformation supportive of innovative ideas based on their own merits rather than on their political expediency in maintaining dominant power relationships is a challenging one. The current surge of interest in innovation is one step towards creating a new awareness and eventual adoption of a societal norm which recognizes, as Carl Jung did in 1928, that " Great innovations never come from above; they come invariably from below, just as trees never grow from the sky downward, but upward from the earth."

REFERENCES

Burns, T., and Stalker, G.M., 1961 "The Management of Innovation," Tavistock Publications, London.

Dvorak, A., Merrick, N.L., Dealey, W.L., and Ford, G.C., 1936, "Typewriting behavior: Psychology applied to teaching and learning typewriting," American Book Company, New York.

Frost, P.J., 1987, Power, politics and influence, in "Handbook of Organizational Communication," F.M. Jablin, L.L. Putnam, K.H. Roberts, and L.W. Porter, eds., Sage Publications Inc., Beverly Hills, California.

Frost, P.J., and Egri, C.P., 1989 (forthcoming) Appreciating executive action, in "The Functioning of Executive Appreciation," S.Srivastva, ed., Jossey-Bass Inc., Publishers, San Francisco, California.

Jung, C.G., 1928, The Spiritual Problem of Modern Man, in "The Portable Jung," J. Campbell, ed., 1971, Penguin Books, New York.

Kanter, R.M., 1988, When a thousand flowers bloom: Structural, collective, and social conditions for innovation in organization, Research in Organizational Behavior, 10:169-211.

Machiavelli, N., 1513, "The Prince," translated by George Bull, 1961, Penguin Books, Baltimore, Maryland.

Nayak, P.R., and Ketteringham, J.M., 1986, "Break-throughs!" Rawson Associates, New York.

Noble, D.F., 1984, "Forces of Production," Knopf, New York.

Parkinson, R., 1972, The Dvorak simplified keyboard: Forty years of frustration, Computers and Automation, 21(11): 18-25.

Pinchot, J. III (1985) "Intrapreneuring," Harper and Row, New York.

Porter, L.W., Allen, R.W., and Angle, H.L., 1981, The Politics of Upward Influence in Organizations, Research in Organizational Behavior, 3:109 149.

Rogers, E.M., 1983, "Diffusion of Innovations, Third edition," Harper and Row, New York.

Wilkinson, B., 1983, "The Shopfloor Politics of New Technology," Heinemann London.

WASTE MANAGEMENT PLANNING IN AN ENGLISH COUNTY:

A CASE STUDY OF O R IN ACTION

R.H.Berry

School of Information Systems
University of East Anglia
Norwich Norfolk NR4 7TJ

INTRODUCTION

In 1978 W'shire County Council's landfill site for domestic waste from the districts of N&b and N&w was almost full. A replacement disposal facility was needed. This paper describes work done with county officers on this problem. Emphasis is placed on the difficulties generated by the context in which W'shire's decision existed. The contention is that OR has looked at waste management as a technical problem which supports complex models and ingored political, organisational and information factors which argue for cruder approaches.

THE PROBLEM CONTEXT

In 1978 local authorities in England were still coming to terms with drastic changes in the organisation of the waste management function. Prior to April 1974 second tier authorities dealt with both waste collection and disposal. Counties had no waste management function. As part of the re-organisation of local government in 1974 counties took on waste disposal while enlarged second tier authorities, called districts, dealt with waste collection. On reorganisation counties inherited existing disposal facilities. These tended to be small scale and close to exhaustion.

In 1978 the two districts of N&w and N&b sent waste to the Blue Lagoon tip site near N&b. N&w also used another site near Cole. Both sites were close to full. The County wanted to identify a long term disposal option for N&w and N&b.

DISCUSSIONS

Early discussions with County staff indicated that officers saw them-selves as choosing between limited alternatives. These consisted of inciner-ation at an existing facility, tipping at a privately owned site, and one of four new landfill locations. A new transfer station could accompany any of these alternatives. It was felt that the options would prove environment-ally acceptable and would not differ significantly in terms of environmental impact. It was agreed that each option would be costed so that the least cost option and cost differences between options could be identified. It was decided that the total of county disposal cost and districts' collection costs should be minimised. This was despite the fact that the county budget would only bear the disposal cost. It was felt that the county would be protected against district requests for compensation if it could be shown that districts' interests had been considered.

One reason for W'shire approaching the author was the author's previous experience in other English counties (Berry,1984). Early discussions confirmed that the W'shire problem was similar to one dealt with previously (Berry,1986). Therefore W'shire staff were introduced to the models used on that occasion.

The first was a non-stochastic simulation model designed to analyse the task of shifting waste between source and disposal site. Given appropriate cost parameters, i.e. a fixed cost of running a vehicle, a variable cost per mile, and a variable cost per loading and unloading operation, this model allowed the costing of a simple haulage operation.

The second model was a simple mathematical programming formulation. The W'shire problem involved a set of waste sources, a transfer station, and a set of possible destinations. The problem was to select a cost minimising destination. The applicability of a programming model, with integer variables to cope with destination selection, and transportation type constraints is obvious. However W'shire staff indicated that they wanted all options costed rather than just the best selected. Therefore a simpler form of model without binary variables to reflect location choice was proposed. This model selected the least cost pattern of routes between waste sources, a transfer station, and a given disposal site.

The decision variables represent number of trips along various routes, and the objective function minimises the sum of operating costs associated with these trips. Constraints ensure that enough trips are made from each source to clear all the waste from that source, and that enough bulk haulage trips leave the transfer station. Further constraints ensure that no more waste is taken to the transfer station than can be coped with.

In the original and revised models integer variables were used to represent numbers of trips along routes. Experience with similar models in other counties suggested that staff lost confidence in a model if output allowed easy criticism. A model allowing a fractional number of trips attracted considerable criticism because officers might have to communicate the results to elected representatives. W'shire staff responded in a similar way.

Both models generate trip related operating costs. There are no capital cost elements involved. The models therefore do not solve the problem of choosing among a set of disposal options where capital costs differ. The third 'model', more properly an approach, the discounted cash flow approach to investment appraisal, does cope with this problem. Compared with the models found in the literature these appear basic (Wilson 1977). However these simple forms were proposed because of experience of data problems in other counties. W'shire staff indicated that data problems would not be found in W'shire and that because the models had such clear implications for the type of data required the search for data would be relatively speedy.

DATA COLLECTION

The first data collection effort concentrated on waste quantity data. County officers produced figures for the amount of waste dumped at both Blue Lagoon and Cole sites based on the number of vehicles arriving at the

592

sites and an estimate of load tipped by the average vehicle. These data underpinned county estimates of waste produced by N&b and N&w.

In order to disaggregate the estimates by area both districts were contacted. N&b provided recent data on sample weighings of waste from each collection round (the route covered by one collection vehicle crew). N&w was unable to provide sample weighings, or data disaggregated by collection round. An informed estimate of total waste was however provided.

Unfortunately county and district estimates of waste generated in N&w differed enormously. The county's estimate was twice that of the district. Further it was clear that it was going to be difficult to split N&w waste according to location. It was decided that the only way to overcome these problems was to provide a method of estimating waste quantities in N&w which would allow identification of quantities with sources. This would also clear up the discrepancy between district and county estimates of waste arising. N&w was asked for a copy of its bin register, a list of streets and the number of bins to be collected from each street. N&b staff provided figures for the average weight of refuse in a dustbin. N&w staff indicated that there was no reason to suppose that N&w bins should be different from N&b bins. Therefore these figures were used to generate an estimate of waste produced in N&w.

The estimated figure was broadly consistent with that provided by N&w staff. To validate the method the exercise was repeated in N&b. The estimate produced was in agreement with the sample weighing based figure previously provided by N&b. It was therefore concluded that the bin count method of estimating waste arising was valid.

County officers were asked to check their estimates of the amount of waste flowing from N&w. Spot checks revealed that the Cole site was being improperly used by vehicles from a district council outside W'shire. Vehicle counts at the Cole tip had simply counted all local authority owned vehicles as belonging to N&w.

It is often claimed that a host organisation benefits as much from the side effects of an OR exercise as from having its problem solved. The removal of the confusion about the amount of waste generated in N&w was a case in point. W'shire was able to curtail unauthorised use of the Cole site, thus conserving scarce tipping capacity. Officers saw this as an early and welcome justification of their decision to involve an outside consultant. However it was worrying that such a mismatch of perceptions could have existed between county and district. It suggested that little communication of substance went on.

The models required waste to be assigned to locations much smaller than a district. Collection rounds were identified as waste sources in N&b. A similar approach could not be applied in N&w. The collection rounds were far from compact because of the rural nature of the district. Therefore the bin register data was used to assign waste to a number of more compact areas.

To get data on collection vehicle costs district officers in N&b and N&w, selected by county staff, were approached by the author. N&w was contacted first. While the relevant staff were helpful, data were largely unavailable. This was because part of the vehicle fleet was maintained by another non-W'shire district – a legacy from pre-organisation days. Only a service charge figure was available for these vehicles. For the rest of the vehicle fleet only aggregate cost data were available and were not accompanied by data on vehicle activity levels. Despite the willingness of district staff to cooperate, it was clear that N&w had little cost data. The management emphasis was not on cost control but on meeting a service level.

Staff in N&b had recently introduced a computerised accounting system which provided monthly cost data on a vehicle by vehicle basis. Associated with the cost data were data on vehicle activity levels. Unfortunately preliminary analysis indicated that there were problems in the data capture procedures underpinning the system. Howevever the data were potentially extremely useful and a request to carry out further analysis was made to district staff. This was refused. District officers indicated that they were unwilling to make district cost data available to the county. The reason given was that they might at some stage wish to hire vehicles out to the county or negotiate about compensation for excessive travel to disposal sites by collection vehicles.

Although it was desirable to use the data held by N&b, there were clearly major problems in the way. Therefore it was decided to ignore this data set. The author had available an existing set of data consisting of, publicity available trip cost parameters for vehicles similar to those used for collection purposes based on the experience of private haulage firms, and tipping and loading cost parameters based on experience with other local authorities. County officers felt that using this data set would allow them to participate in the analysis and would establish estimates of collection costs born by N&b which district staff could then only challenge by making available their own data.

Because there was a possibility that the disposal system selected to replace Blue Lagoon might involve a transfer station, it was necessary to find cost data for bulk haulage vehicles. There was publicly available trip cost data on vehicles similar to those the county would use.

County officers had in mind a very simple transfer station, essentially no more than a covered ramp. On the basis of their previous experience, capital costs were available. Running costs were seen as relatively fixed and related to a requirement to employ a specific number of men on site. Within the capacity limits imposed by the station size, running costs were not seen as variable with respect to throughput.

County officers had carried out a preliminary analysis of four landfill sites. This provided figures for necessary infrastructure investment and cost of void. As far as running costs were concerned, officers saw them as varying with throughput according to a simple step function.

MODEL CHOICE AND USE

The mathematical models described earlier make very crude assumptions about the form of the various cost functions involved. W'shire did not provide adequate data to support refinements. Before finally deciding to make use of the simple models the author asked county staff to confirm that they wished to participate in the analysis and use the models to support a negotiating process between county and district officers. Officers also confirmed that non cost issues would be dealt with outside the models. The possibility of gaining access to the data set held by N&b was discussed. However officers did not seem unduly perturbed by the absence of this data. They indicated that the author's data set offered the advantage of a private sector basis against which public sector costs could be compared.

The possible replacements for Blue Lagoon fell into two distinct groups those involving a transfer station and those that did not. The analysis of those not involving a transfer station required the use of the integer programming models before any discounted cash flow calculations could take place. The analysis of those not involving a transfer station required only the application of the simple haulage model prior to a discounted cash flow comparison.

County staff had no experience of using computer models to analyse waste management problems. To assist them to come to terms with the process the simple haulage model was set up on a programmable calculator and the machine given to county staff to allow them to begin analysing the non transfer systems. For each landfill site the process was first of all carried out by the author and county officers using the programmable calculator. The author and county staff then carried out sensitivity analysis using a FORTRAN version of the haulage model located on a mainframe computer. County staff then carried out a similar analysis for the systems based on the incinerator and the privately owned site near Cole. The transfer based systems were then analysed using a mathematical programming package. Once again the author supported initial analyses leaving county officers to carry out further work when they became comfortable with the process.

Once the trip related costs per period for the various disposal systems had been identified the various systems were compared. Systems based on three landfill sites and not involving transfer generated similar vehicle fleet sizes and investment in infrastructure. They differed only in terms of running cost and could be compared without the use of discounted cash flow analysis. The fourth landfill site involved a higher initial expenditure on site preparation than the others. It also involved a higher trip related cost to the others, though a similar vehicle fleet size. Therefore once again, comparison on cost terms did not require the use of discounted cash flow analysis.

Without exception transfer based systems offered a reduction in trip related costs compared to a system using the same disposal facility but not involving transfer. Without exception transfer reduced the required collection fleet size. However in all cases the saving in trip related costs was outweighted by the period cost involved in operating the transfer station. Further the reduction in collection vehicle fleet size was offset by the need to employ vehicles for bulk haulage. Therefore comparison between transfer and non transfer based systems was also possible without using discounted cash flow.

The analysis clearly demonstrated that on total cost grounds a system based on a particular landfill site without transfer dominated the other alternatives. The author was not involved in the analysis of the systems based on the incinerator or the private site at Cole but county staff reported that neither option challenged the superiority of the preferred landfill site. They also reported a degree of concern that these options meant sacrificing control of part of the disposal system. The conclusion that one particular landfill site was unambiguously best was however misleading. The cost split between districts changed with the choice of site. N&b's costs moved in line with total system costs, a low total cost site generating a low cost collection operation for N&b and a high cost site a high collection cost for N&b. Unfortunately the reverse was true for N&w. County officers found themselves faced with an issue of equity.

The author was not privy to the discussions which preceded a final decision on a replacement for the Blue Lagoon site. Assistance was given with further computer runs, and some model runs were carried out by county officers without the author's involvement. In the event the County's decision was consistent with a choice based on minimising total system cost according to the model outputs.

CONCLUSIONS

The literature on OR in waste management is rich. However the emphasis on theoretically interesting model structures gives a misleading impression of the issues raised in a problem solving situation. Political issues, organisational issues, and personalities have to be dealt with.

The major issue highlighted by this case is that modelling always takes place in a specific context. Here organisational change generating a problem. The waste disposal officers were relative newcomers to the field. They had no data set and no modelling experience, but they wanted both. This lead to the decision to involve a consultant and pressure for the use of simple models that could be taken over for future use.

History had generated a particular organisational context, a split collection and disposal function. As a result project sponsors were not in control of the overall system. This generated debate about the nature of the objective function, and in the final analysis made necessary an explicit political choice. The analysis identified the least cost solution. The debate was then about which district should bear the brunt of the costs.

The organisational structure also affected the data set available and hence the nature of the models and the resulting pattern of use. Lines of communication between county and districts had not been built after reorganisation. This is perhaps not suprising given the degree of dissent there had been nationally about the splitting of the collection and disposal functions. This led to a situation in which county officers held for a significant period of time a false view of how much waste was being produced by a district, and an equally false view about the existence of collection cost data and the likelihood that they would be given access to it.

Finally the organisational structure encouraged competition instead of cooperation between interested parties. The districts saw the county as a potential customer for their services and hence not to be informed about the costs underpinning any price that might be charged.

Despite all this it was possible for an OR activity to be carried out. Sensitivity analysis and crude models coped with the data problems. Analysis allowed the political choices to be made on an informed basis, but did not, and of course never can, replace them.

REFERENCES

Berry, R.H., 1984, "An Examination of the Analysis Process Underlying the Decision to Invest in Reclamation and Disposal Facilities" unpublished Ph.D Thesis, University of Warwick.

Berry, R.H., 1986, "Investing in Waste Disposal Facilities: A Local Authority Case History" in Management Accounting: British Case Studies Volume 2 (Eds. R.Scapens et.al.), CIMA.

Wilson, D.C., 1977, Modelling as an Aid to Solid Waste Management Planning A State of the Art Report, AERE.

CONFLICT MANAGEMENT IN COOPERATIVE GAMES:

SEARCHING FOR CONSENSUS AND CONVERGENCE

L. Valadares Tavares

R. Ferreira dos Santos

Professor of O.R.
Cesur–IST
Technical Univ. of Lisbon

Lecturer
Fac. of Sciences and Technology
New Univ. of Lisbon

1. INTRODUCTION

Life is full of conflicts but, unfortunately, the scientific approach, namely O.R., has not achieved much success to study them. This may be due to the wide variety of multi-attribute phenomena involved by a conflict process as well as to the complexity of the structures of interactions betweeen decision-makers generating conflicts of multiple types.Even the concept of "conflict study" can be rather ambiguous because a pure descriptive attitude is not acceptable, quite often. Then, discussing whose decisions should be supported or which goals should be achieved is a critical issue. A clear option on this matter is made by three different classes of conflict studies:
    *Bargaining Models: in this case, the model is oriented to help one of the two decision-makers and hence "conflict management" means here helping that side to win the conflict.
    *Violence Control Models: violent behaviour or actions with a high potential risk (such as arms race) can be developed when the conflict tension reaches a too high level.The purpose of these models is contributing to bring this tension back to a safe domain. Usually, this purpose implies designing arrangements which can be accepted by all partners and by which they are discouraged to start up agressive actions.This means that the objective is defined in terms of the full set of gains or losses received by all decision-makers.
    *Cooperative Models: these conflicts arise due the existence of cooperation gains (ex. economies of scale) which can be received if decision-makers make coalitions. A coalition implies the unanimous agreement of its members about how such gains are distributed. Usually, the cooperation gain is maximized for the grand coalition which includes all the decision-makers and so the main purpose of these models is helping to reach this state of maximal total benefit.

These models are often applied to the development of multi-purpose and multi-institution projects (ex.constrution of infra-structures, control of environmental systems, implementation of new organizational models)where the cooperation gains are expressed by important cost reductions. Distributing gains is then equivalent to allocating costs, explaining why this area of study is usually called "Cost Allocation" and the models are presented in terms of the distribution of costs instead of gains.

These three types of conflicts have been studied by game theory offering different mathematical solutions but giving less attention to the behaviour of decision-makers, and to the dynamics of the conflict development. This is particularly the case of cooperative games where multiple theoretical solutions have been deduced but always assuming, at least implicitly, that the process of reaching the grand coalition is instantaneous rather than a sequence of iterative steps. Furthermore, the freedom of decision-makers concerning the process of establishing or breaking a coalition has to be acknowledged which means that any gaming solution should contribute to the development of consensus and to the convergence of that coalition sequence for the state maximizing the total cooperation gain. This is the problem studied in this paper, where a new theoretical model for cooperative games is presented searching for further consensus and convergence in this type of conflicts. Empirical analysis confirming the utility of this model is also discussed in section 5.

2. EVALUATION OF COOPERATIVE GAME SOLUTIONS

Let be defined the cooperative (cost) game by (N,C) where $N=\{i: i=1,2,...,n\}$ is the set of decision-makers participating in the game and C is a cost function $2^n \rightarrow R$ assigning to each non-empty coalition $S \subseteq N$ a cost $C(S)$. This function is assumed to be subadditive: $C(S \cup T) \leq C(S)+C(T)$ for all $S,T \subseteq N$ with $S \cap T=0$. Then, the grand coalition, $S=N$, is the state maximizing the cooperation gains, $G(S)= \sum_{i \in S} C(i) - C(S)$ where $C(i)$ is the cost of individual i without joining any coalition. The cooperation gain generated by making a coalition between S and S' is denoted by $G(S,S')$ assuming that $S,S' \subset N$ and $S \cap S'=0$.

Obviously, each player will be interested to minimize its cost, $x(i)$, and the whole system will have a minimal cost if the grand coalition is achieved. Solving this game means finding a vector x with desirable properties, and a long list of mathematical criteria have been proposed by numerous authors, namely:
- Individual rationality: Considering n independent participants, i=1...n, with individual costs $C(i)$, and considering the cost allocation $x(i)$, this criteria imposes that: $x(i) \leq C(i)$, $\forall\ i \in N$.
- Group rationality: Using the same logic and considering that n participants can form (2^n-n-1) different coalitions with associated minimun costs given by $C(S)$, then the verification of this property implies that: $\sum_{i \in S} x(i) \leq C(S)$, $\forall S \subset N$.
- Efficiency: In a Paretian sense, efficiency implies that: $\sum_{i \in N} x(i)=C(N)$. The solutions satisfying these three properties are called stable and their set is the CORE(C).
- Individual marginal cost coverage: It was required by several institutional committees in USA and states the following condition: $x(i) \geq SC(i)$ $\forall i \in N$, with $SC(i)=C(N)-C(N-i)$, denoting by $C(N-i)$ the cost of the set N excluding i.
- Group marginal cost coverage: The same argument for groups, with firm roots in economics and project evaluation literature, requires that: $\sum_{i \in S} x(i) \geq C(N)-C(N-S)=SC(S)$.
- Dummy out property: refered by Tijs and Driessen(1986) this criteria requires that considering $D=\{i \in N: C(S \cup i)-C(S)=C(i)\ \forall S \subset (N-i)\}$ the set of dummy players and C^* the cost function in the game with the players $(N-D)$, being $C^*(S)=C(S)\ \forall S \subset (N-D)$, then must be satisfied the following condition: $x^C(i)=x^{C^*}(i)\ \forall\ i \in (N-D)$.
- Dummy player property: Also refered by Tijs and Driessen it requires that if $i \in D$ then $x(i)=C(i)$.
- Aggregate monotonicity: Considering two cost games C^1 and C^2 being $C^1(S)=C^2(S)$, $\forall S$ excluding N, and $C^1(N)< (>)C^2(N)$ then it is required that $x^{C^1}(i)\leq\ (\geq)x^{C^2}(i)\ \forall i \in N$.
- Strongly monotonicity: Refered by Young(1985) it requires that if a given participant has an increase(decrease) in his separable cost for all coalitions, then he should not be charged less(more). For example, if $SC_S^2(i) > SC_S^1(i)\ \forall S \subset N$ then $x^2(i) \geq x^1(i)$.
- Complementary monotonicity: Refered in Tijs and Driessen(1986) it requires that if $C^2(T)< (>)C^1(T)$, being T a coalition with $1<t<n-1$ participants and $T \subset N$, and if $C^2(S)=C^1(S)$ for all the other coalitions then $x^2(i)\geq\ (\leq)x^1(i)$, $\forall i \in (N-T)$.

These properties are useful to classify the final solution of the game but they ignore its sequential nature. In order that, at each stage, easier consensus between players will be obtained and also to contribute to the convergence of the sequence of decisions to the grand coalition, the following principles are suggested:

A-<u>Cooperativeness</u>: The proposed solution should maximize the cooperation gains which means achieving the grand coalition, according to the adopted assumption of sub-additivity.

B-<u>Progressiveness</u>: The formation of coalitions in real life is always a difficult, interactive and iterative process. Therefore, the proposed game solution will be more easily implemented if it includes explicitly a sequential procedure progressing from the no-coalition state to the desired final state.

C-<u>Equity</u>: At each stage of the game, the proposed allocation of costs should be the result of a clear ethical principle applied to the problem of distributing cooperation gains, so that a concept of fairness can be easily defined, in explicit terms, and shared by all the players.

D-<u>Stability</u>: The allocation solution should avoid that any player or group of players will be tempted to break the solution rules. This implies that the individual and the group rationality must be satisfied.

3. ASSESSING MAJOR SOLUTIONS

Major solutions can be classified into five groups which can be discussed in terms of the four principles presented in section 2:

> A-Solutions based on separable/nonseparable costs;
> B-Solutions strictly based on the game core;
> C-Solutions based on the game nucleolus;
> D-Solutions based on cost gap functions;
> E -Solutions based on marginal costs;

The first group is based on the concept of "separable cost" of player i, $SC(i)$, which is defined by $SC(i)=C(N)-C(N-i)$ and its allocation is given by $x(i)=SC(i)+[C(N)-\sum_{i \in N} SC(i)]$.
.$[W_i/\sum W_i]$, where W_i is a weighting factor equal to $[Min\{C(i), B(i)\}-SC(i)]$ being $B(i)$ the benefit generated by i without any association to other players. Usually, $B(i)>C(i)$ and then it can be shown that the cooperation gains are distributed proportionally to $G(i,N-i)$ what does not help progressiveness. The stability principle can be also violated, because $SC(i) \leq x(i) \leq C(i)$ and these limits can be different from the CORE bounds.

The CORE solutions ignore the sequential nature of a coalition process and are not based on an equity principle. Futhermore, most often there are multiple CORE solutions which stimulates the disagreement between players.

Similar drawbacks are presented by solutions based on the game nucleolus, except what concerns the multiplicity of solutions, which can also violate the monotonicity criterium.

The solutions based on cost gap functions do not guarantee stability with $n \geq 4$, and assume an unrealistic behaviour about the construction of coalitions as it is considered based on separable costs defined for the grand coalition even when just partial coalitions exist. Monotonicity is not fulfilled either.

The last group considers the sequential nature of the coalition process but it assumes an arbitrary order which contradicts most experimental results. Also, it implies an odd ethical principle concerning the allocation of cooperation gains as they are fully received by the player joining a given coalition.

4. PROPOSED MODEL

According to the previous discussion, the proposed model considers the cooperation game as a sequence of stages starting with the no-coalition state and ending when the grand coalition is formed.

At each stage, k, some of the participants can belong to coalitions and hence the set $N=\{i: i=1,...,n\}$ is partitioned into a set of "players", $Y_k=\{j: j=1,...,n_k$ with $1 \leq n_k \leq n\}$ where each j is one individual or a group of individuals(coalition). A new stage, $k+1$, is produced when a new coalition is formed.

The decisions made at each stage concern the formation of coalitions and the distribution of cooperation gains within each coalition, and must satisfy rational and equity requirements. It is assumed that, at each stage k, a function called "coalition attractiveness" is defined for any pair of players, (j,j') belonging to Y_k, $f(j,j')$ and that the coalition to be formed is that one maximizing $f(j,j')$. So, this model is refered to as a "Progressive Allocation Method".

Obviously, the function f implies a metric on Y_k and it describes the behaviour of players. Alternative definitions can be adopted but a simple solution is making $f(j,j')$ equal to the marginal cooperation gain of j and j' to form the coalition (j,j'): $f(j,j')=MG(j,j')= =C(j)+C(j')-C(j,j')=G(j,j')-G(j)-G(j')$, where C and G as previous defined. In other cases, preference can be given to the "marginal cooperation gain by individual", $MG(j,j')/[j+j']$ where j and j' are the number of individuals in j and j' respectively or even to the "relative marginal cooperation gain" given by $MG(j,j')/[C(j)+C(j')]$.

The next problem to be studied is the definition of a criterium to be followed by the distribution of the cooperation gains at each stage, that should express an ethical principle. Two suggestions are given:

1- principle of proportional allocation of incremental gains: the cooperation gain due to the coalition formed by j and j' should be shared by all the members of j and j' in proportional terms to their contribution to the new group (j,j'). An easy and realistic measure of the contribution of j(or j') to the new group is $C(j)$(or $C(j')$). This means that the allocated cost to each individual $i \in (j,j')$ at the end of stage k, $x_k(i)$, is equal to its value at the end of $(k-1)$, $x_{k-1}(i)$, minus its share of $MG(j,j')$. This progressive allocation model is called by the authors the "Proportional Progressive Allocation Method,(PPAM)".

2- principle of expected gains allocation: assuming that in the coalition of j with j' to form (j,j') can exist two distinct and equiprobable sequences, j is joining j' or j' is joining j, then each player will pay the expected marginal cost that result from their association. This is equivalent to share the incremental cooperation gain, $MG(j,j')$, equally by j and j', what corresponds to their expected gain. Each individual $i \in (j,j)$ will receive a share of $MG(j,j')$ through the sequential application(partioning) of this criterium. This progressive allocation model is called by the authors the "Expected Gain Progressive Allocation Method,(EGPAM)".

These Progressive Allocation Methods have common very interesting properties:
- They consider the sequential nature of the game, satisfying the progressiveness principle.
- They guarantee the convergence to the grand coalition, in $(n-1)$ stages for a subaditive game, satisfying the cooperativeness principle.
- They are based on an explicit ethical principle, satisfying the equity principle.
- They guarantee that, at each stage, any player(individual or coalition) is encouraged to cooperate, accepting the new coalition of that stage as it will maximize his additional benefit without violating the adopted ethical principle.
- They satisfy other evaluation criterias, like monotonicity, dummy out and dummy player properties and operacionality.

Obviously, group rationality cannot be guaranted by these methods if reversibility in the process of formation of the grand coalition is adopted.

5. CASE STUDY AND EXPERIMENTAL EVIDENCE

The application of the proposed methods and of other presented methods to a case study is an excellent opportunity to evaluate the behaviour.of the players. The considered case is reported by Young et al.(1982) and Stahl(1980,81), and concerns the implementation of a domestic water supply network extension to six groups of municipalities(A,H,K,L,M,T) of Skane Region in Sweden. Table 1 shows the costs associated to the 63 coalitions that can be formed, $63=2^6-1$, (for more details, see Young et al.(1982)). It refers to a particular case because this is a non-convex game(ex. C(AHK)+C(HKL) < C(AHKL)+C(HK)), although having a CORE, which implies a great difficulty to achieve a stable allocation.

TABLE 1: <u>Total costs for various joint supply systems</u>, in Skr×10^6.
(from Young et al.(1982))

Group	Total cost	Group	Total cost	Group	Total cost	Group	Total cost
A	21.95	KM	31.45	HKM	42.55	A,K,L,T	70.72
H	17.08	K,T	32.89	HK,T	44.94	A,K,MT	72.27
K	10.91	LM	31.10	HL,M	45.81	A,LMT	73.41
L	15.88	L,T	37.86	HL,T	46.98	HKL,M	48.07
M	20.81	MT	39.41	H,MT	56.49	HKL,T	49.24
T	21.98	AHK	40.74	K,LM	42.01	HKMT	59.35
AH	34.69	AHL	43.22	K,L,T	48.77	HL,MT	64.41
A,K	32.86	AH,M	55.50	K,MT	50.32	KLMT	56.61
A,L	37.83	AH,T	56.67	LMT	51.46	AHKL,T	70.93
A,M	42.76	A,K,L	48.74	AHKL	48.95	AHKL,M	69.76
A,T	43.93	A,KM	53.40	AHKM	60.25	AHKMT	77.42
HK	22.96	A,K,T	54.84	AHK,T	62.72	AHLMT	83.00
HL	25.00	A,LM	53.05	AHL,M	64.03	AKLMT	73.97
H,M	37.89	A,L,T	59.81	AHL,T	65.20	HKLMT	66.46
H,T	39.06	A,MT	61.36	AH,MT	74.10		
K,L	26.79	HKL	27.26	A,K,LM	63.96	AHKLMT	83.82

The application of the proposed " Progressive Allocation Methods" led us to the formation of the grand coalition in five stages, giving priority to the coalitions that maximize the marginal cooperation gain, resulting in the following sequence of coalitions and associated marginal gains: (H,L),7.96 -> (HL,K),8.65 -> (M,T),3.38 -> (HLK,A),.26 -> (HKLA,MT),4.54.

Allocating, at each stage, the marginal gains in proportion to players' costs the following final cooperation gains allocation is obtained: A->1.238 ; H->7.8 ; K->3.095 ; L->7.252 ; M-> 2.629 ; T-> 2.776. If the expected gain allocation criterium is applied, then the following final allocation is obtained: A->1.265 ; H->6.459 ; K->4.958 ; L->6.459 ; M->2.825 ; T->2.825.

The correspondent cost allocations, and those obtained by the remaining methods refered in the previous sections, are given in the following table:

	A	H	K	L	M	T
Proport. Prog. Alloc.	20.71	9.28	7.82	8.63	18.18	19.20
Exp. Gain Prog. Alloc.	20.69	10.62	5.95	9.42	17.985	19.155
Sep. Cost Rem.Ben.	19.54	13.28	5.62	10.90	16.66	17.82
Shapley Value*	20.01	10.71	6.61	10.37	16.94	19.18
Nucleolus	20.35	12.06	5.00	8.61	18.32	19.49
Non-Separ.Cost Gap	19.58	13.34	5.69	10.62	16.71	17.88

* value obtained from Young et al.(1982).

Gaming experiments are an useful instrument for testing the realism and soundness of the theoretical allocation methods, because they help us to understand the behaviour of the players and also how close they fit to the proposed solutions.

Using the previous case study, Stahl(1980/81) organized several games obtaining the results given in table 2, where the results of two other games organized by the authors with students of Lisbon Universities are also included.

Stahl and our reports about these games confirms that the process of formation of coalitions is not an instantaneous one but rather a gradual and progressive sequence. This sequence was almost always that one predicted by our model(HL->HKL->MT->HKLA->HKLAMT).

TABLE 2: <u>Gaming Experiments results</u>(in part from Stahl(1981))

	A	H	K	L	M	T
Planners Sweden	21.15	9.70	6.00	9.10	18.37	19.50
Doc. Students IIASA Gr.A	18.15	12.77	8.10	13.25	12.90	18.65
Doc. Students IIASA Gr.B	18.56	13.79	6.75	8.00	17.66	19.05
Scientists IIASA 1	18.65	10.38	6.60	9.18	19.21	19.80
Scientists IIASA 2	21.02	9.85	6.29	9.15	17.83	19.68
Scientists IIASA 3	20.65	10.84	4.67	9.65	18.42	19.59
Scientists Poland	19.02	10.54	5.98	9.94	19.78	18.56
Planners Poland	18.80	10.19	5.80	9.42	19.81	19.80
Planners Bulgarie	19.23	10.63	6.82	9.81	18.10	19.23
Scientists Bulgarie	17.87	12.56	8.10	11.80	15.40	17.99
Students Bulgarie 1*	16.79	14.08	7.90	10.18	18.91	19.50
Students Bulgarie 2*	21.95	13.50	5.50	8.26	20.81	21.98
Students Lisbon 1*	21.95	10.57	6.69	10.00	19.50	19.91
Students Lisbon 2	20.96	9.67	5.84	8.92	18.60	19.84

* games where the grand coalition was not formed

Futhermore, the fitness of each solution to the behaviour of the players is assessed by computing the sum of absolute deviations(SAD), the sum of squared deviations(SSD) and the sum of relative squared deviations(SRSD) for the costs obtained by each method and by each game to all players. The number of games where each method gives a minimal deviation is presented in the following table:

	EGPAM	PPAM	Shapley	Nucleolus	SCRB	NSCG
SAD	8	0	1	2	1	2
SSD	8	0	1	2	2	1
SRSD	7	0	2	2	1	2

The excellency of the proposed method is clearly illustrated by the obtained results. In this particular case the "Expected Gain Progressive Allocation Method" gives the better results because its allocation is in the CORE and it is the closest to games outcomes.

REFERENCES

Hamlen, S. S. , Hamlen, W. A. Jr. , and Tschirhart, John , 1977, "The use of core theory in evaluating joint cost allocation schemes", The Accounting Review, vol.LII, nº3, pp.616-627.

Loehman, E. , Orlando, J. , Tschirhart, J. , and Whinston, A. , 1979, "Cost allocation for a regional wastewater treatment system", Water Resources Research, vol.15, pp.193-202.

Loughlin, J. C. , 1977, "The efficiency and equity of cost allocation methods for multipurpose water projects", Water Resources Research, vol.13, nº1, pp.8-14.

Schmeidler, D. , 1969, "The nucleolus of a characteristic function game", SIAM Journal Appl. Math., vol.17, pp.1163-1170.

Shapley, L. S. , 1953, "A value for n-person games", Ann. of Math. Studies, vol.28, pp. 307-317.

Stahl, I. , 1980, "A gaming experiment on cost allocation in water resources development", IIASA Working Paper-80-38.

Stahl, I. , Assa, I. , and Wasniowski, R. , 1981, "Cost allocation in water resources-six gaming experiments in Poland and Bulgaria", IIASA Working Paper-81-03.

Tijs, S. H. , and Driessen, T. S. H., 1986, "Game theory and cost allocation problems", Management Science, vol.32, nº8, pp.1015-1028.

Young, H. P. , Okada, N. , and Hashimoto, T. ,"Cost allocation in water resouces development", vol.18, nº3, pp.463-475.

CONFLICT RESOLUTION IN GROUP DECISION MAKING:

A RULE-BASED APPROACH TO ANALYSIS AND SUPPORT

David Cray, Gregory E. Kersten and Wojtek Michalowski

Decision Analysis Laboratory
School of Business, Carleton University
Ottawa, Ontario, Canada

CONFLICT IN GROUP DECISION MAKING

A conflict is an interaction of interdependent people who perceive their goals as incompatible and expect interference from each other in achieving these goals (Folger and Poole, 1984). Interdependence means that the actions of one group member impacts on the actions of other members (Putnam, 1986). In this paper we discuss the problem of conflict resolution by groups and how it may be supported. As an example we use conflict resolution by a group of experts who prepare a negotiation strategy. The proposed approach differs from that of Bui (1985), Jarke *et al.* (1987), Raiffa (1982), for example, in that we focus on the support of problem representation, verification of the consistency of this representation, and on the process of restructuring the problem. We assume that the group faces a complex problem and that all group members agree on the overall goal. The conflict exists because group members have different and possibly opposing perceptions of the goal; they attribute different subgoals to the goal and propose different goal/subgoal resolutions.

Artificial intelligence (AI) provides a number of approaches to structure and solve large and complex problems (Charniak and McDermott, 1985, Winston, 1984). AI methods, less restrictive and more flexible than formal mathematical constructs, seem to be an appropriate support tool in the structuring of complex problems. They allow the system to identify inconsistencies and to determine reasons for them. In this paper we present an application of AI methodology, namely rule-based formalism (Kersten *et al.*, 1988) to support conflict resolution in group decision making. In section two we introduce the basic concepts of a rule-based approach to the analysis and support of conflict resolution. In sections three and four we illustrate these concepts, while the implications of the approach are outlined in section five.

RULE-BASED REPRESENTATION OF A PROBLEM DOMAIN

A group of experts from different domains collectively develops a representation of the problem. Expert k, ($k \in K$) prepares model M_k of the sub-problem which is within his/her domain of expertise. We assume that the problem, and its sub-problems are decomposable and can be represented with predicate logic. M_k is a rule-based model with one (possibly artificial) top-level goal. Thus, the problem decomposition may be expressed in terms of rules of the form:

$$\alpha_i \Leftarrow \rho_i([\,\alpha_{ij}\,], j \in J),$$

where α_i, α_{ij} are predicates, $\Leftarrow$ is the connective *if..then..*, and ρ_i is a well-formed formula with permissible connectives *and, or, not.* Each predicate α_i can be written as: PREDICATE_NAME(SAYS_WHO). M is a collection of M_k, ($k \in K$), such that each top-level goal of M_k is a lower-level goal of M which has its own top-level goal.

A graphical representation for an example of a rule-based model *M* problem structure is presented in Figure 1. The top-level goal G links the goals of the two domain experts, a military expert and a foreign affairs expert. Naturally, every domain goal can be further decomposed. As a result, the decomposition defines a hierarchical representation called a *goal representation tree*. At the bottom of this tree we have decision variables which define a position/decision of the group. In negotiation, these variables, here called *facts*, are used to formulate the compromise proposal and to communicate it to the opponent. Figure 1 represents a part of a goal representation tree for the Camp David negotiation problem which will be used to illustrate our considerations. To clarify these concepts with the help of the relationships shown in Figure 1, a predicate's name (a node in this figure) is written as, e.g., MILITARY_STRENGTH, while a rule describing a military strength of Israel can be stated as:

MILITARY_STRENGTH(ISRAEL) ⇐ TERRITORIAL_GAINS(ISRAEL) *and*
[UN_BUFFER_LEBANON(ISRAEL) *or* US_MILITARY_ASSISTANCE(ISRAEL)].

A complete list of goals/subgoals/facts and rules defining relations between them is used to describe the initial problem representation. Next, the consistency of this description must be verified. This can be accomplished when certain logical values are attached to the predicates. Each predicate in our representation may be assigned one of the three values: *true, false, any* (Kersten and Szpakowicz, 1988). A value is assigned through a valuation function τ such that τ(PREDICATE) = *v*, where *v* ∈ {*true, false, any*}. The value *any* of a predicate indicates that it may be either *true*, or *false* without changing a valuation of a goal representation tree into *false*, it also indicates the flexibility of a negotiating position. In our representation τ(DIFFEREN–TIATE_ARAB_STATES(ISRAEL)) = *true* means that the group requires Israel to be successful in dividing the Arab states. We also assume that a goal representation tree is consistent if the valuation of a principal goal (root of the tree) is *true*.

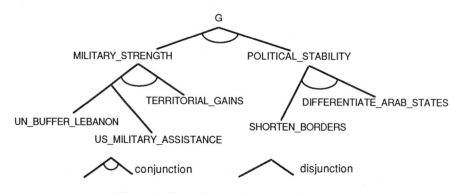

Figure 1. Part of a goal representation tree

The selection of fact valuations which gives a *true* valuation for the principal goal describes a *goal solution*. The selection of a specific goal solution defines a negotiation position. An example of a position for the problem described by the goal representation tree on Figure 1 (assuming that lowest level predicates represent facts) is:

v = {*true, any, true, true, true* },

with the values corresponding to the following predicates:

[UN_BUFFER_LEBANON(ISRAEL), US_MILITARY_ASSISTANCE(ISRAEL), TERRITORIAL_GAINS(ISRAEL), SHORTEN_BORDERS(ISRAEL), DIFFERENTIATE_ARAB_STATES(ISRAEL)].

However, the problem may have no solution, if the valuations of some predicates are contradictory. This situation is not unlikely when experts work independently on problem definition and/or the representation is large and complex. If this is the case, the problem must be redefined by modifying some relationships or introducing new ones.

Having defined the problem, group members consider the possible reactions of the opposing party. Reactions of two types are possible: (i) concessions or rigidity as to particular subgoals/facts, and (ii) modification of the problem definition. Each of these may change the valuations of some predicates as well as relationships between subgoals/facts. These changes are also expressed in the form of rules. Because these rules operate on other rule definitions they are called *meta-rules*.

Reaction of the first type requires no changes in how the problem is understood and defined; the problem domain remains unchanged. What may change are valuations of particular predicates. The player accepts that some subgoals/facts may not be achieved as initially assumed. This type of reaction is governed by meta-rules called *response rules*. A response rule may be a concession related to one or more subgoals/facts, or a requirement that certain subgoals/facts be achieved. This may cause changes in the position; a subgoal/fact, which initially had to exist and had a given value, may change its value or even become irrelevant. Thus, a response rule does not introduce changes in a goal representation tree but it changes the conditions of validation of a particular predicate.

When response rules are applied a problem domain may become empty. Changes of fact values may cause a subgoal (and, consequently, the principal goal) no longer to have *true* valuation. Changing responses without modifying one's own position may not be a sufficient remedy for a negotiator; the problem definition must be modified by reshaping some of the rules. Such modifications are expressed by meta-rules called *restructure rules*. These rules may introduce new predicates, hence they may change the problem representation.

Application of restructure rules corresponds to the incorporation into the problem domain of the expert's knowledge about his/her own side's position, as well as that of his/her opponent. If none of the restructure rules is applicable but the problem representation is consistent and the position is found to be satisfactory, we have arrived at a compromise. If, however, no meta-rules can remove an inconsistency, we have encountered a deadlock in negotiation.

The notions introduced above allow us to describe the development of a position by a group of experts as a process which begins with problem definition and verification. The outcome of the initial phase, an initial position, is then communicated to the opponent. The opponent's response prompts experts' reactions, which may result in a deadlock if the position turns out to be nonmodifiable. If, however, the negotiation is to continue, a new position must be determined, verified and communicated to the opponent. A sequence of positions defines, in turn, a strategy. Using the proposed approach, experts can determine a set of strategies for achieving goals under different conditions dictated by the opponent's changing behavior.

CONFLICT RESOLUTION: AN EXAMPLE

There have been previous attempts to use expert systems methodology to model different aspects of negotiation problems (Michalowski, 1987). However, they did not go beyond theoretical proposals. In this paper we wish to give a practical illustration of some of the artificial intelligence concepts introduced in the previous sections. We shall examine an example loosely based upon the negotiations between Israel and Egypt held at Camp David in September, 1978. The study of the Camp David accords, and the diplomacy surrounding them, has been analyzed as a case study in foreign policy making by the United States (Quandt, 1986). It has also been used as an example of negotiation (Raiffa, 1982), but this rich analytical material has never been employed as the basis for developing negotiation support tools.

We do not intend to structure or describe the complete negotiation process as it was recorded at Camp David. This has already been done by political scientists and economists. What we want to achieve is an illustration, using a high-stakes negotiation process, of the capabilities of our approach to negotiation modelling. In our example, we provide support to one party in the negotiation, support which this party needs to prepare the formulation of consecutive negotiating positions. Such a restriction of the problem implies that negotiation dynamics are not our area of interest. We concentrate on the formulation stage of an initial negotiating position and the resolution of conflict among experts. In our example, we shall assume that Israel is the party whose position we are modelling and supporting in Camp David negotiations.

The root of the goal representation tree (in a case of one goal only), or the nodes next to the root (in a case of multiple goals) corresponds to the overall negotiation goal(s) as perceived by the negotiating party. To give an example, we assume that one of Israel's motives for negotiation was to maintain its military strength. Therefore, this motive represents one of the objectives of Israel's prime minister during the Camp David talks. However, such a goal should be defined in more specific terms. Let us assume that, according to one of the political advisors, maintenance of military strength can be achieved only if Israel keeps the territorial gains of previous Middle East wars. Thus, this expert views a principal negotiation goal as maintaining Israel's military and administrative presence on the West Bank, Golan Heights, Sinai and Gaza Strip. However, the political adviser believes that selective shortening of Israel's border would increase the country's political stability. Therefore, he/she prepares a proposal involving alternative withdrawal plans from the Sinai peninsula. Both the territorial gains and shortening of borders can only be accomplished if a certain number of *true* or *false* assertions (i.e. representing the withdrawal from or keeping of a given territory) is put forward. Some of these assertions are in conflict, as illustrated in Figure 2.

In order to make the goal representation internally coherent, the conflict among the experts involved must be resolved by describing anew the decomposition of their subgoals. In our example this is achieved when one of the advisers accepts necessary changes in his/her perception of the notion of territorial gains, involving, for instance, some kind of withdrawal from the Sinai peninsula.

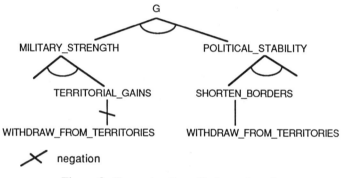

Figure 2. Example of conflicting subgoals

The goal representation tree presented in Figure 1 does not incorporate any elements of knowledge about the possible moves of an opponent. If we only want to find any position consistent with top-level goals, such an omission is acceptable. However, if the initial position is to be further modified, some basic knowledge about the other side's possible behavior is required. In this way we arrive at the other role that experts can play in preparing problem representation. Apart from contributing to a coherent goal representation tree, they are also asked to prepare Israel's responses to the anticipated Egyptian negotiation position.

Knowledge of flexibility on selected issues allows one to model an inherent feature of the negotiation process, which is responses from both sides. Preparation of a response by advisers to Israel's chief negotiator is modelled with the help of the *response rules*. Hence, response rules allow for change in the valuation of a given predicate in response to the anticipated Egyptian negotiation position. Two of the response rules are given below:

τ(WTHDROW_FROM_TERRITORIES(ISRAEL)) = *any* $\Leftarrow$ τ(UN_BUFFER_LEBANON(EGYPT)) = *true*
    and τ(DIPLOMATIC_REL(EGYPT)) = *any*,
τ(TERRITORIAL_GAINS(ISRAEL)) = *any* $\Leftarrow$ τ(AUTONOMY_WEST_BANK(EGYPT)) = *any*.

Please note that neither these rules nor those presented above are historically exact. We do not pretend to historical accuracy. Rather we wish to illustrate the potential offered by our formalism for support systems development.

The party has to prepare a response or the negotiation process ends with a stalemate. However, for a given proposal of the opponent, it may not be possible to formulate a response within a current problem representation. The analysis of the opponents' positions may lead to the introduction of changes in the representation (tree). For example, Israel's firm stand regarding withdrawal from the Sinai, could result in a redefinition of what is considered by Israel's experts as a withdrawal. Instead of simply returning the Sinai, Israel's adviser now suggests the return of part of the Sinai peninsula and the creation of a UN buffer zone there. Hence, the definition of expert's subgoal has changed. To incorporate this kind of change, we introduced the second type of reaction rules: *restructure rules*. An example of a restructure rule is given below:

τ(WTHDRAW_FROM_TERRITORIES(ISRAEL)) = *any* and τ(UN_BUFFER_LEBANON(EGYPT)) = *any*
   $\Rightarrow$
[WTHDRAW_FROM_TERRITORIES(ISRAEL) $\Leftarrow$ RETURN_SINAI(ISRAEL) *or* UN_BUFFER(ISRAEL) *or*
   PHASED_WTHDRAWAL(ISRAEL)]
   $\rightarrow$
[WTHDRAW_FROM_TERRITORIES(ISRAEL) $\Leftarrow$ RETURN_SINAI_PART(ISRAEL) *and* UN_BUFFER(ISRAEL)],

where $\rightarrow$ indicates the replacement of the left-hand side rule with the right-hand side rule.

Restructure rules and response rules are used to prepare a set of positions which the supported party analyzes before the negotiation begins. One of the positions is the initial position, others are reactive positions obtained through projecting the opponent's possible responses to particular negotiating issues. If a position is obtained only with the help of the responses rules, the experts are not required to analyze and verify both the position and the problem representation because the response rules do not change the problem structure. However, if a position is obtained after the restructure rules have been applied the experts should reanalyze the problem and the position since the restructure rule of one expert may affect the representation of the sub-problem for another expert.

CONFLICT DETECTION AND REASONING

The rule-based framework discussed here offers a very useful tool for the detection of the sources of conflict, namely the trace of reasoning. The trace is a chain of rule applications, threading through the goal representation tree and connecting the principal goal to lower subgoals. In the Camp David example we can use the trace to establish that definition of the subgoal of "shortening the borders" is in conflict with the subgoal of "maintaining territorial gains". The trace will also identify the sources of conflict, which in this case is a contradiction in the experts' definition of the subgoals: the requirement to keep the territories and to return them to Egypt. To resolve this conflict, the experts involved would have to modify their definitions of subgoals.

Both response rules and restructure rules include the representation of knowledge about the opponent. The firing of the response rules causes changes in the predicates' valuations within the same representation of the problem, i.e., we search for an alternative position. The restructure rules cause modification of the goal representation tree, i.e., the restructuring of the problem representation so that new goals and facts may be introduced.

As a result, the chief negotiator may wish to check whether a problem domain is nonempty. This can be done using forward chaining (Winston, 1984). In forward chaining we start reasoning from facts and assert all the facts implied by the rules whose "if" clauses are true. In other words, we check what subgoals and ultimately what goals have *true* valuations for a given set of established facts' values. Therefore, forward chaining helps to validate the top-level negotiation goal. Truth valuation of this goal implies accepting the position. Failure can trigger conflict detection through trace.

DISCUSSION

The example we presented here is part of the case we used to test a prototype of a rule-based expert system shell NEGOPLAN (Matwin *et al.*, 1988). The system uses two types of knowledge bases: the rule base, which also includes the facts, and the meta-rule base, which consists of restructure rules and response rules. NEGOPLAN was developed in Prolog on IBM PS/2 and Sun-3 workstations. The case discussed in this paper is represented by a tree with 27 nodes (part of this tree is presented in Figure 1).

We consider NEGOPLAN as an important attempt to provide negotiators with conflict resolution tools based on the knowledge of experts. Iterative use of NEGOPLAN allows the users to build a sequence of positions. Starting from an initial position we can determine consecutive ones which are the result of Egypt's anticipated positions and Israel's reactions to them. It is possible that Israel may prepare several possible positions for a single Egyptian position. Hence we obtain a set of negotiating strategies, based on the opponent's anticipated responses to the proposed negotiating behaviour represented by meta-rules. The meta-rules must be obtained from the experts and/or negotiators through a knowledge acquisition exercise. This parallels the knowledge acquisition phase which precedes the development of an expert system.

Using NEGOPLAN one can develop a large number of strategies, many of which are considered as improbable from the experts' point of view. This is because the system does not make choices in firing meta-rules. This corresponds to the situation in which important changes in the problem domain are made while only minor ones were required to obtain consistent response to the opponent's position.

The set of tools we have proposed in NEGOPLAN enables the user to evaluate the consistency of negotiation goals, and resolve incompatibility of advice given by the negotiator's experts. It also makes it possible to model, in a straightforward fashion, the impact of the opponent's anticipated responses on the negotiator's strategy, and to experiment with a variety of negotiating behaviors.

REFERENCES

Bui, T.X., "N.A.I.: A Consensus Seeking Algorithm for Group Decision Support Systems", *IEEE Conference on Cybernetics and Society*, pp. 380-384, 1985.
Charniak, E., McDermott, D., *Introduction to Artificial Intelligence*. Addison-Wesley, Reading, MA, 1985.
Folger, J.P. and Poole, M.S., *Working Through Conflict*. Scott Foresman, Glenview, IL., 1984.
Jarke, M., Jelassi, M.T. and M.F. Shakun, "MEDIATOR: Towards a Negotiation Support System" *European Journal of Operational Research*, Vol. 31, No. 3, pp. 314-334, 1987.
Kersten, G. E., W. Michalowski, S. Matwin and S. Szpakowicz, "Representing the Negotiation Process with a Rule-based Formalism", *Theory and Decisions*, Vol. 22, 1988.
Kersten, G. E., S. Szpakowicz, "Rule-based formalism and preference Representation: An Extension of NEGOPLAN", Dept. of Computer Science, University of Ottawa, WP 88-14, 1988.
Matwin, S., S. Szpakowicz, Z. Koperczak, G. Kersten and W. Michalowski, "NEGOPLAN: An Expert System Shell for Negotiation Support", *IEEE Expert* (forthcoming).
Michalowski, W., "Multi-person decision support with knowledge base systems", in Y. Sawaragi *et al.* (eds) *Toward Intelligent and Interactive Decision Support Systems*. Springer Verlag, Tokyo, 1987.
Putnam, L.L., "Conflict in Group Decision-Making" in R.Y. Hirokawa and M.S. Pole (eds.), *Communication and Group Decision-Making*. Sage, London, 1986.
Quandt, W., *Camp David: Peacemaking and Politics*. The Brookings Institution, Washington, DC, 1986.
Raiffa, H., *The Art and Science of Negotiation*. Harvard Univ. Press, Cambridge, MA, 1982.
Winston, P.H., *Artificial Intelligence*. Addison-Wesley, Reading, MA, 1984.

O.R. AND THE DECISION MAKER

OPERATIONAL RESEARCH AND THE DECISION MAKER

George Wright

Bristol Business School
Bristol Polytechnic, Coldharbour Lane
Frenchay, Bristol   BS16 1QY

Decision making has strong behavioural components. Decision analysis techniques recognise the primacy of behavioural inputs of subjective probabilities and subjective values, as well as decision problem structures. Implementation of decision analysis in organisational decision making has followed several distinct stages over the last twenty years. Initially, decision analysis was implemented as a normative or prescriptive science which specified optimal solutions to decision problems that the analyst worked through with a client. The underlying decision theory, subjective expected utility, was firmly based on an axiom system that appeared relatively uncontroversial, hence the prescriptions. Psychologists worked on elicitation methods for probability and utility assessment and later on methods to help specify decision problems as decision trees. Later psychological research questioned the ability of human decision makers to provide good enough inputs to decision analysis and a whole literature on heuristics and biases in probability judgements was generated. At the same time, the psychological acceptability of the axiom base was being questioned and issues concerned with the valuation and validation of decision analysis interventions were raised.

Also, decision analysis by the tree approach was being evaluated by practitioners and the focus of academic research began to move to influence diagrams as aids to problem structuring and also to multiattribute utility theory for outcome evaluation. In general, there has been a move away from prescriptions in the practice of decision analysis and a recognition of the limitations of 'optimal' techniques. This is best illustrated by the increasing popularity of decision conferencing in Britain and the United States of America. Decision conferencing utilises decision aiding technologies within a small group setting. The decision analyst now acts as a facilitator for those individuals who have a stake in a key decision. Essentially, the decision problem is 'conferenced' to consensus, with the decision technologies acting in a background manner to answer 'what-if' questions about such issues as resource allocation and outcome evaluation. Since decision conferencing is a relatively new phenomena, many behavioural issues remain uninvestigated. For example, it seems intuitively reasonable that decisions conferenced to consensus stand a good chance of being implemented. But do such decisions show an improved quality over both unaided decisions and prescriptive solutions?

Several of the papers in this stream of the conference are concerned with issues related to the implementation of operational research techniques. OR solutions that have little or no influence on decision making by individuals and organisations are wasted efforts. Themes to be developed in the following chapters include the role of judgement, OR interventions in organisations, the interaction of the OR analyst and the decision maker, and interrelationships between academic knowledge and practitioner experience. Since many OR techniques have less of an inherent behavioural content and emphasis than decision analysis, it follows that these themes and issues are particularly important and non-trivial within this broader perspective on OR and the Social Sciences.

MANAGING THE OR-MANAGEMENT INTERFACE

Rolfe Tomlinson

Institute for Management Research and Development
University of Warwick
Coventry, CV4 7AL. UK

When the first conference on OR and the Social Sciences was held in 1964, there was a general consensus within the OR profession that the subject was concerned with the application of the scientific method to the solution of problems facing decision makers and managers in organisations. Nowadays such statements seem simplistic and misleading. We now know that the scientific method, as it is commonly taught, is seldom actually adopted even by scientists and that they have turned out, under strict scrutiny, to be as subjective and inspirational as anyone else. More importantly, we have also realised that the reductionist methods of science are inappropriate for the serious study of human organisations, and that systems ideas at least have to be incorporated into our methothodology. We have also learnt that although problem solving remains an important part of what OR people do, the most important way that OR achieves success is through altering the process by which decisions are taken, rather than by telling managers what those decisions should be. Thus successful OR requires both an understanding of process in organisations and also in the management of change - subjects commonly thought to be on the private agenda of social scientists.

The changes of understanding indicated in the previous paragraph do not necessarily imply that there has been a volte-face in the way that OR has been conducted over the years. Indeed the changes in the techniques employed are much more apparent than changes in the way that OR studies are undertaken in practice. What has happened is that a small group of practitioners and academics have seriously thought about what they are doing, and have held discussions and published their findings. The self-examination which this has led to within the OR community has been lively and full of controversy. It has been conducted with an awareness of much that has been going on in social science, but there seems to have been relatively little active debate with social scientists. Part of the reason for this may be that much of the social science debate on decision making is perceived to be either at the level of political power-play or, at the other extreme, the decision making behaviour of individuals. Both of these are important, but they are of course very far from being the full story. In my 27 years in the National Coal Board there were several changes of Chairman with consequent differences in style and in their policy objectives. There were major organisational changes; people fell unexpectedly from power and others had spectacular rises. From time to time the unspoken rules of what was acceptable and what was not were substantially altered. BUT the processes leading up to major decisions often remained substantially the same. Policy and investment decisions were made subsequent to organisational processes which had both their formal and informal elements. It would be hard to say if decisions were made rationally (bounded or not) or incrementally but there was certainly a continued conscious effort, if not specifically to improve decisions, then certainly to ensure that identifiably bad decisions were not made. It appears to be the fashion amongst post modernist social scientists to say that this sense of purpose and continuity is an illusion, that the decisions are entirely the consequence of power plays, and that the sophisticated processes that lead to them are simply window dressing. That this is sometimes true is undeniable, but that it is the norm is questionable.

An alternative argument is that the perception that there is a purposeful process leading towards management decisions may have been true in the post-war period up to the mid 70's, but that current changes in management attitudes and organisational practice have made them obsolete It is clear that there have been changes and that these have to some extent affected the practice of OR. Thus, as large centrally managed organisations have been decentralised or broken up altogether, so there has been a tendency for many central OR teams to be broken up and the staff concerned have found themselves operating in smaller groups in relative isolation. On the whole, this does not seem to have done the profession any harm, in term either of numbers or morale. Nor is there so much evidence that this has affected the process or OR, as it was understood in practice by the stronger and most successful groups in the past.

This paper therefore sets out to suggest that there are some important elements in the practice of OR and in its role as a change agent which may be invariant to the changes that are taking place in managing culture and management organisation. This does not mean that the external face of OR is not changing. Indeed, to be a successful change agent OR must continually adapt, both in terms of organisation and in terms of the specific content of its work. The invariance relates to the process by which OR comes into a complex situation and enables change to take place.

It is helpful to develop our basic notions through a practical example. In the late 1960's I was Director of the OR Group of the National Coal Board, at a well established team of about 40 graduates based at Harrow in the suburbs of London. The Coal Board had its main Headquarters in the centre of London; its two main research establishments and all its collieries (which were grouped into 15 Areas) were all 100 miles or more away. Except for the HQ of the Coal Processing and Open-pit mining Divisions our customers were far away from us. Not surprisingly, although we visited the coalfields a great deal, most of our work was actually sponsored by HQ staff. The work was highly regarded - or we would not have continued to grow in numbers over the years - but we felt that we were not being fully effective. So far as staff in the coalfields were concerned (where the real work was done) - we were looking at their problems through the eyes of headquarters staff. So far as HQ staff were concerned we were tackling real problems on a project by project basis. We were conscious that we were missing many of the major problems because we were in no sense identified as part of the overall process of tackling problems; indeed, we often heard about them too late.

We started to tackle this by establishing teams to work specifically on Areas problems, with Area sponsorship - but this did not go far enough. We needed to get much closer to clients, and we did this by breaking the group into a number of teams who each worked full time for one management team - which could be a production Area or a Headquarters branch. The full logic of such a move would have suggested that we should have physically located each group (which could consist of as few as two or as many as six OR staff) with the management concerned. But that would have lost the advantages of our centrally organised group, particularly in control of the work, variety of experience, independence (highly valued by management as well as OR) and the support of senior staff within the OR set up. In other words the OR process would be seriously upset, although the management/OR interaction would become an identifiable process. Our solution was to outstation most of our staff in groups so that any team serving an Area was based within 1 hours call from the Area office. This was perceived by the Area management as having staff 'on the spot', but also retained the unity and integrity of the OR process.

As these procedures became developed I found that, in my annual reviews with team leaders, relatively little time was spent talking about the OR process - our normal control procedures largely took care of that. I found that I was primarily concerned with two things. In the first place, I wanted to know that they understood what was happening in the management area that they were supporting (what were the major problems, who was leading policy on them, how were the decisions going to be taken?) More importantly, however, I was probing to understand how the team was interacting with management in the Area. How they identified problems, teased out the real life complexities, dealt with conflicts and obtained data, discussed intermediate results and prepared for implementation.

The organisational changes introduced at this time have persisted to the present day in a Group that has continued to flourish in the face of continuing contraction in the industry and drastic changes of management structure. There is reason therefore to believe that we were on solid ground when the changes were made. Examining the situation in retrospect, and trying to put a pattern to what was taking place, it is clear that we were placing an increasing emphasis on process. The emphasis was firmly moving away from concentration on projects to a consideration of the wider problems of management and towards a broader understanding of the kind of help that could be provided, i.e. to a consideration of the management process as a whole. This did not mean that projects were unimportant. Over the same period of time we greatly tightened up our project control. We required, among other things, that a sponsor agreed a statement after each project saying what changes had come about as a result of the study. We were thus reinforcing the embedding of the project in the process.

The questions I asked my team leaders did not only focus on the management process, but on the interaction of that process and the OR process. The questions were asking how that interaction was managed. It is my contention that it is that interaction that constitutes the essence of applied OR and that has not changed with changes in the underlying culture. So important is it indeed, that I find it helpful to consider the interaction that goes on during a project as a process in its own right rather than simply as the common ground of the separate processes of management and OR.

One advantage of considering the management - OR interface as a process in its own right is that it makes it easier to study it in balance. As long as it is seen as the intersection of the (limited) OR process and the (almost unlimited) management process, the interaction looks one sided. OR immediately appears to be on the periphery of management, trying to make an impact. It is much more helpful to start with the interaction itself and ask basic questions such as 'What function does the interaction fulfil within the organisation?'

This is a very different question to 'What is OR for' which tends to lead to other questions such as 'Where should OR be placed in the organisation?' Instead we go on to ask questions such as 'What is the organisational need that OR does (should) address?' Answers such as 'to provide a problem solving facility' or 'to ensure effective use of computer technology'; are inadequate, and we must delve more deeply. One answer at this deeper level is to suggest that OR is a necessary part of the organisation's thinking processes i.e. its brain. The concept of an organisation's brain has been put forward by Stafford Beer and the analogy is a powerful one. For survival, any organisation has to sense what is happening outside, correct its behaviour accordingly, and check that the right actions are taken in response to central instructions. Moreover it needs to take a continual overview of what is happening, to ensure that the principles underlying policy changes remain valid. Such behaviour is at least 'brain-like'. If the analogy can be taken seriously, and organisations could be persuaded to a more careful study of how they 'think', then OR should take its place as an important part of the thinking activity.

Thus the first consequence of identifying the management/OR interface as a process is to ask some fundamental questions about the way in which the organisation adapts and controls itself in a changing environment. Only when that question is answered can we seriously examine how to manage and control the interface. If the adaptation process is not explicit and there is no central group responsible for supporting the 'thinking' then the development of ideas will rest with certain key managers, and the OR group might take over some of the 'thinking' role. In other cases, it may require a strong link with planning teams, or with business development teams. What is important is the recognition of management's need to ask the fundamental question.

But the definition of purpose is not sufficient; it is also necessary to manage the process of interaction. It is best to see OR not as actors in their own right, but rather as facilitators of management action. From this viewpoint the interaction process is that of enabling managers to identify factors that need improvement, of determining good policy, and ensuring that an effective solution is implemented. In looking at this it is important again to ensure that the balance between management and OR is correct. Too often the process is thought of as an intervention by OR. More helpfully we should think of this as an element in the whole change process. It should be said however that, even from the OR side, the problem of managing the facilitation process is undervalued. The art of teasing out the relevant parts of the management mess, identifying hidden assumptions, establishing priorities and relationships, sorting out alternatives, and securing a full transfer of solution ownership is only now being brought out of the realm of folk-lore into the open air of critical appraisal, with some foundations in theory. And again what happens appears to be invariant to cultural fads.

In the space available there has only been time to lay out the main issues. The main thrust of the argument has been to try and shift attention away from OR as an intervening process towards management as a process that needs an OR type activity as an integral part of its thinking/ adapting/control activities. This then focuses attention on the management/ OR interface as a process that needs to be managed in its own right.

PSYCHOLOGICAL ASPECTS OF FORECASTING WITH STATISTICAL METHODS

George Wright[1] and Peter Ayton[2]

[1]Bristol Business School
Coldharbour Lane
Frenchay, Bristol, BS16 1QY

[2]Decision Analysis Group
Psychology Dept
City of London Polytechnic
Old Castle St
London E1 7NT

Judgment is implicated in many ways with a variety of forecasting methods. This paper briefly outlines the major methods (pure judgment, extrapolation and econometric modelling) and focuses on the role of judgment in forecasting with statistical models by considering anecdotal reports of the use of these models. We then evaluate studies of the quality of human judgment, most of which have been considered by psychologists using paper-and-pencil tasks with undergraduate students as experimental subjects. Finally we evaluate studies of <u>expert</u> judgment and discuss the implications of these for the quality of judgmental "adjustments" to statistical models in work-a-day situations.

Judgmental Forecasting

Here the information on which the forecast is based is analysed solely by the human mind. Opinions may be proferred from salesmen, executives etc. and these forecasts may be point forecasts or ranges. Since it is often thought that two heads are better than one, methods have been developed to combine individual forecasts with the aim of producing improved forecasts. In the context of behavioral aggregation, Delphi has been the most popular technique (see Parente and Anderson-Parente, 1987 for a review). Delphi involves iterated polling of opinions and feedback of the summarised (anonymous) individual opinions before repolling. The Delphi procedure is specifically designed to avoid any influences resulting from social inter-action (such as pressure to conform to 'dominant' or vociferous individuals). However an alternative strategy – based on the <u>mathematical</u> aggregation of individual forecasts – has also been developed as a means of combining judgments (see Ferrell, 1985 and Lock, 1987 for complementary reviews of this work).

Exploratory prospective analysis, one variant of judgmental forecasting, involves the specification of a set of possible futures as plausible scenarios. Planning involves formulating and specifying a desired future <u>and</u> finding a means of achieving it. Scenarios, once developed, may be <u>forecast</u>

with a degree of judgmental confidence in the scenario occurring. Godet (1983) has advocated the use of such prospective analysis, in which forecasts are generated for scenarios. The contributors to Wright and Ayton (1987) provide reviews of the applications of aggregation, scenario analysis and probability assessment to judgmental forecasting.

Statistical extrapolation

This type of forecasting method is based soley on the variable to be forecast, e.g., historic sales of a product, and time, e.g., on a month-by-month basis. Since only two variables are involved and since elapsed time is not the cause (or at least not the sole cause) of sales, plotted values of the two variables can only indicate the relationship between them. Typical methods of measuring the form of this relationship, and using this measurement to predict the future relationship, include simple linear regression, moving averages and exponential smoothing. Extrapolation methods cannot predict "turning points" where, for example, unique changes in the world have a casual impact on the variable to be forecast.

There are many anecdotal reports that reveal the ubiquity of judgment in extrapolation. For example Lawson (1981) discusses traffic usage forecasting in telecommunications at Bell Telephone Co. of Pennsylvania. The quantities of switching equipment that Bell own today should care for their needs for the subsequent two years. But Bell must buy to serve the traffic loads that they expect for up to five years hence. On the surface, Bell's methodology for forecasting focusses on statistical extrapolation techniques, but as Lawson notes:

"If the projections are acceptable, based on engineering equipment, they are accepted. If not the forecaster decides which historical data points contributed to the bad trend, deletes those historical values, and recycles the process until an acceptable result is obtained. When all else fails, the forecaster abandons the mathematics and substitutes an acceptable forecast based on judgment" (p.19).

And further:

"The eyeball method involves the forecaster in the analysis of data at all steps in the process. It avoids the trap of mathematical projection systems - the tendency to believe the fitted trend without analysis of whether or not the historical data points are representative. In practise it has produced better forecasts than more statistically based systems... The eyeball method is better because it involves the forecaster in analysis of the historical data as the projection is made" (p.24)

Sprague and Bhatt (1978) make a similar point and argue that statistical extrapolation techniques should be combined with judgmental forecasts (perhaps supplied by use of the Delphi technique) by using "subjective weightings". However the precise method of combination is not detailed.

In the context of sales forecasts, Soergl (1983) argues that:

"...while the computer makes it easy to use sophisticated tools, executive judgment will always be needed to assess how the future might differ from the past, and to adjust the forecast accordingly..." (p.43)

Also in sales forecasting, Jenks (1983) points out that:

Quantitatively advanced techniques such as regression modelling,

618

Box-Jenkins, exponential smoothing ... cannot anticipate one-time events such as surprise competitive development nor are they particularly effective for long term planning without significant adjustment by management judgment" (p.82)

Quantitative Forecasting: Econometrics

This is based on a representation of causal relationships between variables. Often, multiple linear regression techniques can be used to model the relationships between changes in precursor variables and their effect on a target variable, where the former are presumed to have a causal impact.

Large econometric models, such as models of a nation's economy, are, in effect, hypotheses about the way in which the economy works. The Wharton model, for example, uses 400 equations which model relationships between variables. Users of such models can simulate the effects of different policy proposals, e.g. the effect of a 2 point cut in the standard rate of income tax on consumer spending. Since the models are hypotheses it follows that alternative models may disagree about fundamentals. For example, the Wharton model links consumer spending closely to changes in disposable personal income whereas the Federal reserve model does not (Malley, 1975).

Sometimes the forecasts of an econometric model may make little sense to the modeller. As Malley (1975) notes:

"... the Wharton model, like just about all others is endlessly subject to 'adjustment'. The notion that models are challenging human judgment has some validity to it some of the time. But a lot of the time the process is reversed - i.e., the models' findings are themselves being challenged and overridden" (p.154)

Irland, Colgan and Lawton (1984) report on economic forecasting by the Maine State Planning Office. Here a 10 year forecast was built around a model of the state's economy. They noted that:

"... the model forecast substantial increases in construction employment. Unfortunately, recent growth (in the period 1975-1980) was almost entirely attributable to massive rebuilding of the paper industry's physical plants and a large spurt in Federally subsidized water, sewer and housing construction. Neither seemed likely to continue ... the model's output had to be 'managed' through the application of operator judgment and experience" (p.13-14)

Glendinning (1975) makes a further point:

"For various understandable reasons, the data on which forecasting is based are often imperfect... This is either because the basic data are not readily available or because those submitting them do not go to the effort or expense of ensuring their accuracy ... the use of forecasts of future levels of national output etc. requires exercise of considerable judgment" (p.409)

McNees and Perna (1981) argue that most judgmental adjustments to econometric models are made for two reasons:

(1) The model has not been performing adequately recently, and/or

(2) some external factor, not incorporated into the model is expected to influence future events.

In summary, the results of statistical extrapolation and econometric modelling are routinely subject to post hoc "adjustment" by managers and modellers alike. These adjustments have face validity to the forecasters themselves. The use of judgment seems unquestioned. However, studies of human judgment in the psychological laboratory are more equivocal. We consider this research next.

Studies of Judgment in the Psychological Laboratory

In the 1960's and 1970's the prevailing view of human judgment was that it was vulnerable to severe biases. Most of the evidence to support this conclusion came from the reports of experiments conducted by Tversky and Kahneman. To illustrate one of their findings consider the following question:
Which of the following causes of death is the more likely of each pair?
(a) Lung cancer or stomach cancer
(b) Murder or suicide

Tversky and Kahneman demonstrated that we often judge the probability of an event by the ease with which relevant instances are imagined. Instances of frequent events are typically easier to recall than instance of less frequent events, thus availability is often a valid cue for the assessment of probability and frequency. However, since availability is also influenced by factors unrelated to likelihood, such as familiarity, recency and emotional saliency, reliance on it may result in systematic biases. People overestimate the relative frequency of diseases or causes of death which are much publicised, such as murder or lung cancer, whereas the relative frequency of less publicised causes of death, such as stomach cancer and suicide, were underestimated. Most of the people in Tversky and Kahneman's experiments thought the first cause of death in each of the pairs cited above were more likely than the second.

Tversky and Kahneman conducted many studies such as the one illustrated here (see Kahneman, Slovic and Tversky, 1982). Paul Slovic (1982) summarised the conclusions to be drawn in an apt expression:

"This work has led to the sobering conclusion that, in the face of uncertainty, man may be an intellectual cripple, whose intuitive judgments and decisions violate many of the fundamental principle of optimal behavior"

However, more recently this somewhat bleak view of human judgment has been challenged. For example, Beach, Christensen-Szalanski and Barnes (1987) argue that issues regarding the quality of human judgment are not settled and that a commonly held belief that, in general, human judgment is poor is not based on convincing data. Beach et al show that empirical studies showing poor judgmental performance tend to be cited in the literature more often than studies reporting good performance results. (Perhaps an availability bias has influenced the judgments of the researchers concerning the quality of the judgments of their subjects!) Furthermore, poor performance by undergraduates on contrived paper and pencil tasks (the basis for many experiments) may not generalise to the world outside the psychological laboratory. Beach et al argue that the quality of human judgment cannot be assessed in isolation from human action. Only by assessing actions which may result from judgment is it reasonable to focus on the quality of judgmental performance.

Perhaps the most commonly quoted article in the forecasting literature which expresses doubt about the capabilities of human judgment is that by Hogarth and Makridakis (1981). They argue that:

"Many of the numerous information processing limitations and biases revealed in the literature apply to tasks performed in forecasting and planning" (p.115)

However the biases quoted have often been identified in undergraduate students' answers to simple paper and pencil tasks completed in the artificial environment of the psychological laboratory. Further, most of the tasks studied have to do with judgment per se, rather than judgment in forecasting.

Studies of Expert Judgment

It might be argued much of the judgmental input to forecasting is made by people who would claim they benefited from the possession of some degree of expertise. A question then naturally arises: How good are experts' judgments? Experts know more than novices but are their judgments less vulnerable to bias? Studies of the quality of 'expert' judgment, when the experts being evaluated are operating realistically in the domain of their expertise are, understandably, somewhat thin on the ground. Even when, individually or collectively, experts can be found who are willing to co-operate to the extent necessary the quality of judgments made by experts is not always readily amenable to analysis. In many cases it may not be possible to collect the kind of data that enable inferences concerning the validity of their judgments to be made. This may be because their judgments are undertaken in complex environments and/or because there is no obvious or available index of judgmental quality.

As to the quality of expert judgement there is a tension in the literature: Some researchers report evidence that experts are vulnerable to the same kinds of errors that might be predicted from Kahneman and Tversky's (and others) studies of students and paper and pencil tasks. Other researchers claim that experts are better judges than the research with paper and pencil tasks would imply.

Likelihood estimation tasks — even when these are not explicitly concerned with probabilities for the future — are of particular interest because forecasts are often ideally expressed via the medium of probability. Christensen-Szalanski, Beck, Christensen-Szalanski and Koepsell (1983) studied the influence of availability on estimates of the risks (mortality rates) of various diseases. They examined the relationship between reporting of diseases in medical journals and the estimates given by a group of experts (physicians) and non experts (college students). To their surprise, they found no statistically significant evidence for a link between the incidence of reporting in the journals influencing the judgments (though as the correlation between journal coverage and mortality is quite high, statistically it is more difficult to identify an effect). But they did find both groups overestimated the risks. In general, physicians were more accurate than students but the estimates of both groups were found to be biased by actual encounters with people with the disease. Christensen-Szalanski et al suggest that experts and non-experts may use similar thought processes but make differently biased judgments because of their different experience of the risky events.

Northcraft and Neale (1987) report a study of real estate agents in Arizona which explored the influence of another heuristically driven bias discussed by Kahneman and Tversky - that of anchoring and adjustment. Students and real estate agents toured properties and made pricing decisions; their judgments confirmed the hypothesis that manipulated valuations apparently made by the property owner would act as an 'anchor' for the values assigned to properties and influence judgments accordingly. Northcraft and Neale conclude:

"   the findings of this study provide strong evidence that previous
laboratory research on decisional heuristics and biases is applicable
to 'real world' information rich interactive estimation and decision
contexts." (p.96)

They further argue that:

"...decision biases and heuristics are more than just parlor tricks
and ... should play an important role in our understanding of everyday
behavior." (p.96)

However other researchers have found evidence that the quality of
judgments made by experts is sometimes far superior to that of students in
the laboratory. Studies of the 'calibration' of subjective probability
judgments have often shown poor performance and typically reveal overconfidence - e.g. when a judge says he is 70% sure of an event occurring he
is less than 70% likely to be correct (cf. Lichtenstein, Fischhoff and
Phillips, 1982). Murphy and Brown (1984) have found that weather forecasters in the United States who, as a matter of routine, produce subjective
probability forecasts for rain are excellently calibrated. Murphy and Brown
suggest that this is because they have considerable experience making several
forecasts a day over many years, receive reliable regular feedback and have
a large amount of relevant data (including objective forecasts). Such
circumstances may well be ideal for the acquisition of judgmental expertise
but may not be found for forecasting in other domains; indeed, good calibration is not always found in experts. For example, Christensen-Szalanski
and Bushyhead (1981) found that physicians expressed confidence in their
diagnoses were badly calibrated and exhibited marked overconfidence. Keren
(1987) has found good calibration in the bids of expert bridge players and
also emphasises the role of feedback. He argues that expert bridge players
resemble weather forecasters in that they get lots of experience with related
items and that the feedback is prompt.

Vertinsky, Kanetkar, Vertinsky and Wilson (1986) found that hockey
players' forecasts of the results of their future games were well calibrated
and concluded:

"Thus while the literature indicates that people are generally weak
intuitive statisticians and are often prey to biases stemming from the
use of heuristics in estimating uncertainties, the performance of field
hockey players denies such a claim, at least for the task in hand.
Furthermore, the study confirms claims of other studies with experts -
that experts can provide high quality probability assessments." (p.403)

Plainly experts can be expected to outperform novices in many judgment
tasks, though even this brief review has shown that experts will not always
be immune to the biases observed in simple laboratory experiments with
naive subjects. Identifying the precise factors which contribute to optimal
judgmental performance remains a task for future research.

Conclusion

We have provided anecdotal evidence that judgment is ubiquitous in a
wide range of forecasting techniques. Research conducted by psychologists
suggests that there are possible systematic biases in human judgmental
processes. The extent to which these biases might be anticipated on an
a priori basis in the judgmental performance of people operating in particular domains remains an unresolved and, for forecasting in particular, a
critical issue.

References

Beach, L. R., Christensen-Szalanski, J. J. J., and Barnes, V., 1987, Assessing human judgment: Has it been done, can it be done, should it be done, in: "Judgmental Forecasting", G. Wright and P. Ayton, eds., Wiley, Chichester.

Christensen-Szalanski, J. J. J., Beck, D. E., Christensen-Szalanski, C. M., and Keopsell, T. D., 1983, Effects of expertise and experience on risk judgments, Journal of Applied Psychology, 68:278-284.

Christensen-Szalanski, J. J. J., and Bushyhead, J. B., 1981, Physicians' use of probabilistic information in a real clinical setting, Journal of Experimental Psychology: Human Perception and Performance, 7:928-935.

Ferrell, W. R., 1985, Combining individual judgments, in: "Behavioral Decision Making", G. Wright, ed., Plenum, New York.

Glendinning, R., 1975, Economic forecasting, Management Accounting, 11:409-411.

Godet, M., 1983, Reducing the blunders in forecasting, Futures, 39:181-192

Hogarth, R. M., and Makridakis, S., 1981, Forecasting and planning: An evaluation, Management Science, 27:115-138.

Irland, L. C., Colgan, C. S., and Lawton, C. T., 1984, Forecasting a state's economy: Maine's experience, Northeast Journal of Business, 11:7-19.

Jenks, J. M., 1983, Non-computer forecasts to use right now, Business Marketing, 68:82-84.

Keren, G., 1987, Facing uncertainty in the game of bridge: A calibration study, Organizational Behavior and Human Decision Processes, 39:98-114.

Kahneman, D., Slovic, P., and Tversky, A., 1982, "Judgment under uncertainty: Heuristics and Biases", Cambridge University Press, Cambridge.

Lawson, R. W., 1981, Traffic usage forecasting: Is it an art of a science?, Telephony, February:19-24.

Lichtenstein, S., Fischhoff, B., and Phillips, L. D., 1982, Calibration of subjective probability: The state of the art to 1980, in: "Judgment under uncertainty: Heuristics and Biases", D. Kahneman, P. Slovic, and A. Tversky, eds., Cambridge University Press, New York.

Lock, A., Integrating group judgments in subjective forecasts, in: "Judgmental Forecasting", G. Wright and P. Ayton, eds., Wiley, Chichester.

Malley, D. D., 1975, Lawrence Klein and his forecasting machine, Fortune, March:152-157.

McNees, S. K., and Perna, K. S., 1981, Forecasting macroeconomic variables: An eclectic approach, New England Economic Journal, May/June:15-30.

Murphy, A. M., and Brown, B. G., 1984, A comparative evaluation of objective and subjective weather forecasts in the United States, Journal of Forecasting, 3:369-393.

Northcraft, M. A., and Neale, G. B., 1987, Experts, amateurs and real estate: An anchoring and adjust perspective on property pricing decisions, Organizational Behavior and Human Decision Processes, 39:84-97.

Parente, F. J., and Anderson-Parente, J. K., 1987, Delphi inquiry systems, in: "Judgmental Forecasting", G. Wright and P. Ayton, eds., Wiley: Chichester.

Slovic, P., 1982, Towards understanding and improving decisions, in: "Human Performance and Productivity: Vol 12, Information Processing and Decision Making", W. C. Howell and E. A. Fleishmann, eds., Erlbaum, Hillsdale, NJ.

Soergl, R. F., 1983, Probing the past for the future, Sales and Marketing Management, 130:39-43.

Sprague, J. C., and Bhatt, D. N., 1978, Design of a telecommunications demand forecasting system, Telephony, 195:46-54.

Vertinsky, P., Kanetkar, V., Vertinsky, I., and Wilson, G., 1986, Predictions of wins and losses in a series of field hockey games: A study of probability assessment quality and cognitive information-processing models of players, Organizational Behavior and Human Decision Processes, 38:392-404.

Wright, G., and Ayton, P., 1987, "Judgmental Forecasting", Wiley, Chichester.

THE USE OF PHILOSOPHY AND PSYCHOLOGY BY THE OPERATIONAL RESEARCH ANALYST IN

INFLUENCING THE DECISION-MAKER

Charles R Leake

Operations Research Division
SHAPE Technical Centre
The Hague, Netherlands

In Operations Research (OR) today there are a number of concerns as to what direction it should take: should it become more abstract or should it become more readily accessible to the decision-maker? Curbing the abstract aspects of the discipline would certainly retard progress in our eventual understanding of the nature of OR problems. Conversely not making OR analyses more accessible to the decision-maker will discourage him from using OR as part of his decision-making process. The latter concern is emphasized by some recent trends in management which avoid using OR in their decision-making process. This problem or dilemma was presented to the Euro VIII Conference on OR in 1986. The theme of this conference as stated in the keynote address was that attention in the OR community needed to be focused on the harmonizing of these 2 aspects of OR. This paper is a follow-up to this theme.

OR is a much more comprehensive discipline than any of the individual sciences which it encompasses. OR may encompass one science or it may bring together several. Historically teams of different scientific disciplines such as psychology, physics, engineering and methematics as well as such disciplines as history and language arts were used to form these teams. These teams worked together as a unit to bring to the decision-maker a clear and thorough understanding of the problem to be solved as well as its solution expressed in a manner that could be understood by the decision-maker. The initial successes of OR are a matter of historical record.

It must be remembered that although science is concerned with the study of facts and generalizations based on facts, OR is concerned with how to use the facts or generalizations of science as well as the theories and facts from other disciplines. Thus, although OR is considered to be a science, it is more an applied science than a theoretical one. The recent over-emphasis on some of the more theoretical aspects of OR, especially the mathematical ones, has been a source of confusion to the decision-maker, the ultimate user of OR.

To see why this could very easily occur let us examine the OR process. The OR team is usually provided with some information by the decision-maker or his subordinate about a condition or state of affairs that is unsatisfactory. Out of this initial guidance, the OR team must determine 3 basic things. They are: 1) identify the problem to be solved; 2) determine data requirements; and 3) determine the connections between the data and the problem to be solved. In addition the team must determine other administrative requirements which have a profound impact on the study such as budget, team organization, equipment, duration of study, etc. All of these determinations require decisions. Decisions require underlying philosophies and strategies that involve the use of psychological techniques and theories.

To illustrate this point, we will consider the problem of identification in greater detail. The same analysis in detail could be done for the other aspects of an OR problems as well. To begin with identifying or formulating the problem is not always a simple task. In many instances the problem as seen by the decision maker may have no solution or one that might be so unreliable as to have little or no meaning. The OR team then has to provide the decision-maker with alternatives to the original problem and devise a strategy to convince the decision maker with. These tasks involve philosophy and psychology.

Let us first consider philosophy. there are 3 philosophies that are generally used in OR. They are idealism, pragmatism and empiricism. Moreover, seldom are they ever used in the pure form intended by their authors. Instead they are used in a more practical sense and in many cases from a subconscious level of thought. An illustration should suffice to make this point.

Many problems which can not be solved directly are modeled. A careful examination of what a model is reminds one of Plato's Republic. Indeed modeling is an example of idealism. The OR team desiring more information or insights into an actual problem study an idealized situation and then use what they have learned from the model to improve their understanding of the actual problem as Plato suggested in his Republic. To illustrate the subliminal aspect, many modelers believe that the solution the model provides is the actual one. However, one who has studied philosophy and understands validity is well aware of the pitfalls of such an approach.

To continue further with this example, consider the problem of data for the model. Here is where some interesting compromises and inconsistencies occur. I for one have yet to encounter a problem where all the data is available and is of the quality desired. Instead, what is encountered is some data of excellent quality, some of a lesser quality, some with which everyone has great misgivings and a huge gap where the most important data requirements are to be found. It is in the putting of this data together and filling the gaps to provide the necessary inputs to the model, that a number of things occur. For example, when data is missing, we might conduct an experiment such as an empiricist would do. Sometimes when the model does not adopt itself to the data in the form in which we find it, the data is truncated, used for interpolation, aggregated, or used to estimate more applicable data. Each one of these activities involves a philosophical approach that can vary from one activity to the next. For example, interpolation of experimental data involves the use of mathematics which is an ideal world. In this ideal world a missing piece of information is obtained that will become part of the set of empirical data which will again serve as input to a model which belongs to an ideal world. And we do all this because it works better than any other system which we know and this approach is pragmatic.

This of course is inconsistent with our normal use of a subject predicate type calculus which does not allow the use of opposites. However, such philosophies as Yin and Yang or Hegel's allow such inconsistencies. Since they are such a part of our everyday practices, it

would imply that our inconsistencies might not be as inconsistent as we
think. It might be the case that our way of examining inconsistencies is
not consistent with reality. Under this premise, the concept of validity
could be re-examined. Instead of using implication as a basis for
validity, the relationship of an entity which we seek to validate in
comparison to the whole of the parts might provide us with a better
guideline for validity.

In spite of inherent inconsistencies we use philosophy in OR. We use
it whenever we make value judgements, or even when we collect data.
Moreover, it is unavoidable, just as are our biases. But we need to
understand these phenomena in order not to mislead ourselves or the
decision-maker. However, philosophy is easy to understand when compared to
psychology which is the next part of our discussion.

Anyone who sat in a meeting that is the least bit perceptive and has
observed the decision-making process understands the psychological aspects
of the process. Aside from some of the more clinical type observations
which one can make, there are several types of psychology that are useful
in decision-making. One has to do with a form of behaviourism known as
anchoring theory and the other is from perception theory which has
applications to learning theory, known as Piaget's developmental theory.

Let us consider anchoring theory first. Anchoring theory is concerned
with least notable differences (lnd) in stimuli. For example, x weighs
less than y, this sound is greater than that, or this solution set contains
more solutions than that. These stimuli are distributed according to
Fechner's Law (E-1).

$$R = a \ln S \qquad\qquad\qquad (E-1)$$

R = response, S = stimulus, a is a constant and ln is the Naperian
logarithm. There are several points of special interest with this law.
They are the threshold limen, the point of saturation and the theoretic
midpoint of the distribution. The threshold limen b is the point along the
stimuli axis at which the first response is recorded. The point of
saturation x occurs when no difference is perceived between greater or
lesser stimuli. The mid point M is given by (E-2).

$$M = \frac{1}{x-b} \int_b^x a \ln t \; dt \qquad (E-2)$$

This function of x is monotically increasing. Hence M increases as x
increases. Or as we move the saturation point out, we also move the
central tendency out. The lesson to be learnt from this is that as you
increase the solution set, instead of aiding the decision-maker you might
be adding to his confusion. A one over the world solution can be as
meaningless as a very narrow solution that tells little or nothing about
the problem to be asked.

This brings us to the next topic in psychology which I have found to
be extremely useful in the teaching of mathematics, that is Piaget's
developmental theory of psychology. For those fans of behavioural
modification, try setting up a behavioural modification program for about
10 varieties of verbal problems. This Augean task is mild, however, in
comparison to the schedule necessary to modify the decision-maker's
behaviour to get him or her to agree with your proposals. Clearly some
alternate approach is suggested. Here is where Piaget's theory offers some
promise.

At this point I wish to digress to a bit of mathematics in order to make Piaget's theory more understandable. Let us consider a set of R of real variables. R is of dimension N and a set C of complex variables where C is of dimension M. On the set R there is a set of functions $\{f_k\}$ defined such that there is a k where:

$$f_k : R \rightarrow C$$

in such a manner that the imaginary components in the M complex variables of C tend to a constant b. When k = 4 and the constant is equal to zero, we have a mathematical representation of Piaget's theory where the functions f_1, f_2, f_3 and f_4 correspond to the 4 development stages of Piaget.

The interesting aspect of Paiget's theory is that when applied to situations it is not strictly limited to chronological age. This is clearly understood in local supermarkets where customers shop for bottles of liquid, not in terms of their content, but in terms of their shape. An illustration of Piaget's rule of invariance for adults. This theory could also be used to explain some problems which analysts have with decision makers. For example, if the analyst is operating on R with f_3 or f_4 while the decision maker is operating with f_2, there will be a real communication problem as f_3 or f_4 implies a different perceptual set than f_2 with very little in common with which to enhance communication. Understanding Piaget can help reduce the accompanying confusion caused by the differences in perception.

Clearly the more one delves into the OR process, the more one becomes convinced that OR is not completely a science, but more of an art. Indeed as one begins to understand this situation, one begins to appreciate both the importance of philosophy and psychology in OR. This is especially true when one chooses a direction to follow. Here clearly the need for understanding the most appropriate philosophies and psychological sets need to be incorporated into the total study data package. They furthermore, need to be brought out into the open and discussed critically in light of the important bearing that they will have on the ultimate results of an OR study.

We have reached the end of our discussion, but before we complete it, let us discuss the main points once again. First and foremost OR is an art as well as a science. Therefore, methods must be selected and are not predetermined as in science. Selection implies philosophical thinking. The model builder's and decision-maker's philosophy or philosophies determine to a large extent whether or not the interface between the analyst and the decision-maker are to be antagonistic or harmonious. When the philosophies are different, the use of psychological theories, in particular Piaget's or anchoring can be useful in channeling the energy involved in a decision-making process in the appropriate direction. The determination of the philosophical bents of the participants in the OR process as well as the choice and use of the appropriate psychological theories are arts. It is in this sense that philosophy and psychology can again play a very important role.

BEHAVIOURAL DECISION THEORY AND OPERATIONAL RESEARCH

Andrew Lock
Dean
Faculty of Management and Business
Manchester Polytechnic

INTRODUCTION

The theme of O.R. and the Social Sciences seems a very appropriate framework in which to offer a brief review of behavioural decision theory and the contribution that it has made to the practice of decision analysis and to give some indications of future potential contributions to O.R., in general, and the broad fields of expert systems and decision support systems.

PROMISE UNREALISED

Decision theory and decision analysis developed in the fields of mathematics, statistics and economics. The field is today about 40 years old, though the real development of practical application is somewhat newer. Most readers will be familiar with the basic paradigms of decision analysis, and probably with the extensions into utility, to incorporate the relative attitude to risk of the decision-maker, and to cover problems with multiple criteria, usually referred to as multi-attribute utility theory or MAUT (Keeney and Raiffa, 1976). In the last ten years, those interested in the field have had to come to terms with a similar crisis to that encountered in Operational Research a decade before (see for example Ackoff 1970a and b), with an effective penetration far lower than earlier messianic predictions. Even today there remains concern about the number of OR/MS groups that are being disbanded (see for example the TIMS president's comments in OR/MS Today, October 1988). Neither volume of applications nor the range of problems to which decision analysis techniques had been applied have anywhere near reached expectations. Indeed many, if not the majority, of applications involved the U.S. government or its agencies, where there was an explicit political need for a defensible methodology to justify particular choices. In addition, in many cases, even the best publicised, little or no notice had been taken of the recommendations when the final decision was actually taken. The number of decision analysts actually practising is quite limited and in almost every case it is quite easy to trace their intellectual genealogy.

SUGGESTED CAUSES

A range of reasons have been put forward in both the OR and decision literatures for the resistance that has been encountered, and the frequent failure to implement recommendations even when a study had been commissioned. Firstly, there is the lack of involvement of the client in the definition of the problem and its solution, sometimes referred to as the 'ownership of the problem' issue. In the most extreme cases, clients have been presented with a single 'optimal' solution based on the analyst's formulation of the problem and selected technique, neither of which they have understood. An interesting piece of evidence from the MIS/DSS literature (Ives and Olson, 1984) suggests that system quality does not necessarily improve with client involvement, though the commitment to the resultant system certainly does. The second issue, which I have examined elsewhere (Lock, 1983), relates to the failure of analysis to take into account the organisational context in which problems arise and solutions have to be implemented. Recent developments in decision analysis have emphasised the importance of client participation in the problem representation, of the process of elicitation of required information and estimates, and of the exploration of the range of outcomes of alternatives. One model that has emerged is that of the decision conference, where, with appropriate computing support, the analyst and participants from the client organisation hammer out a problem representation and estimates for the required data inputs and arrive at a solution to which the participants are then committed (see, for example, Phillips, 1984). This approach raises some concerns about the replicability of particular analyses and the extent to which we are dealing with a genuine decision technology or a technologically sophisticated form of casuistry. It does give a strong emphasis to the consultancy skills of the decision analyst. Fischhoff (1977) usefully brought out these clinical art aspects with an interesting comparison between decision analysis and psychoanalysis.

A PSYCHOLOGICAL VIEW OF JUDGMENT AND DECISION-MAKING

As a consequence of these developments, from having played a comparatively minor role in the development of the decision analytic paradigm, psychologists now play the major role in the development of the field. The recent launch of the Journal of Behavioural Decision Making and the more formal establishment of the Judgement and Decision Making (J/DM) Group has given new focus to this activity. The interest of psychologists in decision theory can be traced back to Edwards' (1954) paper, which indicated the possibility of using the decision theoretic framework, primarily subjective probability and subjective expected utility, as a basis for examining human decision processes. Edwards has remained a major influence, with a strong emphasis (as shown by his comments published with Hogarth's 1975 paper) on ways of improving performance in judgement and decision tasks and developing human potential rather than focussing on how individuals deal with decision-making situations. The decision behaviour field is closely linked with that of the study of human judgement, which is of rather older origin, in which there has been controversy about the quality of human performance in judgmental tasks. Many will be familiar with the work of Kahneman and Tversky (eg Kahneman, Slovic and Tversky, 1982) in the literature that has come to be known under the title of "Heuristics and Biases". This has extensively catalogued the deviations of intuitive judgements from normative models by subjects in laboratory situations. From this many have gained the impression that humans are very poor at a whole range of judgmental tasks. This is not a universally held view. There is substantial empirical work to demonstrate that the identified 'heuristics and biases' often reflect the way problems or questions are presented (frequently called 'framing'). There is also some evidence that results obtained in reported studies with relatively naive undergraduate groups are reversed when they are administered to groups with some training in statistics and decisions analysis. Hogarth, 1981, has also argued that certain identified heuristics, whilst clearly erroneous in specific single instances, can be dynamically efficient. Finally, there are also a significant number of studies which actually show good performance by human judges, and identify the conditions in which this is likely to take place (a useful overview of this alternative view is presented by Beach et al, 1987). These primarily depend on familiarity with the task, the availability of explicit and unambiguous feedback on performance, and the need to examine and learn from such feedback. Interestingly, specialist or expert knowledge per se is not sufficient. A judicious use of this literature provides a great deal of useful material for those who are interested in developing decision aids, decision support systems or any system which requires judgmental input either in its knowledge base (as in an expert system) or in its functioning.

SUBJECTIVE PROBABILITY AND UTILITY

In parallel with the studies of the relative performance of human judgment, there has developed a literature on methods of elicitation of subjective probabilities and preferences from which utility functions for decisions under uncertainty may be derived. Hogarth (1975) presents what is probably the classic summary of the literature on subjective probability and its assessment. The conditions for good performance identified in the preceding paragraph also appear to apply to estimates of subjective probability. It is clear from the literature that considerable difficulty is experienced with events with extremely low probabilities and that it is sometimes hard to separate the estimation of the probability of an event from the evaluation or the utility of the outcome, particularly if outcomes are highly negative. Thus probabilities of outcomes that clearly worry people are overestimated. A good formal summary of utility functions and their assessment is to be found in Keeney and Raiffa (1976), though one should express a little scepticism about the ability of subjects to respond to some of the questions required to construct such functions. A useful and less demanding model for multi-attribute evaluation may be found in Edwards and Newman (1982).

RISK AND MANAGERIAL ATTITUDES

The other contribution of the behavioural sciences to the actual practice of decision support is in improving our understanding of risk and managerial attitudes towards it. The violation of what appear to be compelling utility axioms in actual advice is well documented. Most commentators would go back at least to the Allais paradox (Allais, 1953) (though the interest in apparent paradoxes in rational choice can be traced back to Bernoulli's St Petersburg paradox and beyond). March and Shapira (1987) present an excellent survey of managerial risk perspectives examining both the managerial norms and myths surrounding risk taking and the way in which risks are perceived and defined. The general conclusions of the literature they survey is that managerial decision making rarely conforms to decision theoretic axioms. For decision support practitioners this has a number of possible consequences. The first, mentioned by March and Shapira, is the possibility for educating managers to take a more analytic view of risk. The second concerns the way in which information about outcomes is actually elicited from and presented to decision-makers. These findings indicate directions for the development of aids for decision-making under uncertainty. One ought to observe that the probabilities of events are not wholly exogeneous 'givens'. In most cases there is some human influence on subsequent events, even if it is exaggerated by an 'illusion of control'. In addition, managers seem to prefer options which appear more flexible, in terms of the ability to adapt the strategy to actual outcomes, even when normative models would suggest greater commitment (e.g. building a small plant and having a slow rolling product launch rather than a large plant and national launch). One ought also to observe that clients ought to be encouraged to review the range of original choice alternatives to determine whether new ones can be identified, existing ones altered or hybrids formed from them, which appear better in terms of the identified choice criteria.

PROSPECT

As it has been twenty-five years since the previous OR and the Social Sciences conference, it seems appropriate to offer some thoughts on future prospects. There is certainly considerable scope for the use of the findings from the behavioural decision making literature in the development of intelligent knowledge-based systems. The evidence from the former literature suggests that one should treat with some caution both the validity of overall judgements made by experts and their ability to externalise the set of rules by which they arrive at a specific judgement. The literature on 'bootstrapping' (e.g. Camerer, 1981), the construction of statistical models from multiple judgements by an individual to outperform that individual, suggests that there is considerable scope both for improving and validating expert judgement before its incorporation in a formal system. Whilst there is considerable potential in educating managers in ways of representing decisions, assessing roles and ways of choosing between alternatives with uncertain outcomes, the likelihood of a mass transformation of managerial analytic capability looks slim. The most promising direction looks to be the development of decision aids available to individual decision-makers that help them to develop their own representations of decision situations and then elicit the required judgmental parameters. One of the key issues in the development of successful systems will be the way in which the range of outcomes are displayed to decision-makers and the way in which the integration of these to indicate optional decisions given identified criteria and a preference function is explained to the user. There seems considerable potential in Influence Diagrams as a base methodology for such systems (Howard, 1988; Howard and Matheson, 1983). Influence diagrams are graphical representations of decisions, events, exogenous variable and outcomes which can be manipulated on a screen by a decision-maker. There is considerable similarity with cognitive mapping techniques. Automated influence diagram systems are capable of developing decision tree formulations from the diagrams and eliciting the required parameters from the decision-maker. If decision modelling is to have a significant impact, we have to move away from models which depend on an analyst as an intermediary between them and the decision maker.

REFERENCES

Ackoff, R. L., 1979a, The future of Operational Research is past, Jnl. O. R. Soc., 30:93.
Ackoff, R. L., 1979b, Resurrecting the future of Operational Research, Jnl. O. R. Soc., 30:189.

Allais, M., 1953, Le comportement de l'homme rational devant le risque : critique des postulats et axiomes de l'Ecole Americaine, Econometrica, 21:503.

Beach, L. R.. Christensen-Szalanski, J., and Barnes, V., 1987, Assessing human judgment : has it been done, can it be done, should it be done?, in : "Judgemental Forecasting", G. Wright and P. Ayton, eds., Wiley, Chichester, England.

Camerer, C., 1981, General conditions for the success of bootstrapping models, Org. Behav. and Hum. Perf., 27:411.

Edwards, W., The theory of decision making, Psych. Bull., 51:280.

Edwards, W., and Newman, J. R., 1982, "Multiattribute evaluation", Sage Publications, London.

Fischhoff, B. 1977, Decision analysis : clinical arT of clinical science?, paper presented at 6th Research Conference on Subjective Probability, Utility and Decision Making, Warsaw.

Hogarth, R. M., 1975, Cognitive processes and the assessment of subjective probability distributions, Jnl. Am. Stat. Ass., 70:271.

Hogarth, R. M., 1981, Beyond discreet biases : functional and dysfunctional aspects of judgmental heuristics, Psych. Bull., 90:197.

Howard, R. A., 1988, Decision analysis : practice and promise, Man. Sci., 34:679.

Howard, R. A., and Matheson, J. E., 1983, Influence Diagrams, Working paper No. 195, SRI, Menlo Park, California.

Ives, B., and Olson, M. H., 1984, User involvement and MIS success : a review of research, Man. Sci., 30:586.

Kahneman, D., Slovic, P., and Tversky, A., 1982, "Judgment under uncertainty : heuristics and biases, Cambridge U.P., Cambridge.

Keeney, R. L., and Raiffa, H., 1976, "Decisions with Multiple Objectives", Wiley, New York.

Lock, A. R., 1983, Applying decision analysis in an organisational context, in : "Analysing and aiding decision processes", P. Humphreys, O. Svenson, and A Vari, eds., Hungarian Academy of Sciences, Budapest.

March, J. G., and Shapira, Z., 1987, Managerial perspectives on risk and risk-taking, Man. Sci., 33:1404.

Phillips, L. D., 1984, A theory of requisite decision models, Act. Psych., 56:29.

PROBLEM-PERCEPTIONS AS "REAL" DECISION-INPUTS:

A CASE-STUDY IN MANAGERIAL DECISION-MAKING

Sylvia M. Brown

Dept of Agricultural Sciences
University of Bristol, BS18 9AF
On secondment from
Oxford School of Business
Wheatley, Oxford

ABSTRACT

A study that aims to describe movements in practice in Crop Protection in UK and to discover the reasons for these is being funded jointly by AFRC and ESRC. A major focus of interest is availability and takeup of new technology. Eighty-two organizations have been targeted in various sectors of the industry for the pilot phase.

This paper discusses some methodological issues for Operations Research arising from findings suggesting non-rationality in decision-inputs.

INTRODUCTION

Strategic Decisions in organisations are the outputs of decision-making systems. If Herbert A. Simon (1976) is right and "... the key resource is attention", inputs to decision-systems will be selected and placed in varying orders of priority according to the beliefs, attitudes and perceptions of key decision-makers. It is immaterial whether these are "correct", (e.g. on criteria of verifiability and verification or on a consensus theory of truth). Where decisions are far reaching, it follows that the beliefs, attitudes and perceptions are crucial. Since it is unlikely that decision-makers are in all cases aware of their hierarchy of preferences, some means to make these explicit could form a valuable adjunct to more conventional decision-modelling.

A pilot study of 82 organisations is being sponsored jointly by the Agricultural and Food Research Council and the Economic and Social Research Council. The study examines influences on strategic decisions affecting availability and takeup of Crop Protection Technology in the UK. Perceptions of key decision-makers of their current problem-siutation and future ideal state are being analysed, using Repertory Grids and Cognitive Maps. Results from these are being compared and evaluated against other criteria, e.g. statements of the business mission and declared objectives.

Some implications are discussed, e.g. placement of individuals with particular cognitive and attitudinal structures at strategic decision-nodes in organisations.

"RATIONAL" APPROACHES TO DECISION-MODELLING

Much effort in Management Science has been expended on weighted evaluation models with a strong emphasis on substantive rationality and objectivity, achieved by operationalizability and mathematical processing. Such models are meant to be generally applicable to a range and number of decisions.

There is now a considerable body of work noting the strengths and limitations of such approaches, notably that of Herbert A. Simon (1976), who distinguishes between substantive rationality and procedural rationality and argues that organizational decisions are seldom and seldom should be made according to any of the various Rational Choice models. Especially relevant to the present case are the comments of Abell (1987), that Decision Theory in _any_ of its manifestations cannot cope with the unrecognized (and hence unevaluated) option. Reasons for non-recognition may include the perceptual set of the respondent.

Clearly related to the present study is early work that addressed problem-complexity, e.g. "Interaction" approaches to marketing and buying decisions (Kirsch W. & Kutschker, M., 1982) by characterizing decisions as the outcome of joint decision-processes that are context-dependent. This description is noteworthy; decisions are conceptualized as social events. They should, therefore, be amenable to analysis using appropriate methodology from the Social Sciences. Sociology, however, historically has itself often proceeded by variable-centred methodology, which may be why many Interaction approaches are deterministic in that they treat variables contributing to "potential" as objective and measureable.

Moreover, consumer research is often conducted ante-hoc (i.e. "What would you do if ...?" research) that inevitably biases responses towards a conceptual rationality seldom realised in behaviour.

Thus from their questionnaire results and unstructured interviews, Kirsch & Kutschker (1982) were able to conclude that "... behaviour ... seems only partly rational" from observations of discrepancy between decision-criteria declared by decision-makers and decision-behaviour, but were unable to account for the discrepancy other than by shortfall in use of the criteria. Consequently, they looked for more stable determinants, like traits of the respondent.

Such methods are misguided both on practical and theoretical grounds. Firstly, they can only result in increasingly complex and unwieldly models that fragment the data beyond the limits of significance-testing. Secondly, unwarranted axioms like "Man is a Rational Thinker" are implicit in the model when the psychological evidence suggests no such thing. For example, Newell's (1973) extensive work on simulation of human thought processes strongly suggests "Production System" problem-solving protocols (hence the constant search for decision-paradigms in OR?). Again, Peter Wason's (1977) work reveals many hilarious examples of non-rationality, including self-contradiction, failure to eliminate hypotheses and fallacious inferences, to which most of us are prone most of the time. (See also apropriate references in the S. Brown's paper in "OR as a Social Science" stream, this conference). Thirdly, important inputs to the process, such as perceptions, cognitive style (including empirical bias - Scribner (1977)) and feelings are overlooked.

DECISIONS AS EVENTS, AMENABLE TO ANALYSIS

In contrast to hypothesis-driven, variable-centred research, Ethno-
graphers and other social researchers typically begin with observations of
events and seek to elicit the causes. This is not the forum for yet
another debate about whether there is an objective reality that can be
known; the author accepts the weak anti-naive-perception position that
there is no way of knowing whether there is or not. Thus Brewer & Collins
(1981) assertion that "... theories held in common do affect our constructions of reality" is acceptable whereas their further assertion "... the
problem is to disentangle shared perceptions from shared reality" makes no
sense from this standpoint.

In this study a Social Science comparative approach begins with
particular problem-clarification events addressed by qualitative research.
As many instances as possible of these phenomena are examined, to produce
the set of axioms that fits the majority of cases, i.e. represents the
"theories" held in common by respondents, if any, whether across the whole
sample or within subsets.

To enable this and to assess validity, various forms of triangulation
are necessary entailing multiple methods. As advocated by e.g. Campbell's
(1959) multiple operationalism, the pilot study uses four related techniques. These are:
     Cognitive Mapping
     Influence Diagramming, using a pre-prepared system map
     Rated Repertory Grid
     A pre-prepared rated grid.

In accordance with the arguments of Douglas (1976), the methods
progress from less to more control and response-structuring by the
researcher. The first and most important technique is data-driven. Rather
than hypothesizing a range of factors that might be considered important
and influential and testing for these, respondents' decision-approaches are
being investigated and described first and analyzed for consensus/
discrepancy later, whereafter hypotheses for further investigation may be
generated.

Convergence between methodological individualism and collectivism in
Sociological research has been noted by Knorr-Cetina (1981) (i.e. between
trends deduced from micro-level data or "laws" generalized from macro-level
data). Accordingly, the initial premises of the research are that changes
at industry level result from implemented strategic decisions taken at
organizational level and that some decision-makers and some sectors of the
industry will be more influential than others.

Their decisions will, necessarily, be based in perceptions of the
current situation and predictions about future scenarios. These perceptions and predictions will embody beliefs about systemic structures and
processes.

What the decision-maker believes about the present and the future may
be well or less well-grounded in appropriate beliefs of others. In neither
case can future events verify beliefs. If beliefs are maintained, either
because of or despite lack of reinforcement, and are, consequently, acted
upon, it makes no difference whether or not they are verified (i.e. "true")
or even verifiable. "Good enough" decisions make their own "truth" and
"bad" decisions can always be explained in terms of confounding variables.

Hence it is legitimate to enquire why decision-makers are making and
taking decisions as they are doing, (i.e. what factors they consider

pertinent and in what way) without attempting separation of "objective reality" from "myth" in their accounts of the process.

Since it is also likely that each decision-maker will construe his or her observations differently, will have a different welltanschauung and that the range and number of combinatorial sets of beliefs thus will far outstrip any set of hypotheses that might be generated by a researcher, hypotheses-testing would seem inappropriate, not least at a pilot phase.

Cognitive Mapping requires participants to explore the ramifications of their problem situation. As conducted in the study under discussion, the researcher was able, during the interactive process, to require respondents to explore their perceptions, implicit calculations of risk and "hidden agendae", in addition to the "rational" elements most decision-structuring approaches impose (i.e. how they really were going about making their decisions, warts and all). Many respondents expressed thanks for the insights gained into their own cognitive processes, whilst some became anxious about confidentiality.

AN ILLUSTRATION OF RESULTS OBTAINED

At the time of writing, most of the data were assembled but very little had been analysed. First impressions of the data suggested that one major decision-influencing group of factors was perceptions of the Food and Environment Protection Act (FEPA) and subsequent regulations. This group forms a useful "case-example" whereby to illustrate the argument above.

Who is affected by FEPA?

Sectors of the industry affected directly by legislation include:
Manufacturing companies seeking product registration
Merchants and distributors storing, handling and supplying
products
Consultants and sales representatives advising on Crop Protection
Contractors supplying Crop Protection services
End-users (i.e. growers) storing, handling and using products

Many other sectors of the industry are experiencing "knock-on" effects, for example providers of operator training.

Effect of FEPA on the Crop Protection Industry as a Whole

FEPA and its subsequent regulations aims to affect Crop Protection practices in UK. The degree to which it does so may be consequent on a number of factors, some directly and indirectly related to the legislation, others interacting with the above but of a different conceptual order.

Factors directory related include, e.g.
- the degree to which the law can be enforced
- knowledge of the law by those affected
- heedfulness of the law, if known
- likely interpretations of the law "on the ground'

Factors indirectly connected include, e.g.
- importance of organisational image, i.e. law-abiding/not
environmentally concerned/not
- perceptions of long-term importance of new law
(i.e. "flash in the pan" or "thin end of the wedge")
- availability of 'ways round" the law

        - perceptions of power/lack of it of enforcement agents
        - ability to influence enforcers

    Interacting with these may be, e.g.
        - perceptions of relative costs of compliance or non-compliance
        - congruence of new requirements with existing strategy
        - access to alternative stragegies
        - perceptions of what the competition is doing/will do
        - moral principles, etc. etc.

    If an aspect of the problem-situation is felt to be intractable, one
organisation will devote inordinate amounts of time and attention to
wrestling with it whilst another will "write it off" and pass on to some
more amenable area. For example, some chemical manufacturing companies are
lobbying vigorously to get registration procedures changed and registration
times reduced, whilst some distributors are merely reshuffling their (very
large) product range to meet continuing market needs.

    Some manufacturers are trying to swoop into market niches (gaps in
product-ranges) they see as created by the bottlenecks, others are assuming
"We're all in the same boat" and acting accordingly.

    Clearly, the letter of the law is not some objective, real control
operating on the system to limit its outputs; FEPA only "exists" as a
control in the perceptions of those who feel limited by it.

Implications within the organisation

    Respondents even within the same organisation can display wide
variation in the number and range of factors they take into account.
Particularly striking has been variation in the degree to which decision-
makers feel their enterprise threatened by the Act, its enactors or by
others perceived as less affected. A pessimist produces a different
problem-analysis from an optimist; a high risk-taker devises a different
strategy from a low risk-taker. Both the pessimist and the low risk-taker
may perceive and produce a longer list of adverse consequences.

    It is expected that further analysis will show that variation is
reduced by well formulated and communicated internal Company policy (i.e.
concrete and explicit).

CONCLUSIONS AND PRACTICAL IMPLICATIONS

    Organisational Policy as formulated by senior management will be
translated into strategic decisions by executives whose perceptual-sets may
or may not be congruent with those of their superiors.

    Cognitive Mapping provides a means to assess congruence of perceptual-
set with policy; thereafter corrective action may be needed. It also might
be used to assess attitudinal-set prior to involvement in the decision-
process. A number of participating organisations are also interested in
the technique as a vehicle for team-building.

    Before building any model upon which an organisational decision will
be based, it is important that end-users examine and reflect upon the
perceptions that each will bring to each major strategic decision.

    There is also a case for closer attention to cognitive style and a
wider range of personal characteristics in managerial selection.

REFERENCES

Abell, P., 1987, "The Syntax of Social Life". Clarendon Press, Oxford.

Brewer, M.B. and Collins, B.E. (Eds), 1981, "Scientific Enquiry and the Social Sciences". Jossey Bass, San Francisco.

Campbell, D.T. and Fiske, D.W., 1959, Convergent and Discriminant Validity by the multi-trait, multi-method matrix. Psychological Bulletin 56.

Douglas, J., 1976, Investigative Social Research, Beverley Hills, CA, in: "Linking Data", N.G. Fielding & J.L. Fielding, eds (1986). Sage Publications Inc.

Kirsch, W. and Kutschker, M., Marketing and Buying Decisions in Industrial Markets, in: "Decision-Making", M. Irle, ed. (1982). Walter de Gruyter & Co., Berlin 30.

Knorr-Cetina, K. and Cicourel, A.V. (eds), 1981, "Advances in Social Theory and Methodology". Routledge and Kegan Paul, London.

Newell, A., 1977, On the analysis of human problem-solving protocols, in: "Thinking", P.N. Johnson-Laird & P.C. Wason, eds. Cambridge University Press, Cambridge.

Scribner, S., 1977, Modes of thinking and ways of speaking, in: "Thinking", P.N. Johnson-Laird & P.C. Wason, eds. (1977). Cambridge University Press, Cambridge.

Simon, H.A., 1976, From substantive to procedural rationality, in: "Decision-Making", A.G. McGrew & M.J. Wilson, eds (1982). Manchester University Press.

Wason, P., in: "Thinking", P.N. Johnson-Laird & P.C. Wason, eds (1977). Cambridge University Press, Cambridge.

THE USE OF PROBABILITY AXIOMS FOR

EVALUATING AND IMPROVING FORECASTS

Peter Ayton

George Wright

Psychology Department
City of London
Polytechnic
London E1 7NT

Bristol Business School
Coldharbour Lane
Bristol BS16 1QY

Many management decision-aiding techniques require probabilities as a basic input. In many practical circumstances relevant actuarial data will not be available to permit an estimation of these probabilities by using a relative frequency count. On other occasions actuarial data may be available but it may be perceived to be unreliable; for example, given additional information of other particulars, the actuarial data might reasonably be considered as somewhat inappropriate to serve as a direct input. Consider, for instance, the problem faced by an insurance underwriter who may be able to compute from actuarial tables the likelihood of an oil tanker sinking in any given year, but then has to accommodate the knowledge that the particular tanker she is currently considering will occasionally be travelling through an area that has recently become a war zone. In other contexts, for example sales forecasting, abrupt changes in the world may also invalidate regression and time-series estimates based on techniques of averaging historical data.

In such circumstances the probabilities must be generated by, or at the very least adjusted by, human judgement. With such procedures as decision analysis (cf. Raiffa, 1968), cross-impact analysis (Dalkey 1972) and fault tree analysis (Fischhoff et al., 1978), subjective probabilities have been found to be convenient ways for representing the uncertainty surrounding the possible occurrence of the critical events of interest. The elicitation of subjective probability as a means of gaining access to an individual's internal state of uncertainty is a procedure commonly utilised by decision-aiding techniques (see Wright, 1984). In this context, subjective probabilities usually represent judgemental probabilistic forecasts of the likelihood of occurrence of future events.

A question that is then naturally prompted is how reliable are judgemental forecasts? Human probabilistic judgement has been studied for many years by cognitive

psychologists interested in decision making. Tversky and
Kahneman and others have conducted a number of studies which
have found evidence that the cognitive strategies employed to
produce judgements under uncertainty are vulnerable to
various systematic biases which may result in suboptimal
decision making (cf. Kahneman et al., 1982). Slovic (1982)
summarised the impact of this research as follows:

> "This work has led to the sobering conclusion that man
> may be an intellectual cripple, whose intuitive
> judgements and decisions violate many of the fundamental
> principles of optimal behavior."

However, in recent years this research has become the
subject of a critical reaction. Demonstrations of judgemental
fallibility, it is argued, should not necessarily be
interpreted as having pejorative implications for the
rationality of human reasoning (Cohen, 1981; Beach et al,
1987) and may not always occur in circumstances more akin to
real-world decision making (Ebbeson and Konecni, 1980).

One method for assessing probabilistic judgements is to
test whether or not the judgements conform to the axioms of
probability theory. Judgements expressed as subjective
probabilities must conform to the axioms of probability
theory or else they cannot possibly be valid. For instance,
according to one axiom of probability theory, additivity, the
component probabilities of a set of mutually exclusive and
exhaustive events should add to one. However, Wright and
Whalley (1983) have shown that, when assigning probabilities
to the possible outcomes of a horse race, subjects frequently
produce estimates that sum to greater than one. This
violation was found in circumstances where subjects were
given individuating information concerning the form of each
horse in previous races; but, when this information was not
provided, subjects demonstrated knowledge of probability
theory judging each horse had an equal chance and respected
the axiom. Of course axiom abiding judgements are not
necessarily valid; coherence with respect to the axioms is a
necessary but not sufficient condition for validity (cf.
Ayton and Wright, 1987; Wright and Ayton, 1987).

Another axiom is the intersection law which states that
the probability of event A and event B both happening is the
product of the probability of event A multiplied by the
probability of event B <u>given</u> event A has happened. In terms
of an abstract example, the probability of drawing two
consecutive aces from a pack of fifty-two cards, without
replacing the first card drawn, is:

$$P(2 \text{ Aces}) = \frac{4}{52} \text{ X } \frac{3}{51} = \frac{12}{2652} = .00452$$

Bar-Hillel (1973) asked subjects to choose between
gambles, where the outcome of one gamble depended on a single
elementary event, and the other depended on a complex
compound event occurring. She found evidence that the
subjective probability of a compound event was biased towards
the probability of its components, resulting in
overestimation of conjunctive events and under-estimation of
disjunctive events. These probabilities were for essentially

abstract situations where no knowledge (except that of probability theory) would be available to inform the judgement.

Such psychological findings are consistent with the rationale underlying the use of decision-aiding technologies. Many decision aids assume that, because of a limit to an individual's information processing capacity, better judgements can be obtained by decomposing complex decisions; the argument is that limited capacity information processers can thereby provide more accurate assessments which can then be recomposed, by computer if necessary, according to the normative theories. Cross-impact analysis is based on a modification of the Delphi procedure which allows a forecast for a complex scenario involving the interaction of many events to be computed. Forecasters are required to assess pairwise conditional probabilities to define simple interactions which are then used to compute scenario probabilities.

Recently, we attempted to empirically compare the relative accuracy of probabilistic judgements for decomposed probabilities with judgements for their holistic equivalents. As we have noted, one important factor affecting the elicitation of probabilistic judgements is the extent to which subjects perceive that they posess relevant background knowledge concerning the events of interest. Consequently, instead of presenting an abstract probability problem to our subjects, we asked them for judgements of the likelihood of real events in the world (e.g. that sterling would fall below a certain value within a specified period).

One way of evaluating the accuracy of probabilistic judgements is to measure <u>calibration</u>. Calibration is one method for assessing the veracity of a set of subjective probabilities. A mathematical coefficient of calibration (developed by Murphy, 1973) can be computed to check the degree to which the probabilities given by a subject as a measure of their confidence in a set of propositions correspond to to the actual probabilities of the propositions being correct - as measured by relative frequency. So, for example, if a forecaster expresses 70% confidence in each of a set of twenty events occurring, then fourteen of those events (70%) should actually occur for the forecaster to be perfectly calibrated. The degree to which their subjective probabilities do not correspond to the hit rate will be reflected in the size of the calibration measure. The typical finding from calibration research has been that probabilistic judgements are overconfident - generally for all events assessed as having a 0.XX probability of occurrence less than XX% actually occurr (cf. Lichtenstein et al., 1982).

Following the logic of decomposition and recomposition, we hypothesised that assessment of simpler marginal and conditional probabilities would be better calibrated than holistic assessment of more 'complex' compound probabilities such as intersections, unions and disjunctions. In the earlier card playing example, the intersection of drawing two consecutive aces could be asessed holistically, as could the marginal probability of the first draw being an ace and the conditional probability of a second ace given an ace is drawn first. Our experiment compared the calibration of holistic and normatively recomposed compound probabilities.

Full details of the experiment are reported in Wright et al (1988). Essentially we elicited forecasts from undergraduate students in January 1985 for each of five different events. For each event, forecasts were elicited for five related time periods: February (period A); March (period B); April (period C); 1-15 March (period D); 16-31 March (period E). Subjects also provided judgements for probabilities conditional on the event's occurrence or non-occurrence in the prior time period, i.e. P(B\A), P(C\B), p(E\D), P(B\Ā), P(C\B̄), P(E\D̄). Given these values it is possible to compute the probabilities for the intersections P(A and B), P(B and C), P(D and E). Also values for some disjunctions or unions of two periods are computable i.e. P(A or B), P(B or C), P(D or E), P(B and/or C), P (A and/or B), P(D and/or E).

The complex probabilities were computed from the equations presented below.

Intersection, P(A and B) = P(A) P(B\A)

Disjunction, P(A or B) = P (A)[1-P(B\A][1-P(A)]P(B\Ā)

Union, P(A and/or B) = P(A)P(B\A)+P(A)[1-P(B\A)]+
[1-P(A)]P(B\Ā)

The experimental procedure also required subjects to holistically estimate values for intersections, disjunctions and unions. Given the arithmetical complexity of these equations it might be anticipated that, as far as human judgement is concerned, the two sides of these equalities would seldom balance! However the empirical question of interest is which side, if either, provides the most accurate (best calibrated) judgements?

Against the rationale for decomposition and normative recomposition of judgement we found no evidence that the recomposed complex judgements were better calibrated than the holistically elicited complex judgements. Moreover, even the elicited simple probabilities were no better calibrated than the directly elicted complex probabilities.

A further analysis checked the relationship, across forecasters, between the coherence of the elicited judgements and their calibration. For each individual we measured the discrepancy between their holistic judgements for the complex events and the equivalent value for the same event computed from the recomposed simpler probabilities. It might be expected that more coherent forecasters would tend to be better calibrated since coherence is a necessary (though not sufficient) condition for well calibrated forecasts. However, for the forecasts we examined here, the data provide no evidence that there is a correlation between forecasting coherence and accuracy.

What can be concluded from such results? It might be argued that, if complex probabilistic judgements are as easily elicited holistically as they are by recomposition, then there is no sound basis for practical applications of normative decision theories as decision aids. However such a conclusion relies on the assumption that the complexity of probability judgements is determined by their arithmetic complexity with respect to an essentially arbitrary scheme of decomposition. Consider our earlier example of the probability of drawing two consecutive aces from a pack of cards. It certainly seems reasonable to suppose that given some thought, people could assess the probability of drawing an ace on the first draw and the probability of a second subsequent ace. However it also seems plausible that many people would have difficulty in computing the intersection of the two events. In this example the recomposed simple probabilities might well produce more accurate assessments of the complex probability. However, there is no guarantee that, in real world forecasting tasks, that arithmetically simpler judgements are easier to produce than the arithmetically complex.

An experiment reported by Tversky and Kahneman (1983) illustrates the point. They asked subjects to rate the probabilities of various events, some of which were inclusive of others. Thus the probability that next year there will be an earthquake in San Francisco causing a dam burst resulting in a flood in which at least a thousand people drown <u>cannot</u> be greater than the probability that there will, next year, be a flood somewhere in the USA in which at least one thousand people drown. However their subjects typically rated the logically less likely event as more likely than the general event set of which it is a member. Notice, in this instance, the former event contains a plausible cause for the flood (an earthquake in San Francisco) which, on the face of it, might not have occurred to subjects when judging the latter event. In terms of probability theory, one event represents a more complex conditional than the other.

However, in terms of the forecasters causal understanding of the world, the putatively more complex probability may be easier to judge. It remains to be determined in cases such as Tversky and Kahneman's whether the probability given to the more specific event should be decreased, the less specific likelihood increased or both to some degree. One of these options must be followed in any particular case because, as we have pointed out, incoherent probabilities cannot represent valid judgements.

The reason why our experiment produced no evidence for a correlation between coherence and calibration may well be because, for the forecasters we studied, the level of performance was poor enough such that the more coherent forecasters were as poor judges of the events as the incoherent forecasters necessarily were. Clearly, given this possibility, researchers should now examine the relation between coherence and forecasting for expert forecasters operating in their specific domain of expertise. In addition tests of the principle of decomposition of complex judgements should be performed with expert forecasters. Further consideration should be given to methods for defining the complexity of judgements and, consequently, the general appropriateness of aids to judgemental forecasting.

REFERENCES

Ayton, P. and Wright, G., 1987, Assessing and improving judgmental probability forecasts, Omega, 15:191-196.

Bar-Hillel, M., 1973, On the subjective probability of compound events, Org. Behav. Hum. Perf., 9:396-406.

Beach, L.R., Christensen-Szalanski, J.J.J. and Barnes, V., 1987, Assessing human judgment: has it been done, can it be done, should it be done?, in: "Judgmental Forecasting," G. Wright and P. Ayton, eds., Wiley, New York.

Cohen, L.J., 1981, Can human irrationality be experimentally demonstrated?, Behav. Brain Sciences, 4:317-370.

Dalkey, N., 1972, An elementary cross-impact model, Technological Forecasting and Social Change, 3:341-351.

Ebbeson, E. B. and Konecni, V.J., 1980, On the external validity of decision making research: What do we know about decisions in the real world?, in: "Cognitive Processes in Choice and Decision Making," T.S. Wallsten, ed., Erlbaum, Hillsdale, N.J.

Fischhoff, B., Slovic, P. and Lichtenstein, S., 1978, Fault trees: Sensitivity of estimated failure probabilities to problem representation. J.Expl. Psychol. Human Perception Perf, 4:330-344.

Kahneman, D., Slovic, P. and Tversky, A., 1982, "Judgment under uncertainty: Heuristics and Biases," Cambridge University Press.

Lichtenstein, S., Fischhoff, B. and Phillips, L.D., 1982, Calibration of probabilities: The state of the art to 1980, in: D. Kahneman, P. Slovic, and A. Tversky, eds., "Judgment under uncertainty: Heuristics and Biases," Cambridge University Press.

Murphy, A.H., 1973, A new vector partition of the probability score, J. Appl. Meteor. 12:595-600.

Raiffa, H., 1968, "Decision analysis - Introductory lectures on choices under uncertainty," Addison Wesley, Reading MA.

Slovic, P., 1982, Towards understanding and improving decisions, in: "Human Performance and Productivity: Vol 12, Information Processing and Decision Making," W.C. Howell and E.A. Fleishmann, eds, Erlbaum, Hillsdale, N.J.

Tversky, A. and Kahneman, D., 1983, Extensional versus intuitive reasoning: The conjunction fallacy in probability judgment, Psych Rev, 90:293-314.

Wright, G., 1984, "Behavioural Decision Theory," Penguin Harmondsworth.

Wright, G. and Ayton, P., 1987, The psychology of forecasting, in: "Judgmental Forecasting," G. Wright and P. Ayton, eds., Wiley, New York.

Wright, G., Saunders, C. and Ayton, P., 1988, The consistency, coherence and calibration of holistic, decomposed and recomposed judgemental probability forecasts, J. Forecasting., 7:185-199.

Wright, G. and Whalley, P., 1983, The supra-additivity of subjective probability, in: "Foundations of Utility and risk theory with applications," B.P. Stigum and F. Wenstop, eds., Reidel, Dordrecht.

MODELS AND EXECUTIVE DECISION MAKERS

Tom Hemming and Olle Högberg

Department of Business Administration
University of Stockholm
S-106 91 Stockholm, Sweden

Anders Holvid

ATH Matematik-Konsult AB
Korsfararvägen 140
S-181 40 Lidingö, Sweden

1. INTRODUCTION

Many of the problems faced by analysts could be avoided if OR research had focused more on problem formulation, communication, model building and implementation and less on solution techniques for artificially constrained problems. Recently more effort has been devoted to these largely neglected aspects of OR modeling. A paper by Geoffrion (1987) focuses on these problems and problems of much wider scope related to the possibility of developing general purpose modeling systems integrated with corporate data bases. Of particular concern to Geoffrion are problems of low productivity and poor managerial acceptance of OR models. One reason for low productivity according to Geoffrion is that at least three representations are needed for each model: a "natural" representation, a mathematical representation, and a computer executable representation. Geoffrion wants to replace all these representations by a single representation serving the purpose of all other types.

This paper will discuss model building and communication from another angle than Geoffrion. It is argued that more, not less, model representations are needed. The purpose of additional representations is to improve communication between analysts and managers and other persons concerned in all phases of model building: problem recognition, problem formulation, model building, and implementation. The result will be models of more relevance to real problems, models which are more easily accepted by decision makers and in particular executive decision makers. Focus will be on the use of simple graphical model representations as means for improving communication between analysts and decision makers. Two graphics approaches will be presented. First, the concept of a balloon graph will be introduced as a simple tool for conveying to the decision maker basic characteristics of a decision situation as perceived by the analyst. Because of the simplicity of this approach the decision maker will feel at ease, not being blocked by analytical formulas, remaining able to discuss in a more intuitive way than would otherwise be possible. This approach is particularly useful in the initial problem formulation phase. Second, an approach for graphical representation of large linear programming problems is discussed. The advantage of this approach is its simplicity and suitability for communication as well as model manipulation and documentation. Further, use of this technique for model representation speeds up the model building process to a considerable degree according to an unpublished paper by Holvid (1977).

The problem with decision makers, particularly at the policy level, is that they are busy. In order to anchor an OR project among managers it is necessary to find simple means for improving and speeding up communication between analysts and decision makers. The use of graphics for this purpose is discussed in the second section. One conclusion is that concepts used for communication should not be too involved and analysts should not refrain from using simple informal means of communication. The concept of a balloon graph will be introduced in the third section as one such simple tool but one that has proven to be useful. In the fourth section it is demonstrated how graphics can be used to improve communication and simplify the modeling of complex manufacturing and distribution processes. In the final section remarks are made on the applicability and relevance of this type of graphics.

2. GRAPHICAL MODEL REPRESENTATION

Much has been said about the success and failure of OR. It appears that problems of implementing OR can be found in particular at the policy level. Reasons for OR models being used infrequently can be found in Mintzberg (1979). His findings are well in agreement with the opinion of Geoffrion that OR activities are suffering from low productivity and poor managerial acceptance. This can partly be explained by the inability of the analysts to adopt to the way executive decision makers think and how they approach problems.

As discussed by Mintzberg, executives spend most of their time communicating verbally and relatively little time reading. The analysts are used to stating problem formulations in writing and in a "foreign" language, the language of OR. Analysts have to learn how to communicate with decision makers and other persons concerned. Graphical representations of relationships and data have proven to be important tools for communication. It is striking how easily some OR models are comprehended by decision makers when presented graphically. This is also true for PERT/CPM models.

Consider a construction company. According to OR proponents chances are that it would profit from using the PERT/CPM type of planning tools. The basic ideas of PERT/CPM are relatively easy to explain with the aid of graphics. To illustrate a real world problem, just a few concepts are needed such concepts as nodes and arcs and what they represent. This is probably the reason why this type of technique is used. It may be doubted whether PERT/CPM would be as easily accepted if phrased in structured modeling vocabulary.

What makes PERT/CPM acceptable to managers is its network flow structure, which makes it easy to understand conceptually. The possibility of using graphical representations similar to regular bar charts makes it even more acceptable to managers.

In the next two sections two graphical approaches will be discussed. First, the balloon graph approach which is simple and informal and particularly suitable for overall discussions with executive decision makers. The value of such informal approaches should not be underestimated. Second, a graphics approach suitable for the development of linear programming models of manufacturing and distribution processes is presented. This latter approach is more formal, but has proven to be of value as a means for communication, model formulation, model manipulation and documentation.

3. BALLOON GRAPHICS

Assume that an analyst intends to convince a manager that inventories should be trimmed. In order to do so he may present a genus graph of the same type as in Geoffrion (1987, p. 565) which is supposed to be suitable for managerial communication. It is conjectured that such a graph is not suitable for convincing any manager of the blessings of trimming inventory at an introductory stage. As an alternative the balloon graphics approach is suggested.

The approach is informal and should be adjusted to the problem situation and particular decision makers involved. The balloon graphics approach is best explained by an example. Before resuming the inventory trimming problem, consider part of the problem formulation phase as a discussion stage in which all the phases of OR modeling have to be discussed in order to convey to the decision makers an overall understanding of the problem solving process from problem formulation to implementation. During this stage the analyst also has to acquire a corresponding overall picture of the problem. Consider this discussion as a five step process.

Step 1: In order to be able to carry out a dialogue an appropriate framework for a discussion has to be developed. The role of the analyst is to stimulate the decision maker to think in analytical terms and to express himself verbally. This means that the pros and cons of trimming inventory have to be documented in plain English in the balloons and sand bags as in figure 1 in which the pros are the balloons and the cons are the sand bags.

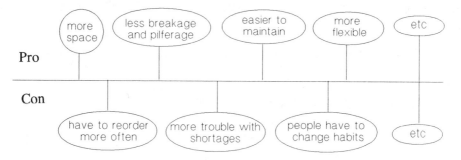

Figure 1. A balloon graph illustrating the pros and the cons for an inventory problem.

Initially the balloons and the sand bags are not drawn when communicating with the decision maker, just the line and the words pro and con or advantages and disadvantages. With the starting point in the line a structured *verbal* dialogue takes place. The result will be a balloon graph as illustrated in figure 1.

Step 2: Further *quantitative* aspects of the verbal information have to be discussed and added to the balloon graph.

Step 3: The *accuracy* of the quantitative information discussed in the previous step has to be judged and related to possible model requirements.

Step 4: The *monetary* implications are discussed in order to give the decision makers an idea of the pros and the cons of a model in monetary terms.

Step 5: Note the balloon regarding *implementation* aspects "persons have to change habits". It is important that such aspects are considered at an early stage and not postponed.

4. A GENERALIZED NETWORK FLOW APPROACH

One way to improve model building is to start with something that has been proven to work and elaborate on it. Chances are then, that managers will accept the result easier than something completely new and unfamiliar. The approach, presented below draws on network flow models and hence called a generalized network flow technique. This technique has been used in Sweden on several occasions by Holvid.

Linear programming models of industrial manufacturing and distribution processes will be regarded as generalized network flow models. The constraints in such models can be divided into two classes: 1. Balance equations and 2. Other constraints such as constraints regarding capacity, blending, and market conditions. In order to represent this type of model graphically a few graphics objects are needed and presented below in figure 2.

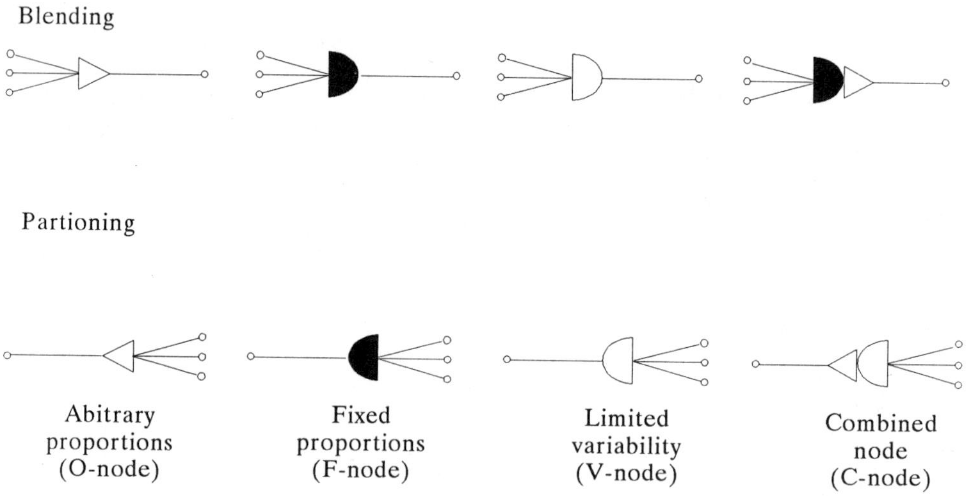

Blending

Partioning

| Abitrary proportions (O-node) | Fixed proportions (F-node) | Limited variability (V-node) | Combined node (C-node) |

Figure 2. Node symbols

Besides the obvious advantage of the generalized network technique for communication purposes it is also convenient for model formulation and manipulation. In this subsection focus will be on this latter property. For a more detailed discussion reference is made to Holvid (1977).

Consider the following simple example of a manufacturing process as illustrated in figure 3.

Example 1.

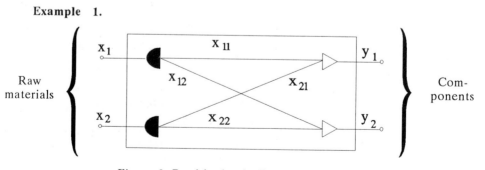

Figure 3. Partitioning in fixed proportions.

The flows x_{11}, x_{12}, x_{21}, and x_{22} between the F- and the O-nodes are determined by the flow variables x_1 and x_2. Thus the process can be described by just two balance equations corresponding to the two O-nodes as below, where a_{ij} and $i,j=1,2$ are the technological coefficients:

$$y_1 = a_{11}x_1 + a_{21}x_2 \qquad y_2 = a_{12}x_1 + a_{22}x_2$$

This kind of partitioning process is common in refinery models. Correspondingly blending processes can be described with the same type of nodes but with the flows in the opposite direction. If the blending proportions are allowed to vary within certain limits then the model will become more complex with additional constraints. It may then be advantageous to use other types of nodes such as the C-nodes in figure 2 and corresponding modeling variants.

Example 1 shows how the flow to an F-node can be represented by a single variable. Substructures within a network composed by more than two F-nodes can also be represented by a single variable. Another possibility to reduce the dimensionality of the model is by removing unnecessary balance variables and restrictions as in example 2.

Example 2:.

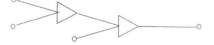 

Figure 4. The structure to the left can be reduced to the one to the right.

As mentioned above one balance equation is needed for each O-node. Thus by reducing the network structure as shown in figure 4 one balance equation and one balance variable can be eliminated. Some additional possibilities to reduce a network are shown in figure 5.

Figure 5. The structures to the left can be reduced to those to the right.

When the network structures are reduced and the corresponding mathemathical representations simplified, care has to be taken in order not to eliminate variables which also are objective function variables and/or used in other constraints.

5. CONCLUDING REMARKS

The graphics approaches outlined here have been used extensively and proven useful. The starting point has been to draw on well-known graphics approaches. The balloon graphics approach is influenced by the mind-mapping technique, although more structured, and the generalized network technique is an extension of network flow techniques.

Of central concern has been to improve communication between analysts and decision makers. However, the generalized network technique is also excellent for communication between analysts and those who provide model input, and a convenient tool for model formulation and documentation. It should also be emphasized that the graphical representation makes it easier to delegate the task of collecting model data to nontechnical persons. Note that relatively little will be gained by eliminating the process of translating a large LP model into a computer executable form in the way intended by Geoffrion (1987) considering the total effort needed.

It is the authors' firm belief that the future success of OR is not so much a matter of developing new solution techniques as developing and using means for communication. This is crucial because of OR claim to provide a set of normative tools defined for general classes of context free problems. In applications the managers and other persons concerned have to provide the analysts with contextual knowledge without having to be confronted with matrix algebra and integrals. Hence, there are good reasons to document and spread the use of informal means for communication and model building.

REFERENCES

Geoffrion, A.M., 1987, An Introduction to Structured Modeling, Management Science, Vol. 33, No. 5 pp 547-588.
Holvid, A., 1977, Towards a Unified Method for Constructing and Implementing LP-Models, unpublished working paper.
Mintzberg, H., 1979, Beyond Implementation: An Analysis of the Resistance to Policy Analysis, in: "Operations Research 78", K.B. Haley, ed., North-Holland and publishing Company, Amsterdam.

A SYNTHESIS OF DESCRIPTIVE AND PRESCRIPTIVE MODELS FOR DECISION MAKING

Osman Coskunoglu

Department of General Engineering
University of Illinois

Elke U. Weber

Center for Decision Research
University of Chicago

INTRODUCTION

Experimental psychology has uncovered many instances where decision makers consistently and persistently violate the postulates of classical normative theories. This paper reviews the main results of this literature, its implications for prescriptive approaches and resulting demands on decision aiding tools. Finally, it reviews artificial intelligence techniques that may help in meeting some of these demands.

The focus of this paper is on individual decision makers in an organizational environment who do not and should not have the liberty of violating normative rules out of personal preference or as a consequence of cognitive limitations but who, usually unknowingly, still are subject to these influences. The goal of the paper is to lay a foundation for the development of flexible decision aids that will apply in such situations.

DESCRIPTIVE MODELS -- BEHAVIORAL DECISION MAKING

The descriptive study of human decision processes originated some 35 years ago when psychologists discovered that newly developed or formalized normative theories of judgment and choice (e.g., expected utility theory or Bayesian probability revision) did not describe human performance in these situations (Edwards, 1954). The field subsequently developed as a dialectic between behavioral and normative theories, a fact that has been deplored by some (Lopes, 1986), but accepted as fruitful and productive by most (March, 1978). There are many ways in which to classify the large body of empirical results [see reviews by Edwards, 1954, 1961; Rapoport & Wallsten, 1972; Slovic, Fischhoff, & Lichtenstein, 1977; Einhorn & Hogarth, 1981; Pitz & Sachs, 1984). For this paper we will discuss the following: (i) Cognitive limitations of human information processing; (ii) Restructuring of the problem representation by the decision maker; (iii) Use of heuristics or simplifying processing algorithms; and (iv) Instability of preference structures.

Cognitive Limitations of Human Information Processing

Several limitation on human information processing capacity can be distinguished (Hogarth, 1987). Because humans cannot process the multitude of incoming information,

653

perception of information is selective. Processing usually occurs in a sequential manner or using simple procedures designed to reduce mental effort which may not always produce optimal results. The final limitation is on memory capacity. Unlike computers which can access all stored information in its original form, much of human memory works by a (less reliable) process of reconstruction. The realization of human information acquisition and processing costs and constraints underlies Simon's advocacy of "bounded rationality" (March, 1978; Simon, 1955). The next two phenomena to be reviewed, namely restructuring of the problem space and the use of heuristics, can be seen as procedures people develop over time and with experience to deal with their cognitive limitations. While adaptive and "boundedly rational" in the context of intuitive judgments, these processing styles can become so habitual or automatic that they will be applied even in important and formal decision situations where they can lead to serious biases.

Restructuring of Problem Representation

One of the basic assumptions of the classical economic model of rational choice (e.g., expected utility theory) is the requirement that choice alternatives be evaluated in terms of their effects on final assets. That is, the outcomes of choice alternatives should be combined with current assets and that alternative should be selected which provides the most desirable final asset position. Continuously updating current asset levels and integrating those with the outcomes of every new choice set requires a significant amount of cognitive effort. Therefore it is perhaps not surprising that people do not encode outcome information in this way, but instead in terms of gains or losses from the status quo or some other reference point. This is one of the central assumptions of Kahneman and Tversky's (1979) prospect theory which (in conjunction with a value function that is concave for gains, convex for losses, and steeper for losses than for gains) accounts for a wide variety of decision behavior that no version of expected utility theory can explain. The reference point used to encode a particular outcome as a gain or a loss can be manipulated by normatively irrelevant changes in context or wording, often leading to reversals of preference between two choice alternatives. These "framing effects" are striking and robust, occur in natural environments, and for experts as well as naive decision makers (McNeil, Pauker, Sox, & Tversky, 1982; Tversky & Kahneman, 1981).

Use of Heuristics

The use of heuristics (i.e., simplified processing rules that provide the correct answer most but not all of the time) has been found extensively in people's judgments of the likelihood of uncertain events (Tversky & Kahneman, 1974). In making such judgments, people over time and experience learn certain regularities in their environment. One such regularity is the fact that similarity is an index of class membership. Estimating the probability that object A belongs to class B on the basis of A's similarity to B is an application of the representative heuristic. Another regularity is the fact that more probable events occur more frequently and thus produce more memories. Estimating the probability of an event by the ease with which instances or occurrences can be brought to mind is an example of the availability heuristic. These heuristics often lead to biased probability judgments because representativeness and availability are diagnostic but not exclusive determinants of probability. Use of the representative heuristic, for example, can lead to the neglect of baserates or prior probabilities. Use of the availability heuristic can lead to overestimation of easily imagined events and to the perception of illusory correlations.

Another apparent regularity of our natural environment is some dependence between the desirability of events and their frequency. Truly wonderful and truly awful events happen rarely, whereas regularly occurring events seem ordinary almost by definition. Thus it is perhaps not surprising that people violate the independence assumption underlying expected utility theory. Violations of the axioms of expected utility theory are extensive and have been reviewed elsewhere (Schoemaker, 1982; Slovic & Tversky, 1974; Keller, 1985). A large and steadily growing literature of alternative models of choice under uncertainty attempts to describe such choice behavior (Weber & Camerer; 1987). Some of these models re-specify the expected utility problem as a multiattribute case with another evaluative dimension such as "regret" (Bell, 1982) or "risk" (Coombs, 1975) in addition to the outcomes' utility. Others weaken one or more of the (subjective) expected utility axioms (Kamarkar, 1978; Machina, 1982). Several of these models explicitly introduce dependencies between subjective probabilities and utilities. In Luce's (1989) rank-dependent subjective expected utility theory, the most general model to yield an interval-scaled utility function, the subjective probability weight functions depend on the rank-order of outcomes. In Becker and Sarin's (1987) lottery-dependent utility theory, the utility function of outcomes depends on the lottery, i.e., on the probability distribution with which the outcomes occur.

Instability of Preference Structures

Another assumption of the classical economic model of rational choice is that of procedure invariance, especially with respect to the elicitation of a decision maker's preference space. That is, his or her preference order for a set of outcomes or alternatives should not depend on the particular method by which it is assessed. However, empirical evidence is beginning to accumulate which suggests that people often do not have stable and well-defined preferences (Fischhoff, Slovic, & Lichtenstein, 1980; Grether & Plott, 1979; Shafer & Tversky, 1985). In such situations, judged or revealed preference is not a reflection of the "true" internal preference structure, but is actually constructed during the elicitation process. Different elicitation procedures highlight different aspects of decision alternatives and may suggest different heuristics or different decision frames, thus giving rise to inconsistent responses (Tversky, Sattath, & Slovic, 1988).

PRESCRIPTIVE MODELS AND TECHNIQUES -- DECISION AIDING

Prescriptive techniques (e.g., decision trees or linear programming) employed in decision aiding are based on normative theories of decision making. Psychological research has raised profound questions about the validity of these normative theories, hence calling "... the foundations of choice theory and decision analysis into question" (Tversky, Sattath, & Slovic, 1988). In particular, a decision aid is developed based on three normative assumptions: Given a decision problem, the decision maker should be (a) able to articulate his preference structure in order to evaluate relative merits of alternative solutions or choice alternatives, (b) be able to assess the probabilities of uncertain events using his or her problem domain knowledge, and (c) select the choice alternative that maximizes (subjective) expected utility. The bad news is that, as reviewed in the previous section, empirical studies of decision making have uncovered human heuristics, biases, and limitation that violate every one of these normative assumption.

Experimental results demonstrating systematic and persistent deviations of human behavior from normative standards have created much controversy (Schoemaker, 1982). From the prescriptive perspective the results are, at one extreme, sharply rejected as "... unwarranted generalizations from unrepresentative experiments" (Phillips, 1983, p. 537), and, at the other extreme, simply ignored (Keeney, 1982). For practioners developing decision aids, there seem to be two relevant points of view. One is that human cognitive limitations and biases, even if they exist outside of the laboratory, do not have any significant effect on the quality of decisions (Christensen-Szalanski, 1986). Little empirical evidence is provided for this position, and the instances cited seem to refer to circumstances with a flat criterion maximum.

In contrast, the other viewpoint acknowledges potential errors and inconsistencies in decision makers and advises the analyst to deal with them prudently (von Winterfeldt &

Edwards, 1986; Watson & Buede, 1987). For example, von Winterfeldt and Edwards view the decision maker's errors and inconsistencies as "... an asset rather than a liability [which] forces both the analyst and the client to think hard and provides them with an opportunity to gain insights into the decision problem" (1986, p. 385). Furthermore, they perceive some non-normative behavior as "creative stress" between the demands of a decision model and human intuition.

This viewpoint seems overly idealistic. It is neither fair nor realistic to expect an analyst to have the expertise and experience that Edwards and von Winterfeldt have in dealing with such conflicting situations without the support of some operational rules from a sound theory. Indeed, the authors of this paper have observed analysts in practice performing blatantly erroneous analyses in order to release the "creative stress" through oversimplistic approaches rather than "thinking hard." After all, practioners, despite their professional training and ethics, operate under their own limitations, biases, and utilities. It has been suggested that decision analysis is a clinical skill even under normal circumstances and one that should only be practiced after an internship with an expert (Brown, Kahr, & Peterson, 1974). Moreover, even if decision makers and analysts judiciously and prudently think hard, how can they detect the violation of, for example, the independence axiom of EU? If they do, how should they proceed? These, and a host of other questions necessitate the development of sound theoretical principles and methodological tools.

Implications of non-normative behavior

What then is the impact of non-normative behavior of a decision maker on decision aiding tools? A prescriptive approach to decision aiding goes through four principal stages: (i) problem formulation, (ii) solution, (iii) post-solution analysis (e.g., sensitivity analysis, reiteration, etc.), and (iv) implementation (i.e., actual execution of the solution). The formulation stage takes into consideration the nature of the problem and may lead to a representation of the problem in form of a decision tree or a linear program. Formulation can further be subdivided into the following three components: identifying the variables, options, parameters, and objectives; establishing the relationships between them (e.g., constraints in a linear program, consequences of options and their probabilities in a decision tree); and determining the preference (value) structure of the decision maker (i.e., the objective function, composed of a multiattribute utility function, to be maximized). Potentially, all of these steps in the prescriptive procedure can be affected by non-normative behavior. Elicitation of probabilities and utility assessment, for example, can be affected by certainty and framing effects (among others). The certainty effect, in turn, violates the independence principle and hence the solution procedure of folding back the decision tree. This suggests two, complementary, approaches to solving the problem: the first one is to change the non-normative behavior and the second to modify the decision tools.

Rectifying non-normative behavior

The key factors in rectifying any non-normative behavior are (a) to anticipate the occurrence of such behavior, (b) to detect it, and (c) to make it explicit to the decision maker and others concerned. In order to appreciate the need for such procedures, one should recognize that an individual operating within an organizational setting cannot knowingly violate rules that are normative from the organizational perspective. For example (adapted from Tversky & Kahneman, 1981), a public health official may choose an immunization program that guarantees to save a particular number of lives in a population at risk over another program that offers some less than certain chance at saving an even larger number of lives in this population when the effects of the two immunization programs are presented in terms of lives saved, but may reverse his or her preference when the identical programs are described in terms of lives lost. From a public policy perspective, such inconsistency is unacceptable, and the only criterion to decide between the two immunization programs (all other things being equal) probably should be the expected number of lifes saved (i.e., the final asset position) which is identical under the two formulations. In their personal choices, decision makers may or may not want to represent alternatives in terms of their final asset position. However, in dealing with an organization's assets they should not have this latitude. In such situations, decision makers need to be reminded and encouraged to use a final results perspective. (Machiavellian decision analysts, aware of the power and mechanisms of "framing effects," can of course also employ decision frames in such a way that the public official or employee will make decisions in line with the policies of his or her organization.)

Cases where a decision maker insists on violating a normative principle knowingly raise the question of whether some important factor has been overlooked in the formulation of the problem. For example, a quality control engineer may find a particular part out of tolerance after the machining operation. The part may not necessarily be defective, but its defectiveness will be revealed only in the actual assembly process. Should the engineer accept the part and send it to the assembly line, taking the risk of an expensive revelation of the defect should one exist, or should she scrap the part? The normative answer would, of course, depend on the probability that a part registering out of tolerance is actually defective as well as on the cost of revealing the defect on the assembly line. However, the quality control engineer will most likely insist on scrapping the part even in situations where this decision has smaller expected utility than the other alternative. Situations where a retained part turns out to be defective are not only costly, but also constitute an identifiable and visible error on her part. Regret theory or some other multi-attribute representation that incorporates the cost of making a wrong decision or accountability could perhaps explain the decision of the quality control engineer. However, the goal may not be to predict or justify her decision by some formal model, but to guide the decisions towards some organizationally acceptable standards. Hence, a decision tool should, in addition to traditionally desired qualities, be able to remove any contextual biasing effects or to make them explicitly known to all concerned if the decision maker insists on his or her non-normative behavior. The latter case may, in fact, reveal a factor not considered in the original problem formulation which is of sufficient normative appeal to be included in future versions of the prescriptive model.

Similar arguments can be made for decision tools designed to elicit probability judgments. Awareness, understanding, and anticipation of the heuristics used to make such judgments and the conditions under which these heuristics will lead to biases, can actually help prevent their occurrence by suggesting effective countermeasures. Thus it has been shown, for example, that base rate neglect as a function of use of the representative heuristic can be significantly reduced by explicitly emphasizing the causal connections between events (Bar-Hillel, 1980).

MODIFICATION OF DECISION TOOLS

Existing decision tools (e.g., procedures or algorithms such as the simplex or decision tree analysis) have at least two shortcomings: they are rigid and they are opaque. If a single data point is missing in a linear program, the program will not run. A structural sensitivity analysis on a decision requires the tedious task of restructuring the complete tree all over. Actually, in existing decision aids that employ a conventional hierarchical programming structure, even a small structural modification is rather tedious (for an example see Lehner, Probus, & Donnell, 1985, p. 471). Furthermore, most tools can not explain, let alone justify, the reasoning process which led to the decision.

If a decision tool is to be used to aid a potentially non-normative decision maker, what type of qualities must it have? The first point is a technical one. Existing decision analysis techniques can not handle the complexities (e.g., dependency between utilities and probabilities or nonlinearities) introduced by some extended utility theories which relax the substitution principle. A preliminary effort towards developing an alternative to the standard decision-tree fold-up procedure has been reported recently (Sarin, 1989). The second point is more encompassing, namely the need to increase the effectiveness of existing tools through increasing their domain of application and their range of functions. The former necessitates relaxing the rigid data and assumption requirements and the latter involves introducing reasoning and interactive explanatory capabilities (the latter providing decision makers with the information necessary to justify the decision taken). Both require a knowledge base, an inference mechanism, and a set of procedural primitives (e.g., modus ponens for symbolic manipulation, a matrix inversion algorithm for numerical computations).

Coupling Symbolic and Numerical Computation

Developing the knowledge base and inference mechanism technology as well as theory necessary for future decision tools is one goal of the field of artificial intelligence (AI). Consequently, the potentials of AI techniques have recently received much attention (von Winterfeldt & Edwards, 1986, pp. 23-25; Lehner, Probus, & Donnell, 1985; Hammond, 1987, Henrion, 1987; Keeney, 1987) amidst some cautionary remarks (Doyle, 1985; Sutherland, 1986) warning against excessive optimism.

The first author is currently developing procedures to couple the symbolic reasoning capabilities offered by AI techniques and languages with the numerical computation techniques of decision tools. Three aspects of this research are briefly discussed. To obtain anticipatory behavior (some form of "intelligence") from an artifact (e.g., a decision tool), the artifact needs to have knowledge and beliefs about its environment. Three types of knowledge can be distinguished: knowledge about the problem domain, about the algorithmic tools, and about behavioral decision theories. These types of knowledge correspond roughly to the three components of decision analysis, namely the decision problem, the decision tools, and the decision maker. The reasoning process employed in the current models is deductive reasoning (Genesereth & Nilsson, 1987). Finally, the models make use of meta-level knowledge and reasoning (Aiello, Cecchi, & Sartini, 1986; Genesereth & Nilsson, 1987, Ch. 10) with reflection capabilities (Genesereth & Nilsson, 1987, pp. 255-261), which can account for formalization of beliefs, default reasoning, inference in changing situations, and reasoning about the reasoning process.

The goal of the research is not to develop a new decision tool per se, at least not in the short run, but to investigate the capabilities of techniques currently available in AI, to create demands for new techniques, and to lay the foundation for a computational theory of decision tools which encompasses both symbolic and numerical processes.

SUMMARY AND CONCLUSIONS

Behavioral scientists have recently been providing experimental evidence on the decision behavior of individuals and have extended classical normative theories of decision making to account for such behavior. This has created a need for decision aids significantly different from existing ones. One way of bridging the gap between necessary (or desirable) and existing decision tools is to leave this task to the skill and imagination of the decision analyst. However, we are more inclined to agree with the statement by the 16th-century Italian philosopher Vico: "Certum quod factum," i.e., one is certain [only] of what one builds. Hence our desire to strengthen and extend the capabilities of existing decision tools. We have been looking for additional tools, techniques, and theories from artificial intelligence, a field currently much interested in discovering how humans deal with certain problems and what problem solution procedures can be automated.

This closes the circle between psychology, operations research, and artificial intelligence. Coordination of research in these three areas has great promise for significantly advancing the following three global research issues in the theory of decision aiding:

(i) How should a decision task been divided between the decision maker and a decision tool (Pitz, 1983)?

(ii) Where in the decision process should the decision maker be aided and how (Humphrey, 1986)?

(iii) What are the fundamentals of a computational theory for developing decision tools?

REFERENCES

Aiello, L., C. Cecchi, & D. Sartini, "Representation and Use of Metaknowledge," Proceedings of the IEEE, 74, 1304-1321, 1986.

Bar-Hillel, M., "The Base-Rate Fallacy in Probability Judgments," Acta Psychologica, 44, pp. 211-233, 1980.

Becker, J. L. & Sarin, R. K., "Lottery dependent utility," Management Science, 33, 1367-1382, 1987.

Bell, D. E., "Regret in Decision Making under Uncertainty," Operations Research, 30, pp. 961-981, 1982.

Brown, R. V., A. S. Kahr, and C. Peterson, Decision Analysis for the Manager, Holt, Rinehart & Winston, New York, 1974.

Christensen-Szalanski, J. J. J., "Improving the Practical Utility of Judgment Research," in: New Directions in Research on Decision Making (Eds.: B. Brehmer, H. Jungerman, P. Lourens, and G. Sevon), pp. 383-410, Elsevier Science, New York, 1986.

Coombs, C. H., "Portfolio Theory and the Measurement of Risk," in: Human Judgment and Decision (Eds.: M. F. Kaplan and S. Schwartz), pp. 63-68, Academic Press, New York, 1975.

Doyle, J., "Expert Systems and the 'Myth' of Symbolic Reasoning," IEEE Transactions on Software Engineering, SMC-11, 1286-1390, 1985.

Edwards, W., "The Theory of Decision Making," Psychological Bulletin, pp. 380-417, 1954.

Edwards, W., "Behavioral Decision Theory," Annual Review of Psychology, 12, pp. 473-498, 1961.

Einhorn, H. J. and R. M. Hogarth, "Behavioral Decision Theory: Processes of Judgment and Choice," Annual Review of Psychology, 32, pp. 53-88, 1981.

Fischhoff, B., P. Slovic, and S. Lichtenstein, "Knowing What You Want: Measuring Labile Values," in: Cognitive processes in choice and decision behavior, (Ed.: T. S. Wallsten), Erlbaum, Hillsdale, NJ, 1980.

Genesereth, M. R. & N. J. Nilsson, Logical Foundations of Artificial Intelligence, Morgan Kaufman, Los Altos, California, 1987.

Grether, D. and C. Plott, " Economic Theory of Choice and the Preference Reversal Phenomenon," American Economic Review, 69, pp. 623-638, 1979.

Hammond, K. R., "Toward a Unified Approach to the Study of Expert Judgment," in: Expert Judgment and Expert Systems, NATO ASI Series, Vol. 35, 1-16, Springer Verlag, Berlin, 1987.

Henrion, M. & D. R. Cooley, "An Experimental Comparison of Knowledge Engineering for Expert Systems and for Decision Analysis," Proceedings of AAAI-87, Seattle, Washington, 1987.

Hogarth, R., Judgement and Choice, 2nd edition, Wiley, New York, 1987.

Humphrey, P., "Intelligence in Decision Support," in: New Directions in Research on Decision Making (Eds.: B. Brehmer, H. Jungerman, P. Lourens, and G. Sevon), pp. 333-361, Elsevier Science, New York, 1986.

Kahneman, D. and A. Tversky, "Prospect Theory: An Analysis of Decisions Under Risk," Econometrica, 47, pp. 263-291, 1979.

Kamarkar U. S., "Subjectively weighted utility: A descriptive extension of the expected utility model," Organizational Behavior and Human Performance, 21, 61-72, 1978.

Keeney, R. L., "Decision Analysis: An Overview," Operations Research, 30, pp, 803-838, 1982.

Keeney, R. L., "Value-Driven Expert Systems for Decision Support," in: Expert Judgment and Expert Systems, NATO ASI Series, Vol. 35, 155-171, Springer Verlag, Berlin, 1987.

Keller, L. R., "The Effects of Problem Representation on the Sure-Thing and Substitution Principles," Management Science, 31, 738-751, 1985.

Lehner, P. L., M. A. Probus, & M. L. Donnell, "Building Decision Aids: Exploiting the Synergy Between Decision Analysis and Artificial Intelligence," IEEE Transactions on Systems, Man, and Cybernetics, SMC-15, 469-474, 1985.

Lopes, L. L., "Aesthetics and the Decision Sciences," IEEE Transactions on Systems, Man, and Cybernetics, SMC-16, pp. 434-438, 1986.

Luce, R. D., "Rank-Dependent, Subjective Expected Utility Representation. Decision Sciences, in press, 1989.

Machina, M. J., "'Expected Utility' Analysis Without the Independence Axiom," Econometrica, 50, pp. 1069-1079, 1982.

March, J. G., "Bounded Rationality, Ambiguity, and the Engineering of Choice," Bell Journal of Economics, 9, pp. 587-608, 1978.

McNeil, B. J., S. G. Pauker, H. C. Sox, and A. Tversky, "On the Elicitation of Preferences for Alternative Therapies," New England Journal of Medicine, 306, pp. 1259-1262, 1982.

Phillips, L. D., "A Theoretical Perspective on Heuristics and Biases in Probabilistic Thinking," in: Analyzing and Aiding Decision Processes (Eds.: P. Humphreys, O. Svenson, and A. Vari), pp. 525-543, Elsevier Science, New York, 1983.

Pitz, G. F., "Human Engineering of Decision Aids," in Analyzing and Aiding Decision Processes (Eds.: P. Humphreys, O. Svenson, and A. Vari), 205-221, Elsevier Science, New York, 1983.

Pitz, G. F. and N. J. Sachs, "Judgment and Decision: Theory and Application," Annual Review of Psychology, 35, pp. 139-163, 1984.

Rapoport, A. and T. S. Wallsten, "Individual Decision Behavior," Annual Review of Psychology, 23, pp. 131-175, 1972.

Sarin, R. K., "Analytical Issues in Decision Methodology," in: Decision and Organizational Theory (Ed.: I. Horovitz), Kluwer-Nijhoff (1989).

Schoemaker, P. J., "The Expected Utility Model: Its Variants, Purposes, Evidence and Limitations," Journal of Economic Literature, 20, pp. 529-563, 1982.

Shafer, G. and A. Tversky, "Languages and Designs for Probability Judgments," Cognitive Science, 9, pp. 309-339, 1985.

Simon, H., "A Behavioral Model of Rational Choice," Quarterly Journal of Economics, 69, pp. 174-183, 1955.

Slovic, P., B. Fischhoff, and S. Lichtenstein, "Behavioral Decision Theory," Annual Review of Psychology, 28, pp. 1-39, 1977.

Slovic, P. and A. Tversky, "Who Accepts Savage's Axioms?" Behavioral Science, 19, 368-373, 1974.

Sutherland, J. W., "Assessing the Artificial Intelligence Contribution to Decision Technology," IEEE Transactions on Systems, Man, and Cybernetics, SMC-16, 3-20, 1986.

Tversky A. and D. Kahneman, "Judgments Under Uncertainty: Heuristics and Biases," Science, 185, pp. 1124-1131, 1974.

Tversky, A. and D. Kahneman, "The Framing of Decisions and the Psychology of Choice," Science, 211, pp. 453-458, 1981.

Tversky, A. S., S. Sattath, and P. Slovic, "Contingent Weighting in Judgment and Choice," Psychological Review, 95, pp. 371-384, 1988.

von Winterfeldt, D. and W. Edwards, Decision Analysis and Behavioral Research, Cambridge University Press, Cambridge, UK, 1986.

Watson, S. R. and D. M. Buede, Decision Synthesis, Cambridge University Press, Cambridge, UK, 1987.

Weber, M. & C. Camerer, "Recent Developments in Modelling Preferences under Risk," OR Spectrum, 9, 129-151, 1987.

WORK

WORK

Vincent Dégot

Research Engineer with the Centre de Recherche en Gestion

Ecole Polytechnique

The papers assembled for this session illustrate various aspects of what might called the "dislocation" of the Operational Research school. This term is intended to convey the idea that OR is no longer the monolithic entity, rooted in specific theoretical concepts and occupying its own territorial reserves, conceived of by its proponents several decades ago; but that it has been taken over in piece-meal fashion by the "vertical" practitioners of business management, each concerned with his own sectorial pre-occupations of a directly relevant kind.

Before going on to describe how the various papers contribute to this view, I would like to make a few observations of my own concerning the concept of Operational Research as I have encountered it in the course of studying, practicing and writing about, as well as teaching, organizational management.

From the very outset, there appears to be some conflict between the words "operational" and "research", which respectively suggest a practical dimension and a more abstract one, leaving us to wonder which of them prevails. The OR manuals dealing with traffic-flow and path-routing optimization formulas and the like incline us to opt for the former; but those explaining graph-theory, linear programming or the "Hungarian" and Ford-Fulkerson algorithms weigh in favour of the latter. This internal contradiction can be seen as one which is central to, and a historical constant of, the OR approach developed during World War Two as an instrument of the military logistics of the Allied invasion of Europe.

This having been said, we must nevertheless remenber that the founders of the discipline gave it the less ambiguous name of "Operations Research" - research to assist operations but not operational in itself - under which it should more properly be discussed. The semantic quibble thus applies solely to the modern, "adulterated" label now given widest currency, but remains relevant to present-day discussion.

It seems to one who, like myself, has for many years past worked outside the strict field of OR, that the latter has flourished mainly along the lines where it has found outlets for this actively operational vocation, i.e. in applications lending themselves to this twist of language. This is typical of all the modern research disciplines - those which are not founded on a body of tried-and-tested theoretical principles confirmed down the ages and of quasi-universal application: their only tenets are those imposed by "scientific fashion" (or, even, the whims of individual practitioners), itself highly sensitive to some so-called sense of social responsibility - whatever that may mean at any given time.

The corporation was seen as comprising a set of problems, contexts and resources particularly relevant to the endeavours of the OR specialist. In the early sixties, this idea was taken up in the areas of production management (optimisation of resources), project management (PERT models) and general business administration.

During that period, whose culmination I experienced at the end of the Sixties, Operational Research was closely associated with the idea of optimisation which is itself associated with a corporate model viewed in micro-economic terms, seeking to apply mathematical devices (investment criteria, input-output tables, etc) to operational ends.

If we return to the theoretical concepts, leaving aside the stochastic tools and the graph-theory adjuncts, we come back to linear programming as OR's bed-rock foundation. Most of its specific algorithms are merely adaptation of LP techniques to match against particular problems. Linear programming forms a further link to the micro-economic and optimisation approaches. We all recall those programming monstrosities, with their thousands of variables and tens of thousands of in-built constraints, the 1000x1000 matrix tables which the computer of the day took many hours to process.

It is clear why this OR approach was consonant with unwieldy and centralised computer systems, and also why the optimisation procedures could not be run frequently: the aim was to define each given paradigm so thoroughly that a stable and durable solution could be produced. This explains two characteristics of Operational Research as thus practiced:
- firstly, it was particularly successful in connection with repetitive processes (refinery scheduling, transport routing, factory lay-out organisation, etc);
- secondly, it implied a relatively static model of the corporation, in which optimisation measure remained effective over fairly long periods of time.

These characteristics are not, of course, inherent to the basic principles of OR, but derived from the computer capabilities available at the time and from the comprehensive formulation of the problems to which it was applied. It can well be imagined that certain computer runs (concerning only inventory movements, as in stock-control systems, and not capital investment decisions) may have been treated on a daily routine, but not very easily if several hours of computer time had to be reserved for them.

One area which lies rather outside this rough definition of OR scope, and which has always been something of a marginal adjunct to it, is that of game theory (taught in France as part of the same curriculum as OR). This technique attracted considerable interest during the Seventues, when strategic planning came to the fore, but its real impact is difficult to evaluate.

Let me now give a few brief example of the limitations met by the attempt to make OR truly operational, as I encountered them some twenty years ago:
- the first example concerns the development of a typological system based on the identification of maximally auto-stable sub-sets within a graph. The accepted method for doing this consisted at the time in discriminating by separation and evaluation of tree-graph components. The alternative developped by a colleague and myself was based on the gradual construction of a graph (with the branches classified in terms of lenght). The result was obtained much more rapidly and, above all, it was possible to produce simultaneously a printout of the graph with a minimum of maximal sub-sets, and to investigate the thresholds where the number of these remained constant, thus indicating the stable components of the whole. However, for a graph comprising some thirty points and despite a corresponding time gain in the ratio of 1 to 100, about 1 hour CPU was required for processing it on an IBM 360/40. This was a relatively abstract research project: although it was initiated in response to a practical problem within a real-life corporation, it was subsequently extended beyond the duration of that particular assignment (Dégot and Hualde, 1975);
- the second example concerns more specifically a request from the French subsidiary of Nestlé to organise the rounds of its representatives. This clearly called for consideration of the route schedulling method according to Little. In the case in point, the duration, the purpose, the timing and the frequency of visits were all variables. This floating demand side, together with the existence of a large number of constraints which were not entirely foreseeable, led us to the conclusion that Little's algorithm was

oversimplified in relation to the facts of the case and could therefore not be applied to it. We were impelled to "cobble together" a method leaving well asside the ideas of optimisation and of OR more generally (apart from the fact that it allowed for the natural tendency of the representatives to optimise their own rounds), but which finally resulted in a mileage gain between 15 and 20 percent as compared with the earlier sitiation (Dégot and Maillard, 1972).

The inference that can be drawn from these two isolated examples are:
- firstly, that OR methods are often designed to resolve only those problems which have very simple and static configurations;
- secondly, that the effectiveness of the OR truly operational activity depends closely on either the capacity or the flexibility (micro-computers) of the computer facilities available.

These factors lead to an attempt to solve the problems posed at a single stroke, meaning that the optimum solution must be based on a vast array of identified constraints. However, the choice of, and the weighting given to, these constraints are often the result of highly "political" options within the corporate context, to the extend that they closely depend on the positions of the deciders, who may place greater or lesser emphasis on such issues as the value of time, the benefits of inventory control, or the costs of delivery shortfalls.

The added flexibility given by decentralised and high-powered data processing facilities assures the decision-makers that they are not entering into long-term commitments whose effects could be to their disadvantage. My experience of the findings of research bodies with which I have worked shows that many of the difficulties involved in the application of OR methods stem from the fact that these are based on accepted models of highly conflictual situations: either these models implied an excessively sharp exposure of the limits of compromise, or they portrayed situations which the people concerned refused to acknowledge as accurate.

These difficulties, seemingly intrinsic to a given stage of OR development, became even more acute with greater decentralisation and flexibility of management control, and as the model of the blindly efficiency-orientated corporation came to be replaced by that of a strategy-concious and "thoughtful" one, Three examples illustrating this evolutionary trend are given detailed treatment in the papers presented in this session.

Rather that attempting to summarise those papers, I prefer to propose a demonstration of their common themes as I see them:
- they on the one hand refer to the problem which, for various reasons such as limlited computer time and the random behaviour of some parameters, cannot be handled by the OR approach to produce solutions in practical terms. In these cases, the alternative adopted consisted in developing new methods - sometimes including OR-derived components but mainly relying on the flexible and decentralised processing facilities offered by micro-computer systems working on a real-time basis. These are the exemples which will be presented under the heading of project management and monitoring;
- on the other hand, they cover management areas which had thus far escaped systematic investigation, such as the management of human, resources, where some researchers have deployed fragment of OR methods and procedures which have now become no more than components of an overall and different system.

It was with these two typical development in mind that the term "dislocation" was used in connection with the OR approach at the beginning of this address: the specific requirements brought out by practical problems were seen as more important than the general principles embodied in theoretical models. Operational research thus became merely an instrument, in the same way as electronic processing, rather than surviving as the basic philosophy of scientific management which some of its early sponsors considered it to be.

In other words, Operational Research could not gain recognition in the corporate world as a specialist function with its own experts (as are the marketing and financial sides), or in the academic world as a separate discipline, due to the failure of the rationalistic corporate model (which alone justified the intensive and successful implementation of OR methods) to survive the winds of management change. By identifying itself to closely with the corporation, which provided not only a field of observation and experiment, but also in many cases the financial support for research projects, the OR school became a tool of corporate pragmatism and was "carved up" to fit into the slots allocated to it by managements.

REFERENCES

Dégot, V. and Hualde, J.M. ,1975, De l'utilisation de la notion de "clique" (sous-graphe complet symétrique) en matiére de typologie de   population,Revue Française d'Automatique, Informatique et Recherche Opérationnelle, janvier 1975

Dégot, V. and Maillard, R., 1972, "L'organisation de la force de vente des produits Nestlé", Publication de l'Ecole des Mines de Paris

QUALITY OF WORKING LIFE AND ORGANISATIONAL EFFECTIVENESS

Reg Sell

Work Research Unit
Advisory, Conciliation and Arbitration Service
London

INTRODUCTION

This paper is written from the standpoint of the Work Research Unit (WRU). This Unit was set up together with a Tripartiate Steering Group within the Department of Employment in 1974 because of its concern that the United Kingdom was not doing enough to combat the spread of boring monotonous jobs such as those in car and electronics assembly.

At that time the prime emphasis was on quality of working life (QWL) and job satisfaction. However no organisation would be interested in pursuing a QWL line which had a neutral or adverse affect on effectiveness. Now the aims of the WRU are broadly put as "to work with organisations to help them increase their effectiveness through paying more attention to QWL."

During the time since the WRU was set up there have been some fundamental changes to its orientation. One has been a move in focus away from the shop floor to cover jobs at all levels in organisations. Expectations of having an 'good' job increase as you move up organisations and so those higher up are more likely to be dissatisfied. The other has been the increased emphasis on the management of change. In the early days it was, to some extent, implicit that you could say to a worker 'I am an expert, you have a rotten job, I will redesign it'. That did not do much for the individual's self respect. There is now a greater realisation that if someone is doing a job without complaint then they have a psychological contract to do it in that way and you should not break contracts unilaterally. This paper looks at the principles and practice of QWL and the management of change.

Following a review of the Unit by Professor Tom Lupton of Manchester Business School the WRU was transferred to the Advisory, Conciliation and Arbitration Service in 1985 and is now part of its Head Office Advisory Services. It works with organisations and provides information on the kinds of activities outlined here. It publishes a bi-monthly Notes, News and Abstracts and a series of Occasional Papers.

From the great number of research studies carried out across the world we can list a number of principles which should be considered in designing jobs.

Variety.   Jobs should consist of a number of different, but related, tasks. Short-cycle or repetative tasks should be avoided. People should not be fixed in one spot or work continually at the same pace.

Autonomy.   People should have some say in how their work is organised and some control over their pace.

Identity.   The tasks an individual does should fit together into a coherent whole. Ideally they should be able to make a complete article or service completely a group of customers.   If that is not possible they should have a good understanding of how what they are doing fits into the whole.

Feedback.   It should be possible to know how well you are doing at your job directly without your boss having to tell you. This avoids the problem of 'only being as good as your last mistake'.

Responsibility.   People should be given an area of work for which they are responsible. If in manufacturing, they are able to do their own inspection, sign off the finished article and despatch it to the customer, this will give more responsibility than passing it to an inspector.

Social contact.   Most people like to be able to inter-act with others as a part of the job and not just during meal breaks.

Achievement.   It is desirable that people go home at the end of the day thinking they have achieved something that is of benefit to the organisation.

Opportunities for learning and development.   Although this is not an easy demand to satisfy directly in many jobs it is often possible to achieve off-line, e.g. using people in task forces or as trainers etc.

Optimum role load.   Both over and underload should be avoided.

Role conflict and ambiguity.   Where these are present stress is more likely. They should therefore be avoided as much as is possible.   They occur when there is uncertainty as to how the individual is being judged (eg by production output, quality or plant safety) and when there are doubts as to the limits of responsibility (e.g. what actions will or will not be supported by the boss).

It is not always possible to design individual jobs such that these principles are fulfilled. For instance it is often difficult to have one person making a complete article because of the complexity. It is, however, often possible to achieve the principle with teams, where people can share tasks so that there is versatility and each person can work in a number of roles. The teams can be semi-autonomous with no appointed leader in which the various leadership roles are rotated or taken on by different people. Alternatively, the team can work with a traditional supervisor. One of the best examples of teamwork is that of Volvo's Kalmar plant in Sweden and this concept will be developed even further in their new Uddevala plant.

If it is not possible to get intrinsic job satisfaction from the job itself, perhaps because of captial expenditure resrictions, then extrinsic measures have to be taken. One way this can be done is by putting people into problem solving groups of the quality circle type or into cross-functional task forces to look at particular problems which relate to several departments. These measures, which can be valuable developmental exercises, can only be really successful if management is prepared to listen and respond to the ideas produced.

It is obvious that there are many jobs in industry which do not fulfil these criteria but there is increasing awareness of the need to make jobs more interesting, to build up commitment and involvement so the knowledge which people have will be released for the benefit of the organisation.

In achieving these ends probably the most important factor which influences what an organiation does is the value system of the top management. If top management really care about the people they employ then they are more likely to create policies which support involvement, participation and good job design. It has to be said, however, that it is easy to expound these principles and exhort managements to take action but it is very difficult to carry the policies through in practice, especially when, for example there are pressures for production.

THE BENEFITS

Traditionally, the benefits of a QWL approach have been seen to be the making of jobs more attractive and so reducing absence, labour turnover, disputes etc. When unemployment is high these types of measures might seem to be less relevant. Even so high levels of absence or turnover continue to give important signals, especially when the turnover is amongst long serving people. Perhaps, with the continuing reduction in the numbers of young people entering the workforce they are likely to become more important in the future.

More interest is now being shown in how a QWL approach can influence an organisation's effectiveness. The problem is that the links are very tenuous. In assembly situations there is a risk that short-term work study based measures will be used. These will almost always come down in favour of fragmented short-cyle tasks. However, in the early 1970's Philips compared group TV assembly methods having 7 people in a group with a traditional line assembly system with around 30 people. Using long term measures including quality, flexibility, training, replacement costs etc. they showed that group assembly was the most effective.

Quality has become a major concern of industry in the last few years and modern Total Quality approaches put great emphasis on the need to involve people to get at their ideas. This requires responsibility to be pushed down the organisation.

One result which can come from pressure to push responsibility down the organisation and to develop more self-management is a reduction in the number of levels of management. This, of course, is easy in a green field site but takes time on an existing one when there is likely to be uncertainty amongst people in middle-management jobs about their future.

FLEXIBILITY AND TEAMWORK

These are two currently fashionable approaches by managements but, if they are to be successful, need to have much attention paid to the QWL needs.

A particular aim of management when introducing flexibility and teamwork is frequently to have a fast response to changing market demands with the minimum number of people. To the employee this can mean feeling that they do not know whether they are coming or going and that they are being pushed from pillar to post.

It is important to build up commitment. Initally this should be to a small team or department. The members of the team need to know what the state of the order book is for their products, to be involved in deciding how the work is to be organised, to know how well they are doing and how their product is being used by the ultimate customer.

Once commitment to the team has been established then this needs to be extended to the whole department and then to the whole organisation. This takes time and requires a lot of effort in the giving of information on departmental and company performance. Only when people have this commitment can they be moved safely between departments and be expected to maintain their performance.

Flexibility also requires a significant input in terms of training and information flow. It is, perhaps, too much to expect complete flexibility from most people. There is a tendency amongst individuals to identify with and to get committed to a particular skill, a group of people, one department. It is much more difficult to get the identity with and commitment to a large organisation.

It is important to consider to what extent flexibility can be forced, even in a small team. Should the members be forced to rotate between jobs? There is a need to ensure that people recruited to work in this kind of organisation are aware of the team work approach and are committed to it and so are willing to take on different tasks. Initially, as part of a total flexibility approach, they must learn and practice the whole range of jobs. As time goes on, however, they may settle into jobs which they like doing. This can be accommodated provided they maintain the whole range of skills so they can take over other tasks when required.

Choice and control over one's work are what we need to aim for. Those people who feel they have some control are more likely to give of their best.

MANAGEMENT OF CHANGE

To have some control is especially important in a period of change in an established organisation when the psycological contracts are all having to be re-negotiated.

This requires there to be a system of participation. This needs to start at the top of the organisation. If the top management do not listen to and involve the next level down then it is unlikely that the middle management will have any thought of involving those below. If this involvement does not take place a lot of value is lost, both in terms of gaining commitment and in achieving an optimum solution.

People who are involved will feed in information which will be of value in getting the new system right. If those who will have to work in the system are a part of the decision making process for the design of the system then they will be more prepared to live with the inevitable short-comings. If they aren't involved there will be continuous objections.

As a part of this process it will be necessary to build up reasonable levels of trust. Unless this is done people will not feel committed and will not be willing to give of their ideas. The development of trust takes time and involves disclosure of information at earlier stages than instinct dictates. It requires the carrying through of low level long requested action such as refurbishing the canteen to demonstrate that ideas will be acted upon and not lost.

CONCLUSION

It can be shown that, a QWL approach can have a major influence on organisational effectiveness. It requires a commitment by management at all levels to involve people and to consider employee needs. It can be shown that inspite of the difficulties, many UK companies, some working with the WRU, are taking a QWL/involvement approach towards increasing their effectiveness.

CAREER PLANING AND MANAGEMENT

Vincent Dégot

Research Engineer with the Centre de Recherche en Gestion

Ecole Polytechnique, Paris, France

This paper attempts to illustrate the fact that certain methods derived from Operational Research theory are sometimes incorporated in a general management study. The implication of this are that:
- the methods in question are employed because, at a given stage in the study, the paradigm took on a configuration reminiscent of the typical OR assignments where problems were first identified as falling within the OR area (transport scheduling and the like) and then approached with a view of finding the most appropriate OR solution to them;
- in consequence, OR can be brought into applications which are new to it ("new", that is to say, in terms of the text-book definition of the OR field): the example taken in this paper is that of career management, not normally seen as an OR province - despite the fact of that discipline's long-established involvement in matters relating to the management of human resources, such as the scientific allocation of task and the organisation of work (plotting the rounds of sales reps, etc).

Career Management: an expanding field

It can be said, in the French context at least, the career management vogue is both a delayed reaction to the economic crisis and an extension of the concept of better quality of working life to include the executive and professionals grades. It thus has a twofold purpose: to benefit the corporation as a whole and at the same time to satisfy the aspiration of the cadre personnel.

In a period of expansion, the corporation is able to give satisfactory short-to-medium-term career prospects to virtually all of its white collar workers, owing to a high rate of upward mobility. But, when the business cycle goes into recession, there are fewer opportunities for career development and these must be more closely administered, given that:
- fewer opportunities mean greater selectivity which, if it is to be accepted, has to be accompanied by greater transparency in management decisions, i.e. by wider notification of evaluation criteria and target profiles;
- the corporation is itself more durably and more irrevocably commited to appointments made when the rate of staff turn-around become slower.

One way in which this turn-around rate can be stimulated is by multiplying the number of horizontal transfers, thus maintaining the cadre personnels's level of "opportunity- awareness" and at the same time building up a body of middle executives with all-round experience of the corporations's business.

However, this more refined and selective management of personnel movements requires new management tools for those who are charged with administering them. A career development policy can succeed only if it is based on mutual confidence between the governing and the governed. For the proper degree of confidence to prevail in times when the very rules of the game are in process of change, with tighter selectivity in promotion procedures, the decisions made at the higher level must be patently consistent at all times and in all places. But, in the case of corporations with several thousands of executive posts, comprising people coming from various background, with greater or lesser seniority, and exercising different disciplines and skills, the number of movement involved is such that consistently strict application of the same standard principles can hardly be feasible. This is all the more true in that the personnel managers are inclined, so as to retain maximum freedom of movement within a more rarified opportunities atmosphere, to deal in terms of increasingly broader areas by grouping together different departments and services which would formely have enjoyed greater autonomy in this respect.

In addition to this underlying trend, there is also a general tendency to treat the components of personnel policy as so many vehicles for passing on corporate themes or values developed at top management level. Career development becomes a "strategic" instrument when it tends to place emphasis on a "dominant profile" of the executive which is representative of the corporate culture and image as projected to the outside world.

When principle-based career management procedures are in force, they help to maintain the desired level of consistency and to stimulate internal mobility, thereby promoting the wider exposure given to the objective principles in question. The procedures described hereunder were developed in response to this concern.

Career Management Principles and procedures

In very rough terms, it can be said that there are three consecutive stages in a career management policy, as follows:
- establishment of job descriptions corresponding to the operating requirement of the corporation concerned. As a general rule, these jobs are hierarchised and represented on an establishment chart, as well as being classified by type of activity. In the case of a large undertaking whose objectives remain fairly stable, most of the individual posts retain the same jobs descriptions for long periods, sometimes with a few slight ajustments when new projects are taken on:
- evaluation of cadre personnel performances, designed to assess not only that performance as displayed in their current posts, but also to judge their potential for future career steps, in the form either of promotion to a higher level of responsibility (managerial promise) or of transfer to another type of activity (latent skills more useful in other parts of the undertaking);
- positive staff administration, which consists in allocating - as far as possible - the right jobs to the right people and satisfying the aspirations of those with good potential (who might otherwise seek employment elsewhere), while at the same time seeing that the undertaking uses its human resources to the best effect.

The first two of these stages are generally carried out through set procedures, sometimes using job analysis and evaluation packages acquired from specialised consultants (Hays International being one of the best known firms in this field).

In all due logic, i.e. if the corporation were as rational as organisation as has often been assumed, all three stages should form a whole: the last one being determined by the other two. This could be so if the corporation was a fossilized structure, within which the characteristics of each post and the "value" of each incumbent could be precisely and definitevely defined. In actual facts, however, it is the third stage which feeds back instructions to the first two, continually causing them to be brought into line with new corporate or social requirements as these arise. For example, by reshaping the profiles (levels of skills, previous career performance) of those appointed to succeed each other in certain posts, it is possible to bring about a gradual change in the basic job descriptions concerned. Also, by making new postings from a different standpoint, the basis of performance evaluation can be given shifts of emphasis more consonant with the new job profiles.

The posts available, the cadre personnel filling them, and the personnel movements policy, together make up what we can call a "career space". Within that space, the cadres follow "trajectories" known as careers. However, the space itself is not fully neutral, but is broken down by factors of interference: functional distinctions, hierarchic levels between which movements are governed by evaluation of "merit" according to performance, or to accademic qualifications, often influenced by a connection with one of several technical or social "families" (computer specialists, graduate of this or that higher educational institute, member of the social establisment, and so on). The order of preference givent to the latter varies according to the dominant culture of particular corporations: technology - or commercial - based companies, the public serviucez, small businesses, etc.

The exemple of the Paris Regional Transport Board

The study referred to in this paper was conducted in liaison with the Paris Regional Transport Board (RATP) which controls, maintains and operates the public bus and underground railway network in Paris and the inner suburbs, employing some 40,000 people of which 3,000 cadre personnel. The study was spread over three years and comprised a number of phases:
- the early stages consisted in determining what significance the cadres and the undertaking attached to the career notion, analysing the local career development systems applying in the different departments of the undertaking, and examining such relevant experiments as were already in hand;
- next, a specific study was engaged upon with one of the departments of the RATP with a view to developping, on the basis of the principles identified earlier, a method for controlling personnel movements. The department in question was responsible for designing, improving and maintaining the Metro system's rolling stock. It employed about 3500 people including 200 in cadre positions;
- a further stage was devoted to analysing the personnel movements (about 85) which had taken place there during the two previous years.

Obviously, this department has an essential technical fuction, so that the cadre personnel were mainly mechanical, electrical, electronic and other engineers. A career development policy there was expected to produce the following results:
- to offer the "promising" cadres a career made up sufficiently rapid and varied steps for them to be fitted to occupy posts of responsibility after a normal period of service (as compared with that corresponding to the careers available to them elsewhere);
- to ensure a sufficient degree of mobility among all the cadres so that none of them would be swallowed up by routine, thus keeping them alert to the openings for their particular skills in the various areas covered by the department (project management, workshop management, engineering design, and so on).

A retrospective review of cadre personnel movements showed that they had been organised so as to fill vacancies which, although to some extend foreseen, were outside the control of the personnel managers: retirement on superannuation, posting to other departments, promotions to senior management level decided by the Board of Directors, and so on.

The overall Personnel Movement process

Given that the filling of a vacancy however caused involves the redeployment of a cadre who in turn leaves a vacancy behind him, it is clear that the administrative problem is not one of organising individual movements, but one of controlling a whole chain of movements. This is a more complex process, but one which gives greater freedom of initiative to the personnel manager.

The latter may, for example, use the appearance of a vacancy as an opportunity for making a series of subsidiary reappointments corres-ponding to his plans for a reshuffle of the cadre establishment. However, such a scheme for making one replacement the occasion for a multi-directionnal cascade, so to speak, of redeployments is a more complex affair than a simple and linear upward process of adjustment.

This complexity calls for the utilisation of management tools to assist with administering on a systematic basis the manifold options which arise whenever a vacancy occurs. It was the absence of such tools which, until recently, condemned personnel managers to thinking on one-for-one replacement lines.

Furthermore, this more complex approach enables a new factor to be considered when a successor to a vacant post has to be selected: how easy will it be to replace the person newly designated in the post he currently occupies? It may turn out that the intended new incumbent is indispensible in his current job or that his immediate superior is not prepared to release him for some reason. In these circumstances, the next-best choice in the normal sequence of selecytion may be side-stepped by resorting to another line of possibilities, as illustrated by the following diagram:

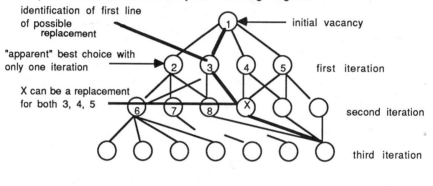

alternative sequences (not all depicted on diagram)

We thus have a typical optimisation process using a tree structure. The software basis described below was developed, in conjunction with the user department, as a means of searching all the possible sequences.

Research algorithm

The above conclusion implied that we define the components of a careers management procedure in different terms from those previously (and often informally) applied:
- the posts themselves are defined using three sets of criteria, i.e. the minimum conditions of eligibility (experience, subject knowledge and current rank, for instance), the corporate strategy options (e.g. to encourage lateral movements as a means of diversifying cadres' experience) and particular aspirations expressed by individuals (such as proximity of workplace to domicile);
- cadre personnel are defined in terms of performance and seniority, but a "role-profile" factor is also considered: a cadre who has been particularly successful as a workshop manager, for example, will be more readily considered for promotion to a more senior post of the same kind.

The basic algorithm consists in an iterative process to identify the best prospect for filling a given vacancy. The outcome is a list of candidates who all meet the basic criteria of eligibility, who are then ranked in order of preference by the careers manager (usualy the deputy departmental head). All those on the short list are starting points for further iteration designed to identify their possible replacements, and so on downwards.

The method does not intend to be a substitute for managerial evaluation, but to assist it in two ways:
- by setting forth a sequence of choices which has been checked out at all the nodal points in the tree structure;
- by providing a systematically constructed field of choice ensuring that none of the possibility is overlooked.

Consequently, the "dynamic programming" process plays only a minor supporting role in the overall procedure of choosing successors to a post falling vacant, and it has priorities other than the strict observance of formal optimisation rules (which are at all events difficult to apply in an area where subjective factors of judgement are always present).

It does however provide career supervisors with a means of broadening their angle of vision, by making it easier for them to cope with the sequence of movements idea - which may seem a superficial one if viewed from a distance, but which in practice represents a considerable new departure in career management attitudes

Also, it presents a public display of the selection criteria, especially those applied to job descriptions and personnel profiles.

Finally it brings some degree of consistency both within and between the successive sequences of investigation: the strategic criteria adopted for a given iteration become imperatives for the other iteration and sequences which follow.

Consequently, in a field where the commitment of the cadre personnel is strong - that on which depends their whole working careers - it introduce a measure of objectivity on which can be fouded, in a later stage, a formal strategy for career development along a number of lines:
- given our ealier assumption that career management implies consistent decision critéria if a climate of mutual confidence is to exist between the governing and the governed (the latter, i.e. the cadres, also need fixed guidelines enabling them to find their way around the "careers space" considered), it is possible to simulate the long-term consequences of various management options concerning horizontal mobility, profile shaping, and so on;
- it also becomes possible to monitor the ongoing changes in cadres profiles, with a view to ascertaining if those profiles essential to proper implementation of corporate aims (including the ability to innovate and to train future senior managers) are effectively perpetuated to a sufficient extent.

Conclusions

The practice of dynamic, or positive, programming depends, we have seen, on an approach to career management problem which incorporates the concept of sequences of cadre parsonnel movement. This may appear to be a commonplace observation, but it in fact implies a thorough overhaul of practical procedures concerning both the formulation of the options and the ways of making them acceptable to both sides. The latter objective cannot be achieved unless the application of the new principles is accompanied by a set of "operating instructions" through which thoses principles can be justified and shown to be workable.

Career management is an area in which many emotional factors intervene. Its greater rationalisation consists not so much in introducing more refined methods of selection, which are bound to retain a large arbitrary element, as in integrating its methodology with an extended perspective in terms of both time and space. Having worked with many organisations in this field, I have reached the conclusion that cadre personnel are more attached to having fixed guidelines enabling them to find their way around the career maze, than to an assurance that the rule applies are fairly and scientifically evolved. But this extended perspective renders the management side's task more complex, due to a multiplication of alternatives, and makes it necessary to employ formulas providing a more systematic means of approching it.

It is in this respect that the procedural formalisation of dynamic programming of cadre careers becomes relevant but, as we have seen, it cannot be introduced, in the first phase at least, otherwise than in controled fashion, using the inter-active mode which micro-computer systems now permit. It can be expected that its role will be more in line with its theoretical definition when it comes to the simulation of management strategies, with the incorporation in the model of formalised and automatic selection criteria which until then remained the prerogative of the managers.

ROUTINIZATION OF WORK AND THE QUALITY OF WORKING LIFE:

A STUDY AMONG CANADIAN NURSES

Vishwanath V. Baba and Muhammad Jamal

Department of Management
Concordia University
1455 de Maisonneuve Blvd. West
Montreal, Quebec, H3G 1M8 Canada

Quality of working life has remained a central concern among social scientists for the past two decades. Though there are disagreements over the precise definition of the concept, there is general agreement over its multidimensional nature and its usefulness as a guiding notion in understanding work (Davis & Cherns, 1975; Hackman & Suttle, 1977; Lawler, 1982; Rice, McFarlin, Hunt & Near, 1985, Walton, 1974). It is defined for the purposes of this research as a way of thinking about work, people and organizations (Nadler & Lawler, 1983). It's distinctive elements include a concern about the impact of work on people as well as on organizational effectiveness and an orientation toward participation in decisions which affect an individual's job in particular and work in general (Nadler & Lawler, 1983; Schuler, 1984). Typical indicators of quality of working life which stem from this definition would include job satisfaction, job involvement, role ambiguity, role conflict, role overload, job stress, job scope, organizational commitment, and turnover intention. This operationalization is also supported by the empirical literature.

Recent theoretical work suggests that work activities, work standards and work outcomes influence perceived quality of work life (Rice et al, 1985). The concept of routinization of work in the context of work design has a long but spotty history. Early writers on management, influenced by the Scientific Management movement and advances in industrial engineering advocated routinization of work as a means to improve efficiency, productivity and organizational effectiveness (Taylor, 1947). However, in the wake of the Human Relations movement, routinization of work began to acquire some negative publicity. Advocates of job enlargement, job enrichment, and job characteristics theory started documenting the ill-effects of task routinization on the quality of working life (Oldham & Hackman, 1980). Especially, the literature on the job characteristics theory is replete with suggestions that routinization of work leads to poor quality of working life because it suppresses innovation, involvement, commitment and other forms of creative expression on the job resulting in poor performance (Oldham & Hackman, 1980). Thus the routinization literature is both ambiguous and inconclusive when it comes to the impact of routinization on quality of working life.

A careful examination of both the early and recent literature on routinization revealed either ambiguous or a narrow and restrictive view of the concept. Routinization in the early literature lacks a precise meaning. It was defined in terms of standardization of work procedure as well as work pattern through the introduction of rest pauses (Taylor, 1947). In other words, routinization of the context of work and routinization of the content of work were intertwined resulting in ambiguity and surplus meaning. In the more recent job and task design literature, the term routinization referred exclusively to the work content. In that sense routinization referred to fragmentation of the task both horizontally and vertically. Influenced by the ideology of the Human Relations movement as well as the writings of humanistic psychologists on the lack of harmony between human nature and organizational realities, the proponents of job enlargement, job enrichment and job characteristics theory argued that routinization of work is essentially bad for both human growth and organizational productivity. In short, the characteristics theory does not include routinization of the job context either in its theoretical articulations or in its empirical verification. The meaning attached to the concept of routinization in this sense is much more restricted and does not allow for speculation on the relationship between routinization of the job context and job content.

Thus, the classical literature suggested that routinization of work has a positive influence on organizational productivity and by implication, on the quality of working life resulting from the overall prosperity generated (Taylor, 1947). On the other hand, the more modern literature with an emphasis on individual well-being, documented the ill effects of routinization on both organizational productivity and the quality of working life. This paper suggests that these mixed findings are due to inadequate theoretical grasp of the concept of routinization. This weakness stems from the failure to separate routinization of the job context from routinization of the job content. Though Emery and Thorsrud (1969) made the observation that separation of work context and content would lead to a better understanding of work life, there has been little empirical research which attempted to verify that observation. The present study argues that routinization in the work context indeed serves to provide the necessary energy, both psychic and otherwise, to innovate and express oneself to the fullest potential in terms of work content with attendant improvements in both quality of working life and work performance.

Therefore this study proposes that routinization of the job context as indicated by employee participation in routine or non-routine shift would have an impact on the individual's perceived quality of working life and job performance. Specifically, this study hypothesizes that people who work routine shifts would demonstrate higher levels of job satisfaction, satisfaction with working hours, job involvement, organizational commitment, job scope, and job performance and lower levels of role ambiguity, role conflict, role overload, stress, and turnover intention compared to those who work non-routine shifts.

METHOD

Sample: The sample consisted of 1148 nurses from eight anglophone hospitals in the greater Montreal area in Canada. The average respondent was 33 years of age, has had a nursing license for 14 years and has held an average of 1.5 jobs in the last five years. There were twice as many anglophone nurses as francophones and 98 percent of the nurses were females.

Procedure: Data were collected in the form of mailback questionnaire with prepaid envelope in two stages. The first stage involved a sample of 200 nurses for pretesting purposes. In the second stage, 2236 questionnaires were distributed among the hospitals and 1148 usable questionnaires were returned yielding a response rate of 51 percent.

Instruments: Demographic information was obtained using appropriate
questions concerning age, gender, education, experience, language of origin
and number of jobs held in the past five years. Information regarding
routinization of the job context was obtained by asking the respondents the
shift or shifts they worked. This information was subsequently recoded into
routine or non-routine shifts. Among the indicators of quality of working
life, job satisfaction and satisfaction with working hours were measured by
a one item 5-point scale each ranging from extremely dissatisfied to
extremely satisfied. Job involvement was measured using a five item short
form of the Lodahl and Kejner (1965) scale. The four item role ambiguity
scale, the six item role conflict scale and the six item role overload scale
were taken from the role factor instrument of Rizzo, House and Lirtzman
(1970). Job stress was measured using a 9 item scale from Parker and
DeCotiis (1983). Organizational commitment was measured using a 15 item
scale developed by Mowday, Steers and Porter (1979). Job scope was measured
by a modified nine item scale developed by Hackman & Oldham (1975). Quality
and quantity of performance were assessed using the two item measure of
Porter and Lawler (1968). Turnover intentions were measured using the one
item measure on future job search behavior by Mobley (1977). Responses to
the above scales were obtained on a 5 item or 7 item Likert format using
appropriate anchors. Cronbach alpha reliability estimates for all the
scales were in the acceptable range. One way analysis of variance was
performed to assess the differences between routine and non routine work
contexts in terms of both the quality of working life and the performance
indicators.

RESULTS

The results clearly showed that nurses who worked routine shifts
perceived a higher quality of working life according to most indicators of
quality of working life compared to those who worked non-routine shifts
(table 1). They also exhibited higher levels of both quantity and quality
of performance compared to those on non-routine shifts (table 1).

Table 1. One Way Analysis of Variance of Routinization, QWL and Performance

Quality of Working Life	Range	Routine Shift (μ) (N = 290)	Non-Routine Shift (μ) (N = 857)	F-value
Job Satisfaction	1 - 5	3.86	3.51	12.08*
Job Involvement	5 - 25	15.47	15.52	0.04
Role Ambiguity	4 - 20	9.37	9.82	3.89**
Role Conflict	6 - 30	19.71	20.06	0.89
Role Overload	6 - 30	15.08	16.36	14.68*
Job Stress	9 - 45	25.64	27.70	14.99*
Satisfaction with hours of working	1 - 5	4.09 '	3.14	95.23*
Job Scope	9 - 63	46.37	44.92	9.10*
Organizational Commitment	15 - 75	51.83	49.84	6.74*
Turnover Intention	1 - 5	1.93	2.22	7.21*
Performance				
Quantity	1 - 5	3.75	3.57	8.13*
Quality	1 - 5	3.88	3.69	8.67*

Note * p < .01
      ** p < .05

Specifically, the one way analysis of variance revealed that nurses who worked routine shifts experienced significantly higher levels of overall job satisfaction, satisfaction with hours worked, job scope, and organizational commitment compared to those who worked non-routine shifts. They also reported significantly lower levels of role ambiguity, role overload, job stress, and turnover intention compared to those on non-routine shifts. Further, they perceived themselves to work harder and turned in higher quality performance compared to those on non-routine shifts. The only indicators of quality of working life which failed to demonstrate significant differences between the two groups are job involvement and role conflict (table 1).

DISCUSSION

The results provided strong support to the hypothesis that routinization of the job context contributes positively toward enhancement of quality of working life. Arguably, this happens because routinization minimizes the need to actively manage the work context so that individuals can channel their energies toward the job itself. If one were to make constant adjustments in one's life in order to meet work demands, even if those adjustments have a certain predictable periodicity, they eventually interfere with the harmony of life. This observation is supported by the empirical literature on shift work and its impact on work and family life (Colquhoun & Rutenfranz, 1980; Coffey, Skipper & Jung, 1988; Dunham, 1977; Frese & Okonek, 1984; Frost & Jamal, 1979; Jamal, 1981; 1986; Jamal & Jamal, 1982; Moore-Ede & Richardson, 1985; Parasuraman, Drake & Zammuto, 1982; Staines & Pleck, 1984; Winget, Hughes & LaDou, 1978). This literature documents the disruptive impact of non-routine shifts on a number of factors including health, family life, stress, work attitudes and behavior, and organizational outcomes. Job characteristic theory suggests that core job dimensions such as skill variety, task identity, task significance, autonomy, and feedback lead to certain critical psychological states such as meaningfulness, responsibility and knowledge of results which in turn bring about positive personal and work outcomes (Hackman & Oldham, 1975). Thus, it is clear that improvement in the core job dimensions or job scope is likely to result in improved quality of working life. This paper argues that routinization of certain aspects of the job context facilitates the above process and increases the probability of a higher quality of working life. In fact, as can be seen from table 1, nurses on routine shifts viewed their core job dimensions or job scope in a more positive light than those on non-routine shifts despite the objective reality that the jobs in both situations were exactly the same. This lends credence to the observation that routinization of work context has a positive impact on quality of working life even when the work content remains the same. Future research should focus on the exact nature of the relationship between routinization on the work context and routinization on the work content so that a more comprehensive theory of routinization can be developed.

In summary, this study provides empirical support to the contention that certain aspects of routinization of work may have a positive influence on the quality of working life. The strengths of this study lie in its attempt to focus on an often ignored aspect of routinization, its inclusion of both context and content relevant factors pertaining to work, a more comprehensive measure of the quality of working life and a large sample. It is hoped that the findings of this study would encourage future researchers to seek a broader understanding of the impact of routinization of work on both job design and quality of working life.

REFERENCES

Coffey, L.C., Skipper, Jr., J.K., & Jung, F.D., 1988. Nurses and shift work: Effects on job performance and job related stress. Journal of Advanced Nursing, 13, 245-254.

Colquhoun, W.P. & Rutenfranz.J., 1980. Studies of Shift Work, Taylor and Francis: London

Davis, L.E. & Cherns, A.B., 1975. The quality of working life. Volume one: Problems, prospects and the state of the art. Free Press: New York.

Dunham, R.B., 1977. Shift work: A review and theoretical analysis. Academy of Management Review, 2, 624-634.

Emery, F.E. & Thorsrud, E., 1969. Form and content in industrial democracy. Tavistock: London

Frese, M. & Okonek, K., 1984. Reasons to leave shift work and psychological and psychosomatic complaints of former shift workers. Journal of Applied Psychology, 69, 509-514.

Frost, P.J. & Jamal, M., 1979. Shift work, attitudes and reported behavior: Some associations between individual characteristics and hours of work and leisure. Journal of Applied Psychology, 64, 77-81.

Hackman, J.R. & Oldham, G.R., 1975. Development of the Job Diagnostic Survey. Journal of Applied Psychology, 60, 159-170.

Hackman, J.R. & Suttle, J.L., 1977. Improving life at work, Scott, Forseman: Glenview, IL.

Jamal, M., & Jamal, S., 1982. Work and nonwork experiences of employees on fixed and rotating shifts: An Empirical assessment. Journal of Vocational Behavior, 29, 282-293.

Jamal, M., 1981. Shift work related to job attitudes. Social participation and withdrawal behavior: A study of nurses and industrial workers. Personnel Psychology, 34, 535-547.

Jamal, M., 1986. Moonlighting: Personal, Social and Organizational Consequences. Human Relations, 39, 977-990.

Lawler, E.E., III, 1982. Strategies for improving the quality of work life. American Psychologist, 37, 486-493.

Lodahl, T., & Kejner, M., 1965. The definition and measurement of job involvement. Journal of Applied Psychology, 49, 24-33.

Mobley, W.H., 1977. Intermediate linkages in the relationship between job satisfaction and employee turnover. Journal of Applied Psychology, 62, 237-240.

Moore-Ede, M.C. & Richardson, G.S., 1985. Medical implications of shift work. Annual Review of Medicine, 36, 607-617.

Mowday, R.T., Steers, R.M. & Porter, L.W., 1979. The measurement of organizational commitment. Journal of Vocational Behavior, 14, 224-247.

Nadler, D.A. & Lawler, E.E., III, 1983. Quality of work life: Perspectives and directions. *Organizational Dynamics*.

Oldham, G.R. & Hackman, J.R., 1980. Work design in the organizational context. In Staw, B. & Cummings, L.L. (eds.). *Research in organizational behavior*, 2, 247-278.

Parasuraman, S., Drake, B.H., & Zammuto, R.F., 1982. The effect of nursing care modalities and shift assignments on nurses' work experiences and job attitudes. *Nursing Research*, 31, 364-367.

Parker, D.F. & DeCotiis, T.A., 1983. Organizational determinants of job stress. *Organizational Behavior and Human Performance*, 32, 160-177.

Porter, L.W. & Lawler, E.E., III., 1968. *Managerial attitudes and performance*. Irwin-Dorsey: Homewood, IL..

Rice, R.W. McFarlin, D.B., Hunt, R.G., & Near, J.P., 1985. Organizational work and the perceived quality of life: Toward a conceptual model. *Academy of Management Review*, 10, 296-310.

Rizzo, J.R., House, R.J., & Lirtzman, S.I., 1970. Role conflict and ambiguity in complex organizations. *Administrative Science Quarterly*, 15, 150-163.

Schuler, R.S., 1984. *Personal and human resource management* (2nd ed.). West: St. Paul. MN.

Staines, G.L. & Pleck, J.H., 1984. Nonstandard work schedules and family life, *Journal of Applied Psychology*, 69, 515-523

Taylor, F.W., 1947. *Scientific Management*. Harper & Brothers: New York.

Walton, R.E., 1974. Improving the quality of work life. *Harvard Business Review*, May-June, 12ff.

Winget, C.M., Hughes, L., & LaDou, J., 1978. Physiological effects of rotational work shifting: A review. *Journal of Occupational Medicine*, 20, 204-210.

ACKNOWLEDGEMENT

This study was supported by a grant (410-87-0106) from the Social Sciences and Humanities Research Council of Canada.

684

OPERATIONAL RESEARCH AND PROJECT CONTROL IN THE COMPUTER INDUSTRY

Steen Rem

Institute of Production Management
and Industrial Engineering
Technical University of Denmark

INTRODUCTION

To me projects that do not involve some form of project control seem difficult to imagine. Equally absurd is the often rigid use of project control systems that cause problems in many projects today.

Lack of results due to lack of project control has been treated by several authors (i.a. R. J. Graham, 1985) but lack of results caused by too much control is seldom described. This may be the result of an inordinate belief in rational control. If the basic assumption is that control is possible (and appropriate), this factor will not be used as an explanatory variable in case of lack of project results.

PROJECT SIZE

Current literature on the subject shows relative agreement that a project can be defined as a task that is unique, complex and of limited duration.

Many firms that organise their work in a project organization seldom differentiate explicitly between types of projects, which may lead to a number of problems and ensuing project failures. The definition of project failure is that the expectations of client/basic organization or supplier/-project (or both) are disappointed.

In this article I shall use the computer industry as an example and a basis of defining different types of tasks/projects.

Characteristic of one type of project - for instance within software production - is that the prototype of the product has been made, but needs adjustments in connection with a specific order. For this purpose a project group is set up to deliver a software package that is compatible with the hardware configuration of the client. The project is small and the project group often comprises a salesman, a programmer or possibly a systems analyst.

Characteristic of another type of project is that <u>the main idea behind</u> <u>the system has been developed</u>, but a number of problems are still outstanding to make the system function. Some modules have not been developed, files cannot be transferred from one module to the other, the system has not been tested, etc. The system is of medium size and the project group comprise a larger group of people and more expertise.

Characteristic of a third type of project is that it involves <u>the develop-</u> <u>ment of an entirely new system/concept</u>, i.e. every aspect of it is new or assumed to be unknown. The project is a large-scale one and the project group very versatile and knowledgeable.

The definition of project size can generally be based on a number of factors; financial aspects, the number of persons involved, the time perspective and the importance of the project to the survival of the firm.

PROJECTS AND UNCERTAINTY

A project can be seen as a combination of 4 basic elements - the project task, the method used, resources and the environment - which ar interdependent (H. Mikkelsen et al., 1985). Naturally such a systems approach is always debatable, but this is not the purpose of this article. Each of these variables involve a degree of uncertainty. R. W. Baker 1986 describes five factors (which) combine to increase the uncertainty inherent in today's project:

(1)   extent to which size (i.e. physical, manpower requirement and financial value) has increased,

(2)   increasing complexity of projects with more disciplines involved

(3)   extent to which technology is "state-of-the-art" or is little tried and requires further research and development,

(4)   increasing involvement of external factors (e.g. government regulations) and environmental concerns (weather and local lobbies),

(5)   internal trading - currency fluctuations, unknown inflation for long-term projects, complexity of financing, e.g. joint ventures.

Here the uncertainty is typically related to the method used, resources and the environment, but not the task itself.

Some projects are characterised by the task being relatively familiar. The method has been used before. Resources are more or less predefined and the environment's perception of and demands on the project are relatively stable. The degree of uncertainty is limited.

At the opposite end we have projects that are characterised by a very uncertain project task and therefore also by a high degree of uncertainty as regards the three other variables.

The normative project literature abounds in directions for the control of projects by project managers (Cleland & King, 1984, Randolph & Posner, 1988), but the directions are only interesting in connection with projects with limited task uncertainty. Where the literature discusses the uncertainty related to projects, the directions are often limited to ways of reducing uncertainty to enable the project manager to control the project by means of the tools available.

THREE MAIN TYPES OF PROJECTS

By combining the two variables - size and project uncertainty - we arrive at three main types of projects.

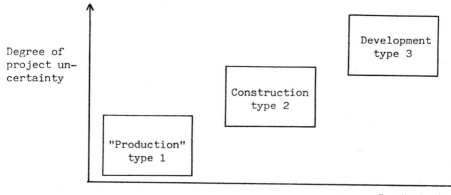

Fig. 1. Types of projects

In principle one of the two variables - project size - can be greatly influenced by the organization as it is a question of allocation resources among projects on the basis of priorities, but such priorities can only be established if the task (goal) is known. If the project task is more or less unknown (types 2 and 3), this type of control will only make sense if management recognises the dichotomy inherent in the relation between goals and means.

The other variable - project uncertainty - is far more difficult for the organization/project manager to influence. On the basis of the uncertainty involved in many projects it is possible to identify "... two complementary developments taking place" (R. W. Baker, 1986):

* use of detailed control procedures for measuring progress and cost, to enable project management to know clearly what work has been done to date and what still needs to be done; this has been aided by use of sophisticated computer systems and a realization that significant effort is required to set up and maintain the procedures

* probabilistic techniques to try to quantify the uncertainty and its effect on a model of the project programme or cost estimate

In my view this development has one main purpose and that is to create models whose application is supposed to make an unpredictable world (more) predictable. By this I mean that many - but not all - project management techniques are developed and applied to "pull" the project towards the point of origin in figure 1, which - in theory - will make it more controllable. This view has been voiced by others (S. Christensen & K. Kreiner, 1987). My point is that the closer to the point of origin - and thereby the smaller the project and the more limited the uncertainty - the less rigid the control necessary.

There is a far greater need for developing techniques/systems that can be used in connection with type 3 projects. Instead of spending resources on developing advanced probabilistic techniques and various computer-based control systems it is much more important to develop methods that will promote recognition that different types of projects involve different control requirements and possibilities and that quite different (e.g. participant-based) types of techniques must be used for some types of projects to be successful.

PERCEIVED CONTROL PROBLEMS

A Danish survey of 10 firms (H. Mikkelsen et al., 1986) shows that the firms are looking for tools that may be of assistance in controlling work, time, finances and documentation and materials as these are considered the most important aspects of project management in the order of priority given. The survey also suggests that lack of project goals and planning is the source of (too many?) changes of plan.

This survey is in line with my own experience with the computer industry. A division manager in a major service bureau said, "We are about to introduce a number of techniques with a view to controlling resources in our projects. We use two tools; one we call a project model, which is a development model, which means that all computer development must be based on the same shell, the same development phases. To this model we add a control tool which has two purposes. One is to ensure that all persons involved in development work do a standardised job. They will when they go by models. The other is to ensure through time planning that projects are delivered on time. This is as far as our ambitions with this system go, but it reveals at an early stage when deviations are in sight".

THE CONSEQUENCES OF A RATIONAL CONTROL PHILOSOPHY

The control problems identifiable in connection with project control are often the result of a basic assumption that the world the projects form part of is rational. The problems originate in the way in which project results are evaluated - when project result and effort are compared. The conclusion very often is that in view of the result too many resources have been used. Within the computer industry it is my experience that a development project (type 3) is evaluated on as short-term basis and is expected to "pay its way". Factors like a redefinition of the project task along the way, lack of the knowledge necessary to carry out the project and spin-off solutions to (different) problems are seldom taken into account. What I mean is that a project must be judged on a basis that takes account of the degree of task uncertainty and project size.

If such projects are to be judged on a financial (rational) basis, which is fair enough, the whole life cycle of the project must be taken into consideration. Over time (successful) type 3 projects become one or more type 1 project(s) and make a financial contribution to the firm that is its long-term basis of survival. Generally speaking type 3 projects are risk investments, which (in the form of type 1 projects) may prove profitable over time.

The consequence of a rational perception of projects is that the two trends presented by R. W. Baker are reenforced; the project manager is confirmed in his belief that the closer he gets to the spending of resources originally budgeted, the more successful he is and therefore he will find that he needs operational (computer-based) control systems that will reveal "overspending" or tell him what to do.

A survey (B. N. Baker et al., 1983) concerning the determinants of project success, noted with concern an increasing emphasis within some corporations on the creation of "... elaborate and detailed reporting and control systems (which led to) excessive delays, red tape, superficial reports, and inadequate information flows". The reports became more important than the drive of the project and the project result suffered.

A manager in the computer department of a major Danish bank said "our goal is that within 6 to 8 months approved changes, i.e. projects initiated, must result in visible improvements for users in the form of partly or entirely new systems". All projects are standardised and follow a given project model and its control systems.

The result will necessarily be that development projects are not experimental, and obtaining a minimum of satisfaction becomes a goal.

PROJECTS AND OPERATIONAL RESEARCH

The question is whether OR can be used in defining and planning the project task and for other project management tasks?

Invoking methodological pluralism, I would say that the more type 1 characteristics a project has, the more OR may be useful because such projects deal with relatively concrete systems that need adjusting. If it is a question of a limited problem where only few variables need to be taken into consideration, it may be reasonable to use and OR model to generate a solution or at least a basis for decision-making.

For type 3 projects a very strict application of OR is less useful as data are a critical factor. Because this type of project often changes in the process - new goals are defined - it is impossible to get a viable data background. I will go so far as to say that if a project at a given time has "sufficient" data, it might destroy the very idea of this type of project, i.e. the development of solutions to problems in a very uncertain environment. The risk is that the model itself becomes all-important and that the goal of the project becomes the provision of problems for the model to solve.

However, I do not deny that OR techniques can be useful in type 3 projects. But when used, they should support creativity and development rather than decision-making, planning and control. OR tools may be useful as a common language or frame of reference for project participants if their ideas can at all be quantified or described. This is doubtful (S. Christensen & K. Kreiner, 1987) as "... experience (shows) that factors that are considered very important by project participants and that may be the key motivation for participating in the project cannot be verbalised precisely. In addition project goals are often symbols that capture people's feelings and stimulate their imagination and visions. But the magic is broken when it is attempted to operationalise the symbols". In any case the problem is that as soon as OR techniques are used, one is trapped by the basic assumption of rationality on which they are based.

The development of computer technology has produced new information systems, which according to some may be used together with OR for more general management tasks in order for project management to be based on scientific, quantitative methods. Again I would maintain that such methods may be useful in connection with type 1 projects, but less so in connection with type 3 projects as it is arguable that the time perspective of the former is so short that it involves a practically static situation. The project manager is in a role culture situation (R. J. Graham, 1985) where the use of OR techniques is not inconsistent with the project culture. Type 3 projects run for longer periods and are more dynamic, and therefore involve a task culture where scientific, quantitative management methods have their natural limitations.

CONCLUSION

If it is at all reasonable to draw a conclusion on this background, it must be that projects can be a great asset to firms if they were to supplement analytical with anarchistic thinking more often. Some (H. Enderud, 1984) have also described the latter as a creative activity that is based on "sequences of thoughts being discontinued, open and never-ending processes. Impulse, intuition and chance are legitimate elements. The method is change-oriented. Focus is on what is different, alternative, opposite, distorted, exaggerated; i.e. anything that may act as a catalyst of new thinking". In other words: rigid control of projects is possible and appropriate in some phases of some types of projects, but impossible in other phases.

REFERENCES

Baker B. N., Murphy, P. C., and Fisher, D., 1983, Factors Affecting Project Success in Project Management Handbook, D. Cleland and King W., Van Nostrand Reinhold Company, New York

Baker R. W., 1986, Handling Uncertainty in International Journal of Project Management, Vol. 4 No. 4 November

Christensen S. and Kreiner K., 1987, Projektledelse i Løst Koblede Systemer-rationalitet i en ufuldkommen verden, København

Graham R. J., 1985, Project Management: Combining Technical and Behavioral Approaches for Effective Implementation, Van Nostrand Reinhold Company, New York

Cleland D. I. and King W. R., 1983, System Analysis and Project Management, McGraw-Hill Book Company

Mikkelsen H. and Riis J. O., 1985, Grundbog i projektledelse, Forlaget Promet

Mikkelsen H, Riis J. O. and Thorsteinsson U., 1986, Mikrodatamat Projektstyring, Driftsteknisk Institut, DtH

Randolph and Posner, 1988, Effective Project Planning and Management - getting the job done, Prentice Hall

POST-INDUSTRIAL SOCIETY

POST-INDUSTRIAL SOCIETY

Steve Cropper

Department of Management Science
University of Strathclyde
Glasgow

OR and the Post-Industrial Society

Operational Research is one of a wider set of organisational services – "Information or knowledge work" which has become the subject of an increasing debate in the social sciences following early, seminal statements about an emerging post-industrial society. The debate has turned around the continuing validity of this vision and its provenance as characterised by Daniel Bell and others. (Bell, 1973).

The post-industrial society was seen as consisting of an economy in which the dominant emphasis and balance of work had shifted from manufacturing to the service sector. Increased use of scientific knowledge in production and the emergence of IT heralded a general rise to power of professional, scientific and technical groups over groups deriving power from wealth or elitism. The nature of employment in the tertiary sector had thereby changed from a low-waged, low-skilled workforce to a highly skilled technocracy. Finally, and associated with these changes, there was a shift in values from material goods to services and in some versions of the theory, toward societal harmony rather than conflict and competition.

Futurists, and some more critical observers too, have come to interpret social, cultural and economic changes since the mid-1970s as reflecting a further shift in dominance toward a fourth sector – information services – and to an "information society" (Lyon, 1986). Certainly, the explosion of IT in both home and workplace has meant that this popular label is now commanding serious consideration by social scientists. But the description of the change process as a formal continuation of the hypothesised "march through the sectors" has also been questioned. Information, it is argued, does not constitute an economic sector in itself, but rather an activity and resource that increasingly pervades and is transforming the other three.

OR and the Consequences of IT.

OR is a prime user and diffuser of IT and its applications; OR is often closely involved in implementing IT processes and products and thereby, it may be argued, in changing organisations and society.

Quite how information technology and the rise of information and knowledge workers are affecting social and economic activity is as yet unclear. Can IT play a progressive role or not? Does it? Does IT deskill workers or require more skilled workers, for example, or both? Does it actually create a new class of worker, of which the operational researcher may be one example? Or, rather, does IT allow information and knowledge work to enter many existing occupations as it becomes increasingly accessible. Certainly, managers seem to be exploiting information made directly accessible to them through spreadsheets and databases. At the root of such questions is a broader issue about the validity of the suggestion that society responds to technological advance – the view of technological determinism. While such a view has been present in the theories of post-industrial and information society, others have recently stressed that technological advance is the result of political choices to invest in IT research, development and application by governments and companies – technological change is socially determined.

An understanding of this continuing debate is important to operational researchers and a clear point of contact between OR and the social sciences. For the social sciences, OR appears as an archetypal knowledge service – characterised by Ian Miles as engaging in the production and interpretation of specialised information, often very client specific, and being a relatively healthy and dynamic sector of the economy, whose employment growth has been far less marked than its output growth. (Miles 1987). For this reason, the history and future of OR bears careful analysis and monitoring - it may be a useful "indicator". For OR, an understanding of the debate in the social sciences is important for two reasons - application and involvement/power.

The Application of Social Science by OR

Not only do the different conceptual frameworks proposed by students of social and economic change provide a sense of the context and impacts of OR work, the also provide material for use by operational workers in their work. Strategic planning activities are increasingly a target for OR. And recent theory and practice of scenario building and forecasting has brought home the importance of articulating multiple, alternative futures. Debates in the social sciences can be used to change the way we think about the structure of economies and society particular. Rather than an ad hoc, atheoretical analysis of future possibilities, application of the explicit and coherent theories of social science as a way of generating possible futures tailored to the needs of the particular organisation and knowledge of the careful empirical work carried out to test or moniter these theories could offer a powerful aid to strategic planning.

The Involvement/Power of OR in Post-Industrial Society

OR is a part of a major wave of change in organisations and society associated in part with the rapid evolution of applicable IT. Does ORs implementation and exploitation of IT - its role as advisor to and information handler for key decision makers mean that it has power to shape wider society? The effect of OR's activities on work, on the flow and availability of cash, information and material goods are neither new, nor particularly related to IT. The control of decisions has always formally eluded OR but its influence on the decision process is changing as it increasingly designs and implements bespoke systems for managers to make use of themselves. Conversely, what have been the changes in OR as a result of changes in technology and society? And what vision does OR have of its future in the context of an information-rich society? It is concerned with the provision of information and knowledge. For the future, what sort of knowledge, in what form, available to whom? Whom should OR serve? Will the skills needed change? Will OR's training needs differ? These questions should be addressed with the clear understanding that the future of OR is bound up at least in part with the wider processes of change that have been mentioned above.

As a final word of introduction, it is worth pursuing this question of the scale of analysis. OR seems often to be concerned with analysis and definition of its role at the micro level - the take up of information services, its positioning amongst the growing set of management information services - and in increasing its impact in organisations. The social sciences tend to look wider, to be concerned at the macro level with the relationship between technology and occupational and societal structures. The two are not mutually exclusive, but only the identification of clear points of mutual concern and an understanding of the different perspectives involved will make for a fruitful debate and for a continuing collaboration. The papers in the section that follows address a variety of topics related to the main theme of OR and post-industrial society and provide the ground for future work and debate.

References

Bell, D (1973)   -   The Coming of Post-Industrial Society, Heinemann, London

Lyon, D (1986)   -   From "post-industrialism" to "information society":
                      a new social transformation? Sociology, 20,4,577-588

Miles, I (1987)   -   From the service economy to the informative society -
                      and back again? Information Services and Use, 7,13-29

IDEAS ON A MEANING-GENERATING WORK AND A SELF-REFLECTING SOCIAL SCIENCE

Gustavo González

Universidad de Los Andes
Bogotá

Ernesto Lleras

Universidad de Los Andes
Bogotá

INTRODUCTION

The UN's Food and Agricultural Organization, FAO, claims that in the latter quarter of this century the youth population of Third World countries would have duplicated its numbers to the point that by the year 2000 there will live in Africa, Asia and Latin America more than 1000 millon young persons, 80% of all the young at that time in the world. This means an educational and employment, not to mention the nutritional, challenge that developed countries have never faced nor will probably ever face. It means too, that the requirements of capital and technology needed to foster development under the assumptions of the conventional political-economical models and sponsored by either of both ideologies: capitalism and socialism, will hardly be met.

Such requirements cannot be met due to the fact that during that same quarter century, Third World countries have faced a strange phenomenon with respect to capital and technology. The more they repay, the more they owe.The more they receive, the less they have. The more they sell, the less they collect. They need ever more working hours to get paid ever less salaries (Galeano, 1988).

At such crossroads there is not much time in looking for scapegoats. One could go way back in time and claim that colonialism was an incubating factor, or side up with friends or foes of the IMF and World Bank, stating as Larosiere (1986) did that a viable external position could hardly be achieved -for developing countries- if large segments of the work force lack the vocational skills -or even worse, the basic nutritional and health standards- to produce goods that are competitive in world markets. Or even argue against high population growth, rapid urbanization, drug traffic, lousy governments, etc...The fact is that, at the end of our century, some nations seem to be getting richer while others become poorer and, within nations too, a select group has access to sumptuous consumption while the rest barely survives. And all this is happening within a framework of ever increasing information and economic interdependency. It is difficult then to enjoy a TV-meal while watching the news on Biafra children starving to death. Furthermore, political economy proves the fact that we all have a small or big portion of responsibility, depending on our power over resources, for such a state of world affairs.

There is no doubt too that only concerted action will take us out of such a situation. But, and this is the question we would like to ponder, is our predominant form of knowledge of reality, "calculative thinking", capable of showing the course for our actions at the onset of such unprecedent human challenges ?

It was Heidegger (1966) who established a very illuminating difference between two forms of thinking: calculative and meditative. The contributions due to calculative thinking are no doubt of great usefulness. In his own words: "...Such thought remains indispensable. But it also remains true that it is thinking of a special kind...Its peculiarity consists in the fact that whenever we plan, research, and organize, we always reckon with conditions that are given. We take them into account with the calculated intention of their serving specific purposes. This calculation is the mark of all thinking that plans and investigates... Such thinking remains calculation even if it neither works with numbers nor uses an adding machine or computer. Calculative thinking computes. It computes ever new, ever more promising and at the same time more economical possibilities. Calculative thinking races from one prospect to the next; never stops, never collects itself. Calculative thinking isn't meditative thinking, not thinking which contemplates the meaning which reigns in everything that is....There are then two kinds of thinking, each justified and needed in its own way..."

Social science in general and management science in particular belong to the first kind of thinking, although one can observe in some of its contributions like in Schumacher (1977) and many others, a thirst for meditative thinking. Why then is this so? And, will this latter type of thinking truly quench our thirst? But, more important still, will it show us the right course for our actions, as needed and stated above?

THE CRUX OF THE NEAR FUTURE

In our view of the present state of affairs we feel that both material and spiritual conditions have to be given in order to attain physical and spiritual viability of persons and communities. That is, not only are physical circumstances: food, shelter, security, etc.. indispensable to survive, but spiritual circumstances, if we are allowed to use the term, are too, indispensable. A concentration camp might comply with the minimum physical requirements, but it surely is far from providing spiritual viability.

For spiritual viability we would expect conditions that would allow us for ways to relate with others, with nature, with the transcendence, and even with oneself, in order to make our lives meaningful as individuals and members of communities. Knowledge, ideologies, philosophies, values, and religions, in one word, culture, all provide for such a viability.

While for physical viability we would expect conditions that would allow for work as a fundamental means to satisfy our basic needs. A work that would not only contribute to making our lives meaningful for self and communities, but to warrant this to future generations.

With this in mind and with all the limitations entailed in a sweeping generalization, nevertheless we will attempt one that hopefully will clarify our argument. The North (a truer description of "developed countries" since it is not value-laden) seem to have attained physical viability, while one could observe a jeopardized spiritual viability in the light of a nuclear hecatombe or in the recent discussions about the poverty of progress (e.g. deterioration of the environment, alienation, anomie). On the other side, the perceptions from international agencies about the physical viability of the

South (a not value-laden description of "Third-World or developing countries") are quite excruciating, although one could argue positively for its spiritual viability but definitely not in the light of physical extinction.

Capitalism and Socialism, as ideologies, have played a crucial role for the spiritual and physical viability of the North. Hostilities based on assumptions about property, surplus value, and who decides on what and how much to produce, and how to produce it, be it the market or the only party, have aided a very dinamic process for economic growth on both sides: East and West. Moreover these hostilities have been dialectical in providing meaning to political and economical actions of imperialism on the part of both sides towards the South. And although this thought and practice has been predominant in the last century, it now seems the tensions about theory (e.g. ideology) due to the original assumptions seem to be loosing force, mainly because such assumptions are being watered down. Perestroika is but one proof of this hypothesis.

So our contention is that a new ideology, which we will call "technological ideology" for the sake of our discussion, is enveloping the preceeding ones. Such an ideology has established now a commom ground between East and West, based on technological competitiveness and similar economical, political and social practices. Calculative thinking has definitely been instrumental to this ever-rolling wave. A wave that, in its wake, is giving birth to new tensions now between North and South.

So the question is, can we provide for the social creation of ideology ? One that will establish a new international order and that will allow not only for the physical viability but for the spiritual viability of both, the North and the South? Heidegger's intuitions have aided us in sketching an agenda that hopefully will lead the course that will provide answers to to such questions.

CALCULATIVE THINKING AND MEDITATIVE THINKING

For Heiddeger the way to what is near is always the longest and thus the hardest for us humans. This way is the way of meditative thinking...it demands of us not to cling one-sidedly to a single idea, nor to run down a one-track course of ideas. Meditative thinking demands of us that we engage ourselves with what at first sight does not go together at all. He characterizes such thinking by, first, proposing an attitude towards technology which says "yes" and at the same time "no", which lets "technical devices enter our daily life, and at the same time leave them outside..as things which are nothing absolute..". This attitude is "releasement toward things". Then, on account of the fact that the "meaning pervading technology hides itself" and since something "which shows itself and at the same time withdraws is the essential trait of what we call the mystery", he proposes a second attitude: that of "openness to the mystery".

This then is quite opposed to calculative thinking although it surely could be made to be complementary. And that is what our philosopher, 33 years ago, suggested with anguish. The atomic hecatombe didn't frighten him as much as the fact that calculative thinking would someday become the only way of thinking. This would rob man from his essential qualities, since man is meditative being too. The fact that today we hail the development of artificial "intelligence", by reducing "intelligence" to its calculative dimension confirms his fears. He was definitely concerned about the spiritual viability of his world.

Although calculative thinking does reckon with conditions that are given and take them into account with the calculated intention of their serving specific purposes, nevertheless, reality is much more complex. We can't always put it to serve our specific purposes. Reality is incomparably more than what can be accommodated in the cubbyholes of logical sequences, as Jaky (1986) so succinctly stated. Its specificity is lost in calculative thinking. Whereby one of the distinguishing features of meditative thinking is precisely its concern with the smallest detail and vastness of reality. Concern on specificity or queerness of things, both when taken singly and when taken together , that is, as a whole, is the means par excellence to restore sensitivity for the real.

The issue of spiritual viability is that of meaning, a joint venture among men. The very unique capacity of human beings for self awareness, makes possible the joint construction of the social world. A world constructed among men out of a permanent process of self-reflection on practice, and practice based on self-reflection. This process is brought together by interaction and thus gives birth to meaning.

The problem then engendered by calculative thinking is that it precludes freedom of will. It shadows down self-reflection by smoothing out the specificity and queerness of things and personalities. Such thinking is inherently tied to necessity and laws. Calculative responses are, so to speak, predetermined by the immanent logic of the calculation that is being performed. This is the antithesis of self-reflection and freedom of will. Freedom of will is not instrumental to calculative thinking, that is why such thinking, left on its own, imposes a heavy yoke.

For freedom of will to grow we necessarily have to excel calculative thinking. As pointed out by Schumacher, the inherent freedom of man relies in his ability to develop his inner space. This inner space has cognitive as well as psychological dimensions. Cognitive in what concerns the development of reflexive abilities. Psychological in that an attitude of autonomy and, at the same time, a sense of identity with self and others have to be developed. Heidegger's releasement towards things and openness to the mistery are definite preconditions for these achievements. That is , they are related to a praxis -of which work is but just one aspect- in which man can reflect and create, and to an interactive environment in which communication is dialogical.

RENEWED SOCIAL PRAXIS

Social, political and economical practices in the North geared only towards technological competitiveness and permanent economical growth, seem ever more inimitable in the future. As mentioned above, the economical and technological gap between North and South is ever increasing. In light of the self defeating economical growth of the North, whereby natural environment is decaying, and a sumptuous standard of living that renders a purposeless life to youth, one wonders then if such a model is worth imitating. It seems as if these facts are confirming Heidegger's anxieties.

Work practices present in industrial and post-industrial settings, go together with calculative thinking. In total absence of meditative thinking, meaning springs out only from these practices, configurating the technological ideology. The main problem with this ideology is that it precludes man from reflecting on his work, from taking an emotional and intellectual "distance" from the practice and product of his work. A liberating social practice then is needed whereby releasement and openness, would help to recuperate autonomy and identity. From this then emanates a work which can become joyful, meaningful, and responsible.

700

This recovery of man's meditative dimension makes possible too, the emergence of a dialogical attitude not only among men themselves but between man and things. Such an attitude is a product of non-exploitative interactions among people. Freire's (1970) contention is that the aforementioned relationship of man with his work produces a psychological condition whereby humility, respectfulness, and hope spring out.

This willed-to social circumstances allow for the permanent construction of the social world and meaning, in a dialectical relationship with meditative thinking.

RENEWED SOCIAL SCIENCE

All that lives no doubt tends towards survival, but only physical survival. Certainly, calculative thinking is instrumental for understanding and manipulating this kind of survival but it doesn't provide the elements for self-reflection and meaning. In the "great chain of being", non-human animals' behaviour is genetically pre-programmed. The laws inherent to this behaviour can be grasped by calculative thinking. This is what makes them space-bounded and ahistorical. Not so with man. His self-reflection and free-will is precisely what allowed for his construction of calculative thinking. But this thinking shouldn't become his tight-shirt.

It is his tight-shirt when we forget about the human dimension of spiritual survivorship. Also, when we claim that culture obeys laws that we can manipulate as we have done with all the rest of non-human beings. In a word, we have ignored meaning. Social science then has quite a difficult but hopeful agenda. From searching for laws, prediction and control, within a frame of equilibrium, which has allowed for the manipulation of man by man, it will now have to explore ways of reaching meaning which will inevitably generate a frame of conflict but new social conditions for a liberatory praxis.

Technological ideology seems to preclude the possibility for a genuine meditative thinking. What we are then proposing here entails the change of practices in several spheres of life. But most fundamentally in the workplace and in education. Since the preconditions for achieving it are related to work practices and thinking practices, in both cases the crux of the matter is to recover the dialogical relationship between self-reflection and work.

REFERENCES

Freire, Paulo, 1970, "Pedagogía del Oprimido", Siglo XXI Editores, México.
Galeano, Eduardo, 1988, Todos pagan lo que pocos gastan, El Tiempo, Octubre 14, Bogotá.
Heidegger, Martin, 1966, "Discourse on Thinking", Harper Torchbooks, New York.
Jaky, Stanley L., 1986, "Chesterton, A Seer of Science", University of Illinois Press, Urbana.
Larosiere, M. de, 1986, IMF Opening Address, Geneva.
Schumacher, E.F., 1977, "A Guide For The Perplexed", Harper and Row, New York.

POST-INDUSTRIALISM, POSTMODERNITY AND O.R.: TOWARD A 'CUSTOM AND PRACTICE' OF RESPONSIBILITY AND POSSIBILITY

Roy Jacques

Dept. of Mgmt., SOM
University of Massachusetts
Amherst, MA 01003 U.S.A.

INTRODUCTION

In his 1986 Inaugural address as President of the Operational Research Society, Society for Long-Range Planning, J. Rosenhead (1986) advocates trans-gressing, "One of the great Unwritten Rules, which is 'Keep politics out of operational research'." Indeed, he reminds us that, "To talk, write and act as if there were no politics in operational research is itself a distinctly political posture". Rosenhead urges us to question the assumptions and method-ology of OR -- which now legitimate excluding the discipline from the larger and messier social context in which OR work is, willingly or not, embedded.

A glance at current research, however, is more supportive of the polar claim implicitly made by Miser (1986), when he praises the contributions of Philip Morse as, "A useful paradigm of how operations research analysts can -- and, as many believe, should -- approach a problem area to which they bring no specialized, contextual knowledge." Can the development of algorithmic virtuosity remain the central concern of OR research? Rosenhead suggests not, as do others (Lenihan, 1986; Raiszadeh & Lingaraj, 1986).

OR, however, like other management disciplines, lacks a critical literature. This absence allows Techology to pose as Science, which then can pose as Truth, rather than as, "Western society's main cultural production". We are thus blinded to "The progressive erosion of the domain of metaphysics, or ultimate Truth, since the time of Descartes". While we seek efficient solutions, more and more disciplines are recognizing that, "The critical question to answer first is: What would count as knowledge" (Calás, 1987).

In the shadow of Sputnik, functionalist grounding of technical work in in a General Systems cosmology may have seemed self-evident. In the shadow of Chernobyl, the ethical immanence of our assumptions is more problematic. Can systems be efficient if they contribute to a devitalized social environment? In such a context, the pretense of objectivity does not leave science socially neutral, but socially indifferent to "structural violence" (Kelly, 1984).

As Rosenhead implies, the scientific subject must come forward, i.e. our work must show Knowers as well as Knowns. If we continue to expect the facts', to 'speak for themselves' they will speak only for our entrenched biases.

POST-INDUSTRIALISM AND POSTMODERNITY

Recognition of a broader social perspective, however, forces us to confront the fact that the meaning of "Knowledge" can no longer be considered self-evident. When the Left (Frankel, 1987) can state that, "It has become somewhat cliche to say that we are living through the dawn of an [sic] new epoch", the Right (Bell, 1976) can suggest that, "Reality is now itself problematic", and even the closed American mind of Allen Bloom (1987) produces a book from the worry that, "Almost every student entering the university believes, or says he [sic] believes that truth is relative" -- one may plausibly suspect that something is going on with the processes relating knowledge formation and social organization.

In order to relate these developments to the problem of grounding OR work in its broader social context, a brief sketch of the post- terrain is necessary. Please bear in mind that in this brief excursus the path traced is that of the argument at hand, not of the general terrain. The interested reader is encouraged to consult the referenced works.

Bell (1976), captures a common theme in suggesting that post-industrialism involves, "A shift in the kinds of work people do, from manufacturing to services." Related to this change is some awareness of a, "Growing crisis which calls into question the viability of present relationships between work, economic production, man [sic] and society, and the ability of organizations to adapt" (Burrell & Morgan, 1979).

This observation leads to divergent prescriptions. Frankel (1987), has suggested three positions, the Neoconservative, the Anti-modern, and the Pro-modern -- using Daniel Bell, Rudolf Bahro, and Alvin Toffler as exemplars. Bell's Neoconservative approach calls for a return to 19th century values of work and family. Bahro advocates an Anti-modern rejection of technocracy for a simple communalism. Toffler, Bahro's polar opposite, welcomes technological development as the mechanism of achieving post-industrial community.

But, is post-industrial society post-modern? To a large extent, no. We We might describe as modern, "Any science that legitimates itself with reference to a metadiscourse" (Lyotard, 1984). For `metadiscourse' one might loosely substitute paradigm, or worldview -- i.e., our assumptions. He continues, "Simplifying to the extreme, I define postmodern as incredulity toward metanarratives". Among other things, this means that a single story describing reality is no longer adequate. Objective `Knowledge of' (an object) ceases to be possible and one must deal with `Knowledge for' (a purpose).

We might look at post-industrialism as pointing toward, without visual- izing, postmodernity. Bell (1976) glimpses that post-industrial society, "Because it centers on services...is a game between persons". In this, he has perceived, but not explored the possibility that the human subject we are/study is "produced" by, "An ensemble of rules for the production of the truth", an ensemble of rules constituting a "Game of Truth" (Foucault, 1984). Burrell & Morgan (1979), characterize this stance as, "essentially a regulative one, concerned to make piecemeal adjustments designed to improve the viability of the technological society characteristic of the present era".

The postmodern, however, may visualize the post-industrial. Lyotard (1984) compares Social post-industrialism with Cultural postmodernism. Frankel (1987) suggests that the modern/postmodern debate, "In many ways has become an explicit debate over the nature of culture and social production in the emerging `post-industrial' society". This awareness leads to awareness of the concentration of information as a sociopolitical problem (Said, 1983). The choice of one's Game of Truth, then, becomes an ethical problem, a "Functional imperative" (Burrell & Morgan). Technical problems do not disappear, but their legitimation becomes problematic if truth only exists within non-universal rule structures -- just as rugby rules are not used to adjudicate chess tournaments.

IMPLICATIONS FOR O.R.:

As Samuels' checker player (Or any other limited-domain success within AI) has hopefully taught us, hubristic General Systems thinking (Miller, 1978) exists within a context that is `computationally intractable'. Even if the universe does consist of Cartesian automata, the borders of our knowledge remain local, incomplete and imperfect.

Numerous recent critiques have argued that the worldview of "neutral" science may be influenced by ethnocentrism (Smircich & Calás, 1986), race (Dill, 1987), sex/gender (Gilligan, 1982), social class (Benson, 1983), as well as privileging a particular framework of interpretation. As Harding (1986) asks, "If problems are necessarily value-laden, if theories are con-structed to explain problems, if methodologies are always theory-laden, and if observations are methodology-laden, can there be a value-neutral design and interpretation of research?"

Smircich & Calás (1986) outline a growing awareness within organizational studies which has increasingly lead to two divergent reactions to this problem. Some, consistent with Habermas and Bell (Harrigan, 1983; Webster & Starbuck, 1988) suggest that increased rigor, multiple-site measurement, and triangulation of methods will carry the day. A diverging approach has been, increasingly, to attempt to understand `knowledge' as a body of embedded assumptions.

That scientific research rests upon assumptions, or that knowledge is incomplete is hardly controversial (Einstein, 1949; Popper 1985; Quine, 1980). In social and applied sciences, however, such recognition has been neither frequent nor thorough. How often do we pass beyond Kuhnian "puzzle solving" (Kuhn, 1970), to discuss the metanarrative within which we position our work -- what Newell & Simon (1976) have termed Laws of Qualitative Structure?

WHAT IS (NOT) TO BE DONE:

How will we `do science' without a single legitimating narrative? The situation is far less hopeless than it may appear. Here are some suggestions for projects that might help us begin to adapt:

Resisting Creating a New Positivity

The failure of one philosophical position to completely explain the world we create does not imply the need to search for `The New Paradigm' as represented by New Age advocates or some superficial readers of Kuhn. A science that produces localized statements of knowledge that are useful because of their internal consistency and instrumental value in no way needs to rest on `real' truth or fit within a grand system. The potential for multiple simultan-eous truths of localized value represents a potential freedom from the domina-tion of a master narrative that claims a much larger domain than it can explain. It also creates the responsibility of addressing the ethical, social, political contexts of our work.

Giving Full Stature to Narrative/Intuitive/Scientific Knowledge Forms

Currently, we privilege statements that legitimate themselves through appeal to scientific grounding. `Science has found...' is preferable to either `I have been told...' or `My feeling is...'. That this is desirable is not self-evident, outside of certain well-specifiable domains. Lyotard (1984)

reminds us that "Science is a subset of learning", in that it imposes two restrictions unique to science -- "objects must be available for repeated access", and objective criteria must exist for judging statements made.

The absence of such qualities from a huge portion of the statements we might make results in a privileging of certain kinds of information at the expense of others. Consider the methodological controversy related to the `case study'. Because the results obtained through such work are not objective, reliable, valid, and significant in a narrowly-defined manner, such work is frequently accorded subordinate status -- useful for exploratory or supportive roles to supplement `hard' research (Hägg & Hedlund, 1979; Benbasat, Goldstein & Mead, 1987). Such a view is thoroughly perspective-dependent (Morgan & Smircich, 1980). A more useful position in a world of multiple meta-narratives would be to accord full status as `research' to work based upon narrative, as well as upon scientific, statements.

A similar argument can be made that devaluing of intuitive knowledge is based on social values, not on objective criteria. Bordo (1986) shows Cartesian objectivism to be historically specific, politically meaningful, and gender based. Again, if the exclusion of such understanding from knowledge production cannot be shown to be based on objective grounds, such exclusion can only be interpreted as a social act that privileges some statements at the expense of others.

A Typology of Contexts for Assessing Scientific Knowledge Use

Providing a typology is dangerous if one is attempting to avoid positive knowledge. This one should be viewed as a representation to further discussion, not as a portrayal of `natural' categories. The purpose of this typology is to illustrate one way in which the failure of grand narrative schemes represents opportunity as well as challenge for researchers.

The basis for this model is computer technology and the positioning of MIS relative to the management disciplines in general. If we imagine the need for an MIS `paradigm', a tension is created. The reference disciplines and journal system for MIS research are strongly skewed toward a technical master narrative. Through time, however, we see more problems moving away from the strictly technical and toward the organizational/managerial (Culnan, 1987; Brancheau & Wetherbe, 1987).

For a scientific/technical approach, this presents a problem. For an approach that allows recognition of multiple views, each competent for specific purposes with specific types of information, this can be an opportunity. One might subdivide MIS crudely into three levels -- Computer Science, Information Systems, and Information Management. Corresponding to these levels, we can propose a dominant means of understanding and an `appropriate' organizational area.

Computer Science understanding can be thought of as close to electrical engineering, thus making it a part of physical science knowledge. In organiz- ations, this area would most closely analogize to production technology problems. In this arena, physical science techniques are most likely to provide answers to the kind of problems that might be raised.

Information Systems understanding most frequently would provide problems amenable to scientific observation of mechanical and social factors. This analogizes well to OR/MS. Whether the question is system load optimization or customer queuing, efficiency criteria will very likely be paramount.

Information Management understanding moves from a reference discipline grounded in science to one grounded in multiple ways of understanding. In this realm, efficiency criteria are very likely insufficient or even inappropriate. Without impugning the ability of scientific analysis to do what it does well -- objectively measurable, instrumental tasks -- we here need to ground such work in a broader context, where subjective, narrative and ethical values will probably supercede objective criteria.

Recognize The Limits of Triangulation

Triangulation is frequently cited by `conservative' social science practitioners as a means of strengthening the results of scientific studies of organizations (Jick, 1979). What rarely, if ever seems to be acknowledged is the incommensurability of results across metanarratives. In other words, results mean different things within different worldviews. Triangulation is only possible where the methods triangulated can be judged from a single reference point.

One distinction we might make in this argument is between questions of technique and questions of philosophy. Where researchers share a philosophy, they may be able to share methods. Within this area we might distinguish two levels of difference, that of organizing metaphor and that of methodology. The former is somewhat problematic, since it may lead one to value criteria differently -- e.g. comparing probabilistic and rule-based expert systems (Henrion, 1987). The latter, since it proposes different methods adjudicable by the same criteria, may be objectively tested -- e.g. comparing a Bayesian problem solution to a fuzzy set approach (Fischer & Henrion, 1987).

Questions of philosophy, however, are inadjudicable and frequent. A classic situation would be the triangulation of a quantitative study with `rich', `thick', or `deep' (to use the inevitable legitimating jargon) qualitative work. This paper has attempted to argue that different researchers bring their social roles into their research -- that the Good, the True and the Real vary from one position to the next. Differences over these values constitute ethical, epistemological, and ontological problems which cannot be glossed over by assuming the neutrality of research.

If I find use of gambling motivational techniques effective, but abhorrent (Evans, 1987) -- I have an ethically incommensurable problem. If I value the detailed understanding of ethnographic work over the probabilistic reliability and validity of hypothetico-deductive designs (Smircich, 1986) -- I have an epistemologically incommensurable problem. If I believe, following Foucault, that the subject I observe is produced by social practices rather than being the rational, free agent producing the practices (Bordo, 1988) -- I have an ontologically incommensurable problem.

What is (not) to be done? One thing that is not to be done is to placidly glide over incommensurable differences. To do so is, knowingly or unknowingly, to attempt to devalue different forms of knowledge -- simply because they are different, not because they are inferior. Also (not) to be done is to develop another grand plan that will allow a `neutral' perspective from which one can adjudicate between science and non-science. To be satisfied with multiple, partial truths that cannot be reduced to one another may not be divinely decreed as our fate, but, given the current state of our knowledge, it seems necessary for at least the foreseeable future.

CONCLUSION

To wrap this argument up into a concise ending would be to contradict it. A critical part of the argument is that we should become more comfortable dealing with fragmentary knowledge. As Foucault (1977) said, "To imagine another system is to extend our participation in the present system" to replace Enlightenment values with another grand narrative would be to extend the devaluation of diverse ways of knowing into future research.

In lieu of a conclusion, let me say this: A world of multiple knowledge frameworks is not valueless; it is a world of possibility and of responsibility. It is a world in which we refuse to use `objectivity' as reason for avoiding personal involvement in our knowledge productions. Many would argue that without a single view of what is Real or True a society will not behave in its own best interest -- that every action is equally good and nothing matters. The epiphets `Nietzschean', `nihilist', or `anarchist' are almost inevitable, as though such labelling precluded further discussion.

Barbara Smith (1987) calls this the Egalitarian Fallacy and argues that there can be value without truth value. If we must make do with localized truths and even ignorance, so be it. As Confucius reminded us in the Analects, "When you do not know something, to know that you do not know it; that is [also] True knowledge".

[I would like to thank Linda Smircich and Marta Calás for]
[valuable suggestions and comments regarding this paper.]

REFERENCE

(A comprehensive bibliography is available from the author)

Bell, D., 1976, "The Cultural Contradictions of Capitalism," Basic Books, New York
Bordo, S., 1986, The Cartesian Masculinization of Thought, Signs, 11:439
Burrell, G. and Morgan, G., 1979, "Sociological Paradigms," Heinemann, London
Calás, M. B., 1987, "Organizational Science/Fiction: The Postmodern in the Management Disciplines, Doctoral Dissertation, University of Massachusetts, Amherst
Culnan, M. J., 1987, Mapping the Intellectual Structure of MIS, 1980-1985: A Co-Citation Analysis, MIS Qtly, 11:341
Foucault, M., 1984, The Ethic of Care for the Self as a Practice of Freedom, Phil. & Social Crit., 12:112
Frankel, B., 1987, "The Post-Industrial Utopians," Polity Press, Cambridge
Gilligan, C., 1982, "In a Different Voice," Harvard University Press, Cambridge, Massachusetts
Hagg, I., and Hedlund, G., 1979, Case Studies in Accounting Research, Acct'g. Orgs. & Soc., 4:135
Harding, S., 1986, "The Science Question in Feminism'" Cornell University Press, London
Lyotard, J. F., 1984, "The Postmodern Condition: A Report on Knowledge", University of Minnesota Press, Minneapolis
Miser, H. J., 1986, in, The Beginnings of Operations Research in the United States, P. E. Morse, ORS Jour., 34:10
Morgan, G., and Smircich, L., 1980, The Case for Qualitative Research, Acad. of Mgmt. Rev., 5:491
Raiszadeh, F. M. E., and Lingaraj, B. P., 1986, Real-World O.R./M.S. Applications in Journals, ORS Jour., 37:937
Rosenhead, J., 1986, Custom and Practice, ORS Jour., 37:335
Smircich, L., 1986, Behind the Debate Over the Validity of Alternative Paradigm Research, prepared for the annual meeting of the American Educational Research Assoc., San Fransisco, California, April, 1986
Smircich, L., and Calás, M. B., 1986, Organizational Culture: A Critical Assessment, in: "The Handbook of Organizational Communication", F. Jablin, L. Putnam, K. Roberts, and L. Porter, eds., Sage, London, 1987
Smith, B., 1987, Value Without Truth-Value, in: "Life After Postmodernism," J. Fekete, ed., St. Martin's Press, New York

PLAY AND ORGANIZATIONS

Jannis Kallinikos

Department of Business Studies
Uppsala University
Sweden

INTRODUCTION

There is a long tradition in western ethics tending to de-
limit all those phenomena that might be clustered under the
rubric of play to childhood. When playing is allowed to enter
the adult world it is, as a rule, delegated to the realms of
leisure and recreation. Work and play, it seems, describe activities and practices which belong to different mental and social
domains. Nowhere else is such a stance reflected more clearly
than in the administrative disciplines where the study of play
is conspicuous by its absence. Apart from a few sporadic
attempts playing as the object of study has remained alien to
organizational analysis. This is regrettable, particularly in
the light of the central importance which the concept of play
has assumed in almost the entire space of the social and human
sciences (Erikson, 1963, 1978; Axelos, 1964; Winnicott, 1971;
Bateson, 1972).

Nevertheless it would be fairly conceivable that if organizational activities conformed closely to that logic of action
described by the rigid sequence of steps embodied in the tools
and techniques of the administrative disciplines, then organizations ought to have grown stiff and become ossified. It is not
surprising that a portion of organizational activities unfold
and need to unfold as a prearranged and predictable sequence of
steps, but such a recognition can by no means obscure the fact
that much of what is going on in organizations is taking place
outside the boundaries of that order incarcerated within the
prearranged and the predictable. Lindblom (1981) forcefully
argued for the necessity of moving beyond the view of decision
making as technology, i.e. a prearranged sequence of steps codified and teachable. According to him all those activities classified as decision making have to be conceived as what he termed
indeterminate practices marked by ambiguity and where improvis-
ation, experimentation and the exercise of intuition and judgement assume a central significance. March (1976) too envisaged
playing and playfulness as central ingredients of those processes by means of which established visions of reality are tran-
scended and novel goals built on different and new constellations

of values are uncovered. Drawing on his analysis March suggested certain normative implications that should bear on the quality of decision making in organizations. Thus, in his view goals have to be treated just as hypotheses, intuition as real and experience as a theory to be revised and changed when needed, whereas memory has to be distrusted and hyprocrisy tolerated as a stage of transition. In another advocacy Dandridge (1986) suggested ceremony as a means for integrating work and play and uppgrading the quality of working life in organizations.

HUMAN COMMUNICATION AND THE ORIGINS OF PLAY

Any estimate of the deep importance which play bears on organizations must trace its multiple and ramified effects on human communication and as a consequence involve an examination of its phylogenetic and ontogenetic origins. Imagination and all those faculties accompanying it unfold upon the same mental and social territory on which play operates. Unfortunately, considerations of space do not allow for other than a fragmentary and largely incomplete treatment of it.

Following Bateson's (1972) phylogenetic account it might be speculated that play has marked an important step in the evolution of communication. Such speculation rests on the assumption that playing cannot occur unless the organism develops certain capacity for simulation or, if you like, vicarious and projective representations of environmental stimuli, allowing for the possibility of pathways along which stimuli might be subject to manipulation and implying that they "can be trusted, distrusted, falsified, denied, amplified, corrected and so forth" (ibid., p. 178). Vicarious and projective representations imply then, among other things, that the organism no longer responds automatically to what he now recognizes as signals in his environment, that he is somehow able to approach, interpret and reconstitute otherness purposefully and, as it is reflected clearly in the case of humans, to rehearse within the realm of fantasy and imagination his own responses. This is easily recognized in the case of players playing a game of chess but even the solitary animal or child acts towards, for instance, inanimate objects as if they were something different. To state it briefly, in that mechanical world of stimulus-response sequences there is no room for play. For play might be said always to involve a step back from the cold facts of objective reality and a consequent attempt to reframe and reconstitute it. The capacity of detaching or dissociating objects, roles and patterns from their original context with the consequent exploration of their potential uses, functions and meanings is one of play's most essential attributes (Miller, 1973). Such is the case of the child striving to explore the possibilities of his blocks or the poet struggling to expand the capacity of language (Bateson, 1972), the artisan-artist exploring the potentialities of his materials (Lévi-Strauss, 1966) and the children or adults playing roles and games respectively (Miller, 1973). No doubt, such a diverse bundle of activities might be differentiated by the emphasis they put on means versus ends, yet all of them presuppose a certain dissociation of means from ends and an experimentation with the alternative uses in which means might be put. Art seems to lie halfway between the instrumentality of work and play's indifference to goals (Lévi-Strauss, 1966). Nevertheless, as noted by Miller (1973, p. 93) "play is not means without the end; it is a crooked line to the end".

The ability of detaching objects, roles and patterns from their original context is clearly manifested in that metacommunicative function of play which Bateson (1972) called framing, i.e. the demarcation and organization of those series of activities to be taken as play. In this regard play establishes a context signifying that the sequence of actions, messages and responses occuring while playing are to be interpreted in terms departing considerably from what might be the rule or the convention. In other words, the detached objects and activities are recasted in a different context within which they are to derive their novel meanings and functions. This is exemplified by animal play and human competitive games where a series of activities bearing a formal resemblance to those of combat are not taken to imply a fight (Bateson, 1972; Miller, 1973; Goffman, 1974). Thus play actions and messages do not stand for what they literally denote leading one to speculate on play as the prototype of such sophisticated human faculties as the ability to make use of metaphors and for symbolic and esthetic creations.

Ontogenetic accounts of play usually assume a development from the solitary, almost hallucinating world of the infant, to the shared reality of games and to cultural experience. Common to such accounts is play's enormous significance for the cognitive and emotional integration of the ego. Erikson (1963) distinguishes a transition from what he calls autocosmic play involving the infant's exploration and sensation of its own and immediate person's body and of closed objects to the microsphere of toys where the small child's preoccupations, themes and interests find expression and therefrom to the macrosphere of rule-regulated games. Winnicott (1971) provides a similar account although casted in somewhat different terms. What is important in the work of these two prominent psychoanalysts is the view that play develops in an intermediate territory extending from the boundaries of the subjective to those of the objectively perceived or shared reality. In Erikson's terms between fantasy and actuality and in Winnicott's terms between the subjective and the actual or shared reality. It is the same space which in the preceding phylogenetic account is opened by the precarious, ambivalent and loosely coupled relationship of the sign and the signified made possible by what we called vicarious representations and the reframing of series of actions.

PLAY AND ORGANIZATIONS

Even though one might reject at the outset that playing has a central place in organizational life, there is still a lot to learn about subtle patterns of human behaviour in organizations by exploiting all those insights provided by the various accounts of play. Erikson (1963, 1978) for instance views planning, experimentation and ritualization in the adult world as extension of action and thinking patterns established in childhood by means of playing, and Freud (1908, 1920) considers fantasy and make-believe in adulthood as the transformation of childhood's play impulses. Winnicott (1971) too assumes creativity broadly viewed as contingent upon the capacity to play, but as this last phrase suggests, Winnicott considers playing to continue throughout life. In this regard such characteristics of play as experimentation, inventiveness and creativity might be viewed as the fundamental modalities of organizational innovation and change. Play may be said to have both a renewing and an adaptive function. For the very act of experimenting with alternative courses of action always introduces randomness, that fundamental requirement of newness, together with the behavioural accompaniments of openness and flexibility.

Nevertheless, there is a more straightforward manner by which play might be thought to contribute to organizational change in particular, and organizational life in general. If we are allowed to define play as the demarcation and particular organization of a series of actions within which a great emphasis is put on the exploration and elaboration of means and where the pattern or organization is derived from more instrumental contexts then we are able to name a number of central organizational activities conforming to such a definition. Practice and training, R & D, scenario planning, decision making in certain circumstances involving, for instance, decision making with symbolic rather than substantive value or withdrawal for the search of solutions, minute patterns of everyday ritualized behaviour and periodical ritual and ceremonies. The list might be extended but what seems important in the present context is the underlying commonalities of such a diverse bundle of activities consisting in their as if character, where a series of results obtained are detached from their instrumental contexts and where the subordination of means to ends appears rather fragile. Obviously in many cases such reframed courses of action with the involved exploration and experimentation result in a solution or reveal an alternative which is to be brought back to the instrumental context. The proposition that planning, decision making, R & D and the other activities mentioned above might be viewed as different forms of play or as activities where the playful element takes on a central role surely runs counter to many commonsense assumptions on the nature of organizations. Yet, it would seem, upon a close examination, that all these diverse practices involve what was referred to as framing, i.e. the demarcation and particular organization of a sequence of activities which means that a series of actions and messages being proposals or trials, rehearsals or manipulations of magnitudes and patterns of action, experiments and so forth have been dissociated from their original context, or to use Goffman's (1974, p. 59) words "decoupled from their usual embedment in consequentiality". Thus manipulation of patterns of action within for instance the frame of planning do not engender, say, the retaliation of competitors. Experimentation in the R & D laboratory is not the same as actual production. These and related cases reveal, and are actually built on, the discrimination of the map from the territory, the sign from the signified. This reasoning however raises the issue of whether those attributes referred to as framing and dissociation are sufficient for characterizing a series of actions as play. For the dissociation of objects, roles and patterns of objects and roles and the consequent experimentation with alternative uses and courses of action might be said to imply, in the case of organizational activities in question, their subordination to the instrumentality of goal accomplishment. The issue of instrumentality and the emphasis put on means versus ends are thereby allowed to become the crucial factors distinguishing play from non-play.

Being released from the tyranny of goals, playful activity is differentied from work by its intrinsically rewarding and pleasurable character and the fact that it is played for its own sake. Nevertheless, this distinction, meaningful as it might be on certain occasions, boils in the present context down to triviality. For as stated, in play the relationship and articulation of means to ends appears to be rather complex. To repeat it, "play is not means without the end; it is a crooked line to the end". Note also that the distinction of means from ends is very difficult to make in many circumstances marked by a considerable element of ambiguity, an altogether common situation in organizations which led Lindblom (1959) to give goals a derived and secondary character arguing that they arise incrementally upon the constant elaboration of means.

The organized character of games provides another region which might be looked upon for tracing play's impact on organizations. For the rule-bound character of both game and ritual might be seen as an attempt to counteract their labile nature. The very fact that both develop in the slim and glittering area between fantasy and actuality, on asserted and simultaneously denied similarity of different domains of action, always threatens to break down the established frame leading to literal interpretations of metaphoric patterns of activity (Bateson, 1972). In this regard, formalization and minute patterns of everyday ritualized behaviour in organizations might be looked upon as manifestations of the attempt to curb and bring the precariousness of human playfulness within tolerable limits. To quote Erikson (1978, pp. 79-82) "ritualization at its best, that is, in a viable cultural setting, represents a creative formalization which helps to avoid impulsive excess and compulsive self restriction ... a mixture of formality and improvization, a rhyming in time".

There is still another way of looking at organizations that should allow playing and playfulness a central place. Reminiscent of Feyerabend's (1975) anarchistic epistemology of "anything goes", Guillet de Monthoux (1983) ardently argued for the necessity of viewing firms and organizations as projects where people invest ideals, sentiments, even their lives. Having declared human actors as "open possibilities" he went on rejecting an analysis of organizations in terms of a set of axioms purporting to detect and delineate the logic of action, an enterprise he called logical and which, according to him, goes back to enlightenment and therefrom to Aristotle, or in terms of behavioural determinism of anthropological and psychological colouring. Even if he sometimes makes it easy for himself, part of his argument is, no doubt, convincing. For even the cold realities of organizational rules might be viewed as embodied intentions gazing forward and being unaccountable by underlying, backward facing psychological factors of the kind alluded to in the preceding paragraph. To refer to Miller (1973) rules can in the last analysis be looked upon as a form of social make-believe. Play and project might in this regard be viewed as overlapping terms for both involve staggering and precarious processes marked by the indeterminacy of human will, being intrinsically rewarding and played or pursued for their own sake. True, there are differences involved since projects seem to be driven by burning and compelling visions (ends?), an assumption that allows a different form of determinism, i.e. of phenomenological colouring, to enter by the backdoor.

CONCLUDING REMARKS

The view of play suggested in the present chapter attributes a central significance to it, both within the context of social life in general, and that of organizations in particular. Seen as involving a step back from the cold facts of what might be termed as objective reality, playing endeavours to reframe and reconstitute it in a ceaseless and committed effort to gain a new and more comprehensive understanding of it. Such a reconstitution has been viewed as a staggering and precarious process involving the exploration of those potentialities inherent to means and the experimentation and testing of alternative courses of action.

The attempt to give play a more central place in the analysis of organizations than it has thus far assumed, might unfold, it seems, along a number of alternative pathways. Certain of these have been suggested but there are surely many others to be uncovered. True also, important problems inherent to such an enterprise have deliberately been left to flow their undercurrent way. The intrinsically rewarding character of play, the often voluntary complication of those processes by which means are elaborated and the more general problem of play's relationship to work and instrumental activity are worth mentioning. Unfortunately, considerations of space alone forced the decision to treat them in passing, leaving their elaborate treatment to a future opportunity.

REFERENCES

Axelos, C., 1964, "Vers la Pensee Planetaire," Minuit, Paris.
Bateson, G., 1972, "Steps to an Ecology of Mind," Ballantine, New York.
Dandrige, T. C., 1986, Ceremony as an Integration of Work and Play, Organization Studies, 7:159.
Erikson, E. H., 1963, "Childhood and Society," W. W. Norton, New York.
    -        , 1978, "Toys and Reasons," Marion Boyars, London.
Feyerabend, P., 1975, "Against Method," New Left Books, London.
Freud, S., 1908, The Relation of the Poet to Day-Dreaming, in "Papers of Sigmund Freud," 4:173, Hogarth, London.
    -        1920, "Beyond the Pleasure Principle," Hogarth, London.
Guillet de Monthoux, P., 1983, "Action and Existence: Anarchism for Business Administration," Wiley, New York.
Goffman, E., 1974, "Frame Analysis," Harper & Row, New York.
Lévi-Strauss, C., 1966, "The Savage Mind," Weidenfeld & Nicolson, London.
Lindblom, C. E., 1959, The Science of Muddling Through, Public Administrative Review, 19:78,
    -            1981, Comments on Decisions in Organisations, in "Perspectives on Organizational Design and Behaviour," Van de Ven, A. H. & Joyce, W. F., eds, Wiley, New York.
March, J. G., 1976, The Technology of Foolishness, in "Ambiguity and Choice in Organizations," March, J. G. & Olsen, J. P. eds, Universitetsförlaget, Oslo.
Miller, S., 1973, Ends, Means, and Galumphing: Some Leitmotifs of Play, American Antropologist, 75:87.
Winnicott, D. W., 1971, "Playing and Reality," Tavistock, London.

AS IF IT MATTERED ANYWAY ....

A REQUIEM AND OVERTURE FOR OPERATIONAL RESEARCH

Jim Bryant

Sheffield City Polytechnic
Pond Street
Sheffield, S1 1WB, U.K.

INTRODUCTION

Operational Research and the Social Sciences is a usefully ambiguous title for a conference. Suggesting, as it does some kind of relationship between two fields of scholarship, it begs the question that significant interaction occurs. At the same time it permits contributions which focus narrowly upon one or other area, by means of a legitimising tokenism: a nod in the direction of 'social problems' or 'modelling approaches' is all that is required. My aim in this paper is no purer than most, but, I believe, at least touches both areas. It is to ask what O.R. must become, if in twenty-five years time we are honestly going to be able to say that the answer to this question really matters to anyone.

ACHIEVEMENT?

Prompted by a wish to look forward at the sort of O.R. that will be needed during the next twenty-five years, I shall naturally begin by looking back over the last twenty-five.

What have been O.R.'s achievements? The frank answer is that no one knows. Moreover, we don't know even now how to make an assessment. Certainly we can disregard the measurement of accounting outcomes of O.R. involvement (e.g. cost saving, efficiency) as worthwhile indicators, for they fail to provide an unequivocal basis for evaluation. More widely, taking O.R. as being concerned "inter alia" with problems, solutions, choices and decisions, the inadequacy of such simplistic yardsticks is even more apparent: as Bevan and Bryer (1978) pointed out, decisions, for example, 'are names not for events but social processes'. Consequently, to investigate the impact of O.R. in terms of changes in these four elements is misguided and unhelpful, precisely because it fails to recognise the nature of the social world in which O.R. operates. We in O.R. need the help of those in the social sciences here – that they might turn their attention to the organizational processes in which O.R. becomes enmeshed – if we are ever to be able to give an authoritative view of the O.R. achievement.

For all the difficulties of assessing what O.R. has done, there is plenty of evidence that many people working in the area are dissatisfied with what has been achieved. This gut feeling is particularly strong

around the impact of O.R. upon major (and indeed minor) social problems.
Thus, in 1967, on the 30th Anniversary of the use of the term 'Operational
Research', Goodeve (1968) sketched the flourishing development of civil
O.R. in the U.K. and then added these remarks:
    'All this is a fairly rosy picture, but I sometimes compare this
picture with our early hopes and anticipations. Despite the setting up
of an Institute for Operational Research especially for the purpose, we
have not had much success in applying scientific disciplines to the social
sciences.'
Similar concerns have been voiced at regular intervals ever since.

    Now before we proceed to consider the O.R. response to this self-
criticism, it is worth reflecting upon it being made at all. Would, say,
accountants or surveyors or engineers or chiropodists be so concerned at
their relative lack of involvement in addressing social problems? Probably
not. We must recall that O.R. owes its origins to the radical science
movement of the Thirties (Rosenhead, 1986), and that something of this
idealistic zeal survives today. There is an undercurrent within O.R. -
stronger at some times than others - which calls us ambitiously to engage
with broader issues than our employers require. This unease with the way
things are, and the nagging belief that we ought, by virtue of our O.R.
perspective, to be able to do something about it, has to underpin any
prospect of rising successfully to the challenge.

    How could the guilt be assuaged? Twenty years ago, the belief was
that O.R. could 'bring new techniques of measurement into the social
sciences' (Goodeve, 1968). I can well recall my own contemporary involvement in work which set out quantitatively to measure the effectiveness of
a police force. Today, such approaches seem rather naive. In any case,
the social scientists have rightly become disenchanted with the promise
of calibration. An alternative hope in the next decade was that a 'systems
approach' would prove viable against the messy problems of society. This
desire to handle the whole problem flowered in such ventures as Ackoff's
Busch Center (Ackoff, 1974) and such magnificent gestures as Beer's Project
Cybersyn (Beer, 1981) in Chile. Again, I found this yearning for holism
reflected in my own work via systems studies of energy and environmental
problems in the mid-Seventies. Today, the evangelising force has drained
from this approach, though it remains as a guiding principle. Instead,
our hopes now attach to a new version of a very old idea: working with
our clients. This role has been explicitly recognised for years. Tomlinson
(1974) urged that 'if O.R. is to be effective, it must be seen as something
which is not done "to", not really done "for", but essentially done "with"
decision-makers'. He added that O.R. needs to be involved in 'the development of an understanding of the situation which requires change' and in
the commitment 'to cause the necessary changes that arise from that
situation'. It is a short step from there to taking seriously the idea
that O.R. is concerned with supporting the unfolding of social processes
in organizations.

    These three movements within O.R. - social measurement, total systems
thinking and partnership in practice - have each had their passing
successes. Further they have trickled through to influence general O.R.
work far beyond the immediate social concerns of the time. Importantly,
they have thereby shaped the thinking, not only of O.R. workers themselves
(and indeed of others in cognate fields) through formal training programmes,
but also of hundreds, if not thousands, of people who have rubbed shoulders
with O.R. practitioners in problem-managing teams or who have been their
clients. In this way, the approach of O.R. has been disseminated through
management, at least in traditional hierarchical organizations. This, if
we could gauge it, is the legacy of the past.

PROSPECTS

In 1977 I acted as discussant in a programme entitled 'Interdisciplinary Research and Social Progress' organised by Steve Cook at the British Association Annual Meeting that year. The basic purpose of the programme was 'to establish the relevance of interdisciplinary research as an aid to solving present and future problems of society'. I am sure that the speakers would not mind my referring to them as a pretty mixed bunch; I will spare their blushes by not reminding them of their contributions, which nevertheless stood out in what I found to be a frankly appalling meeting programme (that king had no clothes!). One talk however made a lasting impression: it was by Mairead Corrigan of the Peace People of Northern Ireland, and from a stance of complete commitment called upon others to pursue similar social goals. Afterwards I wrote (Bryant, 1977) to Steve, that for me Mairead's 'talk was the highlight of the meeting and made me feel rather ashamed at the academic niceties in which it is so easy for one to become involved. If only there was some more positive way in which we could help 'people's groups' of this sort'. Now, twelve years later, following Jonathan Rosenhead's successful initiative for Community O.R. (Rosenhead, 1986), I am happy to find myself with others putting into action my earlier wishes. I tell this story now because I believe that it contains the seeds of O.R.'s future, and of O.R.'s relationship with the social sciences.

Let me return briefly to pick up and embroider two key issues from my retrospective review. First, O.R. people are a restless, uncomfortable crowd, always breaking new ground and at the forefront of change: O.R. is a subversive activity which paints a moving target for others to aim at. Second, O.R. is concerned with mess management, and mess management is a collective responsibility: O.R. supports people's thinking through complex problems in a social context. These features suggest first, that the content of O.R. work will lead unpredictably into virgin territory over the next 25 years; second that it must rely upon a command over the handling of social processes within a client group context. Thus we cannot know what problems the O.R. people of tomorrow will be caught up in, but we can assume that they will be present, if they are there at all, in order to affect how people think about them.

An O.R. praxis that relies upon process support - and support moreover to client groups - as its fundamental strength, will need practitioners with special skills. These skills centre upon the ability to help problem-owners to articulate and explore the meanings which situations have for them, and to negotiate with others new meanings that are consistent with their intentions for deliberate action. This package of abilities lies to one side of that which most present-day practitioners would lay formal claim to (though if their initial modesty were dispelled, many would no doubt suggest that the new abilities were just what they do anyway in the everyday commonplace of their working lives). But such facilities cannot be developed upon nothing. They require a substrate of sound theory: theory about the social world; thoery which the social sciences must provide. It is only with such an underpinning body of theory that a practitioner can effectively reflect-in-action (Schon, 1983), and it is only through reflection-in-action that professional growth can occur, through a better understanding of the way in which both practitioners and their clients construct the realities which they address - including any apparent problems.

The whole direction which I have suggested is one which relies upon a number of extremely delicate assumptions. First, it assumes that social institutions will be prepared to support the challenge to orthodoxy which good O.R. represents. Second, it assumes that the O.R. profession will show a greater preparedness to work as collaborators with their clients, and to see their strength as being to aid expression rather than to provide 'clever' analysis. Third, it assumes that the theory of social processes in organizations will be seen as centrally relevant to effective O.R. practice, and moreover, will be extended and developed through reflective practice. What it does not do is to assume any particular configuration for society in 2014: we must aim for something far more robust than that.

CODA

The last conference on 'O.R. and the Social Sciences' has, during the intervening years, been variously seen as a symbol of hope (Tomlinson, 1974) and as a blighted vision (Sadler, 1978). Let us hope that the present Conference will be viewed more positively in twenty-five years from now as at least cementing a new and closer relationship between O.R. and the social sciences. It doesn't matter if O.R. doesn't matter twenty-five years hence, but it does matter if what it potentially offered to society remains just a hollow promise.

REFERENCES

Ackoff, R.L., (1974), The social responsibility of operational research, Opl.Res.Q., 25: 361
Beer, S., (1981) "Brain of the Firm" (2nd Edn.), Wiley, Chichester
Bevan, R.G. and Bryer, R.A., (1978), On measuring the contribution of O.R., J.Opl.Res.Soc., 29: 409
Bryant, J.W., (1977), Personal Communication, 5th September 1977
Goodeve, C., (1968), The growth of operational research in the civil sector in the United Kingdom, Opl.Res.Q., 19: 113
Rosenhead, J., (1986), Custom and Practice, J.Opl.Res.Soc., 37: 335
Sadler, P., (1978), O.R. and the transition to a post-industrial society, J.Opl.Res.Soc., 29: 1
Schon, D.A., (1983), "The Reflective Practitioner: how professionals think in action", Temple Smith, London
Tomlinson, R.C. (1974), O.R. is, J.Opl.Res.Soc., 25: 347

AUTHOR INDEX

SUBJECT INDEX

Knowledge utilisation, 154

Labour, 486
Leadership, 537
Learning processes, 298
Life cycle
  of project work groups, 313
Linguistics, 259

Management
  education, 513
  and operational research, 613
  women in, 207
Managerial communication, 207
Market
  segmented, 554
Master system, 268
Materials requirement planning, 333
Measurement, 482
Messes, 225, 238, 241
Methodologies
  critical, 73
  hard and soft, 66
Methodology, 165
  choice, 320
  scientific, 177
  systems, 171
Modelling Social Behaviour, 503
Moral behaviour, 547
Multi-attribute decision-making, 386

National Coal Board
  operational research in, 614
Negotiation modelling, 605
Network, 650
Norms
  modelling social, 531

Operational research
  and behavioural decision theory,
    629
  and education, 14, 17, 102, 134
  and information systems, 446
  and managerialism, 91
  nature of, 4, 90, 579, 623, 716
  and organisation theory, 117
  and philosophy, 626
  and post-modernism, 61, 705
  and psychology, 627
  and science, 36
Operational Research and the
    Decision-Maker, 611
Operational Research as a
    Social Science, 107
Optimised production timetable, 331
Organisational
  democracy, 26
  design, 118, 339, 397
  interactions, 525
  structure, 305

Organisations
  circular, 26
  multidimensional, 21
  as negotiated order, 44

Personnel management process, 675
Planning, 367
Play, 709
Policy-making, 441, 594
Positivist operational research, 289
Post-Industrial Society, 691
Power, Conflict and Control, 559
Pragmatic structures, 391
Premise setting, 573
Prisoner's dilemma, 547
Problems of Measurement, 469
Process management, 43
Production planning, 319, 325, 331
Productivity, 198, 485
Professionalisation, 54
Project management, 312, 480, 614,
    648, 685

Quality of life, 272
Quality of working life, 667, 679

Rationality, 152
  collective, 584
Responsible scheming, 371

Schools
  location of, 275
Science and technology, 123
Scoring scale, 497
Simulation of social systems, 514,
    533, 538
Social action, 392
Social praxis, 700
Social psychology of organising, 350
Social responsibility
  and operational research, 131, 717
Social science
  nature of, 38, 89
  and intervention, 294
  in operational research 139
  renewed, 701
Socio-technical systems theory, 118
Soft systems methodology, 40, 172,
    232, 253, 507
  critique of, 69
Solidarity, 351
Stakeholder, 486
Strategic choice, 370
Strategic options development
    and analysis, 48
Strategy, 434
Systems dynamics, 190, 215
Systems and Operational Research, 219
Systems thinking, 171

Technological change, 337, 587

722